ATLAS OF CRUSTACEAN LARVAE

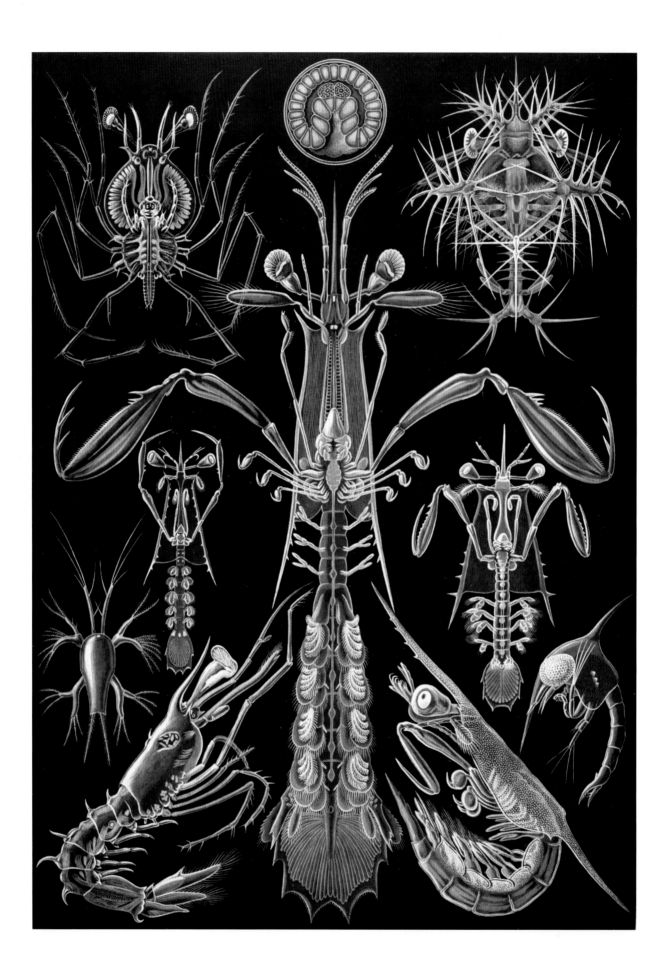

Atlas of Crustacean Larvae

EDITED BY JOEL W. MARTIN, JØRGEN OLESEN, AND JENS T. HØEG

JOHNS HOPKINS UNIVERSITY PRESS | BALTIMORE

This book was brought to publication with the generous assistance of the Carlsberg Foundation.

Johns Hopkins University Press
2715 North Charles Street
Baltimore, Maryland 21218-4363
www.press.jhu.edu

Library of Congress Cataloging-in-Publication Data

Atlas of crustacean larvae / edited by Joel W. Martin, Jørgen Olesen, and Jens T. Høeg.
      pages cm
 Includes bibliographical references.
  ISBN 978-1-4214-1197-2 (hardcover : alk. paper) — ISBN 978-1-4214-1198-9 (electronic) — ISBN 1-4214-1197-0 (hardcover : alk. paper) — ISBN 1-4214-1198-9 (electronic) 1. Crustacea—Larvae. I. Martin, Joel W. II. Olesen, Jørgen III. Høeg, Jens.
  QL435.A85 2014
  595.3—dc23   2013017982

A catalog record for this book is available from the British Library.

Frontispiece: Plate number 76 from the 1904 publication *Kunstformen der Natur* (*Artforms of Nature*) by Ernst Haeckel (1834–1919), depicting various forms of crustacean larvae (classified by Haeckel as Thoracostraca).

Special discounts are available for bulk purchases of this book. For more information, please contact Special Sales at 410-516-6936 or specialsales@press.jhu.edu.

Johns Hopkins University Press uses environmentally friendly book materials, including recycled text paper that is composed of at least 30 percent post-consumer waste, whenever possible.

To my beautiful wife Sue, who made all things possible for me, and to
Alex and Paul, my own wonderful and constantly metamorphosing larvae
JWM

To Signe, Alma, Ida Marie, and Nora
JO

To Marianne and Julie, to Graham Walker, and to Dieter and Kriemhild
Waloszek for friendship and invaluable inspiration in studying crustacean
larvae, and in fond memory of Alexey V. Rybakov
JTH

CONTENTS

CONTRIBUTORS

Addis, Alberto
Dipartimento di Scienze della Natura e
    del Territorio
Università di Sassari
Sassari, Italy

Ahyong, Shane T.
Marine Invertebrates
Australian Museum
NSW, Australia

Boesgaard, Tom
Copenhagen, Denmark

Boyko, Christopher B.
Department of Biology
Dowling College
Oakdale, NY, USA

Carcupino, Marcella
Dipartimento di Scienze della Natura e
    del Territorio
Università di Sassari
Sassari, Italy

Chan, Benny K. K.
Biodiversity Research Center
Academia Sinica
Taipei, Taiwan

Clark, Paul F.
Department of Life Sciences
Natural History Museum
London, UK

Criales, Maria M.
Division of Marine Biology and
    Fisheries
University of Miami
Miami, FL, USA

Cuesta, José A.
Instituto de Ciencias Marinas de
    Andalucia CSIC
Cadiz, Spain

Dos Santos, Antonina
Instituto Português do Mar e da
    Atmosfera
Lisboa, Portugal

Gerken, Sarah
Biological Sciences and Liberal Studies
University of Alaska
Anchorage, AK, USA

Gómez-Gutiérrez, Jaime
Departamento de Plancton y Ecología
    Marina
Centro Interdisciplinario de Ciencias
    Marinas
La Paz, Baja California Sur, Mexico

Goy, Joseph W.
Department of Biology
Harding University
Searcy, AR, USA

Grygier, Mark J.
Lake Biwa Museum
Shiga, Japan

Guerao, Guillermo
Unitat Operativa de Cultius
    Experimentals
Institut de Recerca i Tecnologia
    Agroalimentàries
Tarragona, Spain

Hampson, George
WHOI Oceanographer Emeritus
North Falmouth, MA, USA

Haney, Todd A.
Science Department
Sage Hill School
Newport Coast, CA, USA

Harvey, Alan
Department of Biology
Georgia Southern University
Statesboro, GA, USA

Haug, Carolin
Biocenter—Department of Biology II
    and GeoBio-Center
LMU Munich
Planegg-Martinsried, Germany

Haug, Joachim T.
Biocenter—Department of Biology II
    and GeoBio-Center
LMU Munich
Planegg-Martinsried, Germany

Hiruta, Shimpei
Division of Natural History Sciences
Hokkaido University
Sapporo, Hokkaido, Japan

Hiruta, Shin-ichi
Hokkaido University of Education
Kushiro, Hokkaido, Japan

Høeg, Jens T.
Department of Biology
University of Copenhagen
Copenhagen, Denmark

Huys, Rony
Department of Life Sciences
Natural History Museum
London, UK

Iliffe, Thomas M.
Department of Marine Biology
Texas A&M University
Galveston, TX, USA

Kado, Ryusuke
School of Marine Biosciences
Kitasato University
Kanagawa, Japan

Koenemann, Stefan
Montessori Bildungshaus Hannover
Hannover, Germany

Kolbasov, Gregory A.
Invertebrate Zoology
Moscow State University
Moscow, Russia

Kutschera, Verena
WG Biosystematic Documentation
University of Ulm
Ulm, Germany

Maas, Andreas
WG Biosystematic Documentation
University of Ulm
Ulm, Germany

Martin, Joel W.
Research & Collections Branch
Natural History Museum of Los
    Angeles County
Los Angeles, CA, USA

Martinsen, Søren Varbek
Copenhagen, Denmark

†McLaughlin, Patsy

Møller, Ole Sten
Institute of Veterinary Disease Biology
University of Copenhagen
Copenhagen, Denmark

Olesen, Jørgen
Natural History Museum of Denmark
University of Copenhagen
Copenhagen, Denmark

Palero, Ferran
Centro Superior de Investigación en
    Salud Pública
Valencia, Spain

Petrunina, Alexandra S.
Department of Invertebrate Zoology
Moscow State University
Leninskie Gory, Russia

Pohle, Gerhard
Atlantic Reference Centre
Huntsman Marine Science Centre
St. Andrews, New Brunswick, Canada

Richter, Stefan
Institut fuer Biowissenschaften
Universitaet Rostock
Rostock, Germany

†Rybakov, Alexey V.

Santana, William
Pró-Reitoria de Pesquisa e Pós-
    Graduação
Universidade Sagrado Coração
Bauru, Brazil

San Vicente, Carlos
Creixell, Tarragona, Spain

Schminke, Horst Kurt
Oldenburg, Germany

Smith, Robin J.
Lake Biwa Museum
Kusatsu, Shiga, Japan

Waloszek, Dieter
WG Biosystematic Documentation
University of Ulm
Ulm, Germany

Watling, Les
Department of Biology
University of Hawaii at Manoa
Honolulu, HI, USA

Wolff, Carsten
Department of Biology
Humboldt University
Berlin, Germany

ACKNOWLEDGMENTS

We are grateful to a large number of people and organizations who made this volume possible. Generous funding to help offset the cost of publication was provided by the Danish Carlsberg Foundation through a grant to Jørgen Olesen. We are extremely grateful for that crucial support, without which this volume could not have been produced. In Los Angeles, we sincerely thank Madelyn and Jerry Jackrel and Miriam Schulman for generous personal donations that also helped fund the publication of the book. We also acknowledge several grants from the U.S. National Science Foundation, including the "Tree of Life: Decapod Crustacea" grant (DEB 0531616) to Joel W. Martin. Partial support for J. W. Martin was also provided by a contract from the National Oceanic and Atmospheric Administration in support of his ongoing studies of decapod crustaceans from the Hawaiian Islands.

Several colleagues, some of whom are contributors to the volume, devoted time and effort toward improving draft versions of the chapters, and we are indebted to them. In particular we are grateful to Marnie Knight and Annie Townsend (Euphausiacea); Steven G. Morgan, Paul Clark, Jose Paula, and Joe Goy (Brachyura); Buz Wilson, Kim Larsen, and Gary Poore (Isopoda and Tanaidacea); Gary Poore (Pentastomida); Charles Fransen and Joe Goy (Amphionidacea); Ray Bauer and Joe Goy (Caridea); Ray Bauer (Dendrobranchiata); Dieter Waloszek, Hans-Uwe Dahms, Ryusuke Kado, and Andreas Maas (the nauplius larva); Simone M. Helluy (Astacidea); Todd Haney (Amphipoda); Dieter Waloszek and Andreas Maas (Euphausiacea); and Geoff Boxshall (Thermosbaenacea, Spelaeogriphacea, and "Mictacea"). Although all authors contributed in some way to the creation of the glossary (and not all of them agree with our final definitions), we especially thank Ole Sten Møller for his helpful comments and insight. In the Los Angeles laboratory of J. W. Martin, we are grateful to the following for help and support at various stages: Regina Wetzer, Dean Pentcheff, Adam Wall, Jonathan Sepulveda, Jenessa Wall, and Phyllis Sun. We also appreciate the assistance of Richard Hulser of the Natural History Museum of Los Angeles County's library in locating sometimes obscure literature.

Several workers provided us with previously unseen photographs and/or permission to reproduce these figures. Among them we particularly thank Jean-François Cart for permission to use some of his spectacular images of various branchiopods (chapters 5, 8, 9, 11), David Wrobel for allowing us to use his remarkable photograph of an Eryoneicus larval stage in life (chapter 52), Hsiu-Lin Chin and Arthur Anker for their beautiful color photographs of crab megalopae (chapter 54), and Roland Meyer and Roland Melzer for permission to use their outstanding scanning electron microscope images of crab zoeae (also in chapter 54).

We sincerely thank senior editor Vincent Burke at Johns Hopkins University Press for his interest, encouragement, and help from the beginning of the project through to completion; we also are grateful to copyeditor extraordinaire Kathleen Capels, JHU Press editorial assistant Jennifer Malat, and production editor Debby Bors for their attention to detail and for keeping us on track. In addition, we thank Rick Harrison for his encouragement many years ago, when we first approached him with the possibility of producing such a volume.

Finally, we are deeply grateful to our families. J. W. Martin thanks his beloved late wife, Sue, for her ever-present support, love, interest, and encouragement. This book would not have been possible without her. He also thanks his children, Alex and Paul, for their interest, inspiration, strength, and humor. J. Olesen is thankful to Signe, Alma, Ida Marie, and Nora for their support and patience, especially during the last period where too much family time in the evenings was spent on setting up photo plates of "weird, small animals" for this book. He also deeply appreciates the pioneering and inspiring teaching at the University of Copenhagen about larvae and their development in an evolutionary context by Claus Nielsen, Reinhardt M. Kristensen, and Jens T. Høeg. Last but not least, he is much indebted to Dieter Waloszek for his unique and infectious enthusiasm for crustacean larvae. J. T. Høeg thanks his Marianne and Julie for always accepting how much a scientist loves his work. He also wants Ib Svane, Graham Walker, Jørgen Olesen, Peter Gram Jensen, Henrik Glenner, Anders Garm, and Benny Chan to know that they all mattered more than they can possibly imagine.

ATLAS OF CRUSTACEAN LARVAE

1

Introduction

JOEL W. MARTIN
JØRGEN OLESEN
JENS T. HØEG

OVERVIEW

The great biodiversity of the Crustacea, both in terms of the number of species and in terms of their tremendous morphological range, continues to fascinate and challenge us. The most recent compilation (Ahyong et al. 2011) estimates over 1,000 extant families and nearly 70,000 described species. But morphological diversity is where the group is truly staggering. In the opening line to their updated classification of the extant Crustacea, J. W. Martin and Davis (2001) wrote, "no group of plants or animals on the planet exhibits the range of morphological diversity seen among the extant Crustacea." This remains true today. And the larval forms of the Crustacea are no less diverse. Indeed, they are spectacular in their variety of sizes and forms, elegant in their function, and aesthetically simply beautiful. To many of us, these larval stages are even more fascinating biologically than are the adults.

For many years now it has been a goal of ours to compile the best available images of the wonderful and diverse larval forms of the Crustacea into a single volume. This has not been possible in the past, primarily because, until relatively recently, images of crustacean larvae were restricted to line drawings, many of which were inaccurate. Additionally, these drawings (and paintings, in some cases) were widely scattered in the primary literature, rather than gathered into one volume. Although some of these historical illustrations were, and are, quite good (in fact, we have used many of them in the current book), advances in digital photography and scanning electron microscope (SEM) techniques have provided us with unprecedented detail and spectacular images of a wide variety of larval forms. At the same time, advances in our understanding of basic crustacean biology have led to recent revelations in answer to long-standing questions, such as the mode of larval development in the Remipedia (Koenemann et al. 2007a, 2009), the parentage of such long-standing "larval genera" as *Cerataspis* (Bracken-Grissom et al. 2012), and (at last!) the

developmental stage that follows the mysterious larvae that constitute the Facetotecta (Glenner et al. 2008; Scholtz 2008). Knowledge of crustacean larval biology has also increased tremendously. For example, larvae at deep-sea hydrothermal vents are now known to be affected by weather patterns at the sea surface (Adams et al. 2011). Crustacean larvae are known to possess a different visual apparatus from that of the adult (e.g., see Jutte et al. 1998 for stomatopod larvae). Some crustacean larvae have recently been shown to respond to sound, which might help them orient to coral reefs when settling (e.g., Montgomery et al. 2006; Stanley et al. 2010, 2011, 2012; Kessler 2011; Simpson et al. 2011). The role of crustacean larvae in biological invasions is enormous (e.g., Galil et al. 2011). Our understanding of chemical communication and larval settlement cues, including the important role of bacterial cues and quantitative proteomics in metamorphosis and settling, continues to improve (e.g., see Clare 2011; Hadfield 2011, 2012; Huang et al. 2012). New findings are announced regularly at annual meetings of the Crustacean Society and the biennial International Larval Biology Symposium. New methods of preparing larval forms for morphological study continue to appear (e.g., see Michels and Büntzow 2010 for new staining methods; Ivanenko et al. 2011 and Fritsch et al. 2013 for confocal microscopy of copepod and branchiopod larvae). This combination of new biological information, new techniques, and newly available images suggested to us that the timing was right, and the material available, to compile all of this information and these images into a single volume.

In addition to using recently published photographs and SEM images, we encouraged all authors to produce, wherever possible, new figures and photographs specifically for this book, images that had not been published previously. Many authors accepted this challenge, and the atlas you are now holding contains drawings, photographs, and SEM images of crustacean larvae never seen before. The overall result is, we feel, well worth the efforts we have expended.

METHODOLOGY, TERMINOLOGY, AND GLOSSARY

This atlas is the result of the work of 46 authors, all of whom are world experts on crustacean larval development, representing 14 countries. Authors were invited to contribute to the atlas based on their special knowledge of the subject, and each was asked to follow a standard format and provide basic information on larvae of a given taxon. Within those constraints, however, authors were allowed to expand where necessary. As noted above, they were also encouraged to include (or to produce) the best possible images of larval development for the group of crustaceans under their purview. Interested readers will note that the chapters that follow are not in proportion to the diversity of taxa. That is, for some taxa that are relatively species-poor, such as the Cephalocarida and Remipedia, considerable room has been allowed because of the morphological and phylogenetic importance of the larvae of these groups. Conversely, some taxa that are megadiverse, such as the Amphipoda, are given relatively brief treatments, either because larvae per se are not known for the taxon or because larval diversity is relatively low.

For each chapter, we also asked authors to list up to five of the most relevant studies that would serve to introduce students to the larvae described in that chapter. Because many of these references also are cited in other chapters, we have included the selected references (along with all other references cited in the text) in the full references section at the end of the volume. Thus there is a small amount of duplication, which we felt was necessary to allow each chapter to serve as a stand-alone work on the morphology of larvae in any given taxon.

The diversity of authors, and the vast amount of literature on larval crustaceans, has led, not surprisingly, to the realization that some terms are used and defined differently, both historically and by various authors today. Terms such as larva, post-larva, juvenile, and the like have been used inconsistently in the literature, depending on who was using them, and even today there is a lack of standardization for many of these terms. We have included a glossary of terms used in the descriptions of crustacean larvae, but it is important to note that our definitions may not be accepted by all workers; indeed, many of our authors disagreed among themselves as to the correct definitions. For terms that overlap with the field of crustacean neuroanatomy (compound eye, median eye, nauplius eye), we used definitions provided by Richter et al. (2010). Many additional definitions can be found on the Crustacea glossary associated with the Assembling the Tree of Life: Decapoda site at http://decapoda.nhm.org. Definitions we have used here are rather broad, in an attempt to adequately reflect the use of these terms in different taxa. Several terms are used often in the atlas, and these we felt were in need of a more general discussion.

Larva

The simple question "What is a larva?" turns out to be not so simple. The term larva (from the Latin *larua*, meaning "ghost, specter, or mask") was first used in a biological sense by Linnaeus (1768) for insects in which the larvae are so different from the adults that they mask the adult morphology. Since then, the term has come to be used for a variety of developing or immature forms of many different phyla. It has been defined extremely broadly, as, for example, by D. Smith (1977) as "an intermediate developmental form between eggs and adults," a definition so broad as to be limited in its usefulness (since it would also include embryos and juveniles). The word has also been defined along more functional lines, as in Young's (1999) definition of a larva as "a post-embryonic stage of the life cycle which differs from the adult morphologically and is capable of independent locomotion," although his phrasing might rule out some parasitic forms among the Crustacea. In Young's (2001) valuable historical overview, he notes that "no single definition is useful for all approaches to the study of larvae," a statement with which we fully agree. In this book, with its emphasis on morphology, we also define larva very broadly, as "any immature, post-embryonic form of an animal that differs morphologically from the adult and often develops into the adult either gradually (via anamorphosis) or by more abrupt changes in morphology (metamorphosis)." This definition depends to some degree on the definition of the word metamorphosis, which also has been used broadly and inconsistently in the past and has been applied to plants as well as animals. We define metamorphosis very broadly: "originally referring to any change in form, the term in biological applications now more commonly refers to a profound change in morphology (typically accompanied by behavioral and functional changes) during the life cycle of an organism." By our broad definition of larva, several groups of crustaceans do not have true larvae, as the embryos develop into small versions of the adults, a process often referred to as direct development. This pertains, for example, to nearly every group within the diverse Peracarida, which accounts for a large percentage of overall crustacean biodiversity. In an attempt to be as inclusive as possible, however, we have also included chapters on development within those groups, to give readers an idea of what transpires in the ontogeny and development of each major group of crustaceans.

Embryology

Embryology per se is not the subject of this volume, and we discouraged authors from including information on purely embryological development. For discussions of crustacean embryology, we refer the interested reader to the works of, for example, D. Anderson (1982) and Scholtz (2004). The line between embryology and larval development is sometimes crossed in this volume, however, particularly for those crustacean groups in which true larvae are not seen. Thus, for example, in many of the peracarid groups, our authors have included images of late embryos to give the reader an idea of how development proceeds in groups lacking true larvae. One reason for this is that in some cases the embryos can provide traces of the original developmental patterns that would otherwise be lost. In some of these late embryos, the vestiges (or precursors) of larval morphology can be seen, and we thought it important to include that information here.

The Nauplius

The nauplius larval phase is such a common component of crustacean development, and is seen in so many groups, that we have devoted a separate chapter to its morphology and diversity. Today, more than 300 years after the first observation of a nauplius (by Antonie van Leeuwenhoek in 1699), there is still debate over the evolutionary importance of this phase (and its component stages), the fossil record of naupliar-like larvae, the terminology used, the role of the nauplius in phylogeny, and even the basic biology and natural history of these larval forms. Chapter 2 is a summary of current knowledge of the nauplius, with subsequent chapters presenting more detailed information for groups that exhibit this larval form.

Phases, Stages, and Instars

The literature on crustacean larvae contains many adjectives to describe the different morphological forms that make up a larval developmental series. We have encouraged our authors to employ the word phase to describe broad larval categories: a phase is a morphologically distinct form that is part of the crustacean life cycle. Examples of larval phases are nauplius, zoea, mysid, megalopa, and the like. In contrast, the word stage is used to denote an ontogenetic subset (or subsets) of a particular phase. Thus, for example, most crabs go through a zoeal phase during their development. Within the zoeal phase there are anywhere from one to nine (or more) zoeal stages, often referred to as zoea I, zoea II, zoea III, and so forth. This usage follows most other treatments of crustacean larvae (e.g., Harvey et al. 2001). Exceptions to this usage are frequent in the literature, however, and we have pointed out areas in the atlas where the lines between phases and stages are not quite as clear as in the above example. An instar is usually defined as the period between molts (ecdyses). Thus it can be used synonymously with our use of the word stage, but it can also have a slightly different meaning. The difference depends on whether we know the history of the organism being studied. For example, if a given larva is collected from the plankton, we might employ the word stage in a less defined sense, when we are not aware of the exact developmental sequence. Another example is in the field of fossil larvae, where the word stage is more appropriately used than instar, since the exact developmental sequence (and thus the instar number) cannot be known with certainty.

Post-Larvae and Juveniles

As with the term larva, the words post-larva and juvenile have been used often and inconsistently in the past. Occasionally, life forms that are clearly larval by our earlier definition have been referred to as post-larvae in the literature. A common example is the use of post-larva to describe the brachyuran megalopa, which is clearly different from the adult. In this atlas, we employ the terms post-larva and juvenile somewhat interchangeably. A post-larva or juvenile crustacean is clearly beyond (older than) the classically defined larval series, lacks the larval characters, and is (in most ways) a miniature version of the adult. The slight difference between the terms post-larva and juvenile is usually construed as a matter of degree, with post-larva referring to those stages immediately following larval development, and with juvenile referring to all subsequent stages up until the last molt before adult onset, which typically is marked by maturation of the reproductive organs and appendages. In some crustacean groups, there are only slight, if any, metamorphoses from larval phases all the way through to adult phases, so some flexibility has been allowed in the terminology used in the chapters.

GENERALITIES OF CRUSTACEAN DEVELOPMENT AND LARVAE

Just as it is impossible to generalize about the Crustacea as a whole, it is also impossible to generalize about their wonderful and diverse forms of larvae. What follows is a very brief introduction to some of the basic patterns seen in crustacean larval development. All crustaceans develop from eggs, which may be shed directly into the water, borne by the parental female on modified appendages, or carried in some kind of brood chamber (formed in various ways), depending on the group. Embryonic development varies greatly (see, e.g., Nielsen 2012), but the cleavage pattern in crustaceans appears to be basically determinate. It can appear to be spiral-like, and many earlier workers have referred to arthropod development as being spiral, but differing from the typical spiral cleavage known in, for example, polychaete annelids, sipunculans, and most mollusks (e.g., R. Brusca and Brusca 2003). More recent workers, however, note that the similarity to spiral cleavage is superficial. Nielsen (2012) states that "early stages may resemble spiral-cleaving embryos, but the cleavage pattern is not spiral," and in some groups it can appear nearly radial (G. Brusca 1975; D. Anderson 1982). Cleavage can be (and is usually) holoblastic (typically in groups with very little egg yolk, such as copepods and barnacles), but it also can be meroblastic and superficial in groups with more yolk-rich eggs. Presumptive endoderm arises from the 4D cell, and presumptive mesoderm from the 3A, 3B, and 3C cells; ectoderm arises from the 3d and 4d cells, with the stomadeum arising from the 2b cell independent of gastrulation (see D. Anderson 1973, 1982; G. Brusca 1975; F. Schram 1978, 1986; Nielsen 2012). These features set crustaceans apart from other arthropod groups (F. Schram 1986).

Development can be direct, with the eggs hatching as miniature versions of the adult (as in the Peracarida), or it can be indirect, with several distinct types of larvae. Development following embryology is assumed to be primitively anamorphic (a long series of larval stages, with each subsequent stage differing only slightly from the previous one), although just a few extant crustacean groups display that type of growth today. Notable for their anamorphic development are the extant Anostraca (fairy shrimps) and the related (extinct) †*Rehbachiella* from the Cambrian (e.g., Walossek 1993; also see chapters 3, 4, and 5). As in all arthropods, growth is teloblastic, with the addition of somites coming from a posterior growth zone just anterior to the terminal (anal) segment. The majority of crustaceans exhibit some type of metamorphic change during their development, and in some cases (e.g., the Cirripedia and Tantulocarida) the change in form is particularly striking. Larval development in crustaceans typically includes a planktonic dispersal phase that can last from a few days to several weeks or even months. Some larvae, particularly in groups that exhibit abbreviated development, are benthic or spend only a brief amount of time in the plankton. In species that have planktonic larvae and benthic adults, there is often a distinct phase that serves as a transitional form (both morphologically and ecologically) from the planktonic mode of life to the benthic mode, such as the decapodid phase (megalopa) in decapods or the cyprid larval phase of barnacles. Some of these transitional settlement forms do not feed, which would appear to set a limit on the time spent before their metamorphosis to a post-larval form. Crustacean larvae can be lecithotrophic, planktotrophic, or parasitic; many begin as lecithotrophs but become planktotrophs as they develop.

HISTORY OF STUDIES ON CRUSTACEAN LARVAE

The study of crustacean larvae has a rich history that in many ways reflects and encapsulates the history of all research on the Crustacea. The nauplius larva is shared by a large number of crustaceans and is considered by many to be a basic, defining feature of the group (see chapter 2). The study of this larval form can be traced all the way back to Antonie van Leeuwenhoek, who, according to Gurney (1942), observed the hatching eggs of *Cyclops* (Copepoda) in 1699. Nauplii played an even more important historical role later, when J. Thompson (1830) discovered that barnacles were, in fact, crustaceans, based on the presence of nauplii in the mantle cavity. A few years later (J. Thompson 1836), he added to this by recognizing that the larvae of the parasitic (and highly modified) Rhizocephala not only were crustaceans, but were definitely cirripedes, based on the presence of frontolateral horns on the nauplii. As far as we can determine, this finding marked the first application of the use of crustacean larval characters in elucidating phylogenetic relationships, and it was among the first cases of larval characters used to clarify evolutionary relationships in any group of marine invertebrates. Since that time, the study of crustacean larvae has virtually exploded, and today the literature is vast and rich, with newsletters and even entire journals devoted to the roles that crustacean larvae play in plankton ecology, dispersal, and biogeography; arthropod phylogenetics; crustacean aquaculture; and more.

We have asked each author to comment briefly on some of the more important historical publications and events in the study of the larvae of each taxon. For other discussions of the importance of crustacean larvae as part of the overall history of crustacean research, see Anger (2001), Harvey et al. (2001), J. W. Martin and Davis (2001), and the chapters that follow.

CRUSTACEAN LARVAE, PHYLOGENY, AND CLASSIFICATION

As noted by J. W. Martin and Davis (2001), "for many groups of crustaceans, a study of the systematic relationships *is* a study of the larvae." The reason for this is twofold. First, larval characters are often the best, or the only, characters that we have available for deducing phylogenetic relationships. This is especially obvious for those groups, such as the Facetotecta, in which the larvae are known but the definitive adults are not (but see Glenner et al. 2008). Second, characters of the larvae, which for the most part exist in a relatively uniform environment (the plankton) compared with that of the adults, may be less subject to the selective pressures that are unique to their post-larval life (e.g., Felder et al. 1985). Thus, the argument goes, characters of the larvae could more accurately reflect genealogy than would characters of the adults, which might be more strongly acted on by selection as an adaptation to various adult habitats. For this reason, a large number of studies on crustacean phylogeny have employed (or have been based exclusively on) characters of the larvae. For a list of such studies, see J. W. Martin and Davis (2001).

Ideally, any classification should reflect current knowledge of phylogeny. Although tremendous progress has been made in our understanding of the relationships within the Crustacea, and of the Crustacea to other groups of arthropods, there is still no universally accepted, phylogenetically based classification of the entire Crustacea. One of the most recent attempts, by Regier et al. (2010), is reproduced here (fig. 1.1). Despite the enormous amount of molecular data on which that phylogeny is based, there are still debates about the implied relationships, and not all crustacean groups are represented in Regier et al.'s phylogenetic tree.

Fortunately, classification is not a serious problem for the purposes of this atlas. Although we sincerely hope that our book is helpful toward the study of crustacean phylogeny, it is primarily an atlas of larval morphology. Regardless of the eventual results of ongoing studies on crustacean phylogeny, the morphological details presented here should not change significantly, other than the addition of anticipated (and hoped for) increases in our knowledge. For the purpose of organizing this volume, we have followed the rather conservative classifications of J. W. Martin and Davis (2001) and Ahyong et al. (2011).

There are some differences between our organization and that of J. W. Martin and Davis (2001) and Ahyong et al. (2011). The most obvious exception is that we no longer recognize the class Maxillopoda, about which even J. W. Martin and Davis (2001) expressed doubts, and which has subsequently been shown to be an artificial grouping based on morphological and (predominantly) molecular analyses (e.g., Regier

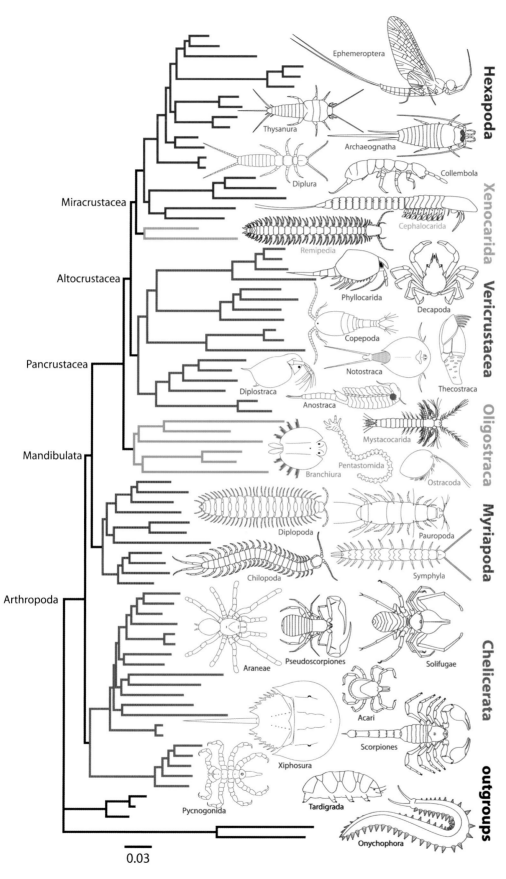

Fig. 1.1 A recently proposed phylogenetic tree of the Arthropoda, including major branches of the Crustacea, based on extensive molecular sequence data (from Regier et al. 2010, reproduced with permission from *Nature*). Scale bar indicates nucleotide changes per site.

et al. 2010), although Ahyong et al. (2011) employed it. Thus, for organizational purposes only, the major groups of crustaceans treated here are arranged in roughly the same order as in the J. W. Martin and Davis (2001) and Ahyong et al. (2011) classifications, with the following exceptions: (1) within the Branchiopoda we recognize the taxa Onychocaudata, Cladoceromorpha, and Gymnomera (following molecular and morphological evidence summarized in Olesen and Richter 2013), although this creates some currently unnamed taxonomic ranks (one of the limitations of the Linnean system of classification); (2) the former class Maxillopoda has been abandoned, and former subclasses within that class have now been elevated to the level of class; (3) all taxa at or below the level of superfamily shown in J. W. Martin and Davis (2001) have been removed for brevity; (4) names and dates of authorities for the taxa have been removed; (5) within the Acrothoracica, the orders Apygophora and Pygophora in J. W. Martin and Davis (2001) have been replaced by Cryptophialida and Lythoglyptida (following Ayhong et al. 2011); (6) the former order Pedunculata is no longer recognized within the Thoracica, and some former suborders are now orders (Ahyong et al. 2011); (7) within the Ostracoda: Podocopa, the order Palaeocopida is recognized for the family Punciidae; (8) within the Peracarida, we have retained the former subdivisions of the Amphipoda (as opposed to listing all families alphabetically, as was done by Ahyong et al. 2011), and several orders of isopods are newly recognized (with the Anthuridea and Flabellifera no longer recognized), following Ahyong et al. (2011); and (9) within the Decapoda, the Glypheidea and the Procaridoida are recognized as pleocyematan infraorders (following De Grave et al. 2009; De Grave and Fransen 2011), the name Palinura has been replaced by Achelata, the Polychelida (formerly placed in the Palinura by J. W. Martin and Davis 2001) is recognized at the level of infraclass, and the former Thalassinidea has been replaced by the 2 infraorders Axiidea and Gebiidea (following papers in De Grave et al. 2009; J. W. Martin et al. 2009). We again note that this is not intended to be a purely phylogenetic arrangement. Taxon names appearing in **bold** below are those treated in the book.

Subphylum Crustacea
    Class **Branchiopoda**
        Subclass Sarsostraca
            Order **Anostraca**
        Subclass Phyllopoda
            Order **Notostraca**
            Order Diplostraca
                Suborder **Laevicaudata**
                Suborder **Onychocaudata**
                    Infraorder **Spinicaudata**
                    Infraorder Cladoceromorpha
                        Subinfraorder **Cyclestherida**
                        Subinfraorder Cladocera
                          Ctenopoda
                          Anomopoda

                          Gymnomera
                          Onychopoda
                          Haplopoda
    Class **Remipedia**
    Class **Cephalocarida**
    Class **Thecostraca**
        Subclass **Facetotecta**
        Subclass **Ascothoracida**
            Order Laurida
            Order Dendrogastrida
        Subclass Cirripedia
            Superorder **Acrothoracica**
                Order Cryptophialida
                Order Lythoglyptida
            Superorder **Rhizocephala**
                Order Kentrogonida
                Order Akentrogonida
            Superorder **Thoracica**
                Order Ibliformes
                    Suborder Iblomorpha
                Order Lepadiformes
                    Suborder Heteralepadomorpha
                    Suborder Lepadomorpha
                Order Scalpelliformes
                    Suborder Scalpellomorpha
                Order Sessilia
                    Suborder Brachylepadomorpha
                    Suborder Verrucomorpha
                    Suborder Balanomorpha
    Class **Tantulocarida**
    Class **Branchiura**
            Order Arguloida
    Class **Pentastomida**
            Order Cephalobaenida
             Order Porocephalida
    Class **Mystacocarida**
            Order Mystacocaridida
    Class **Copepoda**
        Subclass Progymnoplea
            Order Platycopioida
        Subclass Neocopepoda
            Superorder Gymnoplea
                Order Calanoida
            Superorder Podoplea
                Order Misophrioida
                Order Cyclopoida
                Order Gelyelloida
                Order Mormonilloida
                Order Harpacticoida
                Order Poecilostomatoida
                Order Siphonostomatoida
                Order Monstrilloida
    Class **Ostracoda**
        Subclass **Myodocopa**
            Order Myodocopida
                Suborder Myodocopina

Order Halocyprida
  Suborder Cladocopina
  Suborder Halocypridina
Subclass **Podocopa**
  Order Platycopida
  Order Paleocopida
  Order Podocopida
    Suborder Bairdiocopina
    Suborder Cytherocopina
    Suborder Darwinulocopina
    Suborder Cypridocopina
    Suborder Sigilliocopina
Class **Malacostraca**
Subclass Phyllocarida
  Order **Leptostraca**
Subclass Hoplocarida
  Order **Stomatopoda**
Subclass Eumalacostraca
  Superorder **Syncarida**
    Order Bathynellacea
    Order Anaspidacea
  Superorder **Peracarida**
    Order **Spelaeogriphacea**
    Order **Thermosbaenacea**
    Order **Lophogastrida**
    Order **Mysida**
    Order **Mictacea**
    Order **Amphipoda**
      Suborder Gammaridea
      Suborder Caprellidea
        Infraorder Caprellida
        Infraorder Cyamida
      Suborder Hyperiidea
        Infraorder Physosomata
        Infraorder Physocephalata
      Suborder Ingolfiellidea
    Order **Isopoda**
      Suborder Phreatoicidea
      Suborder Microcerberidea

Suborder Asellota
Suborder Calabozoida
Suborder Cymothoida
Suborder Limnoriidea
Suborder Phoratopidea
Suborder Sphaeromatidea
Suborder Tainisopidea
Suborder Valvifera
Suborder Epicaridea
Suborder Oniscidea
  Infraorder Tylomorpha
  Infraorder Ligiamorpha
Order **Tanaidacea**
  Suborder Tanaidomorpha
  Suborder Neotanaidomorpha
  Suborder Apseudomorpha
Order **Cumacea**
Superorder **Eucarida**
  Order **Euphausiacea**
  Order **Amphionidacea**
  Order **Decapoda**
    Suborder **Dendrobranchiata**
    Suborder Pleocyemata
      Infraorder **Stenopodidea**
      Infraorder Procaridoida
      Infraorder **Caridea**
      Infraorder Glypheidea
      Infraorder **Astacidea**
      Infrorder **Axiidea**
      Infraorder **Gebiidea**
      Infraorder **Achelata**
      Infraorder **Polychelida**
      Infraorder **Anomura**
      Infraorder **Brachyura**

2

Joel W. Martin
Jørgen Olesen
Jens T. Høeg

The Crustacean Nauplius

GENERAL: Because crustaceans are so morphologically diverse, it is difficult to find characters that are shared by all, or even by most, members of the group. Indeed, there are few characters that we can point to as being specifically and uniquely crustacean. One such character is the nauplius larva, a developmental phase that immediately identifies its owner as a crustacean and that has been used widely as a feature that unites them (e.g., Cisne 1982; F. Schram 1986; Walossek and Müller 1990; Walossek 1993; T. Williams 1994b; Harvey et al. 2001). It is also important to note that a series of papers by Waloszek and colleagues (e.g., Walossek and Müller 1990; Walossek 1993) has revealed larval stages in the fossil record (some of which are referred to as head larvae) that occurred well before the nauplius stage of the true Crustacea (their "Eucrustacea"). Thus in their view the ancient arthropodan head larva, with four pairs of limbs, is a precursor to the shorter nauplius larvae, with their characteristic three pairs of limbs (the nauplius appendages) found only in the crown group of the Crustacea (Eucrustacea) (see chapter 3).

The nauplius appears to be a developmental phase that nearly all crustaceans pass through at some point in their development, whether or not it is obvious (T. Williams 1994b; but also see Scholtz 2000). It is the earliest free-living phase in the development of a large number of crustaceans. Even so, many groups have bypassed this phase during their development, such that eclosion from the egg occurs at a later (postnaupliar) stage of development. Other groups have evolved such bizarre larval modifications that the original naupliar characters are scarcely (if at all) evident. But it is widely assumed that all crustaceans, from anostracans to barnacles to decapods, originally had, as their primary form of larva, a nauplius. This is supported in part by amazingly preserved larval arthropods from the Lower Cambrian, about 500 million years old, some of which are identifiable as being naupliar or metanaupliar (X. Zhang et al. 2010; also see chapter 3). According to Walossek (1999), the locomotory and feeding apparatus of larval crustaceans "was so advanced by the Up-

per Cambrian that virtually no differences exist" between the early nauplii of †*Bredocaris admirabilis* or †*Rehbachiella kinnekullensis* (both Cambrian) and extant nauplii of barnacles, copepods, or branchiopods (also see Walossek 1993, 1995; Walossek et al. 1996).

Because so many crustaceans, extant and extinct, have in common a naupliar phase in their development, we present here a general overview of the nauplius larva, so that other authors need not repeat all of these features in the chapters that follow and can instead focus on characters unique to their groups.

MORPHOLOGY: The crustacean nauplius larva (plural = nauplii), at its simplest (earliest) stage, consists of an oval, pear-shaped, or rounded body that bears three pairs of functional appendages, destined to become the antennules, antennae, and mandibles of the adult. However, in the nauplius these appendages have different functions from their adult counterparts. Because the nauplius does not benefit from having multiple pairs of appendages with different specializations, the three naupliar appendages must address its combined locomotory, sensory, and feeding needs. Although most figures of nauplius larvae depict them flattened out (e.g., figs. 2.1A, 2.2), to show the appendages, this belies their true shape and orientation in life, where the limbs extend laterally but function in three dimensions (exopods up, endopods down; fig. 2.1B, C, D) and are therefore in a different plane from that of the larval body and shield (also see Glenner et al. 1995; as well as models of fossil forms in Walossek 1993; X. Zhang et al. 2010).

The antennules (first antennae) are uniramous (unbranched), lacking a flagellum. They may (but do not always) bear tubular, blunt-ended sensory aesthetasc setae (fig. 2.1B), and it is assumed that they play a primarily sensory role. Originally, however, they were probably used in locomotion and feeding (e.g., Walossek 1993). The antennae and mandibles are almost always biramous, with the two branches termed an exopod and an endopod (fig. 2.1A, B). These appendages are usually natatory and feeding limbs (although some nauplii are benthic rather than swimming). The antennae and mandibles typically bear a few plumose setae to aid in propulsion, and they may be quite complex (fig. 2.1D; also see Moyse 1987, his fig. 7A–C, for the naupliar antenna of *Lepas*). In many groups, one or both of these appendages bear a naupliar process— sometimes referred to as the masticatory spine, gnathobase, coxal spine, coxal endite, arthritic process, or arthrite (the latter term usually restricted to copepods) by some authors and for some taxa—at the base of the appendage (on the coxal article). Indeed, some authors have defined the naupliar larval phase operationally as those stages bearing a naupliar process on the second antenna (Dahms et al. 2006, as "arthritic process"). The process is not morphologically the same in all taxa, which is one reason why more descriptive terms (such as masticatory spine) are sometimes used.

The nauplius larva also usually bears, at the anterior end, a single eye that is referred to as the nauplius eye or naupliar eye (fig. 2.1A), even in those cases (e.g., notostracans, cladocerans, and copepods) where it persists in the adult. The nauplius eye actually consists of a cluster of three or four median eyes (or eye cups; see Richter et al. 2010 for definitions). This relatively simple eye is typical but not universal (it is lacking, for example, in the nauplii of euphausiids, cephalocarids, mystacocarids, and remipedes; Mauchline and Fisher 1969; Koenemann et al. 2007a; Richter et al. 2010). Where present, it is not morphologically identical among all crustacean larvae. The number of median eyes making up the nauplius eye can be three or four (the latter only in branchiopods), and in some taxa a tapetal layer is reported, in addition to the light-absorbing pigment layer. In his classic studies of the nauplius eye in adult crustaceans, Elofsson (1963, 1965, 1966) argued that it appears to have been derived independently at least four times (although this was well before the advent of cladistic thinking and methodology).

The body is often covered with a dorsal cephalic (naupliar) shield, sometimes also called a head shield, that is usually widest anteriorly. When expanded over the head and thoracic segments (more often in post-nauplius stages) it is often called a carapace. This shield may be present at hatching or may develop in later stages (see below), and its appearance can therefore be gradual or abrupt. A prominent oval or circular raised area—termed the dorsal organ (reviewed by J. W. Martin and Laverack 1992), or sometimes referred to as the sensory dorsal organ (Laverack et al. 1996)—is a characteristic component of the head shield in extant and extinct branchiopods (Walossek 1993; Olesen 2005) and may be a plesiomorphic feature in that clade. A similar organ is seen in post-nauplius larvae in some other taxa (see J. W. Martin and Laverack 1992; Laverack et al. 1996).

The body of the nauplius, at least initially, shows no external signs of segmentation. During subsequent molts, the body adds new segments posteriorly, and new appendages (both cephalic and thoracic) begin to appear as buds. These may include additional head appendages, which can be rudimentary or functional, and rudimentary thoracic appendages.

Once these additional somites and appendage buds appear, the larva is usually referred to as a metanauplius. Often the term orthonauplius is used to designate the earlier, simpler nauplius that lacks these limb buds. In some taxa, the egg hatches as an orthonauplius, with the metanauplius stage reached in one to several molts; in many groups, however, the eggs hatch directly as metanauplii. Late metanauplius stages can also show the beginnings of paired compound eyes. The labrum (a small, posteriorly directed cuticular flap arising from between the bases of the antennules and covering the mouth) is also typically present. A pair of caudal setae, one on each side of the anal area, is also common.

Many crustacean groups (the Cephalocarida, Branchiopoda, Ostracoda, Mystacocarida, Copepoda, Thecostraca, Euphausiacea, and Dendrobranchiata) produce free-living nauplii; other groups, with much-modified or direct development (e.g., crayfish, caridean shrimps, mysids, anaspidaceans, and branchiopods), show at least some vestige of an egg-nauplius stage (an embryo with naupliar appendages that are more developed than the more posterior limbs) during development (Scholtz 2000). In other direct-developing groups, egg-nauplii are absent (e.g., in peracarids such as amphipods, isopods, tanaidaceans, and cumaceans).

In some groups there are several morphologically distinct nauplius stages, separated by molts (e.g., most free-living Copepoda go through six nauplius and five copepodid stages; also see chapter 27); these stages are usually referred to simply as nauplius I, II, III, and so forth. Cirripedes also pass through a phase with six quite-similar nauplius stages before metamorphosing to the cypris larva (see chapter 22). In other taxa, such as cephalocarids, mystacocarids, and branchiopods (e.g., H. Sanders 1963; Olesen 2001; Olesen 2004), it is not as straightforward to distinguish between a nauplius phase and later stages or phases, because the development is more gradual and the differences between stages are less distinct. The same applies to the Cambrian microcrustacean †*Rehbachiella* from the Orsten lagerstätte of Sweden, which exhibits extremely gradual development with only minor differences between the stages; this mode of development has been suggested to approach the ancestral developmental mode in the Crustacea. Taxa with a distinct nauplius phase, such as copepods and cirripedes, are therefore probably specialized.

The term pseudometanauplius is sometimes used, especially in literature on krill (euphausiid) development, to describe a nauplius stage that is between the ortho- and metanauplius stages, a stage that lacks limbs but has an extended and more slender trunk than that seen in a typical orthonauplius. Terminology within the Crustacea is not consistent, however, since basically the same thing is seen in branchiopod taxa, such as anostracans and spinicaudatans, where the posterior part of the hatching stage is extended but still lacks limb rudiments (e.g., see chapter 9).

In some taxa (e.g., mysids), development takes place within a brood chamber, and the larva, although bearing the three nauplius limbs, does not swim. In this case, the developing larva is often called a nauplioid or, if further limbs are evident,

Key to the Nauplii of Major Crustacean Taxa

1 Cephalic shield composed of platelets and a caudal process .... **Facetotecta**
   Cephalic shield, if present, not composed of platelets:
   —Body visibly segmented in the early stages. Predominantly benthic...2
   —Body not visibly segmented in the early stages 3

2 Antennules with six segments. Found in interstitial waters of subtidal soft substrates .... **Cephalocarida**
   —Antennules with eight segments. Found in interstitial waters of sandy substrates ....**Mystacocarida**
   —Antennules with four segments. Trunk eight-segmented and covered by a cephalic shield. Found in
     plankton ....**Ostracoda** family **Punciidae**

3 With frontal filaments, frontolateral horns, and a caudal process ....**Cirripedia: Thoracica**
   —With frontal filaments and frontolateral horns; caudal process and labrum absent or greatly reduced; no
     real gnathobase on the antenna (at best a small spine or seta); mouth and anal opening absent (related to
     non-feeding) ....**Rhizocephala**
   —With frontal filaments and a caudal process, but no frontolateral horns (possibly with several pairs of horn-like
     lateral processes)..**Ascothoracida**
   —Not as above 4

4 With a bivalved shell ....**Ostracoda** (except family **Punciidae**)
   Without a bivalved shell:
   —With a telson and rostral helmet-like projection (hood); antennules segmented.............................**Euphausiacea**
   —With a telson, but lacking a rostral hood; antennules segmented.......................................**Decapoda: Penaeoidea**
   —Antennules not segmented. Protopod of the antenna more than half the length of the limb. Mandible
     uniramous, shorter than the antenna. Primarily found in inland waters ....**Branchiopoda** (7)
   —Antennules not segmented. Mandible biramous, clearly longer than the antenna. Known only from anchialine
     cave waters... **Remipedia**
   —With five-segmented antennules. Antennal exopodal segments increase from six at N I to a final number of
     nine at N VI. A pair of caudal setae at N I is reduced in size at N II................................**Copepoda: Polyarthra** (5)

5 Caudal process present. Body pear-shaped ....**Longipediidae**
   —Body shape rectangular or broadly oval. If a caudal process is present, then only as a spiniform
     protuberance..**Canuellidae**
   —Not as above 6

6 Body shape ovoid. Caudal armature symmetrical. Antennules uniramous and three-segmented. Antenna
   biramous, consisting of a two-segmented protopod, one-segmented endopod, and five-segmented exopod.
   Mandible biramous, consisting of a two-segmented protopod, two-segmented endopod, and four-segmented
   exopod without a gnathobase..**Copepoda: Cyclopoida**
   —Body elongate. Caudal armature often asymmetrical. Body flexed ventrally in many taxa. Antennule and
     antenna held forward. Antennules three-segmented, with a broad and elongate distal segment. Antenna
     biramous, consisting of a two-segmented protopod, two-segmented endopod, and six-segmented exopod.
     Mandible biramous, consisting of a two-segmented protopod, one-segmented endopod, and four-segmented
     exopod. Labrum large and spinulose..**Copepoda: Calanoida**
   —Body shape broad and more or less flattened. Antennule held perpendicular to the longitudinal axis of
     the body, one- to three-segmented. Post-maxillar limbs widely spaced. Antenna and mandibular endopod
     elongate. Mandibular endopod with one or two stout setae terminally on the inner process ....
     ....**Copepoda: Harpacticoida: Oligoarthra**

7 Antennules as lateral horns. Carapace forms a dorsal shield covering the entire body, with a large dorsal organ
   in the center. Labrum large, rounded, and flattened (at least in *Lynceus brachyurus*). Caudal rami small triangular
   projections, non-articulated to the body .... **Branchiopoda: Laevicaudata**
   —Antennules as small, sensillae-bearing buds. Carapace in the later naupliar stages small, with a bivalved
     appearance covering the early buds of the trunk limbs. Labrum large, terminating in either one (species
     of Limnadiidae) or three (species of Cyzicidae) large spines, or sometimes without spines (species of
     Leptestheridae). Caudal rami elongate and pointed, with claw-like projections non-articulated to the body
     ....**Branchiopoda: Spinicaudata**
   —Antennules tubular and relatively short (length is 3–4 times width). Carapace in the later stages forms a
     small dorsal shield covering a small part of body. Labrum in the later stages is relatively large and fleshy,
     with small marginal spines. Caudal rami elongate, wide basally and pointed distally, non-articulated to the
     body..**Branchiopoda: Notostraca**
   —Antennules tubular and mostly rather long (in *Eubranchipus*, length is 10 times width). Carapace absent.
     Labrum large, ellipsoid in shape, with a dense row of marginal sensillae, but spines are absent. Caudal end
     terminates in a pair of caudal setae...**Branchiopoda: Anostraca**
   —Antennal protopods extremely long (about as long as the entire trunk). Mandibular palps long, unsegmented
     tubular structures, with distal setation only .... **Branchiopoda: Cladocera: Haplopoda**

a post-nauplioid (see chapter 38), names that are similar to the word nauplius (and of course derived from that word), but these developmental stages should not be confused with true naupliar larvae. In these nauplioids, the yolk is extensive, filling most of the head and abdomen of the developing larva. It appears that the term nauplioid basically covers the same developmental stage as the term egg-nauplius, which is a term preferred by some authors (e.g., Scholtz 2000) for direct-developing crustacean (malacostracan) embryos with distinctly visible nauplitar appendages (antennule, antenna, and mandibles).

Other names for nauplitar larvae are specific to the different crustacean groups. For example, the Facetotecta have a unique y-nauplius (or nauplius y) larval stage. See the individual taxonomic chapters for these variations and their terminology. The somewhat confusing term naupliosoma refers to a pre-zoeal stage in the Achelata, rather than to a naupliar stage.

MORPHOLOGICAL DIVERSITY: Despite their incredible numbers and widespread occurrence, nauplii exhibit a surprisingly conservative morphology (Dahms 2000). At the same time, there are "many variations on this relatively simple theme" (fig. 2.3) (Harvey et al. 2001), and crustacean nauplii have "undergone remarkable adaptive radiation within the limits of their organizational constraints" (Dahms et al. 2006) and have displayed "functional plasticity" over time (T. Williams 1994b). One obvious example is in the Cirripedia (barnacles), where nauplii are easily recognizable as barnacle larvae by their possession of frontolateral horns (fig. 2.2F). The curious y-larvae of the Facetotecta are another easily recognized variant on the basic nauplius shape; these larvae are all small (250–620 μm long) and possess a uniquely faceted cephalic shield (fig. 2.2G). Yet another variant is the unusual, flattened nauplii of laevicaudatan branchiopods, where the naupliar appearance is barely recognizable, since the naupliar appendages are concealed between a dorsal shield (carapace) and an extremely enlarged ventral labrum (see chapter 8). The nauplii of the only two groups of malacostracans that exhibit naupliar eclosion (the Euphausiacea and Dendrobranchiata) differ from those of almost all other groups in being completely lecithotrophic, thus lacking the full complement of feeding mechanisms, a difference thought by Scholtz (2000) to be of phylogenetic significance. Not all forms of nauplii are mentioned here; other variations on the basic naupliar morphology can be found in the individual taxonomic treatments (also see fig. 2.3).

Opposite is an identification key to the naupliar larvae of some of the major groups of crustaceans, modified after Dahms et al. (2006), Koenemann et al. (2007a), and Olesen (pers. comm.). For a somewhat similar key that also includes some non-crustacean arthropod taxa and non-naupliar crustacean larvae, see Fornshell (2012). A brief comparison of the groups is given in table 2.1.

NATURAL HISTORY: In terms of numbers, the crustacean nauplius may well be "the most abundant type of multi-cellular animal on earth" (Fryer 1987b; Harvey et al. 2001). Because of their abundance, nauplii play critically important roles ecologically (e.g., Alekseev 2002; Dahms and Qian 2004). The nauplius is the earliest free-swimming phase of any crustacean. Swimming and feeding in the nauplius (see Gauld 1959; T. Williams 1994a) are accomplished with the cephalic appendages, although some nauplii (e.g., euphausiids, dendrobranchs, remipedes, and early branchiopod larvae) are lecithotrophic (non-feeding and depending on their yolk reserves). In addition to lecithotrophy developing secondarily (from ancestral feeding nauplii) in the above taxa, it appears that it has also developed from feeding nauplii multiple times *within* certain taxa, such as the Cirripedia (D. Anderson 1994; also see chapters 17–22). Where feeding occurs, it appears to be triggered by chemical cues, at least in barnacle nauplii (D. Anderson 1994). Feeding is typically either by suspension feeding or feeding on coarse particles. Nauplii are known to eat small (or large, in the case of some barnacle nauplii; see Moyse 1963) phytoflagellates, diatoms, and other microplankton (e.g., Moyse 1963; M. Barker 1976; Stone 1989; Harvey et al. 2001). Swimming in most non-copepod groups is accomplished mainly by a rowing motion of the second antennae, while most copepod nauplii are capable of swimming in rapid, jerky movements (Gauld 1959; T. Williams 1994a), making them rather difficult to catch with a fine pipette. A few nauplii are benthic rather than free-swimming (e.g., in the Mystacocarida, Cephalocarida, Ostracoda, and some copepods; see Dahms 2000). Nauplius larvae often tow long threads of fecal material behind them. Nauplii tend to lack pigmentation, although the median nauplius eye and darker oil droplets are sometimes visible inside them (fig. 2.1C). In some taxa, the nauplii are retained, such as in the mantle cavity of barnacles (D. Anderson 1986) or in the marsupium of peracarids (e.g., see chapter 38).

PHYLOGENETIC SIGNIFICANCE: Naupliar characters have a long history of use in crustacean and arthropod phylogenetics, and the very presence of naupliar larvae is a unique and defining feature of the Crustacea. Within each taxonomic group where naupliar larvae are found, morphological naupliar characters have been used as indicators of relationships, and these historical studies are mentioned briefly in each of the chapters that follow. The presence of nauplii or nauplius-like larvae in the fossil record, and the phylogenetic significance of those forms, is discussed in chapter 3. The difference between two major functional forms of nauplii was proposed by Scholtz (2000) as being of phylogenetic significance: free-living nauplii of the Euphausiacea and Dendrobranchiata (both Malacostraca) are all lecithotrophic, whereas in most other groups the nauplii are feeding (discussed also by Koenemann et al. 2007a).

HISTORICAL STUDIES: The study of the crustacean nauplius can be traced all the way back to Antoine van Leeuwenhoek, who, according to Gurney (1942), observed the hatching eggs of *Cyclops* (Copepoda) in 1699. The name nauplius was

Table 2.1 Brief comparison of major extant crustacean groups hatching as a nauplius

Taxon[1]	Hatch as	Distinguishing features	No. of naupliar stages	Locomotion and feeding	Metamorphosis at end of naupliar phase	Characters indicating end of phase
Cephalocarida	metanauplius	well-developed head shield	15	benthic, non-swimming; probably feeding	no distinct metamorphosis; gradual, sequential addition of limbs	loss of NP on A2 coxa, loss of md palp
Remipedia	orthonauplius	no head shield, no nauplius eye, very large md; A2 and md unusually flexible despite being unsegmented	8 known, probably incomplete	swimming; lecithotrophic	apparently no distinct metamorphosis known; gradual, as far as is known	unknown; first known post-naupliar stage (pre-juvenile) has head shield, frontal filaments, anterior thoracic limbs
Branchiopoda	orthonauplius or metanauplius	A1 small, unsegmented; md uniramous	varies according to group[2]	swimming; feeding (planktotrophic), except stages 1–3 in some groups (non-feeding)	varies by group; extremely gradual in anostracans, more metamorphic in other taxa[3]	loss of NP on A2 coxa, loss of md palp (but palp rudiment can be retained in juveniles)
Branchiura	metanauplius-like or juvenile-like[4]	functional post-md appendages; max with bipartite hook; carapace with setal fringe	1 (debatable as to whether it is a metanauplius)[4]	swimming or non-swimming; pre-adapted for parasitism	varies by genus[4]	loss of A2 and md
Ostracoda	orthonauplius-like or later; enclosed in bivalved shell	bivalved shield (except punciids); often with caudal rami	4–7 or 8 instars[5]	swimming, crawling, or attached to substrate; feeding varies[6]	no distinct metamorphosis; gradual, with sequential addition of limbs	see column at left
Thoracica	orthonauplius	frontolateral horns	6	swimming; planktotrophic, lecithotrophic	distinct metamorphosis into a cyprid larva	see column at left
Acrothoracica	orthonauplius	frontolateral horns	usually 4	swimming; lecithotrophic	as above	as above
Rhizocephala	orthonauplius	frontolateral horns; no NP on A2	usually 4 (5 in *Briarosaccus*)	swimming; lecithotrophic	as above	as above
Ascothoracida	orthonauplius	no frontolateral horns	up to 6	swimming; planktotrophic, lecithotrophic	as above[7]	as above
Facetotecta	orthonauplius (nauplius y)	no frontolateral horns; faceted shield of many plates	up to 6	swimming; planktotrophic, lecithotrophic	as above	as above
Mystacocarida	metanauplius	prominent supra-anal process; A1 of 8 segments; no nauplius eye	6	benthic; feeding	no distinct metamorphosis; gradual	loss of NP on A2 coxa
Copepoda	orthonauplius, metanauplius, or (rarely) copepodid	extremely variable[8]	4–(more typically) 6 (1 in *Misophria*)	swimming; planktotrophic, with many exceptions[8]	distinct metamorphosis into a copepodid	loss of NP on A2 coxa; naupliar somites separated
Malacostraca (Euphausiacea and Dendrobranchiata)	orthonauplius or metanauplius (or later)	no NP on A2, no mandibular gnathobase	up to 3 (in krill) or 6 (in dendrobranchs)	swimming; lecithotrophic	distinct metamorphosis into a calyptopis (krill) or a protozoea (dendrobranchs)	thoracopodal swimming

Source: Based on Walossek 1993; Dahms et al. 2006; Addis et al. 2007; Koenemann et al. 2007a, 2009; Høeg et al. 2009a, 2009b; Haug et al. 2011a; Olesen et al. 2011.

Abbreviations: A1 = antennule, A2 = antenna, max = maxillule, NP = naupliar process (= coxal spine), md = mandible.

[1] In taxa such as the Cephalocarida, some Branchiopoda, and the Mystacocarida (where there is no distinct metamorphosis between a nauplius phase and a juvenile phase), the exact number of naupliar stages can be debated. We have generally used the loss of the naupliar process (masticatory spine) on the antennal coxa as indicating the end of a naupliar phase, but we are aware that trunk-limb development at that stage in some cases is far beyond what is normally described as naupliar (e.g., in the Cephalocarida and Anostraca). Another criterion for distinguishing between a nauplius and a juvenile phase is the loss of the mandibular palp in taxa such as the Cephalocarida and Branchiopoda. This is problematic, however, since the palp is lost gradually and sometimes even persists long into the juvenile phase (as a rudiment). The same is seen in the Cambrian crustacean †*Rehbachiella kinnekullensis*, where development is extremely gradual, which probably is ancestral within the Crustacea. Hence no clear criteria for distinguishing between a naupliar phase and later developmental phases exist that can be used across all crustacean taxa, with their diverse developmental styles.

[2] Anostraca: many (more than 20 in some species); Notostraca: probably 7–10; Laevicaudata: 4–6; Spinicaudata: 6–7; Haplopoda: unknown, but the literature depicts a minimum of 4 different stages. See chapters 4–14.

[3] Anostraca: gradual; Notostraca: less gradual; Spinicaudata and Laevicaudata: with a more distinct shift (metamorphosis) between the naupliar and juvenile phases. See chapters 4–14.

[4] Varies by genus. The hatching stage of some species of *Argulus* has been called a metanauplius, but since additional thoracic appendages are present, we refer to them as metanauplius-like. Additionally, *Chonopeltis* larvae have no naupliar swimming appendages and may be non-motile. See chapter 24.

[5] No distinction is made between larvae and juveniles. See chapters 28 and 29.

[6] If swimming, then without the specialized swimming setae on A2. Feeding may be planktotrophic, lecithotrophic, or benthic. See chapters 28–30.

[7] The antenna and mandible may be present as specialized oral appendages in this group.

[8] Variability is great, and there are many exceptions, especially among the parasitic groups. See chapter 27.

originally published (posthumously) by O. Müller (1785) (1776, according to Holthuis 1993), for creatures he assumed were adults (but were actually the larvae of copepods), although crustacean metamorphosis was not universally accepted for nearly another half century. J. Thompson (1830), by observing barnacle nauplii, is usually credited with proving that crustacean larvae undergo a radical metamorphosis. The term nauplius was subsequently used for all such larvae, regardless of the identity of the adult. Charles Darwin famously referred to the nauplius in *On the Origin of Species* (1859), noting that "even the illustrious Cuvier did not perceive that the barnacle was, as it certainly is, a crustacean; but a glance at the larva shows this to be the case in an unmistakable manner." It is likely that the name originally referred to one of two characters in Greek mythology, both named Nauplius, one of whom was the son of Poseidon and Amymone (Holthuis 1993) and for whom the city of Nauplia (today Nafplion, a seaport in Greece) is named. Whether the original (ancestral) nauplius was pelagic and dispersal-adapted (as argued by de Beer 1958), benthic and not dispersal-adapted (H. Sanders 1963), or demersal (Mileikovsky 1971; also see Cisne 1982) has been a subject of some debate. For more historical perspectives on the study of crustacean nauplii, see Dahms (2000) and Scholtz (2008). On fossil crustacean nauplii, see Walossek (1993); also see chapter 3.

Selected References

Dahms, H-U., J. A. Fornshell, and B. J. Fornshell. 2006. Key for the identification of crustacean nauplii. Organisms, Diversity & Evolution 6: 47–56.

Gauld, D. T. 1959. Swimming and feeding in crustacean larvae: the nauplius larva. Proceedings of the Zoological Society of London 132: 31–50.

Sanders, H. L. 1963. The Cephalocarida: Functional Morphology, Larval Development, Comparative External Anatomy. Memoirs of the Connecticut Academy of Arts and Sciences No. 15. New Haven: Connecticut Academy of Arts and Sciences.

Scholtz, G. 2000. Evolution of the nauplius stage in malacostracan crustaceans. Journal of Zoological Systematics and Evolutionary Research 38: 175–187.

Walossek, D. 1993. The Upper Cambrian *Rehbachiella* and the phylogeny of Branchiopoda and Crustacea. Fossils and Strata 32: 1–202.

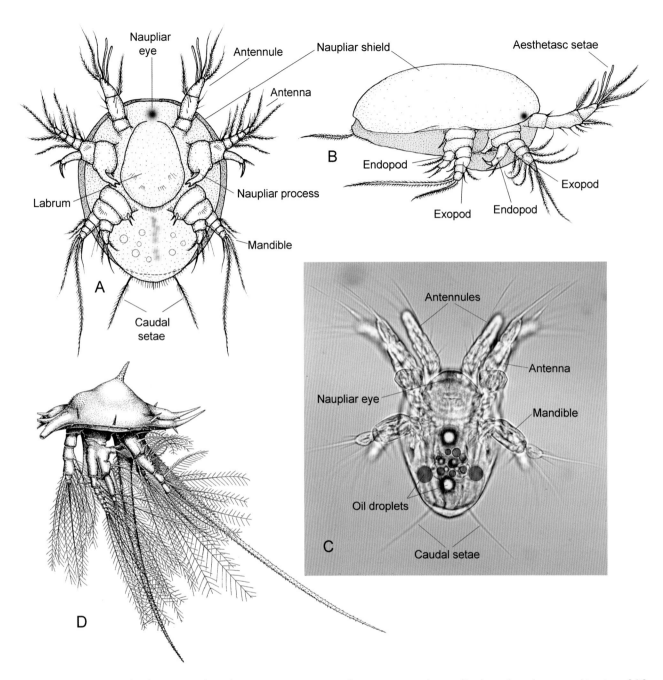

Fig. 2.1 Crustacean nauplius larvae. A and B: schematic representations of a crustacean orthonauplius larva (based on a combination of different copepod nauplii characters). A: ventral view. B: lateral view. C: copepod nauplius, light microscopy. D: drawing of a cirripede nauplius, showing the complex three-dimensionality in life, lateral view. A and B original, partly following Dahms et al. (2006); C from Micrographia, www.micrographia.com, courtesy of John Walsh; D modified after Glenner et al. 1995b.

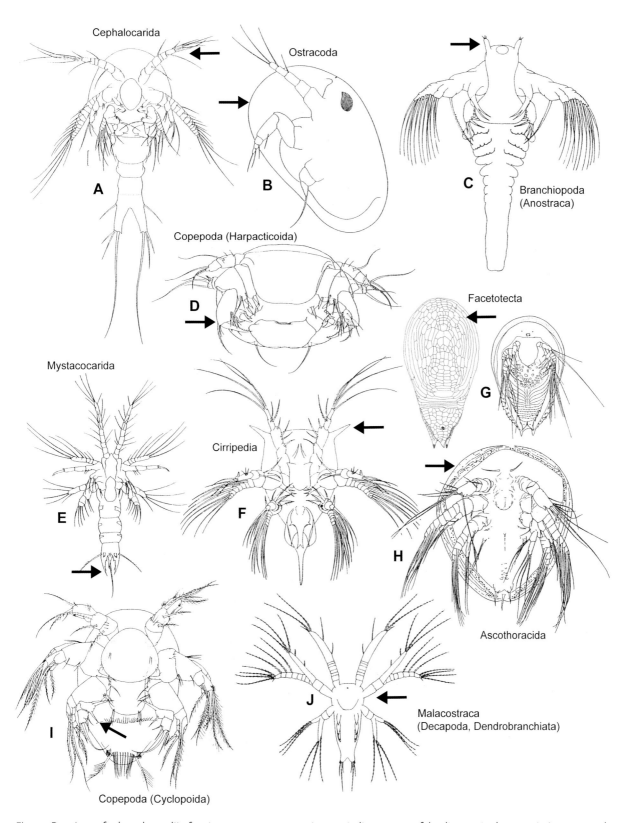

Fig. 2.2 Drawings of selected nauplii of major crustacean groups (arrows indicate some of the diagnostic characters). A: metanauplius of *Hutchinsoniella macracantha* (Cephalocarida), ventral view. B: nauplius I of *Cypris fasciata* (Ostracoda), lateral view. C: metanauplius of *Artemia* cf. *franciscana* (Branchiopoda: Anostraca), ventral view. D: nauplius II of *Paramphiascella fulvofasciata* (Copepoda: Harpacticoida), dorsal view. E: metanauplius of *Derocheilocaris remanei* (Mystacocarida), dorsal view. F: metanauplius of *Semibalanus balanoides* (Cirripedia), ventral view. G: nauplius y-larva I (Facetotecta), dorsal (*left*) and ventral (*right*) views. H: nauplius VI of *Baccalaureus falsiramus* (Ascothoracida), ventral view. I: nauplius of *Macrocyclops fuscus* (Copepoda: Cyclopoida), ventral view. J: metanauplius of *Penaeus duorarum* (Decapoda: Dendrobranchiata), ventral view. A–J modified after Dahms et al. 2006 (which see for original attribution of all images; this figure reproduced by permission of H-U. Dahms).

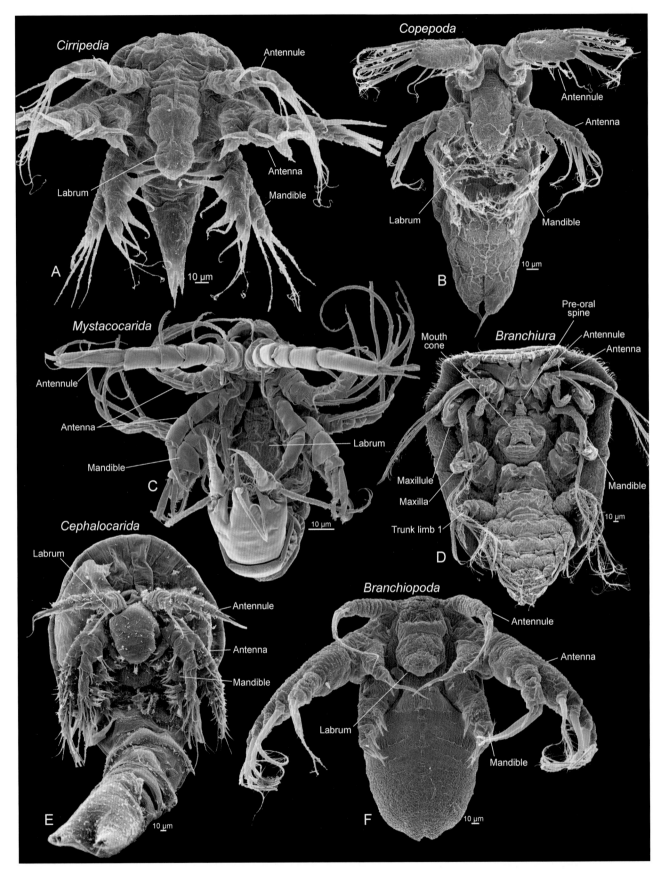

Fig. 2.3 Selected crustacean nauplii, ventral views, SEMs. A: Cirripedia. B: Copepoda. C: Mystacocarida. D: Branchiura. E: Cephalocarida. F: Branchiopoda. A–F original.

3

Carolin Haug
Joachim T. Haug
Andreas Maas
Dieter Waloszek

Fossil Larvae (Head Larvae, Nauplii, and Others) from the Cambrian in Orsten Preservation

GENERAL: The Orsten is a special type of phosphatic preservation of mainly arthropod fossils from the Cambrian (approximately 525 to 488 million years old) that provides an exceptional view on the early evolution of the Crustacea and their larval types. Many workers (e.g., Stein et al. 2008; J. Haug et al. 2009b, 2010a, 2010b) use the term Crustacea *sensu lato* (abbreviated *s. l.*) to encompass extant crustaceans and all earliest representatives of the Crustacea (the "Pan-Crustacea" of Walossek and Müller 1990), and that designation is also used here. Among these fossils are many crustacean larvae. The fossils are preserved in an uncompressed state, that is, in three dimensions and in extremely fine detail, with preservation of microstructures such as setae and setules, membranous areas (e.g., limb joints, mouth, and anus), and even faceted compound eyes (Maas et al. 2006). The full life cycle for most of these animals is unknown, but at least in some groups it included a very long series of gradually changing stages (anamorphic development).

LARVAL TYPES: Apart from a few putative adult forms (e.g., species of †*Skara*; see K. Müller and Walossek 1985; Walossek and Müller 1998a, 1998b; Liu and Dong 2007), the majority of Orsten fossils are different larval and immature stages of crustacean species for which the adult most often is unknown (a putative exception is †*Bredocaris admirabilis*; K. Müller and Walossek 1988; Walossek and Müller 1998a).

Head Larvae: Head larvae have an ocular somite, four appendage-bearing somites, and an undifferentiated trunk bud. The mouth is located at the rear of a prominent ventral sclerotization (the hypostome), another plesiomorphic feature. The anterior three appendages mainly work in conjunction for locomotion and feeding, a character new to the head larva of Crustacea *s. l.*, in contrast to the euarthropod condition. This functional change was facilitated by the evolution of an exopod on the two post-antennular limbs, both with inwardly directed setation (figs. 3.1F, J; 3.2A, C; 3.3 C). This is one of the autapomorphies of Crustacea *s. l.* (J. Haug

et al. 2009b, 2010a). This setation must have been very efficient, since it has been retained in the early larvae of all extant crustaceans. The fourth limb of the head larva (last head limb of the adult, shaped like the trunk limbs) was also functional and may have assisted in locomotion.

Another limb-associated autapomorphy of Crustacea *s. l.*, developed first in post–head larvae, is a special setiferous endite, the proximal endite, located medioproximally to the basipod (e.g., fig. 3.2A, C). The proximal endite might have facilitated better food transport toward the mouth along the ventral body surface. Subsequently, during crustacean evolution, the proximal endite became pre-displaced, appearing as a developed feature in the hatching head larvae, such as in †*Martinssonia elongata* (fig. 3.2K; J. Haug et al. 2010a). Development itself was strictly anamorphic, where new segments and appendages appeared progressively during ontogeny at the anterior end of the budding zone in front of the telson. Such head larvae are retained among Crustacea *s. l.* until the level of the Labrophora, the taxon containing the exclusively Cambrian Phosphatocopina and the crustacean crown group, the Eucrustacea (*sensu* Walossek 1999, corresponding to "Crustacea" of various authors; also see J. Haug et al. 2010a; Olesen et al. 2011).

Nauplius (Including Orthonauplius and Metanauplius): In our view, the Eucrustacea—not Crustacea *s. l.*—is characterized by a new type of hatching larva, the nauplius or orthonauplius. The orthonauplius possesses only four segments—the ocular segment and three appendage-bearing segments (Walossek and Müller 1990; Maas et al. 2003)—and a short hindbody with a dorsocaudal spine and initial furcal spines. This stage is followed by a number of metanauplius stages, where further segments develop progressively and the primordia of further limbs appear. This phase is then followed by a juvenile phase, leading to the adult. Larval and immature stages can be distinguished by their number of segments and appendages and by the progressive development of the adult morphology. The demarcation between the metanaupliar and juvenile phases is not always clear and depends on the taxon (and author). Additional differences between the plesiomorphic head larva and the eucrustacean orthonauplius are present in the morphology of specific limb- and mouth-associated structures. Many characters of the nauplius, however—such as the labrum (a posterior outgrowth from the hypostome), sternum (a fusion product of at least two post-oral sternites), paragnaths (a pair of humps located on the mandibular part of the sternum), sternitic setulation, and a coxa on the antennae and mandibles—developed earlier (i.e., in the ground pattern of the Labrophora). These features thus characterize the head larva of the Labrophora. This foreshadowed, and in a way enabled, the evolution of the fully functional feeding and locomotory orthonauplius.

All these larvae, including (1) the head larvae of Crustacea *s. l.* and the Labrophora, (2) post–head larvae of these levels, and (3) orthonauplii and metanauplii of the Eucrustacea, are found in the Orsten and are described in more detail below.

MORPHOLOGY: Because of the diversity of the Orsten larvae, it is not possible to address the detailed morphology of all taxa. Therefore, only certain taxon-specific details are mentioned.

Early Crustacea *s. l.*: A number of Orsten taxa represent derivatives of the stem-lineage toward the Labrophora. Their detailed relationships are given in J. Haug et al. (2010b) but are not particularly relevant for the description of their larvae. The order in which we describe the taxa reflects the order in which we believe the species evolved.

Of the 3 species (†*Oelandocaris oelandica*, †*Henningsmoenicaris scutula*, and †*Sandtorpia vestrogothiensis*) representing the monophyletic Oelandocarididae, true head larva stages are known only for †*H. scutula*. For this species, a phase with four stages of the head larva type is known (fig. 3.1A–C). Additionally, the species has a set of post–head larva stages, with five appendage-bearing head segments and more stages with progressively more trunk segments (fig. 3.1D–F; J. Haug et al. 2010b). From stage 2 onward, a large bowl-shaped head shield with marginal extensions is present dorsally (fig. 3.1B). One of the most striking features of post–head larva stages of †*H. scutula* is a pair of anterolateral stalked compound eyes. Eye development can be followed in the sequence. The eyes become apparent first as anterolateral elevations below the anteroventral shield margin near the hypostome in stage 3 (fig. 3.1G). In the next stages, these elevations develop into bubble-like blisters (stage 5; fig. 3.1H). Finally, the eyes become stalked in stage 8 (fig. 3.1I). In †*H. scutula*, new segments (trunk segments 2 and 3) are, at their first appearance, incorporated into a posterior tagma (the pygidium), dorsally forming a uniform shield. Later in ontogeny, these segments become liberated (jointed against the posterior body region). This mechanism of freeing the thoracic segments is also known from other Paleozoic euarthropod taxa, such as the Trilobita and the Agnostina. It is, therefore, assumed to represent the plesiomorphic condition of development for Crustacea *s. l.*, and thus must have been modified at least once in crustacean in-group taxa (J. Haug et al. 2010b).

†*Oelandocaris oelandica* and †*Sandtorpia vestrogothiensis* are represented only by later stages. †*O. oelandica* is known from one specimen with five appendage-bearing head segments and one trunk segment, and six other specimens having more trunk segments (fig. 3.1J, K; Stein et al. 2005, 2008). Similar to †*H. scutula*, †*O. oelandica* possesses a prominent head shield (fig. 3.1J). †*S. vestrogothiensis* is represented by a single specimen with five appendage-bearing head segments and a prominent bowl-shaped head shield, which extends posteriorly into a long tail spine (fig. 3.1L–N; J. Haug et al. 2010b).

A larva of uncertain systematic affinity is "larva C" (fig. 3.1O, P). It possesses an egg-shaped body, and it lacks a dorsal shield, eyes, mouth, anus, or any other ventral structures, as well as a distinct trunk bud, but it has five pairs of appendages (K. Müller and Walossek 1986a). These appendages grossly resemble those of remipede larvae (Koenemann et al. 2009; also see chapter 15).

†*Cambropachycope clarksoni* (fig. 3.2A, B) and †*Goticaris longispinosa* (fig. 3.2C–E) form the Cambropachycopidae. This taxon is characterized by a single large anterior compound eye that is dorsoposteriorly drawn out into a cone-shaped, backward-pointing tip (fig. 3.2A–E) and uniramous paddle-shaped trunk limbs (fig. 3.2A, C). †*G. longispinosa* is known from two head larva stages and several later stages, which bear trunk segments. Two blisters behind the compound eye may represent external median eyes (fig. 3.2C). A true head shield does not occur in †*G. longispinosa*. Plesiomorphically, the head retains only four appendage-bearing segments in the Cambropachycopidae. †*C. clarksoni* is known only from later stages, of which the earliest possesses a head composed of four appendage-bearing segments and one trunk segment. A head shield, lacking marginal wings, is developed (fig. 3.2A, B). Later stages retain the head tagma but possess more trunk segments (J. Haug et al. 2009b). Development of the eyes can be followed during ontogeny. The compound eye is preserved with its facets still visible (fig. 3.2E), and new facets are added during development (J. Haug et al. 2009b).

†*Musacaris gerdgeyeri* (fig. 3.2F–I) and †*Martinssonia elongata* (fig. 3.2J–M) are possibly more closely related to the Labrophora than are other Orsten taxa. A head larva stage is known from both species. While further development of †*Martinssonia elongata* is known only from stages having five and seven trunk segments and a long tail end with a bifid rear (K. Müller and Walossek 1986b; J. Haug et al. 2010a), the five known stages of †*Musacaris gerdgeyeri* are head larvae, although the largest is more than three times as large as the smallest one (fig. 3.2F–I). Feeding setation (on the endopod and basipod) and exopod setation change only slightly, compared with the condition in the earliest preserved stage (J. Haug et al. 2010a), and body segmentation is lacking throughout. A head shield with slight lateral margins is evident only in the later stages of †*Martinssonia elongata*.

Phosphatocopina: In terms of specimen numbers, the phosphatocopines (fig. 3.3A–P; Maas et al. 2003) are the most abundant Orsten crustaceans, with more than 50,000 specimens known. Various species are known, and all specimens with preserved cuticular remains are immature stages. Phosphatocopines have a large shield, which encompasses the entire body (or at least most of it), resulting in a bivalved appearance (fig. 3.3A–P). In †*Klausmuelleria salopensis*, most likely a representative of the Dabashanellidae (the suggested sister-group to all remaining phosphatocopines or Euphosphatocopina; Maas et al. 2003), the shield is univalved in the supposed earliest stage (Siveter et al. 2001, 2003b). Other representatives of the Dabashanellidae also bear a univalve shield, but their assignment to a certain stage is difficult, as soft parts are missing (Zhao 1989; Shu 1990). Many phosphatocopine specimens are head larvae consisting of only the head, with four appendage-bearing segments (Maas et al. 2003).

Phosphatocopines are important in that they demonstrate the labrophoran level in crustacean evolution. They show this in their possession of a complete cephalic feeding system, which is also present (but mostly modified) in the sister taxon Eucrustacea. This system includes the labrum arising behind the hypostome (fig. 3.3A, N), the sternum (fig. 3.3A), the pair of paragnaths (fig. 3.3L), a fine sternitic setulation extending to the flanks of the labrum, and coxal portions on the antennae and mandibles (as modifications of the proximal endite in fig. 3.3A, K; while the proximal endite is retained on posterior limbs in fig. 3.3G, H, K, O). The Phosphatocopina, however, lack the nauplius and the specialization of the fourth cephalic appendage into a feeding aid (the maxillule).

The phosphatocopine head larva was clearly a mobile and feeding larva. This is important to note, as the hatching orthonauplius of the crown-group level, the first larva of the ontogenetic sequence of the Eucrustacea, also uses this equipment for feeding and locomotion but lacks the fourth pair of appendages. With the metanauplius, the maxillule (an autapomorphy of the Eucrustacea) appears, but it is still an undeveloped bud (see the metanauplius of †*Bredocaris admirabilis*, fig. 3.4A–G, or †*Wujicaris muelleri*, fig. 3.5H). The head composition of the adult (four post-antennular segments, with their biramous limbs included) is achieved one or more stages later (a feature retained from the Labrophora). This implies that head larvae and metanauplii are not equivalent, but that the metanauplii are subsequent stages of a new series of larval stages (Maas et al. 2003), and the complete series together is likewise a new feature, characteristic of the Eucrustacea. It is therefore necessary to consider these morphologies in full detail to understand crustacean evolution and phylogeny.

Eucrustacean Larvae: Orthonauplii and subsequent larvae are also known from the Orsten fossil assemblages. The Cambrian eucrustacean larvae are, in general appearance, similar to their extant relatives. This is best seen in the metanauplius of †*Wujicaris muelleri* from the early Cambrian of China (about 525 million years ago), the oldest reported fossil eucrustacean larva (fig. 3.5H; X. Zhang et al. 2010). This stage is characterized by a large, laterally expanding head shield that is drawn out posterodorsally into a prominent spine, which resembles that of larval cirripedes (see chapter 22). Anteriorly, the hypostome bears a pair of ovoid, putatively median eyes, and medially it is drawn out into a long spine, in a position comparable

only with the pre-oral spine anterior to the mouth cone in the Branchiura (X. Zhang et al. 2010; also see chapter 24).

A sequence of five successive stages, starting with a metanauplius carrying an undeveloped limb bud (the maxillule) on its hindbody, is known from †*Bredocaris admirabilis* (fig. 3.4A–G; K. Müller and Walossek 1988; Walossek and Müller 1998a). This species has a bowl-shaped head shield throughout its development, with a posterior V-shaped indentation, while the trunk is only faintly segmented in the last stage, the putative adult. The ontogeny of †*B. admirabilis* is characterized by a progressive limb addition on the hindbody, but there is no external segmentation of the entire hindbody (no tergites), and limb development is delayed in this region (fig. 3.4A–G). This means that the post-maxillulary appendages—the maxillae and five thoracopods—appear gradually, but they keep the same initial form as when they first appeared as buds (fig. 3.4C). Only in the putative adult are all trunk limbs (now numbering seven) fully developed (K. Müller and Walossek 1988). A very similar ontogenetic pattern is found in extant thecostracans (e.g., the Cirripedia). Among the Thecostraca, however, the developmental jump in appendage development occurs during the molt to a stage called the cypris larva, or cypridoid (which molts further to the adult; Høeg et al. 2009a).

Unassigned Eucrustacean Larvae: Two additional, still unassigned, eucrustacean larval forms are present in the Orsten. Both larva A1 and larva A2 are apparently orthonauplii (fig. 3.4H–L). They possess an ovoid body (110–120 μm long), three pairs of appendages with only weakly developed setation (short, weakly sclerotized spinules), and a short trunk bud with a dorsocaudal spine and initial furcal spines. Larva A1 has a labrum, which is lacking in larva A2, and both lack a mouth and an anus (K. Müller and Walossek 1986a; Roy and Fåhræus 1989; Walossek and Müller 1989; Walossek et al. 1993). This morphology superficially resembles that of malacostracan nauplii, but malacostracan nauplii have a well-developed swimming setation and no indication of feeding equipment. These differences make affinities of these two fossil larvae to the Malacostraca unlikely. No later stages corresponding to larvae A1 and A2 have been found to date. Larva A1 is the longest-ranging crustacean larva in terms of geological and geographical range, found in deposits from the Middle Cambrian to the Lower Ordovician, and from Sweden to Australia and Newfoundland (Maas et al. 2006).

†*Rehbachiella kinnekullensis*: The most complete ontogenetic sequence of any Orsten crustacean, which is also the longest sequence known for any crustacean, is that of †*Rehbachiella kinnekullensis*, the sister species to all remaining branchiopods (fig. 3.5A–G; Walossek 1993; Olesen 2007). No fewer than 30 successive stages—the largest still immature—of that species are known. Ontogeny starts with a 160 μm orthonauplius, followed by a long series of gradually developing stages that add segments and appendages. Trunk segments develop in two steps, and limbs require several steps to attain a fully functional appearance (with more than 200 setae medially). This strictly anamorphic development of †*R. kinnekullensis* can be used as a reference scheme for all other eucrustaceans

(Walossek 1993, his fig. 43, his table 4). It provides very detailed insights into the morphogenesis of a variety of structures. For example, the comprehensive head shield of the late stages begins as a small bowl-shaped structure (similar to the shield of †*Bredocaris admirabilis*) and progressively elongates posteriorly, eventually covering at least seven trunk segments and most of the proximal portion of the thoracic limbs. Again, the development of the limbs and their substructures (endites, coxae on the mandibles) can be monitored, as can features such as the mandibular palp and the dorsal organ. Nevertheless, this species also has certain unique specializations (autapomorphies), such as the development of one segment taking two molts (Walossek 1993).

MORPHOLOGICAL DIVERSITY: Orsten-type larvae belong to a variety of taxa and exhibit substantial morphological disparity. Morphological differences, pointing to different life habits, occur even between larvae of the same taxon, such as the one-eyed †*Goticaris longispinosa* and †*Cambropachycope clarksoni*. †*G. longispinosa* is (in general) rather slender, which is indicative of a life as an active hunter of prey during swimming. †*C. clarksoni*, with its huge pair of broad, paddle-shaped trunk appendages, may have been an ambush predator (J. Haug et al. 2009b). Later larval stages of the Phosphatocopina display a relatively high morphological disparity (fig. 3.3), which appears less in earlier stages. This disparity precluded earlier researchers from assigning the earliest larval stages to a specific species (Maas et al. 2003).

NATURAL HISTORY: All Orsten crustacean taxa are marine and generally thought to be part of the meiofauna (small-sized animals living on or in the bottom in a flocculent layer). This layer was a soft bottom of segregated inorganic and organic (biogenic) material that was progressively more compacted from the water-sediment interface downward and was progressively more anoxic along a redox cline (K. Müller and Walossek 1991; Waloszek 2003a). Like their systematic affinities, the ecological niches inhabited by these Orsten forms were similarly diverse. Several of their larvae performed sweep-net feeding (e.g., those of †*Rehbachiella kinnekullensis*, †*Wujicaris muelleri*, or †*Bredocaris admirabilis*). Such larvae presumably produced water currents by moving their appendages, particularly the long setae-bearing exopods of the antennae and mandibles, and the resulting water currents not only propelled the larvae forward, but also dragged food particles toward the mouth (K. Müller and Walossek 1988; Maas et al. 2003). This life at a low Reynolds number cannot be discussed in detail here, but see, for example, Purcell (1977) and Strickler (1984). Late stages of †*R. kinnekullensis*, which developed progressively more trunk limbs, changed to post-maxillulary suspension feeding by developing special filter-feeding setae and more structures associated with this mode of feeding, similar to living anostracan and diplostracan branchiopods. In these taxa, the metachronous beat of the maxillae and trunk appendages induces a water current running along the ventromedian interlimb space; food particles dragged along passively in this current are caught by sieving setae. Retention setae sort out unsuitable particles, and pusher setae on the proximal endites in the sternitic food groove move the nutrients toward the mouth (Walossek 1993). The water current leaving the limbs produces the forward movement of the animal (described for *Artemia* sp. by Barlow and Sleigh 1980).

Certain crustacean larvae from the Orsten were undoubtedly active predators, such as the Cambropachycopidae, with their huge single compound eye (J. Haug et al. 2009b). Other early larval stages known from the Orsten did not feed, as the mouth and/or anus were not open yet. These larvae were much inflated, indicative of a lecithotrophic mode of living. Examples are larvae A1 and A2 (fig. 3.4H–L) and "larva C" (fig. 3.1P), as well as those of †*Martinssonia elongata* (fig. 3.2J–M) and †*Musacaris gerdgeyeri* (fig. 3.2F–I). Additionally, the very early stages of †*Henningsmoenicaris scutula* (fig. 3.1A, B) were lecithotrophic (K. Müller and Walossek 1986a; J. Haug et al. 2010a, 2010b), although they had a tiny mouth.

PHYLOGENETIC SIGNIFICANCE: The importance of the Orsten fauna lies in their exceptional contribution to understanding crustacean evolution. The Orsten crustaceans are a paraphyletic assemblage of taxa that branched off from different nodes within Crustacea *s. l.*; the Orsten also contains early representatives of crown-group taxa, or Eucrustacea, the group containing all extant crustaceans and some fossil forms that are clearly crustaceans (for a detailed discussion, see J. Haug et al. 2009b). We posit that the hatching stage in the ground pattern of Crustacea *s. l.* was a larva possessing an ocular and four appendage-bearing segments and an undifferentiated trunk bud. The anterior tagma of early larvae matches the euarthropod adult head condition, and this larval type has therefore been called a head larva (Walossek and Müller 1990). A head larva is also present in fossil euarthropods, such as trilobites. Thus we regard the head larva as a feature of the Euarthropoda that was retained in Crustacea *s. l.* The Orsten larvae have been identified as representatives of different evolutionary levels within the Crustacea, which makes them extremely valuable for comparative morphology and evolutionary considerations. In general, larval characters of the Orsten crustaceans have been of high importance for comparative morphological or phylogenetic interpretations, as well as for the understanding of the early evolution of the Crustacea (for the discovery of epipodites in Cambrian crustaceans and the evolution of these, see X. Zhang et al. 2007 and Maas et al. 2009; for the evolution of the crustacean feeding apparatus, see Waloszek et al. 2007; for the evolution of crustacean appendages, see J. Haug et al. 2013; and for a discussion of the phylogeny of this taxon, see particularly Walossek 1993, 1999; Maas et al. 2003; Waloszek 2003b; Haug et al. 2010b).

HISTORICAL STUDIES: In 1983, Klaus J. Müller, the discoverer of the Orsten, described several crustacean species and mentioned that larval stages of different species were present in this material (K. Müller 1983). Later, the fauna was found to include rather complete ontogenetic sequences. Subsequently, extensive monographic studies were published on different crustacean species from the Orsten, including the description of ontogenetic stages and sequences (e.g., †*Martinssonia elongata* in K. Müller and Walossek 1986b; †*Bredocaris admirabilis* in K. Müller and Walossek 1988). These studies were amended by detailed descriptions of developmental sequences of several phosphatocopines (Maas et al. 2003) and stem crustaceans (Stein et al. 2008; J. Haug et al. 2010a, 2010b). The accumulation of the many details recognizable on Orsten taxa allowed the determination of systematic affinities of certain species (e.g., of †*B. admirabilis* to the Thecostraca in Walossek and Müller 1998a, 1998b). The study of †*R. kinnekullensis* by Walossek (1993) was especially crucial for understanding the evolution and developmental mode of eucrustaceans, as this species provides a reference scheme for staging developmental sequences. More recently, several Orsten crustaceans and their ontogenies have been reexamined. New species were described, based only on larval stages (†*Yicaris dianensis* [not illustrated here; see X. Zhang et al. 2007]; †*Sandtorpia vestrogothiensis* in Haug et al. 2010b; †*Musacaris gerdgeyeri* in J. Haug

et al. 2010a; †*Wujicaris muelleri* in X. Zhang et al. 2010). Additional material awaits closer examination.

Selected References

Haug, J. T., A. Maas, and D. Waloszek. 2010b. †*Henningsmoenicaris scutula,* †*Sandtorpia vestrogothiensis* gen. et sp. nov. and heterochronic events in early crustacean evolution. Transactions of the Royal Society of Edinburgh, Earth and Environmental Science 100: 311–350.

Maas, A., D. Waloszek, and K. J. Müller. 2003. Morphology, Ontogeny and Phylogeny of the Phosphatocopina (Crustacea) from the Upper Cambrian "Orsten" of Sweden. Fossils and Strata No. 49. Oslo: Taylor & Francis.

Müller, K. J., and D. Walossek. 1986a. Arthropod larvae from the Upper Cambrian of Sweden. Transactions of the Royal Society of Edinburgh, Earth Sciences 77: 157–179.

Walossek, D. 1993. The Upper Cambrian *Rehbachiella* and the phylogeny of Branchiopoda and Crustacea. Fossils and Strata 32: 1–202.

Walossek, D., and K. J. Müller. 1998b. Cambrian "Orsten"-type arthropods and the phylogeny of Crustacea. *In* R. A. Fortey and R. H. Thomas, eds. Arthropod Relationships, 139–153. Systematics Association Special Volume No. 55. London: Chapman & Hall.

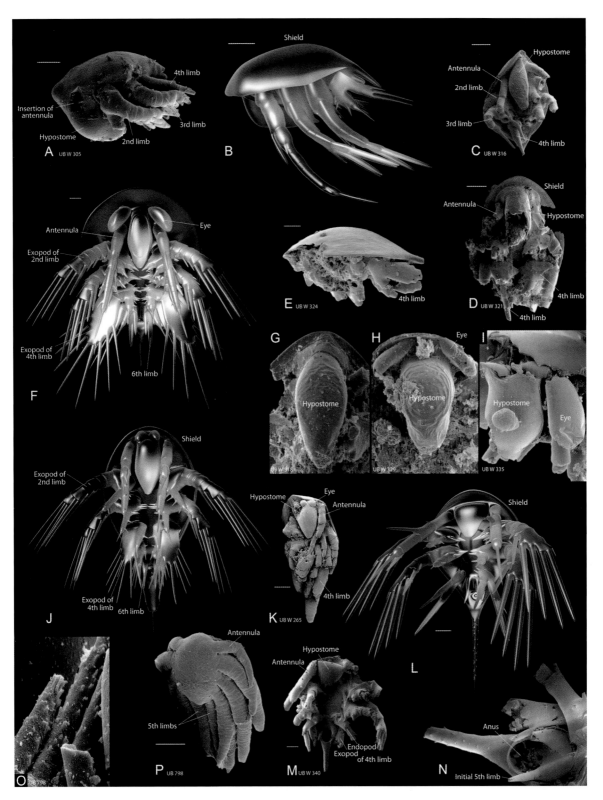

Fig. 3.1 Crustacea *s. l.*, derivatives of the lineage toward the Eucrustacea, SEMs (unless otherwise indicated). A–I: †*Henningsmoenicaris scutula*. A: stage 1, lateral view. B: 3D model of stage 2, lateral view. C: stage 3, ventral view. D: stage 5, ventral view. E: stage 5, lateral view. F: 3D model of stage 7, ventral view. G–I: details of the developing lateral eyes. G: stage 3. H: stage 5. I: stage 8. J and K: †*Oelandocaris oelandica*. J: 3D model of the earliest-known stage, ventral view. K: single specimen of the earliest-known stage, ventral view. L–N: †*Sandtorpia vestrogothiensis*. L: 3D model, lateroventral view. M: single specimen and holotype, representing an early larval stage. N: closeup of the tail spine, with its scale-like ornamentation. O and P: "larva C" (of uncertain affinity). O: closeup of the exopod setae; note the fine setules; image flipped vertically. P: dorsal view. Scale bars: 50 μm. A–P original.

Fig. 3.2 Crustacea *s. l.*, derivatives of the lineage toward the Eucrustacea (*continued*), SEMs (unless otherwise indicated). A and B: †*Cambropachycope clarksoni*, lateral views. A: 3D model of stage 2. B: stage 1. C–E: †*Goticaris longispinosa*. C: 3D model of stage 3, lateral view. D: stage 2, lateral view. E: closeup of the single anterior compound eye of the specimen in D, with the facets preserved. F–I: †*Musacaris gerdgeyeri*. F: stage 1, ventral view. G: 3D model of stage 2, lateroventral view. H: stage 3, lateroventral view. I: 3D model of stage 5 (oldest-known stage), still appearing very larval-like, lateroventral view. J–M: †*Martinssonia elongata*. J: 3D model of stage 1, lateroventral view. K: stage 1, ventral view. L: stage 2, dorsal view. M: 3D model of stage 2, lateroventral view. Scale bars: 50 µm. A–M original.

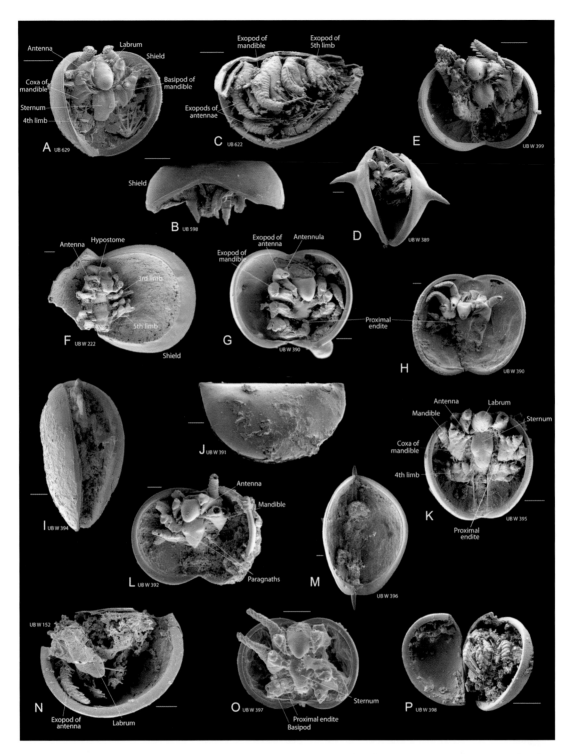

Fig. 3.3 Larvae of the Phosphatocopina, SEMs. A–D: †*Vestrogothia* sp. A: earliest-known stage, ventral view. B: earliest-known stage, anterior view. C: slightly older stage, with the left valve broken off, lateral view. D: older stage, ventral view. E: early stage of †*Falites* sp., ventral view. F: relatively early stage of †*Trapezilites minimus*, ventral view. G–P: representatives of †*Hesslandona*. G and H: †*H. suecica*. G: early stage, ventral view. H: slightly older stage, ventral view. I: early stage of †*H. toreborgensis*, with at least some ventral details preserved, lateroventral view. J: young stage of †*H. trituberculata*, lateral view. K: early stage of †*H. necopina*, ventral view. L: early stage of †*H. angustata*, ventral view. M: smallest known specimen of †*H. curvispina*, ventral view. N: stage 1 of †*H. unisulcata*, lateral view. O and P: †*Hesslandona* sp. O: early larval stage, ventral view. P: early stage of a "cracked" specimen, lateroventral view. Scale bars: 50 μm. All specimens are UB W, referring to the collection in the Steinmann Institute, University of Bonn, Germany; collection numbers for previously unpublished specimens are in parentheses. A–C, F, and N modified after Maas et al. 2003; D (389), E (399), G (390), H (391), I (394), J (393), K (395), L (392), M (396), O (397), and P (398) original.

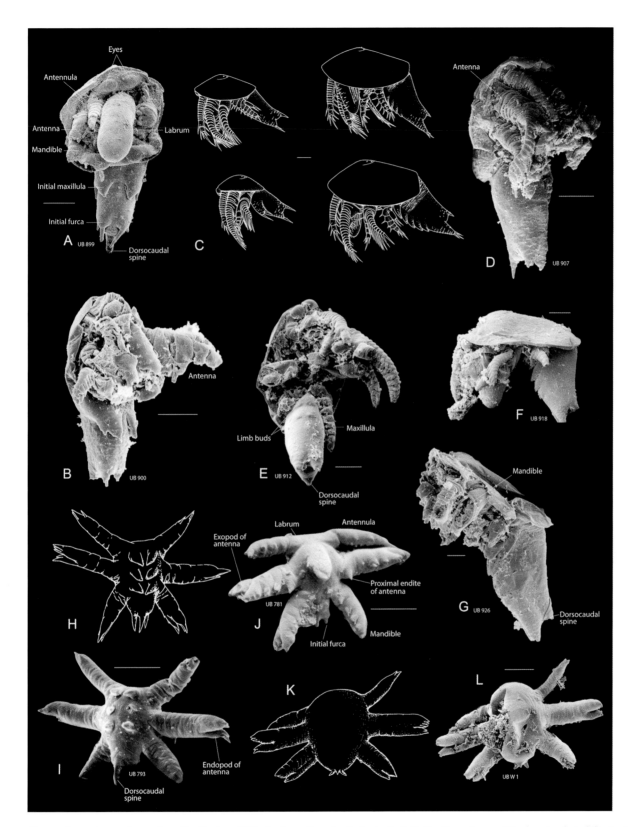

Fig. 3.4 Eucrustacean larvae from the Orsten, SEMs (unless otherwise indicated). A–G: the thecostracan †*Bredocaris admirabilis*. A: stage 1, ventral view. B: stage 1, ventrolateral view. C: drawing of the ontogenetic sequence of metanauplii I, II, IV, and V (*clockwise, starting from the lower left*), lateral view. D: stage 2, ventral view. E: stage 2, ventral view. F: stage 4, lateral view; image flipped horizontally. G: stage 5, lateral view. H–J: "larva A1" (as designated in the original publication). H: drawing, dorsal view. I: specimen, dorsal view. J: specimen, ventral view. K and L: "larva A2." K: drawing, dorsal view. L: specimen, dorsal view. Scale bars: 50 μm. A, B, D–G, I, J and L original; C modified after K. Müller and Walossek (1988); H modified after K. Müller and Walossek (1986a); K modified after Walossek and Müller (1989).

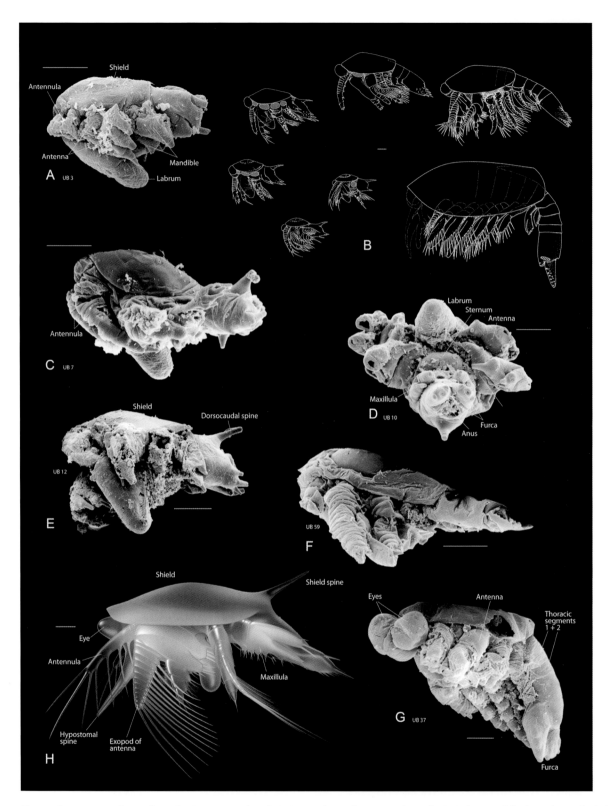

Fig. 3.5 Eucrustacean larvae from the Orsten (*continued*), SEMs (unless otherwise indicated). A–G: larvae of the branchiopod †*Rehbachiella kinnekullensis*. A: larval stage 1 of series A, orthonauplius, lateral view; image flipped horizontally. B: drawing of parts of the ontogenetic sequence, larval stages 1–4, trunk-segment stages 2, 5, and 8 (*starting clockwise from the center*), lateral view. C: larval stage 2 of series A, metanauplius, lateral view; image flipped horizontally. D: larval stage 3 of series A, posteroventral view. E: larval stage 3 of series A, lateral view. F: larval stage 4 of series B, lateral view. G: trunk-segment stage of series A, with two trunk segments, lateral view. H: 3D model of a metanauplius of †*Wujicaris muelleri*, lateral view. Scale bars: 50 μm. A and C–H original; B modified after Walossek (1996), also see Walossek (1993, his figs. 6 and 7, his table 4).

4

Joel W. Martin
Jørgen Olesen

Introduction to the Branchiopoda

The Branchiopoda is a relatively species-poor assemblage—estimated at approximately 1,000 extant species (J. W. Martin and Davis 2006; Brendonck et al. 2008)—of small, typically phyllopodous (leaf-limbed) crustaceans. Somewhat unusually among crustaceans, they are known almost exclusively from ephemeral ponds, lakes, and other bodies of fresh water on all continents; only a few species are marine (Brendonck et al. 2008; Rogers 2009). Most are small (5–30 mm long as adults), but some predatory anostracans (e.g., *Branchinecta gigas*) can reach adult lengths of 150 mm. Although traditionally divided into four main groups—the Anostraca, Notostraca, Conchostraca, and Cladocera—it is now known that this simple division masks a far greater diversity than was formerly recognized. The group today is treated as the three groups Anostraca, Notostraca, and Diplostraca. The Diplostraca contains the Laevicaudata, Spinicaudata, and Cladoceromorpha (clam shrimps and water fleas). The Spinicaudata and Cladoceromorpha have recently been suggested to constitute the Onychocaudata (Olesen and Richter 2013). The Cladoceromorpha, in turn, contains the Cyclestherida [monotypic] and the cladoceran groups Anomopoda, Ctenopoda, Haplopoda, and Onychopoda (the latter two united in the Gymnomera) (following Olesen and Richter 2013). Branchiopods are also known from the fossil record, including the Devonian †*Lepidocaris rhyniensis* from the Rhynie chert and the fascinating Cambrian †*Rehbachiella* from the Orsten fauna (Walossek 1993; also see chapter 3).

Anostracans (commonly called fairy shrimps or brine shrimps) are easily recognizable crustaceans that are archaic looking in some respects. They are graceful upside-down swimmers that lack a carapace and bear paired, stalked compound eyes; both characters are unique in the Branchiopoda. Notostracans, commonly called shield shrimps or tadpole shrimps (the latter because of their superficial resemblance to frog larvae), are large (up to 100 mm long in some cases, although most species are in the 30–50 mm range as adults)

predatory and scavenging branchiopods easily recognizable by their sizeable, horseshoe-shaped dorsal shield and long multisegmented abdomen bearing filamentous caudal rami (cercopods). "Conchostracans" (now partitioned among the Spinicaudata, Cyclestherida, and Laevicaudata) are commonly called clam shrimps, because of their superficial resemblance to bivalved mollusks; all species have a bivalved carapace. Cladocerans (sometimes called water fleas) are recognized today as encompassing many disparate morphological forms with unclear relationships. Most species have a recognizable head and trunk, a bivalved carapace that does not enclose the head, a single compound eye, four to six trunk limbs, and a limbless recurved post-abdomen with a pair of strong terminal claws.

Branchiopods have long played an important role in discussions of crustacean phylogeny, with various workers stressing their presumed primitive morphology. Monophyly of the group has rarely been questioned, but relationships within the Branchiopoda are still not entirely understood and are the subject of much study (e.g., Fryer 1987a; J. W. Martin 1992; J. W. Martin and Davis 2001; Walossek 1993; Olesen et al. 1996; Olesen 2000, 2004, 2007, 2009; Olesen and Grygier 2004; Stenderup et al. 2006; Richter et al. 2007; Regier et al. 2010). Enough is known, however, to state that, whereas the Cladocera appears to be a natural group, the grouping of large branchiopods is not.

Larval developmental modes range from nearly completely anamorphic development (e.g., in some anostracans) to extreme metamorphosis (e.g., in the Laevicaudata). Nauplii (where known) are distinct from those of all other crustaceans in having short unsegmented antennules, large antennae in which the length of the protopod is more than half the length of the entire limb, and uniramous mandibles (Olesen 2007). All groups are treated in the ten chapters that follow, with the exception of the Cambrian †*Rehbachiella*, which is mentioned in chapter 3.

Among the chapters on extant branchiopods, we have

included a chapter on some fascinating Devonian fossils from the Rhynie and Windyfield cherts (see chapter 6). Our reason for doing so is that three of these Devonian fossils—†*Lepidocaris rhyniensis*, †*Castracollis wilsonae*, and †*Ebullitiocaris oviformis*—have been identified as branchiopods (Scourfield 1926; Fayers and Trewin 2003; L. Anderson et al. 2004). Additionally, the unnamed fossils of orthonauplius larvae from the Windyfield chert were originally considered to have branchiopod affinities (C. Haug et al. 2012). Although †*Lepidocaris* is still considered a branchiopod, the branchiopod affinities of the Windyfield nauplii are now under discussion (see chapter 6).

5

Anostraca

Jørgen Olesen

GENERAL: Anostracan branchiopods are commonly called fairy shrimps or brine shrimps. They are a group of mostly colorful freshwater crustaceans, characterized by their peculiar habit of swimming upside down while using their phyllopodous thoracopods to filter organic particles from the water (although a few species scrape algae from surfaces). Anostracans are among the most diverse groups of branchiopods, with about 300 species divided among 8 families (Brendonck et al. 2008). Different from other extant branchiopods, they lack a dorsal shield (carapace). In the majority of species the body is composed of 19 body segments, the anterior 11 of which each carry a pair of thoracopods, followed by 2 genital segments, 6 limbless abdominal segments, and a telson. Some atypical species may bear as many as 17 or 19 thoracopods (in the subarctic genera *Polyartemiella* and *Polyartemia*). Anostracans are medium-sized crustaceans, usually 1–3 cm long, but a few raptorial species (e.g., *Branchinecta raptor* or *B. gigas*) can grow significantly larger, the latter up to 10 cm long (e.g., Fryer 1966; Rogers et al. 2006). Anostracans occur worldwide, and are even found in Antarctica, but their distribution is scattered, since all of the species are adapted to special types of water bodies, such as snowmelt ponds or desert pools after rain. The life cycle consists of dioecious (male and female) adults, but some strains of *Artemia* are parthenogenetic. Eggs are carried in a midventral brood pouch, located behind the last pair of thoracopods (fig. 5.1A). Two types of eggs are laid: (1) thin-shelled summer eggs that continue developing and hatch quickly, or (2) thick-shelled winter eggs (resting eggs) in which development is arrested at about the gastrula stage. Larvae hatch as orthonauplii and—at least in some species—their development is long and gradual (anamorphic).

LARVAL TYPES

Nauplius (Including Orthonauplius and Metanauplius) and Post-Larval Stages: Some anostracans have a very gradual (anamorphic) development, with only minor differences between the stages (Benesch 1969; Fryer 1983; Schrehardt 1987).

As many as 25 pre-adult stages have been reported for *Artemia salina* (Benesch 1969; also see Walossek 1993), and Fryer (1983) identified 20 instars of *Branchinecta ferox* before the larvae acquired a full complement of functional trunk limbs. Weisz (1947) and Hentschel (1967) also reported relatively many stages. Other authors have reported fewer stages for various species (e.g., Claus 1873, 1886; Oehmichen 1921; Heath 1924; Cannon 1926; Hsü 1933; Pai 1958; D. Anderson 1967; Baqai 1983; Jurasz et al. 1983), but many of these descriptions may be incomplete (see Fryer 1983). There have been attempts to name various phases of the development of *Artemia*, but since its development does not naturally fall in distinct phases—in contrast to many other crustaceans—many of these attempts fall short or appear arbitrary. Schrehardt (1987), based on Kaestner (1967), used a quite detailed naming scheme for various phases (orthonaupliar, metanaupliar, post-metanaupliar, post-larval) which has been adopted here for convenience, although any attempt to divide the development of anostracans into discrete phases conflicts with its gradual nature. The hatching stage in *Artemia salina*, and in other taxa where this stage is known (e.g., fig. 5.1B, D), is a lecithotrophic (containing yolk and carrying non-functional mouthparts) larva with no clear external anlagen of the post-mandibular limbs; this stage can therefore be termed an orthonauplius (Schrehardt 1987). The orthonaupliar phase (which lasts for only one stage in *A. salina*) ends with the formation of post-mandibular limb buds. The metanaupliar phase (which lasts for four stages in *A. salina*) ends when the first pair of trunk limbs has become functional. The post-metanaupliar phase (which lasts for seven stages in *A. salina*) ends when complete body segmentation has been achieved and the natatory function of the antennae has been lost. The post-larval phase (which lasts for five stages in *A. salina*) is characterized by the development of the antennae and genital structures.

What complicates attempts to set up an unequivocal division of anostracan ontogeny into distinct and well-defined phases is not only that development is gradual, but also that the naupliar (cephalic) feeding system in many intermediate developmental stages operates concomitantly with the adult (thoracic) feeding system (e.g., larvae in fig. 5.3). Many larval aspects, such as the natatory antennae with naupliar processes and the mandibular palps, are retained well into anostracan development, alongside a fully functional adult thoracopodal system (fig. 5.1D–K). Hence there is no clear distinction in time between larval and adult phases. Other attempts to summarize anostracan development include that of Walossek (1993), who used the even more gradually developing Cambrian †*Rehbachiella kinnekullensis* as a reference ontogeny when referring to larval stages of anostracans and other crustaceans.

MORPHOLOGY: The morphological description presented here is based mainly on three different stages of *Eubranchipus grubii*, each of which represent three different phases of anostracan development as categorized by Schrehardt (1987), based on his work with *Artemia salina* (orthonaupliar, metanaupliar, and post-metanaupliar). The same material of *Eubranchipus grubii* was used by Møller et al. (2004). These three stages are very similar to the first three stages described for the same species by Oehmichen (1921), so it cannot be ruled out that they represent nearly the full early development of *Eubranchipus grubii*; if so, its development is significantly more abbreviated than that of species such as *Artemia salina* and *Branchinecta ferox*.

Hatching Stage / Orthonauplius of *Eubranchipus grubii*: The orthonauplius in fig. 5.1B was the earliest larva found by Møller et al. (2004). It is similar to the earliest larva described by Oehmichen (1921), which he cultured from sampled mud, and which he considered to be the hatching stage. It is a rather elongate lecithotrophic-appearing larva, which is still much expanded by yolk (both body and limbs). The only appendages present at this stage are the antennules, antennae, and mandibles (the naupliar appendages), but the feeding structures of the two latter appendages are not yet functional. The morphology of the naupliar appendages remains relatively unchanged during the entire developmental phase (but see below), and this morphology is virtually identical to that of other anostracan species. The antennules are slender tubular appendages with distal setation only. The antennae are very large; the large protopod of each antenna is divided into a rather short proximal coxal part, carrying a not-yet-functional naupliar process later used for food manipulation, and a long basipodal part, carrying a characteristic long seta that is also later involved in food manipulation. The exopod is long and carries a dense row of natatory setae along the posterior margin. The endopod is short and weakly segmented, and it carries distal setation only. The mandible is weakly subdivided into four segments: a proximal coxa with a not-yet fully developed gnathal edge medially and a large posteriorly directed seta; a rather large basis carrying two median setae; and two more distal segments that Møller et al. (2004) interpreted as endopodal, but which most likely are exopodal (see homologies of very similar larval mandible in *Lynceus biformis*, a laevicaudatan branchiopod; Fritsch et al. 2013). In the

earliest-known stage of *Eubranchipus grubii*, the surface of the bean-shaped hindbody is subdivided into about 12 segments, of which the three anteriormost are the most developed. The hindbody terminates in a pair of undeveloped setae.

The hatching stage (orthonauplius) in other anostracans is quite similar to that of *E. grubii*, but some differences exist with respect to the size and shape of the hindbody. In *Artemia salina* the hindbody is elongate and cone-shaped, with no signs of limb buds externally (Schrehardt 1987), whereas in *Branchinecta ferox* (Fryer 1983) and *Branchinecta occidentalis* (fig. 5.1D) it is more ellipsoid-shaped.

Metanauplius of *Eubranchipus grubii*: The metanauplius in fig. 5.2 (also described by Møller et al. 2004) is similar to Oehmichen's (1921) stage 2 of the same species; he called it the "stage after the first molt." When compared with the gradually developing *Artemia salina*, however, the degree of development in this metanauplius of *E. grubii* corresponds approximately to metanauplius 5 of *A. salina*. It is uncertain whether the described stage is the true stage 2 of *E. grubii*. The early buds of the compound eyes are triangular structures, which have not yet developed into the stalked structures seen in adults (fig. 5.2B). Frontally, there are external signs of a pair of frontal organs (fig. 5.2B). Dorsally, there is a large dorsal organ on the cephalon (fig. 5.2F, H). The antennules are still slender tubular appendages, with distal setation and no clear segmentation. The antennae are now fully developed and involved in both swimming and feeding (fig. 5.2A, F). The coxal portion of the antenna is distinctly set off from the basis and is drawn out into a long setule-bearing naupliar process (masticatory spine), and both this process and a characteristic long feeding seta located more distally at the basis are directed toward the mouth region and assist in food handling (fig. 5.2A, E). The mandibles have the same basic morphology as in the hatching stage (coxa, basis, and two exopodal[?] segments), but the articulations between the segments have become clearer, the setation is more developed, and the gnathal edge of the coxa has become functional (fig. 5.2A). The labrum is a large, swollen, U-shaped structure covering the mandibular gnathal edges; it carries long setules along the distal edge (fig. 5.2A, E). The maxillules are a pair of elongate, laterally placed limb buds, while the maxillae are a pair of rounded buds placed closer to the midline of the larva (fig. 5.2A, D). Two rows of thoracopodal limb buds are present at varying degrees of development (fig. 5.2A, C, I–K). The anterior buds are the most developed and have all the future limb parts present (exopod, epipod, etc.); the posterior ones are merely undifferentiated lobes. The limb buds occupy the entire ventrolateral corner of the trunk, resulting in a laterally directed proximal-distal limb axis (characteristic for branchiopod larvae; Olesen 2007). The most developed limb buds are divided into a row of six primordial ventral endites, terminating distally (laterally) in a portion that is interpreted as the endopod, a more distinct lateral portion that is the exopod, and three dorsolateral portions that are the three epipods (fig. 5.2C, I–K) (for alternative terminology, see Ferrari and Grygier 2003; Pabst and Scholtz 2009). The hindbody terminates in a pair of setule-bearing caudal setae (fig. 5.2G). Dorsally, the trunk bears rows of paired setae, one pair for each body segment (fig. 5.2F, J).

Post-Metanauplius of *Eubranchipus grubii*: The post-metanauplius in fig. 5.3 (see Møller et al. 2004 for more details) is a bit earlier in development than Oehmichen's (1921) stage 3 of the same species, so clearly Oehmichen (1921) missed at least this stage. When compared with the gradually developing *Artemia salina*, the degree of development in this post-metanauplius of *E. grubii* corresponds approximately to Schrehardt's (1987) post-metanauplius 4 or 5 of *A. salina*. The buds of the compound eyes have become more rounded, but the eyes are still not stalked (fig. 5.3A, H, I, M). Dorsally, there is a large dorsal organ on the cephalon (fig. 5.3H, M). The antennules are still slender tubular appendages, with distal setation and no clear segmentation (fig. 5.3A, I, M). The morphology of the antennae is much like that of the metanauplius described above, with one important exception: the tip of the coxal naupliar process has branched into two setule-bearing spines (fig. 5.3A, E, F). The morphology of the mandibles and the labrum is the same as in the previously described stage (fig. 5.3A). The maxillules and maxillae (not shown here, but see Møller et al. 2004, their fig. 7B) are very similar to those in the previously described stage. Two rows of thoracopods are present at varying degrees of development (fig. 5.3A, H, K, J–M). The anterior four to six pairs of thoracopods are quite well developed, with some setation. These anterior thoracopods have begun to attain a more adult-like orientation by shifting the proximodistal axes of the limbs from pointing laterally to pointing ventrally; they have become functional and assist in swimming and feeding, while the remaining limbs are undeveloped buds. During this phase of development at least six median endites are present, and perhaps even a seventh endite makes a transitory appearance and disappears later in development. Later (not shown), endites 1 and 2 fuse to become one larger endite in the adults (Møller et al. 2004), a pattern also seen in other anostracans (e.g., Fryer 1983; T. Williams 2007). Posterior to the limb buds of the eleventh pair of thoracopods, the anlagen to at least eight additional body segments can be seen externally (fig. 5.3A, B). The caudal setae have become longer than in the previously described stage (fig. 5.3A–C). The trunk segments still bear a pair of setae dorsally (fig. 5.3J, M). A post-metanaupliar stage of another species (*Chirocephalus diaphanous*) is illustrated in fig. 5.1C.

Development from Post-Metanauplius to Adult: Several changes occur from the post-metanaupliar stage described above to the full adult morphology. These changes have not been examined in detail for *Eubranchipus grubii*, but they can be compiled from other species. The cephalic naupliar feeding system (antennae and mandibular palps) operates for a very long time hand in hand with the thoracopodal adult feeding system. For example, in *Artemia salina* the larval antennae (including the naupliar process) and the mandibles are present and active in swimming and feeding as late as post-metanauplii 5 and 6, but after that they either atrophy (mandibular palp) or transform significantly and attain another function. In males, the naupliar swimming antennae become modified

into large clasping organs, which are used to hold the females during mating. Both Schrehardt (1987) and R. Cohen et al. (1998), who studied all larval stages of 2 different species of *Artemia*, were able to trace many details of how the transformation takes place during the post-larval phase. The essence is that the powerful clasper is (for the most part) formed from the original larval antennal exopod, in which the segments have fused and a secondary joint has appeared, dividing the clasper into a basal and a distal part. Both the endopod and the protopod (in part) degenerate. The frontal knob of the clasper—which is a structure playing a role in pre-copulatory and copulatory activities (Wolfe 1980)—is formed as a new structure at the median side of the former exopod.

MORPHOLOGICAL DIVERSITY: Based on what is known, larval morphology within the Anostraca is relatively uniform. Most differences concern the general shape of the larva, such as the hindbody, and the relative size of the appendages, such as the antennae.

NATURAL HISTORY: Most species live in temporary water bodies of various kinds, so their development is quite accelerated in time, particularly in desert species where water dries out quickly. Due to the works of Barlow and Sleigh (1980), Fryer (1983), and T. Williams (1994a), who studied 2 different anostracan species (*Branchinecta ferox* and *Artemia* sp.), the mechanics of locomotion and feeding in both larval and adult anostracans are known in much detail. This makes the Anostraca better understood in this respect than any of the other branchiopods, and even better than most other crustaceans. Since the basic composition of the anostracan naupliar feeding apparatus is quite similar to that of other large branchiopods (the Notostraca, Laevicaudata, and Spinicaudata), its mechanics will be treated in some detail here, as many of the principles can transferred to other larval branchiopods.

Both swimming and feeding in the early nauplii (orthonauplius and metanauplii) start out being accomplished by the naupliar appendages. Then very gradually, in small steps from stage to stage, these roles are transferred to the thoracopods, of which more and more are formed during ontogeny. When adulthood is reached, the original locomotory and food-collecting functions of the larval appendages are entirely lost and are now handled by the thoracopods. A swimming nauplius generally lies with its ventral surface uppermost (Fryer 1983). In early nauplii, all three pairs of naupliar appendages (the antennules, antennae, and mandibles) show coordinated action during swimming, but only the antennae contribute significantly to locomotion (Barlow and Sleigh 1980; Fryer 1983). A typical beat cycle of the antennae includes an effective stroke, where they move through an arc of about 180° in which the limbs are relatively stretched and the setae spread out like a fan, followed by a recovery stroke where the limbs are bent and the setal fans are folded to reduce water resistance. Hence the mechanics of anostracan naupliar swimming are relatively simple, but numerous subtle refinements of the basic swimming pattern, such as limb rotation and variation in the amplitude that each antenna travels, allow for changes in both direction and speed. Because of the small size of these (and other) crustacean larvae, to them water is a viscous medium (a low Reynolds number medium), which means that the larvae do not really swim in the water, but rather lever themselves through it. Fryer (1983) reported that the backward swing (power stroke) of the antennae propels the animal forward only during the period of the swing, and that forward movement virtually stops at the end of the swing. During the recovery stroke, despite the fact that the limbs are orientated to reduce water resistance (drag), the nauplius is actually pushed slightly backward (but less so in later and more streamlined larvae). Feeding, which happens as the larvae swim, is also rather well understood, although Barlow and Sleigh (1980) and Fryer (1983) are in disagreement on some central points. Barlow and Sleigh (1980; also see Gauld 1959) reported that the natatory setae of the antennal endopod and exopod are responsible for the collection of food particles, while Fryer (1983) thought that their function is solely concerned with locomotion. Fryer (1983) believed that the large basipodal feeding setae (figs. 5.2A; 5.3A, E–G) (his "distal masticatory spines"), which are equipped with pairs of long hooked setules (fig. 5.3G), are conveying particles to the mouth. The mandibular palps, which are about a half cycle out of phase with the antennae, play an important role in cleaning food particles off the basipodal feeding setae (and perhaps, in part, the antennal natatory setae), carrying them toward the labral gland secretions and within the range of the very mobile coxal naupliar process of the antennae. These processes convey the food particles farther, within the vicinity of the mandibular gnathobases. During larval development the responsibility for both locomotion and feeding is gradually transferred from the naupliar appendages to the thoracopods. In many intermediate larval stages the larval and adult systems operate hand in hand. In the later larval stages, while the naupliar appendages are still present and active, their contribution to swimming and feeding is limited (except for the mandibular gnathobase).

PHYLOGENETIC SIGNIFICANCE: It is uncertain when the anostracan crown group first appeared in the fossil record, although it might have been as early as the Silurian (see Belk and Schram 2001). A well-preserved extinct anostracan from the Miocene—†*Branchinecta barstowensis*—was reported by Belk and Schram (2001). Chapter 3 discussed two older and remarkably well-preserved extinct taxa with similarities to the Anostraca. The Cambrian †*Rehbachiella kinnekullensis* was originally treated as an early anostracan by Walossek (1993), but it is better placed outside the Branchiopoda, probably as a stem-lineage branchiopod (Olesen 2007, 2009; also see chapter 3). The Devonian †*Lepidocaris rhyniensis* was also originally treated as an early anostracan by Scourfield (1926), a placement for which there are still strong arguments (Olesen 2007, 2009; also see chapter 6). The phylogenetic position of the Anostraca is now rather well established as a sister-group to

the remaining branchiopods, based on both morphological and molecular data (Stenderup et al. 2006; Richter et al. 2007; Olesen 2009; Regier et al. 2010).

Aspects of larval development and larval morphology in the Anostraca have played an important role in discussions of the phylogeny and evolution of the Branchiopoda and the Crustacea. First of all, the very gradual (anamorphic) development of at least *Artemia* spp. and *Branchinecta ferox* (e.g., Fryer 1983; Schrehardt 1987) has been discussed as a primitive crustacean attribute, based on the assumption that a gradual larval development, starting with an orthonauplius, is primitive. Among Recent crustaceans, only the Cephalocarida comes close to the Anostraca with respect to the number of stages (H. Sanders 1963; Addis et al. 2007; also see chapter 16). The Cambrian stem-lineage branchiopod †*Rehbachiella kinnekullensis* has an even more gradual larval development (Walossek 1993; also see chapter 3). It remains to be seen whether this gradual type of development applies to all anostracans. For example, in two studies of *Eubranchipus grubii*, not many larvae were found, indicating that development in this species may not be particularly gradual (Oehmichen 1921; Møller et al. 2004). The general morphology of early anostracan larvae is in some ways different from that of other crustaceans, but in many respects it is shared with that of other branchiopods and therefore recognized as an evolutionary novelty (synapomorphy) for the Branchiopoda (Olesen 2007, 2009). The larval similarities to most other large branchiopods include (1) a pair of very large swimming antennae; (2) antennal coxal naupliar processes of the same morphology; (3) a uniramous, three-segmented mandibular palp with almost identical setation; and (4) rows of primordial endites of thoracopod limb buds facing ventrally.

HISTORICAL STUDIES: Anostracan larval development received attention quite early (e.g., Shaw 1791), but the earliest works containing high-quality illustrations were those of Claus (1873, 1886) and Sars (1896a, 1896–1899).

Selected References

Benesch, R. 1969. Zur Ontogenie und Morphologie von *Artemia salina* L. Zoologische Jahrbücher, Abteilung für Anatomie und Ontogenie der Tiere 86: 307–458.

Fryer, G. 1983. Functional ontogenetic changes in *Branchinecta ferox* (Milne-Edwards) (Crustacea, Anostraca). Philosophical Transactions of the Royal Society of London, B 303: 229–343.

Møller, O. S., J. Olesen, and J. T. Høeg. 2004. Study on the larval development of *Eubranchipus grubii* (Crustacea, Branchiopoda, Anostraca), with notes on the basal phylogeny of the Branchiopoda. Zoomorphology 123: 107–123.

Schrehardt, A. 1987. A scanning electron-microscope study of the post-embryonic development of *Artemia*. In P. Sorgeloos, D. A. Bengtson, W. Declair, and E. Jasper, eds. *Artemia* Research and Its Applications. Vol. 1, Morphology, Genetics, Strain Characterization, Toxicology, 5–32. Wetteren, Belgium: Universa Press.

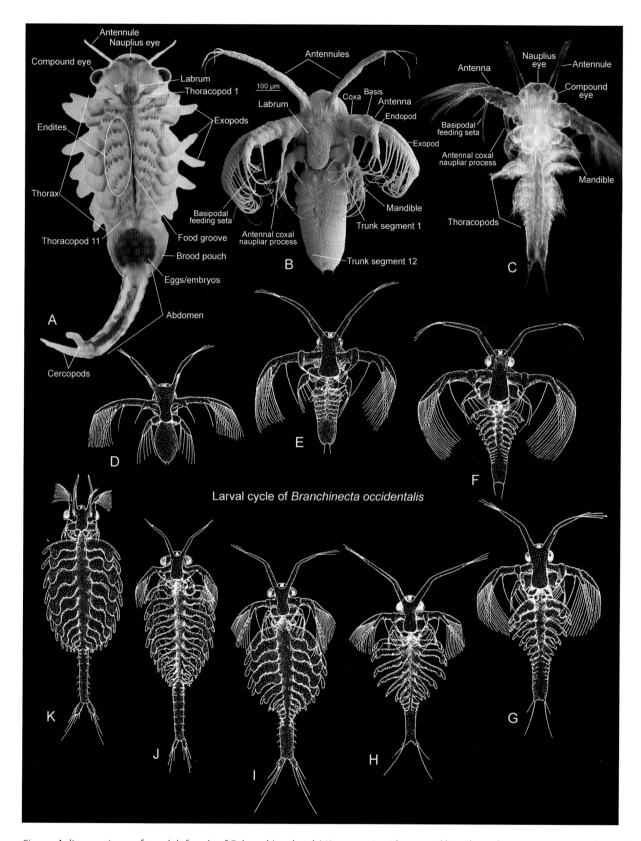

Fig. 5.1 A: live specimen of an adult female of *Eubranchipus bundyi* (Anostraca), with a ventral brood pouch containing eggs/embryos, light microscopy, dorsal view. B: orthonauplius of *E. grubii* (Anostraca), probably the hatching stage, SEM, ventral view C: post-metanaupliar stage of *Chirocephalus diaphanous* (Anostraca), SEM, ventral view. D–K: drawings of the larval cycle of *Branchinecta occidentalis*, estimated to be the eight first stages, dorsal view. A of material from California, USA, collected in April 2010, courtesy of J. Olesen and D. C. Rogers; B modified after Møller et al. (2004); C courtesy of Jean-François Cart; D–K modified after Heath 1924.

Fig. 5.2 Metanauplius of *Eubranchipus grubii* (Anostraca), similar to Oehmichen's (1921) stage 2 (see text), SEMs. A: ventral view. B: closeup showing the eye lobes (compound eye) and frontal organ, frontal view. C: thoracopodal limb buds of the right side, ventral view. D: closeup of the buds of the maxillules and maxillae. E: labrum and antennal coxal naupliar process. F: dorsal view. G: caudal setae. H: dorsal organ. I: thoracopodal limb buds of the right side, lateral view. J: lateral view. K: anterior thoracopodal limb buds, lateral view. A–K modified after Møller et al. (2004).

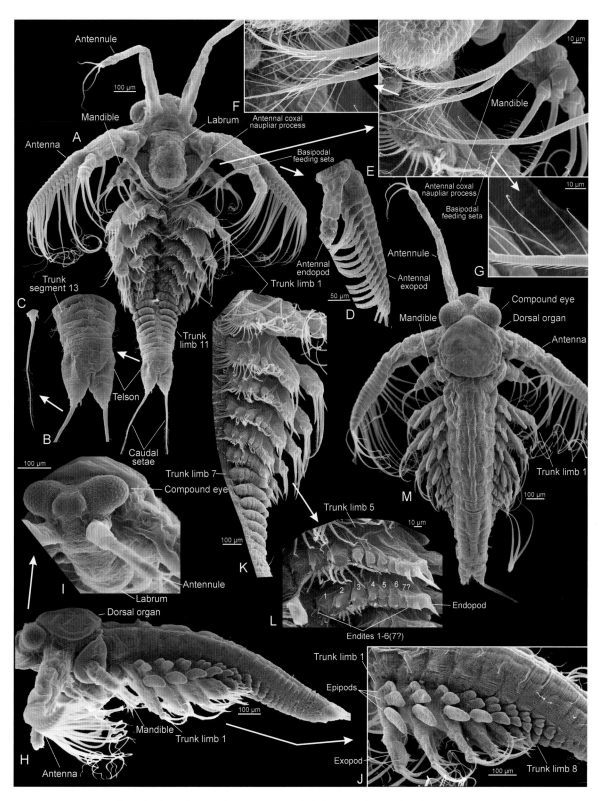

Fig. 5.3 Post-metanauplius of *Eubranchipus grubii* (Anostraca), slightly earlier than Oehmichen's (1921) stage 3 (see text), SEMs. A: ventral view. B: telson and posterior rudimentary trunk segments. C: closeup of the caudal seta. D: endopod and exopod of the antenna. E: naupliar process and basipodal feeding seta of the antenna and mandibular palp. F: closeup of the sweeping setules of the antennal naupliar process. G: closeup of the paired hooked setules of the antennal basipodal feeding seta. H: lateral view, with the antennules and caudal setae omitted. I: closeup showing the compound eyes, labrum, and left-side antennule, frontal view. J: trunk limbs, lateral view. K: row of left-side thoracopods, ventral view. L: thoracopods 5 and 6, ventral view. M: dorsal view, with the right-side antennule omitted. A–M modified after Møller et al. (2004).

6

Carolin Haug
Joachim T. Haug
Jørgen Olesen

Uniquely Preserved Fossil Larvae, Some with Branchiopod Affinities, from the Devonian

The Rhynie and Windyfield Cherts

GENERAL: The Rhynie chert, together with the relatively close Windyfield chert, is a fossil deposit of Devonian age (ca. 400 million years ago) that provides a unique view into the early evolution of a proposed terrestrial and freshwater biota (but see Channing and Edwards 2009). Although the finds are dominated by exceptionally well-preserved plants, there are also faunal components, mainly arthropods such as arachnids, collembolans, myriapods, crustaceans, and even a pterygote insect (the earliest one known). Three-dimensionally preserved crustacean larvae of the branchiopod †*Lepidocaris rhyniensis* have long been known from the Rhynie chert fauna (Scourfield 1926, 1940), and another type of crustacean larva (an orthonauplius) of yet-uncertain (possibly non-eubranchiopod) affinity has been discovered recently in the Windyfield chert (Fayers and Trewin 2004; C. Haug et al. 2012). The preservation of these larvae is exquisite and even allows for the examination of limb formation and setal morphology, in detail challenged only by the older (Cambrian) crustacean larvae from the Orsten fauna (see chapter 3). Based in part on published data (Scourfield 1926, 1940) and on new examinations of Scourfield's original material, we discuss the larvae of †*Lepidocaris rhyniensis* and present some highlights of the yet-unnamed orthonauplii from the Windyfield chert.

LARVAL TYPES

Branchiopod-Like Metanauplii: Two types of crustacean larvae occur in the Rhynie and Windyfield cherts. The first are larval stages of †*Lepidocaris rhyniensis* (fig. 6.1A–I). This species has been interpreted as the possible sister species to the Anostraca (Scourfield 1926, 1940; Walossek 1993; Olesen 2004, 2007, 2009) or to the Eubranchiopoda (Branchiopoda *s. str.*) (F. Schram and Koenemann 2001; Olesen 2004). Cephalocarid affinities also have been proposed (e.g., H. Sanders 1957). The smallest more or less entire larva of †*L. rhyniensis* has 9–10 trunk segments. Proposed earlier stages reported by Scourfield (1926) are too poorly preserved to judge whether these are entire animals or just assemblages of fragments.

Unnamed Larval Type: The other type of crustacean larva, which is yet unnamed, is known exclusively from the Windyfield chert. It is an orthonauplius (fig. 6.1J) not yet known from older developmental stages. This larva was provisionally interpreted as a eubranchiopod (Fayers and Trewin 2004; C. Haug et al. 2009), but possible branchiopod affinities are still under discussion (see below and C. Haug et al. 2012).

MORPHOLOGY: Detailed investigations of the morphology of both larval types, based on the original material, are in progress. Yet certain details are known from earlier investigations and can be given here. The following brief description of the early larval stages of †*Lepidocaris rhyniensis* is based mainly on the same two specimens examined by Scourfield (1940) (fig. 6.1A–F), since these are the most complete ones, but Scourfield (1926) treated many aspects of the development of †*L. rhyniensis* (fig. 6.1G, H). In general, fragments of the trunk are quite common among the material, while complete larvae have been found only rarely.

Larvae of †*Lepidocaris rhyniensis* are characterized by an enormous labrum with a broad, almost straight posterior margin. The antennules consist of a single segment with only distal setation (fig. 6.1A, B, I) and appear rather similar to eubranchiopod larvae, such as notostracans and some anostracans (see chapters 5 and 7). The antennae were originally described as consisting of a two-segmented protopod carrying a large naupliar process basally, a three-segmented endopod, and a five-segmented exopod (fig. 6.1A–D, I). The general morphology of the antennae is quite similar to that of larvae of extant branchiopods (see chapters 5, 7, 8, and 9; also see Olesen 2004). For example, the tip of the naupliar process is divided in two branches in a way comparable with that of anostracan larvae (figs. 5.3E; figs. 6.1A, D). In adults the antennal naupliar processes are lost, but the antennae are still retained as swimming appendages (see Scourfield 1926) and are not reduced or modified, as they are in anostracans and notostracans. The mandibles in the larvae consist of a large coxa with a gnathobase medially. Scourfield (1940) reported an unbranched three-segmented palp distally (fig. 6.1C), but this palp is not clearly observable in the specimens examined by us. The palp is lost in adults, so that only the coxa, which bears the gnathal edge, is left; such a loss is also seen in extant branchiopods (and in the Cambrian †*Rehbachiella kinnekullensis*). The preserved larval stages of †*L. rhyniensis* also provide quite extraordinary information on the development of the trunk limbs. While adult limb morphology of †*L. rhyniensis* is, in some respects, similar to that of extant branchiopods (phyllopodous limbs with six endites, an unsegmented endopod, and an exopod) (Olesen 2007), early development is quite different. In †*L. rhyniensis* the future distal tips of the endopod and the exopod of the trunk limbs point ventroposteriorly, and their lateral sides are free from the body (fig. 6.1H). This is in contrast to the larvae of all extant branchiopods, where the future distal tips of the endopod and exopod point laterally. The orientation of the early limb buds in †*L. rhyniensis* most likely is plesiomorphic; this was used as an argument for placing the species as a sister-group to the Eubranchiopoda (Branchiopoda *s. str.*) by F. Schram and Koenemann (2001), while Scourfield (1926), Walossek (1993), and Olesen (2007, 2009) argued for an in-group position as a sister-group to the Anostraca, based on other morphological aspects. Scourfield (1926) described a special pattern of development in the furca. The original caudal furcae (primary furcae) that are present in the larvae (fig. 6.1A, B, E) are gradually overgrown by another pair of caudal processes (secondary furcae) (Scourfield 1926). In adults, the secondary furcae are long, rod-like terminal structures, while the primary furcae consist of two small knobs, with a characteristic conical seta in the center. This replacement of the original larval furcae during development is unseen in other branchiopods, but it can be compared (convergently) with eumalacostracan crustaceans, where modified abdominal appendages (uropods) replace the caudal furcae as the most terminal appendages.

The yet-unnamed orthonauplii from the Windyfield chert are often preserved as isolated limbs (mostly antennae and mandibles) or fragments of these, especially parts of the exopods. Complete specimens, however, can also be found (fig. 6.1J). A prominent feature of these larvae is the strong similarity between the antennae and the mandibles. The exopods of both limbs are especially very similar. Both are composed of ring-shaped annuli, which are armed with many fine setules. Three crustaceans from the Rhynie chert fauna—†*Lepidocaris rhyniensis*, †*Castracollis wilsonae*, and †*Ebullitiocaris oviformis*—have been identified as branchiopods (Scourfield 1926; Fayers and Trewin 2003; L. Anderson et al. 2004). The yet-unnamed orthonauplii from the Windyfield chert were suggested not to be branchiopods in the strict sense (C. Haug et al. 2012), but this was based on a specific character (apomorphy) suggested by Olesen (2007, 2009), which later work has shown needs reinterpretation. It is well known that the mandible in larval branchiopods in the strict sense is uniramous (e.g., see chapters 5, 7, 8 and 9). Olesen (2007, 2009) assumed that this was the result of a lost exopod. It was recently shown in a pa-

per by Fritsch et al. (2012) on *Lynceus biformis* (a laevicaudatan branchiopod) that it is, however, completely the opposite: the exopod is retained, but the endopod is lost. Hence the presence of mandibular exopods in the unnamed orthonauplii does not exclude a eubranchiopod affinity. Further studies will focus on where the new type of crustacean should be placed phylogenetically.

MORPHOLOGICAL DIVERSITY: The larvae of †*Lepidocaris rhyniensis* differ significantly from the tiny nauplii from the Windyfield chert, which is due to their different systematic relationships. Within each type of larva, however, the material is very uniform.

NATURAL HISTORY: The Rhynie and Windyfield cherts are interpreted as representing a special biotope (e.g., L. Anderson and Trewin 2003; Fayers and Trewin 2004). As an extant analog, Yellowstone National Park, with its strong geothermal activity, is often cited, and a freshwater environment is assumed, but recent investigations suggest that higher salinity levels might have occurred (see the review by Channing and Edwards 2009). The crustacean larvae are usually found in plant-rich chert areas, which makes it difficult to take proper images, due to the low transparency of the chert (fig. 6.1F). Swimming and feeding of the larvae has, of course, not been studied, but considering how similar the naupliar apparatus appears to be to those of anostracans or laevicaudatans (and other branchiopod larvae), swimming and feeding probably were similar (see Fryer 1983; Olesen 2005).

PHYLOGENETIC SIGNIFICANCE: The Rhynie and Windyfield chert fossils and their surroundings provide a unique view into the early evolution of a proposed terrestrial and freshwater biota. The †*Lepidocaris* fossils of the Rhynie chert shed light on evolution within the branchiopods. Reinterpretations of certain morphological characters of †*L. rhyniensis* were offered by F. Schram (1986), Walossek (1993), and F. Schram and Koenemann (2001). All of these authors interpreted Scourfield's first trunk limb as a limb-shaped second maxilla, which is therefore not reduced, as it is in all extant branchiopods. F. Schram and Koenemann (2001) also found support for positioning †*L. rhyniensis* on the branchiopod stem-lineage. Olesen (2007, 2009), however, based on Cannon and Leak (1933), again found support for Scourfield's (1926) original interpretation of the second maxillae as being reduced. Based on a detailed comparison with numerous eubranchiopods, Olesen (2007, 2009) supported an in-group eubranchiopod affinity of †*L. rhyniensis*. A reinvestigation of the complete type material and additional material is in progress (C. Haug et al., in prep.). The affinities of the Windyfield chert orthonauplii are as yet unclear (C. Haug et al. 2012).

HISTORICAL STUDIES: The main work on †*Lepidocaris rhyniensis* was provided by Scourfield (1926), who described two sexual forms of the adult, as well as some of the developmental stages. Scourfield (1940) later amended his earlier work, based on the two well-preserved larvae that are shown here. Another work, also based on original material (although only a single specimen), was presented in the larger comparative work on branchiopod feeding mechanisms by Cannon and Leak (1933). Several authors have contributed to our phylogenetic understanding. The Windyfield chert was discovered only recently (see the review by Fayers and Trewin 2004). In his unpublished PhD dissertation, Fayers (2003) attempted to describe the tiny orthonauplii from the Windyfield chert in detail. New investigative techniques for the Rhynie chert in general (Kamenz et al. 2008), and especially of these nauplii (C. Haug et al. 2009, 2012), as well as the presence of significantly more material, are the basis for an ongoing investigation of the crustacean larvae from these deposits.

Selected References

Fayers, S. R., and N. H. Trewin. 2004. A review of the palaeoenvironments and biota of the Windyfield chert. Transactions of the Royal Society of Edinburgh, Earth Sciences 94: 325–339.

Haug, C., J. T. Haug, S. R. Fayers, N. H. Trewin, C. Castellani, D. Waloszek, and A. Maas. 2012. Exceptionally preserved nauplius larvae from the Devonian Windyfield chert, Rhynie, Aberdeenshire, Scotland. Paleontologia Electronica 15(2): 24A.

Olesen, J. 2007. Monophyly and phylogeny of Branchiopoda, with focus on morphology and homologies of branchiopod phyllopodous limbs. Journal of Crustacean Biology 27: 165–183.

Scourfield, D. J. 1926. On a new type of crustacean from the Old Red Sandstone (Rhynie chert bed, Aberdeenshire)—*Lepidocaris rhyniensis* gen. et sp. nov. Philosophical Transactions of the Royal Society of London, B 214: 153–187.

Scourfield, D. J. 1940. Two new and nearly complete specimens of young stages of the Devonian fossil crustacean *Lepidocaris rhyniensis*. Proceedings of the Linnean Society 152: 290–298.

Fig. 6.1 (*opposite*) Fossil larvae from both formations (Devonian). A–I: †*Lepidocaris rhyniensis*. A: very well-preserved and rather complete specimen, with details of the appendages visible, ventral view; total length ca. 710 μm. B: same as A, original photograph of the specimen while still embedded in the chert. C: larval mandible (coxa and palp). D: branched antennal naupliar process, photograph. E: another well-preserved specimen; total length ca. 560 μm. F: same as E, photograph of the specimen while still embedded in the chert; dirt conceals parts of the body (white area). G: anterior part (ventral) of a young male, showing the larval palps on the mandibles and the developing claspers. H: cross-section through the body segment, showing three different stages of the development of the posterior trunk limbs; H^I = early stage, H^{II} = a later stage, H^{III} = a still-later stage. I: 3D model, lateroventral view. J: well-preserved specimen of a yet-unnamed orthonauplius from the Windyfield chert, with many details of the appendages visible, photograph; total length ca. 600 μm. A, B, and D = Scourfield's specimen 1 (NHM IC 38487); E and F = Scourfield's specimen 2 (NHM IC 38486); J = NMS G.2010.16.2.1. Abbreviations: NHM = Natural History Museum, London; NMS = National Museums of Scotland, Edinburgh. A–F modified after Scourfield (1940); G and H modified after Scourfield (1926); I based mainly on drawings in Scourfield (1940); B, D, F, J originals.

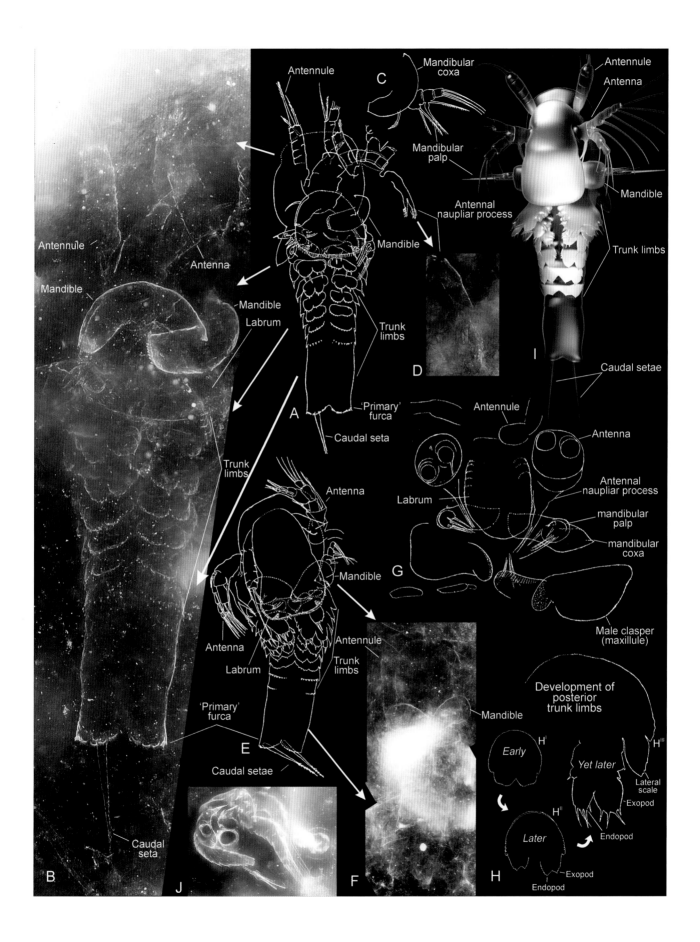

C Mandibular coxa

Mandibular palp

Antennule

Antenna

Antennal naupliar process

Mandible

Antennule

Antenna

Mandible

Trunk limbs

I

Caudal setae

A Antennule

Antenna

Mandible

Labrum

Mandible

Trunk limbs

'Primary' furca

Caudal seta

D

Antennule

Antenna

Labrum

Antennal naupliar process

mandibular palp

mandibular coxa

G

Male clasper (maxillule)

Antenna

Mandible

Antenna

Labrum

Antennule

Trunk limbs

'Primary' furca

E

Caudal setae

Mandible

F

Development of posterior trunk limbs

Early H[I]

Yet later H[III]

Lateral scale

Exopod

Later H[II]

Endopod

Exopod

Endopod

H

J

B Antennule

Antenna

Mandible

Labrum

Trunk limbs

Caudal seta

7

Notostraca

JØRGEN OLESEN
OLE STEN MØLLER

GENERAL: Notostracans are commonly called tadpole shrimps. They are a species-poor group of distinctly archaic-looking branchiopod crustaceans. The number of species is much debated because of morphological plasticity, and cryptic species may be involved (King and Hanner 1998; Rogers 2001; M. Korn and Hundsdoerfer 2006). Traditionally, 11 species divided in 2 genera (*Triops* and *Lepidurus*) have been recognized (Linder 1952; Longhurst 1955; Brendonck et al. 2008). Notostracans, like most other large branchiopods, inhabit inland freshwater (occasionally brackish) sites, mostly temporary pools on all continents except Antarctica. Notostracans are rare in most places, but they can be locally very abundant in most parts of the world. Some species can reach 100 mm long (J. W. Martin 1992) but most are around 30–50 mm. Superficially, notostracans are characterized by a shield-like dorsal carapace and a long limbless tail section, the latter composed of a variable number of cylindrical trunk segments (often termed body rings) mostly extending beyond the carapace and ending in a pair of long, thin, multiarticulate caudal rami. The endites of trunk limb 1 are long and antenna-like. The compound eyes are internalized. In some taxa, the trunk may bear as many as 72 pairs of limbs (J. W. Martin 1992), the posterior-most of which are very small and densely placed, with (unseen in other crustaceans) several limb pairs attached to each trunk segment (polypody) (e.g., fig. 7.3K, showing a juvenile; also see Linder 1952). The trunk limbs vary greatly in morphology along the body axis (reduced serial similarity), but at least the posterior trunk limbs are clearly phyllopodous, while the anterior ones have various sclerotized parts and appear almost segmented. Notostracans are omnivorous (consuming detritus and a variety of small organisms) and predominantly benthic, although they can swim well (Fryer 1988). The monophyly of the Notostraca has never been questioned, but their phylogenetic position within the Branchiopoda is not entirely settled. Morphological data support the Notostraca as a sister-group to the "bivalved branchiopods" (the Diplostraca) (Olesen 2007, 2009; Richter et al. 2007), while molecular data often

place them at various positions within the Diplostraca (e.g., Taylor et al. 1999; Spears and Abele 2000; Braband et al. 2002; Stenderup et al. 2006).

Reproduction is complicated and known to vary among populations (see the summary in Zierold et al. 2009). In *Triops cancriformis*, populations can be gonochoric, hermaphoditic, or androdioecious. The latter is a very rare reproductive mode in animals, in which populations are made of hermaphrodites, with a small proportion of males (e.g., Zierold et al. 2009). This lack of males long suggested that *Triops* was parthenogenetic, but the presence of testicular lobes seems to suggest hermaphroditism (Zaffagnini and Trentini 1980; Zierold et al. 2009). The eggs are released by the female and then carried for a while in pouches on the eleventh pair of trunk limbs (Fryer 1988). Various authors have reported that *Lepidurus arcticus* attaches its eggs to moss fronds (see Fryer 1988). When dry, the shed eggs enter into a diapause phase before hatching. Their life cycle after hatching is very fast, and individuals of *Triops cancriformis* become mature about two weeks after hatching.

LARVAL TYPES

Metanauplius and Later Stages: Larval development is known in most detail for *Triops cancriformis* (Claus 1873; Fryer 1988; Møller et al. 2003) and for *Lepidurus arcticus* (Borgstrøm and Larsson 1974), but the development of *L. apus* and *T. longicaudatus* is also partly known (Brauer 1874; Fryer 1988), and, at least between *Triops* and *Lepidurus*, there are significant differences (see below). Here we will refer to *T. cancriformis*, the early instars of which have been restudied during the preparation of this chapter. The development of *T. cancriformis* is essentially anamorphic (gradual), but, in contrast to the extreme gradual development of anostracans (at least in some species) (see chapter 5), large steps are taken at each molt, resulting in substantial morphological differences from stage to stage. For example, at stage 3 of *T. cancriformis*, the rudiments of about 10 trunk limbs are already present (fig. 7.2), while in stage 3 of *Artemia salina* only a few anterior limbs are present as rudimentary limb buds (Schrehardt 1987). As in anostracans, the development of notostracans does not naturally fall into distinct phases. In the first two stages of *T. cancriformis*, only the naupliar appendages are functional and active (fig. 7.1A–C), while the maxillules are merely small undeveloped buds, so these two stages can be termed metanauplii. But in stage 3—while the naupliar appendages are still active—the anterior trunk limbs buds are rather well developed, so this stage can hardly be termed a nauplius. The naupliar locomotion/feeding system is retained for a very long time in development and operates concomitantly in many developmental stages with the adult (thoracic) feeding system (e.g., the late stage in fig. 7.3F).

MORPHOLOGY: A number of early instars of *Triops cancriformis* have been reinvestigated in the process of preparing this chapter, based on the same material used in Møller et al. (2003). Also see the summary of early development in Fritsch and Richter (2010).

Stage 1 (Metanauplius) (fig. 7.1A–C): This is the hatching stage, which has a plump appearance and is expanded with yolk. It is clearly a non-feeding stage, as the mouth appendages are not developed. The antennules are small, tubular, densely annulated appendages with two distal setae: a long seta with a broad basal part, and a shorter one. The antennae are very large, almost reaching the caudal end of the trunk. Basally, there is a pair of unbranched naupliar processes that are coxal feeding structures, which apparently are still inactive at this stage (they appear to be soft). The basis is long, densely annulated with rows of small spines, and carries a short mediodistal seta (the basipodal feeding seta in anostracans) (see chapter 5). The endopod apparently is two-segmented, with three distal setae (one short, two longer). The exopod is not truly segmented, as the dorsal and ventral sides are segmented in two different ways: (1) the dorsal side is annulated in a way similar to the basis, and (2) the ventral side is subdivided into about five larger sclerites, each corresponding to a long seta. The labrum is situated posterior to the antennae and is a relatively large rounded structure with no spines. The mandibles are uniramous and subdivided into four portions: (1) a proximal coxa with a small, undeveloped seta, but with no gnathobase for feeding; (2) a basis, medially extending into a rudimentary endopod carrying two weakly developed median setae; (3) an exopod segment with two median setae; and (4) another exopod segment with two distal setae (terminology based on Fritsch et al. 2013). The maxillules are a pair of undifferentiated buds. The maxillae were not seen. Posterior to the maxillules, there is a relatively long hindbody, with a few anterior limbs weakly indicated externally, that ends terminally in a pair of rounded caudal lobes. Dorsally, the head section contains a large dorsal organ. The primordial carapace is seen as a pair of dorsal swellings behind the dorsal organ.

Stage 2 (Metanauplius) (fig. 7.1D–H): This stage is significantly more advanced than stage 1, and these larvae are probably feeding (at least according to Claus 1873; Fryer 1988). The general shape is more elongate and differs from stage 1 in a variety of respects. A pair of frontal filaments is now present (fig. 7.1I). The antennae have generally become larger; the coxal naupliar processes of the antennae appear more robust, with scattered denticles; and the endopod of the antenna is more clearly two-segmented, with the distal three setae now being approximately the same size. The labrum is now more lifted, providing room for an atrium oris, and it carries a total of eight small spines: two at the laterodistal margin, two at the inner side of the labrum but still close to its margin, and four at the inner side, in a row farther away from the margin and closer to the presumed mouth opening (fig. 7.1G). The gnathobases of the mandibles have developed considerably and carry some primordial spines, but they still seem incapable of meeting in the midline of the animal. The maxillulary limb buds have become more prominent and are divided into an anterior, smaller part (carrying one primodial seta) and a posterior, more globular part (fig. 7.1H). The maxillae are now present as a pair of small lateral limb buds (fig. 7.1G). About seven pairs of trunk limb buds are present externally, with the four anterior ones differentiating early into endites. The trunk is extended significantly posteriorly and ends in a pair of conical protrusions, which will become the caudal rami. Finally, a short carapace lobe is present as a posterior extension of the head shield.

Stage 3 (fig. 7.2): This stage is significantly more advanced than stage 2, and the larvae are certainly feeding (see Fryer 1988). It differs from stage 2 in several respects. The antennal coxal naupliar processes have become branched, with the posterior branch being smaller than the anterior one (fig. 7.2G, H). The margins of the labrum have become plate-like (fig. 7.2B). The mandibular gnathobases now lie closer together and have attained an armature of short, sharp denticles. The setae of the mandibular palp are more developed and bear more denticles (fig. 7.2C, D). The maxillular buds have become larger, with short spines (fig. 7.2A); and 9–10 trunk limb buds are now present, the anterior 3–4 being divided ventrally into distinct endites (1–5) and endopod lobes carrying anlagen of setation (fig. 7.2J). The limb-bearing part continues into a narrow hindbody, continuing further into an expanded telsonic part carrying a pair of caudal rami, which are much longer than in stage 2 but still not articulated to the telson (fig. 7.2A, I, L). The telson extends into a pair of small triangular spine-like plates both ventrally and dorsally between the caudal rami (fig. 7.2A, I, L). A pair of telsonic setae is now clearly present dorsally (fig. 7.2L). The carapace, which is continuous with the head shield, has become larger, with anterior and lateral folds overhanging the basal parts of the limbs, and with the posterior fold approximately reaching trunk limb 4 (fig. 7.2A, I).

Stage 4 (fig. 7.3A–E): This is the assumed stage 4, which in some ways is earlier than Møller et al.'s (2003) stage 4 (which probably is a true stage 5). It differs from stage 3 in a variety of ways. The posterior branch of the antennal coxal naupliar process is almost reduced (fig. 7.3C), seemingly later than Møller et al.'s (2003) stage 4, where this branch is more prominent. The maxillules are more developed, with longer spines pointing toward the mouth region (fig. 7.3C). The anterior trunk limbs are more developed, with longer and more setose endites and an endopod (fig. 7.3D). A total of 13–14 limb buds are seen, at decreasing degrees of development (fig. 7.3A), earlier than stage 4 of Møller et al. (2003). The hindbody (the part with only weakly developed limbs buds) is more slender (fig. 7.3A); and the caudal rami are longer and more slender.

Stages 5–6: This development has not been followed in detail, neither here nor in Møller et al. (2003). It appears that this still remains to be done for these remarkable crustaceans. Based on Fryer (1988), however, who studied the first 4–5 stages in detail and the following stages more briefly, the outline of the development of these stages is known. As in some anostracans (see chapter 5), the naupliar locomotory/feeding apparatus is gradually replaced by the thoracopodal apparatus, which eventually means that the characteristic naupliar swimming antennae (with their naupliar processes) and the mandibular palp are either lost or reduced in the adults, although the small sensory antennae still remain. For a number of stages, the two systems operate concomitantly. One stage, where both systems are active, has been examined in the course of this work, the results of which are presented here.

Stage 7 (fig. 7.3F–I): This staging is based on Fryer's (1988) work on *T. cancriformis*. The setation of various trunk-limb endopods and endites depicted by Fryer (1988, his figs. 112–114) for what he termed "stage 7" is identical to what is seen in the examined stage. Much of the adult's general morphology is attained by stage 7. When compared with the previously examined stage (stage 4), a number of changes have taken place. The antennules are more slender and shorter, relative to the length of the animal (fig. 7.3F). The antennae are much smaller, relative to the size of the other limbs, and the naupliar processes have become reduced to small knobs (fig. 7.3F). The labrum has developed a flattened, plate-like margin that covers a pair of robust mandibular gnathobases (fig. 7.3F–H). The mandibular palps have been reduced to small knobs (fig. 7.3G). A pair of paragnaths are present (fig. 7.3F–H). Limb buds of more than 25 pairs of limbs can be recognized; The anterior-most buds have well-developed, distinct, setose endopod and endites, while the endites at the posterior-most buds cannot be distinguished (fig. 7.3F, I). The posterior trunk limb buds have become increasingly narrow posteriorly. Uniquely among the crustaceans, each trunk segment corresponds to more than one pair of trunk limb buds (fig. 7.3I), a developmental trend that is continued so that, later in development and in the adults, 3–4 (or more) trunk limbs are inserted on a trunk segment (polypody) (fig. 7.3K, late juvenile). The caudal rami have become very long, and, according to Fryer (1988), superficial articulations appear for the first time (fig. 7.3F). The carapace has become very large, covering nearly all appendages and much of the limbless abdomen (fig. 7.3F).

MORPHOLOGICAL DIVERSITY: Not much is known about the development of other species in the Notostraca. Fryer (1988) studied *Triops longicaudatus* for comparison with

T. cancriformis and noted that the development of the former involved more early instars. The early development of *Lepidurus arcticus* was studied by Borgstrøm and Larsson (1974) and Fryer (1988), who found that its development is speeded up, compared with that of *T. cancriformis*. *L. arcticus* hatches as a non-feeding nauplius. By stage 2, it already looks like a miniature adult, but it still does not feed, despite (degenerated) mandibular palps being present. At stage 3 the larva is essentially a miniature adult, and it starts to feed by the adult mechanism. At no stage do the antennae (which are devoid of a naupliar process and basipodal setae) or the mandibular palps play any role in feeding, which is in strong contrast to *T. cancriformis*.

NATURAL HISTORY: The basics of the larval swimming and feeding apparatuses in notostracans are similar to those of anostracans (see chapter 5), which is not too surprising, considering the many similarities between the larvae of these two taxa (Fryer 1983, 1988; Olesen 2007). But there are also important differences. In the early stages of *Triops cancriformis*, both swimming and feeding are accomplished by the naupliar appendages; and these two functions are gradually (but less gradually than in anostracans) transferred to the thoracopodal apparatus (Fryer 1988). The mechanics of early larval swimming were described in much detail by Fryer (1988). Stage 1 and several succeeding stages row themselves through the water by using the antennae as oars, essentially similar to what was described for the Anostraca (see chapter 5). Over a series of molts, this mechanism is gradually replaced by the adult mechanism as the trunk limbs develop and the antennae atrophy. Stage 1 of *Triops cancriformis* does not feed, but from stage 2 until the thoracopodal feeding system takes over, the food-collecting element in the naupliar feeding apparatus is the mandibular palp (Fryer 1988), from which the food particles are transferred to the antennal naupliar processes that again pass the collected material forward to the mouth. This is not exactly identical to the situation in anostracan, spinicaudatan, and (probably) laevicaudatan larvae, where Fryer (1983, 1988) assumed that a characteristic, long, setulated seta (basipodal feeding setae) (see chapter 5), located mediodistally at the basis, assisted in the collection of food particles. Setae in this position are present in *T. cancriformis* (basipodal setae) (figs. 7.1A, D; 7.2A, F; 7.3A), but they are short and lack the long setules that are present on the corresponding setae in anostracans, so Fryer (1988) assumed that they play no major feeding role in notostracan larvae. Another difference between notostracan larvae and other branchiopod larvae is that the setae of the mandibular palps (fig. 7.2C, D) in the former are shorter and more robust. Fryer (1988) related this difference to presumed differences in their feeding strategies, and assumed that the stouter setation of the mandibular palp in *Triops* reflected their assumed feeding on relatively coarse particles, which may even be swept off the substratum. This is in contrast to anostracan larvae, which may be referred to more accurately as suspension feeders.

PHYLOGENETIC SIGNIFICANCE: Based on the generally accepted phylogeny of the Branchiopoda, where the Notostraca are placed at least above the level of the Anostraca, it is simplest to assume that the notostracan naupliar feeding apparatus, with its robust setation adapted for sweeping coarse particles, has evolved from an anostracan/spinicaudatan filter-feeding type of naupliar apparatus. Despite these specializations in the naupliar apparatus, however, it still essentially belongs to the branchiopod type, as outlined by Olesen (2004, 2007, 2009).

HISTORICAL STUDIES: Pioneering observations on hatching eggs, nauplii, and several of the early instars of *Triops cancriformis* were made by Schaeffer as early as 1756. Other early (and sometimes remarkably good) contributions are those of Zaddach (1841), Claus (1873), Brauer (1874), and Sars (1896a). The single outstanding modern treatment is that of Fryer (1988), from which we have taken much information. Nonetheless, a complete treatment of all stages from hatching to maturity remains to be done.

Selected References

Borgstrøm, R., and P. Larsson. 1974. The first three instars of *Lepidurus arcticus* (Pallas), (Crustacea: Notostraca). Norwegian Journal of Zoology 22: 45–52.

Claus, C. 1873. Zur Kenntniss des Baues und der Entwicklung von *Branchipus stagnalis* und *Apus cancriformis*. Abhandlungen der Königlichen Gesellschaft der Wissenschaften in Göttingen 18: 93–140.

Fryer, G. 1988. Studies on the functional morphology and biology of the Notostraca (Crustacea: Branchiopoda). Philosophical Transactions of the Royal Society of London, B 321: 27–124.

Møller, O. S., J. Olesen, and J. T. Høeg. 2003. SEM studies on the early development of *Triops cancriformis* (Bosc) (Crustacea: Branchiopoda: Notostraca). Acta Zoologica (Stockholm) 84: 267–284.

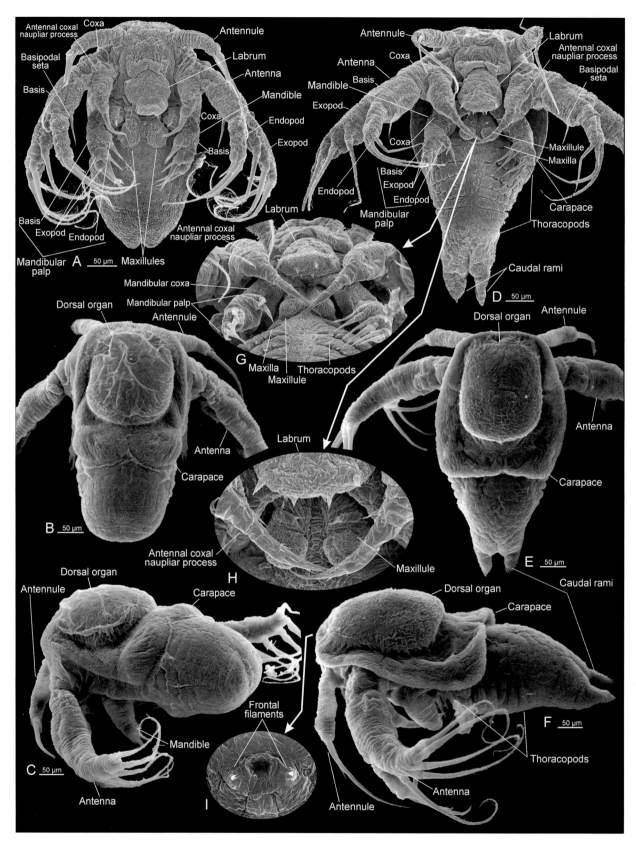

Fig. 7.1 Stages 1 and 2 of *Triops cancriformis* (Notostraca), SEMs. A: stage 1, ventral view. B: stage 1, dorsal view. C: stage 1, lateral view. D: stage 2, ventral view. E: stage 2, dorsal view. F: stage 2, lateral view. G: closeup of the naupliar feeding apparatus of stage 2. H: closeup of the naupliar processes (antennae) and maxillular buds of stage 2. I: closeup of the frontal filaments of stage 2. Most images original, but the same material used in Møller et al. (2003).

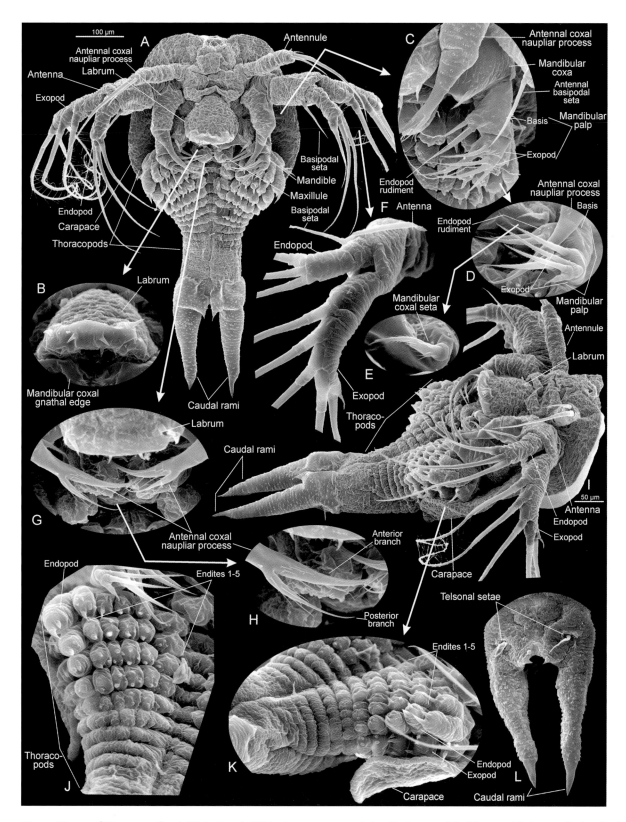

Fig. 7.2 Stage 3 of *Triops cancriformis* (Notostraca), SEMs. A: stage 3, ventral view. B: closeup of the labrum, with the marginal and sublabral spines. C: mandible and naupliar process of the antenna. D: closeup, showing the setation of the mandible. E: closeup of the mandibular coxal seta. F: antenna, showing segmentation. G: closeup, showing the branched naupliar processes (antennae). H: closeup of the right-side branched naupliar process. I: stage 3, lateral view. J: right-side thoracopodal limb buds, ventral view. K: left-side thoracopodal limb buds, lateral view. L: caudal rami and part of the telson, with telsonal setae, dorsal view. Most images original, but the same material used in Møller et al. (2003).

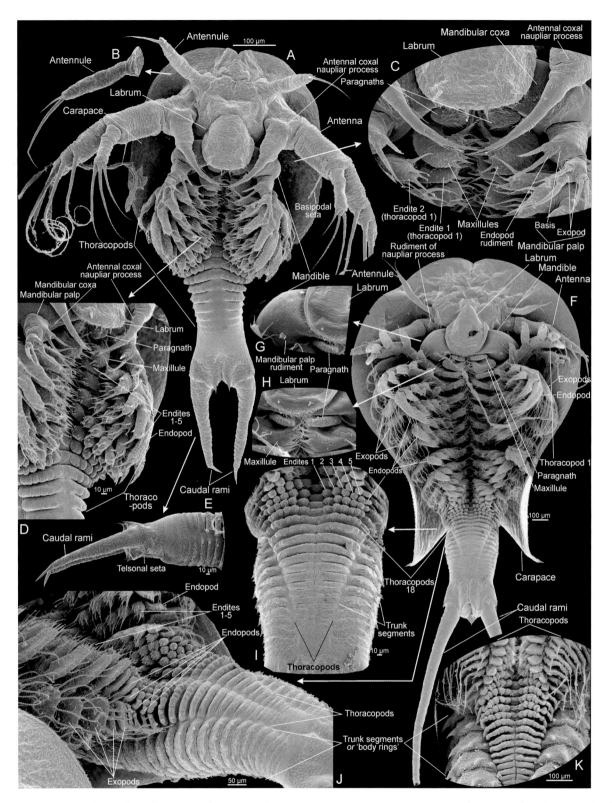

Fig. 7.3 Stages 4?, 7, and one later stage of *Triops cancriformis* (Notostraca), SEMs. A: stage 4, ventral view. B: right-side antennule of stage 4. C: labrum, naupliar processes (antennae), and mandible of stage 4. D: left-side trunk limbs, mandibles, paragnaths, maxillules, and naupliar processes (antennae) of stage 4. E: caudal region of stage 4, lateral view. F: stage 7, ventral view. G: closeup of the mandibular coxa, with rudiment of the palp, and paragnath of stage 7. H: closeup of the paragnaths and maxillules of stage 7. I: posterior limb buds of stage 7, showing the early developmental "mismatch" between the thoracopods and trunk segments (polypody). J: posterior limb buds of a late juvenile, showing the well-developed "mismatch" between the thoracopods and trunk segments (each trunk segments carries 3–4 thoracopodal buds). K: trunk segments of a late juvenile, ventral view. Most images original, of the same material used in Møller et al. (2003).

JØRGEN OLESEN
JOEL W. MARTIN

Laevicaudata

GENERAL: Laevicaudatans (Lynceidae) are small bivalved branchiopods sometimes (along with spinicaudatans and cyclestheriids) called clam shrimps. Among clam shrimps, the Laevicaudata are easily recognized by the globular form of the carapace, which can enclose the entire animal, including the very large head (fig. 8.2H). Growth lines of the carapace—such as those seen in other clam shrimps—are absent. Caudal claws are also lacking. Only 3 genera and approximately 36 species are known, all in one family, the Lynceidae (J. W. Martin and Belk 1988; Brendonck et al. 2008). Laevicaudatans inhabit ephemeral (temporary) freshwater pools, ponds, and streams. In terms of larval development, the best-known species are *Lynceus brachyurus* and *L. biformis* (Olesen 2005; Olesen et al. 2013); most of this chapter is based on the former. Their life cycle consists of four to six larval (naupliar) stages, juvenile instars, and the adult, all of which are free-living.

LARVAL TYPES

Nauplius and Heilophore: *Lynceus brachyurus* probably has five or six larval stages (the number is not known precisely; see Olesen 2005), considered to be modified nauplii; *L. biformis* has four larval stages (Olesen et al. 2013). Nauplii of *L. brachyurus* are remarkable for their large and nearly circular dorsal shield, huge plate-like labrum, and well-developed lateral horns (modified antennules) (figs. 8.1A–C; 8.2A–C). All larval stages of *L. brachyurus* are basically similar, except for an overall increase in size and limb development, and a change in the morphology of the antennal coxal naupliar process (masticatory spine), which becomes bifid after the first or second molt (fig. 8.1C, J, N, Q–S). The last larval stage of *L. brachyurus* and *L. andronachensis* was termed a "heilophore larva" by Botnariuc and Orghidan (1953), a term that has only been used occasionally since then for laevicaudatan larvae.

MORPHOLOGY: The large dorsal shield, plate-like labrum, and lateral horns (antennules) immediately distinguish larvae of *Lynceus brachyurus* from all other branchiopod (and indeed all other crustacean) larval forms (figs. 8.1A–C; 8.2A–C). Additionally, larvae of *L. brachyurus* have a large, rounded dorsal organ (fig. 8.2B) in the center of the dorsal shield. A pair of small frontal organs (one per side) area located on the cone-shaped protrusion of the head region (fig. 8.1C, F, I). The trunk limbs are visible as rows of immovable lateral limb buds, but they do not appear until late in the larval sequence (figs. 8.1C, E; 8.2G). The morphology of the antennae is—with an important exception (the naupliar process)—relatively unchanged during naupliar development. The basic components, from proximal to distal (partly shown in fig. 8.1J; see Olesen 2005), are a short coxa carrying a huge (relative to the size of the limb) naupliar process; a short segment that is probably a part of the basis; the basis proper; an exopod composed of two small segments and four additional larger segments (of which the most proximal is by far the largest), with two medially directed natatory setae carried on the distal segment and one per segment on each of the three preceding segments; and a two-segmented endopod with three (early stage) or four (late stage) distal setae. In the first nauplius stage(s) the large coxal naupliar processes are long curved spines, which are directed toward the mouth region, each carrying 10–15 long posteriorly directed setules and a loosely defined row of shorter anteriorly directed setules (fig. 8.1L). In later stages the naupliar processes on each side have developed a characteristic articulation approximately midway, allowing more flexibility (fig. 8.1J, N, Q–S), and the tip has becomes branched (fig. 8.1C, J, Q–S). A small pair of characteristic spines is found both distally at the tip of the unbranched naupliar process in the early nauplii and at the anterior branch of the naupliar process in the later nauplii, which indicates homology (fig. 8.1M, O). This suggests that the posterior branches become added during development. The setules on the anterior branch are arranged in a dense comb-like row directed toward the mouth; the setules at the posterior branch are arranged more loosely, similar to the hairs on a bottle-brush. The morphology of the mandibles remains largely the same during the entire larval development period. In *L. brachyurus* the mandibles are divided into four segments: coxa, basis, and two exopodal segments (the endopod is reduced); see Fritsch et al. (2013) for *L. biformis*. The coxa has not yet developed a gnathobasic process and a gnathal edge, but instead carries a large seta (fig. 8.1J, K), which is directed toward the mouth and is involved in feeding.

After the last larval stage, the morphology changes abruptly, by just one molt, into a juvenile/adult morphology (fig. 8.2G, which shows the exuvium of the last nauplius and the first juvenile stage a few minutes after molting). This change involves a shift of the dorsal shield into a bivalved carapace, which, when situated inside the shield of the last nauplius, does not fill it out completely anteriorly (fig. 8.1D; see discussion in Olesen 2005). By the first juvenile instar, the trunk limbs become more adult-like and movable (fig. 8.2D, G). Other characteristic larval structures, such as the plate-like labrum and the antennular lateral horns, are either lost or modified. The juvenile labrum appears much smaller than the larval labrum, of which it fills only a fraction, as can be seen when examining a first-stage juvenile through the transparent cuticle of a last-stage larva just about to molt (fig. 8.1B, P). Additionally, the antennular lateral horns become modified into smaller appendages, with a morphology more typical of branchiopod antennules. The outline of the future juvenile antennules can also be seen through the cuticle of the basal part of the lateral horns of the last larval stage (fig. 8.1B, P).

MORPHOLOGICAL DIVERSITY: Despite the apparent uniformity in descriptions of larvae of *Lynceus brachyurus*, there is some variation in the group. Larvae are known only from a few taxa, and all differ from each other in a number of respects, such as labrum morphology and the extension of the dorsal shield. *Lynceus andronachensis*, in contrast to *L. brachyurus*, has a more normal-sized labrum (Botnariuc 1947). *L. biformis* also has a smaller labrum, one that is uniquely equipped with four robust spines at its distal margin (fig. 8.2K) (Olesen et al. 2013). Another distinctive character of *L. biformis* is its lack of a dorsal shield in the first larval stage. Despite the many peculiarities of the larvae of *Lynceus* species, the morphology of the naupliar feeding apparatus falls well within the branchiopod pattern recognized by Olesen (2004, 2007, 2009).

NATURAL HISTORY: Larvae of *Lynceus brachyurus* are somewhat inefficient swimmers. They swim by moving the second antennae; no other limbs actively move except for the mandibles, which are thought to be used only in feeding (Olesen 2005). Feeding, although not directly observed for lynceid larvae, is assumed to be similar to that seen in other branchiopod larvae, with particulate matter brought into the

mouthfield under the labrum by a sweeping motion of the antennae (Olesen 2005). During feeding the relatively flexible naupliar processes (coxal masticatory spines) move actively under the labrum (fig. 8.1Q–S) (Olesen 2005). The mandibles of the larvae have no functional gnathal edge. Instead, a large gnathobasic seta is present close to where the gnathal edge will later appear (fig. 8.1J, K). These setae probably assist the highly mobile naupliar processes in sweeping food particles toward the mouth region.

PHYLOGENETIC SIGNIFICANCE: Laevicaudatans are clearly branchiopods, but the relationship of the group to other branchiopods has been the subject of some debate. The larvae of species of *Lynceus* display many peculiarities not seen in other branchiopods. The morphology of the naupliar feeding apparatus, however, is clearly branchiopodan (Olesen 2004, 2007, 2009). For discussions of possible relationships of laevicaudatans to other branchiopods, see Walossek (1993); Olesen et al. (1996); Olesen (2000, 2004, 2007, 2009); Olesen and Grygier (2004); Stenderup et al. (2006); Richter et al. (2007); Regier et al. (2010); also see chapter 4.

HISTORICAL STUDIES: Surprisingly detailed studies (for the time) of larvae of the genus *Lynceus* were published by Grube (1853), Gurney (1926), Botnariuc (1947), and Monakov and Dobrynina (1977). For more modern coverage see Olesen (2004, 2005), Fritsch et al. (2013), and Olesen et al. (2013).

Selected References

Fritsch, M., T. Kaiji, J. Olesen and S. Richter. 2013. The development of the nervous system in Laevicaudata (Crustacea, Branchiopoda): insights into the evolution and homologies of branchiopod limbs and "frontal organs." Zoomorphology 132: 163–181.

Martin, J. W., and D. Belk. 1988. A review of the clam shrimp family Lynceidae Stebbing, 1902 (Branchiopoda, Conchostraca) in the Americas. Journal of Crustacean Biology 8: 451–482.

Olesen, J. 2004. On the ontogeny of the Branchiopoda (Crustacea): contribution of development to phylogeny and classification. *In* G. Scholtz, ed. Evolutionary Developmental Biology of Crustacea, 217–269. Crustacean Issues No. 15. Lisse, Netherlands: A. A. Balkema.

Olesen, J. 2005. Larval development of *Lynceus brachyurus* (Crustacea, Branchiopoda, Laevicaudata): redescription of unusual crustacean nauplii, with special attention to the molt between last nauplius and first juvenile. Journal of Morphology 264(2): 131–148.

Olesen, J., M. Fritsch, and M. J. Grygier. 2013. Larval development of Japanese "conchostracans": pt. 3, larval development of *Lynceus biformis* (Crustacea, Branchiopoda, Laevicaudata) based on scanning electron microscopy and fluorescence microscopy. Journal of Morphology 274(2): 229–242.

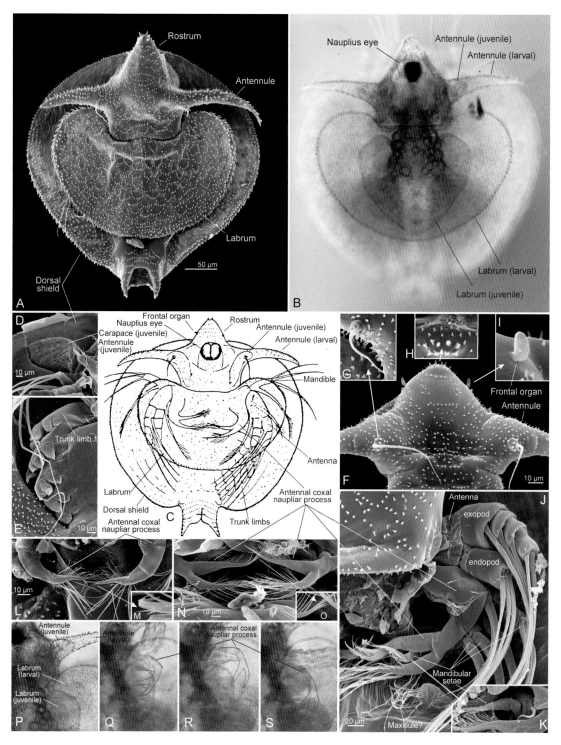

Fig. 8.1 Larval (naupliar) stages of *Lynceus brachyurus* (Laevicaudata), with selected appendages and other structures, SEMs (unless otherwise indicated). A: intermediate stage, ventral view. B: last larval (naupliar) stage, with juvenile tissue seen inside, light microscopy, ventral view. C: drawing, same as B. D: closeup of the left-side head region of the last larval stage, with the cuticle removed to show the next-stage (first juvenile) antennule and carapace inside. E: rudiments of the trunk limbs of the last larval stage. F: cone-shaped protrusion of the head region and basal parts of the lateral horns (modified antennules). G: closeup of the antennular seta, inserted in a small socket made up of spines. H: closeup of the small pore at the tip of the cone-shaped protrusion of the head region (arrow). I: closeup of the left-side frontolateral organ. J: right-side second antenna, showing the exopod, endopod, and bifid antennal coxal naupliar process. K: closeup of the gnathobasic seta of the mandible. L: unbranched antennal naupliar processes of the early naupliar stage. M: closeup of the tip of the naupliar process, showing a pair of spines. N: branched antennal naupliar process of the intermediate naupliar stage. O: closeup of the tip of the anterior branch of the naupliar process, showing a pair of spines. P: closeup of the last-stage nauplius, showing the juvenile labrum and carapace through a transparent cuticle, light microscopy. Q–S: three different positions of the antennal naupliar process during feeding, from video recordings. A, B, and D–S of material from Dyrehaven, north of Copenhagen in Denmark, used in Olesen (2005); C modified after Gurney (1926).

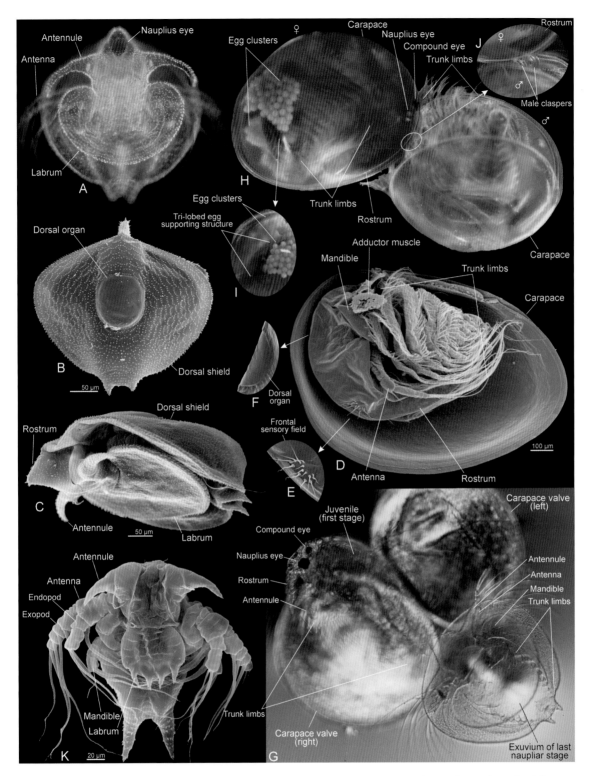

Fig. 8.2 Larval (naupliar) stages, early juveniles, and adults of *Lynceus brachyurus* (A–J) and an early nauplius of *L. biformis* (K) (Laevicau-data), SEMs (unless otherwise indicated). A: early larval stage, ventral view, light microscopy. B: early larval stage, dorsal view. C: early larval stage, lateral view. D: early juvenile with left-side carapace valve removed, lateral view. E: closeup of the left-side frontal sensory field. F: closeup of the dorsal organ seen from the left side. G: exuvium of the last nauplius (*right*) and first juvenile (*left*), a few minutes after molting (metamorphosis), light microscopy; the bivalved carapace of the juvenile is in an unnatural open position; the juvenile body is in the left carapace half. H: adult female (*left*) and male (*right*) in the process of mating, light microscopy. I: closeup of the egg cluster of the female showing the trilobed support structure behind the eggs. J: closeup of the male clasping organ (first pair of trunk limbs) while attached to the rim of the female carapace. K: early nauplius of *L. biformis*; note the small labrum with four terminal spines and the absence of a large dorsal shield (in contrast to *L. brachyurus*). A of *L. brachyurus* specimen from France, courtesy of Jean-François Cart; B–H of *L. brachyurus* material from Dyrehaven, north of Copenhagen in Denmark, used in Olesen (2005); K of *L. biformis* material from Japan, used in Olesen et al. (2013).

9

Spinicaudata

Jørgen Olesen
Mark J. Grygier

GENERAL: Spinicaudatans are a group of branchiopods that (along with the laevicaudatans and cyclestheriids) are commonly known as clam shrimps (see chapters 8 and 10). They are a relatively species-poor group of branchiopods (ca. 150 species; Brendonck et al. 2008), characterized by their bivalved and growth-line-bearing shells (carapaces), somewhat resembling those of bivalved mollusks. They are further characterized by their small and unsegmented antennules, biramous swimming antennae, up to about 32 pairs of serially similar foliaceus trunk limbs (decreasing in size and complexity from front to back), and laterally compressed body end, which bears a pair of large claws. All trunk segments bear appendages (the abdomen is absent), which is unusual for crustaceans. Spinicaudatans occur worldwide, except in Antarctica, but their distribution is scattered, since all species are adapted to live in temporary waters of various kinds (e.g., desert pools after rain, snowmelt ponds, and rice paddies). A diverse fossil record is present, mostly of the carapace only (Tasch 1969), but sometimes limb details are preserved (W. Zhang et al. 1990; Orr and Briggs 1999). Currently 3 extant families—the Cyzicidae, Limnadiidae, and Leptestheriidae—are recognized by most authors (Olesen et al. 1996; J. W. Martin and Davis 2001; Olesen 2007; Rogers et al. 2012), but this may change, as relationships within the Spinicaudata are controversial (Naganawa 2001; Hoeh et al. 2006; Schwentner et al. 2009). Spinicaudata used to be grouped with the other clam shrimps (the Laevicaudata and Cyclestherida) in the "Conchostraca," but this latter, traditional taxon has proven to be artificial (paraphyletic) (Olesen 1998; Braband 2002; Richter et al. 2007). The life cycle begins with eggs carried beneath the carapace, attached to trunk limb exopods of the female, before deposition in the sediment (figs. 9.1A, B; 9.2L; 9.3K). After hatching, there are five to seven larval (naupliar) stages, but possibly a different number of true instars, before metamorphosis to the bivalved juvenile (example shown in fig. 9.3J). The reproductive pattern of the Spinicaudata exhibits a wide range of breeding

systems, including dioecy, androdioecy, parthenogenesis, and hermaphroditism (Sassaman 1995; Weeks et al. 2009).

LARVAL TYPES

Nauplius: The first part of the post-embryonic development of spinicaudatans is the naupliar phase, which consists of five to seven stages, not all of which clearly represent molt-separated instars. The nauplii are characterized externally by a generally dorsoventrally flattened overall appearance, bud-like antennules, the presence of naupliar processes on the antennae, uniramous mandibular palps, non-articulated caudal claws, and (only in later larvae) non-functional lobate trunk limbs and a small dorsal carapace covering only the anterior pairs of these limbs (figs. 9.1; 9.2; 9.3). Several naming schemes have been applied to the relatively few larval stages of the Spinicaudata. In a number of species—such as *Leptestheria saetosa*, *L. dahalacensis*, and *Imnadia yeyetta*—some authors have recognized five larval stages, which have been termed nauplius I and II, metanauplius, peltatulus, and heilophore (Botnariuc 1947; Petrov 1992; Eder 2002). Other authors have not followed this terminology, instead referring more generally to larvae in the post-embryonic (but pre-juvenile) phase as larvae, stages, or nauplii (D. Anderson 1967; Olesen and Grygier 2003, 2004; Pabst and Richter 2004). Here we use the second set of terms, a nauplius being defined by the presence of a naupliar feeding apparatus, including a pair of distinct antennal naupliar processes. The first stage(s) of all spinicaudatan taxa can be termed orthonauplii, since no post-mandibular limb buds are present externally. There is no clear taxon-related pattern in the number of naupliar stages. Five, six, and seven stages have been reported for both the Limnadiidae and Cyzicidae, while five is the highest number reported for the Leptestheriidae (table 9.1).

MORPHOLOGY: The state of development of a nauplius within a spinicaudatan larval series is typically distinguished by combinations of characteristics: (1) degree of lecithotrophy

Table 9.1 Overview of most complete descriptions of spinicaudatan development

Family	7 larval stages	6 larval stages	5 larval stages	Fewer than 5 stages (probably incompletely known)
Limnadiidae	*Limnadia lenticularis* (Lereboullet 1866; Sars 1896a; Zaffagnini 1971), *Eulimnadia braueriana* (Olesen and Grygier 2003)	*Eulimnadia texana* (Strenth and Sissom 1975), *Limnadopsis parvispinus* (Pabst and Richter 2004)	*Limnadia stanleyana* (D. Anderson 1967), *Imnadia yeyetta* (D. Anderson 1967)	*Eulimnadia stoningtonensis* (W. Berry 1926), *Eulimnadia diversa* (Mattox 1937), *Imnadia voitestii* (Botnariuc 1947)
Leptestheriidae			*Eoleptestheria ticinensis* (Ficker 1876), *Leptestheria saetosa* (Petrov 1992), *Leptestheria dahalacensis* (Kapler 1960; Monakov et al. 1980; Eder 2002)	*Leptestheria intermedia* (Botnariuc 1947), *Eoleptestheria variabilis* (Botnariuc 1947)
Cyzicidae	*Caenestheriella gifuensis* (Olesen and Grygier 2004)	*Caenestheriella packardi* (Sars 1896b)	*Caenestheria* sp. (Bratcik 1980)	*Caenestheriella gynecia* (Mattox 1950), *Eoleptestheria variabilis* (Botnariuc 1947)

(yolk and undeveloped feeding parts), (2) number of small sensilla on the antennular limb buds, (3) morphology (unbranched or branched) of the antennal naupliar processes, (4) number of pairs of limb buds and their state of development, (5) presence and size of the early carapace, (6) shape and size of caudal lobes/claws, and (6) overall size. Most differences in larval development between various spinicaudatans reflect variability in the timing of the development of the structures mentioned above (heterochrony), such as changes in the stage at which feeding structures become active, the naupliar processes become branched, the carapace anlagen appear, and so forth. A major difference between some representatives of the spinicaudatan families is the shape of the labrum.

Limnadiidae (fig. 9.1): The morphology of the seven naupliar stages described for *Eulimnadia braueriana* by Olesen and Grygier (2003) is briefly outlined as an example here. Nauplius I has a swollen appearance and seems to be lecithotrophic (filled with yolk). It has weakly segmented antennae and mandibles, and a non-functional feeding apparatus (undeveloped naupliar process and mandibular coxal process) (fig. 9.1C, D). The labrum is relatively short, with a triangular distal part. A large dorsal organ is located dorsally on the head region. Nauplii II and III are similar to each other in morphology but differ greatly in size. Their shape is more slender and more elongate than in nauplius I, and the feeding apparatus appears to be functional: a naupliar process with setules (fig. 9.1E), and a gnathal edge on the mandibular coxal process, with teeth. The labrum is more distinctly pointed than before. In nauplius IV the tips of the antennal naupliar processes have become bifid and the caudal lobes more pointed (fig. 9.1F), but otherwise it is very similar to the previous stage. In nauplius V the anlagen of the carapace have appeared as a pair of dorsolateral swellings between the mandibles and the first pair of trunk limbs, the early anlagen of the first five or six pairs of which are present. In nauplius VI the small carapace has become free from the body. The labrum has become significantly wider basally, and four pairs of trunk limbs are developed enough so that all limb parts—five endites, the endopod, the exopod, and the epipod—can be recognized as small separate lobes.

The anlagen of three more pairs of limbs are also present. All seven limbs are still in a horizontal position. In nauplius VII the size of the free carapace has increased and its future bivalved shape can be recognized. The limb buds of the first five pairs of limbs are now much developed, and the anlagen of three or four more pairs are present (fig. 9.1I–K). Seen from the side, the posterior part of the body is curved ventrally.

A pointed, elongate, almost lanceolate larval labrum (as in *Eulimnadia braueriana*) is not diagnostic for the Limnadiidae, since it is not present in all limnadiids. A largely similar labrum is found in larvae of *Limnadia lenticularis*, *E. texana*, *E. diversa*, *E. stoningtonensis*, and even in free-living larvae of the non-spinicaudatan *Cyclestheria hislopi* (see chapter 10), but not in *Limnadia stanleyana*, *Limnadopsis parvispinus*, and *Imnadia yeyetta*, which have a more regular, rounded labrum.

Seven naupliar stages have also been described for *Limnadia lenticularis* (Sars 1896a); the sequence is quite similar overall to that of *Eulimnadia braueriana*, but not exactly the same. The two first stages are very similar, including the lecithotrophic appearance of nauplius I. The antennal naupliar process becomes branched in the molt to nauplius IV in both species. Apart from this, there are many minor differences in the timing of the development of the carapace and trunk limbs, and of stages three to seven; none is exactly equivalent in the 2 species (Olesen and Grygier 2003). In two other limnadiids—*E. texana* and *Limnadopsis parvispinus*—only six larval stages have been described. In these 2 species the naupliar processes become branched at the molt to nauplius III (one stage earlier than in the above 2 species). Use of this (and other structures) as a homology marker to align various naupliar series suggests that either stage II or stage III is extra in *E. braueriana* and *Limnadia lenticularis*, or missing in *E. texana* and *Limnadopsis parvispinus*. Pabst and Richter (2004) suggested that the Limnadiidae originally had six naupliar stages, basing this idea on the many similarities between the limnadiid larval cycle and that of a non-limnadiid, the cyzicid *Caenestheriella packardi* (Sars 1896b). In some other limnadiids—*Limnadia stanleyana* and *Imnadia yeyetta*—only five stages have been described.

Cyzicidae (fig. 9.2): Seven naupliar stages were described for *Caenestheriella gifuensis* by Olesen and Grygier (2004), summarized here as an example. Nauplius I appears lecithotrophic, since the body, labrum, and appendages are inflated with yolk. Feeding structures (antennal naupliar processes and mandibular coxae) are undeveloped (fig. 9.2A, B). The hindbody is distinctly globular. The antennules are small buds. The antennal basis and both antennal rami are divided into more segments (at least superficially) than in later stages (fig. 9.2C). The mandibles are only weakly segmented. Nauplii II and III (fig. 9.2D, E) are largely similar to each other (except in size) but differ significantly from nauplius I. The lecithotrophic appearance is lost and the feeding structures (naupliar processes and mandibular coxae) are fully developed (fig. 9.2E, F); this does not necessarily mean that feeding has started, however. The labrum has become more flattened and has developed three spines on the distal margin, the median of which is the largest. The segmentation of the antennal basis and rami has changed by the fusion of some of the primordial segments in nauplius I (compare figs. 9.2C and 9.2G). The mandible is clearly divided into four segments: coxa, basis, and two exopodal segments (following terminology by Fritsch et al. 2013). The hindbody has become more elongate, with two long caudal spines (fig. 9.2D, E). Nauplius IV is larger than the previous stage, from which it also differs in two significant respects: the antennal naupliar process has become branched, and the carapace anlagen have appeared. Nauplius V differs from the preceding stage in being larger, and in having a relatively larger carapace anlagen and a number of weakly developed limb buds. In nauplius VI the small carapace has become free from the body and the buds of the trunk limbs are more distinctly developed. In nauplius VII the carapace has become larger and the limb buds are more developed (fig. 9.2J), so that the different parts (the endites, endopod, exopod, and epipod) can be clearly recognized in them. The labrum is very large, and the median spine reaches the caudal spines. In lateral view the body is curved slightly ventrally, foreshadowing the adult form.

In *Caenestheriella gifuensis* the labrum has three large characteristic spines: a median spine, flanked by two smaller lateral spines. This is seen in some other species of the Cyzicidae as well, including *Cyzicus tetracerus* (Botnariuc 1947; Botnariuc and Orghidan 1953) and *Caenestheria* sp. (Bratcik 1980). These spines are absent in *Caenestheriella packardi*, however (Sars 1896b); therefore their presence is not diagnostic for the Cyzicidae. *Caenestheriella gifuensis* is the only cyzicid for which seven larval stages have been described. Five stages were described for *Caenestheria* sp. by Bratcik (1980), and six for *Caenestheriella packardi* by Sars (1896b). Olesen and Grygier (2004) noted that nauplii II and III of *C. gifuensis* are difficult to separate, and also that nauplii VI and VII are quite similar to each other; these may not be distinct instars. In the larval series of *Caenestheria* sp. described by Bratcik (1980), both an early and a late nauplius are missing in *Caenestheria* sp. When comparing the larval series of *E. braueriana* to the six described stages of *C. packardi*, it appears that either nauplius II or III

of *E. braueriana* is missing in *C. packardi*. It remains uncertain, however, whether these differences among the abovementioned species are true, or whether too few or too many stages have been described in some species.

Leptestheriidae (fig. 9.3): Those leptestheriids for which development has been treated in most detail are *Eoleptestheria ticinensis*, *Leptestheria dahalacensis*, and *L. saetosa* (considered a synonym of *L. dahalacensis*) (Ficker 1876; Kapler 1960; Monakov et al. 1980; Petrov 1992; Eder 2002). A few stages of *L. intermedia* and *E. variabilis* were described by Botnariuc (1947). SEM photos of a few stages of *Leptestheria kawachiensis* are included in this chapter.

Five larval stages have been described for *Eoleptestheria ticinensis*, *Leptestheria dahalacensis*, and *L. saetosa*. The five stages of *L. dahalencensis* and the five stages of *L. saetosa* seem to correspond closely to each other. As in other spinicaudatans, nauplius I of these 2 species has a lecithotrophic appearance, but the hindbody of nauplius II also appears to be filled with yolk. The antennal naupliar process becomes branched at the molt to nauplius III. The anlagen of both the carapace and the anterior trunk limbs become visible externally in nauplius IV (termed the "peltatulus" stage by Petrov 1992 and Eder 2002). In nauplius V the carapace becomes free and the parts of the limb buds (the endites, endopod, exopod, and epipod) become visible; this stage was called the "heilophore" by Petrov (1992) and Eder (2002). The figures for this chapter illustrate one early stage of *E. ticinensis* (fig. 9.3A) and two stages of *L. kawachiensis* (fig. 9.3B–I). The earlier stage of *L. kawachiensis* (fig. 9.3B) corresponds approximately to nauplius III of *L. dahalacensis*, and the older one (fig. 9.3D–I) corresponds approximately to nauplius V of the latter. Five stages were also described for *E. ticinensis* by Ficker as early as 1876; according to him the antennal naupliar processes were already branched in the second stage. In all other known spinicaudatans this does not occur before the third or fourth larval stage, so probably one of the earlier stages was missed by Ficker.

MORPHOLOGICAL DIVERSITY: Most variation is treated above.

NATURAL HISTORY: Egg laying has rarely been studied in spinicaudatans, but some of the trunk-limb epipods are used for egg storage/transfer in adult females of *Leptestheria dahalacensis* (Tommasini et al. 1989; also see J. W. Martin 1992). The eggs pass from the gonoduct into the dilated epipods of certain trunk limbs, from which they are later discharged through a distal median slit on the epipod and adhere externally (by mucus) to the exopods of the middle trunk limbs (figs. 9.1A; 9.2L; 9.3K). In many cases the eggs bear characteristic ornamentations externally (fig. 9.1B) (Thiéry and Gasc 1991; Shen and Huang 2008; Rabet 2010). Swimming in all spinicaudatan nauplii is accomplished primarily by means of the large antennae, possibly with a minor contribution from the mandibular palps. Sars (1896a) remarked that the movements of the hatching stage of *Limnadia lenticularis* are rather awkward, and that movement becomes more regular and energetic only after the larva has undergone its first molt and developed a more complete setal armature. In *L. lenticularis* the trunk limbs remain immovable, even in the last larval stage. For the later larval stages in this species, Sars (1896a) remarked that movement is brought about chiefly by the antennae, whose rhythmical strokes from front to back propel the body through the water in a peculiar jerky manner. At each stroke their "basal, incurved projection" (antennal naupliar process) is brought in between the labrum and the body, so that it is brought into contact with the oral region. Sars (1896a) speculated that this projection was of importance in the feeding process, something that was later confirmed for other branchiopod larvae with a very similar feeding apparatus. For example, in anostracan larvae the swimming setae of the antennae also have been suggested to act as a food-collecting fan, while those of the mandible play a part in food transfer from the antennal fan to the sublabral space, where food can be further manipulated by the antennal process (Cannon 1928; Gauld 1959; Fryer 1983; for more details, also see chapter 5). The same has been mentioned for the spinicaudatan *Limnadia stanleyana*, in which the feeding apparatus seems ready by nauplius II, although, according to D. Anderson (1967), feeding does not start before naupliar stage IV.

PHYLOGENETIC SIGNIFICANCE: The naupliar feeding apparatus in all examined spinicaudatans is very similar to that of larvae of other large branchiopods; its morphology has been conservative in branchiopod evolution. Its features have been suggested as a supporting character (synapomorphy) for the Branchiopoda (Olesen 2007, 2009). In the Spinicaudata (as in the Anostraca, Notostraca, and Laevicaudata), the feeding apparatus consists of large antennae (for swimming) with a coxal naupliar process that becomes branched after a few molts, and a mandibular coxa with a three-segmented palp that has a practically identical setation among these taxa. The use of larval characters for the elucidation of spinicaudatan phylogeny is difficult, since larvae are only known from a few species, and the conservativeness of branchiopod larval morphology may also limit its use in this regard, but at least labrum morphology seems promising. The characteristic pointed and almost lanceolate shape of the labrum of some limnadiids may indicate a close relationship among these taxa; it may also be the case for those cyzicids in which the larval labrum bears three characteristic spines.

HISTORICAL STUDIES: Classical studies are those of Ficker (1876), Sars (1896a, 1896b), and Botnariuc (1947), which were all very detailed and were not matched in their detail until SEM was introduced.

Selected References

Anderson, D. T. 1967. Larval development and segment formation in the branchiopod crustaceans *Limnadia stanleyana* King (Conchostraca) and *Artemia salina* (L.) (Anostraca). Australian Journal of Zoology 15: 47–91.

Botnariuc, N. 1947. Contributions à la connaissance des phyllopodes conchostracés de Roumanie. Notationes Biologicae 5: 68–158.

Olesen, J., and M. J. Grygier. 2003. Larval development of Japanese "conchostracans": 1, larval development of *Eulimnadia braueriana* (Crustacea, Branchiopoda, Spinicaudata, Limnadiidae) compared to that of other limnadiids. Acta Zoologica (Stockholm) 84: 41–61.

Pabst, T., and S. Richter. 2004. The larval development of an Australian limnadiid clam shrimp (Crustacea, Branchiopoda, Spinicaudata), and a comparison with other Limnadiidae. Zoologischer Anzeiger 243: 99–115.

Fig. 9.1 (*opposite*) Adult and nauplii of the Spinicaudata (Limnadiidae), SEMs. A and B: *Limnadia lenticularis*. A: ovigerous female. B: closeup of the ornamented eggs. C–M: *Eulimnadia braueriana*. C: nauplius I, ventral view. D: left-side antennule, antenna, and mandible of nauplius I. E: closeup of the antennal naupliar process (unbranched) and mandibular gnathal edges of nauplius II. F: nauplius IV, ventral view. G: closeup of the labrum of nauplius IV. H: closeup of the mandibular gnathal edges of nauplius V. I: nauplius VII, ventral view. J: left-side trunk limbs of nauplius VII, ventral view. K: trunk limbs of nauplius VII, lateral view. L: antennal naupliar process and mandible of nauplius VII. M: closeup of the antennal naupliar process of nauplius VII. A and B of material from France, courtesy of Jean-François Cart; C–M of material from Japan, used in Olesen and Grygier (2003).

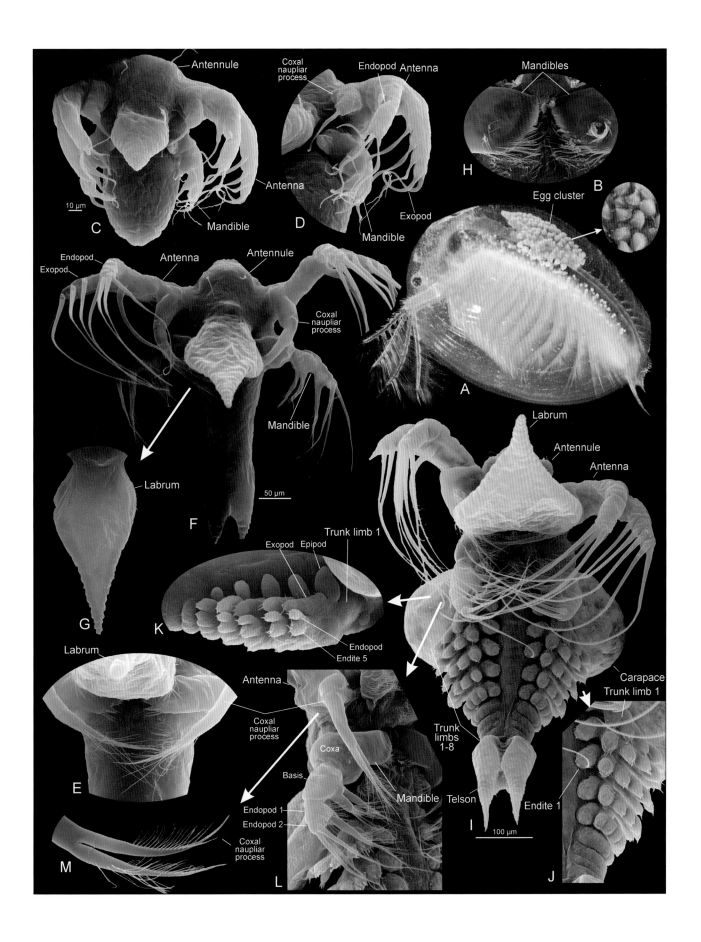

Antennule

Coxal
naupliar
process

Endopod Antenna

Mandibles

Egg cluster

B

Antenna

Exopod

Mandible

D

Mandible

C

10 µm

A

Endopod

Antenna

Antennule

Exopod

Coxal
naupliar
process

Labrum

Labrum

Antennule

Antenna

Mandible

Labrum

G

50 µm

F

Trunk limb 1

Exopod Epipod

Carapace

Trunk limb 1

Labrum

K

Endopod

Endite 5

Trunk
limbs
1-8

Antenna

Coxal
naupliar
process

Coxa

Basis

Mandible

E

Endopod 1

Endopod 2

Telson

Endite 1

J

Coxal
naupliar
process

M

L

I

100 µm

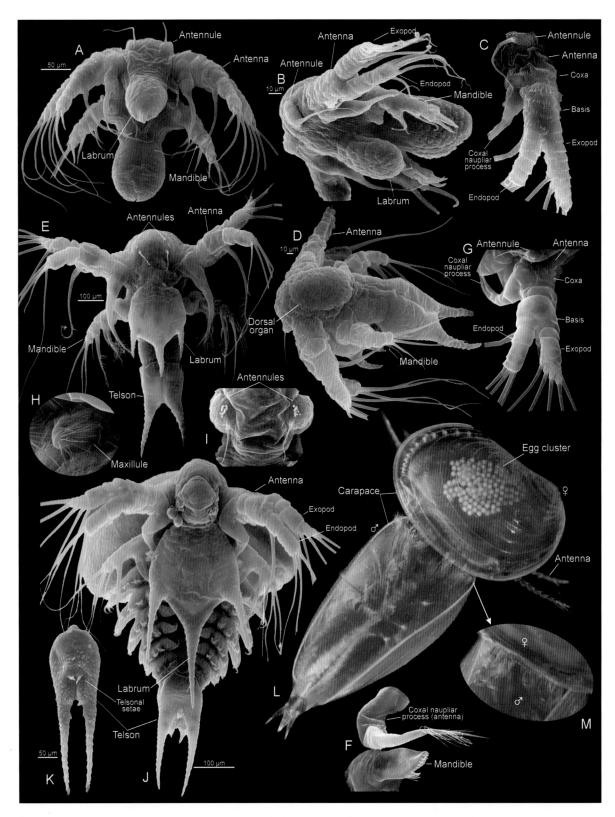

Fig. 9.2 Spinicaudata (Cyzicidae), SEMs. A–K: nauplii of *Caenestheriella gifuensis*. A: nauplius I, ventral view. B: nauplius I, lateral view. C: antenna of nauplius I. D: nauplius II, dorsolateral view. E: nauplius III, ventral view. F: closeup of the antennal naupliar process and gnathal edge of the mandible of nauplius II or III. G: antenna of nauplius IV. H: closeup of maxilla 1 of nauplius IV. I: closeup of the antennules of nauplius VI. J: nauplius VII, ventral view. K: telsonal region of nauplius VII, dorsal view. L and M: adults of *Cyzicus tetracerus*. L: mating couple; female with egg clusters and male with claspers. M: closeup of the male claspers attached to the carapace margin of the female. A–K of material from Japan, used in Olesen and Grygier (2004); L and M of material from France, courtesy of Jean-François Cart.

Fig. 9.3 Spinicaudata (Leptestheriidae), SEMs. A: early stage of nauplius of *Eoleptestheria ticinensis*. B–J: nauplii and juvenile of *Leptestheria kawachiensis*; the larval sequence is not known in detail, and the stages are therefore not numbered. B: early larva, with a branched antennal naupliar process, ventral view. C: mandible and maxillule of an intermediate larva. D: late larva, lateral view. E: late larva, ventral view. F: antennule and antenna, with the naupliar process of a late larva. G: closeup of the basipodal feeding seta of the antenna of a late larva, with rows of setules. H: row of early trunk limbs of a late larva. I: telsonal region of a late larva, dorsal view. J: first juvenile stage, lateral view. K: ovigerous adult female of *E. ticinensis*. A and K of material from France, courtesy of Jean-François Cart; B–J original, of material collected by Mark J. Grygier in September 1999 from rainwater-filled ruts in harvested rice paddies in Kataoka-cho, Kusatsu City, Shiga Prefecture, Japan.

10

JØRGEN OLESEN

Cyclestherida

GENERAL: Cyclestheridans are small bivalved branchiopods commonly called clam shrimps (together with spinicaudatans and laevicaudatans; see chapters 8 and 9). Traditionally they have been grouped with the Spinicaudata, but they are now placed in the monotypic order Cyclestherida. Among clam shrimps, cyclestheridans are easily recognized by the rounded shape of the growth-line–bearing carapace, the rounded and serrate tip of the rostrum, the characteristic row of large dorsal spines in the caudal region, and, not least, the direct-developing embryos located dorsally between the carapace valves (fig. 10.1A). Commonly only 1 species, *Cyclestheria hislopi*, is recognized, but another has been reported briefly from China (see Olesen et al. 1996). A recent study suggests that *Cyclestheria hislopi* may be a species complex of several cryptic species worldwide (Schwentner et al. 2013). Adults inhabit temporary freshwater pools but, in contrast to other clam shrimps, they sometimes also inhabit smaller permanent freshwater bodies, such as lakes or reservoirs (Roessler 1995). Their distribution is restricted to tropical or subtropical parts of Australia, Asia, Africa, and the Americas (J. W. Martin et al. 2003; Olesen et al. 1996). *Cyclestheria hislopi* is a phylogenetically important taxon. Despite its general similarities to the other clam shrimps, both molecular and morphological evidence suggest that it is closely related (the sister-group) to the Cladocera (water fleas) and therefore can be viewed as a kind of living "link species" between spinicaudatans and cladocerans (e.g., J. W. Martin and Cash-Clark 1995; Olesen et al. 1996; Richter et al. 2007).

The life cycle alternates between parthenogenetic and sexual reproduction (heterogony, or cyclic parthenogenesis). Embryos are brooded in the parthenogenetic part of life cycle. Males are known from only a few populations worldwide, and resting eggs (ephippia), from which juveniles hatch, are known from Colombia (Roessler 1995). Putative free-living larvae have been reported from a single population in Cuba (Botnariuc and Viña Bayés 1977), but the status of these larvae needs confirmation.

LARVAL TYPES

Embryos, or Larval Stages: The developing stages in the brood chamber have variously been termed "embryos" (e.g., Sars 1887), "larvae" (Olesen 1999), or "embryonized larvae" (Olesen 2009). Fritsch and Richter (2012) distinguished between an early "embryonic phase" (before shedding the egg membrane) and a later phase of "embryo-like larvae" (after shedding the egg membrane, but while still in the brood chamber). During the parthenogenetic part of the life cycle, the embryos/larvae go through all developmental stages while occupying a dorsal brood chamber between the female parent's carapace valves, while in the sexual part of the cycle, juveniles hatch directly from ephippium-like resting eggs (direct development) (Roessler 1995; Olesen 1999). From most parts of the world, only the parthenogenetic part of the life cycle has been reported, and males are very rare (Sassaman 1995; Olesen et al. 1996). At least seven developmental stages are passed before a juvenile morphology is achieved (Sars 1887; Olesen 1999; Fritsch and Richter 2012), but "the precise number may not known"; this aspect of their development needs to be examined in more detail. At least the first four stages are passed within an egg membrane, but it is not known in detail whether that development is gradual or occurs in discrete stages (instars). Until at least stage 7, long filaments connect the embryonized larvae to upper prolongations of the exopods of the middle pairs of trunk limbs of the female parent (fig. 10.1A–C). In the earlier stages the filament is attached to the eggshell, but in later stages it is attached to the forehead of the embryonized larvae. the entire development of the embryonized larvae / juveniles can be divided into three phases: (1) an embryonic phase (enclosed in the egg membrane), (2) a post-embryonic phase (free of the egg membrane), and (3) a juvenile phase (limbs that are capable of movements).

MORPHOLOGY: Four developmental stages from the parthogenetic part of the life cycle are illustrated and described here, which roughly represent the entire pre-juvenile development of the cyclestheridans.

Stage 2 (following the stage designation of Olesen 1999): Stage 2 is still enclosed in the egg membrane (fig. 10.1D–G). It is nearly globular and contains much yolk. Only a few structures can be seen externally (after removal of the egg membrane). The naupliar appendages (the antennules, antennae, and mandibles) are present as distinct undeveloped buds and form a separate head section. The outline of more posterior appendages also can be seen. Dorsally, a yet-undeveloped dorsal organ is at the rear of the head.

Stage 4: Stage 4 is the last stage still enclosed in an egg membrane (fig. 10.1H–L). It still contains much yolk, giving it an embryonic appearance. The antennules and mandibles are small undifferentiated buds, while the antennae have developed some early segmentation. The maxillules and maxillae are small flattened buds. In the buds of the maxillae, the anlagen to the openings of the future maxillary glands can be seen (fig. 10.1L). The buds of approximately 10 trunk limbs are present in various degrees of development. The anterior buds are the most developed, and an early endopod/exopod bifucation is present, as well as weakly developed endites and an epipod. The posterior limb buds are just undifferentiated lobes. A distinct carapace fold covers a small part of the body dorsally. The carapace originates ontogenetically in the region of the maxillules and maxillae.

Stage 6: Stage 6 is free of the egg membrane (fig. 10.2A–E) (but see stage 6 in Fritsch and Richter 2012). It is more distinctly dorsoventrally compressed than in the previous stages. At the head, is a distinctly protruding globular dorsal organ dorsally. The embryonized larvae are still connected to the exopods of the female parent via long embryonic filaments. The attachment zones of these filaments are the foreheads of the larvae (see the scar on the forehead and the remains of the filament/membrane in fig. 10.2E). The antennules are still small globular buds. The general morphology of the antennae has become more like that of the adult, with more distinctly developed early segmentation. The mandibles have begun to attain their adult triangular shape (seen from a lateral view) and have started to develop the median part, which will eventually become the gnathal edge of the mandible in the adult. Both maxillules and maxillae are still small undifferentiated buds. Approximately 12 pairs of trunk limbs are developed to varying degrees, with their prospective tips still pointing laterally. The anterior trunk limbs are the most developed and bear the anlagen to all major adult limb parts: endites 1–5, the endopod, the exopod, and the epipod. A short non-bivalved dorsal carapace lobe, covering the anterior part of the trunk, originates in the region of the maxillules and maxillae.

Stage 7: This stage is very different from stage 6 (fig. 10.2F–M). It is still somewhat embryonic looking, being much expanded by the yolk inside, but the entire body shape now approaches that of the adults. The body has become more laterally compressed (instead of dorsoventrally compressed) and has started to bend ventrally, and the anterior trunk limbs have attained a vertical orientation, approaching the orien-

tation in adults. The morphology of the various limb parts also approaches that of the adults, with, for example, a large elongate exopod. The bivalved appearance of the carapace is now clear. The labrum and the mandibles have grown significantly larger. The dorsal organ has become even larger than in the previous stage. The forehead has a scar where the embryonic filament was attached. Anlagen of setation are visible at the antennae and at various parts of the trunk limbs, such as the endites. The anlagen of the furcal claws and the paired telson setae can also be seen.

None of the later stages have been illustrated here. Sars (1887) showed that a subsequent stage is much more developed and essentially looks like a small adult animal, but it still exhibits some minor differences. Various internal-organ systems are now active (e.g., the circulatory system), and the trunk legs, present now in full number, have commenced their rhythmical movements. The mouthparts also seem fully developed and are active in conveying food to the mouth. When the young have attained an even fuller development, agreeing with the adult animal in every aspect except size, they leave the brood pouch of the mother (Sars 1887). According to Roessler (1995), two to three molts are needed before this takes place.

Not much is known about the putative free-living larvae of *Cyclestheria hislopi* reported by Botnariuc and Viña Bayés (1977). Based on the single drawing they presented, it more-or-less looks like a general spinicaudatan free-living heilophore larva (see chapter 9). It has bud-like antennules, typical spinicaudatan swimming antennae with a naupliar process, a mandible with a uniramous palp, and a large pointed labrum (much like the larvae of certain limnadiid spinicaudatans).

MORPHOLOGICAL DIVERSITY: Development is known only from populations of *Cyclestheria hislopi*, so there is no interspecific variation.

NATURAL HISTORY: The life cycle in at least some populations alternates between sexual and parthenogenetic reproduction, as in cladocerans (Roessler 1995). The life cycle of *Cyclestheria hislopi*, based on studies in Colombia, involves the presence of sexual females and males, which is related to particular environmental conditions. Resting eggs are also produced and are stored in a kind of primitive ephippium, as is well known from many cladocerans. In *C. hislopi* the entire carapace (not just a part, as in some cladocerans) constitutes the ephippium, and its deposition results in the death of the female (Roessler 1995). Some confusion still exists concerning the full life cycle. First, the putative presence of free-living larvae in a population in Cuba, but not found anywhere else, conflicts with the existence of the life cycle described above. Second, in most places where *C. hislopi* is known, males are absent, despite detailed searches (K. K. Nair 1968; Paul and Nayar 1977; Olesen et al. 1996), so it seems as though the life cycle is not identical in all populations.

The parthenogenetic part of reproduction, with its many succeeding generations, occupies the major part of the life cycle and is therefore the best studied. One peculiar aspect is the way in which, during development, the parthenogenetic embryonized larvae are attached to the middle trunk limbs (exopods) of the female parent by long embryonic filaments that resemble umbilical cords (fig. 10.1A–C). As far as is known, no nutritional transfer takes place. Spinicaudate clam shrimps also have long exopodal filaments, to which the eggs are attached in clusters under the carapace, but the embryos are not brooded (see chapter 9). The development of the first seven stages in *Cyclestheria hislopi* seems to rely only on yolk, but in the following stage, while the larvae are still within the brood chamber of the female, mouthparts, trunk limbs, and certain internal organs are already active, and this stage has been reported to start feeding (Sars 1887).

PHYLOGENETIC SIGNIFICANCE: *Cyclestheria hislopi* exhibits morphological similarities to other clam shrimps, but both molecular and morphological evidence suggest that it is closely related to the Cladocera (water fleas), making it a phylogenetically important species. Developmental aspects of *C. hislopi* have played a major role in discussions of its evolution and phylogenetic position. The heterogonic reproduction (cyclic parthenogenesis) of the species and its direct development have often been discussed as linking *Cyclestheria hislopi* to the Cladocera (e.g., Sars 1887; Schminke 1981; J. W. Martin and Cash-Clark 1995; Olesen et al. 1996; Olesen 1999; Fritsch and Richter 2012), and molecular data have further confirmed that the species represents the sister-group to the Cladocera (e.g., Taylor et al. 1999; Spears and Abele 2000; Stenderup et al. 2006; Richter et al. 2007).

HISTORICAL STUDIES: The first description of the development of *Cyclestheria hislopi* was by the well-known Norwegian carcinologist G. O. Sars in his seminal work from 1887, which is still one of the best sources of information. Sars raised a culture of *C. hislopi* in Norway, based on Australian mud, and he studied many of its morphological, behavioral, and developmental aspects. The parthenogenetic part of the life cycle and the embryonized larvae were later reported by a number of authors (e.g., Dodds 1926; K. K. Nair 1968; Paul and Nayar 1977; Egborge and Ozoro 1989). The three most detailed recent treatments are those of Roessler (1995), Olesen (1999), and Fritsch and Richter (2012).

Selected References

Fritsch, M., and S. Richter. 2012. Nervous system development in Spinicaudata and Cyclestherida (Crustacea, Branchiopoda)—comparing two different modes of indirect development by using an event pairing approach. Journal of Morphology 273: 672–695.

Olesen, J. 1999. Larval and post-larval development of the branchiopod clam shrimp *Cyclestheria hislopi* (Baird, 1859) (Crustacea, Branchiopoda, Conchostraca, Spinicaudata). Acta Zoologica (Stockholm) 80: 163–184.

Olesen, J., J. W. Martin, and E. W. Roessler. 1996. External morphology of the male of *Cyclestheria hislopi* (Baird, 1859) (Crustacea, Branchiopoda, Spinicaudata), with comparison of male claspers among the Conchostraca and Cladocera and its bearing on phylogeny of the "bivalved" Branchiopoda. Zoologica Scripta 25: 291–316.

Roessler, E. W. 1995. Review of Colombian Conchostraca (Crustacea): ecological aspects and life cycles—family Cyclestheriidae. Hydrobiologia 298: 113–124.

Sars, G. O. 1887. On *Cyclestheria hislopi* (Baird), a new generic type of bivalve Phyllopoda, raised from dried Australian mud. Forhandlinger i Videnskabsselskabet i Kristiania 1887: 1–65.

Fig. 10.1 (opposite) Adult female and embryonized larvae of *Cyclestheria hislopi* (Cyclestherida), SEMs. A: parthenogenetic female, with the left-side valve of the carapace removed, showing a cluster of embryonized stage 7 larvae, ventral view. B: closeup of larvae from A. C: closeup showing the forehead of the larvae, where the embryonic filament attaches. D: embryonized stage 2 larva, ventral view. E: stage 2, with the egg membrane only partly removed, dorsal view. F: stage 2, with the egg membrane completely removed, dorsal view. G: stage 2, frontal view. H: embryonized stage 4 larvae, lateral view. I: stage 4, ventral view. J: stage 4, dorsal view. K: anterior part of the body of stage 4, lateral view. L: anterior part of the body of stage 4, ventral view. A–L of material collected in Colombia in 1994, used in Olesen (1999).

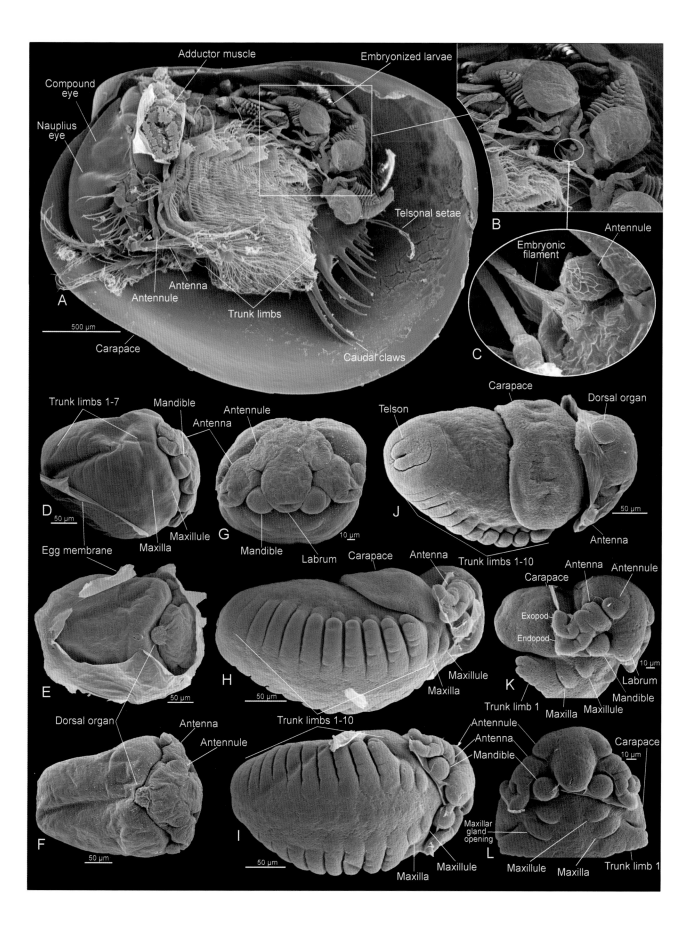

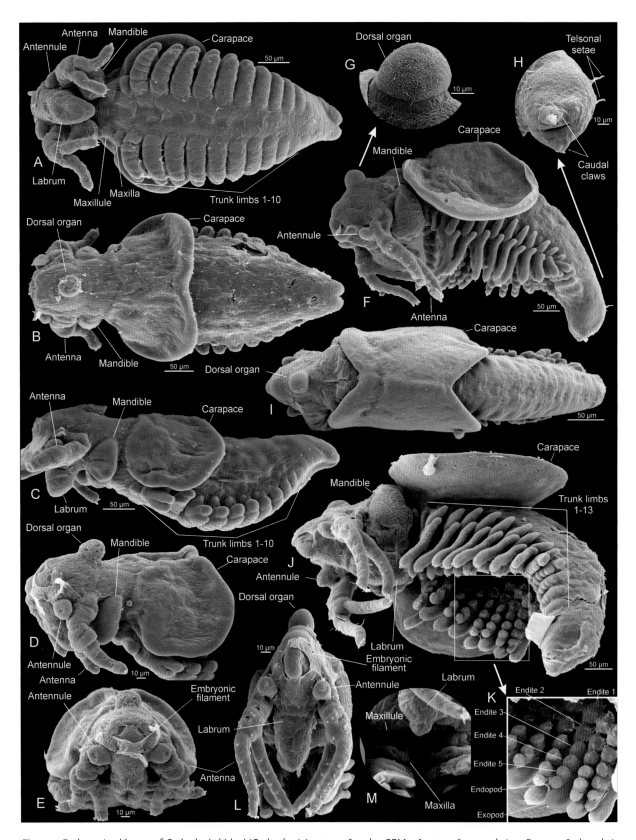

Fig. 10.2 Embryonized larvae of *Cyclestheria hislopi* (Cyclestheria), stages 6 and 7, SEMs. A: stage 6, ventral view. B: stage 6, dorsal view. C: stage 6, lateral view. D: anterior part of the body of stage 6, lateral view. E: stage 6, frontal view. F: stage 7, lateral view. G: closeup of the dorsal organ of stage 7. H: closeup of the caudal region of stage 7. I: stage 7, dorsal view. J: stage 7, lateroventral view. K: closeup of the trunk limbs of stage 7. L: stage 7, frontal view. M: closeup of the maxillules and maxillae of stage 7. A–M of material collected in Colombia in 1994, used in Olesen (1999).

11

JØRGEN OLESEN

Cladocera: Anomopoda

GENERAL: The Anomopoda is a diverse taxon of water fleas (Cladocera). They are commonly found in most freshwater habitats, and even interstitially or in moist habitats (Frey 1980; J. W. Martin 1992; Dumont and Negrea 2002). Forró et al. (2008) estimated 537 currently described species and recognized 11 families. Sizes range from less than 0.3 mm to about 6 mm, and males are smaller than females. The Daphniidae contains the "typical" *Daphnia* (fig. 11.1A), and most members of this family are open-water forms (or even pelagic), while other families, like the Eurycercidae (fig. 11.2A), Chydoridae (fig. 11.2H), Macrothricidae (fig. 11.2E), and Acantholeberidae, contain lesser known bottom-dwelling or vegetation-associated species. As do most other cladocerans, anomopods have a pair of biramous swimming antennae, small sensory antennules, a carapace covering only the trunk and appendages, and a post-abdomen ending in a pair of claws (e.g., Fryer 1987a; J. W. Martin 1992; Dumont and Negrea 2002). The five to six pairs of trunk limbs in the Anomopoda are not serially similar, as they are in the Ctenopoda; rather, each limb pair is modified to perform various functions in different taxa. These modifications are linked to a diversity in feeding strategies. Most species feed by filtration or scraping, but others are scavengers on other crustaceans (*Pseudochydorus*), and 1 genus, *Anchistropus*, is ectoparasitic on freshwater hydras (Fryer 1968, 1974, 1991). The monophyly of the Anomopoda has been supported by most workers (Fryer 1995; Olesen 1998; Negrea et al. 1999; Stenderup et al. 2006; Richter et al. 2007). A variety of phylogenetic positions of the Anomopoda within the Cladocera have been suggested, so this question must be considered unresolved. As in other cladocerans, the life cycle of anomopods is heterogonic (alternating parthenogenesis and sexual reproduction).

LARVAL TYPES

Stages (Instars): There are no true larvae, but the stages of development (instars) are described here. The life cycle is heterogonic. In temperate regions many cycles of parthenogenetic generations normally occur from spring to late summer or early autumn, followed (after mating) by the release of resting eggs that hatch in the spring and start a new parthenogenetic cycle (e.g., Dumont and Negrea 2002), essentially similar to what is summarized for ctenopod cladocerans in chapter 12. As in ctenopods, development is direct and takes place in a brood pouch formed by the carapace valves. Development from resting eggs is also direct (as opposed to *Leptodora kindtii*; see chapter 13). Most earlier treatments have dealt with early embryogenesis or single stages (e.g., Samassa 1893; Cannon 1921; Baldass 1941), while important aspects of later development have escaped attention. The most detailed recent treatment is that of Kotov and Boikova (2001), who studied the late embryogenesis of *Daphnia galeata* and *D. hyalina* by observing live embryos removed from the brood chambers of females. As for the Ctenopoda (see chapter 12), Kotov and Boikova (2001) recognized four stages (instars) taking place in the female's brood chamber, separated by the shedding of membranes (molts). An important difference between the Ctenopoda and Anomopoda is that one more instar is incorporated into the egg in the latter, which therefore hatches as instar IV, not instar III (still inside the female's brood chamber). Hence the development of the Anomopoda appears to be shortened, or more embryonized, compared with the Ctenopoda. In the resting eggs of anomopods (and ctenopods and onychopods) all embryonic stages are incorporated into the egg membrane (Kotov and Boikova 2001).

MORPHOLOGY: The development of *Daphnia hyalina*, about which many details are known, is briefly summarized here, based mostly on Kotov and Boikova (2001). Surprisingly little is known about the late development of other species of the diverse Anomopoda, so my summary of *D. hyalina* is supplemented by new SEM examinations of the developmental stages of 6 selected anomopod species, representing 5 different families.

Daphnia hyalina (Daphniidae) (fig. 11.1B): The laying of parthenogenetic eggs in the dorsal brood chamber in *Daph-*

nia has often been described (Weismann 1876–1879; Baldass 1941; Rossi 1980; Zaffagnini 1987). The first instar lasts until the (first) outer egg membrane is shed, after approximately 16 hours (Kotov and Boikova 2001). It covers early embryogenesis and the early appearance of the antennules, antennae, labrum, maxillary zone, and first thoracic segment. The second instar lasts until the second membrane is shed, after approximately 20 hours (Kotov and Boikova (2001). During this instar the maxillules and maxillae start to differentiate, the last thoracic segment and the eye pigmentation appear, and the swimming antennae perform their first movements. The third instar lasts until the third membrane is shed, after approximately 5 hours (Kotov and Boikova (2001). During this instar the first movements of the post-abdomen and the labrum occur, and gut peristalsis and the first heartbeats can be seen. The fourth instar lasts until the final molt, which takes place outside the female's brood chamber after approximately 16 hours (Kotov and Boikova 2001). During this instar the thoracic limbs become free and full fusion of the eyes take place. The first movements of the mandibles, the maxillules, the thoracic limbs, and the eye capsule can be seen. The fourth instar is released from the female's brood chamber as a neonata approximately half an hour before the final molt.

Simocephalus vetulus (Daphniidae) (fig. 11.1C–E): Two stages have been examined, a relatively early stage (fig. 11.1D), developmentally corresponding to approximately a 20–25 hour stage of *Daphnia hyalina*, and a late stage (fig. 11.1E), roughly corresponding to a 40–50 hour stage. Both have been removed from the brood chambers of females. The earliest stage (fig. 11.1D) has a rather globular overall shape and is very embryo-like. Anlagen of various structures can be recognized: small antennular buds with some distance between them, large biramous antennae that have become free of the body, maxillulary and maxillary buds, anlagen to five pairs of trunk limbs, and carapace and post-abdominal anlagen. In the later stage that was examined (fig. 11.1E), much of the adult shape and morphology has been achieved. This includes a small rostrum, on which a pair of small tubular antennules are inserted; an anterior head shield partly covering the bases of the antennae; and well-developed carapace valves that cover the trunk limbs and (in part) the post-abdomen, which now carries a pair of post-abdominal setae.

Moina brachiata (Moinidae) (fig. 11.1G–J): Three stages have been examined: an early stage (fig. 11.1G, H), an intermediate stage (fig. 11.1I), and a late stage (fig. 11.1J), which, based on Grobben's (1879) comprehensive study of the development of *Moina rectirostris*, approximately covers the entire development of this species. All have been removed from the brood chambers of females. The earliest examined stage of *M. brachiata* (fig. 11.1G, H) has a peculiar gross morphology that seems characteristic for *Moina* (identical to what Grobben described for *M. rectirostris*). This includes the facts that the embryo is strongly curved, so that both ends bend dorsally; the antennular buds are still placed laterally; no labrum covers the mouth opening; the trunk limbs constitute large undifferentiated ventrolateral lobes on the body; and, unlike other

cladoceran embryos examined by SEM, the cellular pattern is visible on the surface, which is probably related to the embryo's direct exposure to nutritional fluid in the female's brood chamber (see below). The intermediate stage (fig. 11.1I) looks more like that of other anomopods. The antennular buds have moved slightly medially and have become rather large, which reflect the relatively large size of adult antennules in moinids. A labrum has appeared. Relatively large maxillulary buds and smaller, more laterally placed maxillary buds are now present, the latter with a small pore (maxillary gland opening). The buds of five pairs of limbs are now present, all already different from each other, foreshadowing the situation in adults, where each pair of limbs acts in a separate fashion from the other limbs in the filtration cycle (e.g., Fryer 1991). The late stage that was examined (fig. 11.1J) has much of the adult shape and morphology. This includes a pair of rather long tubular antennules, a pair of very long antennae, and a bivalved carapace enclosing the trunk limbs and the post-abdomen.

Eurycercus glacialis (Eurycercidae) (fig. 11.2A–D): Three stages have been examined: an early stage (fig. 11.2B), an intermediate stage (fig. 11.2C), and a late stage (fig. 11.2D), representing three distinctly different phases of development. All have been removed from the brood chambers of females. The earliest examined stage (fig. 11.2B) is globular in shape. A membrane covering most of the embryo has been removed (but probably at least the antennae are free at this stage), which reveals various aspects of gross morphology. The antennules are small buds, with some distance between them. The antennae are rather small and biramous. The mandibles are small buds, and the maxillules and maxillae (with maxillary gland pores) are yet smaller buds, aligned in position with the gnathobase of the trunk limbs. Five pairs of trunk limb buds are present, all being slightly different in morphology from each other (no serial similarity). An early carapace fold is present on each side of the body, originating behind the maxillary limb buds. In the morphology of the intermediate stage (fig. 11.2C), a head shield has appeared. The antennular buds have become tubular and have migrated toward the midline. The labrum has become larger and more pointed. The mandibular buds have become larger. The antennae have also become larger, with weakly developed setation, The rest of the body is covered by a membrane. In the late stage that was examined (fig. 11.2D), much of the adult shape and morphology has been achieved. This includes a distinct head shield, a pair of small tubular antennules with distal sensillae, a bivalved carapace, five to six trunk limbs with weakly developed setation, and a large post-abdomen that extends well beyond the carapace. Aspects of late embryogenesis (e.g., development of the maxillae) were studied by Kotov (1996) in *E. lamellatus*, a closely related species.

Macrothrix hirsuticornis (Macrothricidae) (fig. 11.2E–G): Two stages have been examined: a relatively early stage (fig. 11.2F) and an intermediate stage (fig. 11.2G). Both have been removed from the brood chambers of females. The earliest examined stage (fig. 11.2) was surrounded by a membrane, which was removed before examination. Several rudimentary structures are developed externally at this stage: a rather broad labrum, a pair of fleshy antennular limb buds with some distance between them, a pair of mandibular buds, a pair of relatively small biramous antennae placed laterally on the body, small maxillulary and maxillary buds (the latter with maxillary gland pores) aligned in position with the proximal portions of the trunk limbs, five pairs of largely undifferentiated trunk limb buds, and rudimentary lateral folds of the carapace placed laterally to the maxillules and maxillae. The intermediate stage (fig. 11.2G) was also surrounded by a membrane that was partly removed before examination. This stage is modified from the previously described stage in various ways. The labrum has become larger and more pointed posteriorly. The antennules have migrated toward each other medially and have become elongate and club-shaped, foreshadowing the situation in adults, The mandibular buds have become slightly larger. The antennae have become much larger, again foreshadowing the situation in adults. The buds of the maxillules and maxillae are still present, although the maxillulary buds have moved toward the midline and have developed some primordial setation. The maxillary buds are smaller and still have the pores of the future maxillary glands. Five trunk limb buds are present, which are slightly different from each other morphologically; trunk limbs 1–4 have weak setation along the future median limb margin.

Acantholeberis curvirostris (Acantholeberidae) (fig. 11.2I): One stage—a relatively early one—has been examined. Prior to examination it was covered by a membrane, which has been removed. It is quite similar to equivalent stages of other anomopods. In gross morphology it has a short and broad labrum, a pair of short broad antennular buds, a pair of relatively short biramous antennae, and a pair of small mandibular buds. The buds of the maxillules and maxillae are only weakly indicated and are aligned in position with the gnathobases of the trunk limbs. No maxillary gland pores can be seen. Weakly indicated buds of five pairs of trunk limbs are present, with the proximal gnathobases being the most distinct part.

Lathonura rectirostris (Macrothricidae) (fig. 11.2J, K): One stage—a relatively early one—has been examined (fig. 11.2K). It is quite similar to equivalent stages of other anomopods. Its gross morphology has a short and broad labrum, and a pair of relatively long antennular buds that are widest distally. The buds of the maxillules and maxillae are only weakly indicated. The maxillae buds are slightly laterally displaced and have maxillar gland pores. Four pairs of weakly indicated trunk limb buds are present, which are all are different from each other.

MORPHOLOGICAL DIVERSITY: The morphological diversity of the developmental stages of the Anomopoda is less than in taxa with free-living larvae. There are many differences in details, however, such as the specific morphology and position of various embryonic limb buds. Also, the overall shape of the embryos is varies greatly among taxa. Embryos of *Moina*, due to their nährboden-supplied development, have a quite different general morphology from embryos of other

taxa. The Anomopoda is the most diverse high-level taxon within the Branchiopoda, and the development of a very large number of its species remains unknown.

NATURAL HISTORY: In the majority of anomopod species, the development of embryos in the brood chamber relies on yolk supplies. Only in moinids does a nährboden produce a substance that nourishes the embryos within a closed brood chamber (Weismann 1876–1879), such as in *Penilia* (Ctenopoda) and the Onychopoda. Patt (1947) stated that when the total volume of a fully formed embryo is compared with that of a newly laid egg, the Cladocera that possess a nährboden exhibit a greater increase during this period than those Cladocera in which the nährboden is lacking. This clearly applies to the moinids examined here, since the earliest examined stage is significantly smaller (devoid of yolk) than the later stages (fig. 11.1G–J). The characteristic exposed cell surfaces found in early embryos (fig. 11.1G–I) of *Moina brachiata* can probably be explained by the fact that the embryos in the female's brood chamber are surrounded by nutritional fluid. Various limbs and organ systems start functioning before the embryos are released from the female's brood chamber, so that when juveniles leave the chamber, they are ready to move and feed. Little information is available on this process, with the detailed study by Kotov and Boikova (2001) on species of *Daphnia* being an important exception. These authors found that the development of *Daphnia*, and therefore probably other anomopods, is more embryonized than that of the relatively similar ctenopods, since an extra stage (stage 3) is passed within the egg membrane (see chapter 12).

PHYLOGENETIC SIGNIFICANCE: Most workers agree that the Anomopoda is monophyletic and nested within the Cladocera (e.g., Fryer 1995; Olesen 1998; Negrea et al. 1999; Stenderup et al. 2006; Richter et al. 2007), but the group's relationship to other cladocerans is unsettled. A small SEM study of cladoceran embryos conducted in preparing this chapter suggests that studies of the limb development of a variety of anomopod species could be phylogenetically informative. Many of the examined species pass through a relatively similar phase (a phylotypic stage) with rather small limb buds (except the antennae), such as *Simocephalus*, *Eurycercus*, *Macrothrix*, *Acantholeberis*, and *Lathonura* (figs. 11.1D; 11.2B, F, K, I). Major differentiations occur later in development and most notably include modifications in the size and position of the antennules (which migrate medially and sometimes become very large), the position of the maxillules and maxillae, and the morphology of the trunk limbs. The latter is especially significant for anomopods, which, in contrast to other cladocerans (the Ctenopoda, Onychopoda, and Haplopoda), is characterized by the marked non-seriality of the trunk limbs. This differentiation is intimately linked to the various feeding strategies in different families and has been central to anomopod evolution (Fryer 1968, 1974, 1991, 1995). The non-seriality of the trunk limbs is seen quite early in development, and intermediate embryos clearly indicate which limbs and which limb parts will come to play a prominent role in the adults. For example, in *Moina* and *Daphnia* the gnathobases of trunk limbs 3 and 4 are already much developed (fig. 11.1B, I), which foreshadows their important role in the adult feeding system of these taxa (Fryer 1991).

HISTORICAL STUDIES: The development of anomopod cladocerans, as well as other cladocerans, received a significant amount of early attention, probably due to the easy accessibility of water fleas (such as anomopods) in many kinds of fresh water. Their morphology was the favorite object of study by seventeenth-century naturalists, such as Jan Swammerdam and Antonie van Leeuwenhoek. Among the most detailed early treatments of anomopod development are those of Weismann (1876–1879) and Grobben (1879), which laid most of the basis for later studies. The most comprehensive recent study is that of Kotov and Boikova (2001).

Selected References

Grobben, C. 1879. Die Entwicklungsgeschichte der *Moina rectirostris*: zugleich ein Beitrag zur Kentniss der Anatomie der Phyllopoden. Arbeiten aus dem Zoologischen Institute der Universität Wien und der Zoologischen Station in Triest 2: 203–268.

Kotov, A. A., and O. S. Boikova. 2001. Study of the late embryogenesis of *Daphnia* (Anomopoda, "Cladocera," Branchiopoda) and a comparison of development in Anomopoda and Ctenopoda. Hydrobiologia 442: 127–143.

Weismann, A. 1876–1879. Beiträge zur Naturgeschichte der Daphnoiden. Zeitschrift für Wissenschaftliche Zoologie, vols. 27–33: 1–486.

Fig. 11.1 (opposite) Embryos and adults of various anomopods (Cladocera). A: adult female and embryos of *Daphnia magna* (Daphniidae), light microscopy, lateral view. B: drawing of embryos of *D. hyalina*, ventral view. C: adult female of *Simocephalus expinosus* (Daphniidae), light microscopy, lateral view. D and E: *S. vetulus*, SEMs. D: relatively early embryonic stage, ventral view. E: late stage, lateral view. F: adult female of *Moina salina* (Moinidae), light microscopy, lateral view. G–J: *M. brachiata*, SEMs. G: early stage, ventral view. H: early stage, dorsal view. I: intermediate stage, ventral view. J: late stage, ventral view. A, C, and F of material from France, courtesy of Jean-François Cart; B modified after Kotov and Boikova (2001); D and E original, of material from Dyrehaven, north of Copenhagen in Denmark; G–J original, of material from Jutland in Denmark.

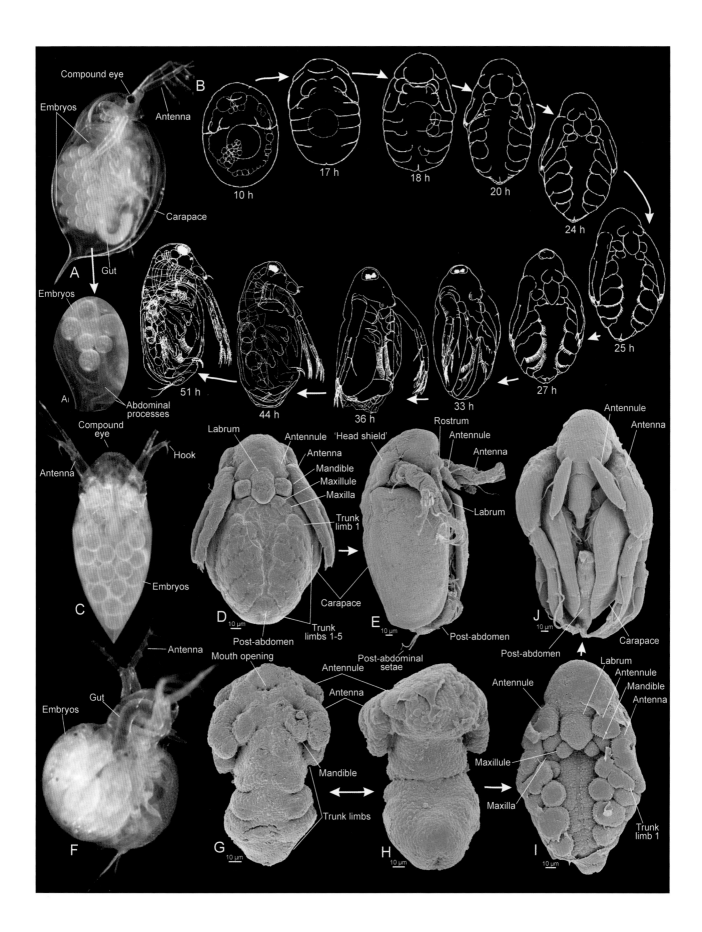

Compound eye

Embryos

Antenna

Carapace

A

Gut

B

10 h   17 h   18 h   20 h

24 h

25 h

27 h

51 h   44 h   36 h   33 h

Embryos

Abdominal
processes

Compound eye

Antenna

Hook

Embryos

C

Antenna

Gut

Embryos

F

Labrum

Antennule

Antenna

Mandible

Maxillule

Maxilla

Trunk
limb 1

D

Carapace

Trunk
limbs 1-5

Post-abdomen

10 μm

Mouth opening

Rostrum

'Head shield'

Antennule

Antenna

Labrum

E

Carapace

Post-abdomen

10 μm

Antennule

Antenna

Post-abdominal
setae

Antennule

Antenna

Mandible

Trunk limbs

G

10 μm

H

10 μm

Antennule

Antenna

Maxillule

Maxilla

Trunk
limb 1

I

10 μm

Antennule

Antenna

J

Post-abdomen

Carapace

Labrum

Antennule

Mandible

Antenna

10 μm

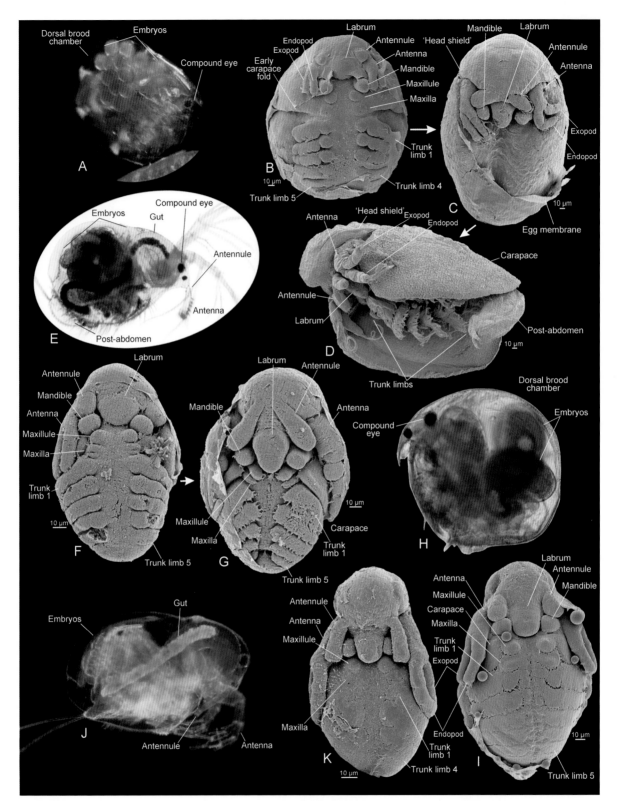

Fig. 11.2 Embryos and adults of various anomopods (Cladocera) *(continued)*, SEMs (unless otherwise indicated). A–D: *Eurycercus glacialis* (Eurycercidae). A: adult female, light microscopy, lateral view. B: early stage, ventral view. C: intermediate stage, ventral view. D: late stage, lateral view. E–G: *Macrothrix hirsuticornis* (Macrothricidae). E: adult female, light microscopy, lateral view. F: relatively early stage, ventral view. G: intermediate stage, ventral view. H: adult female of *Chydorus sphaericus* (Chydoridae), light microscopy, lateral view. I: relatively early stage of *Acantholeberis curvirostris*, ventral view. J and K: *Lathonura rectirostris* (Macrothricidae). J: adult female, light microscopy, lateral view. K: relatively early stage, ventral view. A, E, H, and J of material from France, courtesy of Jean-François Cart; B–D original, of material from Qeqertarsuaq, Greenland; F and G original, of material from Greenland; I original, of material from Læsø in Denmark; K original, of material from Bornholm in Denmark.

12

JØRGEN OLESEN

Cladocera: Ctenopoda

GENERAL: Ctenopods are small, often hyaline water fleas (Cladocera). The species are divided among 3 families (the Sididae, Holopediidae, and Pseudopenilidae), which are found most commonly in open water, although some are benthic or associated with vegetation. Forró et al. (2008) estimated 50 currently described species. Most are freshwater species, but there are a couple of marine species. Their size is typically from 1 to 3 mm. As is true for most other cladocerans (anomopods), they have a pair of large biramous swimming antennae, small sensory antennules, a carapace covering only the trunk and appendages, and a post-abdomen ending in a pair of claws (e.g., Fryer 1987a; J. W. Martin 1992; Dumont and Negrea 2002). Characteristics for ctenopods include six pairs of serially similar, metachronally beating phyllopodous legs, which in most species are used for filter feeding. Species of *Holopedium* (Holopediidae) develop a gelatinous dorsal mantle (jelly coat), which most likely is a floating device (Montvilo et al. 1987). *Sida crystallina* (Sididae) has a unique type of neck organ that it uses to attach itself to vegetation while filtration feeding (Günzl 1978). *Penilia* (Sididae) is marine and is widespread in the world's oceans (e.g., Johns et al. 2005), while *Pseudopenilia* (Pseudopenilidae) is found only in the deep anaerobic zone of the Black Sea (Korovchinsky and Sergeeva 2008). The monophyly of the Ctenopoda is agreed upon by most workers (e.g., Fryer 1987a; Korovchinsky 1990; Olesen 1998), but the position of ctenopods within the Cladocera is not yet clarified, and different treatments have lead to different results (e.g., Olesen 1998, 2009; Negrea et al. 1999; Braband et al. 2002; Stenderup et al. 2006; Richter et al. 2007). As in other cladocerans, the life cycle of ctenopods is heterogonic (alternating parthenogenesis and sexual reproduction).

LARVAL TYPES

Stages (Instars): There are no true larvae, but stages of development are described here. The life cycle is heterogonic. In temperate regions many cycles of parthenogenetic generations normally occur from spring to late summer or early autumn (and sometimes to midwinter), followed (after mating) by the release of resting eggs that hatch in the spring and start a new parthenogenetic cycle (Korovchinsky and Boikova 1996; Dumont and Negrea 2002). In the parthenogenetic part of the life cycle, development takes place in a dorsal brood chamber, formed by the carapace valves (fig. 12.1L). Development is direct, since the offspring are released from the brood chamber as miniature adults. Development from the hatching eggs is also direct (as opposed to *Leptodora kindtii*; see chapter 13). Most earlier studies of the development of the parthenogenetic generations of the Ctenopoda dealt with the early embryogenesis of various species (Samassa 1893; Agar 1908; Dejdar 1930; Baldass 1937; Löpmann 1937), while only a few treat their late development—the topic of this chapter. The later development of species of the marine genus *Penilia*—which (unusually among cladocerans) have a closed brood chamber with a maternal placenta-like structure (e.g., Egloff et al. 1997), similar to moiniids and onychopod cladocerans (see chapters 11 and 14)—has received some attention (Sudler 1899; Della Croce and Bettanin 1965; Onbé 1978; Atienza et al. 2008; Fritsch et al. 2013). The most recent treatment of the late development of the parthenogenetic part of the life cycle in ctenopods was by Kotov and Boikova (1998), who studied live embryos removed from female brood chambers of *Sida crystallina* and *Diaphanosoma brachyurum*. They suggested a new staging system, involving four instars demarcated by the shedding of membranes, which they interpreted as embryonic molts (instars). The development of *Diaphanosoma brachyurum* summarized in this chapter is based on Kotov and Boikova (1998) and on new SEM data presented here (fig. 12.1).

MORPHOLOGY

First Instar: The first instar of *Diaphanosoma brachyurum* lasts from the entry of the egg into the brood chamber until the outer egg membrane is shed (first molt) (Kotov and Boikova 1998). It lasts 15 hours and covers two phases: early embryogenesis (egg organization, and blastula and gastrula formation), and the beginning of the later embryogenesis.

Second Instar (fig. 12.1A–G): The second instar lasts until the second egg membrane is shed (second molt). It lasts 12–13 hours. The entire body is enclosed in a membrane (partly removed in the specimens shown in fig. 12.1). This instar is characterized by large antennal limb buds that now approximately reach the first trunk limbs but still are not free of the body. Body segmentation takes place In the early phase of this instar, followed by a long period in which the formation of the trunk limbs take place. The examples of embryos illustrated here represent two phases of instar II, an early phase (fig. 12.1A–C) and a late phase (fig. 12.1D–F), with a number of differences between embryos belonging to these two phases. In the early phase the embryos have more laterally placed antennular buds, which have migrated more ventrally in the late phase. The labrum has become more pointed in the late phase; the antennae have become longer, with more straight rami; and the mandibular buds have become more distinct. In the early phase it is not possible to distinguish between the maxillular and maxillar limb buds, which can be done in the late phase. The anlagen to the carapace have enlarged significantly in the late phase, and a dorsal organ has appeared.

Third Instar: The third instar lasts until the third egg membrane has been shed (third molt). It has free and movable swimming antennae, while the trunk limbs and post-abdomen are still enclosed in a membrane. This instar lasts 6–7 hours. Kotov and Boikova (1998) noted two phases for this instar: one where eye pigmentation is absent, and another where the embryo has small red-brown eyes.

Fourth Instar (fig. 12.1H, I): The fourth instar lasts until the fourth egg membrane is shed (fourth molt), which takes place shortly after the embryo has been released from the female's brood chamber. This instar lasts 18–19 hours. It is characterized by long swimming antennae, which grow in size during the duration of this instar, and by free trunk limbs. A stage examined in relation to this chapter (fig. 12.1H, I) seems to belong to the early phase of the fourth instar (not to the third), since no membrane surrounds the trunk limbs and post-abdomen. This stage has basically developed all of the major structures seen in the adult, but in simpler, less setose versions. The antennules are small and tubular, with a distal seta. The antennae are very large, especially the protopods. The labrum is distinctly pointed. The mandibles have started to curve toward each other, but the future gnathal edges do not yet meet in the midline. Both the maxillules and maxillae are present as small bud-like limbs, the maxillule with primordial setation, and the maxillae with the opening to the future maxillary glands. There are six pairs of trunk limbs, with both setose exopods and setose median limb margins; the latter is weakly subdivided into apparently five rudimentary endites. Epipods are also present (but not shown). Posteriorly, the post-abdomen has developed a pair of weak claws.

The fourth instar is the one that leaves the brood chamber; when newly released, it is called the neonata (Kotov and Boikova 1998). This is not a special instar, since it cannot be distinguished from the late embryos in the brood chamber and no molt separates them. The neonata (fig. 12.1J) molts into the first juvenile (fig. 12.1K) within a few minutes, which marks the end of what Kotov and Boikova (1998) defined as the embryonic period. The molted first juvenile instar differs from the neonata in being generally more graceful. In addition, the antennae bear longer and setulated setae, the carapace covers all the limbs, and the post-abdominal claws are more developed. The post-embryonic development of *Diaphanosoma brachyurum* was studied by Boikova (2005).

MORPHOLOGICAL DIVERSITY: The known differences in the late embryogenesis of species of the Ctenopoda are rather small. Kotov and Boikova (1998) examined both *Sida crystallina* and *Diaphanosoma brachyurum* and found four instars in both, but with a slightly different duration in the 2 species (especially instar IV, which endures longer in *S. crystallina*). The timing of the onset of the function and early movements of various appendages and organ systems also vary between the 2 species.

NATURAL HISTORY: In the majority of ctenopod species, the development of embryos in the brood chamber relies on yolk supplies. Only in *Penilia* does a nährboden produce a substance that nourishes the embryos within the closed brood chamber (Sudler 1899; Potts and Durning 1980). Kotov and Boikova (1998) reported that in embryos of *Diaphanosoma brachyurum* and *Sida crystallina*, movements of various limbs and organs start long before the embryos leave the brood chamber. For example, in *D. brachyurum*, the first movements of the swimming antennae start during instar II, and the first heartbeats and gut peristalsis can be seen during instar III, while the first movements of the mandibles, labrum, and trunk limbs cannot be seen before instar IV. Kotov and Boikova (2001) compared the embryogenesis of the Ctenopoda and Anomopoda and found that the development of ctenopods is less embryonized than that of anomopods (at least of *Daphnia*). In both *Diaphanosoma brachyurum* and *S. crystallina*, two instars (III and IV) are passed after hatching, but this still occurs within the female's brood chamber, while in anomopods, instar III is passed within the egg membrane. Hence the development of ctenopods is less embryonized than that of anomopods, and in this respect their development constitutes an intermediate between the Cyclestherida, where larvae have also become embryonized (see chapter 10), and the Anomopoda.

PHYLOGENETIC SIGNIFICANCE: Nearly all workers agree that the ctenopods are a monophyletic group, but their relationship to other cladocerans is unsettled, with several different hypotheses put forward (Olesen 1998, 2009; Negrea et al. 1999; Braband et al. 2002; Stenderup et al. 2006; Richter et al. 2007). Additionally, aspects of their development appear to be intermediate between the Cyclestherida and Anomopoda.

HISTORICAL STUDIES: Most early studies deal only with early embryogenesis (Samassa 1893; Agar 1908; Dejdar 1930; Baldass 1937; Löpmann 1937), with Sudler (1899), who treated the late embryogenesis of *Penilia*, as a significant exception. The most important recent studies are that of Kotov and Boikova (1998) and Fritsch et al. (2013).

Selected References

Baldass, F. 1937. Entwicklung von *Holopedium gibberum*. Zoologische Jarhbücher, Abteilung für Anatomie und Ontogenie der Tiere 63: 399–454.

Fritsch, M., O. R. P. Bininda-Emonds, and S. Richter. 2013. Unraveling the origin of Cladocera by identifying heterochrony in the developmental sequences of Branchiopoda. Frontiers in Zoology 10: 35.

Kotov, A. A., and O. S. Boikova. 1998. Comparative analysis of the late embryogenesis of *Sida crystallina* (O. F. Müller, 1776) and *Diaphanosoma brachyurum* (Lievin, 1848) (Crustacea: Branchiopoda: Ctenopoda). Hydrobiologia 380: 103–125.

Sudler, M. T. 1899. The development of *Penilia schmaeckeri* Richard. Proceedings of the Boston Society of Natural History 29: 107–133.

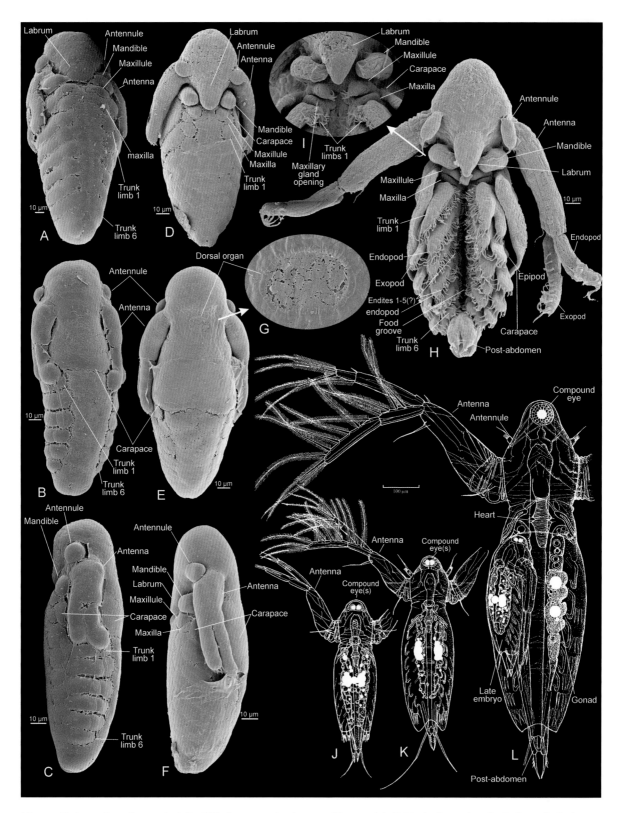

Fig. 12.1 Embryos, juveniles, and adult of *Diaphanosoma brachyurum* (Ctenopoda), SEMs (unless otherwise indicated). Embryos staged according to Kotov and Boikova (1998). A–C: embryos from the same brood chamber, approximately 17–18 hours old, representing the early phase of instar II. A: ventral view. B: dorsal view. C: lateral view. D–F: embryos from the same brood chamber, approximately 21–25 hours old, representing the late phase of instar II, with only partly free antennae. D: ventral view. E: dorsal view. F: lateral view. G: closeup of the dorsal organ of the specimen in E. H: embryo, at approximately 34–35 hours old, representing an early instar IV, ventral view. I: closeup of the mouth region of an embryo approximately 34–35 hours old. J: drawing of a newly released embryo (neonata), before the final embryonic molt. K: drawing of a first juvenile instar, after this molt. L: drawing of an adult female, with one embryo shortly before its release. A–I original, from material collected in Denmark in 1993 by Jørgen Olesen; J–L modified after Kotov and Boikova (1998).

13

Cladocera: Haplopoda

Jørgen Olesen

GENERAL: The Haplopoda is a taxonomically small group of large (typically about 12 mm), nearly transparent predatory water fleas (cladocerans). It consists of only 1 genus, *Leptodora*, which contains 1 widespread species, *L. kindtii*, described more than 150 years ago in an 1844 newspaper article by Focke (see Dumont and Hollwedel 2009), and another species, *L. richardi*, described recently (Korovchinsky 2009). *Leptodora kindtii* is fairly widespread in fresh water, and sometimes occurs in brackish water (e.g., parts of Chesapeake Bay) (W. Johnson and Allen 2012), but it is most often found in the deeper parts of lakes across the Northern Hemisphere. The species was recently shown to have had a rather complicated evolutionary history (Xu et al. 2010). Unlike most other cladocerans, *Leptodora* is an obligate carnivore, feeding on other planktonic crustaceans such as smaller water fleas and copepods, and much of its peculiar morphology is explainable as adaptations to this lifestyle. This includes its nearly hyaline body; large eye; two rows of segmented legs, forming a large feeding basket; and modified mouth region, housing a pair of styliform mandibles and a peculiar trilobed lower lip. The large abdomen is also possibly (at least in part) an adaptation to a predatory lifestyle. Live prey (e.g., *Diaphanosoma*) are seized by the long first pair of legs and maneuvered into the feeding basket, while the long limbless abdomen is temporarily bent forward to assist in trapping the prey. The styliform mandibles are pushed deeply into the tissue of the prey, and the pharynx starts suctioning movements (Sebestyén 1931; Herzig and Auer 1990; Browman et al. 1989). Another specialization in *Leptodora*, unseen in other cladocerans, is the presence of free-living metanauplii in the sexual part of the life cycle. The unusual appearance of *Leptodora* has resulted in much attention being paid to its phylogenetic position. Most of the recent morphological and molecular evidence suggest that *Leptodora* (Haplopoda) is the sister-group to the Onychopoda, another group of water fleas to which this species bears some resemblance in its morphology (segmented trunk legs) and feeding behavior (raptorial feeding, in contrast

to filtration) (J. W. Martin and Cash-Clark 1995; Olesen 1998; Richter et al. 2001, 2007, Swain and Taylor 2003). As in other cladocerans, the life cycle of *Leptodora* is heterogonic (alternating parthenogenesis and sexual reproduction).

LARVAL TYPES

Metanauplius and Direct Development: The life cycle is heterogonic. It begins in the early spring, when free-living metanauplii hatch from the resting eggs (winter eggs). The first worker to report free-living larvae (metanauplii) of *Leptodora* was Sars (1873), and apparently there have been only two later treatments (Warren 1901; Sebestyén 1949). In all cases the larvae were collected in early spring (February–April). According to Sebestyén (1949), who studied their development based on specimens kept in petri dishes, the larval phase is passed exceedingly fast. By the third day after hatching the legs assume their final position, being ready to act as a feeding basket. This short duration may explain the few reports of free-living larvae. Based on Sebestyén (1949), three larval stages seem to be passed before the feeding basket has been achieved. Sars (1873), however, depicted a larval stage apparently not accounted for by Sebestyén (1949), so the number of stages seems to be at least four, or even higher. This topic needs more study. When the larvae have become adults, parthenogenesis, which is by far the dominant part of the life cycle, starts, and it lasts from spring to autumn. Varying numbers of eggs are deposited in the dorsal brood pouch (modified carapace) via the dorsally extending oviduct. T. Andrews (1948) reported that egg deposition required 5–10 minutes, starting immediately following ecdysis at the beginning of the first adult instar. All early developmental stages (embryos) are passed within the brood pouch (direct development), and the offspring are not released before they assume the shape of the adults. Both T. Andrews (1948) and Boikova (2008) reported that the brooding period is around 70–90 hours (at room temperature). According to T. Andrews (1948), the free-swimming juveniles/adults are released a few hours prior to the following ecdysis of the mother. At the end of the parthenogenetic part of the life cycle, when conditions become unfavorable (in autumn), males appear and resting eggs are produced, which serve to carry the species through the winter. Resting eggs were studied briefly by Sebestyén (1949), who found that the newly released eggs were surrounded by a gelatinous envelope not present in material collected in the early spring. She suggested that this coating enables the eggs to float in different layers in the open water, but sooner or later they reach the bottom, where they spend the winter and where she assumed hatching takes place.

Few workers have attempted to divide the embryonic development of *Leptodora* into discrete stages, something that is complicated in taxa with embryonized larvae. Boikova's (2008) identification of four stages, which is a modification of a previous suggestion by Olesen et al. (2003), is followed here. Boikova's (2008) scheme takes into consideration not only morphological change, but also the shedding of egg membranes and embryonic molts.

MORPHOLOGY

Metanaupliar Phase (fig. 13.2I–O): Sebestyén (1949) studied—in "classical natural history style"—the development of several free-living metanauplii kept in petri dishes. She managed to hatch larvae from 21 resting eggs that could be kept alive for no longer than 6–7 days. By the third day the adult shape is attained and the characteristic adult predatory feeding begins. Sebestyén (1949) roughly divided development into three phases before adult shape was achieved—a nauplius phase, a metanauplius, and a transitional state—but no molting was directly observed and no complete exuviae (only fragments) were found, even after careful investigation. Sars (1873) reported a larval stage that falls clearly between Sebestyén's (1949) nauplar and metanaupliar phases, so there is some indication that the true pre-adult sequence of free-living leptodoran larvae consists of four stages. Metanauplius III may actually consist of more than one instar, since the larvae grouped in this category differ in some minor aspects, most significantly in the degree of reduction of the mandibular palps. Sebestyén (1949) found some indication that only a part of the exuvium is shed between certain stages, so this topic is complicated and needs more study. The first three stages will be termed metanauplii here, since rudimentary post-mandibular limbs are present. The largest known pre-adult stage is too developed to be termed a metanauplius, and may more appropriately be referred to as a pre-juvenile. Sebestyén (1949) called it a "transitional stage," since it combines larval and adult morphology.

Metanauplius I (fig. 13.2I): This is the hatching stage (Sebestyén 1949). The stage has been illustrated in greatest detail by Warren (1901, his fig. 1), and its general morphology was confirmed by Sebestyén (1949). It is most significantly characterized by (1) a pair of enormous swimming antennae, which at this stage are relatively much longer than in any other period of life; (2) a pair of unique, tubular, unsegmented, posteriorly directed mandibular palps, each of which carries three setae

distally and the rudiment of the styliform gnathobase proximally; and (3) a globular, yolk-containing, post-mandibular body region still lacking external rudiments of the caudal rami. The body is only slightly longer than wide. The post-mandibular part of the body is shorter than the mandibular palps. The antennules are small rounded buds. Rudiments of the six pairs of thoracopods are present as conspicuous transverse limb buds positioned laterally on the globular body, and, according to Warren (1901), the maxillules and maxillae are represented by tiny limb buds between the mandibles and the rudiments of the first pair of thoracopods. The lower lip is not yet developed. On three occasions Sebestyén (1949) observed the hatching of this stage, which takes place on the fourth day from the beginning of the differentiation of the resting egg and lasts for about 15–20 minutes. The stage has a duration of only 1–2 hours. This metanaupliar stage swims immediately, and occasionally the eggshell—still covering the main part of the body—is carried with it for a while (possibly a partial molt).

Metanauplius II (fig. 13.2J): This stage, described by Sars (1873, his fig. 9), differs from the previous stage in being more slender / less globular. The body is approximately twice as long as wide. The post-mandibular part of the body is now longer than the mandibular palps and terminates in a pair of pointed rudiments of the caudal rami. The mandibular palps now carry four setae (as opposed to three). The rudiments of the six thoracopods are now present as large, tube-like, transverse limb buds.

Metanauplius III (fig. 13.2K–M): Larvae categorized as belonging to this stage were described by Sars (1873, his figs. 1, 2), Warren (1901, his fig. 2), and Sebestyén (1949, her figs. 4, 5). This stage differs from the previous one in being more slender, about three times longer than wide. Sebestyén (1949) depicted a fifth, small distal setae of the mandibular palp, not shown on the specimen drawn by Sars (1873). The six pairs of rudimentary limb buds appear larger than in the previous stage. The rudiments of the caudal rami are more distinctly set off from the remaining hindbody. Warren (1901) described a very similar larval specimen, but with the mandibular palp having atrophied to a small triangular protrusion that appeared to lack setation (fig. 13.2K). The larval specimens of Sars (1873) and Sebestyén (1949) may represent a stage different from that of Warren (1901), but this needs more study.

Pre-juvenile / "transitional stage" / metanauplius VI (fig. 13.2N, O): This stage, described by Sars (1873, his figs. 5, 6) and Sebestyén (1949, her fig. 5), is significantly different from the previous one and has started to approach the morphology of the adults in several respects, which is why it may be better termed a pre-juvenile stage rather than a metanauplius. Sebestyén (1949) called it a transitional stage. This stage is more slender than the previous one, 4–5 times longer than wide. Immediately behind the thorax region, the hindbody bends weakly ventrally, foreshadowing the condition in adults. The hindbody has started to become segmented, but it has not yet attained full adult segmentation. The six pairs of thoracopods are no longer attached to the body along their length, but have become free and hang down from the body, still non-functional but ready to form the typical catch-basket of the adult. The carapace / brood pouch is present as a small dorsal flap immediate behind the head. The caudal rami are long and pointed, bearing rows of denticles, as in adults. The mandibular palps are still present but are small triangular protrusions, as they are in some specimens belonging to the previous stage.

Embryonic Development: Boikova's (2008) division of the embryonic development of *Leptodora* into four stages is followed here, with the main aspects of development building on Boikova (2008) and Olesen et al. (2003), unless specified otherwise.

Stage 1: Stage 1 starts when the egg is placed in the brood pouch and continues until the embryo possesses rudiments of all head and thoracic appendages in the ventral germ band. This stage, as recognized by Boikova (2008), lasts for 40 hours, during which the embryo retains the shape and size of the egg. After approximately 19 hours, the antennae appear and segmentation of the head begins. By 23 hours, the incipient furrows of the labrum and antennules appear. By 31 hours, the furrows of the first pair of trunk limbs appear. By 26 hours, the outer egg membrane cracks longitudinally and is shed as two hemispheres. The small rudiments of the maxillules and maxillae appear almost simultaneously after 33 hours, at the same time as the furrows of the second trunk limbs.

Stage 2 (figs. 13.1A; 13.2A, G, H): Stage 2 starts when the embryo begins changing its shape and becomes more elongate. This stage lasts until the embryo is released from the inner egg membrane, in about 8 hours. During stage 2, the embryo loses its egg shape, but it is still somewhat globular, with much yolk. The limbless part of the hindbody appears posteriorly in the germ band. The buds of the labrum and antennules become rounded and much more prominent. The antennae elongate significantly, but they are still attached to the remaining embryo along their full length. Two dorsolateral swellings immediately behind the mandibles represent paired rudiments of the carapace. No segmentation of the buds of the trunk limbs is seen externally, but methods such as distalless staining show its presence (fig. 13.2G, H). Specimens belonging to what is here recognized as stage 2 were categorized as stage 1 by Olesen et al. (2003), who did not study early development.

Stage 3 (figs 13.1B–E; 13.2B–F): Stage 3 starts when the embryo is released from the inner egg membrane and becomes a free larva (although still in the brood pouch), and lasts until the first larval molt, a period of about 9 hours. The antennae become free and are no longer attached to the embryo along their full length, and the rami are equipped with ventral rudimentary setae. The overall shape of the embryo is now much more larva-like, with an elongate, cone-shaped, limbless hindbody. During this stage the hindbody grows longer and gradually becomes segmented. The rudiments of the antennules, the mandibles, and the labrum are now distinctly more developed and have become partly free from the body. The trunk is still enveloped in a membrane, so segmentation is not clearly revealed externally by SEM. Other methods, such as fluorescence staining, show that the trunk limbs during

this stage are divided into at least five portions, which are precursors of the future segments (fig. 13.2F). The two dorsolateral swellings (the anlagen to the carapace) (fig. 13.1E) are united and form a narrow dorsal stripe posterior to the head (fig. 13.1G). The maxillary processes (possibly displaced proximal endites of the first trunk limbs) are present, with 2–3 setae near the base of the first pairs of trunk limbs (fig. 13.2C). During this stage the buds of the maxillules enlarge and fuse with the median sternitic plate under the labrum, forming the precursor of the future three-lobed labium (lower lip) (fig. 13.2E). A stage described by Olesen et al. (2003) as a separate stage (their stage 3) may represent specimens fixed in the process of molting between stages 3 and 4 (fig. 13.1G, H), as suggested by Boikova (2008).

Stage 4 (fig. 13.1I–P): Stage 4 starts with a molt and ends with a final larval molt. It lasts about 21 hours. It is a very movable larva, characterized by an adult-like appearance. The whole body is long and adult-like, terminating in a segmented abdomen (but possessing fewer segments than in adults), with a pair of long caudal claws posteriorly. In the earliest part of this stage, the carapace is present as a small dorsal flap behind the head, which gradually enlarges and moves posteriorly during development (compare the carapace development in fig. 13.1I, K, P) (Samter 1895; Olesen et al. 2003). A distinct globular yolk portion is seen in the earliest part of this stage (fig. 13.1P), but it gradually becomes smaller. The head extends anteriorly into a large eye section and carries a large saddle-shaped dorsal organ (fig. 13.1N). The mandibles now approach the S-shaped appearance in the adults, but they have not yet become clearly denticulated. The labrum is large and spatula-shaped, with a pair of papillae at its inner surface (fig. 13.1O). Under the labrum a clearly trilobed labium now appears, with the broad median lobe composed of an elevated sternitic part and the two smaller lateral lobes composed of the modified maxillules (Olesen et al. 2003). The former maxillular setation can still be partly recognized (fig. 13.1L). The trunk limbs hang down freely from the body, and the full adult segmentation and rudimentary setation are present.

MORPHOLOGICAL DIVERSITY: No variation is known.

NATURAL HISTORY: Sebestyén (1949) reported on various aspects of the free-living larvae (metanauplii) of *Leptodora kindtii*, based on observations of live larvae in petri dishes, and this section is based on her work. Swimming is achieved by vigorous oar-like strokes of the antennae, resembling the movement of adults; it begins immediately after the larvae hatch from the resting eggs. During swimming the larvae are mainly horizontal, with the ventral side down, and they change direction frequently. Despite being offered young copepods by Sebestyén (1949), the larvae of *Leptodora kindtii* did not feed during the first 1–3 days after hatching, which fits well with the absence of well-developed feeding structures in this stage. Until at least the third day the larvae rely entirely on yolk (lecithotrophic). Sebestyén (1949) found that the feeding basket formed by the trunk limbs was already developed by the third day, and copepod chitin remnants at the bottom of the containers suggested that feeding took place. By the fifth day food was observed in the gut. The long, tubular, unsegmented mandibular palps are unique larval organs in the metanauplii of *Leptodora kindtii*. They persist only in the earliest phase of development and then degenerate slowly. Free-living larvae of large branchiopods also have mandibular palps, which are three-segmented and well known to be involved in feeding (see chapters 5–10). The function of the mandibular palp in metanauplii of *Leptodora* is not yet understood, but various suggestions have been given: (1) it is involved in swimming, (2) it acts as a rudder/stabilizer during swimming, (3) its internal granules may assist in activating the yolk, and (4) it functions as a sensory organ (Sars 1873; Warren 1901; Gerschler 1911; Sebestyén 1949). Suggestion (2) appears reasonable from a functional perspective. In the earliest larvae, the huge swimming antennae are so dominating in size that it makes sense that the mandibular palps may act as a pair of rudders/stabilizers in order to maintain swimming direction. The mandibular palps degenerate concurrently with the development of the hindbody, which may take the place of the mandibular palps as swimming stabilizers.

PHYLOGENETIC SIGNIFICANCE: *Leptodora* (Haplopoda) is considered the sister-group to the Onychopoda, another group of water fleas to which this genus bears some resemblance in its morphology and feeding behavior. Various developmental aspects of *Leptodora* have played a role in evolutionary questions. *Leptodora kindtii*—together with onychopod cladocerans—is unique among branchiopods in having tubular segmented trunk limbs (like many non-branchiopod crustaceans) modified for raptorial feeding. All other branchiopods generally have flattened (phyllopodous) trunk limbs that are used for filtration. Therefore, when considering that these raptorial cladocerans are nested deeply in the phylogenetic tree of the Branchiopoda (e.g., Olesen 1998; Richter et al. 2007), the tubular, segmented, raptorial limbs of *Leptodora* must have evolved from phyllopodous limbs—but how? Olesen et al. (2001) found that the early limb development in *Leptodora* is similar to that of other branchiopods: rows of transverse ventral limb buds, subdivided in some portions (fig. 13.2F–H; also see chapters 5–10). In the adults, however, these lobes have different fates. In *Leptodora* they end up constituting limb segments, while in other branchiopods (except onychopods) they largely constitute endites (lobes at the inner limb margin). Therefore, in an evolutionary context, the true segments of the trunk limbs of *Leptodora* must be homologous to the endites of the trunk limbs of other branchiopods, and must have evolved from these (see details in Olesen et al. 2001).

Leptodora has a complicated mouth region, which is an adaption to its predatory lifestyle. The unique styliform mandibles are used for puncturing the prey. The trilobed lower lip does not appear in other crustaceans, and it has been difficult to homologize it to the mouthparts of other crustaceans. A developmental study by Olesen et al. (2003), however, followed the fate of the early maxillular limb buds and showed that they end up forming the two lateral lobes of the trilobed lower lip. The maxillar limb buds, which, according to Samter (1900), make a fleeting appearance very early in development, become displaced laterally during later development, as they do in a number of other cladocerans, but they reach a more extreme lateral position in *Leptodora* (fig. 13.1K; Olesen et al. 2003).

Evolutionary understanding of the unique, posteriorly placed dorsal brood pouch, which is a modified carapace, also benefits from being evaluated in a developmental context. The free part of the carapace in the adults is attached to the posterior part of the thorax, overhangs the abdomen, and forms the brood pouch. How did the carapace in *Leptodora* end up in such an unusual position? The early anlagen to the carapace (brood pouch) in *Leptodora* are a pair of dorsolateral swellings immediately behind the head (fig. 13.1E, G), which are similar to what is seen in early stages of other carapace-bearing branchiopods (see chapters 7 and 9). When the carapace grows posteriorly, the anterior part of the carapace gradually fuses to the thorax, leaving only the posterior parts of the valves free, which then constitute the brood pouch (Olesen et al. 2003).

One of the most striking aspects in the development of *Leptodora* is the presence of two different types of development—direct and with free larvae—in two different parts of the life cycle. In all other cladocerans, which basically have the same type of life cycle (cyclic parthenogenesis, or heterogony), development is direct from both resting eggs and in the brood pouch. Are the free-living larvae in *Leptodora* primitive for cladocerans, or have they reappeared secondarily? An assumption of its primitive nature led Eriksson (1934) to separate *Leptodora* (Haplopoda) from the remaining cladocerans; he combined the latter in his Eucladocera. Most evidence, however, suggests that the Haplopoda and Onychopoda are sister taxa, which necessitates a more complicated interpretation of the evolution of free-living larvae versus direct development within the Branchiopoda. Either the free-living larvae (metanauplii) of *Leptodora* have reappeared secondarily within the Cladocera, as was suggested for taxa with free nauplii within the Malacostraca (Scholtz 2000), or direct development has appeared more often within the Cladocera. The latter is preferred here.

HISTORICAL STUDIES: Development in the parthenogenetic part of the life cycle of *Leptodora* was studied in detail by P. Müller (1868), but this work was published in Danish and has therefore been partly overlooked. Another early and important work (published in Norwegian-Danish) was by Sars (1873), who was the first to find and illustrate metanauplii. Other early studies on larvae and development include those of Samter (1895, 1900), Warren (1901), T. Andrews (1948), and Sebestyén (1949).

Selected References

Boikova, O. S. 2008. Comparative investigation of the late embryogenesis of *Leptodora kindtii* (Focke, 1844) (Crustacea: Branchiopoda), with notes on types of embryonic development and larvae in Cladocera. Journal of Natural History 42: 2389–2416.

Olesen, J., S. Richter, and G. Scholtz. 2003. On the ontogeny of *Leptodora kindtii* (Crustacea, Branchiopoda, Cladocera), with notes on the phylogeny of the Cladocera. Journal of Morphology 256: 235–259.

Sebestyén, O. 1949. On the life-method of the larva of *Leptodora kindtii* (Focke), (Cladocera, Crustacea). Acta Biologica Hungarica 1: 71–81.

Warren, E. 1901. A preliminary account of the development of the free-living nauplius of *Leptodora hyalina* (Lillj). Proceedings of the Royal Society of London, B 68: 210–218.

Fig. 13.1 Embryos removed from the dorsal brood chamber of *Leptodora kindtii* (Cladocera: Haplopoda), SEMs. A: stage 2, lateroventral view. B: stage 3, ventral view. C: stage 3, lateral view. D: stage 3, dorsal view. E: closeup of left-side anlagen of the stage 3 carapace / brood pouch. F: caudal region of stage 3, with the dorsal side showing the anlagen to the post-abdominal setae. G: embryo in the process of molting between stages 3 and 4, dorsal view. H: same as G, ventral view. I: stage 4, lateral view. J: anlagen to the post-abdominal setae of stage 4. K: stage 4 head and thorax, lateral view. L: stage 4, closeup of the trilobed lower lip and left-side styliform mandible. M: stage 4 head, ventral view. N: stage 4 dorsal organ. O: stage 4, underside of the labrum. P: stage 4, lateral view of part of the thorax and abdomen, showing the developing carapace / brood pouch and expanded yolk portion. A–O of material used in Olesen et al. (2003).

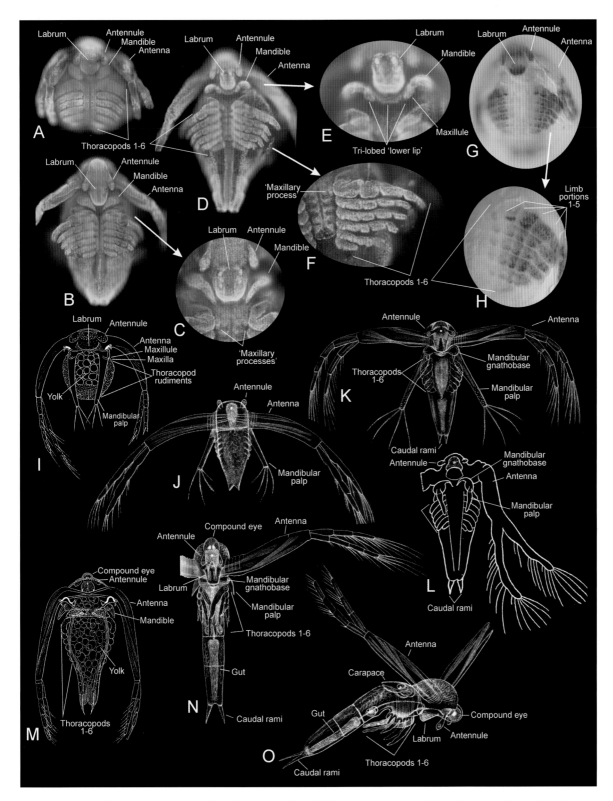

Fig. 13.2 *Leptodora kindtii* (Cladocera: Haplopoda). A–H embryos, light microscopy . A–F: blue color stems from DAPI fluorescent staining. A: stage 2, ventral view. B: early phase of stage 3, ventral view. C: closeup of the mouth region of an early stage 3 specimen. D: late phase of stage 3, ventral view. E: closeup of the mouth region of a late stage 3 specimen. F: closeup of the left-side thoracopods of a late stage 3 specimen. G and H: brown color stems from the expression of the distalless gene. G: late phase of stage 2, ventral view. H: closeup of the limb portions of G. I–O: free-living larvae (metanauplii). I: metanauplius I, light microscopy, ventral view. J: metanauplius II, light microscopy, dorsal view. K: early metanauplius III, light microscopy, ventral view. L: drawing of an intermediate metanauplius III, ventral view. M: late metanauplius III, light microscopy, dorsal view. N: pre-juvenile (transitional stage), light microscopy, ventral view. O: pre-juvenile, light microscopy, lateral view. A–H modified after Olesen et al. (2003); I and M modified after Warren (1901); J, K, N, and O modified after Sars (1873); L modified after Sebestyén (1949).

14

JØRGEN OLESEN
STEFAN RICHTER

Cladocera: Onychopoda

GENERAL: The Onychopoda is a species-poor taxon of pe-culiar water fleas, characterized by an enormous compound eye and four pairs of segmented (non-phyllopodous) trunk limbs, both of which are related to their raptorial/predatory lifestyle (fig. 14.1A). Another special character is the closed brood chamber (modified carapace) in which the embryos are nourished by a substance produced by the female. Currently, the Onychopoda consist of approximately 33 species divided into 3 families: the Polyphemidae (fig. 14.1A), Cercopagididae, and Podonidae (Mordukhai-Boltovskoi 1968; Rivier 1998). Their size ranges from less than 1 mm in polyphemids (fig. 14.1A) and podonids to about 1 cm in cercopagidids (when the long caudal appendage is included). In contrast to other branchiopods, which are mainly freshwater species, onycho-pod cladocerans are found in all major types of water bodies. There are 2 species in fresh water, 7 species in the world's oceans, and approximately 24 species in brackish water in the Ponto-Caspian region. The monophyly of the Onychopoda has been supported by both morphological and molecular evidence (J. W. Martin and Cash-Clark 1995; Olesen 1998; Richter et al. 2001; deWaard et al. 2006), and their sister-group appears to be the Haplopoda, another distinct taxon of pred-atory water fleas (Stenderup et al. 2006; Richter et al. 2007). In all cases where this has been studied, the Polyphemidae, with the well-known species *Polyphemus pediculus* (fig. 14.1A), found commonly in many freshwater habitats, has come out as the sister-group to the remaining onychopods (J. W. Martin and Cash-Clark 1995; Richter et al. 2001; Cristescu and Hebert 2002). As in other cladocerans, the life cycle of onychopods is heterogonic (alternating parthenogenesis and sexual repro-duction).

LARVAL TYPES

Heterogonic Life Cycle: The heterogonic life cycle of onychopods was reviewed by Rivier (1998), who reported much variation among both species and populations in the timing and duration of the two parts of the life cycle, depend-ing on factors such as temperature and food availability. In *Polyphemus pediculus*, the length of each parthenogenetic cycle (from embryos appearing in brood chamber to the release of juveniles) even depends on the generation number, being longest in the first generation (2–6 days). Clutch size is also highly variable and can, in some populations of *P. pediculus*, reach as many as 40–60 embryos early in the season.

Few workers have made an effort to divide the develop-ment of onychopods into stages, but attempts have been made for *Evadne nordmanni* and *Podon leuckarti* (Gieskes 1970; Platt and Yamamura 1986). Platt and Yamamura (1986) divided the development of *E. nordmanni* into six arbitrary stages, but no system based on embryonic molting has been put forth, such as those proposed for other cladocerans (e.g., Kotov and Boikova 1998, 2001), and it may not be possible to establish such a system, since onychopod embryos cannot be studied in vitro. It is unknown if embryonic molting is present at all. In podonids the release of neonates from the mother's brood chamber is synchronized with a molt. In *Polyphemus pediculus* and *Bythotrephes*, however, the release of neonates appears not to be synchronized with molting (Egloff et al. 1997). The process of release of neonates in *P. pediculus* was studied in much detail by Butorina in several papers (reviewed in Buto-rina 2000), who found that neonates, through a complicated process, are released via a single transverse slit lying dorsal to the rectum and normally covered by the caudal appendage (see normal orientation in fig. 14.1A, B), which, during release, swings ventrally to expose the slit.

MORPHOLOGY: Three different developmental stages of *Polyphemus pediculus* (Polyphemidae), representing the major span of development, have been examined by SEM as an ex-ample: an early stage (fig. 14.1C–E), an intermediate stage (fig. 14.1F, G), and a late stage (fig. 14.1H, I). All specimens have been removed from the brood chambers of females.

The earliest stage (fig. 14.1C–E) is roughly divided in two parts: a large globular head section (mostly the compound eye), and a large trunk section with lateral rudiments of four pairs of trunk limbs. An oval labrum anlagen lies between the head and the trunk. On each side of the labrum there is a pair of widely separated antennular buds, and, posterior to these, a pair of mandibular buds; external anlagen of maxillules or maxillae are absent. Laterally, a pair of long biramous antennae are still attached to the trunk along their full length.

In the intermediate stage (fig. 14.1F, G), the antennules have become tubular and have migrated toward the midline of the animal. The labrum has become partly free of the body. The antennae have become longer and are free of the trunk. The mandibles have become partly free of the body, and the gnathal edge is weakly indicated. The buds of four pairs of trunk limbs are present, with a row of setae at the posterior margin. The three anterior limbs have a distinct gnathobasic portion. A weak anal slit is present posteriorly. Dorsally, there is a large trilobed dorsal organ. More posteriorly, there is a short carapace lobe.

In the last stage that was examined (fig. 14.1H, I), much of the adult morphology has been achieved. A pair of small tubular antennules are placed closely together and directed anteriorly. A pair of large biramous swimming antennae have a four-segmented exopod, with a rather short proximal segment, as in adults. The endopod has three distinct segments, similar to adults, plus a possible rudimentary segment that is absent in adults. Four pairs of setae-bearing trunk limbs have become free of the trunk. Trunk limbs 1–3 each have a distinct exopod and endopod, with the latter only weakly subdivided into three segments. The caudal appendage is well-developed.

MORPHOLOGICAL DIVERSITY: Not much is known about the variation in gross morphology of the developing stages in the 3 onychopod families, but a preliminary study of embryos of *Cornigerius maeoticus* (Podonidae) and *Bythotrephes longimanus* (Cercopagididae) performed by the authors shows that the basic components of the embryos are the same. For example, in all taxa the head is dominated by the large compound eye early in development. In *Bythotrephes*, the caudal appendage, which in adults is a dominant feature, characteristically appears early in development (e.g., Claus 1877). As development progresses in the female's brood pouch, and space gets tight, the very long and slender caudal processes end up being curled around the juveniles before the young are released (e.g., Rivier 1998).

NATURAL HISTORY: As in other cladocerans, development in the parthenogenetic part of the life cycle is direct, as the offspring are released from the brood chamber with what is essentially an adult morphology, sometimes (in podonids) even being sexually mature at this early stage, with small four-celled embryos in their own embryonic brood chambers (Egloff et al. 1997; Rivier 1998). This adaptation probably speeds up their reproductive potential. In contrast to other cladocerans (except *Moina* and *Penilia avirostris*), the embryos have little yolk and undergo total cleavage (e.g., *Polyphemus pediculus*; Kühn 1913). Again in contrast to other cladocerans

(except *Moina* and *Penilia avirostris*), development takes place in a closed brood chamber, functioning as a kind of uterus, in which a placenta-like organ, a nährboden, supplies the necessary nutrition for growth (Egloff et al. 1997). In some podonids (e.g., *Evadne anonyx* and *Cercopagis pengoi*) the length of the embryos is increased by a factor of 10 during development, and in *Polyphemus pediculus* by a factor of 5–6, while the embryonic size in other cladocerans (e.g., *Bosmina*) is only doubled, due to the presence of yolk from the onset (Patt 1947; Rivier 1998). As far as is known, the development of those cladocerans that possess a nährboden in a closed brood chamber can take place only inside this chamber, and not, for example, in an isotonic Ringer's solution, as is the case for other cladocerans (e.g., Obreshkove and Fraser 1940; Kotov and Boikova 1998, 2001; Boikova 2008).

PHYLOGENETIC SIGNIFICANCE: As is well known, onychopod cladocerans, together with the Haplopoda (*Leptodora kindtii*), are unique in having stenopodous segmented trunk limbs used for raptorial feeding, as opposed to the phyllopodous filtratory trunk limbs found in most other branchiopods. The stenopodous limbs of the Haplopoda have arisen secondarily within the Branchiopoda (Olesen et al. 2001; also see chapter 13), and, since the Onychopoda is the sistergroup of the Haplopoda (e.g., Richter et al. 2007), it may be assumed that the stenopodous limbs of the Onychopoda also appeared secondarily, probably even in the common ancestor to the Haplopoda and Onychopoda. As in the Haplopoda, the early development of the trunk limbs in *Polyphemus pediculus* is essentially similar to that seen in embryos of cladocerans with phyllopodous limbs in the adults. For example, the gross morphology of the trunk limb buds in an intermediate stage of *P. pediculus* (fig. 14.1F) is very similar to that of an intermediate stage of *Macrothrix hirsuticornis* in the Anomopoda (fig. 11.2G), where an almost similar division into primordial limb portions and a similar type of primordial setation is seen. This embryological similarity is in marked contrast to the limb differences seen in the adults of these groups, but it suggests a deeply founded relationship between the stenopodous segmented raptorial limbs in onychopods like *P. pediculus* and the more phyllopodous filtratory limbs in anomopods such as *M. hirsuticornis*. Probably the segments in onychopod limbs were derived from median lobes (endites) in filtratory limbs, as in *Leptodora kindtii* (although detailed homologies are not yet worked out).

HISTORICAL STUDIES: Kühn (1913) provided a detailed treatment of the early embryogenesis of *Polyphemus pediculus*, but there is very little information about the early development of other groups (see reviews by Egloff et al. 1997; Rivier 1998). Scattered information about morphological aspects of the later parts of development was provided by Weismann (1876–1879) and Claus (1877), and subsequently by Gieskes (1979) and Platt and Yamamura (1986).

Selected References

Egloff, D. A., P. W. Fofonoff, and T. Onbé. 1997. Reproductive biology of marine cladocerans. Advances in Marine Biology 31: 79–167.

Rivier, I. K. 1998. The Predatory Cladocera (Onychopoda: Podonidae, Polyphemidae, Cercopagidae) and Leptodoridae of the World. Leiden, Netherlands: Backhuys.

Weismann, A. 1876–1879. Beiträge zur Naturgeschichte der Daphnoiden. Zeitschrift für Wissenschaftliche Zoologie, vols. 27–33: 1–486.

Fig. 14.1 (opposite) Embryos and adult of *Polyphemus pediculus* (Onychopoda), SEMs (unless otherwise indicated). A: adult female, with late embryos in the brood chamber, light microscopy, lateral view. B: drawing of a brood chamber, with nährboden and eggs, lateral view. C: five early embryos from the same clutch, removed from the female's brood chamber, dorsolateral view. D: early embryo, ventral view. E: early embryo, lateral view. F: intermediate embryo, ventral view. G: intermediate embryo, dorsal view. H: late embryo, ventral view. I: late embryo, lateral view. A and C–I original, of material collected in Denmark; B redrawn from Butorina (1998).

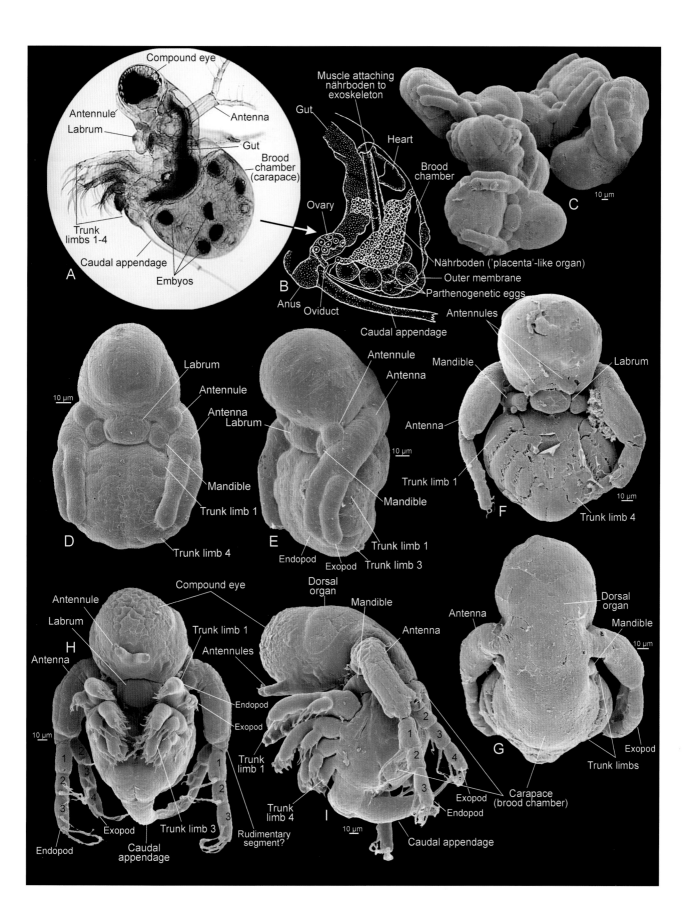

A

Compound eye
Antennule
Antenna
Labrum
Gut
Brood chamber (carapace)
Trunk limbs 1-4
Caudal appendage
Embyos

B

Muscle attaching nährboden to exoskeleton
Gut
Heart
Brood chamber
Ovary
Nährboden ('placenta'-like organ)
Outer membrane
Parthenogenetic eggs
Anus
Oviduct
Caudal appendage

C

10 μm

D

Labrum
Antennule
Antenna
Mandible
Trunk limb 1
Trunk limb 4
10 μm

E

Antennule
Antenna
Labrum
Mandible
Endopod
Exopod
Trunk limb 3
Trunk limb 1
10 μm

F

Antennules
Mandible
Labrum
Antenna
Trunk limb 1
Trunk limb 4
10 μm

H

Antennule
Compound eye
Labrum
Trunk limb 1
Antennules
Antenna
Endopod
Exopod
Trunk limb 1
1 2 3 4
1 2 3
Endopod
Exopod
Caudal appendage
Trunk limb 3
Rudimentary segment?
10 μm

I

Dorsal organ
Mandible
Antenna
1 2
1 3
2 4
3
Exopod
Endopod
Trunk limb 4
Caudal appendage
10 μm

G

Dorsal organ
Mandible
Antenna
Exopod
Trunk limbs
Carapace (brood chamber)
10 μm

15 Remipedia

Jørgen Olesen
Søren Varbek Martinsen
Thomas M. Iliffe
Stefan Koenemann

GENERAL: Remipedes are a relatively small group of blind, unpigmented, chilopod-like crustaceans that includes 24 described extant species. The body length of adults ranges from 9 to 45 mm among species. All extant remipedes inhabit difficult-to-access underwater caves with hypoxic groundwater (Koenemann and Iliffe 2013). They were discovered as late as 1979 and described shortly thereafter as a new crustacean class of uncertain affinity (Yager 1981). Most species have been described from the Caribbean region, but 2 species are known from the island of Lanzarote (Canary Islands) and one from Western Australia. Little is known about behavioral aspects of the Remipedia, such as feeding, but observations of individuals attacking and grabbing a mysid, an *Artemia*, and a cave shrimp (*Typhlatya*) indicate that they are at least partially predatory (Koenemann et al. 2007b). Remipedes are characterized by an unusual—for crustaceans—undivided homonomous trunk, consisting of 16–42 segments (depending on the species). Each trunk segment bears a pair of laterally directed paddle-shaped appendages, which are moved metachronally during swimming (Kohlhage and Yager 1994). Several features of the head and its armament are also unusual for the Crustacea. The anteroventral head region bears a pair of bifurcate pre-antennular frontal filaments that probably have a sensory function. The sensory antennules each consist of two branches: a long dorsal ramus with intrinsic muscles, and a short ventral flagellum. The antennae are biramous, each with an outwardly curved three-segmented endopod and a flap-like undivided exopod. The antennae are ceaselessly beating during swimming, which made Carpenter (1999) and Koenemann et al. (2007b) suggest that they are involved in filtration as a feeding mode supplementary to predation. The mandibles lack a palp, and their masticatory parts are covered by the posterior lobes of the labrum and extend into the atrium oris. Posterior to the mandibles, the remipede cephalon is armed with three pairs of powerful, relatively similar uniramous appendages that are modified as prehensile raptorial mouthparts. The first two pairs are the maxillules and the maxillae, while the third pair, the maxillipeds, are probably derived from the first trunk segment (Koenemann and Iliffe 2013).

Since their discovery only about 30 years ago, the Remipedia have played an important role in discussions of the evolution of arthropods in general and crustaceans in particular. Because of their homonomous trunk, lacking tagmosis, they were initially considered a primitive sister-group to all other crustaceans (e.g., F. Schram 1983, 1986; R. Brusca and Brusca 2003), but later authors interpreted the morphology of at least the post-mandibular predatory head appendages as being derived (Hessler 1992a; Walossek 1993). On morphological grounds, the Remipedia have often been suggested as being closely related to the Malacostraca, sometimes with the Hexapoda as a third key player (Moura and Christoffersen 1996; Fanenbruck et al. 2004; Fanenbruck and Harzsch 2005; Koenemann et al. 2007a, 2009), but other crustacean subtaxa have also been suggested as their closest relatives (Jenner 2010). Most recently, the Remipedia and Cephalocarida surprisingly appeared as sister-groups in two studies based on molecular data sets (Koenemann et al. 2009; Regier et al. 2010). The life cycle includes orthonauplii, metanauplii, and at least one pre-juvenile stage in what appears to be an anamorphic series.

LARVAL TYPES

Nauplius and Pre-Juvenile: Eight naupliar stages (two orthonauplii and six metanauplii) and one pre-juvenile stage are reported, based on specimens identified as post-embryonic stages of *Pleomothra apletocheles* (Koenemann et al. 2007a, 2009). All stages were free-living and lecithotrophic (nonfeeding). The description in this chapter is based on those specimens. The fact that the limited number of larvae (14) found to date represents as many as nine different developmental stages suggests that more stages remain to be found. Based on what is known, it seems that post-embryonic development is anamorphic, with only small changes between the stages.

MORPHOLOGY

Nauplii: Eight naupliar stages are currently recognized. All are swollen by yolk, have no feeding structures, and therefore appear lecithotrophic. The color of the live specimens is whitish-yellow (unpigmented), and a nauplius eye is absent (fig. 15.1A–D). The body of the earliest-known larva (orthonauplius) is ovoid in shape, and the anterior part of the body, which bears the naupliar appendages, is approximately similar in length to the limbless hindbody (fig. 15.1A). During the metanaupliar phase the hindbody grows longer, relative to the naupliar region, and the overall shape of the body becomes gradually more elongate (fig. 15.1B–D). In all nauplii, the naupliar limbs are the only fully developed appendages. None of the studied naupliar specimens exhibited any clear body segmentation, but in at least one larva there are some segment-like indentations on each side of the hindbody (fig. 15.1B). The antennules are relatively long and slender, and remain uniramous during most of the naupliar development (fig. 15.1E, L), but in late metanauplii, the anlagen to an additional ventral antennular branch that is present in adults can be seen (fig. 15.1L). Distally, there are two naked setae and the anlagen to a third seta (fig. 15.1E, orthonauplius). The antennae are biramous, consisting of a long undivided protopodal part and two rami bearing some large naked setae distally (or subdistally). The endopod bears two setae distally, while the exopod is equipped with three setae and the anlagen to a fourth (fig. 15.1E, orthonauplius). The mandibles are biramous and almost twice the length of the antennae; they consist of an extremely long undivided protopodal part and two rami that bear either two (endopod) or four (exopod) large naked setae distally (fig. 15.1F). Remarkably, none of the mandibular appendages are clearly subdivided into segments, such as a coxa and a basis; not even the rami exhibit any external traces of being articulated to the protopod. The antennae and mandibles bear cuticular ornamentation arranged as densely-spaced rows on one side (anterior) of the limbs (fig. 15.1E–G). In the orthonauplii, there are (by definition) no external signs of post-mandibular limbs. Fluorescence microscopy (Koenemann et al. 2009), however, revealed that early limb buds of the maxillules, maxillae, maxillipeds, and a couple of trunk limbs are already present. These findings are corroborated by SEM photography of an orthonauplius, from which the outer body membrane has been peeled off (fig. 15.1I). At this early stage, the limb buds already show signs of being either uniramous (the maxillules, maxillae, and maxillipeds) or biramous (the trunk limbs). During the known part of the metanaupliar phase, an increasing number of post-mandibular limb buds gradually appear (Koenemann et al. 2009). Our preliminary interpretation is that in this phase, a pair of limb buds is added during each molt, but this needs to be confirmed by studying a larger sample of larvae. In one of the specimens, which Koenemann et al. (2009) recognized as a metanauplius IV, we can observe that the trunk limbs start their development as pairs of weakly marked wide buds, located posteriorly on the lateroventral corners of the body (fig. 15.1J). These early flap-like structures develop into larger buds, consisting of the well-known elements of the adults: a protopod and two rami of approximately the same size, without any clear subdivision into segments. During the entire naupliar development, the tips of the trunk limb buds are directed ventrally, which is in contrast to the orientation in adults, where the trunk limbs are directed laterally, in a very characteristic manner for the Remipedia.

Pre-Juvenile: Only one known specimen represents the developmental phase between the metanaupliar and juvenile phases (which are known from a number of remipede species). This specimen was referred to as "post-naupliar" by Koenemann et al. (2007a) or as "pre-juvenile" by Koenemann et al. (2009) (fig. 15.2). There is a considerable morphological gap between the oldest known metanauplius and the pre-juvenile, and also between the pre-juvenile and juveniles/adults; therefore, it is likely that a number of stages from this developmental phase are yet to be found. The pre-juvenile represents a curious combination of larval and adult anatomy. Despite the generally progressed development of this stage, it is probably still lecithotrophic, as feeding structures are undeveloped and there is no clear labrum and mouth opening (fig. 15.2C). The general form of the stage is much like that of adults, but with fewer developed body segments, as well as a number of weakly developed structures or persisting larval rudiments. As in adults, there is a clear separation into a cephalic region, which includes a maxillipedal segment, and a trunk region, which carries the laterally directed swimming appendages (fig. 15.2A, B). The cephalon has various structures (fig. 15.2C–G), including a pair of large bifid frontal filaments; a pair of long antennules with an additional small ventral branch; and a pair of apparently non-functional antennae, which are bent outward in a characteristic way, as in adults. The endopod is three-segmented, without any further visible segmentation. A pair of mandibles consist of the rudiments of the long biramous larval palps, now obviously non-functional (and previously used for locomotion), which are bent together ventrally to the body. The undeveloped basal (coxal) process appears to be the anlagen of the gnathobasic part of the adult mandible. The undeveloped (probably non-functional) maxillules, maxillae, and maxillipeds are all subdivided into a number of segments (portions) that lack setation. The dorsal side of the cephalon is equipped with a head shield that ends posterolaterally in a pair of triangular processes slightly overlapping the first trunk segment (fig. 15.2A). The trunk is composed of at least eight distinguishable segments, each equipped with a broad dorsal tergite and transverse cuticular rims (termed sternal bars in adults) on the ventral side. The sternal bars connect the limbs of each side and possibly act as support structures for them (fig. 15.2B). The presence of 10 pairs of limbs denotes additional segments in the posterior trunk. The anterior 6–7 pairs of the trunk limbs are directed laterally (in an adult-like manner), with an accompanying segmentation pattern, while the posterior appendages, which are merely small buds, are directed ventrally, like the limb buds of the metanauplii (fig. 15.2J). A pair of caudal rami articulated to the remaining part of the body is present posteriorly (fig. 15.2I, J).

MORPHOLOGICAL DIVERSITY: Variation, where known, is included above.

NATURAL HISTORY: Notes on swimming by nauplii were given by Koenemann et al. (2009), based on video recordings of several specimens made in the lab a few hours after collection (fig. 15.1M, N). Because the larvae are highly active swimmers, it was necessary to chill individuals in a petri dish in a refrigerator to slow down swimming movements for video recordings. It appears that the antennules are mostly directed forward of a lateral position or perform feeble asynchronous strokes, adding little to locomotion. The antennae and mandibles, in contrast, are very active, and both limbs contribute to locomotion, despite bearing only a few bare distal setae (fig. 15.1M). The antennae perform more or less constant strokes that cover a range of about 180°, lateral to the body axis. The mandibles cover a range of movements from a position about 45° laterally to the main body axis to a position stretched out along the body in a posterior direction. During the power stroke, the mandibles are almost stretched out, while they are slightly bent during the recovery stroke. Both the antennae and the mandibles are unusually flexible. Apart from the insertion point of the mandibles on the trunk, there are no specific joints or distinct articulations; rather, the limbs appear to be flexible along their whole length. In the video recordings used in Koenemann et al. (2009), there are several examples that demonstrate how the antenna's flexibility enables it to clean both the antennules and the mandibles. The extraordinarily elongate mandibles are equally flexible and often bend ventrally over the trunk, with the rami meeting in a "folded-hands" position. The two rami can even move independently, as they occasionally move toward each other like a pair of forceps (fig. 15.1N). The latter type of movement may have been at least partly triggered by the unnaturally cool conditions under which the larvae were kept during the video recordings. Nonetheless, these movements still illustrate the unusual flexibility of the remipede naupliar limbs.

Due to the presence of large amounts of yolk in the body (at least during the naupliar phase), the lack of developed feeding structures, and the lack of a mouth opening, the larvae (including the pre-juvenile specimen) are assumed to be nonfeeding. Other crustacean taxa are characterized by an early lecithotrophic development (e.g., rhizocephalan, penaeid, and euphausid larvae), but the lecithotrophic phase in the Remipedia seems to last an unusually long time. Koenemann and Iliffe (2013) remarked that although the orthonauplii contain large amounts of yolk, it did not seem as though the yolk is the sole energy provider for their remarkable growth and swimming movements in the water column. The authors further suggested that if the yolk of the larvae indeed is their exclusive source of energy, then it may be assumed that development from lecithotrophic to feeding forms proceeds in a relatively short period of time (to reduce energy consumed by locomotion). These aspects need to be studied in much more detail, however.

PHYLOGENETIC SIGNIFICANCE: The discovery and description of developmental stages of the Remipedia was a long-awaited event in crustacean research. We might have expected that the discovery of remipede larvae would shed some light on the much-debated evolutionary history of this group of crustaceans (e.g., Koenemann and Iliffe in press). Although some of the larval features have contributed to discussions concerning the phylogenetic position of the Remipedia, expectations fell short of the success story of crustacean larvae that have sometimes provided the final clue for the affinity of certain modified taxa (e.g., that barnacles are crustaceans, rather than mollusks; J. Thompson 1830).

The finding that the earliest stage of the Remipedia is an orthonauplius is important, since this larval type is found only in the Crustacea. The growing evidence for crustacean paraphyly with respect to the Hexapoda (e.g., Regier et al. 2010) means that an orthonauplius as a hatching stage would have been lost in the Hexapoda clade, as well as in several crustacean subtaxa. Moreover, in the scheme put forward by Walossek 1999, which takes crustacean fossils into account, an orthonauplius type of larva as the hatching stage is considered an autapomorphy at the level of Crustacea *sensu stricto* (Eucrustacea). Apart from this, it is difficult to use larval characteristics for determining the phylogenetic position of the Remipedia with great confidence. Koenemann et al. (2007a, 2009) discussed the presence of an extended lecithotrophic larval stage of the Remipedia as an indication of affinity to the Malacostraca. Certain malacostracan nauplii (dendrobranchiate Decapoda and euphausiaceans) and remipede larvae share the presence of a large amount of yolk; the lack of a mouth, anus, and gut; the lack of gnathobasic structures of the antennae and mandibles; and the presence of a late-occurring labrum. Most of these features, however, are characteristic for lecithotrophic nauplii in subgroups of various crustacean taxa, such as the Cirripedia, Branchiopoda, and Copepoda (Dahms 1989; Walossek et al. 1996; Olesen and Grygier 2004). Koenemann et al. (2009) also noted other similarities between the Remipedia and Malacostraca that were related to development, such as in the formation of the nervous system and the morphogenesis of antennules, which in both groups have two branches, with the additional branch occurring quite late in ontogeny. Yet another developmental character in the Remipedia is the loss of the mandibular palp during ontogeny. A similar loss is seen during development in the Branchiopoda and Cephalocarida. In some other crustaceans (e.g., some malacostracans) and mandibulates in general (hexapods and myriapods), a palp is entirely lacking, so we consider this character to be of little value in resolving remipede affinities. In sum, there is no conclusive development-based evidence for the phylogenetic position of the Remipedia. The available information mostly suggests a malacostracan affinity, but that is contradicted by at least some molecular evidence (e.g., Regier et al. 2010).

HISTORICAL STUDIES: The early development of the Remipedia remained unknown until 2006, when two diving explorations of anchialine cave systems on Abaco Island in the Bahamas yielded 14 crustacean larvae that could be identified as post-embryonic stages of *Pleomothra apletocheles* (Koenemann et al. 2007a, 2009). Those papers are the only studies on remipede larval development.

Selected References

Koenemann, S., J. Olesen, F. Alwes, T. M. Iliffe, M. Hoenemann, P. Ungerer, C. Wolff, and G. Scholtz. 2009. The post-embryonic development of Remipedia (Crustacea): additional results and new insights. Development Genes and Evolution 219: 131–145.

Koenemann, S., F. R. Schram, A. Bloechl, M. Hoenemann, T. M. Iliffe, and C. Held. 2007a. Post-embryonic development of remipede crustaceans. Evolution & Development 9: 117–121.

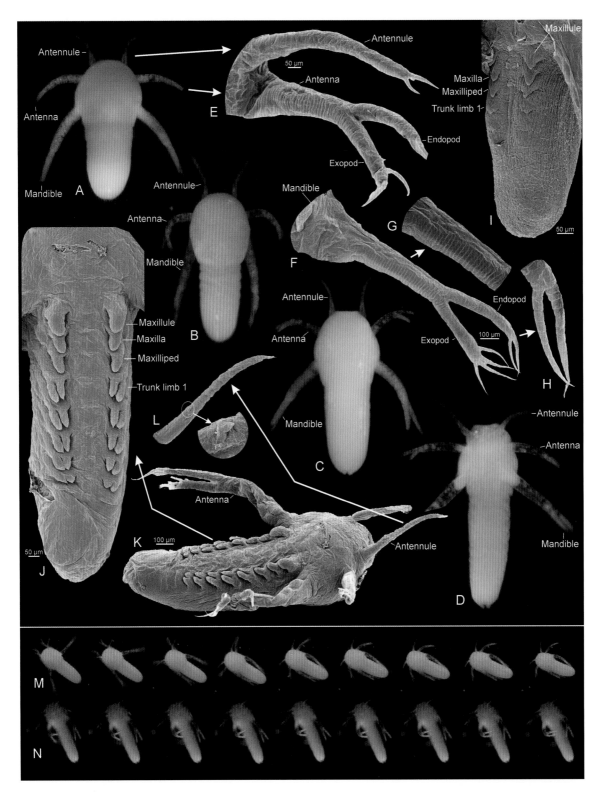

Fig. 15.1 Nauplii of the remipede *Pleomothra apletocheles*. A–D: live nauplii, light microscopy. A: orthonauplius, dorsal view. B: metanauplius, dorsal view. C: metanauplius, dorsal view. D: metanauplius, ventral view. E–I: SEMs of the same specimen. E: orthonauplius antennule and antenna, dorsal view. F: orthonauplius mandible, dorsal view. G: protopod of an orthonaupliar mandible, showing the cuticular ornamentation, arranged as densely spaced rows. H: two naked distal setae of an orthonaupliar mandibular endopod. I: orthonauplius trunk, showing the early limb buds of the maxillules, maxillae, maxillipeds, and a couple of trunk limbs, ventral view. J–M: SEMs of the same specimen. J: metanauplius, with the limb buds of the maxillules, maxillae, and maxillipeds and five pairs of trunk limbs, ventral view. K: metanauplius trunk, ventrolateral view. L: metanauplius antennules, and anlagen to the ventral branch of the antennules. M: swimming sequence of a metanauplius, showing the movability of the antennae and mandibles, video recording. N: sequence showing the unusual flexibility of the mandibles, video recording. A–M of material from Abaco Island in the Bahamas, used in Koenemann et al. (2007a, 2009).

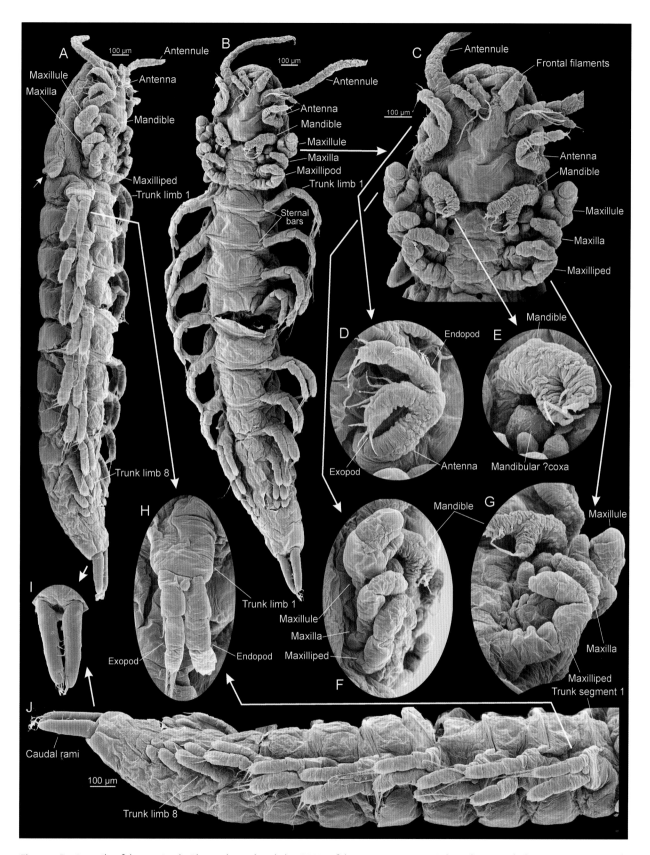

Fig. 15.2 Pre-juvenile of the remipede *Pleomothra apletocheles*, SEMs of the same specimen. A: lateral view, with the arrow pointing at the small triangular posterior protrusion of the head shield. B: ventral view. C: cephalon, ventral view. D: antenna. E: mandibular palp. F: maxillule, maxilla, and maxilliped, lateral view. G: mandibular palp, maxillule, maxilla, and maxilliped. H: trunk limb 1. I: caudal rami. J: trunk, lateral view. A–J of material from Abaco Island in the Bahamas, used in Koenemann et al. (2007a, 2009).

16

Cephalocarida

Jørgen Olesen
Søren Varbek Martinsen
George Hampson
Alberto Addis
Marcella Carcupino

GENERAL: Cephalocarids are a species-poor group of small (less than 4 mm), marine, ancient-looking crustaceans (fig. 16.1A–C). They were first described by H. Sanders (1955), based on specimens from the muddy bottom of Long Island Sound, USA, but they have been found subsequently in the North and South Atlantic Ocean, the North and South Pacific Ocean, and (recently) in the Mediterranean Sea, at depths ranging from shallow subtidal to bathyal (Hessler and Elofsson 1992; J. W. Martin et al. 2002; Carcupino et al. 2006). To date, 12 species have been described, belonging to 5 genera, with *Lightiella* and *Sandersiella* being the most species rich (5 and 4 species, respectively) (Olesen et al. 2011). Adult cephalocarids are easily distinguished from other crustaceans by the number of body segments (20, including the telson) and their characteristic body tagmosis. The body (posterior to the cephalon) consists of nine limb-bearing thoracic segments (in *Lightiella* thoracic limb 8 is lacking), 10 limbless abdominal segments, and a telson. Because of their primitive appearance, cephalocarids have played an important role in discussions on crustacean phylogeny (e.g., H. Sanders 1957, 1963; Hessler 1992a; Walossek 1993). Cephalocarids are hermaphroditic. Their life cycle includes one to two large embryos borne on specialized egg-carrier limbs, and anamorphic development with at least 14 (*Hutchinsoniella macracantha*) or 15 (*Lightiella magdalenina*) metanaupliar stages, followed by a low number of juveniles.

LARVAL TYPES

Embryos: In *Hutchinsoniella macracantha*, the best-studied cephalocarid, the gonopores are at the posterior surface of thoracopod 6. When the large eggs emerge, they are directed to the posterior egg carriers (specialized ninth thoracopods), to which they are cemented and carried for a while (fig. 16.1A–D) (Hessler et al. 1995). *Hutchinsoniella* and *Lightiella* appear to be different with respect to the number of embryos being carried. In *Hutchinsoniella* two embryos are usually present, one on each egg carrier (fig. 16.1A–C), while 16 of 17 exam-

ined specimens of *Lightiella incisa* carried only a single embryo (H. Sanders and Hessler 1964), a tendency later confirmed in other samples of *Lightiella* (J. W. Martin et al. 2002; Addis et al. 2007; Olesen et al. 2011). In *H. macracantha* the bodies of the brooding embryos are tightly flexed within the eggs, so that the posterior part of the body is pressed under the cephalon, and they are always oriented the same way within the egg membrane (fig. 16.1D) (Hessler et al. 1995). Cephalocarids, and in particular their larvae, are extremely rare, so their development has been examined only in a few species; in some cases just a few stages are known. The 2 best-studied species are *Hutchinsoniella macracantha* and *Lightiella magdalenina* (H. Sanders 1963; Addis et al. 2007). In addition, several stages are known for *Lightiella incisa* (H. Sanders and Hessler 1964), while only a few stages have been reported for *L. serendipita* and *L. monniotae* (Jones 1961; Olesen et al. 2011).

Metanauplius: In the metanauplius, development appears to be rather gradual (anamorphic), with many instars separated only by small changes. H. Sanders (1963) described 18 post-embryonic stages for *Hutchinsoniella macracantha*, 13 of which carried an antennal naupliar process on the coxa (termed a masticatory spine in some other crustaceans), which are therefore, by definition, metanauplii. Recent (August 2011) sampling of *H. macracantha* by Jørgen Olesen, Søren Varbek Martinsen, and George Hampson yielded several larval stages (fig. 16.1E–H), at least one of which was not described by H. Sanders (1963). This undescribed stage has 12 trunk segments (fig. 16.1G) and falls between H. Sanders's stages 5 and 6. Hence the number of metanauplii of *H. macracantha* is higher than was reported by H. Sanders (1963), but a complete restudy of its larval development remains to be done. In *Lightiella magdalenina*, 15 metanaupliar larval stages and two juveniles have been described, but an earlier stage than this species' previously known stage 1 is described in this chapter (figs. 16.2A; 16.3A); the exact number of instars of *L. magdalenina* may still be unknown (Addis et al. 2007). The earliest-known stages of *H. macracantha* and *L. magdalenina* are already metanauplii, as they have, in addition to the naupliar append-ages (the antennules, antennae, and mandibles), relatively well-developed maxillules and early limb buds of the maxillae. The shortest stage 1 larva known for the Cephalocarida is that of *H. macracantha*, which has a trunk region composed of two incipiently jointed segments plus the telson; in the stage 1 specimen of *L. magdalenina* described by Addis et al. (2007), the trunk region is composed of five well-developed segments plus the telson. Additional sampling of the same species, how-ever, has yielded an even earlier specimen, also with five trunk segments (plus the telson), but with trunk segments 4 and 5 being only weakly developed (figs. 16.2A; 16.3A). In the early part of the metanaupliar development, two body segments seem to be added for each molt (approximately stages 1–6), while only one segment is added per molt (with some varia-tion) in the later part of development (Addis et al. 2007). In this respect the development of cephalocarids is less gradual than in *Artemia* and the Cambrian Orsten crustacean †*Rehbachiella kinnekullensis* (Walossek 1993), for example. A striking char-acteristic of cephalocarid development—different from other gradually developing crustaceans—is that the body segments develop much faster than the trunk limbs. As remarked by Addis et al. (2007) and also treated by Olesen et al. (2011), the addition, development, and functionality of the trunk limbs are extremely delayed. When the formation of thoracic and abdominal segments is finished and their adult number has been achieved, only three trunk limbs are present in *L. mag-dalenina*.

MORPHOLOGY: Egg (embryo)-carrying adults of *Hutchin-soniella macracantha*, as well as four different larvae, are illus-trated here (fig. 16.1), based on recent (August 2011) sampling. Three early stages of *Lightiella magdalenina*, two of which are described by Addis et al. (2007; their stages 2 and 4), are il-lustrated here (figs. 16.2; 16.3). An additional stage, which is apparently earlier than stage 1 of Addis et al. (2007), and there-fore referred to as "stage 1," is illustrated here as a supplement to their description (figs. 16.2A–E; 16.3A, B). In addition, a few illustrations of a late metanauplius of *L. monniotae* are in-cluded (figs. 16.2N–P; 16.3H–J), corresponding approximately to stage 10 of *L. magdalenina*, as well as an illustration of a late juvenile (fig. 16.2O). The general morphology of stage 1 of *Hutchinsoniella macracantha* and *Lightiella magdalenina* will be outlined first, followed by a summary of the development of the major morphological aspects (based on H. Sanders 1963; Addis et al. 2007; observations made when preparing this chapter).

Stage 1 of *Hutchinsoniella macracantha* is a metanauplius with a short, broad, flattened dorsal head shield; uniramous antennules; biramous antennae bearing naupliar processes (coxal masticatory spines in other crustaceans); biramous mandibles; weakly developed maxillules; two post-cephalic body segments; and a telson. Both stage 1 of *Lightiella mag-dalenina*, described by Addis et al. (2007), and the even ear-lier "stage 1" illustrated here (figs. 16.2A–E; 16.3A, B), are in general very similar to stage 1 of *H. macracantha*, but with

more post-cephalic body segments: five in stage 1 of Addis et al. (2007), while, in the "stage 1" illustrated here, the two posterior of these are incompletely developed (fig. 16.2A).

The uniramous antennules undergo only minor changes during their entire development. In the first stages of both *Hutchinsoniella macracantha* and *Lightiella magdalenina*, they are six-segmented and bear one seta on each of segments 2, 3, and 5, as well as a cluster of setae and a distinctly superficially segmented aesthetasc distally at the sixth segment. This same basic morphology is retained throughout development, with only some minor rearrangements of the setal distribution. An-other minor change concerns the presence of a few clusters of small sausage-shaped sensillae on segments 5 and 6 in the "stage 1" of *L. magdalenina* (fig. 16.2B), which appear to be lacking in the later stages examined by SEM.

The antennae remain large and biramous during develop-ment and retain much of their original structure. One major change is that the naupliar processes are lost at the end of the metanaupliar phase. The antennal exopods are multiannulate in all known stages of the Cephalocarida (as in some other crustaceans, such as mystacocarids and Orsten crustaceans). In both *Hutchinsoniella macracantha* and *Lightiella magdalenina*, the exopod is divided into 13 short segments in "stage 1" (fig. 16.2C), a number that is retained in *L. magdalenina* in stage 2 (fig. 16.3E) and, in both species, is increased during further development (15 segments in stage 4) (fig. 16.2J) to a total number of 19 in the adults. The endopod is two-segmented and rather short in all developmental stages. In *H. macracan-tha*, *L. magdalenina*, and *L. monniotae*, the antennal naupliar process is a finger-like coxal endite, with some distal setation (fig. 16.3B, D, H). In *L. magdalenina* there are three distal setae and one lateral seta, all pointing toward the mouth region. In *L. magdalenina* the naupliar process is lost after 15 stages, whereas in *H. macracantha* it is lost after 13 stages. This loss, by definition, marks the end of the metanaupliar phase.

The mandibles undergo significant changes during devel-opment, as the palp (basis and two rami) is gradually lost. In the earliest stages the mandible consists of the same compo-nents as the antennae: a protopod divided into a coxa and a basis. The coxa bears a broad endite with an already-formed gnathal endite; an endopod and an exopod are also present. For both *Hutchinsoniella macracantha* and *Lightiella magda-lenina*, the gnathal edge of the mandibular coxal process is already well developed in stage 1 (fig. 16.3B) and basically is similar to that of the adult. In the earliest larval stages, the mandibular exopod has an annulate-like segmentation, as does the antennal exopod, but the mandibular exopod is by far the smaller of the two (only six to seven segments) (fig. 16.2P). The mandibular endopod in larval stages has often been described as bilobate, but it actually consists of two clear segments that are inserted on a rather large basis (fig. 16.3E, G, I). The bilobed appearance is caused by a large median pro-trusion of the proximal endopodal segment, making it almost a broad as the basis. In *H. macracantha* the palp (basis plus two rami) is reduced in stage 14 (the first juvenile stage), but atro-phication of the exopod starts at least as early as stage 11 (still a metanauplius; but note the uncertainty with respect to the true number of larval stages). In a late larva of *L. monniotae*, corresponding approximately to stage 10 of *L. magdalena*, the palp is still present (endopod, fig. 16.3I; exopod, fig. 16.2P) (also see Olesen et al. 2011), and a rudiment of the palp is present long after the metanauplar phase in *L. monniotae* (fig. 16.2O).

The maxillules undergo rather significant changes during development. In *Hutchinsoniella macracantha*, the maxillule starts as a relatively undifferentiated biramous limb. A short cone-shaped unsegmented median part constitutes the future endopod, and the exopod is a lateral broad flap-like structure, with some marginal setation. The early endopod segmentation appears to be clearer in the first stage of *Lightiella magdalena* than in the first stage of *H. macracantha*. The endopod becomes significantly longer during development, and by stage 2 of *H. macracantha* it already has become four-segmented (also see fig. 16.3I, a late metanauplius of *L. monniotae*). In all adult cephalocarids the endopod is three-segmented, a condition that was shown to take place by a reduction of the fourth segment during ontogeny in *L. monniotae* (Olesen et al. 2011). During development, the exopod becomes larger and acquires more marginal setae, but its basic morphology is the same. The protopod undergoes much change during development. In the early part of development its median margin is loosely arranged into endites (fig. 16.3I, a late metanauplius of *L. monniotae*), but in stage 11 of *L. magdalena* the proximal endite has started to enlarge into finger-like processes that are extended even farther in juveniles and adults, reaching into the atrium oris (fig. 16.2O, a late juvenile of *L. monniotae*).

The maxillae and thoracopods undergo much change during development. They are treated together here, because they have a very similar development and are almost identical in the adults. The maxillae are present as rudimentary lobed structures in stage 1 of both *Hutchinsoniella macracantha* and *Lightiella magdalena*. Thoracopod 1 appears in stage 3 in *H. macracantha* and in stage 5 of *L. magdalena*. Development from rudimentary lobes to adult morphology occurs in roughly three to four steps (H. Sanders 1963; Addis et al. 2007). All stages of limb development, because of serial displacement in the degree of development, are roughly exhibited in a single late-larval specimen of *Lightiella monniotae* (fig. 16.3J). Early in development the limb is a ventrally pointing bifurcate lobe, devoid of functional setae. The exopod and pseudepipod (part of exopod) then become developed, but the endopod is a small unsegmented conical structure. Next, the endopod is segmented but still small. Finally, the limb is fully developed (e.g., see Olesen et al. 2011). The development of the thoracopods is much delayed, compared with the trunk limbs. Development of the left and right side limbs is asynchronous in the examined specimen of *L. monniotae* (fig. 16.3J).

A consistent correlation exists between the presence of tergopleura and trunk limbs. Tergopleura are broad lateral extensions of each tergite that cover the basal parts of limbs (fig. 16.2N). In the adult, only the thoracic segments have tergopleura, while the abdomen consists of ring-like segments, most of which have lateral spines at the posterior margin. In the larvae, all trunk segments start as ring-like abdominal segments (but are more dorsoventrally flattened than in the adults). The anterior larval segments are transformed into thoracic segments with tergopleura, while the posterior segments are retained as ring-like abdominal segments. This transformation in morphology is displayed in a single late larval specimen of *Lightiella monniotae*, in which the anterior three to four segments bear tergopleura (fig. 16.2N) (Olesen et al. 2011).

MORPHOLOGICAL DIVERSITY: Larvae are known only from *Hutchinsoniella macracantha*, *Lightiella serendipita*, *L. incisa*, *L. magdalena*, and *L. monniotae* (Jones 1961; H. Sanders 1963; H. Sanders and Hessler 1964; Addis et al. 2007; Olesen 2011), and they differ from each other only in minor aspects.

NATURAL HISTORY: Aspects of larval locomotion and feeding have been studied only for *Hutchinsoniella macracantha* (H. Sanders 1963). Larvae of *H. macracantha* are wholly benthonic (as are the adults), move over the substratum ventral side down, and feed on deposited detritus. In larvae (and adults), the antennules and antennae are responsible for locomotion, but in adults the leaf-like metachronally beating thoracopods play a major role in locomotion. When larvae of *H. macracantha* move around, the antennules are mostly beating constantly, while they are often held motionless in adults, indicating that they play a larger role for locomotion in larvae than they do in adults (observations by Jørgen Olesen in August 2011). Feeding in early larvae mainly involves the cephalic appendages and is therefore very different from that of the adults, where food is handled by the post-mandibular appendages and brought forward to the mouth region (atrium oris) (H. Sanders 1963). In larvae, food particles are swept inward by the exopods of the antennae and the mandibles and are then pushed to the posterior end of the labrum by means of spines and processes on the antennules, antennae, and mandibles. From here the food is passed to the molar process of the mandible by the endites of the first maxilla and the naupliar process of the antennae. H. Sanders (1963) called the larval method of feeding "cephalic feeding." He noted that feeding is initiated in stage 2 in *H. macracantha*. In "stage 1" of *Lightiella magdalena*, both the molar process of the mandible and the naupliar process of the antenna are well developed (fig. 16.3B), which suggests that the larva is feeding.

As thoracopods are added in the later larval stages, the adult (post-mandibular) method of feeding becomes predominant. Several structures important in larval cephalic feeding are gradually lost, such as the naupliar processes of the antennae and the palps of the mandibles. The maxillules play a central role in the cephalic (larval) and post-mandibular (adult) feeding systems, since food items from both the larval and the adult feeding systems are passed to these limbs before they are moved into the atrium oris.

PHYLOGENETIC SIGNIFICANCE: Cephalocarids have played a crucial role in debates over crustacean evolution and phylogeny since their discovery (H. Sanders 1957, 1963; Hessler 1992a; Walossek 1993). Very early, it was assumed that the primitive mode of development in the Crustacea probably was by a regular addition of segments and limbs at each molt (anamorphic development), somewhat similar to what is seen in annelids (e.g., Calman 1909). H. Sanders (1963) suggested that the Cephalocarida showed the closest approximation to such an assumed ancestral development. Cephalocarid development is indeed gradual (see H. Sanders 1963; Addis et al. 2007), but in some ways it seems modified, compared with that of certain branchiopods (anostracans) and the Orsten crustacean †*Rehbachiella kinnekullensis* (Walossek 1993; Olesen et al. 2011). First, the development of the limbs is much delayed, compared with the development of the trunk segments. Second, there seem to be some jumps in the early phase of larval development, compared with some other gradually developing crustaceans. Based on what is known about cephalocarid development, two trunk segments appear to be added in one molt in the early part of their development, which is in contrast to some anostracans, where only one seg-

ment is added per molt, or to †*R. kinnekullensis*, which perhaps had an even more gradual development, with two molts per segment (Walossek 1993).

HISTORICAL STUDIES: The classic study of cephalocarid development is that of H. Sanders (1963), which is still the most important source for information about larval morphology and functional morphology.

Selected References

Addis, A., F. Biagi, A. Floris, E. Puddu, and M. Carcupino. 2007. Larval development of *Lightiella magdalenina* (Crustacea, Cephalocarida). Marine Biology (Berlin) 152: 1432–1793.

Olesen, J., J. T. Haug, A. Maas, and D. Waloszek. 2011. External morphology of *Lightiella monniotae* (Crustacea, Cephalocarida) in the light of Cambrian "Orsten" crustaceans. Arthropod Structure and Development 40: 449–478.

Sanders, H. L. 1963. The Cephalocarida: Functional Morphology, Larval Development, Comparative External Anatomy. Memoirs of the Connecticut Academy of Arts and Sciences No. 15. New Haven: Connecticut Academy of Arts and Sciences.

Sanders, H. L., and R. R. Hessler. 1964. The larval development of *Lightiella incisa* Gooding (Cephalocarida). Crustaceana 7: 81–97.

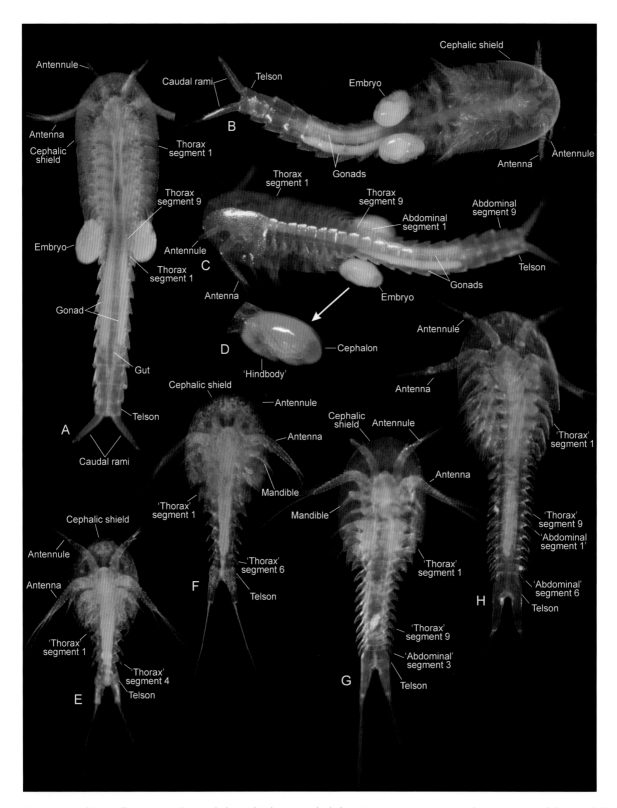

Fig. 16.1 *Hutchinsoniella macracantha* (Cephalocarida), live animals, light microscopy. A–D: egg (embryo)-carrying adults. A: adult, dorsal view. B: adult, ventral view. C: adult, laterodorsal view. D: closeup of the left-side egg (embryo). E–H: four different larval stages. E: stage with four trunk segments (plus telson), corresponding to stage 2 of H. Sanders (1963), ventral view. F: stage with six trunk segments (plus telson), corresponding to stage 3 of H. Sanders (1963), dorsal view. G: stage with 12 trunk segments (plus telson), between stages 5 and 6 of H. Sanders (1963) but not corresponding to either of these, ventral view. H: stage with 15 trunk segments (plus telson), corresponding to stage 8 of H. Sanders (1963). A–H original, of material collected in Buzzards Bay, Massachusetts, USA, in August 2011 by Jørgen Olesen, Søren Varbek Martinsen, and George Hampson.

Fig. 16.2 Metanauplii of the Cephalocarida, SEMs. A–M: stages "1," 2, and 4 of *Lightiella magdalenina*. A: "stage 1" (earlier than the stage 1 described by Addis et al. 2007), with five trunk segments (plus telson), lateral view. B: antennule of "stage 1," left side. C: antenna of "stage 1," right side. D: closeup of the antennal exopod segments of "stage 1." E: telson of "stage 1." F: stage 2, with seven trunk segments (plus telson), lateral view. G: cephalic shield of stage 2. H: telson of stage 2. I: stage 4, with 11 trunk segments (plus telson), lateral view. J: antenna of stage 4, right side. K: stage 4, frontal view. L: trunk segments of stage 4. M: closeup of the lateral spine at the posterior margin of abdominal segment 2 of stage 4. N–P: late stages of *L. monniotae*. N: pleuroterga of a late metanauplius. O: closeup of the mandible, with rudimentary palp, and maxilla 1, with enlarged coxal endite, of a juvenile. P: closeup of the mandibular exopod of a late metanauplius. A–M original, of material from the Mediterranean Sea, collected at the same locality as the material used in Addis et al. (2007); N–P of material from New Caledonia, used in Olesen et al. (2011).

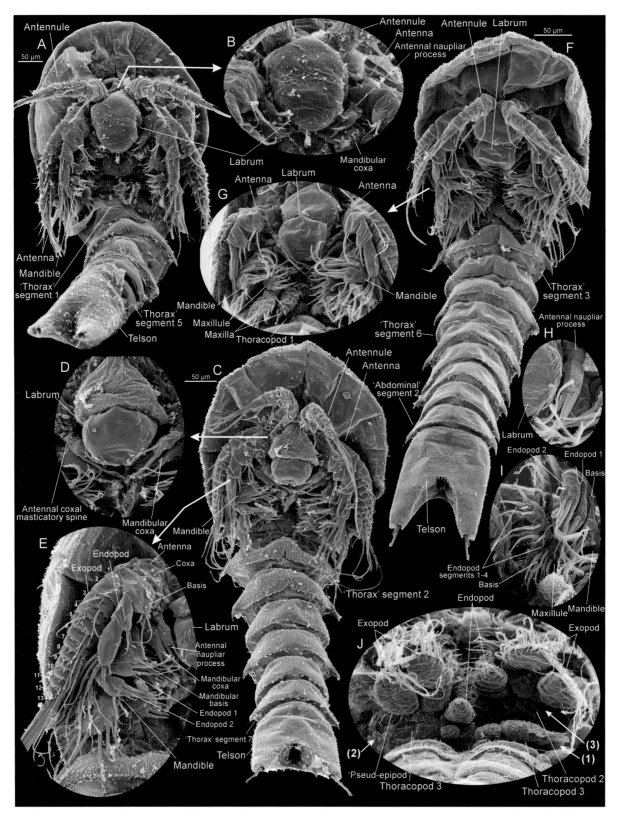

Fig. 16.3 Metanauplii of the Cephalocarida, SEMs. A–G: *Lightiella magdalenina*. A: "stage 1" (earlier than the stage 1 described by Addis et al. 2007), with five trunk segments (plus telson), ventral view. B: labrum and part of the feeding apparatus of "stage 1." C: stage 2, with seven trunk segments (plus telson), ventral view. D: labrum and part of the feeding apparatus of stage 2. E: antenna and mandible of stage 2, right side. F: stage 4, with 11 trunk segments (plus telson), ventral view. G: labrum and part of the feeding apparatus of stage 4. H–J: late stage of *L. monniotae*. H: antennal naupliar process, left side. I: mandible and maxilla 1, right side. J: thoracopods 2 and 3 from the posterior of the metanauplius, in three different phases of development (indicated by the numbers 1–3 in parentheses); development of the left and right limbs is asynchronous. A–G original, of material from the Mediterranean Sea, collected at the same locality as the material used in Addis et al. (2007).

17

Introduction to the Thecostraca

Jens T. Høeg
Benny K. K. Chan
Joel W. Martin

The Thecostraca, which includes the well-known Cirripedia (barnacles), is a large assemblage of diverse crustaceans in which parasitism and adaptation to unusual habitats have resulted in a wide range of unusual morphologies and lifestyles. Larval morphology has always played a critical role in studies of their systematics and evolution, and until the advent of molecular evidence, the affinities among the thecostracan taxa rested exclusively on the morphology of the larvae (Høeg 1992; Høeg and Møller 2006; Høeg et al. 2009a). Thecostracan larvae share some important characteristics. Development typically consists of a series of nauplius instars, followed by a terminal phase that is called the cypridoid larva (Høeg et al. 2004). The cypridoid larva is specialized to initiate the permanently attached phase found in nearly all juvenile and adult Thecostraca (table 17.1). Historically, the classification of the Thecostraca has changed frequently. Today the taxon thought to consist of 3 major subgroups (subclasses): the Facetotecta, Ascothoracida, and Cirripedia.

Thecostracan nauplii gradually develop a trunk, but they never develop any post-cephalic (thoracic) appendages. These limbs all appear in one step, at the molt to the cypridoid stage. The latter is called y-cyprid (or cypris y) in the Facetotecta, a-cyprid in the Ascothoracida, and cyprid (or c-cyprid), in the Cirripedia. The body of the cypridoid is wholly or partially covered by a large carapace (head shield), which in a-cyprids and cyprids is drawn down on either side, with the two halves connected by an adductor muscle. The cypridoid carries a pair of prehensile antennules and six pairs of natatory appendages on the segmented thorax, enabling this larva to both swim through the water and attach securely to the substratum. In facetotectans and ascothoracidans, the cypridoid body terminates with a large segmented abdomen and a telson furnished with unsegmented furcal rami. Cirripede cyprids retain the rami, but both the abdomen and telson are either rudimentary or absent. The cypridoid larva is very well equipped with sensory organs, including five pairs of lattice organs on the carapace, a nauplius eye, a pair of compound eyes, conspicuous frontal filaments, and one or two antennular aesthetascs.

The Facetotecta is an incompletely understood group (to say the least), but it is nevertheless widespread and can be very abundant locally. For more than 100 years the facetotectans were known only from their larvae (referred to as y-larvae): a series of five y-nauplius (or nauplius y) stages, followed by a single y-cyprid. Recent work (Glenner et al. 2008; also see Scholtz 2008), using a crustacean molting hormone (20-HE) to induce metamorphosis, has revealed that the subsequent stage is a reduced limbless slug-like juvenile, called an ypsigon. The adults, although still unknown, are very likely advanced parasites in still-to-be-identified hosts. Most of the species or "forms" characterized to date are known from the western Pacific Ocean (off Japan), the North Atlantic Ocean, and the Mediterranean Sea, although that distribution might reflect search intensity, and large unstudied collections, such as those from the Indian Ocean, exist. In some locations more than 40 different types of y-nauplii, all representing undescribed species, can be present in the plankton, indicating a rich biodiversity of adults that is yet to be discovered. The name Facetotecta comes from the unique ornamentation of the cephalic shield in both y-nauplii and y-cyprids. Molecular evidence (Pérez-Losada et al. 2009) indicates both the monophyly of the Facetotecta and their sister-group status to the remaining thecostracans (the Ascothoracida and Cirripedia).

The Ascothoracida is a small group containing approximately 100 known species, all of which are ecto- or endoparasitic on cnidarians (the Anthozoa) and echinoderms. Today, 2 subgroups are recognized: the Laurida and Dendrogastrida. In many species the nauplii are brooded, with the larvae released as a-cyprids (previously called ascothoracid larvae), but planktonic nauplii are known from representatives of 3 of the 6 families.

The Cirripedia is by far the best known of the 3 subclasses and exhibits an extreme diversity in both body forms and

Table 17.1 Summary of larval development in the Thecostraca

Taxon	Hatching instar	Naupliar development[1]	Settlement (cypridoid) instar	Metamorphic instar	Chapter
Facetotecta	nauplius	5 known instars, planktotrophic or lecithotrophic	y-cyprid	ypsigon	18
Ascothoracida	nauplius or a-cyprid	up to 6 instars, planktotrophic or lecithotrophic	a-cyprid[2]	unknown[3]	19
Cirripedia					
Acrothoracica	nauplius or cyprid	6 instars, lecithotrophic	cyprid	intermediary stage[4]	20
Rhizocephala					
Kentrogonida[5]	nauplius[6]	4–5 instars,[7] lecithotrophic	cyprid	kentrogon or trichogon	21
Akentrogonida	cyprid	absent	cyprid	none[8]	21
Thoracica	nauplius or cyprid	6 instars, planktotrophic or lecithotriophic[9]	cyprid	juvenile barnacle[10]	22

[1] When nauplii are present. All thecostracan taxa contain species that hatch as cypridoid larvae.

[2] Some Ascothoracida have 2 consecutive a-cyprid instars, separated by a molt.

[3] In the ground pattern, ascothoracidans seem to change very little between the a-cyprid and the attached juvenile. Early-metamorphosed females of *Ulophysema* and *Dendrogaster*, however, are drastically altered.

[4] Not well described, but an instar seems to be intercalated between the cyprid and the true juvenile.

[5] The Kentrogonida is not monophyletic, since the Akentrogonida is nested within the taxon.

[6] A few species hatch as cyprids.

[7] Not well established and may include 6 in some species.

[8] Akentrogonids seem to skip the kentrogon and trichogon instars and inject the parasite or male cells directly into the host or the virgin female parasite.

[9] Most balanomorphans have planktotrophic nauplii, whereas the majority of pedunculated barnacles have lecithotrophic nauplii.

[10] The cyprid molts into a juvenile, which is a feeding barnacle with shell plates, but deviations occur in species with dwarf males.

lifestyles. The group includes the familiar suspension-feeding stalked and acorn barnacles (the Thoracica), the parasitic Rhizocephala, and the lesser-known Acrothoracica (D. Anderson 1994). Morphologically, the cirripedes are united only by their very characteristic larvae. All cirripede nauplii, including the parasitic forms, bear a pair of frontolateral horns anteriorly on the head shield. The tips of these horns bear sensory setae and a complex exit pore for two large unicellular gland cells (Walker 1973; Semmler et al. 2008). Strictly homologous structures are unknown elsewhere in the Crustacea, including the Facetotecta and Ascothoracida. The last naupliar instar heralds the formation of the terminal cypris stage. Its head shield is enlarged, partially overhanging the body, and the distal part of the antennule is pronouncedly swollen, indicating the development of the cypris attachment organ. The cirripede cyprid (c-cyprid) is much more specialized for substrate location and attachment than are the cypridoid larvae of the Facetotecta and Ascothoracida. The cyprid can traverse long distances over the substratum by walking on its complexly shaped antennules, and it employs an impressive array of sensory organs (Bielecki et al. 2009; Maruzzo et al. 2011). Final attachment is not mechanical, as in facetotectans and ascothoracidans, but by secretion from a cement gland, which produces one of the strongest known bioadhesives (Walker et al. 1987). Until settlement of the cypris, development is very similar in all cirripedes, irrespective of the habitat and biology of the adult. But the metamorphosis that follows proceeds very differently in the 3 subgroups (Høeg and Møller 2006; Høeg et al. 2012). The cyprid endows cirripedes with an immense advantage in locating a suitable site for attachment. The complex morphology and biology of this larva goes a long way toward explaining the success of the Cirripedia, which today are dominant organisms in many marine habitats, either as suspension feeders or parasites in crab populations.

Thoracican cirripedes total more than 1,000 species worldwide and include the familiar stalked and acorn barnacles. They inhabit almost all marine environments, including rocky intertidal zones, mangroves, subtidal reefs, corals, hydrothermal vents, cold seeps, and other deep-sea habitats. In most species, the adults are armed with calcified shell plates, which caused them to be classified as mollusks until the discovery of their larvae. Thoracicans are often epibiotic on other marine organisms, such as decapod crustaceans, sea turtles, whales, corals, sponges, and hydroids. Despite this range of life forms and habitats, all thoracican cirripedes share a common life-history pattern, consisting of planktonic larval stages and a benthic sessile adult stage that obtains nourishment by suspension feeding. The only exception is a handful of parasitic species (unrelated to the Rhizocephala).

The Acrothoracica consists of small burrowing barnacles found in deep-sea carbonate substrata and in the calcareous shells or skeletons of marine invertebrates, notably gastropod shells inhabited by hermit crabs. They have separate sexes, and the larger, suspension-feeding females bear tiny dwarf males attached externally to the mantle sac. Females lack calcareous plates and live inside burrows (Tomlinson 1969; Kolbasov 2009). Currently, the group contains some 80 species in 11 genera and 2 orders (the Lithoglyptida and Cryptophialida) (Kolbasov 2009). Although first discovered at relatively high latitudes, their greatest diversity is in the tropics.

The approximately 250 species of the Rhizocephala are another entirely parasitic group of barnacles, perhaps the most modified of all the Crustacea (Høeg 1995). The sexes are separate, with the female parasite hosting cryptic dwarf males deep within its body. All known species infest other crustaceans, typically decapods. The female parasite lacks all appendages, segmentation, and internal organs, other than gonads. The parasite consists of root-like extensions (called the interna) that penetrate and pervade the body of the host. The interna is connected to the sac-like externa, normally situated on the abdomen of the host. The larval stages are tiny, and they provide the only morphological clue that these parasites are highly modified cirripedes. The nauplii bear distinctive frontolateral horns, as do the nauplii of all of the other Cirripedia (J. Thompson 1836), and the terminal stage is a typical cirripede cyprid (Høeg et al. 2004).

In some thecostracan taxa the nauplii can be either planktotrophic or lecithotrophic. Planktotrophy predominates in acorn barnacles (Thoracica: Balanomorpha), whereas many pedunculated species have lecithotrophic nauplii. Species of the genus *Ibla* may be on the evolutionary borderline, being

planktotrophic but sometimes able to complete development without feeding (Høeg et al. 2009b). Planktotrophic species also exist in the Facetotecta and Ascothoracida, although lecithotrophy is probably more common. The Acrothoracica and Rhizocephala have exclusively lecithotrophic nauplii. The cypridoid larva is always lecithotrophic, being specialized to find a settlement site using the energy accumulated by the nauplius (via planktotrophy) or handed over from the egg (lecithotrophy). The number of naupliar instars is maximally six, but that number has been reduced to five or even four in some species (and to only two in *Ulophysema oresundense*). Finally, a considerable number of thecostracans have an abbreviated planktonic development, with no free nauplius stage. Normally, the larva will simply hatch as a cypridoid, but cases exist in which the hatching stage is a nauplius that is simply brooded until it molts and is released as a cyprid.

All of the above groups are treated in the chapters that follow.

18

Facetotecta

Jens T. Høeg
Benny K. K. Chan
Gregory A. Kolbasov
Mark J. Grygier

GENERAL: The Facetotecta (or enigmatic y-larvae) have been found in the plankton in many seas. Adult stages of the facetotectans are still unknown in situ. Their larval development consists of at least five naupliar instars and a cypridoid (y-cypris) larva (Itô 1990; Grygier 1996; Kolbasov and Høeg 2003). The naupliar head shield and the carapace of the y-cyprid have a surface pattern of reticulated cuticular ridges, forming plates. Itô (1985) proposed the new genus *Hansenocaris* for his 3 new species, described on the basis of their respective y-cyprids. Currently, the facetotectans encompass 7 species, established on the basis of y-cypris morphology and assigned to a single genus, *Hansenocaris* (Itô 1990; Kolbasov et al. 2007), but their number will have to be very significantly extended (Glenner et al. 2008). An additional 4 species of *Hansenocaris* were described on the basis of naupliar stages (Belmonte 2005), but these are dubious, because they were not established on a full naupliar series (five instars), and the corresponding (following) y-cyprids are unknown. The y-cyprid metamorphoses into a slug-like stage, called the ypsigon (Glenner et al. 2008). The morphology of both the y-cyprid and the ypsigon strongly indicates that adult stages are advanced parasites in still-to-be-identified hosts. The life cycle includes at least five free-swimming naupliar stages, a y-cypris larva, and a slug-like ypsigon that is probably a juvenile. Larval stages are free living; adults are unknown and are probably parasites.

LARVAL TYPES

Nauplius and Y-Cypris: Only lecithotrophic forms have been reared through all stages. The nauplius stage is called a y-nauplius. The y-nauplii (fig. 18.1) are planktotrophic or lecithotrophic. All stages are semitransparent, with a pigmented nauplius eye and, in feeding forms, a brownish gut colored by food particles (fig. 18.1A–D). The y-cyprid is always nonfeeding. Its carapace is univalved and does not cover the whole body of the larva (fig. 18.2). The y-cyprid is the last larval stage and probably serves for infestation of a host.

MORPHOLOGY

Y-Nauplius: The naupliar body consists of a cephalic anterior part, covered by the head shield, and a posteriorly projecting hindbody (fig. 18.1F, J, M–O). Cuticular ridges divide the head shield and hindbody into numerous dorsal polygonal plates, which are all arranged in a symmetrical pattern (fig. 18.1I–L). Some dorsal plates of the head shield have symmetrical pores and setae (fig. 18.1L). The dorsocaudal organ (fig. 18.1L), which has a round knob shape, lies posteriorly on the dorsal side of the hindbody. The hindbody terminates in a conspicuous spine, and a pair of smaller ventral spines represents the furcal rudiment (fig. 18.1F, N, O). A pair of pores projects a little anteriorly to the labrum (fig. 18.1E). The prominent labrum consists of a big proximal rounded part and a smaller distal elongated part (fig. 18.1E, H, G, M, N). Three pairs of naupliar appendages surround the labrum (fig. 18.1E, F, H, N, O). The uniramous antennules consist of a basal portion, with several circular annulations, and an elongated distal portion that bears setae. The biramous antennae and mandibles bear setae and gnathobases on the inner margins. The area behind the mandibles may bear the setiform maxillulary rudiments, and more posteriorly the ventral surface is ornamented by several closely spaced, quite prominent ridges armed with minute denticles (fig. 18.1M–O).

Y-Cypris: The y-cyprid has a univalved carapace that only partially covers the larval body (fig. 18.2A–E). The carapace resembles an inverted boat, but with elongated sharp posterior ends (fig. 18.2A, B). Long longitudinal cuticular ridges may ornament the carapace (fig. 18.2A–E), and its surface also bears numerous pores, forming a symmetrical pattern (fig. 18.2D, E). Five pairs of lattice organs (without a pore field) are grouped around big unpaired anterior (lattice organs 1 and 2) and posterior (lattice organs 3–5) pores (fig. 18.2E). Paired compound eyes lie anteriorly in the body (fig. 18.2A). A complex of structures—including the antennules, labrum, paraocular processes, post-ocular filamentary tufts, and two pairs of rudiments of antennae and mandibles—is situated ventrally under the compound eyes (fig. 18.2C, F). The antennules consist of four segments (fig. 18.2A, C, F). The second segment may be armed with a conspicuous curved hook (fig. 18.2F). The fourth (small) segment carries a well-developed aesthetasc (fig. 18.2A, F). A pair of bifurcate paraocular processes are connected to the compound eyes; these processes represent the frontal filaments. A pair of post-ocular filamentary tufts is situated behind the paraocular processes (fig. 18.2F). The conspicuous labrum consists of a wider basal part and a protruded distal part, with prominent curved hooks (fig. 18.2F). At metamorphosis the ypsigon exits from the y-cypris cuticle, just behind the labrum (Glenner et al. 2008). The hooks may therefore function to rip open the (unknown) host animal and so provide an entrance for the ypsigon. The thorax carries six pairs of biramous natatory thoracopods (fig. 18.2A–C). The abdomen consists of two or four segments, including a long telson that is covered by cuticular ridges and terminates in short unsegmented furcal rami (fig. 18.2A, B).

MORPHOLOGICAL DIVERSITY: Facetotectan nauplii exhibit diversity in the form and size of the head shield and the position and number of its cuticular plates. There are 13 naupliar morphotypes known, but only some of these have been correlated with cyprids. Y-cyprids differ in the size and form of the carapace and the paraocular processes, in the number of abdominal segments (2 or 4), and in the presence or absence of antennular hooks (Kolbasov et al. 2007). There are several distinct types of ypsigons that can be correlated with the preceding type of y-cyprid (Glenner et al. 2008).

NATURAL HISTORY: Y-nauplii can be both planktotrophic and lecithotrophic, and they have a characteristic bobbing mode of swimming, making them rather easy to spot when alive. The function and homology of the naupliar horn pores and dorsocaudal organs are unknown. Grygier (1987) interpreted horn pores as incipient frontal filaments, but Itô and Takenaka (1988) considered this interpretation to be uncertain, and Itô (1990) showed that the papillae each have a pore and probably represent the exit ducts of the head gland, which was described by Elofsson (1971). The main part of the dorsocaudal organ lies internally, and only its knob is exposed externally. Unlike the intermittent swimming seen in cirripede cyprids, the y-cyprids (which are non-feeding) swim almost constantly, with the thoracopods swinging out laterally during their beat. The bifurcate paraocular process in y-cyprids may be homologous to the frontal filaments of other cypridoid larvae, and the post-ocular filamentary tuft may represent the external portion of a gland (Itô and Takenaka 1988). After treatment with a crustacean molting hormone, y-cypris larvae molt into a special slug-shaped ypsigon stage of indefinite nature, indicating that this is the first stage in a parasitic mode of life (Glenner et al. 2008; Pérez-Losada et al. 2009).

PHYLOGENETIC SIGNIFICANCE: Much remains unknown about this enigmatic group and its relationships with other crustaceans. Molecular evidence (Pérez-Losada et al. 2009) supports the monophyly of the Facetotecta and their sister-group status to the Ascothoracida and Cirripedia. Based on the morphology of the y-cyprid and the subsequent ypsigon stage, facetotectans are assumed to be the larvae of adults that are parasitic on yet-to-be-identified hosts.

HISTORICAL STUDIES: Facetotectan larvae were discovered more than 100 year ago by Hansen (1899). Thorough studies of the larval development using SEM were published by Itô (1990) and Kolbasov and Høeg (2003). The anatomy of some features in y-larvae were studied with a transmission electron microscope (TEM) by Elofsson (1971) and Itô and Takenaka (1988). The presumably juvenile stage—the ypsigon—following the y-cypris was reared by Glenner et al. (2008).

Selected References

Glenner, H., J. T. Høeg, M. J. Grygier, and Y. Fujita. 2008. Induced metamorphosis in crustacean y-larvae: towards a solution to a 100-year-old riddle. BMC Biology 6: 21.

Grygier, M. J. 1991. Towards a diagnosis of the Facetotecta (Crustacea: Maxillopoda: Thecostraca). Zoological Science 8(6): 1196.

Itô, T. 1990. Naupliar development of *Hansenocaris furcifera* Itô (Crustacea: Maxillopoda: Facetotecta) from Tanabe Bay, Japan. Publications of the Seto Marine Biological Laboratory 34(4/6): 201–224.

Kolbasov, G. A., M. J. Grygier, V. N. Ivanenko, and A. A. Vagelli. 2007. A new species of the y-larva genus *Hansenocaris* Itô, 1985 (Crustacea: Thecostraca: Facetotecta) from Indonesia, with a review of y-cyprids and a key to all their described species. Raffles Bulletin of Zoology 55: 343–353.

Kolbasov, G. A., and J. T. Høeg. 2003. Facetotectan larvae from the White Sea with the description of a new species (Crustacea: Thecostraca). Sarsia 88: 1–15.

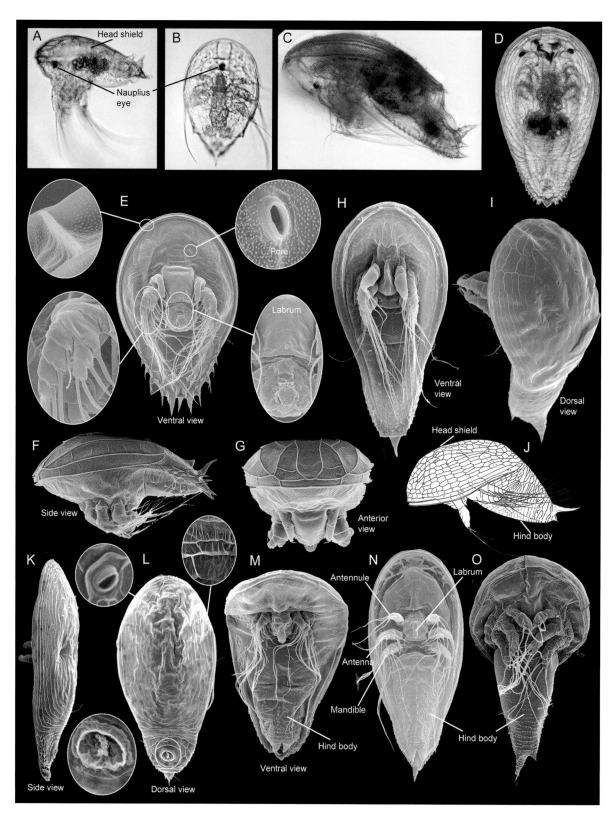

Fig. 18.1 Facetotectan y-nauplius larvae, SEMs (unless otherwise indicated). A: y-nauplius, light microscopy, lateral view. B: planktotrophic y-nauplius, light microscopy, dorsal view. C and D: last y-nauplius instars, with developing y-cyprids, light microscopy. C: lateral view. D: ventral view. E–G: y-nauplii. H and I: y-nauplii of *Hansenocaris mediterraneus*. J: schematic drawing of a y-nauplius of *H. mediterraneus* (comparable to I), lateral view. K–N: different y-nauplius instars of *H. itoi*. O: y-nauplius. A, B, E–G, and O original, of material from Okinawa, Japan; C and D original, of material from the English Channel; H and I original, of material from the Mediterranean Sea; K–N original, of material from the White Sea.

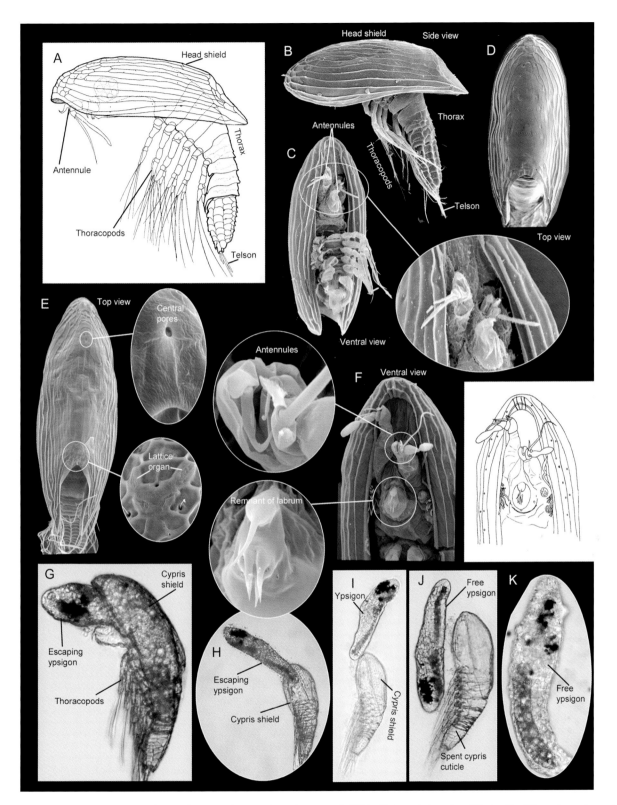

Fig. 18.2 Facetotectan y-cyprids. A: schematic drawing, lateral view. B–F: SEMs. B: lateral view. C and F: details of the antennules and labrum, ventral view. D: dorsal view. E and F: *Hansenocaris itoi*. E: dorsal view, with insets of the pores and lattice organs. F: (*left*) ventral view, with insets showing the antennules and remnant of the labrum; (*right*) schematic drawing of the same specimen. G–K: live y-cyprids and ypsigons, based on treatment with 20-Hydroxyecdysone (see Glenner et al. 2008), light microscopy. G: ypsigon emerging from a y-cyprid, just in front of the labrum. H: ypsigon almost escaped from a y-cyprid. I and J: free ypsigons, close to their now-empty and discarded y-cypris larvae. K: free ypsigon, with pigmented material and oil cells. A–D original, y-cyprids of unknown affiliation; E and F original, of material from the White Sea; G–K original, of material from Okinawa, Japan.

19

Ascothoracida

Jens T. Høeg
Benny K. K. Chan
Gregory A. Kolbasov
Mark J. Grygier

GENERAL: The Ascothoracida are ecto-, meso- or endoparasites in echinoderms and anthozoans (Grygier and Høeg 2005). Except for the secondarily hermaphroditic species of the Petrarcidae, they have separate sexes. The larger female is accompanied by dwarf cypridiform-like males, often living inside her mantle cavity. Ascothoracidans use their piercing mouthparts for feeding on their hosts, although in some advanced forms nutrient absorption may also take place through the integument. The approximately 100 described species are classified in 2 orders, the Laurida and Dendrogastrida (Grygier 1987; Kolbasov et al. 2008b). The life cycle includes up to six free-swimming naupliar instars (fig. 19.1A–C, F, G), an a-cypris larva (fig. 19.1D, E, H–J), a juvenile, and an adult. The larval stages are free living or brooded. Juveniles and adults are parasitic on different taxa of cnidarians and echinoderms.

LARVAL TYPES:

Nauplius and A-Cyprid: Some Ascothoracida retain the thecostracan ground pattern, with six instars of planktotrophic nauplii, but most species either release lecithotrophic nauplii or brood their offspring inside the mantle cavity until their release as a-cyprids, also called ascothoracid larvae (Itô and Grygier 1990). The a-cyprid locates and attaches to the host, using its prehensile antennules.

MORPHOLOGY

Nauplii: The nauplii have an oval bowl-shaped head shield that is broader anteriorly. The head shield can carry short setae, small pores, and feeble ridges (Itô and Grygier 1990; Grygier 1992). Nauplii have a nauplius eye and a pair of setiform frontal filaments (fig. 19.1A, B). The labrum is small and slightly protruding (fig. 19.1G). There are three pairs of natatory limbs: uniramous antennules, and biramous antennae and mandibles, all armed with long setae (fig. 19.1A, B, K, L). Beginning with nauplius instar II, there are also setiform vestiges of the maxillules. Under the hindbody, the last nau-

pliar instar stages have small vestiges of the thoracopods of the ensuing a-cyprid. The hindbody ends in an unpaired terminal spine and a pair of caudal setae or spines (fig. 19.1A, B).

A-Cyprid: The a-cyprid (or ascothoracid larva) has a bivalved carapace that surrounds the body, although the hindbody will normally protrude posteriorly (fig. 19.1D, E, I, J). The carapace surface can be almost smooth (fig. 19.1I) or covered by prominent polygonal cuticular ridges (fig. 19.1J). There are five pairs of lattice organs (without a pore field) along the dorsal hinge line. The prehensile antennules are Z-shaped and composed of four to six segments, whose precise homology to the antennular segments in facetotectan y-cyprids and cirripede cyprids is uncertain. The distal segment bears a hooked claw (for mechanically grasping the host) and different types of setae, including long aesthetascs (fig. 19.1D, M, N). The compound eyes are often reduced, but their vestiges may be present (Hallberg et al. 1985; Grygier 1992). Frontal filaments are attached at the borderline between the anterior part of the larval body and the inner surface of the carapace. Behind the antennules, the mantle cavity houses a conspicuous oral pyramid, with vestiges of piercing mouthparts (fig. 19.1D, E). The thorax carries six biramous and natatory thoracopods (fig. 19.1D, E, H, N). The long abdomen consists of five segments, including an elongated telson that terminates with unsegmented furcal rami (fig. 19.1D, E, H, I).

MORPHOLOGICAL DIVERSITY: The a-cypris larvae of the Ascothoracida differ in the shape and ornamentation of the carapace and in the segmentation of the antennules. The a-cyprids of the Laurida (fig. 19.1J) normally have a more sculptured (reticular ornamentation) carapace, with five pairs of the lattice organs, while those of the Dendrogastrida (fig. 19.1I) have less carapace ornamentation and only four pairs of lattice organs (Kolbasov et al. 2008b). In addition, the laurid a-cyprids normally have six-segmented antennules, as opposed to the four-segmented antennules in dendrogastrid a-cyprids. The Dendrogastrida brood their larvae until the a-cypris stage, whereas the Laurida release their larvae as free-swimming planktotrophic or lecithotrophic nauplii.

NATURAL HISTORY: Free-swimming lecithotrophic nauplii may be reared in culture to the a-cypris stage (Itô and Grygier 1990; Grygier 1992). Development from nauplius I to the a-cypris seems to last approximately one month in warm waters (Itô and Grygier 1990). Species of the Dendrogastrida brood their larvae inside the female's mantle cavity until the cypridoid stage, which consists of two a-cypris instars (Brattström 1948; Wagin 1976; Kolbasov et al. 2008b). The first a-cypris instar has an infantile morphology. The carapace is smooth, without pores, setae, and lattice organs. The antennules have only a rudimentary armament, and the abdomen has only four of the five segments present in the second a-cypris instar. The first a-cypris instar is normally retained within the mantle cavity, while the second instar has the definitive morphology for host location and attachment and is released into the plankton. The a-cyprids of the Laurida (called Tessmann's larvae) can be found in the plankton. Settlement and metamorphosis of the a-cyprid have never been observed, so a full life cycle is not known for any species, but a tentative life cycle is proposed for *Ulophysema oeresendense* in Grygier and Høeg (2005). Primitively, there seems to be little change from an a-cyprid to an adult. This is especially true for the males, where it is questionable whether any molt separates the a-cyprid from the sexually mature form. In advanced species (dendrogastrids), the adult females can grow into elaborate forms, with bizarre extensions from the body.

PHYLOGENETIC SIGNIFICANCE: Ascothoracidans constitute one of the 3 subclasses of the Thecostraca. Historically, the study of relationships within this group has relied heavily on developmental features (e.g., see Høeg 1992; Høeg and Møller 2006; Høeg et al. 2009a). Modifications to accommodate their parasitic lifestyle separate most ascothoracidan larvae from those of other thecostracans.

HISTORICAL STUDIES: The larval development of *Ulophysema öresundense* was studied in detail by Brattström (1948). Thorough studies of the larval development of laurid ascothoracidans, using SEM, were published by Itô and Grygier (1990) and Grygier (1992). The morphology of both the first and second a-cyprid larvae were studied with SEM by Kolbasov et al. (2008b). Wagin (1976) published a monograph on the Ascothoracida, including their development.

Selected References

Brattström, H. 1948. On the Larval Development of the Ascothoracid *Ulophysema öresundense* Brattström. Vol. 2, Studies on *Ulophysema öresundense*. Undersökningar över Öresund No. 33. Kungliga Fysiografiska Sällskapets i Lund Förhandlingar, n.s., 59, No. 5. Lund, Sweden: C. W. K. Gleerup.

Grygier, M. J. 1992. Laboratory rearing of ascothoracidan nauplii (Crustacea: Maxillopoda) from plankton at Okinawa, Japan. Publications of the Seto Marine Biological Laboratory 35(4/5): 235–251.

Itô, T., and M. J. Grygier. 1990. Description and complete larval development of a new species of *Baccalaureus* (Crustacea: Ascothoracida) parasitic in a zoanthid from Tanabe Bay, Honshu, Japan. Zoological Science 7: 485–515.

Kolbasov, G. A., M. J. Grygier, J. T. Høeg, and W. Klepal. 2008b. External morphology of ascothoracid-larvae of the genus *Dendrogaster* (Crustacea, Thecostraca, Ascothoracida), with remarks on the ontogeny of the lattice organs. Zoologischer Anzeiger 247: 159–183.

Wagin, V. L. 1976. Ascothoracida. Kazan, Russia: Kazan University Press [in Russian].

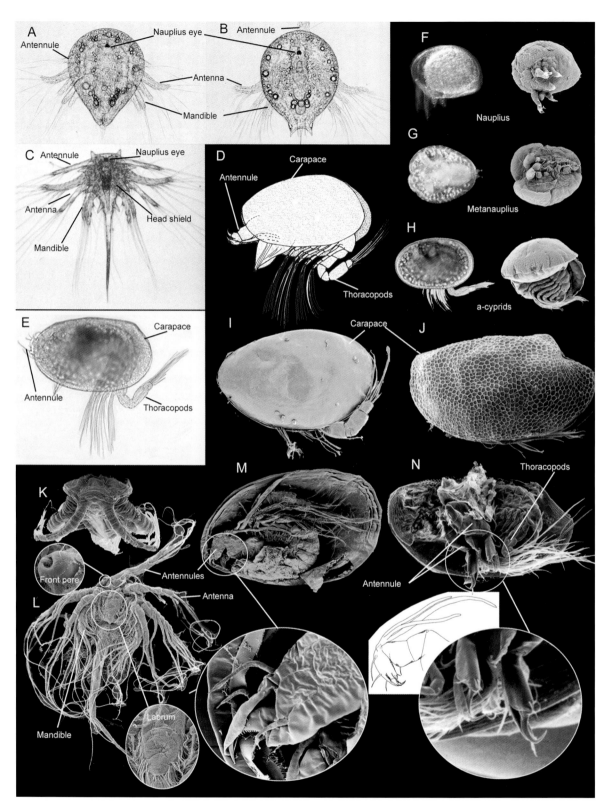

Fig. 19.1 Nauplii and a-cyprids of the Ascothoracida. A and B: lecithotrophic nauplii of *Baccalaureus falsiramus*, light microscopy, dorsal view. C: planktotrophic nauplius of *Zibrowia* sp., light microscopy, ventral view. D: schematic drawing of an a-cyprid. E: a-cyprid of *Dendrogaster rimskykorsakowi*, light microscopy, lateral view; note the mouth cone protruding ventrally. F–H: *D. rimskykorsakowi*. F and G: nauplii, light microscopy (*left*) and SEMs (*right*). H: a-cyprids, light microscopy (*left*) and SEM (*right*). I: a-cyprid of *Ulophysema oeresundense*, SEM, lateral view. J: a-cyprid of *Baccalaureus falsiramus*, SEM, lateral view. K and L: nauplius I of *Zibrowia* sp., SEMs. K: frontal view. L: ventral view. M and N: a-cyprid of *U. oeresundense*, SEMs; note the claw, used for mechanical attachment, at the tip of the antennules in the closeups and the drawing. M: right half of the carapace removed to reveal the body and appendages (ventral side is up). N: left half of the carapace removed (ventral side is down). A–H, K original, of material from Japan and Okinawa; I, M, and N original, of material from the Sound (Øresund) in Denmark.

20 Acrothoracica

GREGORY A. KOLBASOV
BENNY K. K. CHAN
JENS T. HØEG

GENERAL: The Acrothoracica is a group of small burrowing epibiotic barnacles, found largely in carbonate sediments and the skeletons of marine mollusks. Acrothoracicans are dioecious, with suspension-feeding females bearing dwarf males attached externally to the mantle sac. Females lack calcareous plates and live inside burrows (Tomlinson 1969; Kolbasov 2009). Currently, the acrothoracicans encompass approximately 67 species, assigned to 11 genera and 2 orders, the Lithoglyptida and Cryptophialida (Kolbasov 2009). The acrothoracicans were first discovered at relatively high latitudes, but their greatest diversity is found in the tropical seas of the world. The life cycle consists of at least four free-swimming naupliar stages (when present), a cypris larva, a juvenile, and an adult. Larval stages are free living; juvenile and adults are epibiotic, living in burrows. There are no known morphological differences between male and female larvae, and sex may not be determined until the time of attachment.

LARVAL TYPES

Nauplius: Free-swimming nauplii of the type typical for cirripedes (fig. 20.1A–D) were found for several genera in the Lithoglyptida: *Berndtia, Armatoglyptes, Lithoglyptes,* and *Trypetesa* (Kühnert 1934; Utinomi 1961; Turquier 1967; Tomlinson 1969; plus our own unpublished data).

Cypris: Most acrothoracicans, including the Cryptophialida, are supposed to hatch as cyprids, while the naupliar phase is passed in embryonic form within the egg coat (Kolbasov et al. 1999; Kolbasov 2009). The terminal larval stage is always a typical cyprid (figs. 20.1E–G; 20.2).

MORPHOLOGY

Nauplius: The nauplii have a developed head shield, with long frontolateral horns and a long dorsocaudal spine (fig. 20.1A, B, D). Frontolateral horns are applied to the shield in stage 1 (fig. 20.1A) and then extended from the body in subsequent stages (fig. 20.1B, D). A nauplius eye and frontal fila-

ments are located in the anterior part of the body (fig. 20.1A, B). The thoracoabdominal process terminates with a pair of sharp ventral spines (fig. 20.1A, B). Nauplii have three pairs of typical natatory limbs: uniramous antennules, biramous antennae, and mandibles covered with long setae (fig. 20.1A, B). The labrum is either rudimentary (fig. 20.1A, D) or well developed, long, and triangular (fig. 20.1B, C). There seem to be no clear-cut criteria for separating acrothoracian nauplii from those of other cirripedes (which explains their absence from the key to naupliar larvae in chapter 2).

Cypris: Cypris larvae of the Lithoglyptida (figs. 20.1E; 20.2A, B, F) and Cryptophialida (figs. 20.1F, G; 20.2C, D) differ significantly in their morphology. The lithoglyptids have typical spindle-shaped cirripede cypris larvae, with a well-developed carapace that completely encloses the body. As in thoracican cirripedes, they have frontolateral pores (fig. 20.2A, B, H), lattice organs with a conspicuous terminal pore (figs. 20.1E; 20.2G), and the setae on the fourth antennular segment arranged in subterminal and terminal groups (fig. 20.2L). Similarly, they have six pairs of natatory thoracopods and a distinct, although very small, abdomen and telson (figs. 20.1E; 20.2B, E). The cypris larvae of the Cryptophialidae have a much smaller carapace, leaving parts of the body directly exposed (figs. 20.1F, G; 20.2C, D). The carapace is conspicuously ornamented by deep pits and hexagonally arranged ridges. It bears a few very long setae but lacks frontolateral pores (figs. 20.1F). The lattice organs are plate-shaped, with a thick cuticular border, but they lack a terminal pore (fig. 20.2I). The fourth antennular segment has all the setae clustered in one terminal group (fig. 20.2J, K). The thorax and thoracopods are rudimentary and not suitable for swimming (figs. 20.1F, G; 20.2C).

MORPHOLOGICAL DIVERSITY: The cyprids of the Lithoglyptida differ in the shape of the carapace (i.e., the ratio between length and height), in details of their antennular setation, in the armament of the thoracopods, and in the degree of telson cleavage. The cyprids of the Cryptophialida differ in carapace ornamentation, size, and the number of carapace setae, as well as in the number of setae of the fourth antennular segment (Kolbasov and Høeg 2007; Kolbasov 2009).

NATURAL HISTORY: Nauplii of *Trypetesa* are lecithotrophic, with a reduced labrum, mouth, and anus, and they do not increase in size during subsequent molts (Turquier 1967), whereas nauplii of *Armatoglyptes*, with a developed labrum, are planktotrophic (our own data). The cypris larvae of the Cryptophialida are unable to swim and can disperse only by antennular walking, resulting in highly gregarious populations of adults (Turquier 1985; Kolbasov and Høeg 2007).

PHYLOGENETIC SIGNIFICANCE: Larval characters have firmly established the position of the Acrothoracica within the Cirripedia. Their nauplii bear the distinctive frontolateral horns, with sensory setae, and the associated gland cells that characterize the group (Walker 1973; Semmler et al. 2008), and the final larval stage is a typical cyprid.

HISTORICAL STUDIES: A thorough study of the larval development of the genus *Trypetesa*, using light microscopy, was published by Turquier (1967). The morphology of nauplii of *Berndtia* was studied with light microscopy by Utinomi (1961). The morphology of cypris larvae of different acrothoracican taxa was studied with SEM by Kolbasov et al. (1999) and Kolbasov and Høeg (2007). Information on all known larvae of the Acrothoracica was summarized by Kolbasov (2009).

Selected References

Kolbasov, G. A. 2009. Acrothoracica, Burrowing Crustaceans. Moscow: KMK Scientific Press.
Kolbasov, G. A., and J. T. Høeg. 2007. Cypris larvae of the acrothoracican barnacles (Thecostraca, Cirripedia, Acrothoracica). Zoologischer Anzeiger 246: 127–151.
Turquier, Y. 1967. Le développement larvaire de *Trypetesa nassarioides* Turquier, cirripède acrothoracique. Archives de Zoologie Expérimentale et Générale 108: 33–47.
Utinomi, H. 1961. Studies on the Cirripedia Acrothoracica, 3: development of the female and male of *Berndtia purpurea* Utinomi. Publications of the Seto Marine Biology Laboratory 9: 413–446.

Fig. 20.1 *(opposite)* Nauplii and cypris stages. A–E: Lithoglyptida. A: drawing of nauplius I of *Berndtia purpurea*, ventral view. B: nauplius II of *Armatoglyptes taiwanus*, showing the frontolateral horns and appendages, light microscopy, ventral view. C: magnified view of the terminal part of the labrum in B. D: nauplius IV of *Trypetesa lampas*, SEM, ventral view. E: drawing of a cyprid of *A. taiwanus*, lateral view. F and G: Cryptophialida. F: drawing of a cyprid of *Cryptophialus heterodontus*, dorsolateral view. G: drawing of a cyprid of *Australophialus melampygos*, dorsal view. A modified after Utinomi (1961); B–D original; E–G modified after Kolbasov and Høeg (2007).

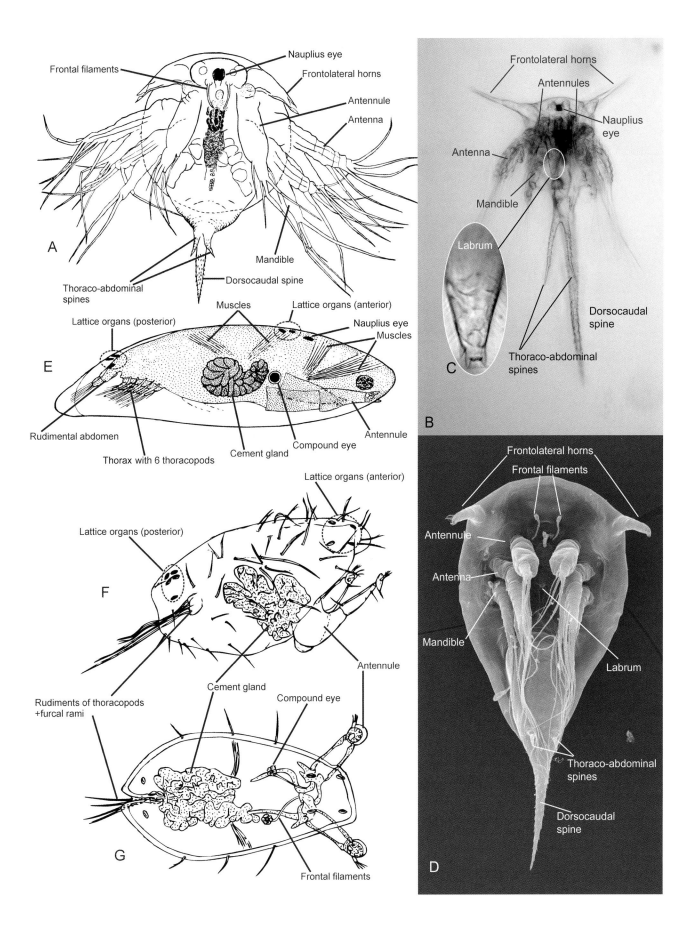

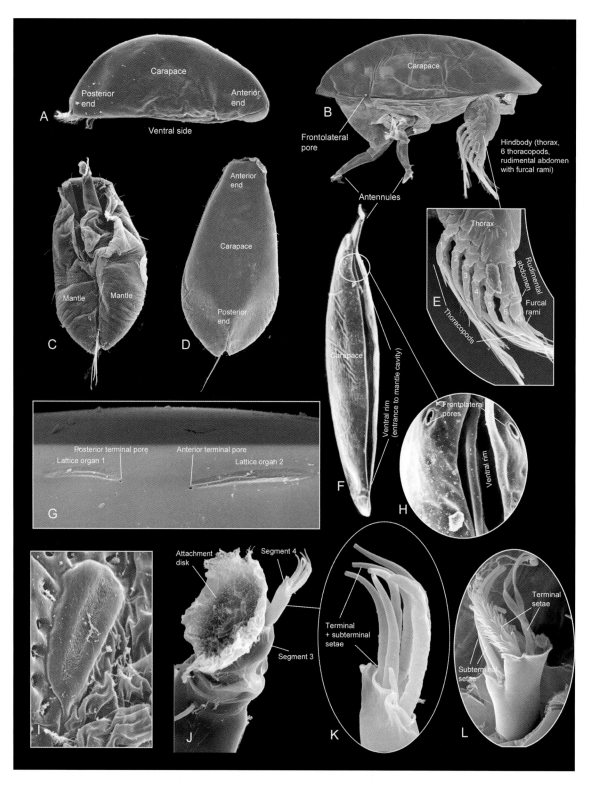

Fig. 20.2 Cypris larvae. Lithoglyptida (A, B, E–G, H, and L) and Cryptophialida (C, D, I–K). A: *Weltneria spinosa*, lateral view. B: *Kochlorine grebelnii*, with the antennules and hindbody stretched out from the carapace, lateral view. C and D: *Australophialus melampygos*. C: ventral view. D: dorsal view. E: thorax of *K. grebelnii* with the thoracopods, and rudimental abdomen with the furcal rami, lateral view. F: *Armatoglyptes mitis*, ventrolateral view. G: closeup of the anterior lattice organs of *Trypetesa lateralis*, lateral view. H: closeup of the frontolateral pores of *A. mitis*, ventral view. I: lattice organ 4 of *Cryptophialus wainwrighti*. J and K: *A. melampygos*. J: distal part of the antennule. K: closeup of the terminal part of the fourth antennular segment, showing the union of the terminal and subterminal setae. L: fourth antennular segment of *W. spinosa*, with separate subterminal and terminal setae. A, B, E, G, and I–K modified after Kolbasov (2009); C, D, and L modified after Kolbasov and Høeg (2007); F and H modified after Kolbasov et al. (1999).

Rhizocephala

Jens T. Høeg
Benny K. K. Chan
Alexey V. Rybakov

GENERAL: Rhizocephalans are very successful and entirely parasitic members of the Cirripedia. With more than 250 species, they comprise about one quarter of all cirripede species, and their parasitic mode of life is therefore a major component in the success of the subclass (Høeg et al. 2005). Rhizocephalans parasitize other Crustacea, with most species infesting brachyuran and anomuran decapods, but they also occur on carideans, peracarids, stomatopods, and even balanomorphan cirripedes (Høeg and Lützen 1995, 1996). The majority of species belongs to the suborder Kentrogonida and are found exclusively on decapods, whereas species of the highly specialized suborder Akentrogonida also infest other crustaceans (Høeg and Rybakov 1992). Rhizocephalans occur in all marine environments, including the deep sea. Remarkably, they also infest semiterrestrial true crabs and hermit crabs, and they even occur on freshwater crabs. All rhizocephalans have separate sexes, and this is reflected in both the size and detailed morphology of their larvae (Høeg 1984; Walker 1985; Høeg and Lützen 1995, 1996). Larval development deviates little (if at all) from that found in other Cirripedia (Rybakov et al. 2002), but the metamorphosis following attachment involves highly specialized stages, unlike any found elsewhere in the subclass (Høeg 1985a, 1987a). The parasite is female and has an extremely simplified morphology. The parasite consists of an internal nutrient-absorbing system of rootlets (interna) and an external reproductive sac (externa) that contains an ovary, a mantle cavity, and organs specialized to host dwarf males, but it lacks appendages, segmentation, a gut, and sensory organs (Høeg and Lützen 1995, 1996). The life cycle consists of 4–6 naupliar instars, a cyprid, a kentrogon and vermigon (females), a trichogon and sperm-producing adult (males), and (in females) an endoparasitic phase and an adult consisting of interna and externa hosting cryptic dwarf males.

LARVAL TYPES

Nauplius: Most species hatch as a nauplius. Few species have been cultured through all stages, so the number of nau-

pliar instars is uncertain, but it seemingly can vary from four to six.

Cyprid: The nauplius phase is followed by a typical cyprid. Some deep-sea species, all freshwater species, and all members of the Akentrogonida have an abbreviated development and hatch as cyprids. Following cypris settlement, the ensuing metamorphosis involves highly specialized stages (kentrogon, vermigon, and trichogon) unlike any seen elsewhere in the Cirripedia. Male and female larvae differ in size (fig. 21.2F). This is most apparent at the cypris stage, where the sexes also differ in clear-cut details of their morphology (fig. 21.2C–E).

MORPHOLOGY

Nauplius: The naupliar larvae are always lecithotrophic and have three pairs of limbs: uniramous antennules, biramous antennae, and mandibles (fig. 21.1). Unlike non-feeding nauplii in the Thoracica, those in the Rhizocephala are very small, rarely exceeding 360 μm in length, and contain only limited amounts of nutrients that are stored in oil cells. The nauplii have only the merest rudiment of a labrum, and they lack both plumose setae for suspension feeding and antennal and mandibular endites (fig. 21.1B). Limb setation changes little (if at all) during development and therefore offers no clue to distinguish between instars. The instars differ principally by a gradual increase in their length, due to the development of the post-cephalic trunk. As in other cirripedes, the morphology of the last nauplius instar heralds the ensuing cypris by possessing an enlarged head shield and a distinct distal distension of the antennule, indicating the development of the cypris attachment organ. The nauplii always have a pair of frontolateral horns, a pair of frontal filaments, and (normally) a pigmented naupliar eye (fig. 21.1A, B).

Cyprid: The non-feeding cyprids, in almost all details, are quite similar to those of thoracican and acrothoracican cirripedes, but they differ in the sensory armature on the antennules (fig. 21.2A). In the Kentrogonida, both male and female cyprids have an olfactory aesthetasc seta on the small fourth antennular segment, but it is relatively much longer in males (Walker 1985). Males carry an additional, even longer aesthetasc on the third segment (the attachment organ). These and other differences, seen easily with a light microscope, offer a reliable means of determining the sex of individual larvae (fig. 21.2C–E). Cypris length varies with both sex and species and ranges from a mere 70 μm to 360 μm. In the Kentrogonida, male cyprids are larger than females, with the size ranges either completely separated (fig. 22.2F) or overlapping to some degree (Høeg 1984; Høeg and Lützen 1995). Most rhizocephalan cyprids have lattice organs with a pore field (fig. 21.2B), frontal horn pores, frontal filaments, and a nauplius eye, while compound eyes are present only in a few species (Høeg and Lützen 1993).

MORPHOLOGICAL DIVERSITY: Most species of the Peltogastridae and Lernaeodiscidae have a flotation collar encircling the naupliar body (fig. 21.1C–G, I), consisting of an exceedingly thin cuticle reinforced by a lattice of ridges (fig. 21.1J, K). The collar, absent from the first instar (compare fig. 21.H and I), provides extra flotation for the nauplius. Some species lack the nauplius eye (fig. 21.1C). Variation in cypris morphology has considerable systematic value and concerns the body outline, the antennular shape, and the armament on the sensory organs. Most akentrogonid species lack antennular aesthetascs, and in some the fourth segment carries only a single bifid seta.

NATURAL HISTORY: The small size of the larvae reflects the parasites' need to produce large broods, and it may also partially explain why rhizocephalan larvae are only rarely identified in plankton samples. Because larvae are small and non-feeding, larval development is normally short (3–7 days), but it can last up to one month in cold-water species (Høeg 1995; Høeg and Lützen 1995). In the Kentrogonida, female cyprids settle only on prospective host animals. Here they metamorphose into a kentrogon stage and produce a hollow injection stylet that passes through the host's underlying integument (fig. 21.2H, J, K). The attachment site is always where the cuticle is thin (gills, or the base of setae), thus facilitating penetration of the stylet (fig. 21.2J). The parasite is injected as a worm or slug-shaped vermigon stage, whose exceedingly thin cuticle allows it to change form while passing down through the stylet (fig. 21.2H, J). The injected vermigon migrates through the blood system to the abdomen of the host, where the root system (interna) develops, and eventually emerges under the abdomen as a virginal externa. Male cyprids settle only at the mantle opening of such virginal females, where they metamorphose into a trichogon stage (fig. 21.2I, L). The trichogon resembles the vermigon, except for being conspicuously armed with cuticular spines (fig. 21.2L, M). It becomes implanted into the female, where it develops into a cryptic dwarf male. Species of the Akentrogonida use the cypris antennule to penetrate the host or the virginal female without forming kentrogons or trichogons.

Some sacculinid species infest freshwater or semiterres-

trial crabs, but release their larvae into salt water when the host migrates to a river mouth (Høeg and Lützen 1995). A few sacculinids occur on true freshwater crabs, and they are unique among cirripedes in releasing their larvae (as cyprids) into the limnic environment. The cyprids of the Chthamalophilidae (only 70 μm long) are unique in lacking a thorax and thoracopods and can therefore disperse only by walking on the antennules. This situation is also found in some acrothoracicans, and in both cases this limited capacity for dispersal results in very dense populations of adults.

Rhizocephalans are parasitic castrators, so many host species have evolved elaborate means for grooming away the larvae before they become internal and gain control of the host. The specialized rhizocephalan metamorphosis is therefore largely an adaptation to circumvent host defenses against successful infection (Ritchie and Høeg 1981; Høeg et al. 2005).

PHYLOGENETIC SIGNIFICANCE: The simplified nature of adult rhizocephalans offers no clues to their affinities, but J. Thompson (1836) relegated them to the Cirripedia, based on the presence of nauplii with frontolateral horns. This conclusion has never been questioned and offers one of the most convincing cases for using larval morphology in animal systematics (Høeg 1995; Høeg and Møller 2006). Both larval and molecular datasets are in agreement that the Akentrogonida is monophyletic and nested within the Kentrogonida, rendering the latter a paraphyletic assemblage (Glenner and Høeg 1994; Glenner et al. 2010).

HISTORICAL STUDIES: J. Thompson's early (1836) work is noted above. In the 1940s and 50s André Veillet published several insightful but largely neglected studies on rhizocephalan larvae (reviewed in Høeg 1985a, 1987a; Høeg et al. 2004). Important modern studies on naupliar development are Collis and Walker (1994), Walossek et al. (1996), Rybakov et al. (2002, 2003), and Høeg et al. (2004). Glenner et al. (1989) used SEM to provide the first comparative account of rhizocephalan cyprids.

The benchmark paper of Delage (1884) described development from the nauplius through the formation of the kentrogon stage, but observation of the actual infection of the host had to await discovery of the vermigon, more than 100 years later (Glenner and Høeg 1995a; Glenner et al. 2000). The metamorphosis from cyprid to kentrogon was described in detail using TEM by Høeg (1985a) and Glenner (2001); Høeg et al. (2012) studied this process by video microscopy. Yanagimachi (1961) found that rhizocephalans can have separate sexes, and his discovery of sexually bimorphic larvae was later extended to all kentrogonid Rhizocephala (Ritchie and Høeg 1981; Høeg 1984; Walker 1985; Glenner et al. 1989). The trichogon stage and its implantation as a cryptic dwarf male into the female parasite was described and compared with the kentrogon by Høeg (1987a, 1987b). Cypris morphology and metamorphosis in akentrogonid species was studied by Høeg (1985a, 1990), Lützen et al. (1996), and Glenner et al. (2010). Glenner and Høeg (1994) discussed the evolution of metamorphosis in the Rhizocephala and considered the kentrogon to be homologous to the first juvenile instar of thoracican barnacles.

Selected References

Glenner, H. 2001. Cypris metamorphosis, injection and earliest internal development of the kentrogonid rhizocephalan *Loxothylacus panopaei* (Gissler), Crustacea: Cirripedia: Rhizocephala: Sacculinidae. Journal of Morphology 249: 43–75.

Glenner, H., J. T. Høeg, A. Klysner, and B. Brodin Larsen. 1989. Cypris ultra-structure, metamorphosis and sex in seven families of parasitic barnacles (Crustacea: Cirripedia: Rhizocephala). Acta Zoological (Stockholm) 70: 229–242.

Høeg, J. T. 1985a. Cypris settlement, kentrogon formation and host invasion in the parasitic barnacle *Lernaeodiscus porcellanae* (Müller) (Crustacea: Cirripedia: Rhizocephala). Acta Zoologica (Stockholm) 66: 1–45.

Høeg, J. T. 1987a. Male cypris metamorphosis and a new male larval form, the trichogon, in the parasitic barnacle *Sacculina carcini* (Crustacea: Cirripedia: Rhizocephala). Philosophical Transactions of the Royal Society of London, B 317: 47–63.

Rybakov, A. V., O. M. Korn, J. T. Høeg, and D. Walossek. 2002. Larval development in *Peltogasterella* studied by scanning electron microscopy (Crustacea: Cirripedia: Rhizocephala). Zoologischer Anzeiger 241: 199–221.

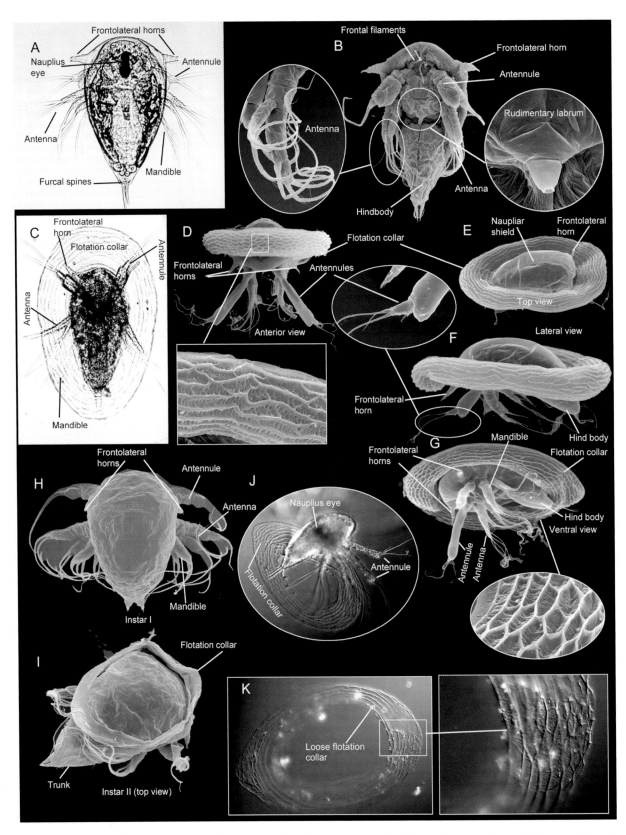

Fig. 21.1 A: nauplius of *Sacculina carcini*, light microscopy, dorsal view. B: nauplius II of *Lernaodiscus porcellanae*, SEM, ventral view. C: nauplius of *Peltogaster paguri*, showing the flotation collar, light microscopy, dorsal view. D–G: nauplii of *P. paguri*, SEMs. D: anterior view. E: dorsal view. F: oblique lateral view. G: oblique ventral view. H: nauplius I of *L. porcellanae*, with no flotation collar at this instar, SEM. I: nauplius II of *L. porcellanare*, with the small flotation collar now developed, SEM. J and K: nauplii of *Briarosaccus tenellus*, light microscopy. J: loosening of the flotation collar before molting to the next nauplius. K: loose flotation collar after the molt. A–K original.

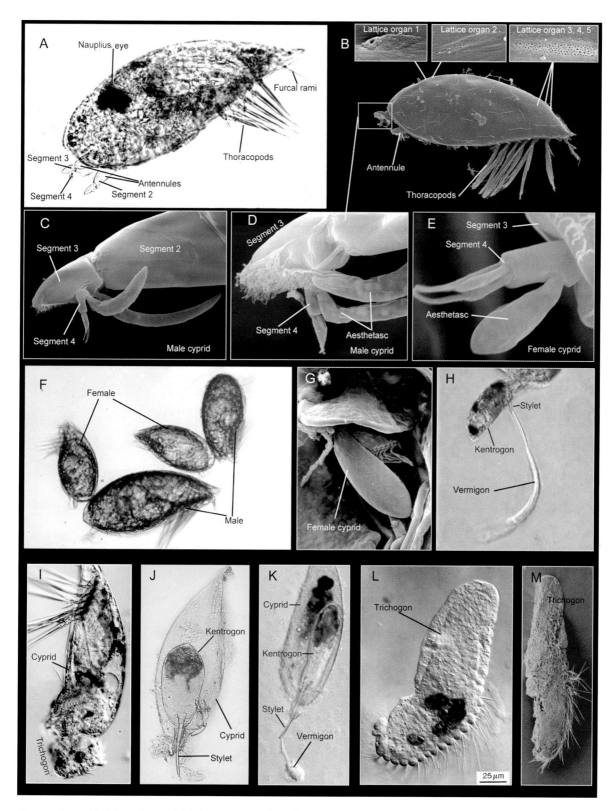

Fig. 21.2 A: cyprid of *Sacculina carcini*, light microscopy, lateral view. B: female cypris of *Lernaeodiscus porcellanae*, showing the lattice organs, SEM, lateral view. C–E: *S. carcini*. C and D: antennules of male cyprids; note the long, male-specific aesthetasc on segment 3, which is absent in females, SEMs. E: fourth antennular segment in a female cyprid (compare with C and D), SEM. F: live large male and small female cyprids of *Peltogaster* sp., light microscopy, lateral view. G: cyprid of *L. porcellanae*, settled on a crab gill, SEM. H: live kentrogon of *Loxothylacus panopaei* on a crab gill, injecting the vermigon, light-microscopy. I–M: *S. carcini*, lateral views. I: trichogon emerging through the antennule of a cypris, light microscopy. J and K: kentrogons, dissected free from their attachment on a crab limb, light microscopy. J: kentrogon encased in the cypris carapace, with its stylet extended. K: kentrogon encased in the cypris carapace, with a vermigon emerging from the tip of its stylet. L: free trichogon, armed with spines, light microscopy. M: free trichogon, SEM. A–M original.

22

Thoracica

BENNY K. K. CHAN
JENS T. HØEG
RYUSUKE KADO

GENERAL: Thoracican cirripedes include the well-known crustaceans referred to as barnacles. The group contains more than 1,000 species worldwide and exhibits an extreme diversity in body forms, including stalked and acorn barnacles, cup-shaped sponge barnacles, and the asymmetrical Verrucomorpha (D. Anderson 1994). Thoracican cirripedes inhabit almost all marine environments, including rocky intertidal zones, mangroves, subtidal reefs, corals, sponges, deep-sea hydrothermal vents, and cold seeps, and they can be epibiotic on a wide range of other marine organisms. Despite this range of life forms and habitats, all thoracicans share a common life-history pattern, consisting of planktonic larval stages that eventually settle and metamorphose into sessile adults (Moyse 1987; Walker et al. 1987; Høeg and Møller 2006). The adults are suspension feeders, except for a handful of parasitic species. The life cycle consists of 4–6 naupliar instars, a cyprid, and metamorphosis into sessile juveniles and adults.

LARVAL TYPES

Nauplius: Naupliar larvae (figs. 22.1; 22.2) can be planktotrophic or lecithotrophic. Lecithotrophic larvae are larger in size when compared with planktotrophic larvae, and their bodies contain many lipid reserves (fig. 22.1E, F). They also have a rudimentary labrum (fig. 22.1E), a structure otherwise essential for naupliar suspension feeding (Walossek 1993). The setae on the swimming limbs in lecithotrophic larvae are often simple (fig. 22.1 D, G), instead of the plumose setae seen in planktotrophic larvae (fig. 22.2A, B).

Cyprid: In some species, the embryos can hatch directly as cyprids (most scalpellids), or the naupliar phase can be spent as brooded larvae inside the mantle cavity until the cyprid stage is reached (intertidal *Tetraclitella*). Such brooded nauplii also have greatly reduced limbs and a rudimentary labrum (fig. 22.1F). As in all other cirripedes, cyprids of thoracican barnacles (figs. 22.2C; 22.3) are non-feeding. The body is composed of a carapace, which develops from the naupliar head

shield and flexes down to cover both sides of the larva (Høeg et al. 2004).

MORPHOLOGY

Nauplius: Thoracican naupliar larvae are often pear-shaped and have, like all other cirripede nauplii, a pair of frontolateral horns. Instar I larvae are globular, with the limbs and labrum poorly developed (fig. 22.1B). These larvae have short frontolateral horns pointed posteriorly. The frontolateral horns have a horizontal array of pores on the top surfaces, in addition to a large terminal gland pore. All naupliar instars carry a pair of frontal filaments, and most species (except deep-sea hydrothermal vent species) also have a single medial nauplier eye from instars I to VI (figs. 22.1A; 22.2A). Paired compound eyes are also visible late in instar VI, where the larvae also have a much more pronounced globular shape (fig. 22.1D), but the compound eyes are not fully developed until the succeeding cypris instar. The nauplii have three pairs of limbs (figs. 22.1A, C; 22.2A): uniramous antennules, and biramous antennae and mandibles. The antennules, antennae, and mandibles are equipped with a variety of setae (figs. 22.1C; 22.2A, B). On the ventral side, the nauplier labrum covers the pre-oral chamber and the mouth (fig. 22.1C). The hindbody of the nauplius consists of a trunk (figs. 22.1A, C; 22.2A). It represents the developing thorax, which only becomes segmented in the cyprid, and it accordingly changes in shape through the nauplier instars.

Cyprid: The cyprid has an elongated torpedo-shaped body that is covered by the carapace, which is drawn down on both sides (fig. 22.3), but it only rarely has a middorsal ridge (fig. 22.3C) and therefore is not truly bivalved. All cyprid appendages can be completely withdrawn into the mantle cavity formed by the carapace (fig. 22.3A). On its surface the carapace bears setae, lattice organs, and various types of gland pores. As in all other thecostracans, there are five pairs of lattice organs, two situated anteriorly and three situated posteriorly (fig.

22.3B, C). They are elongated chemosensory organs, perforated by pores (fig. 22.3B), and develop from setae in the head shield of the nauplius (P. Jensen et al. 1994a; Rybakov et al. 2003). Anteroventrally, the carapace has a pair of large pores on each side, which are the exits for the frontolateral glands (fig. 22.3B). Cyprids of a few species (e.g., *Lepas*) can have these pores on horns, just as in the nauplii. A pair of multicellular cement glands are located in the middle of the body, but they exit on the third antennular segment (the attachment organ) (fig. 22.3A). The appendages consist of antennules, six pairs of natatory thoracopods, and a pair of unsegmented furcal rami (caudal rami) (fig. 22.3B). The furcal rami sit on a small telson with a deep medial cleft, and the telson halves have therefore been mistaken for basal segments in the rami. The abdomen, if developed at all, is (at best) a tiny disc inserted between the thorax and the telson. Antennae and mandibles are absent. The four-segmented antennules are highly specialized and used for exploratory walking in search of a suitable habitat for settlement (fig. 22.3B) (Lagersson and Høeg 2002; Maruzzo et al. 2011). They carry an array of sensory setae, as well as the gland pores for the cement gland, which is used in attachment (Walker et al. 1987; Moyse et al. 1995; Bielecki et al. 2009). The third segment is elaborated as the attachment organ (fig. 22.3B). The fourth segment contains four short subterminal setae and five terminal setae, some of which sweep over the bottom during the exploratory phase (fig. 22.3B).

MORPHOLOGICAL DIVERSITY: The morphology of thoracican nauplii differs markedly among families and genera. In general, the naupliar larvae of *Ibla* and chthamalid, scalpellid, and verrucid barnacles are more globular and have short frontolateral horns pointing ventrally (fig. 22.1G). Balanid nauplii are pear-shaped and have frontolateral horns pointing anterolaterally (fig. 22.1C, D). Iblid nauplii are globular in shape, with a very short trunk (fig. 22.1G, H). Lepadid and poecilasmatid barnacles have much longer frontolateral horns, an elongated body form, and a trunk with extremely long dorsal thoracic spines (figs. 22.1J, K, L; 22.2B, D, E, F). The stalked barnacle *Capitulum mitella* has very long posterior-shield spines (fig. 22.1I). Regrettably, the terminology of the naupliar spines has never been consistently defined (Walossek et al. 1996). The labrum of stalked barnacles and of verrucid and chthamalid species is unilobed (figs. 22.1J, K; 22.2D), whereas the balanoid labrum is trilobed (fig. 22.1C). The number, form, and arrangement of setae on the naupliar limbs are taxonomically important (Newman and Ross 2001). Functionally, they are coupled to the mode of feeding in ways that are still only poorly understood. Planktotrophic larvae have a well-developed labrum and setose basal enditic spines (or gnathobases) on the antenna and mandibles, to aid in feeding.

The carapace of thoracican cyprids is often rice-shaped (fig. 22.3A). The surface sculpturing of the carapace varies among species (Elfimov 1995), but the functional significance of this is wholly unknown. *Chthamalus malayensis* and *Capitulum mitella* have honeycombed sculpturing (fig. 22.3D), whereas

the carapace of *Tetraclita* is smooth (fig. 22.3B). The shape of the attachment disc is quite variable (from bell- to hand- or hoof-shaped) (fig. 22.3B, D, F, G), and this seems to be related to the type of substrate used by the adult (Chen et al. 2013). In general, the attachment discs of non-commensal barnacles are circular or oval in shape (fig. 22.3B, D). In some coral barnacles, the attachment disc is modified to be spear-shaped and is believed to be inserted into the coral tissue for settlement (fig. 22.3E) (Brickner and Høeg 2010).

NATURAL HISTORY: As in all other cirripedes, the functions of the frontolateral horns and gland are still unknown. The frontal filaments are probably specialized sensory organs (Walker et al. 1987). The three pairs of naupliar limbs are involved in swimming, sensing, and food collection. The uniramous antennules are principally used in sensing, and perhaps for balancing. Both the antennal and mandibular exopods can swing through a wide arc at the sides and are the main swimming limbs. The setae of the antenna are spread during the propulsive stroke and feathered in the recovery stroke. The antennae are also used for collecting food, such as diatoms and flagellates. Endopods and the basal enditic spines of the antennae and mandibles are used to transfer the collected food particles to the pre-oral chamber under the labrum and then into the mouth. The nauplius eye enables the larva to respond positively to light. The function of the image-forming compound eyes is unknown, but they can be seen to move actively on their short stalks when the cyprid is exploring a surface. During surface exploration, the cyprid walks over the surface on its antennules and can change direction or turn around almost on the spot (Lagersson and Høeg 2002; Maruzzo et al. 2011). It can also decide to return to swimming in search of a more suitable locality elsewhere. After their final cementation, the cyprids pass through a complex metamorphic molt (Walley 1969; Høeg et al. 2012; Maruzzo et al. 2012). The carapace is shed and the resulting juvenile begins to develop mineralized shell plates, while the natatory thoracopods of the cyprid are transformed to a basket of food-collecting cirri (fig. 22.2H). The compound eyes are shed in metamorphosis, but the nauplius eye is retained in a transformed state.

PHYLOGENETIC SIGNIFICANCE: Although a wide range of life forms and habitats is found in the group, all thoracians share a life-history pattern that includes planktonic larval stages that eventually settle and metamorphose into sessile adults (Moyse 1987; Walker et al. 1987). The naupliar larvae share the distinctive frontolateral horns (with sensory setae) and exit pores for gland cells with other cirripedes.

HISTORICAL STUDIES: The first barnacle nauplii were found by J. Thompson in 1823 (Winsor 1969). Walley (1969) provided the classic histological study of larval development, while Høeg et al. (2012) and Maruzzo et al. (2012) used video recordings to compare thoracican metamorphosis with that of other cirripedes. Nott (1969), Nott and Foster (1969), Moyse

(1987), Elfimov (1995), Glenner and Høeg (1995b), O. Korn (1995), Moyse et al. (1995), and Bielecki et al. (2009) provided comprehensive comparative studies on thoracican nauplii and cyprids. The development of nervous and muscular tissues in the nauplius was studied by Semmler et al. (2008, 2009), using a confocal laser scanning microscope. Newman and Ross (2001) developed a revised setation formula to record the diagnostic setation of the limbs for species identification of naupliar larvae. Lagersson et al. (2003), Bielecki et al. (2009), and Maruzzo et al. (2011) documented the function and structure of the cyprid antennules in a very detailed manner. P. Jensen et al. (1994a, 1994b) and Høeg et al. (1998, 2004) described the lattice organs and the structure of cyprids in thecostracan evolution. Finally, the settlement biology of thoracican cyprids was reviewed by Clare (1995), Walker (1995), and Aldred and Clare (2008).

Selected References

Bielecki, J., B. K. K. Chan, J. T. Høeg, and A. Sari. 2009. Antennular sensory organs in cyprids of balanomorphan cirripedes: standardizing terminology using *Megabalanus rosa*. Biofouling 25(3): 203–214.

Chan, B. K. K. 2003. Studies on *Tetraclita squamosa* and *Tetraclita japonica* (Cirripedia: Thoracica), 2: larval morphology and development. Journal of Crustacean Biology 23: 522–547.

Glenner, H., and J. T. Høeg. 1995b. Scanning electron microscopy of cyprid larvae in *Balanus amphitrite amphitrite* (Crustacea: Cirripedia: Thoracica: Balanomorpha). Journal of Crustacean Biology 15: 523–536.

Moyse, J. 1987. Larvae of lepadomorph barnacles. *In* A. J. Southward, ed. Barnacle Biology, 329–362. Rotterdam: A. A. Balkema.

Newman, W. A., and A. Ross. 2001. Prospectus on larval cirriped setation formulae, revisited. Journal of Crustacean Biology 21: 56–77.

Fig. 22.1 (*opposite*) Naupliar larvae of the Thoracica, SEMs (except for A). A: naupliar instar II of *Balanus improvisus*, showing the morphological parts of a typical thoracican nauplius, light microscopy, dorsal view. B: naupliar instar I of *Fistulobalanus albicostatus*, showing the poorly developed or rudimentary labrum, limbs, and trunk, frontal view. C: naupliar instar II of *Tetraclita japonica*, showing the trilobed labrum, setose limbs, and basal enditic spines, ventral view. D: naupliar instar VI of *T. japonica*, showing the globular-shaped body and the shield assuming the shape of the ensuing cyprid, lateral view. E and F: naupliar instar IV of *T. divisa*, brooding the nauplii in the mantle cavity until they reach the cypris stage; such nauplii have a reduced labrum and limbs. E: ventral view. F: lateral view. G and H: naupliar instar VI of the stalked barnacle *Ibla cumingi*. G: ventral view. H: honeycombed sculpturing on the head shield, dorsal view. I: naupliar instar VI of *Capitulum mitella*, showing the pronounced posterior shield spines, dorsal view. J: naupliar instar II of *Octolamis cor*, showing the unilobed labrum and the long setose dorsal thoracic spines, ventral view. K: naupliar instar II of *Lepas anatifera*, showing the setose unilobed labrum, ventral view. L: late naupliar instar of *L. anatifera*, showing the long dorsal thoracic spine and caudal spine, dorsal view. A–L: original.

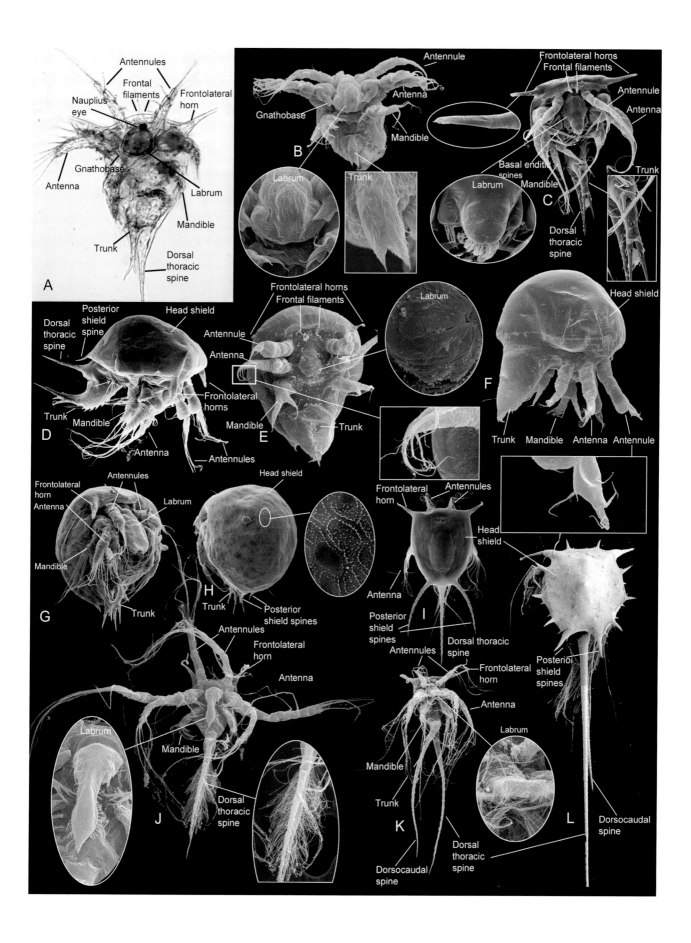

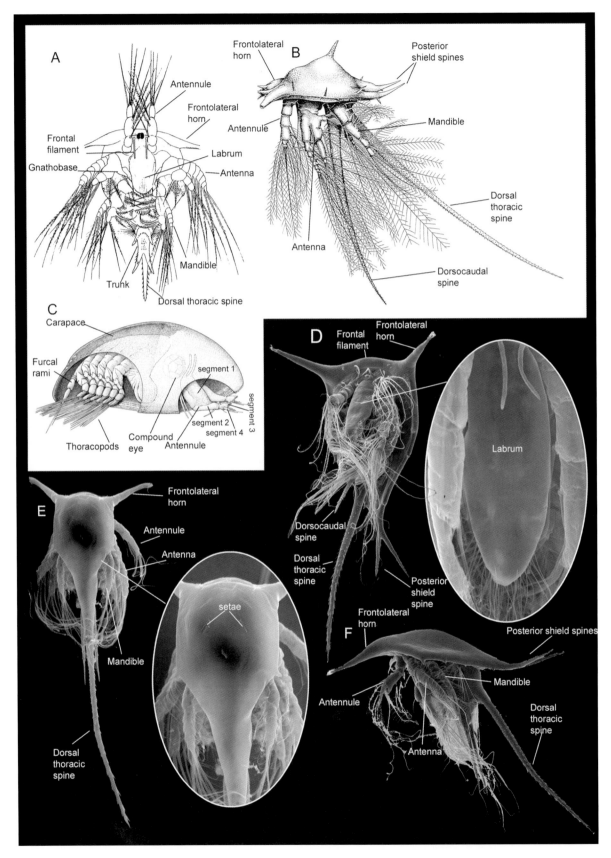

Fig. 22.2 Naupliar and cypris larvae of the Thoracica. A: schematic drawing of a naupliar larva of *Balanus improvisus*, ventral view. B: schematic drawing of a nauplius of *Lepas*, lateral view. C: schematic drawing of a cyprid of *Scalpellum scalpellum*, lateral view. D–F: nauplius of *Lepas*, SEMs. D: ventral view. E: dorsal view. F: lateral view. A–F original.

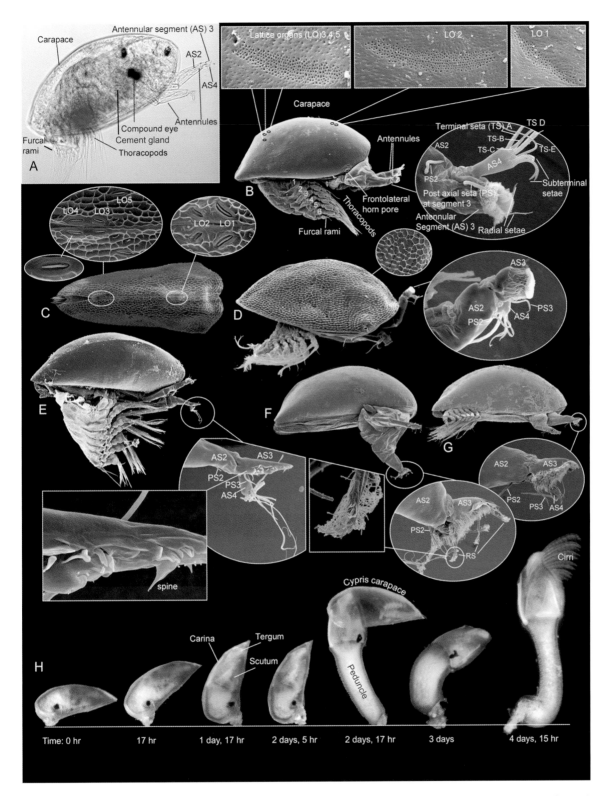

Fig. 22.3 Cypris larvae of the Thoracica. A: cyprid of *Amphibalanus amphitrite*, light microscopy, dorsal view. B: cyprid of *Tetraclita japonica*, showing details of the third and fourth antennular segments, the thoracopods (1–6), and the lattice organs on the carapace surface, lateral view. C and D: cyprid of *Capitulum mitella*, showing the honeycombed carapace sculpturing. C: five pairs of lattice organs, dorsal view. D: structure of the antennule, lateral view. E: cyprid of the coral barnacle *Savignium crenatum*, showing the pointed attachment disc, lateral view. F: cyprid of the stalked barnacle *Ornatoscalpellum stroemi*, showing the hoof-shaped attachment disc, with curiously brush-shaped radial setae, lateral view. G: cyprid of *Scalpellum scalpellum*, showing the hoof-shaped attachment disc, lateral view. H: metamorphosis in a single specimen of *Lepas anatifera*, followed from cypris settlement to the feeding juvenile stage, showing the shedding of the cypris carapace (at 2 days, 17 hours) and the development of the peduncle and shell plates (carina, terga, scuta), light microscopy. Abbreviations: AS = antennular segment, LO = lattice organs, PS = post-axial seta, RS = radial seta, TS = terminal seta. A–H original.

23

Tantulocarida

Rony Huys
Jørgen Olesen
Alexandra S. Petrunina
Joel W. Martin

GENERAL: The Tantulocarida is a small group of micro-crustaceans that exhibits a unique protelean life cycle, composed of obligatory ectoparasitic larvae and free-swimming non-feeding adults. Larval tantulocaridans utilize other marine crustaceans as hosts, including copepods, amphipods, tanaidaceans, isopods, cumaceans, and ostracods (Huys 1991; Boxshall and Vader 1993). The Tantulocarida was established as a new class of Crustacea by Boxshall and Lincoln (1983), but the roots of its taxonomic history reach back to the beginning of the twentieth century. The first genera to be formally described were *Cumoniscus* and *Microdajus*, both of which were originally classified as epicaridean isopods (Bonnier 1903; Greve 1965). Currently, the group accommodates 34 described species assigned to 22 genera (Knudsen et al. 2009; Kolbasov and Savchenko 2009; Savchenko and Kolbasov 2009; Mohrbeck et al. 2010). The most widely used classification (Huys 1990a) recognizes 5 families, but recent contributions suggest that the Deoterthridae and Basipodellidae are possibly paraphyletic or polyphyletic (Kolbasov et al. 2008a; Savchenko and Kolbasov 2009; Petrunina and Kolbasov 2012). Tantulocaridans have been recorded from abyssal depths to the intertidal zone and from polar to tropical waters, including unusual habitats like anchialine cave pools (Boxshall and Huys 1989), coral reefs (Huys 1990b) and hydrothermal vents (Huys and Conroy-Dalton 1997). Recent contributions (Gutzmann et al. 2004; Mohrbeck et al. 2010) suggest that the current number of described species greatly underestimates their actual diversity. Tantulocaridans exhibit no recognizable cephalic limbs at any stage of their intricate life cycle (except for the antennules in the sexual female), but the shared position of the female copulatory pore (ventrally on the cephalothorax, at about the level of the incorporated first thoracic segment) and the male genital apertures (on the seventh post-cephalic trunk segment) is interpreted as evidence that their affinities lie with the Thecostraca (the Facetotecta, Ascothoracida, and Cirripedia) (Huys et al. 1993; also see Petrunina et al. 2013).

The known life-history stages appear to form a complex dual life cycle, combining a sexual phase (with free-swimming adults) and a presumed parthenogenetic multiplicative phase that takes place on the host (fig. 23.1) (Huys et al. 1993).

LARVAL TYPES

Tantulus Larva: The infective stage, or tantulus larva (figs 23.2C; 23.3A, B), is the only stage that occurs in both cycles. Prior to mate location, the larva goes through a benthic non-feeding phase—probably at the sediment-water interface—as part of the temporary meiofauna (Huys 1991). Attached larvae undergo development on the host without conventional molting. In the parthenogenetic pathway, the tantulus forms a large dorsal trunk sac immediately behind the cephalon, causing the ventral deflection of the larval trunk, which is subsequently sloughed (fig. 23.2A, B). The contents of the sac differentiate into eggs that are later released as fully developed tantulus larvae (fig. 23.3I). The sexual cycle involves a unique type of metamorphosis in which a free-swimming adult is formed within the expanded trunk sac of the preceding tantulus larva. In the female pathway the trunk sac forms immediately posterior to the cephalic shield, and the larval trunk is sloughed (fig. 23.3F–H, K). In the male pathway the site of trunk sac formation is at or near the back of the larval thorax, causing the ventral deflection of the urosome and thoracopods (figs. 23.2A, D, E; 23.3J). Throughout metamorphosis the developing sexual females and males are supplied with nutrients via an umbilical cord (fig. 23.3F, G).

Nauplius: A benthic non-feeding stage that can be recognized as a nauplius has been reported recently but not formally described (Martínez Arbizu 2005; Mohrbeck et al. 2010), suggesting that eggs produced by the sexual female hatch as nauplii rather than as tantulus larvae.

MORPHOLOGY: The tantulus larva (body size 75–195 μm) (figs. 23.2A–C; 23.3A, B) is composed of an anterior tagma (prosome), consisting of the cephalon and a thorax of six pedigerous segments, and a limbless tagma (urosome), consisting of the last thoracic segment and a one-segmented abdomen that bears paired caudal rami (fig. 23.2F, J). The cephalon is covered with a dorsal shield bearing pores, sensilla, and (usually) epicuticular lamellae (figs. 23.2C; 23.3A, B). The tantulus has no cephalic appendages and is permanently attached to the host by a frontal oral disc (figs. 23.2C; 23.3A); in free-swimming larvae this disc is often accompanied by several pairs of sensilla and a funnel-shaped organ or proboscis (fig. 23.3B, E). The larva has six pairs of thoracopods (fig. 23.2G, H). The first five pairs are biramous and typically have a well-developed precoxal endite (figs. 23.2I; 23.3C, D); the sixth pair is uniramous and lacks an endite. The endites carry palmate spines apically and function as coupling devices, holding the two members of a leg pair together. Each endopod typically has a grappling apparatus at its tip, which may either act as an additional coupling apparatus for synchronizing thoracopod movements (Huys 1991), or as an attachment device, enabling the tantulus to hold onto the host prior to cementing down the oral disc (Boxshall 1991). Near the center of the disc is a tiny (0.5–2.0 µm) aperture (the stylet pore), through which the tantulus punctures the host's integument with its cephalic stylet. The stylet is operated by paired protractor and retractor muscles (fig. 23.3E) (Huys 1991). Host penetration may be purely mechanical, but large secretory glands inside the cephalon may assist this process. Little of the internal anatomy of the tantulus can be discerned once it has attached itself to the host.

MORPHOLOGICAL DIVERSITY: Tantulocarid taxonomy is largely based on larval characters, such as the cephalic pore pattern, stylet shape, thoracopod morphology and armature, cephalic and abdominal surface ornamentation (fig. 23.2C, F, J), and caudal ramus structure. The mode of male trunk-sac formation in the preceding tantulus larva is diagnostic at the family level, with swelling occurring either posterior to the sixth thoracic tergite and additional expansion between the cephalon and the first tergite (the Basipodellidae); or exclusively posterior to tergite 6 (the Onceroxenidae, Microdajidae, and Deoterthridae); or between tergites 5 and 6, with secondary expansion between tergites 4 and 5 (the Doryphallophoridae) (Huys 1991). Microdajid tantulus larvae that develop into parthenogenetic females produce a long drawn-out neck region, which has not yet been recorded in females of other families (Ohtsuka 1993).

NATURAL HISTORY: The infective tantulus larva attaches permanently to a crustacean host, which supplies all the nutrients it requires throughout its entire life. Nutrients are absorbed and transported back into the larval cephalon by a tubular rootlet system that penetrates the host and appears to be continuous with the umbilical cord that supplies a sexual adult developing within its trunk sac. The non-feeding sexual male must store sufficient food reserves during development to maintain it during the free-swimming mate-location phase.

The sexual female must obtain sufficient nutrients to allow the fertilized eggs to complete their development. Reproduction is semelparous, with each female (sexual or parthenogenetic) producing a single batch of larvae that are probably all released simultaneously. Thus heavily infested hosts may have encountered a swarm of newly released larvae (Huys 1990b; Boxshall and Vader 1993). The larvae do not appear to suppress host molting hormonally, but they may slow host growth by diverting resources into their own development (Huys 1991). Mating between free-swimming adults, as well as the hatching and development of larvae produced by the sexual female, have never been observed. Similarly, it is not known whether there is a simple alternation between the sexual and parthenogenetic cycles, or whether there is some kind of genetic switch mechanism. The presence of sexual males, parthenogenetic females, and tantulus larvae all on the same host individual (fig. 23.2A) (e.g., see Huys 1990b; Knudsen et al. 2009) indicates that it is not a simple seasonal switch from one cycle to the other.

PHYLOGENETIC SIGNIFICANCE: Tantulocarids are assumed to be allied to the thecostracans (the Facetotecta, Ascothoracida, and Cirripedia), based largely on the placement of the adult genital openings (Huys et al. 1993) and 18S rDNA sequence data (Petrunina et al. 2013). Larval morphology plays an important role in tantulocarid taxonomy and systematics.

HISTORICAL STUDIES: Boxshall and Lincoln (1987), Huys (1991), and Huys et al. (1993) described aspects of the dual life cycle. Huys (1991) described the internal anatomy of the cephalon, as well as the musculature of the trunk and legs of free tantulus larvae of *Xenalytus scotophilus* prior to their attachment to the host. Tantulus larvae of a number of genera have been examined with SEM, including *Amphitantulus* (Boxshall and Vader 1993), *Arcticotantulus* (Kolbasov et al. 2008a; Knudsen et al. 2009), *Coralliotantulus* (Huys 1990b), *Doryphallophora* (Boxshall and Lincoln 1987), *Itoitantulus* (Huys et al. 1992b), *Microdajus* (Boxshall and Lincoln 1987; Boxshall et al. 1989; Kolbasov and Savchenko 2009), *Onceroxenus* (Boxshall and Lincoln 1987), *Serratotantulus* (Savchenko and Kolbasov 2009), and *Tantulacus* (Huys et al. 1992a). Knudsen et al. (2009; Petrunina et al. 2013) also published the first photographs of live tantulus larvae, developing males, and parthenogenetic females of *Arcticotantulus kristenseni*, and they presented an overview of the subclass. Petrunina et al. (2013) published TEM sections of newly settled tantulus larvae of *Arcticotantulus pertzovi* and also provided complete 18S rDNA sequences for this species and *Microdajus tchesunovi*.

Selected References

Boxshall, G. A., and R. J. Lincoln. 1987. The life cycle of the Tantulocarida (Crustacea). Philosophical Transactions of the Royal Society of London, B 315: 267–303.

Huys, R. 1991. Tantulocarida (Crustacea: Maxillopoda): a new taxon from the temporary meiobenthos. P.S.Z.N. [Pubblicazioni della Stazione Zoologica di Napoli] 1, Marine Ecology 12: 1–34.

Huys, R., G. A. Boxshall, and R. J. Lincoln. 1993. The tantulocarid life cycle: the circle closed? Journal of Crustacean Biology 13: 432–442.

Knudsen, S. W., M. Kirkegaard, and J. Olesen. 2009. The tantulocarid genus *Arcticotantalus* [*sic*] removed from Basipodellidae into Deoterthridae (Crustacea: Maxillopoda) after the description of a new species from Greenland, with first live photographs and an overview of the class. Zootaxa 2035: 41–68.

Kolbasov, G. A., A. Yu. Sinev, and A. V. Tschesunov. 2008a. External morphology of *Arcticotantulus pertzovi* (Tantulocarida, Basipodellidae), a microscopic crustacean parasite from the White Sea. Entomological Review 88: 1192–1207 [original Russian text published in Zoologicheskii Zhurnal 87: 1437–1452].

Fig. 23.1 (*opposite*) Schematic summary of the presumed dual life cycle of the Tantulocarida. A solid star indicates the presumed position of free-living nauplius stage(s) in the sexual life cycle. The sediment-water interface is indicated by a dotted line (modified after Huys et al. 1993). Drawing based on published data derived from several taxa and a variety of host groups (Boxshall and Lincoln 1987; Boxshall et al. 1989; Huys 1991; Huys et al. 1992b, 1993).

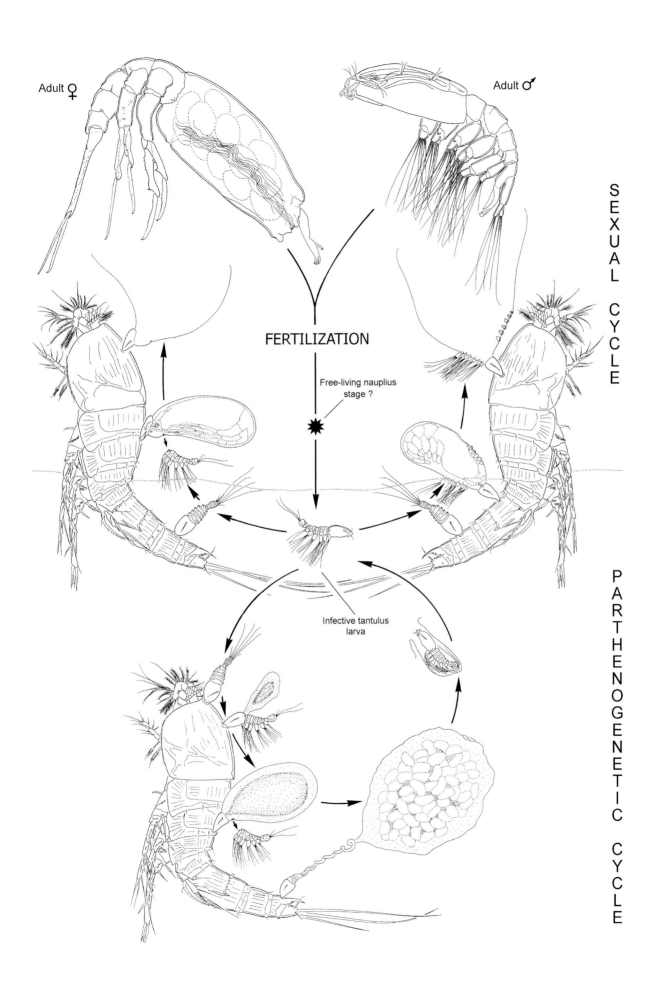

Adult ♀

Adult ♂

SEXUAL CYCLE

FERTILIZATION

Free-living nauplius stage ?

Infective tantulus larva

PARTHENOGENETIC CYCLE

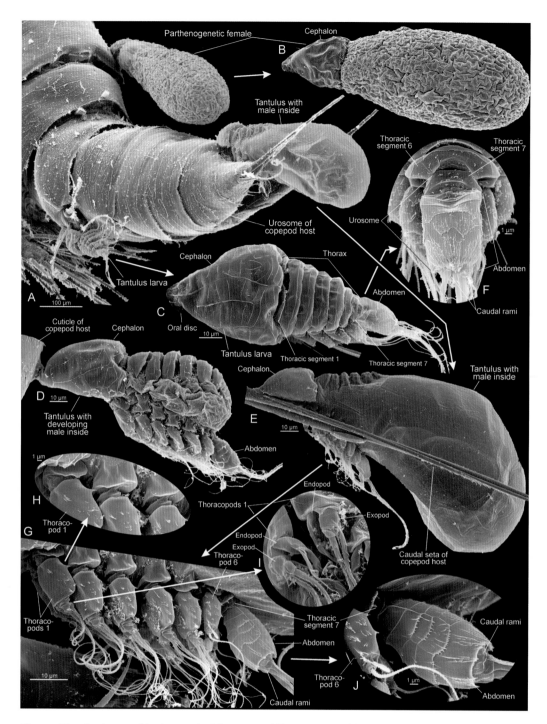

Fig. 23.2 Tantulus larvae of *Arcticotantulus kristenseni* at different stages of development, SEMs. A: recently attached, unmodified tantulus larva (prior to metamorphosis), young parthenogenetic female, and tantulus with a developing male inside an expanding trunk sac; stages are attached to a harpacticoid copepod host (*Bradya* sp.). B: parthenogenetic female at an early stage of trunk-sac formation; note the absence of the postcephalic trunk and thoracopods of the tantulus, and the highly wrinkled integument (which will stretch as the trunk sac enlarges and the eggs develop inside), lateral view. C: premetamorphic tantulus larva, showing the oral disc and basic tagmosis, dorsolateral view; note that the first thoracic tergite is largely concealed beneath the posterior margin of the cephalic shield. D: tantulus, with a developing male inside, showing the initial onset of trunk-sac formation posterior to the sixth thoracic tergite, causing the downward deflection of the urosome and thoracopods, lateral view. E: large metamorphosed tantulus containing a male at an advanced stage of development, attached to the caudal ramus seta of a harpacticoid host, lateral view. F: posterior end of the tantulus, showing the position of the major body articulation between the sixth thoracic segment and the urosome (encompassing thoracic segment 7 and the one-segmented abdomen), dorsal view; note the surface sculpturing on the abdomen. G: thoracopods 1–6 and the abdomen of a metamorphosed tantulus (containing a developing male; see D), lateral view. H–J = closeups of G. H: protopods of thoracopods 1–3, lateral view. I: first pair of thoracopods, showing the biramous structure, ventral view. J: abdomen and caudal rami, lateral view; note the surface ornamentation. A and C–E modified after Knudsen et al. (2009); B and F–J original.

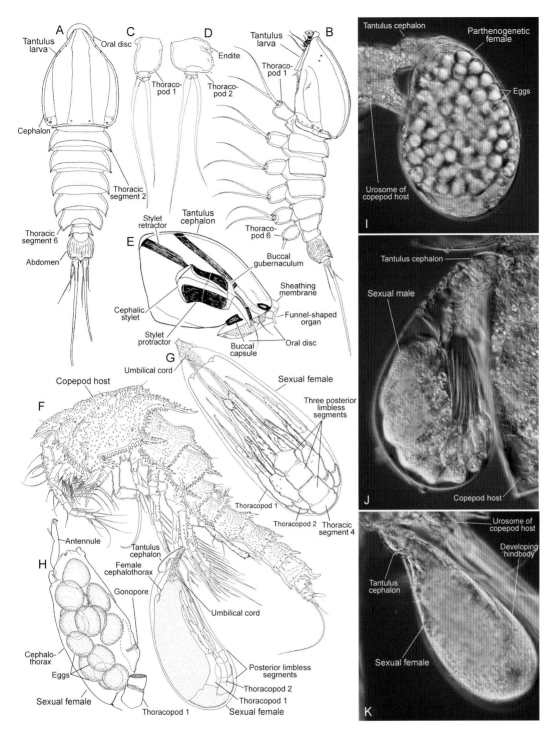

Fig. 23.3 Morphological details of tantulus larvae and other aspects of the life cycle of the Tantulocarida. A–E: drawings of *Xenalytus scotophilus*. A: tantulus larva, dorsal view. B: tantulus larva, lateral view. C: thoracopod 1 of the tantulus, anterior view. D: thoracopod 2 of the tantulus, anterior view. E: sagittal section through the cephalon of a free-living tantulus (prior to infection), revealing the cephalic stylet and associated musculature. F–H: drawings of *Itoitantulus misophricola*. F: harpacticoid host (*Styracothorax gladiator*) with an attached metamorphosed tantulus containing the sexual female stage of *I. misophricola* at an advanced state of development, lateral view. G: sexual female of *I. misophricola* developing within the trunk sac of the preceding tantulus stage, ventral view. H: cephalothorax of a developing sexual female of *I. misophricola*, showing the eggs and gonopore position. I: live parthenogenetic female of *Arcticotantulus kristenseni*, containing eggs, attached to the urosome of a copepod host, light microscopy. J: live tantulus of *A. kristenseni*, containing a male at a late stage of development, attached to the cephalothorax of a harpacticoid host, light microscopy. K: live tantulus of *A. kristenseni*, containing a sexual female at an early stage of development, light microscopy; note the developing postcephalothoracic trunk (hindbody). A–E modified after Huys (1991); F modified after Huys (1993); G and H modified after Huys et al. (1993); I–K modified after Knudsen et al. (2009).

24 Branchiura

OLE STEN MØLLER
JØRGEN OLESEN

GENERAL: Branchiurans are commonly known as fish or carp lice. They are a relatively small group of mostly limnic fish ectoparasites, consisting of 4 genera and approximately 230 species. They can also occur in marine habitats. *Argulus* is widespread in fresh waters, where it feeds on a wide variety of hosts, and it is a significant problem in aquaculture. Branchiuran adults are approximately 5–15 mm long, characteristically dorsoventrally flattened, and always equipped with a carapace (of varied shape, sometimes completely two-lobed) covering four pairs of thoracopods used for swimming (figs. 24.1D, 24.3N). The abdomen is unsegmented and carries two rudimentary furcal rami, which are more clearly visible in larvae and juveniles (figs. 24.1C, 24.2R; 24.3F, O, R). Two large and well-developed compound eyes are always present, as well as a smaller three-cupped nauplius eye (fig. 24.1C, D) (Thiele 1904; Møller 2009). The host attachment always involves the maxillules, which are modified into muscular suction cups (in *Argulus*, *Chonopeltis*, and the monotypic *Dipteropeltis*) (fig. 24.3N, P) or strong hooks (in *Dolops*) (fig. 24.1E–G) (Møller et al. 2008; Kaji et al. 2011). The attachment is non-permanent, and adults, which are excellent swimmers, often leave the host (*Chonopeltis* excepted). They feed on their hosts by biting into the integument with their mandibles, situated at the tip of the mouth cone (figs. 24.2A, F; 24.3A, E, F, K, N), and feed on the blood and tissue fluids from the wound. *Argulus* and *Dipteropeltis* have a stylet-like pre-oral spine (figs. 24.2A, F, G; 24.3A, B, K). This structure is not connected to the esophagus or alimentary canal, but it presumably is used to puncture and penetrate the host's integument and thus physically promote hemorrhaging (Debaisieux 1953; Shimura 1983). Glands are associated with this pre-oral spine and are suggested to secrete anticoagulants, but this is not completely clarified (Shimura and Inoue 1984; Swanepoel and Avenant-Oldewage 1992; Gresty et al. 1993).

Our knowledge of branchiuran larvae is unfortunately not nearly as complete as that of the adults, and hatching-stage larvae are known from fewer than 20 species thus far, most of them from the genus *Argulus*. Unfortunately, larvae of some

of the more aberrant *Argulus* species, such as *A. paranensis* or *A. brachypeltis*, have not been described. Until recently, the larvae from the single African species of *Dolops* (*D. ranarum*) were the only ones known from this genus. Gomes and Malta (2002) mentioned the general shape of the hatching stage of the South American *D. carvalhoi*, and the larva of this species was recently described using SEM and light microscopy by Møller and Olesen (2012). *Chonopeltis* larvae are known only from the literature, and nothing is known about the larvae of *Dipteropeltis*. The life cycle varies among genera.

LARVAL TYPES: Thus far, three larval types have been described from the Branchiura (table 2.1). The first type is often referred to as a metanauplius (but is more progressed in development), such as in *Argulus foliaceus* (figs. 24.1C; 24.2A). The second type is referred to as juvenile-like and is found in at least 2 species of *Dolops* (fig. 24.1E, G), as well as in *Argulus*, where it can occur as either the hatching stage or as stage 2 following a metanauplius-like hatching stage (Fryer 1961; Avenant et al. 1989; Rushton-Mellor and Boxshall 1994; Gomes and Malta 2002; Møller and Olesen 2012). The final type is an odd-looking non-swimming type found in *Chonopeltis* (fig. 24.1H).

Metanauplius-Like Larvae: The metanauplius-like larvae of *Argulus foliaceus* resemble other crustacean nauplii by hatching with a naupliar swimming apparatus (exopods of the antennae and mandibular palps) (figs. 24.1C; 24.2A, C–E) and by lacking functional thoracopods (figs. 24.1C, 24.2A, K, O, Q) (Rushton-Mellor and Boxshall 1994; Møller et al. 2007). Having functional post-mandibular appendages and even differentiated first thoracopods, however, by definition disqualifies this kind of larva from being considered a metanauplius, and thus we employ the term metanauplius-like (see chapter 2).

Juvenile-Like Larvae: The juvenile-like larvae of *Argulus megalops* hatch with all four pairs of thoracopods developed and active. They resemble the adults in most aspects, except for a lack of suction discs and a few other details of setal and spinal armature; hence the term juvenile-like (C. Wilson 1902).

Non-Swimming Larvae: *Chonopeltis* larvae have no naupliar swimming apparatus, and, as the thoracopods are not completely developed, it is still not known how these larvae move about, if at all (fig. 24.1H). The maxillule of the *Chonopeltis* larva displays a strong bipartite distal hook (fig. 24.1H) (Fryer 1956, 1961), similar to the condition in *Argulus* larvae (of both metanauplius-like and juvenile-like types). In *Dolops* the larvae of only 2 species (*D. ranarum* and *D. carvalhoi*) are known, and both of them are of the juvenile-like type (fig. 24.1E, G) (Avenant et al. 1989; Gomes and Malta 2002; Møller and Olesen 2012).

The metanauplius-like stage found in some species of *Argulus* is a single instar, and it molts into a juvenile-like larva. The metanauplius-like stage is often caught free-swimming in lakes (pers. obs. for *A. foliaceus*), but it probably quickly seeks out a host. The exact number of juvenile-like stages probably varies both inter- and intragenerically, but typically between 8 and 12 stages are reported, all of them from fish hosts (five stages are presented here for *Argulus foliaceus* in figs. 24.2A; 24.3A, F, H, N) (Claus 1875; Tokioka 1936; Rushton-Mellor and Boxshall 1994). All evidence points toward the juvenile-like larvae of both *Argulus* and *Dolops* being parasitic from the onset; thus all larval stages can be considered parasitic. Similar to the adults, the larvae can and probably do leave their hosts relatively frequently. Juvenile-like larvae have also been reported from lake and estuary plankton (C. Wilson 1902, 1904). Copulation can take place on or off the host (observed only in *A. japonicus* by Avenant-Oldewage and Everts 2010), but all known branchiuran eggs are deposited away from the host, so a fully pelagic life cycle is not possible (Piasecki and Avenant-Oldewage 2008). The larval development of *Chonopeltis* (fig. 24.1H) seemingly is composed of 8–10 larval stages, and stage 1 larvae have been found on host fishes. Generally, however, the life cycle of *Chonopeltis* is poorly known and is compounded by their aberrant morphology (Fryer 1959, 1961; van As and van As 1996). No reports of *Dipteropeltis hirundo* larvae or life-cycle data can be found in the literature (Møller and Olesen 2009).

MORPHOLOGY

Metanauplius-Like Larvae: The intrageneric variability of the metanauplius-like larvae in *Argulus* is small, and the

Table 24.1 Overview of the known hatching-stage morphologies of Branchiura larvae

Larval (hatching-stage) type	Taxon
metanauplius	*Argulus americanus, A. foliaceus, A. catostomi, A. coregoni, A. japonicus, A. lepidostei, A. maculosus*
juvenile-like	*A. funduli, A. megalops, A. stizostethii, A. varians, Dolops carvalhoi, D. ranarum*
non-swimming larvae	*Chonopeltis australis, C. brevis, C. inermis, C. minutus*

Source: Refer to text; Kellicott 1880; Bouchet 1985.

highly studied *A. foliaceus* larvae will serve as a good example (also see Shimura 1981; Shafir and van As 1986; Lutsch and Avenant-Oldewage 1995). The hatching-stage larvae in *A. foliaceus* are approximately 600–800 µm long and generally ovoid in shape (figs. 24.1C; 24.2A, S). Most characteristically, they are equipped with long exopods on the antennae and mandibles (mandibular palp; see Møller et al. 2007), in other words, a naupliar swimming apparatus (figs. 24.1C; 24.2A–E). The antennal endopods are large, but mandibular endopods are lacking and have probably been integrated into the mouth cone (24.2A, C) (Gresty et al. 1993). The mandibles (the mandibular coxal endites) are already specialized for cutting and are situated within the fully developed mouth cone (fig. 24.2F) (Møller et al. 2007). The pre-oral spine is also already present at this stage (fig. 24.2A, F, G). The maxillules are equipped with strong bipartite distal hooks, and the maxillae have two or three smaller distal hooks for attachment purposes (fig. 24.2A, C, H–J). The carapace is richly fringed with short setal-like structures for a close seal to the host's integument (fig. 24.2L, M). Thus the larvae have all the morphological prerequisites for a parasitic lifestyle. The four thoracopods are differentiated, but only the first pair is beginning to demonstrate movement, and in general they only show weak setation (fig. 24.1C; 24.2A, K).

Juvenile-Like Larvae: Stage 2, the first juvenile-like stage (fig. 24.3A), and the following stages of *A. foliaceus* are in several ways significantly different from stage 1. The naupliar swimming apparatus has degenerated, and only vestiges of the antennal and mandibular exopods are present (fig. 24.3C–E, G). Swimming is now fully taken over by the four pairs of thoracopods, an excellent example of a posterior shift of propulsion (Møller et al. 2007). One of the most significant changes during ontogeny is the shift in the host-attachment mode of the maxillules. In stage 2 (as in stage 1), the distal bipartite hook is responsible for attachment (figs. 24.1C; 24.2H), but in the following stages this function is gradually transferred to the characteristic proximal suction discs, while the distal podomeres of the limbs degenerate (compare four different instars in fig. 24.3). It was recently shown that the suction disc is a fusion product of the first and second podomere, and the discs can truly be considered evolutionary novelties (Kaji et al. 2011). In stage 6 the maxillular suction discs are fully formed and probably functional (fig. 24.3N, O). Another clear indicator of the stage the animal has reached is the number of posteriorly directed stout teeth of the proximal podomere of the maxilla (fig. 24.2A, I). In *A. foliaceus*, the hatching metanauplius-like larva carries one tooth on this podomere (fig. 24.2A), stages 2–4 carry two (fig. 24.3A, F), and from stage 5 and onward the adult number of three teeth is reached (fig. 24.3H, N). The setation of the thoracopods increases gradually through the stages, and the dorsally positioned flagellae of the appendages develop after stage 5 (not shown). The number of larger spines in the anterior cephalic area also increases gradually to the full adult complement (fig. 24.3A, N). The only major morphological development taking place in postnaupliar ontogeny is the development of the suction discs in the maxillules, and a general increase in setation and in the number of spines.

Some *Argulus* species skip the metanauplius-like stage and hatch as juvenile-like larvae (see table 2.1). In these cases the hatching larvae closely resemble the second or third stage of the metanauplius-type larvae, so the *A. foliaceus* stages 2–3 larvae (figs. 24.3A = stage 2 and 24.1D = stage 3) would be morphologically similar, for example, to an *A. megalops* hatching-stage (juvenile-like) larva. The maxilla stage indicator from *A. foliaceus* shows two teeth on the proximal podomere in the *A. funduli* juvenile-like hatching stage, for example, thus indicating that at hatching this larva is equivalent to stages 2–4 *A. foliaceus* larvae. Remaining development in the species hatching as juvenile-like larvae is not very well documented, but it seems to be identical to that of the metanauplius-like hatchers (C. Wilson 1902, 1907b, 1912; Rushton-Mellor and Boxshall 1994).

In *Dolops*, the hatching-stage larvae are known only from 2 species: *D. ranarum* and *D. carvalhoi* (fig. 24.1E–G). The larvae resemble each other closely and belong to the juvenile-like larval type, with a length of approximately 0.7–1.0 mm (see Møller and Olesen 2012 for details; or Avenant et al. 1989). The maxillule already has approximately the adult shape, terminating in a strong recurved hook, with a smaller spine-like structure slightly anterior to this (fig. 24.1E, F). In the juvenile-like larvae in *Dolops*, just as in *Argulus*, the setation of the thoracopods is not fully developed at hatching. Thus far, the development from the hatching stage to the adult has not been described in detail for any *Dolops* species, but very probably it resembles the above-mentioned *Argulus* species that hatch as juvenile-like larvae.

Non-Swimming Larvae: All of the known *Chonopeltis* larvae show a peculiar mix of traits, in that they possess neither naupliar appendages nor usable thoracopods (fig. 24.1H) (Fryer 1956, 1961). *Chonopeltis* larvae are generally smaller than other branchiuran larvae (ca. 4–500 μm), and they have only one pair of antennae developed, presumed to be the second pair (Møller 2009; Møller and Olesen 2009). These antennae, however, are uniramous, do not carry any major setation, and in general are unlike any antennae known from the Branchiura (Fryer 1956, 1961; Møller et al. 2008). Both pairs of maxillae are well developed and generally resemble those of the metanauplius-like larvae of *Argulus*, such as the maxillule having a terminal barbed two-partite hook (K. Barnard 1955; Fryer 1956, his fig. 78). The development of the suction disc of the maxillule has not been documented in detail, but Møller et al. (2008) suggested that it develops in the same way as in *Argulus*, based on the persistence of the distal part of the limb in late juvenile specimens in several *Chonopeltis* species, such as *C. inermis* (K. Barnard 1955), *C. brevis* (Fryer 1961), and *C. lisikili* (van As and van As 1996). The pronounced trilobate outline of the adult carapace is not visible in the larvae, and the cephalic-lobe support rods characteristic of the genus are not developed. The thoracopods are differentiated but clearly do not show adult morphology (fig. 24.1H) (Fryer 1964; Avenant-Oldewage and Knight 1994). The final number

of larval stages remains unclear, but 8–10 were reported by Fryer (1961), and generally development in *Chonopeltis* seems more gradual than in the other genera.

MORPHOLOGICAL DIVERSITY: Morphological diversity of larvae in the Branchiura is low, based on our current knowledge. In all *Argulus* and *Dolops* species from which larvae are known, the juvenile-like larva occurs at some time during larval development. One group of *Argulus* species hatch as metanauplius-like larvae, but the following stage is always juvenile-like (table 2.1). Another group of *Argulus* species hatch as juvenile-like larvae, as does the known *Dolops* species. Only *Chonopeltis* larvae differ markedly in their morphology. No evidence points toward the presence of gender-based size differences in branchiuran larvae. Only a very small number of species (all of them in the genus *Argulus*) have been investigated in enough detail, however, to reveal any such differences.

NATURAL HISTORY: The metanauplius-like larvae of *A. foliaceus* or *A. coregoni* are frequently found (sometimes in large numbers) in plankton samples from infected lakes (pers. obs.), and the reproductive potential of these species is very large (Hakalahti and Valtonen 2003; Hakalahti et al. 2003, 2004). The development time for the eggs is strongly correlated with water temperature, and the more widespread species (*A. japonicus*, *A. foliaceus*, and *A. coregoni*) typically hatch after about three weeks at 20°C–25°C. At low temperatures, development time is extended, even to the point of the eggs being able to overwinter (Shafir and van As 1986; Mikheev et al. 2001; Hakalahti et al. 2003, 2004). In *Dolops ranarum*, a development time of 27–35 days at approximately 24°C was reported (Fryer 1964), but these findings were contradicted by Avenant et al. (1989), who reported the first hatches after 57 days, at a slightly lower temperature. The only other data on hatching in a South American *Dolops* species were given by Gomes and Malta (2002) on *D. carvalhoi*, which hatched after 16 days when kept at 26°C–28°C. Fryer (1961) reported that *Chonopeltis inermis* eggs began hatching after 22 days at room temperature. All known branchiuran eggs are deposited away from the host, in a single layer on smooth surfaces, and are commonly ordered in rows. The quantity of eggs in a single egg batch is highly variable, but one batch can contain more than 500 eggs (*A. foliaceus* and *A. coregoni*), and the eggs are very heavily cemented onto the surfaces (fig. 24.1A, B) (Mikheev et al. 2001; Møller 2009). *Chonopeltis inermis* females generally produce smaller egg batches (ca. 130 at most) (Fryer 1956). The known hatching rates in *Argulus* are typically high, reported at 70–90 percent in the case of *A. coregoni* (Hakalahti et al. 2004) and approximately 90 percent (and above) in *A. foliaceus* under lab conditions (pers. obs.), but experimental data on the *Argulus* species hatching as juvenile-like larvae are lacking. Gomes and Malta (2002) found highly variable hatching rates in *D. carvalhoi*, varying from 0 to 60 percent.

Similar to the non-feeding nauplius stages of parasitic siphonostomatoid copepods (to which the Branchiura were thought to belong until the 1930s; see Møller 2009), the metanauplius-like stage of *Argulus* was thought to be non-feeding by C. Wilson (1902). Although *Argulus* metanauplius-like larvae are able to survive for up to six days and even molt in the absence of a host (pers. obs.), if a host is present, the larvae will attack it successfully (Shimura 1981; Møller et al. 2007). The fully formed mouth cone and the morphology of the mandibular coxal processes clearly enable *Argulus* metanauplii to feed as adults do, by ripping and cutting the host's integument. Thus a parasitic lifestyle is possible, although not mandatory, for the metanauplius-like larvae (Lutsch and Avenant-Oldewage 1995; Møller et al. 2007). The feeding mechanism of the juvenile-like larvae of *Dolops* and *Argulus* is expected to be identical to that of the adults, but the feeding mechanism of the *Chonopeltis* larva is still unknown. Reportedly, however, *Chonopeltis* larvae and adults are unable to swim freely, and the host change / finding mechanism remains mysterious (Fryer 1956, 1961; van Niekerk and Kok 1989; Avenant-Oldewage and Knight 1994).

PHYLOGENETIC SIGNIFICANCE: In trying to resolve the internal phylogeny of the Branchiura, the main characters have been the different morphologies of the larval maxillules. The fact that the adult morphology in *Dolops* (terminal hooks) is similar (but not identical) to the larval morphology in *Argulus* led authors to speculate that the *Dolops* condition was explainable as a neotenous condition (e.g., Overstreet et al. 1992). This was also supported by the fact that *Dolops* and *Argulus* share several derived traits, such as in the antennules and antennae. Møller et al. (2008), however, found *Dolops* to be the sister-group of the remaining Branchiura, based on molecular evidence. If this topology is accepted, there is no practical way of testing whether the adult hooks in *Dolops* are secondary (the neotenous event taking place on the *Dolops* branch) or belong to the Branchiura ground plan (synplesiomorphy). The latter implies that the adult suction-disc condition of the maxillules in *Argulus*, *Chonopeltis*, and *Dipteropeltis* is synapomorphic. Undoubtedly the early common ancestor of the Branchiura had neither suction discs nor hooks on the maxillules, but there is no possible out-group comparison, as the closest relative of the Branchiura (except the highly derived Pentastomida) still needs to be identified. If the ontogeny of the maxillules in *A. foliaceus* reflects their evolutionary direction, however, then the relatively simple maxillulary hooks pre-dated the more complicated maxillulary suction discs (for discussion, see Møller et al. 2008; Møller and Olesen 2009).

HISTORICAL STUDIES: The body of literature on the Branchiura is not as comprehensive as that of other crustacean groups, but it does contain a significant amount of older works. One of the first illustrations of an *Argulus* larva (*A. catostomi*) was given in Dana and Herrick (1837), and Claus (1875) beautifully illustrated the *A. foliaceus* metanauplius-like larva. C. Wilson (1904, 1907b) provided some of the first experimental larval data, as well as descriptions of some juvenile-like larvae, while M. Martin (1932) was the first to provide data on late embryonic development ("*A. viridis*" = *A. foliaceus*). Tokioka (1936) described *A. japonicus* development. Major later works on *Argulus* development include those of Shimura (1981) on *A. coregoni*; Rushton-Mellor and Boxshall (1994), who presented a very detailed description of the larval development of *A. foliaceus*; and Lutsch and Avenant-Oldewage (1995) on *A. japonicus*. In *Dolops*, the development of the larvae is still unknown, but the morphology of the hatching stage of *D. ranarum* was described by Fryer (1964) and Avenant et al. (1989). A hatching-stage larva of *D. carvalhoi* was recently described by Møller and Olesen (2012). Larval development in *Chonopeltis* was described first by Fryer in three papers (Fryer 1956, 1961, 1977), and later as a short note in van As and van As (1996). For a review of the more recent literature on the Branchiura in general, see Piasecki and Avenant-Oldewage (2008), Møller (2009), and Møller and Olesen (2009).

Selected References

Avenant, A., J. G. van As, and G. C. Loots. 1989. On the hatching and morphology of *Dolops ranarum* larvae (Crustacea: Branchiura). Journal of Zoology (London) 217: 511–519.

Fryer, G. 1961. Larval development in the genus *Chonopeltis* (Crustacea: Branchiura). Proceedings of the Zoological Society of London 137: 61–69.

Martin, M. F. 1932. On the morphology and classification of *Argulus* (Crustacea). Proceedings of the Zoological Society of London 102: 771–806.

Møller, O. S., J. Olesen, A. Avenant-Oldewage, P. F. Thomsen, and H. Glenner. 2008. First maxillae suction discs in Branchiura (Crustacea): development and evolution in light of the first molecular phylogeny of Branchiura, Pentastomida, and other "Maxillopoda." Arthropod Structure and Development 37: 333–346.

Rushton-Mellor, S. K., and G. A. Boxshall. 1994. The developmental sequence of *Argulus foliaceus* (Crustacea: Branchiura). Journal of Natural History 28: 763–785.

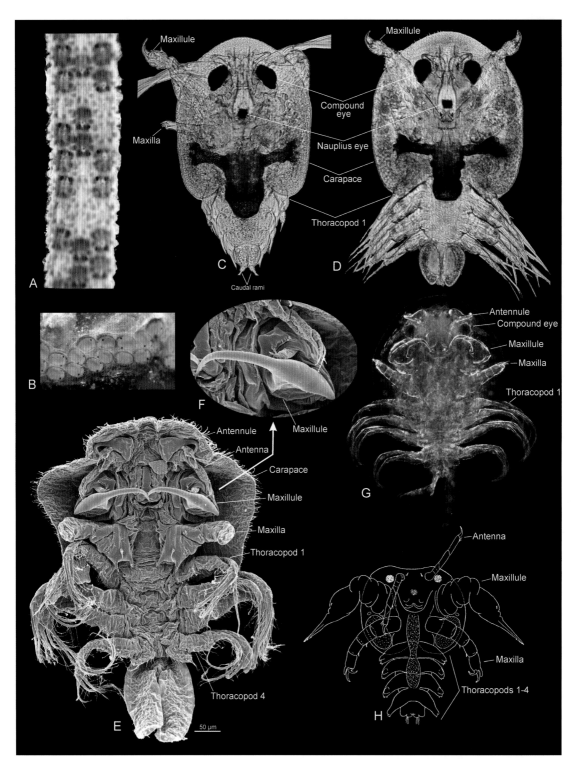

Fig. 24.1 Early developmental stages of branchiurans, light microscopy and SEMs. A: egg string of *Argulus foliaceus*, as found *in vivo*; note the ordered rows and the visible developing compound eyes, light microscopy. B: egg string of *A. japonicus*; the embryos are significantly younger than those in A, since the eyes are less developed and the eggs contain more yolk, light microscopy. C and D: *A. foliaceus*. C: stage 1 larva (metanauplius), SEM, Nomarski-contrast, dorsal view. D: stage 3 larva, EDF, Nomarski-contrast, ventral view. E–G: *Dolops carvalhoi*. E. larval stage 1, EDF, ventral view. F: maxillule distal hooks (detail from E). G: larval stage 1, light microscopy, ventral view. H: drawing of larval stage 1 of *Chonopeltis inermis*, ventral view. A , C, and D original, of material from the public Danmarks Akvarium in Copenhagen, Denmark (for details, see Møller et al. 2007); B original, of material collected in South Africa and donated by Professor A. Avenant-Oldewage, University of Johannesburg (for collection details, see Møller et al. 2008); E–G collected by T. Kaji, University of Shizuoka, Japan, used in Møller and Olesen (2012); H modified after Fryer (1956).

Fig. 24.2 Metanauplius-like (stage 1) larva of the branchiuran *Argulus foliaceus*, SEMs. A: ventral view; the full extent of the naupliar append-age is not visible. B: antennule; note the large hook of the first podomere and the smaller distal part. C: antenna, with the proximal part showing both the endo- and exopods; the mandibular exopod and maxillule are also visible. D: antenna exopod, with the characteristic step-like arrangement of the distal setal shafts. E: mandibular exopod, showing the same step-like arrangement of the setae as the exopod in D. F: mouth cone and pre-oral spine. G: distal part of the pre-oral spine. H: maxillule terminal hooks; note the two separate parts, as well as the barbs. I: maxilla, also showing the distal hooks, though significantly smaller than those of the maxillule. J: detail from I. K: thorax and abdomen/telson, ventral view; thoracopod 1 is probably moveable, and the setation is not fully developed. L: dorsal carapace, anterior view. M: brush border of the carapace, with dense setation. N: circled area from L, showing the weak setation. O: thoracic segments 3 and 4, and telson with caudal rami, dorsal view. P: dorsocaudal spine of the telson. Q: posterior part of the thorax and telson, dorsal view; note the lobed carapace. R: caudal rami. S: dorsal view; note the lobed posterior of the carapace, the dorsal organ, and the brush border. T: dorsal midline setae of the trunk segments. U: small rounded structure with a characteristic folded cuticle, always found on the dorsal surface of the posterior part of the carapace; its function and origin are still unknown. A–S original, of material from the public Danmarks Akvarium in Copenhagen, Denmark (for details, see Møller et al. 2007).

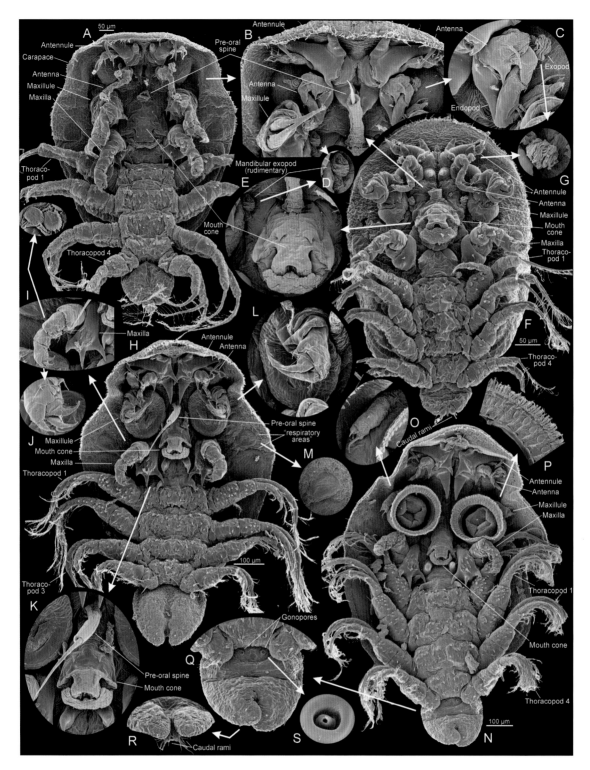

Fig. 24.3 Larval stages of the branchiuran *Argulus foliaceus*, SEMs. A: stage 2 larva, ventral view. B: stage 2 larva, anterior cephalic area (a different specimen from A; here the pre-oral spine is very prominent), ventral view. C: antenna, with the endopod and vestigial exopod. D: vestigial mandibular exopod (see fig. 24.2C). E: mouth cone of larval stage 3; the proximal part of the pre-oral spine is visible, along with the vestigial mandibular exopod. F: larval stage 3, ventral view. G: vestigial antennal exopod. H: larval stage 5, ventral view. I: maxilla, showing the proximal podomere, with posteriorly directed teeth. J: distal hooks of the maxilla. K: mouth cone and prominent pre-oral spine. L: distal hooks of the maxillule; these hooks are probably still somewhat active in this stage. M: cuticular folding of the carapace, due to shrinkage, but probably indicating the outline of the respiratory area. N: larval stage 6; note the nearly completed development of the maxillule suction discs. O: small remaining vestige of the distal podomeres of the maxillule. P: suction-disc rim, showing the central part, with support rods and the peripheral brush border. Q: two telsonic lobes, ventral view. R: two telsonic lobes, viewed directly from the posterior (ventral is up); note the small caudal rami protruding dorsally. S: gonopore closeup. A–S original, of material from the public Danmarks Akvarium in Copenhagen, Denmark (for details, see Møller et al. 2007).

25 Pentastomida

JOEL W. MARTIN

GENERAL: Pentastomids are a small group of parasites of the respiratory tract (trachea, lungs, and nasal passages) of reptiles, amphibians, birds, and mammals. On occasion their hosts can include humans (Riley 1986), where they can cause dermatitis and visceral pentastomiasis (Paré 2008). They are sometimes called tongue worms, because of the resemblance of some species of the genus *Linguatula* to a vertebrate tongue. The name Pentastomida (meaning "five mouths") comes from the distinguishing characteristic in some species, which have five similar outgrowths on the head. One of these is indeed the mouth, and the other four are short protuberances, each of which typically bears two terminal chitinous hooks. The adult body is wormlike and sheathed in a porous cuticle. There are approximately 130 species known from 6 families (Heymons 1935; Self 1969; Riley 1986; Almeida and Christofferson 1999; J. W. Martin and Davis 2001). Extant members are often treated as a subclass that is divided into 2 (Riley 1986; J. W. Martin and Davis 2001) or 4 (Almeida and Christofferson 1999) orders; neither classification is widely accepted (see Poore and Spratt 2011; Poore 2012; Christoffersen and De Assis 2013).

Inclusion of the Pentastomida in the Crustacea is controversial (see J. W. Martin and Davis 2001). Early morphological (e.g., Wingstrand 1972; Riley et al. 1978; Storch and Jamieson 1992) and more recent molecular studies (e.g., Abele et al. 1989; Zrzavý 2001; Lavrov et al. 2004; Møller et al. 2008; Regier et al. 2010; K. Sanders and Lee 2010) place them firmly within the Crustacea and allied to the Branchiura. Fascinating fossils of creatures identified as pentastomids, however, are known from the Late Cambrian to Early Ordovician (e.g., Walossek and Müller 1994; Walossek et al. 1994; Maas and Waloszek 2001; Waloszek et al. 2006; Castellani et al. 2011), long before any of the current vertebrate hosts existed. These authors argue that pentastomids are not crustaceans.

The life cycle consists of an egg, a primary larva, a nymph, and an adult. Most species have a diheteroxenous (two hosts) life cycle (fig. 25.1A). At least 1 species uses a single host, her-

ring gulls (Banaja et al. 1975), and 1 species has a free-living larva (Winch and Riley 1986).

LARVAL TYPES

Primary Larva and Nymph: A primary larva hatches from the egg in a rounded or elongate oval form that bears four to six short legs. As it grows it molts (up to nine times in some species) and becomes more wormlike, at which stage it is referred to as a nymph (which is now infective to the secondary host). The adult form is achieved after a series of molts, during which some of the legs are lost.

MORPHOLOGY

Primary Larva (fig. 25.1A, C, E): The primary larva, sometimes called simply the larva, can be seen through the egg prior to hatching (fig. 25.1A, B, D). The body of the larva is short, somewhat tardigrade-like, and usually bears two pairs of stubby appendages with terminal hooks (fig. 25.1C, E). Annulations in the cuticle are not yet evident. The short tail may be strongly hooked in some species (Riley 1986). A penetration apparatus, including a chitinous penetration spine used to penetrate the gut wall of the host, is present within the mouth circle (fig. 25.1E). An assortment of glands and sensillae, some located on cuticular sensory papillae, have been described from various species (e.g., Esslinger 1962, 1968; Winch and Riley 1986).

Nymph (fig. 25.1F, G): The primary larva is separated from the infective stage (nymph) by a series of molts, usually six to eight or nine (Riley 1986). Nymphs are elongate (vermiform), annulated, and develop gradually into the adult form. They are distinguished by having single or double rows of uniform penetration spines (Riley 1986). Vestigial trunk limbs have been reported from some fossil forms (e.g., fig. 25.1H, I).

MORPHOLOGICAL DIVERSITY: Pentastomid development in general is well known, because of their medical significance. Larval strategies, however, can vary widely, from indirect development with successive nymphal stages to direct development (see Castellani et al. 2011). Larval morphology is not known for all groups. Developing cephalobaenids and porocephalids (the two orders recognized by Riley 1986) differ in the details of the penetration apparatus and the hook fulcra (Riley 1986, his table 2). Hooks apparently are retained by infective instars of some species and not others. Because life cycles of many species are unknown, it is highly likely that larval diversity is underappreciated (see Riley 1986 for further details).

NATURAL HISTORY: Pentastomids are dioecious, with the females larger than the males. Adults feed on blood or mucous, depending on their location in the host's respiratory system. Eggs laid in the respiratory tract are coughed out or are passed through the digestive tract and then eaten by an intermediate host, often a fish or small mammal. The egg is infective to the primary (intermediate) host and is acquired as a contaminant of water or food. The primary larva then hatches from the egg, burrows through the intestine, and forms a cyst, usually in visceral tissue (including the lymph nodes, liver, intestine, swim bladder, and reproductive organs). After several molts, the larva within the cyst attains a more adult-like morphology and is now called a nymph. When an intermediate host is eaten by what will become the definitive host, the death of the intermediate host is the apparent cue for excystment. When the intermediate host is eaten, the nymph exits the cyst and either burrows into or crawls up the esophagus and enters the respiratory system of the definitive host, where it matures. Definitive hosts are typically reptiles, but they may also be amphibians, birds, and mammals. Intermediate hosts can be species of these groups or fishes or insects. The varying life cycles of species of each of the then-known families were reviewed in detail by Riley (1986). Fossil forms, if indeed parasitic during that period, might have parasitized conodonts.

PHYLOGENETIC SIGNIFICANCE: Treatment of pentastomids as crustaceans is considered controversial by some workers. Larval morphology has played an important role for in-group systematics (e.g., Riley 1986), but it has shed little or no light on relationships to other crustaceans, as the larval stages bear no resemblance to those of other crustaceans. Molecular and spermatological evidence place pentastomids close to the extant branchiurans (fish lice), another predominantly parasitic group.

HISTORICAL STUDIES: Important historical treatments of the Pentastomida include papers by Sambon (1922), Heymons (1926, 1935), Self (1969), and Riley (1986), as well as literature reviewed (but in some case overlooked) by each of these authors, especially the works (in German) of Osche, Böckeler, and von Haffner. For fossil pentastomids, see, for example, Walossek and Müller (1994), Walossek et al. (1994), Maas and Waloszek (2001), Waloszek et al. (2006), and Castellani et al.

(2011). Studies focusing more on extant pentastomid larvae include works by Fain (1961, 1964), Esslinger (1962, 1968), Riley and Self (1979), and Riley (1981, 1983). Riley (1986) remains the definitive reference for nearly all aspects of pentastomid biology. A large body of literature exists on the medical and veterinary significance of pentastomids (e.g., see Paré 2008).

Selected References

Castellani, C., A. Maas, D. Waloszek, and J. T. Haug. 2011. New pentastomids from the Late Cambrian of Sweden: deeper insight of the ontogeny of fossil tongue worms. Palaeontographica: Beiträge zur Naturgeschichte der Vorzeit, A, Paleozoology-Stratigraphy 293: 95–145.

Poore, G. C. B., and D. M. Spratt. 2011. Pentastomida. Australian Faunal Directory. Australian Biological Resources Study. Online at www.environment.gov.au/biodiversity/abrs/online-resources/fauna/afd/taxa/PENTASTOMIDA.

Riley, J. 1986. The biology of pentastomids. Advances in Parasitology 25: 45–128.

Self, J. T. 1969. Biological relationships of the Pentastomida: a bibliography on the Pentastomida. Experimental Parasitology 21: 63–119.

Waloszek, D., J. E. Repetski, and A. Maas. 2006. A new Late Cambrian pentastomid and a review of the relationships of this parasitic group. Transactions of the Royal Society of Edinburgh, Earth Sciences 96: 163–176.

Fig. 25.1 (*opposite*) Life cycle and developmental stages of pentastomids. A: schematic drawing of the generalized life cycle of a typical pentastomid. B and C: drawings of *Raillietiella furcocerca* (Cephalobaenidae). B: egg. C: primary larva, dorsal view. D and E: drawings of *Porocephalus crotali* (Porocephalidae). D: egg. E: primary larva, dorsal view. F: drawing of a nymph of *Sebekia oxycephala* (Sebekidae), from a poeciliid fish, dorsolateral view. G: outer hook of a nymph of *Waddycephalus* sp. (Sambonidae), from the abdominal wall of an Australian gecko, light microscopy. H and I: Upper Cambrian fossil of †*Heymonsicambria scandica*. H: SEM image. I: reconstruction. A modeled loosely after a figure of the life cycle of *Raillietiella frenatus* in Mehlhorn (2008); B and C modified after Esslinger (1968); D and E modified after Esslinger (1962); F modified after a SEM image in Almeida et al. (2010); G modified after Barton (2007); H and I modified after Castellani et al. (2011) and Waloszek et al. (2006).

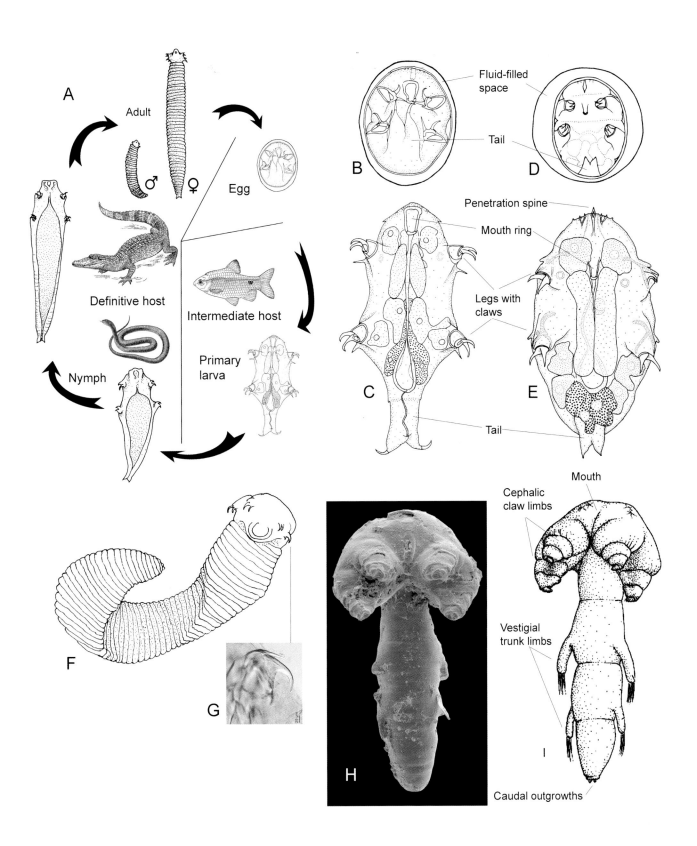

A

Adult

♂

♀

Egg

Definitive host

Intermediate host

Nymph

Primary larva

Fluid-filled space

Tail

B

D

Penetration spine

Mouth ring

Legs with claws

Tail

C

E

Mouth

Cephalic claw limbs

Vestigial trunk limbs

F

G

H

I

Caudal outgrowths

26 | Mystacocarida

Jørgen Olesen
Joachim T. Haug

GENERAL: Mystacocaridans are tiny, interstitial, somewhat worm-like crustaceans. They are a species-poor group (13 species) found interstitially in the intertidal zones of sandy beaches in many parts of the world, such as the Mediterranean area, Africa, the western coast of North America, South America, and Western Australia (Hessler 1992b; Boxshall and Defaye 1996). Despite being so widespread geographically, the taxon is characterized by an extreme conservatism in its morphology. A number of features—some also applicable to the known larvae—are characteristic for all mystacocarids and are probably modifications to a life among sand grains: a pair of very long antennules; a characteristic flexing zone in the cephalon (fig. 26.1F); a rather long tubular and flexible body, with the segments bearing characteristic toothed furrows laterally (fig. 26.1A, D, F); and four pairs of small, setae-bearing lobate thoracopods (fig. 26.2K) (Pennak and Zinn 1943; Delamare-Deboutteville 1953; Hessler and Sanders 1966; Olesen 2001). Another characteristic is that, in many respects, adults retain a larval appearance. Mystacocarids crawl among sand grains, and locomotion is brought about by alternating cyclic movements of the biramous antennae and mandibles (Lombardi and Ruppert 1982). The Mystacocarida are divided into only 2 genera, *Derocheilocaris* (8 species) and *Ctenocheilocaris* (5 species). The phylogenetic position of the Mystacocarida within the Crustacea is not entirely settled. Traditionally, they have been grouped in the now-abandoned Maxillipoda, along with other crustacean taxa such as the Copepoda and Branchiura (Dahl 1956; Walossek and Müller 1998a, 1998b). A close relationship with the Orsten fossil †*Skara* and with the Copepoda (Walossek and Müller 1998a, 1998b) was recently supported (J. Haug et al. 2011a). A recent comprehensive molecular phylogeny placed the Mystacocarida closest to the Branchiura and Pentastomida (Regier et al. 2010).

The life cycle is assumed to start with freely laid eggs, from which the metanauplii (stage 1) hatch. Stage 1 is followed by seven additional immature stages. McLachlan (1977) found that reproduction for *Derocheilocaris algoensis* occurs through-

out the year and that growth from an egg to a reproductive adult appears to take two months. In *D. typicus*, about 40 days is required (J. Hall and Hessler 1971). These authors also noted that in *D. typicus*, some young individuals are present at all times of the year, but they are most abundant in the summer months.

LARVAL TYPES: Larval development has been examined in 4 species: *Derocheilocaris remanei*, *D. typicus*, *D. algoensis*, and *D. katesae*, and a few stages were treated briefly for *D. delamarei* (Pennak and Zinn 1943; Delamare-Deboutteville 1954; Hessler and Sanders 1966; McLachlan 1977; Cals and Cals-Usciati 1982; Olesen 2001; J. Haug et al. 2011a). Nothing is known about development in *Ctenocheilocaris*. It was long believed that the 2 well-studied species, *Derocheilocaris typicus* from the Atlantic coast of the United States and *D. remanei* from the Mediterranean area, which are quite similar in adult morphology, differed significantly with respect to their larval sequences (summarized by Hessler and Sanders 1966). Reinvestigations of *D. remanei*, however, have corrected the original description and have shown that larval developments in *D. remanei* and *D. typicus* are, in fact, very similar, including the exact same number of stages, with largely the same morphology (Olesen 2001; J. Haug et al. 2011a). This provides further evidence for the general view of the mystacocarids being evolutionarily conservative. The development of *D. algoensis* follows the same course (McLachlan 1977). The treatment of mystacocarid development given below is based on *D. remanei* but, as far as is known, it can be considered general for mystacocarids.

Metanauplius: The earliest-known stage of *Derocheilocaris remanei* can be termed a metanauplius, as it has three distinct post-cephalic segments and a pair of undeveloped maxillules (figs. 26.1A; 26.2A, D). Delamare-Deboutteville (1954) originally described the earliest stage as an orthonauplius lacking maxillular limb buds, but further investigations have strongly suggested that such a stage is non-existent (Cals and Cals-Usciati 1982; Olesen 2001; J. Haug et al. 2011a). Based on an analysis of size categories and morphology, J. Haug et al. (2011a) identified nine developmental stages of *D. remanei*: eight immature stages, and the adult. Since the development from larvae to adults is rather gradual, and since the adults in many respects are larva-like (e.g., larval morphology of the antennae and mandible are retained in adults), there are few distinct landmarks during development to distinguish between a larval (or naupliar) and a juvenile/adult phase. One such landmark is the loss of the naupliar processes between stages 6 and 7, which Hessler and Sanders (1966) used to distinguish between a naupliar and a juvenile/adult developmental phase (see chapter 2). But since this developmental change is not accompanied by other significant changes, we refer to the eight immature stages of *D. remanei* as stages 1–8 (although we refer to stage 1 as a metanauplius).

MORPHOLOGY: It is characteristic for mystacocarids that, in some important respects, the early larvae are largely similar to the adults. This includes the antennules, which retain their large size during the larva's entire development, and the larval feeding apparatus (the antennae and mandibles), which are largely similar when comparing early larvae and the adult (except for the loss of the naupliar process). In other respects the larvae change significantly during development. This involves general body growth (with the addition of body segments), the development of post-cephalic limbs, the specific morphology of the lateral toothed furrows of the body segments (see J. Haug et al. 2011a), and some modifications of the telson and its ornamentation.

Here we summarize the development of *Derocheilocaris remanei*, starting with a description of the major aspects of stage 1 (metanauplius). The subsequent stages highlight important changes from the previous stage. The description is based on Olesen (2001) and J. Haug et al. (2011a), as well as on new images produced for this chapter (figs. 26.1, 26.2). More details can be found in these two papers, and in Hessler and Sanders (1966) for the very similar *D. typicus*.

Stage 1 (Metanauplius) (figs. 26.1A–C; 26.2A–D): Stage 1 has antennules, antennae, mandibles, weakly developed maxillules, and three fully formed post-cephalic segments (not including the telson), which include a maxilliped segment and two thoracoabdominal segments. The cephalic shield is divided into a small anterior lobate part, corresponding approximately to the antennules, and a larger posterior part (fig. 26.1A). The labrum of the hypostome-labrum complex is large and spoon-shaped. Both the posterior part of the head shield and the maxilliped segments laterally carry pairs of X-shaped toothed furrows, and the two following trunk segments (post-cephalic segments 2 and 3) bear a pair of more simple, elongate toothed furrows (see J. Haug et al. 2011a for a precise description and ontogenetic change). The uniramous antennules are eight-segmented and of considerable length. The biramous antennae have a four-segmented endopod, terminating in a characteristic hooked structure (fig. 26.2I, J, although this represents stage 4), a nine-segmented exopod where most segments bear one seta each (fig. 26.1A,

B), and a complex coxal naupliar process divided into four separate branches (see J. Haug et al. 2011a, their figs. 4, 5). The biramous mandibles have a three-segmented endopod; a seven-segmented exopod distally, with long locomotory setae; and a characteristic shoe-shaped coxal gnathobase (see J. Haug et al. 2011a, their fig. 11A). The maxillules are rudimentary; each consists of a flattened lobe with five finger-like medial processes and a distolateral seta (fig. 26.2D). No other limbs are present. The telson is a short conical structure, which continues posteriorly in a pair of large pointed caudal rami (figs. 26.1A, C; 26.2A, C). A weak flexing zone between the telson and the caudal rami is seen dorsally (fig. 26.2A). The caudal rami bear rows of setules (figs. 26.1A, C; 26.2A, C). Dorsomedially between the caudal rami, the telson bears a conical supra-anal process, which continues into a robust spine ventrally and a long seta dorsally (figs. 26.1A; 26.2A). On each side, there is an additional long seta at the basis of the caudal rami, yielding a total number of three long telsonal setae. A triangular flattened cuticular process is located on each side, between the supra-anal process and the caudal rami.

Stage 2 (figs. 26.1D, E; 26.2E–J): Stage 2 has four fully formed post-cephalic body segments (not including the telson) and the anlagen to a fifth segment, present externally as a pair of lateral toothed furrows at the telson (fig. 26.2D, E). The development of the maxillular limb buds has progressed further, with them having some weak segmentation and some median setation (Olesen 2001; J. Haug et al. 2011a, their figs. 6B, 9). None of the more posterior limbs are developed. The telson region has become modified in at least four major ways: (1) the articulation between the telson and caudal rami has become more distinct but is not yet fully formed ventrally (fig. 26.1E); (2) an additional long seta has appeared laterally at the basis of each caudal ramus, resulting in two setae on each side and a total number of five long setae in the telsonal region; (3) a characteristic seta with a long broad socket has appeared dorsally at the basis of each caudal ramus (fig. 26.2H); and (4) the caudal rami have become divided into a proximal broad tubular part and a distal claw-like part, with the latter forming a 90° angle with the former (fig. 26.1E).

Stage 3 (J. Haug et al. 2011a, their fig. 2C): Stage 3 has six fully formed post-cephalic body segments (not including the telson) and the anlagen to a seventh segment, present as a pair of lateral toothed furrows at the telson. Segmentation of the maxillule is adult-like, but setation is only weakly developed (J. Haug et al. 2011a, their fig. 6C). This stage has a weak indication of the maxilla. A small triangular bud of post-cephalic limb 4 is present. The basal part of the caudal rami has grown longer, and an additional seta has appeared at each ramus, close to where the rami articulate with the telson (J. Haug et al. 2011a, but similar to stage 4 shown in fig. 26.2L), yielding a total number of seven long setae in the telsonal region.

Stage 4 (figs. 26.1F–H; 26.2K, L): Stage 4 has eight fully formed post-cephalic body segments (not including the telson) and the anlagen to a ninth segment, present as a pair of lateral toothed furrows at the telson. The maxillule now has more advanced setation than in stage 3, and it does not undergo significant further change (J. Haug et al. 2011a). The maxilla is present as a limb bud, which is weakly subdivided into three portions (J. Haug. et al. 2011a, their figs. 8A, 9). Limb buds of the maxilliped segment and post-cephalic segments 2 and 3 were mentioned for stage 4 of *D. typicus* by Hessler and Sanders (1966), but they were not noted for *D. remanei* by Olesen (2001) and J. Haug et al. (2011a).

Stage 5 (J. Haug et al. 2011a, their fig. 2E): Stage 5 has nine fully formed post-cephalic body segments (not including the telson) and the anlagen to a tenth segment, present as a pair of lateral toothed furrows at the telson. The maxilla is still not fully developed, and it is differentiated into three portions that bear a number of endites and setae medially (see J. Haug et al. 2011a, their figs. 8B, 9). The maxilliped appears as a limb bud, divided into three portions (J. Haug et al. 2011a, their figs. 8D, 9), with a general aspect much like that of the maxilla when it appeared in the previous stage. The limbs of post-cephalic trunk segment 4 are now essentially fully developed and have the appearance of an unsegmented small limb with three setae. Post-cephalic limbs 2, 3, and 5 are also present, but less developed.

Stage 6 (J. Haug et al. 2011a, their fig. 2F): Stage 6 has 10 fully formed post-cephalic segments (not including the telson). This is similar to the body segmentation in adults. The maxilla has essentially reached its definite form, and its morphology does not undergo significant change in the following stages (J. Haug et al. 2011a, their figs. 8C, 9). The maxilliped is still roughly subdivided into three portions, but it now has more primordial setae than in the previous stage (J. Haug et al. 2011a, their figs. 8E, 9). Otherwise it is very similar to stage 5.

Stage 7 (J. Haug et al. 2011a, their fig. 3A): Stage 7 has 10 fully formed body segments (not including the telson). The naupliar process of the antenna is reduced. The maxilliped reaches its definitive form and does not undergo further significant change. The bud-like trunk limbs have achieved their adult setation.

Stages 8 and 9 (Adult) (J. Haug et al. 2011a, their fig. 3B, C): Based on detailed measurements and on differences in the morphology of the toothed furrows, J. Haug et al. (2011a) distinguished two almost identical stages, 8 and 9, of which the latter represents the adult.

MORPHOLOGICAL DIVERSITY: Larval stages are known only for 4 species of *Derocheilocaris* (*D. typicus, D. remanei, D. alogensis,* and *D. katesae*); there is basically no variation among them.

NATURAL HISTORY: Essentially no information is available concerning functional (feeding and locomotion) aspects of mystacocarid larvae. Lombardi and Ruppert (1982) treated feeding and locomotion for adults of *Derocheilocaris typicus,* and much of this information can be transferred to larvae, since the morphology of the locomotory/feeding apparatus in larvae and adults is identical in some respects. Both larvae and adults have very long antennules, a pair of biramous antennae, and mandibles with dorsally directed exopods and ventrally di-

rected endopods. Many researchers (e.g., Hessler 1969; J. Hall 1972; pers. obs. by Jørgen Olesen in 1996) have reported that when mystacocarids are removed from their interstitial habitat, they are incapable of effective locomotion. Lombardi and Ruppert (1982) found that the antennae and mandibles are the principle organs of locomotion in adults. The exopods and endopods of the antennae and mandibles (or their setae) are pressed against the dorsal and ventral substrate, which causes the animal to move forward when these appendages perform their power strokes during the coordinated cyclic movements these limbs undergo (Lombardi and Ruppert 1982). Due to a morphological similarity between larvae and adults, and because larvae and adults are found in the same habitat, it is almost certain that the basic locomotion of larvae is exactly the same as that described for adults. Feeding and locomotion are—as in many other Crustacea—intimately linked. Lombardi and Ruppert (1982) considered adult mystacocarids to be substrate scrapers, using the medial setae on the maxillules, maxillae, and maxillipeds to scour and trap diatoms and other unicellular algae from the surface of the substrate. The feeding mechanism of early mystacocarid larvae has not been studied, but since functional maxillules, maxillae, and maxillipeds are absent, the process must be different from that of adults.

PHYLOGENETIC SIGNIFICANCE: One striking feature of the Mystacocarida is that the adults are so larvae-like, having retained the naupliar locomotion/feeding apparatus (biramous mandibles and antennae). For this reason the Mystacocarida has often been suggested as having had a neotenic origin (e.g., Hessler and Newman 1975; Hessler 1992b). This is probably an oversimplification, however, since many aspects of mystacocarid morphology certainly are derived (e.g., the toothed furrows on the body segment). Another possibility regarding the larval-like branched mandibles and antennae of adults would be simply to interpret them as retained primitive features (plesiomorphies), as is partly seen in at least some of the Orsten crustaceans (e.g., †*Skara,* †*Rehbachiella,* and †*Bredocaris*) and the Cephalocarida.

HISTORICAL STUDIES: The Mystacocarida was described as a new crustacean order by Pennak and Zinn (1943). Their description was rather comprehensive and even included preliminary descriptions of a number of larval stages. The description contained various mistakes, however, which were corrected by Hessler and Sanders (1966). Another early contributor was Delamare-Deboutteville, who, in a series of works, added significantly to our knowledge of the development of *Derocheilocaris remanei* (e.g., Delamare-Deboutteville 1953, 1954). The most recent treatments are by Olesen (2001) and J. Haug et al. (2011a).

Selected References

Cals, P., and J. Cals-Usciati. 1982. Développment postembryonnaire des crustacés Mystacocarides: le problème des différences présumées entre les deux espèces voisines *Derocheilocaris remanei*

Delamare-Deboutteville et Chappuis et *Derocheilocaris typicus* Pennak et Zinn. Comptes Rendus Hebdomadaires des Séances de l'Académie des Sciences, ser. 3, 294: 505–510.

Haug, J. T., J. Olesen, A. Maas, and D. Waloszek. 2011a. External morphology and post-embryonic development of *Derocheilocaris remanei* (Mystacocarida) revisited, with a comparison to the Cambrian taxon *Skara*. Journal of Crustacean Biology 32: 668–692.

Hessler, R. R., and H. L. Sanders. 1966. *Derocheilocaris typicus* Pennak & Zinn (Mystacocarida) revisited. Crustaceana 11: 141–155.

Olesen, J. 2001. External morphology and larval development of *Derocheilocaris remanei* Delamare-Deboutteville and Chappuis, 1951 (Crustacea, Mystacocarida), with a comparison of crustacean segmentation and tagmosis patterns. Det Kongelige Danske Videnskabernes Selskab, Biologiske Skrifter 53: 1–59.

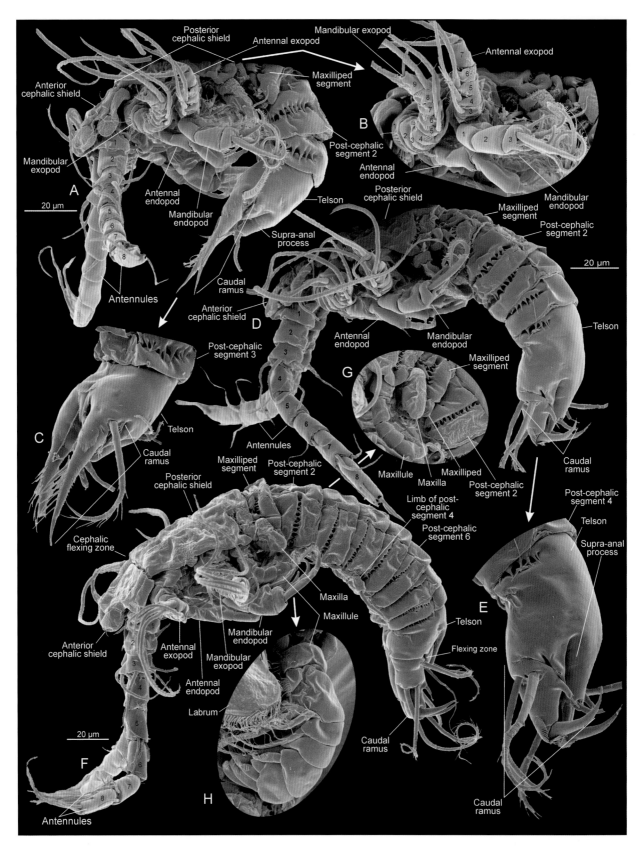

Fig. 26.1 Larvae of *Derocheilocaris remanei* (Mystacocarida), SEMs. A: stage 1 metanauplius, lateral view. B: left antenna and mandible of stage 1. C: telson of stage 1. D: stage 2, lateral view. E: telson of stage 2. F: stage 4, lateral view. G: maxillule, maxilla, and maxilliped of stage 4. H: left maxillule of stage 4. A–H of material from Canet Plage (Beach) in southern France (the type locality of the species), used in Olesen (2001) and J. Haug et al. (2011a).

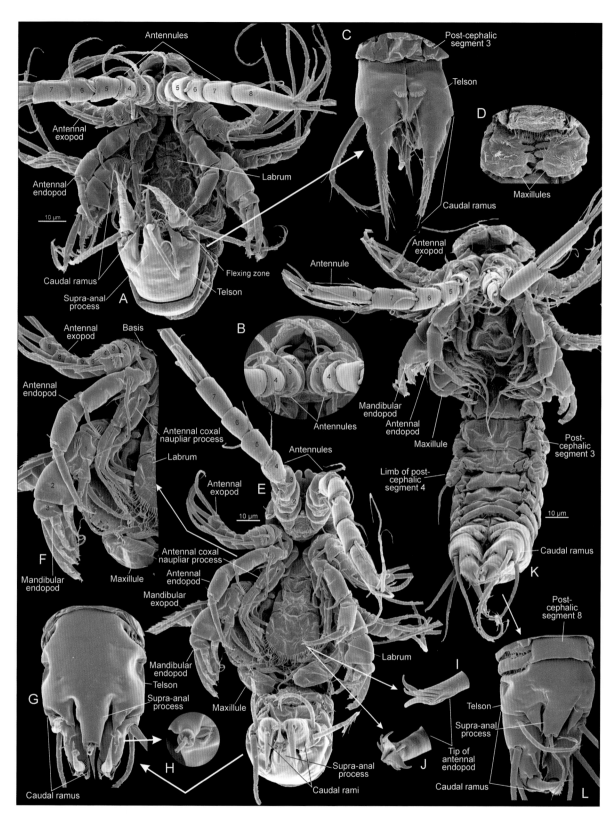

Fig. 26.2 Larvae of *Derocheilocaris remanei* (Mystacocarida), SEMs. A: stage 1 metanauplius (same specimen as fig. 26.1A), ventral view. B: closeup of stage 1, frontal view. C: telson of stage 1, ventral view. D: closeup of the maxillular limb buds of stage 1, ventral view. E: stage 2 (same specimen as fig. 26.1D), ventral view. F: left antenna and mandible of stage 2. G: telson of stage 2, dorsal view. H: closeup of the characteristic seta of the right caudal ramus of stage 2. I and J: two different views of the distal hooks of the antennal endopod. K: stage 4, ventral view. L: telson of stage 4, dorsal view. A–K of material from Canet Plage (Beach) in southern France (the type locality of the species), used in Olesen (2001) and J. Haug et al. (2011a).

27 | Copepoda

RONY HUYS

GENERAL: Copepods outnumber every other group of multicellular animals on earth, including the hyperabundant insects and nematode worms (Hardy 1970; Huys and Boxshall 1991). These small crustaceans are found throughout the world's natural and man-made aquatic environments, spanning the entire salinity range from fresh water to hypersaline water, including the most unusual continental habitats (Reid 2001). Copepods, with more than 12,500 described species, and outnumbering the insects in terms of individuals by up to three orders of magnitude (Schminke 2007), carry a global biological importance that is belied by their generally small size. In the pelagic realm, which encompasses a volume of 1,347 million km^3 and is the largest biome on the planet, copepods are the dominant members of the holozooplankton, both numerically and in terms of biomass (Harris et al. 2000). In addition to life strategies that encompass free-living, substrate-associated, and interstitial habits, copepods also have extensive impacts in their role as associates or parasites of the majority of aquatic metazoan phyla, from sponges to chordates, including reptiles and marine mammals. This variety of life strategies has generated an incredible morphological plasticity and disparity in body form and shape that are arguably unrivalled among the Crustacea. Copepods underpin the world's freshwater and marine ecosystems (Costanza et al. 1997). They are sensitive bioindicators of local and global climate change (Richardson 2008), key ecosystem-service providers (Frangoulis et al. 2005; Falk-Petersen et al. 2007), and parasites of economically important aquatic animals (M. Costello 2009). They sustain the majority of the world's fisheries (Costanza et al. 1997) and, through their roles as vectors of disease (Huq et al. 1983; Dick et al. 1991), also have a number of direct and indirect effects on human health and our quality of life.

Copepods are typically small. In planktonic and benthic forms, total body length is usually between 0.2 and 5.0 mm, although some species of *Valdiviella* (Calanoida) can reach 28 mm in length. The real giants amongst the copepods are the parasites, with the largest being members of the siphono-stomatoid family Pennellidae (*Pennella balaenopterae* reaches about 250 mm in length and carries egg sacs that may exceed 350 mm). Most classifications (e.g., Huys and Boxshall 1991; J. W. Martin and Davis 2001; Boxshall and Halsey 2004) recognize 2 infraclasses: the Progymnoplea (containing the single order Platycopioida) and the far more speciose Neocopepoda, which is divided into the superorders Gymnoplea (order Calanoida) and Podoplea (orders Misophrioida, Cyclopoida, Mormonilloida, Harpacticoida, Siphonostomatoida, and Gelyelloida). The ecological adaptability displayed by copepods is reflected in their tremendous morphological plasticity, which makes it difficult to formulate a rigorous diagnosis of the subclass Copepoda that is both informative yet sufficiently comprehensive to cover the bizarre parasites as well as the free-living forms. Virtually all copepods have a stage in their life cycle—either the adult or one of the copepodid instars—exhibiting a cephalosome into which the maxilliped-bearing first thoracic somite is incorporated and possessing at least two pairs of swimming legs, the members of which are linked by an intercoxal sclerite. The life cycle typically consists of nauplii (0–6) and copepodids (1–6), the last copepodid stage being equivalent to the adult (fig. 27.1A).

LARVAL TYPES: Post-embryonic development of copepods is divided into a naupliar phase and a copepodid phase. Primitively, each phase consists of six stages (fig. 27.1A).

Nauplius: Larval development and life cycles are highly variable and can be significantly abbreviated (fig. 27.6A), although most copepods hatch at the (ortho)nauplius stage, a simple larval form described earlier (see chapter 2). Typically, after six naupliar instars (often designated NI, NII, NIII, etc.), the final nauplius stage molts into a copepodid stage.

Metanauplius: Occasionally the first stage in the life cycle is a metanauplius, and some parasitic species are known to hatch as a copepodid.

Copepodid: Copepodid stages are often referred to as CoI, CoII, and so forth (or CI, CII, etc.). The first copepodid resembles the adult but has a simple, unsegmented abdomen and only three pairs of thoracic limbs (maxillipeds and legs 1–2). There are significant changes in body size and shape, as well as in the appendages, in the molt from NVI to CoI, collectively known as metamorphosis (e.g., Gurney 1942; Dahms 1992). Intermolt stages are important for tracing the origin and homology of larval structures between naupliar and copepodid stages; examples are provided by Hulsemann (1991) for calanoids and by Dahms (1992) for harpacticoids.

Chalimus: In some parasitic groups a stage following the infective copepodid (or one of the copepodid stages) is called the chalimus; it differs from the copepodid in its possession of a frontal filament that aids in attachment to the host (e.g., I.-H. Kim 1993; Ohtsuka et al. 2009).

Other Stages: Some parasitic copepods have an onychopodium or (transient) pupal stages in their life cycle.

MORPHOLOGY

Nauplius: Copepods lack any external expression of somites during the naupliar phase of development. The naupliar body usually increases in size from one stage to the next. Early nauplii have three well-developed limbs (antennules, antennae, and mandibles), and the setose buds of the caudal rami. Buds of other limbs between the mandible and caudal ramus may be added during the naupliar phase. During the molt to CoI, the naupliar limb buds, including the caudal rami, appear as functional ("transformed" *sensu* Ferrari and Dahms 2007) limbs. Copepods possess, at most, six naupliar stages (NI–NVI) (virtually all of the Calanoida, the Harpacticoida, and free-living Cyclopoida) (fig. 27.6A), with stage reductions primarily in parasitic taxa (e.g., Izawa 1987) but also in free-living orders (e.g., Gurney 1933a; Matthews 1964). According to Izawa (1987), abbreviated naupliar phases with only five, four, three, or two stages result from the suppression of NII, NII–NIII, NII–NIV, and NII–NV, respectively, but this formula does not seem to be applicable to all parasitic copepods (Ferrari and Dahms 2007). In the lernaeopodid *Salmincola californiensis* (Kabata and Cousens 1973) and some Nicothoidae (*Hansenulus trebax*, see Heron and Damkaer 1986; *Neomysidion rahotsu*, see Ohtsuka et al. 2007), the short-lived nauplius remains inside the egg membrane and hatches direct at the first copepodid stage (figs. 27.5D; 27.9C). Some Pennellidae, such as *Cardiodectes medusaeus* (P. Perkins 1983), *Peniculisa shiinoi* (Izawa 1997), *Peniculus minuticaudae* (Ismail et al. 2013), and *Peroderma cylindricum* (Samia 1993) have lost the naupliar phase completely; the embryo develops directly into the first copepodid stage (fig. 27.9A). In the Cucumaricolidae and

some Chordeumiidae (e.g., *Parachordeumium amphiurae*), the life cycle is even more abbreviated, hatching from the egg at the second copepodid stage, with no intervening nauplius stage (fig. 27.6A) (Paterson 1958; Goudey-Perrière 1979). The key below (after Ferrari and Dahms 2007) is useful for copepods in which all six naupliar stages are expressed, particularly for the free-living species.

Metanauplius: Some copepods have a naupliar stage with more than three pairs of functional appendages. The life cycle of thaumatopsyllids includes a parasitic metanauplius that lives in the stomach of its ophiuroid host (Dojiri et al. 2008); it possesses bilobate maxillules and limb buds of legs 1–2 (figs. 27.3A, B; 27.7A–D). Thaumatopsyllid metanauplii usually show sexual dimorphism in their body shape, pigmentation, nauplius-eye morphology, and (occasionally) gonadal structure (Dojiri et al. 2008; Hendler and Dojiri 2009; Hendler and Kim 2010). At least some representatives of the Chordeumiidae have functional maxillules that are subsequently lost in the copepodid stages and in adults (fig. 27.3E, F) (Jungersen 1914). Adult micrallectids have a complete set of functional cephalothoracic appendages but lack all posterior limbs, and their organization is comparable with a metanauplius (fig. 27.3C, D) (Huys 2001).

Copepodid: Copepodid stages usually have their thoracic and abdominal somites separated by an arthrodial membrane, lack the naupliar endite on the antennary coxa, exhibit well-developed post-mandibular appendages, and have their swimming legs united by an intercoxal sclerite (interpodal bar). The antenna shifts from a naupliar paroral to a copepodid pre-oral position and loses its masticatory function. During the copepodid phase, body size and the number of somites usually increase (fig. 27.1A, B), but many exceptions exist. During each molt to a new copepodid stage, one new somite is added from a growth zone that is located in the anterior part of the posterior abdominal (anal) somite. First copepodid stages have five post-cephalosomic trunk somites, and there seems to be a functional tagma boundary between the third and fourth somites in gymnopleans, but not in podopleans (fig. 27.1C). This tagma boundary is not morphologically specialized, but instead is a flexure point defined by the behavior of the animal. At CoII the functional flexure point is located between the fourth and fifth post-cephalosomic somites and remains the definitive prosome-urosome boundary in podopleans; in gymnopleans this boundary is positioned between the fifth and sixth somites and is not attained until CoIII. There are up to nine pairs of well-developed limbs at the first copepodid stage: antennules, antennae, mandibles, maxillules, maxillae,

Three functional limbs (antennule, antenna, mandible); bud of the caudal ramus with one pair of setae .... NI	
Bud of the maxillule a simple lobe, with one seta or the posterior part of the body distinctly narrower than the anterior part .... NII	
Bud of the caudal ramus with more than one pair of setae .... NIII	
Mandibular gnathobase present and/or bud of the maxillule multilobate, with no more than six setae .... NIV	
Bud of the maxilla present or bud of the maxillule multilobate, with at least seven setae .... NV	
Bud of swimming legs 1 and 2 present .... NVI	

Table 27.1 Common pattern of development for legs 1–4 during copepodid phase

	Leg 1	Leg 2	Leg 3	Leg 4
N	$1^{\wedge}B$	$1^{\wedge}B$		
CoI	1 + 1	1 + 1	$1^{\wedge}B$	
CoII	2 + 2	2 + 2	1 + 1	$1^{\wedge}B$
CoIII	2 + 2	2 + 2	2 + 2	1 + 1
CoIV	2 + 2	2 + 2	2 + 2	2 + 2
CoV	3 + 3	3 + 3	3 + 3	3 + 3
Adult	3 + 3	3 + 3	3 + 3	3 + 3

Note: 1 + 1 = reorganized leg with 1-segmented exopod and endopod; 2 + 2 = leg with 2-segmented exopod and endopod, etc. Abbreviations: N = pre-metamorphic nauplius, Co = copepodid instar, $1^{\wedge}B$ = primary setose leg bud (Ferrari 1988).

maxillipeds, swimming legs 1–2, and the caudal rami, plus a setose bud of swimming leg 3. The remaining limbs are added as buds during the copepodid phase, one stage later than that in which their respective somites are expressed. Most limbs develop segments during the copepodid phase. Ferrari (1988) identified a common pattern of development for legs 1–4 during the copepodid phase (table 27.1), but he found that there are 23 additional patterns that produce an adult leg with three-segmented rami. Published studies suggest that all copepods have at least one copepodid stage in their life cycle (usually copepodid I), the only known exceptions being *Parachordeumium amphiurae* and *Cucumaricola notabilis*, which hatch from the egg at the copepodid II stage (see Paterson 1958; Goudey-Perrière 1979). Copepodid I (or, rarely, CoII) is often the infective stage in the life cycle of symbiotic copepods. In some Chitonophilidae, extreme transformation and a gross increase in size (as a result of hypermorphosis) take place at the final molt (fig. 27.11C, D) (Huys et al. 2002). Goudey-Perrière (1979) distinguished up to eight post-naupliar instars in the life cycle of the chordeumiid *Amphiurophilus amphiurae*, which she related to copepodids II–VI (instar 1 = CoII, instar 2 = CoIII, instar 3 = CoIV, instars 4–5 = CoIV; instars 6–8 in females or 6–7 in males = CoVI). The process of setal formation during copepodid molting was described by B. Dexter (1981).

Chalimus: The chalimus is one of up to four stages in the copepodid phase of development that attach themselves to the host by means of a frontal filament (figs. 27.4B, C; 27.9A, B). The possession of a frontal filament is a feature of several families within the large fish-parasitic clade of siphonostomatoid copepods, including members of the Caligidae, Pandaridae, Cecropidae, Pennellidae, and Lernaeopodidae (C. Wilson 1907a; Sproston 1942; Kabata and Cousens 1973; Grabda 1974; T. Schram 1979); its presence in the Hatschekiidae was inferred by T. Schram and Aspholm (1997), who recorded a frontal filament–secreting organ in *Hatschekia hippoglossi*. The pre-formed frontal filament carried within the frontal region of the infective copepodid is everted and attached to the host before the molt to chalimus I. This filament is a discrete structure and remains permanently attached to the host. At the subsequent molt to chalimus I, an additional bulb of material is secreted at the origin of the filament, around its base. At

each of the next three molts, a further bulb of material (an extension lobe) is secreted at the origin of the filament, so chalimus II has two lobes at the base of its filament, chalimus III has three lobes, and chalimus IV has four lobes (fig. 27.4A, C). The nature of development between first copepodid and adult caligids has caused considerable confusion in determining the number of true instars (stages separated by true molts). Ohtsuka et al. (2007) reviewed the dissenting views on *Lepeophtheirus*, which is the only example in the entire Copepoda where the number of stages in copepodid-phase development has been reported to exceed five (four chalimus stages and two pre-adults) before the adult. Their reinterpretation suggests that *Lepeophtheirus* conforms to the basic caligid life cycle, consisting of two naupliar, one copepodid, and four chalimus stages (corresponding to the second to fifth copepodid stages) preceding the adult (see also Venmathi Maran et al. 2013).

Pupal Stages: In the Nicothoidae there are three basic types of post-larval development (Hansen 1897; Heron and Damkaer 1986): (1) direct metamorphosis from copepodids to adults in both sexes; (2) indirect metamorphosis from copepodids via 1–3 intermediate pupal stage(s) to both sexes of adults; and (3) direct metamorphosis in one sex, but indirect in the other. Females of *Hansenulus trebax* appear to pass through two pupal stages of post-larval development before transformation to a small adult (fig. 27.9C) (Heron and Damkaer 1986). In the first pupal stage, the body is divided into a prosome and a hirsute trunk; the antennules bear a few vestigial setae; the antennae, maxillules, and maxillae are rudimentary; and the maxillipeds are one-segmented. The emerging second pupal stage has fully developed oral appendages and two pairs of swimming legs, similar in details and size to those of the adult female. Both pupal stages are attached to the marsupium of the mysid host by paired dorsal filaments extruding from a middorsal vent. Ohtsuka et al. (2007) inferred from the presence of a thin membranous structure within the copepodid exuvium of *Neomysidion rahotsu* that only one transient pupal stage may be passed through (as a double molt) as the copepodid enters the host (fig. 27.5D). In some Lernaeopodidae (e.g., *Nectobrachia indivisa*), the infective copepod becomes attached to its fish host by the frontal filament and molts into a stage that Heegaard (1947) referred to as the "pupa" and Kabata (1981) equated with the pre-adult, since no subsequent molting takes place. Pupal stages also feature in the life cycle of *Alella macrotrachelus* (Lernaeopodidae), where they are followed by either four (female pathway) or two (male pathway) chalimus stages (fig. 27.9B) (Raibaut 1985). The homologies between these pupae and traditional copepodid stages are as yet unknown.

Onychopodid: The life cycle and bizarre sexual biology of *Gonophysema gullmarensis* was elucidated by Bresciani and Lützen (1961) (fig. 27.5A). This species has a single nauplius stage, which is lecithotrophic and molts into a free infective copepodid stage. After settlement on the ascidian host, the copepodid undergoes a metamorphosis to the onychopodid larva, a simple elongate sac-like stage provided with paired grasping antennae that are used to attach the larva to the skin of the host. The conical head end of the onychopodid larva penetrates the host epithelium and enters the underlying tissues, leaving the shed copepodid exuvium behind. The onychopodid migrates within the vascular tissues of the host before finally settling just below the epithelium of the peribranchial cavity, where the transformation into the adult female form takes place. These parasites lack any trace of a mouth, digestive tract, and anus in the adult, feeding instead by the uptake of nutrients across the specialized body integument (Bresciani 1986). The young stage in the transformation of the adult female is penetrated by one or more onychopodids. These male onychopodids enter the female via the atrial pore and atrium and pass into the testicular vesicle of the female. This testicular vesicle is of ectodermal origin and represents an invagination of the external surface (fig. 27.5B). Once in position, the males undergo a metamorphic reduction so that little tissue except the gonads remain (fig. 27.5C). *Gonophysema gullmarensis* is not hermaphroditic, as originally described (Bresciani and Lützen 1960); instead it exhibits cryptogonochorism, where males are reduced to little more than germinal layers and are housed within specialized invaginated vesicles within the body of the female.

Post-Mating Metamorphosis: Once attached to, or embedded in, the host fish, mated females of many species belonging to the Pennellidae and Sphyriidae (both Siphonostomatoida) and Lernaeidae (Cyclopoida) undergo a profound metamorphosis (without molting) to produce the final body form (figs. 27.6B–D; 27.9A). The metamorphosis involves a considerable increase in volume, especially in the trunk region. This is achieved partly by expansion of the highly folded integument, and partly by production of a new integument (J. Smith and Whitfield 1988). In the pennellid *Lernaeocera branchialis* (fig. 27.6D), straightening out these folds generates an approximately 6-fold increase in length, but this mechanism provides only part of the overall 10- to 20-fold increase in length that takes place.

MORPHOLOGICAL DIVERSITY

Nauplii: Differences among the six stages of a typical naupliar phase include changes in the number of limb buds, in segments of limb rami (exopod, endopod), and in setation elements on limb segments and the caudal rami. Nauplii are typically oval in shape, but they can be elongate in species leading a pelagic lifestyle (fig. 27.2K) (e.g., *Macrosetella gracilis*, see Tokioka and Bieri 1966; *Microsetella norvegica*, see W. Diaz and Evans 1983) or interstitial mode of life (e.g., *Paraleptastacus brevicaudatus*, see Dahms 1990a). The nauplius stages of *Rhincalanus* spp. and several members of the Pontellidae are characterized by extreme elongation and by a slender body that tapers posteriorly to an acute point (fig. 27.2G, I) (Gurney 1934; Björnberg 1972). Harpacticoids belonging to the superfamily Thalestrioidea typically have nauplii with very wide bodies (fig. 27.2F, H. L). The nauplius dorsal shield is usually smooth, although in some harpacticoids it can be ornamented with tiny spinules (fig. 27.2J). All nauplius stages of *Longipedia* are characterized by the presence of a long median caudal process, which becomes progressively shorter during the planktonic naupliar phase (fig. 27.2B) (Onbé 1984). Caudal

spines have also been reported in the early instars (NI–NII) of planktonic *Microsetella* spp. (W. Diaz and Evans 1983), and in the lecithotrophic nauplii of some Lichomolgidae, Taeniacanthidae, and Bomolochidae (Kabata 1976; Izawa 1987). Caudal spines (in planktonic harpacticoids) and enlarged caudal setae (in calanoids, e.g., *Euchaeta, Rhincalanus*, etc.) (fig. 27.2O) are assumed to play a role in swimming or maintaining buoyancy in pelagic nauplii. Left-right asymmetry is often expressed in the caudal setae of calanoid nauplii, with either the left side (fig. 27.2I) (in the Acartiidae, Candaciidae, Centropagidae, Fosshageniidae, Pontellidae, and Pseudodiaptomidae) or the right side (fig. 27.2G) (in the Eucalanidae and Rhincalanidae) being better developed. The nauplius stages of the planktonic *Euterpina acutifrons* lack a caudal spine but exhibit a cluster of long caudal spinules (fig. 27.2N) (Haq 1965; Dahms 1990c).

Copepod nauplii typically have a centrally placed nauplius eye toward the front of the dorsal shield (fig. 27.2A), consisting of a median ventral and two dorsolateral cup-shaped ocelli, with a common backing of two shielding pigment cells. The nauplius eye is absent in *Misophria pallida* (Gurney 1933b), and in 2 families of highly transformed parasites of polychaetes, the Herpyllobiidae (Lützen 1968) and the Xenocoelomatidae (Bresciani and Lützen 1974). Naupliar stages of the tisbid genus *Scutellidium* typically display a midventral oral sucker, which is derived from the labrum and used for adhesion to the surface of algal fronds (Brian 1919; Gurney 1933a; Branch 1974), but see Dahms (1990c) for an exception.

Nauplii of most copepods are planktotrophic, but lecithotrophy has evolved repeatedly. Lecithotrophic nauplii are often characterized by a yolk-rich body (fig. 27.10F), the absence of a labrum, a weakly invaginated stomodeum and proctodeum, and the absence of the antennary masticatory process (fig. 27.2P) and mandibular gnathobase. In the Calanoida, the members of the family Euchaetidae have lecithotrophic nauplii (fig. 27.2O), whereas in closely related families they are planktotrophic. Nauplii in the calanoid families Aetidaeidae (*Chiridius armatus, Aetideus armatus*—both with only four naupliar instars) and Phaennidae (*Xanthocalanus fallax*) rely entirely on their yolk supply until they reach the first copepodid stage (Matthews 1964). In the Harpacticoida, species of the genera *Pseudotachidius* and *Leptocaris* have lecithotrophic nauplii (Gurney 1932; Dahms 1989), and within the orders Cyclopoida and Siphonostomatoida many parasitic groups possess them (fig. 27.10E, F) (Izawa 1987; Boxshall and Halsey 2004). The antennary arthrite may be absent in the early naupliar stages of copepods that display a mixotrophic naupliar phase and lack a functional mouth at NI (e.g., *Calanus finmarchicus*, see Ferrari and Dahms 2007; *Pseudodiaptomus marinus*, see Uye et al. 1983). The planktotrophic nauplii in the family Ergasilidae carry a characteristic spatulate element on the distal segment of the mandibular endopod (e.g., Huys and Boxshall 1991, their fig. 2.10.24C, D). Sexual dimorphism in naupliar stages has not been recorded thus far (but see, e.g., Hendler and Dojiri 2009; Hendler and Kim 2010 for metanauplii). Both thaumatopsyllids (fig. 27.7A) (Dojiri et al. 2008) and monstrillids (fig. 27.8) (Malaquin 1901) have a protelean life cycle, combining an endoparasitic naupliar phase and free-living non-feeding adults. Monstrillid eggs hatch into nauplii that locate a host and burrow into its tissues. After undergoing a considerable metamorphosis in the host's blood system, the endoparasitic sac-like naupliar stages develop root-like absorptive processes and bear virtually no resemblance to other crustacean larvae. Once development is complete, the monstrillid leaves its host as a last copepodid stage and undertakes a single molt into the adult. The infective nauplius stage of *Monstrilla hamatapex* was described in detail by Grygier and Ohtsuka (1995); the antennary arthrite and piercing mandibles in this lecithotrophic stage presumably assist in the attachment to and/or penetration of the host.

Copepodids: Ferrari and Dahms (2007) described the fundamental differences in post-naupliar body architecture between the gymnoplean *Ridgewayia klausruetzleri*, the podoplean *Dioithona oculata*, and the thaumatopsyllid *Caribeopsyllus amphiodiae*. Their paper should be consulted for a further discussion on the remaining variation in the association of somites along the anteroposterior axis of the body. This variation results from one of two processes: the formation of somite complexes that result from the failure of an arthrodial membrane to form between two somites, or the suspension of the addition of somites to the body. Readers interested in the evolutionary reduction of body somites in poecilostome Cyclopoida are referred to Izawa (1991). The morphology of the infective copepodid shows little variation in symbiotic copepods. In some families it attaches itself to the host by a cephalic frontal filament prior to molting into a chalimus stage (figs. 27.4B, C; 27.9A) (e.g., the Pennellidae, and caligiform families) or a pupal stage (fig. 27.9C) (some Nicothoidae). Sexual dimorphism is usually expressed in later copepodid stages of harpacticoids, most free-living cyclopoids, and many calanoids, particularly in body size, antennule segmentation and armature, and legs 1–5. In many harpacticoid families, males form a spinous process (apophysis) on a swimming-leg segment, which may originate as a produced segmental margin, or may be derived by modification of a setal element present in females (Lang 1948; Huys 1990c). Most poecilostome cyclopoids exhibit sexual dimorphism in the maxillipeds, which is typically first expressed at the molt to CoV (e.g., Itoh and Nishida 1995). According to Izawa (1986), sexual dimorphism in *Acanthochondria* is apparent as early as CoII, and this is probably characteristic of the speciose fish-parasitic family Chondracanthidae, with its tiny males and giant females (Boxshall and Halsey 2004). Dimorphic males (with different functions/lifestyles) have been recognized in the notodelphyid genus *Pachypygus* (Dudley 1966; Hipeau-Jacquotte 1978, 1987), the planktonic harpacticoid *Euterpina acutifrons* (Haq 1965) and the myicolid *Pseudomyicola spinosus* (Do et al. 1984). In all of these cases the expression of male dimorphism during development commenced at CoIV. Several planktonic families (e.g., the Aegisthidae, Euchaetidae, Lubbockiidae, Mormonillidae, and Pontoeciellidae) and deepwater harpacticoid lineages have non-feeding adult males with atrophied mouthparts, and male copepodid stages with fully functional mouthparts. Little information is available on the precise developmental onset of this transformation in the feeding mode.

NATURAL HISTORY: Egg sacs (fig. 27.10A–E) are a typical attribute for most major copepod orders (the Cyclopoida [including the Poecilostomatoida], Harpacticoida, and Siphonostomatoida), and their presence has been confirmed in the Mormonilloida (Huys et al. 1992c). In most calanoid families, eggs are freely released into the water column. Exceptions—where eggs are either contained in single or paired multiseriate (or uniseriate) sac(s)—are found in the Aetideidae, Arietellidae, Clausocalanidae, Diaptomidae, Pseudodiaptomidae, and Temoridae, although it is not always clear whether these are true egg sacs, with an enclosing sac membrane (Huys and Boxshall 1991; Mazzocchi and Paffenhöfer 1998). Eggs are retained in a mass on the ventral side of the urosome in the Centropagidae and Euchaetidae and in some members of the Clausocalanidae and Temoridae. Gurney's (1933b) observations on *Misophria pallida* showed that its eggs are loosely attached to the female urosome and not contained in sacs; it is unknown if this is the typical condition for all Misophrioida. No information is available for the Gelyelloida or Platycopioida. Monstrillid females lack egg sacs; instead, the eggs are attached to the paired ovigerous spines by means of a mucous substance secreted by the terminal part of the oviduct (fig. 27.12C). Huys and Boxshall (1991) showed that egg masses are produced iteratively, the ovigerous spines growing accordingly when a new batch is being spawned (fig. 27.12B). The complex of caligiform and dichelesthiiform families within the Siphonostomatoida are characterized by (sometimes coiled) linear egg strings containing a single column of closely packed disc-shaped eggs (figs. 27.4B; 27.10G). The presence of caudal balancers (fig. 27.3G–I) in the nauplii of the Caligidae, Dissonidae, Kroyeriidae, and Pandaridae appears to be correlated with the possession of uniseriate egg sacs (G. A. Boxshall, pers. comm.). In the highly modified Phyllodicolidae, eggs are extruded in elongate masses that break down, with the eggs attached separately to an axial filament originating at the genital aperture (fig. 27.12A) (Laubier 1961). Similarly, in some members of the Chitonophilidae, eggs are attached to the genital area by individual filaments (fig. 27.11I, J) (Huys et al. 2002). The number of eggs contained in a single sac can range from one—such as in the siphonostomatoid families Calverocheridae (Stock 1968), Micropontiidae (Gooding 1957), and Stellicomitidae (Humes and Cressey 1958)—to 2,000–3,000 in the Chordeumiidae (Bartsch 1996). Within the benthic Harpacticoida, members of the Darcythompsoniidae (Lang 1948) and Phyllognathopodidae (Chappuis 1916) reportedly lack egg sacs and release their eggs directly into the environment, while parastenocaridids carry them for a short period before attaching them to the substratum (Schminke 1982). In some symbiotic families associated with invertebrates, eggs are either laid free inside the host (the Lamippidae, see Bouligand 1960; the Sponginticolidae, see Silén 1963), deposited in masses inside a gall on the host (the Mesoglicolidae, see Taton 1934), attached to the inside of the tunic of the host ascidian (the Intramolgidae, see Marchenkov and Boxshall 1995), or laid inside a capsule enclosing the adult female (the Codobidae, see Heegaard 1951).

Specialized brooding in (semi)enclosed chambers formed from body somites and/or appendages has evolved independently many times in podoplean copepods (Grygier and Ohtsuka 2008). In several harpacticoids (*Phyllopodopsyllus*, *Eudactylopus*, *Phyllothalestris*, *Paramenophia*, and the Tegastidae), the eggs are retained in a single ventral sac, enclosed in a brood pouch formed by the modified foliaceous fifth legs (fig. 27.11E, F) (Lang 1948; Gamô 1969a, 1969b; Huys et al. 1996). Similarly, the paired multiseriate egg sacs in the Ascidicolidae (Cyclopoida) are partly or completely covered by the expanded fifth legs (Illg and Dudley 1980). The monstrillid genus *Maemonstrilla* represents the only example of subthoracic brooding among planktonic copepods (Grygier and Ohtsuka 2008). An incubatory pouch that is formed dorsally or dorsolaterally within one or more pedigerous somites (fig. 27.11A, B) is present in the cyclopoid families Buproridae (Illg and Dudley 1980), Gastrodelphyidae (Dudley 1964), and Notodelphyidae (e.g., Sars 1921). Similar brooding has been recorded in the mytilicolid *Pectenophilus ornatus*, although the origin of the brood pouch in this highly modified cyclopoid remains unknown (fig. 27.12E–H) (Nagasawa et al. 1988; Huys et al. 2006). Some Chordeumiidae maintain their loose egg masses in a subthoracic cage that is formed from modified and ventrally downturned cephalic appendages and thoracic outgrowths (fig. 27.11G, H) (Stephensen 1935; Goudey-Perrière 1979). The vermiform female of *Nucellicola holmanae*, a chitonophilid endoparasite of gastropods, is enveloped in a membranous tube that is possibly of host origin; the tube becomes filled with eggs and developing nauplii (Lamb et al. 1996). In members of the Micrallectidae, the lecithotrophic nauplii develop within the eggs retained in the genital tract of the female; the naupliar maxillae develop early and are visible through the body wall of the female (Huys 2001).

In many copepod species, adult males clasp subadult females for an extended period before transferring spermatophores. Males clasping juvenile females (CoI–CoV, inclusive) is interpreted as mate guarding and is widespread among podoplean copepods. It differs from copulation, which takes place only between adults, and is often distinguishable from mate guarding by a difference in the clasping posture (Boxshall 1990). Studies reporting pre-copulatory mate guarding in harpacticoid copepods have been summarized by Kern et al. (1984). Males may grasp female copepodids around their caudal setae, caudal rami, anal somite, or fourth leg, or by the posterolateral margins of the dorsal cephalothoracic shield (Lang 1948). Within the family Harpacticidae, adult males typically clasp all juvenile stages from CoI to CoV (e.g., Itô 1970). Fiers (1998) suggested that the atypical development of leg 4 in female copepodids of many of the Laophontidae is a juvenile adaptation to pre-copulatory mating guarding. Observations of adult males attaching themselves to the dorsal surface of female CoV stages have been recorded in both free-living cyclopoids (the Cyclopidae, see Hill and

Coker 1930) and symbiotic cyclopoids (the Notodelphyidae, see Thorell 1859; Giesbrecht 1882). Do et al. (1984) recorded adult males of the poecilostome species *Pseudomyicola spinosus* clasping juvenile stages from the third copepodid onward. Pre-copulatory mate guarding is a common phenomenon in the fish-parasitic Pennellidae and Caligidae, where males frequently clasp attached chalimus stages around the frontal filament (Ho 1966; Boxshall 1974, 1990).

Many freshwater cyclopoids enter diapause in the later copepodid stages (typically CoIV–V), although some may enter a state of complete torpor or active diapause as early as CoII (e.g., *Cyclops scutifer*). Marine calanoids that have dormant stages either produce diapause eggs (the Acartiidae, Centropagidae, Pontellidae, Temoridae, and Tortanidae) or diapause during the copepodid phase (the Calanidae). Freshwater calanoids (the Diaptomidae) can produce resting eggs that can lie in the sediment and remain viable for up to 300 years (Hairston et al. 1995); fossil eggs of *Diaptomus* were reported from Late Quaternary lake sediments (Bennike 1998). Borutzky (1929) reported encysted nauplii in the freshwater harpacticoid *Bryocamptus arcticus*. The only known example of juvenile marine harpacticoids going into dormancy is that by Dahms et al. (1990), who recorded copepodids of *Drescheriella* sp. in a non-encysted dormant stage within the ice in the Antarctic. Delayed naupliar development has been reported for some marine harpacticoids (Coull and Dudley 1976) and calanoids (Uye 1980), but it is not clear whether such prolongation of the naupliar phase represents a genuine form of dormancy. Reports of diapause for parasitic copepods are unknown. For an excellent review of copepod dormancy, see Williams-Howze (1997).

The nauplii of polyarthran harpacticoids (the Longipediidae and Canuellidae) are planktonic suspension feeders with good swimming abilities, while the copepodids remain close to the substratum (in sediment or on an invertebrate host). Oligoarthran harpacticoids typically have benthic nauplii (except *Microsetella* spp., see W. Diaz and Evans 1983) and even those species that secondarily became holoplanktonic complete an essentially substratum-bound life cycle. Björnberg (1965) noted that all developmental stages of *Macrosetella gracilis* are usually found in association with cyanobacteria (*Trichodesmium*), and it seems likely that all planktonic miraciids exhibit a similar specialized lifestyle (Huys and Böttger-Schnack 1994; O'Neil and Roman 1994). Ovigerous females of the widespread pelagic thalestrid species *Parathalestris croni* use floating macroalgal clumps as nests for their non-swimming nauplii (Ingólfsson and Ólafsson 1997). Members of the genus *Balaenophilus* live attached to their cetacean hosts (Bannister and Grindley 1966) or chelonian (juvenile loggerhead turtles) hosts (Ogawa et al. 1997) throughout their life cycles.

Various species of the Thalestridae (*Amenophia*, *Parathalestris*, and *Thalestris*) and Dactylopusiidae (*Dactylopusioides* and *Diarthrodes*) are obligatorily endophagous in macroalgae during most of all of their lives (Brady 1894; Bocquet 1953; Harding 1954; J. Green 1958; Fahrenbach 1962; Ho and Hong 1988; Shimono et al. 2004, 2007). Nauplii and copepodids live in excavated burrows and galleries or in newly formed capsules or galls (for a discussion on trends in reduction and specialization in frond-mining nauplii, see Dahms 1990b). Stenheliinid nauplii move in a sideways crab-like crawl (Bresciani 1961) and, like all other developmental stages, build mucoid tubes (almost immediately after hatching) that extend into the sediment (Lorenzen 1969; Chandler and Fleeger 1984; Williams-Howze and Fleeger 1987). Their sideways motility and strongly ellipsoid body shape may be adaptations to tube-dwelling.

Very little information is available on the diets of planktotrophic nauplii, but they are recorded as feeding on phytoplankton and naupliar fecal pellets (E. Green et al. 1992). The antennae and mandibular palps are used for swimming and creating a weak feeding current (Paffenhöfer and Lewis 1989). According to Sekiguchi (1974), most calanoids have mixotrophic nauplii, which survive on their yolk reserves during the early stages of development before becoming planktotrophic. The first nauplius stage to feed varies among species, but NIII or NIV appears to be the most common (Mauchline 1998).

Copepodid I (rarely CoII) acts as the infective stage in most symbiotic copepods. Exceptions include species with a protelean life cycle and members of the Ergasilidae, where only the adult females are parasitic (NI–NVI or NI–NIII) and CoI–CoV and adult males are free-living (fig. 27.6A). Several pennellids have an unusual life cycle, involving two different hosts and, hence, two infections (fig. 27.6A). After a brief planktonic phase, the infective copepodid stage locates the first host, either a fish, as in *Lernaeocera* (Scott and Scott 1913; Sproston 1942; Slinn 1970) or a pelagic gastropod mollusk, as in *Cardiodectes* (Ho 1966; P. Perkins 1983). Development from the attached chalimus stages through to sexually mature adults takes place on the gills of the first host. Mating also occurs on this host, after which the mated female leaves the first host and finds a second host, usually a fish, but occasionally a marine mammal, where it completes its metamorphosis (fig. 27.9A). The naupliar and infective copepodid stages of symbiotic copepods can be temporary members of the plankton. A few genera, such as *Saphirella*, were established to accommodate unusual forms that are now known to be the copepodid stages of symbiotic adults (e.g., the Clausidiidae, see Itoh and Nishida 1995). The nauplii and first copepodid larvae of sea-lice (the Caligidae) function as the dispersal and infective stages of the life cycle, respectively, and even adult caligids (e.g., *Caligus elongates*) are not infrequently taken in coastal plankton samples. Pre-metamorphic adult females of the Pennellidae can be found in plankton samples, since it is this stage that is responsible for locating and infecting the final host. The majority of copepods that are parasitic on fishes use more than one type of attachment during their life cycles. The sequence of attachment devices (antennae, cephalothoracic suction cup, holdfast, frontal filament, bulla) used during the post-naupliar phase was reviewed by Kabata (1981), who recognized eight different types of attachment succession. As in *Gonophysema*, members of the highly modified Xenocoelomatidae (endoparasitic in terebellid polychaetes) exhibit cryptogonochorism (Bocquet et al. 1970; Bresciani and Lützen

1974), but no onychopodid is involved. The male copepodid penetrates the atrium of the female, molts, and passes into a special receptacle (*receptaculum masculinum*) formed by a modification of the spermatic ducts of the female. It then develops into a functional testis, resulting in a pseudohermaphroditic condition.

Metridinid calanoids are strongly bioluminescent from the nauplius stage through to the adults of both sexes, while all species of *Lucicutia* (Lucicutiidae) are probably bioluminescent, even in the copepodid stage (Herring 1988).

PHYLOGENETIC SIGNIFICANCE

Ordinal Level: Björnberg (1972) inferred ancestor-descendant relationships of free-living cyclopoid, harpacticoid, and calanoid copepods, using naupliar characters of a large number of planktonic species. Her analysis challenged the widely accepted view that calanoids are close to the base of the Copepoda and placed the Cyclopoida at this position instead, leaving the Calanoida as the most derived order. The gymnoplean type of tagmosis (displayed in the Platycopioida and Calanoida), in which articulation between the prosome and urosome lies between the fifth pedigerous and genital somites, is generally regarded as the plesiomorphic condition (Huys and Boxshall 1991). Ferrari et al.'s (2010) phylogenetic analysis, however—using naupliar and post-naupliar characters, with the Mystacocarida as the sister-taxon of the Copepoda—supported the controversial hypothesis that the highly specialized thaumatopsylloid tagmosis is the most ancestral one, while the gymnoplean architecture is the youngest. Dahms (1990c, 2004a) compared naupliar characters between oligoarthran and polyarthran harpacticoids and, in a subsequent paper (Dahms 2004b), suggested removing the Polyarthra from the Harpacticoida and placing it as the sister-group of all remaining copepods, because neither a larval nor an adult synapomorphy uniting oligoarthrans and polyarthrans could be identified (Tiemann 1984). Dahms (2004a) also used post-embryonic characters to confirm the monophyly of the Copepoda and hypothesized a sister-group relationship based on naupliar characters for the Copepoda/Thecostraca. A recent molecular analysis by Huys et al. (2007) suggested that the order Monstrilloida is nested within a fish-parasitic clade of the Siphonostomatoida, sharing a common ancestor with the stem species of the caligiform families (sea lice); this unforeseen relationship was shown to be congruent with both antennulary and caudal ramus ontogeny. Boxshall and Huys (1998) analyzed the development of antennulary segmentation and setation patterns across 6 orders of copepods and produced a hypothetical general model for antennulary development in the Copepoda as a whole.

Family Level: Dudley (1966) employed attributes of the development of naupliar appendages and concluded that the Notodelphyidae should be placed in the gnathostome cyclopoids, rather than the poecilostome cyclopoids. Based on naupliar morphology, Dahms and Hicks (1996) concluded that the Parastenheliidae are related to the Thalestridae, which is in agreement with a recent analysis using adult characters (Willen 2000). Dahms (1990b) noted that several of the assumed derived naupliar states in the Thalestridae are shared with some species of the Harpacticidae, hinting at a relationship between both families; since then, however, a close alliance between harpacticids and thalestridimorphs has been rejected (Willen 2000). Dahms (1993b) investigated comparative copepodid development in the Tisbidae and related tisbidimorph families and suggested a close relationship between the Tegastidae and Peltidiidae. The presence of three aesthetascs derived from ancestral segments XXI, XXV, and XXVIII on the antennule of copepodid I (or II) provides a useful signature for the poecilostome Cyclopoida and has recently been used to place taxa exhibiting a highly modified adult morphology (e.g., the Chordeumiidae, Chitonophilidae, and Herpyllobiidae) in this order (López-González and Bresciani 2001; Huys et al. 2002; Boxshall and Halsey 2004).

Genus and Species Levels: Dahms et al. (1991) inferred phylogenetic relationships for 6 species of *Tisbe*, based on the morphology of NVI; their final analysis, however, in which adult characters were integrated with naupliar ones, resulted in significant discrepancies. Ferrari (1991) abstracted segmentation patterns from the development of legs 1–6 to group species of the calanoid genus *Labidocera* and genera within the Diaptomidae (Calanoida) and Cyclopidae (Cyclopoida). Dahms (1993a) provided a phylogeny for 3 genera in the Tisbidae, based on naupliar character states. Dahms and Bresciani (1993) described the naupliar development of the stenheliinid species *Delavalia palustris* and discovered several apomorphies, warranting the removal of this species from the Miraciidae. In a later paper, Dahms et al. (2005) discussed the naupliar morphology within the Stenheliinae and the evolutionary novelties displayed by the nauplii of *Stenhelia peniculata*. Schutze et al. (2000) placed 35 species from 29 genera of the Cyclopidae into groups that were based on ten developmental patterns of the female antennule, while Ferrari (1998) and Ferrari and Ivanenko (2005) used developmental data on the maxilliped and legs 1–7 to derive ancestor-descendant relationships among genera within this family. Groups defined by antennulary developmental patterns (Schutze et al. 2000) are not comparable with the lineages derived from the development of thoracopods (Ferrari 1998; Ferrari and Ivanenko 2005). Ferrari and Ueda (2005) examined the development of the female leg 5 and the genital complex in the Centropagoidea and used them as attributes to group species into this calanoid superfamily. Dahms et al. (2009) proposed a phylogeny of 8 *Tisbe* species, using exclusively naupliar characters.

HISTORICAL STUDIES: Post-embryonic development in the Copepoda has been studied for over 250 years and has produced an impressive body of literature. For entry into this literature, interested readers should consult the comprehensive bibliography compiled by Ferrari and Dahms (2007), who also provided an overview of the early history of copepod developmental studies and a chronology of the important descriptive observations and conceptual discoveries (also see Damkaer 2002). Only a few are repeated here. J. Lange (1756)

depicted both nauplii and copepodids of a freshwater cyclopid, and his illustrations are also the earliest for a crustacean nauplius. Ramdohr (1805) described the complete life history of a free-living cyclopid. Surriray (1819) illustrated a nauplius that hatched from the egg of a transformed parasitic copepod. Burmeister (1835) described a chalimus, which Krøyer (1838) later identified as an immature stage of a parasitic copepod. C. Wilson (1905) illustrated the complete development of a caligid, including the nauplius, copepodid, chalimus, and adult stages. Nordmann (1832) compared the nauplius and the first copepodid of highly modified parasites (Lernaeopodidae) to similar instars of free-living copepods and concluded that both categories belonged to the same group of Crustacea. Oberg (1906) determined homologies of antennulary setae between NVI and CoI of *Temora longicornis* by studying intermolt stages. Giesbrecht (1913) proposed that during copepodid development, one new somite is added immediately anterior to the anal somite during each molt. Significant conceptual studies on the timing of setal additions during development were added by Illg (1949) and Dudley (1966). Björnberg (1972) used naupliar morphology to present the first phylogeny of copepods based on developmental data. Izawa (1987) studied the development of several parasitic poecilostome Cyclopoida with an abbreviated naupliar phase.

Harding (1954) pointed out that the new harpacticoid genus *Fucitrogus* described by Brady (1894) was, in reality, based on a nauplius stage of a *Thalestris* species. Various authors have proposed generic names based on copepodid stages. Some have been confidently synonymized with existing generic names (*Aphelura = Pontella; Euchaetopsis = Euchaeta; Pseudolovenula = Megacalanus; Pseudocletopsyllus = Cletopsyllus*), while others remained genera inquirenda: *Specilligus; Hessia; Centromma; Microcryobius; Mawsonella; Plagiopus; Faurea;* and *Nogagella*. The genera *Paurocope, Saphirella,* and *Lanowia* represent juvenile copepodid stages of clausidiids. The early copepodid stages of clausidiids, particularly the first copepodid, are commonly found in coastal plankton. The genus *Saphirella* is still used in the literature as a collective name for clausidiid juveniles.

Selected References

Dahms, H.-U. 1992. Metamorphosis between naupliar and copepodid phases in the Harpacticoida. Philosophical Transactions of the Royal Society of London, B 333: 221–236.
Dudley, P. L. 1966. Development and Systematics of Some Pacific Marine Symbiotic Copepods: A Study of the Biology of the Notodelphyidae, Associates of Ascidians. University of Washington Publications in Biology No. 21. Seattle: University of Washington Press.
Ferrari, F. D., and H.-U. Dahms. 2007. Post-embryonic development of the Copepoda. Crustaceana Monographs 8: 1–232.
Izawa, K. 1987. Studies on the phylogenetic implications of ontogenetic features in the poecilostome nauplii (Copepoda: Cyclopoida). Publications of the Seto Marine Biology Laboratory 32(4/6): 151–217.
Raibaut, A. 1985. Les cycles évolutifs des copépodes parasites et les modalités de l'infestation. L'Année Biologique, ser. 4, 24: 233–274.

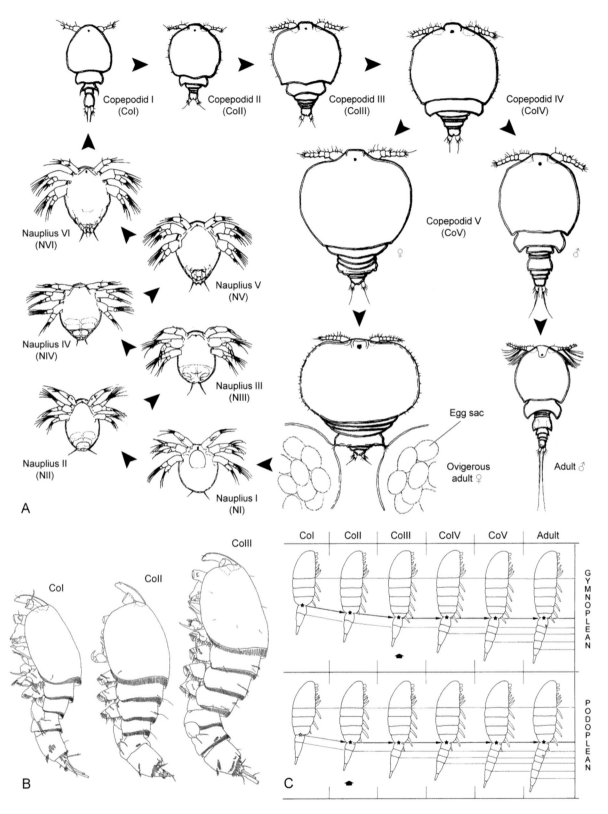

Fig. 27.1 A: drawing of the basic copepod life cycle, consisting of six naupliar and six copepodid stages, as exemplified by *Cancerilla tubulata* (Siphonostomatoida: Cancerillidae), dorsal view. B: drawing of copepodids I–III of *Parastenhelia megarostrum* (Harpacticoida: Parastenheliidae), lateral view. C: comparison of the developmental pattern in podoplean and gymnoplean copepods, lateral view; solid stars indicate the position of the major body articulation, a hollow star indicates the poorly defined flexure plane, and arrows indicate the stage at which definitive tagmosis is attained and specialization of the joint commences. A modified after Carton (1968); B modified after Dahms (1993c); C modified after Huys and Boxshall (1991).

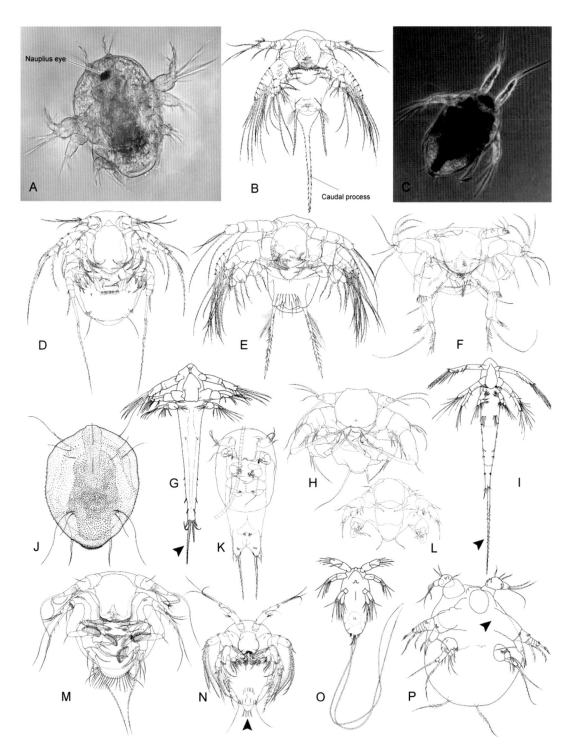

Fig. 27.2 Examples of naupliar diversity, drawings (unless otherwise indicated). A: unidentified *Cyclops* sp. (Cyclopoida: Cyclopidae), light microscopy, dorsal view. B: nauplius I of *Longipedia minor* (Longipediidae), ventral view. C: unidentified instar of *Ergasilus sieboldi* (Cyclopoida: Ergasilidae), light microscopy (note the cyan pigment inside), ventral view. D: nauplius II of *Phyllognathopus viguieri* (Phyllognathopodidae), ventral view. E: nauplius I of *Canuella perplexa* (Canuellidae), ventral view. F: nauplius II of *Delavalia palustris* (Miraciidae), ventral view. G: nauplius IV of *Rhincalanus cornutus* (Rhincalanidae), ventral view; the black arrow shows the asymmetry of the caudal setae. H: nauplius I of *Alteutha oblonga* (Peltidiidae), ventral view. I: nauplius IV of *Pontellopsis brevis* (Pontellidae), ventral view; the black arrow shows the asymmetry of the caudal setae. J: nauplius I of *Rhizothrix minuta* (Rhizothricidae), dorsal view. K: nauplius VI of *Macrosetella gracilis* (Miraciidae), holding a *Trichodesmium* filament, ventral view. L: nauplius I of *Parategastes sphaericus* (Tegastidae), ventral view. M: nauplius I of *Zaus spinatus* (Harpacticidae), ventral view. N: nauplius I of *Euterpina acutifrons* (Tachidiidae), ventral view; the black arrow indicates a cluster of enlarged spinules. O: lecithotrophic nauplius VI of *Euchaeta marina* (Euchaetidae), ventral view. P: lecithotrophic nauplius I of *Pseudotachidius* sp. (Pseudotachidiidae), ventral view; the black arrow indicates the absence of the antennary masticatory process. A courtesy of Sam Brutcher; B, D–F, H, J–N, and P (harpacticoid nauplii) modified after Dahms (1990c); C original; G, I, and O (calanoid nauplii) modified after Björnberg (1972).

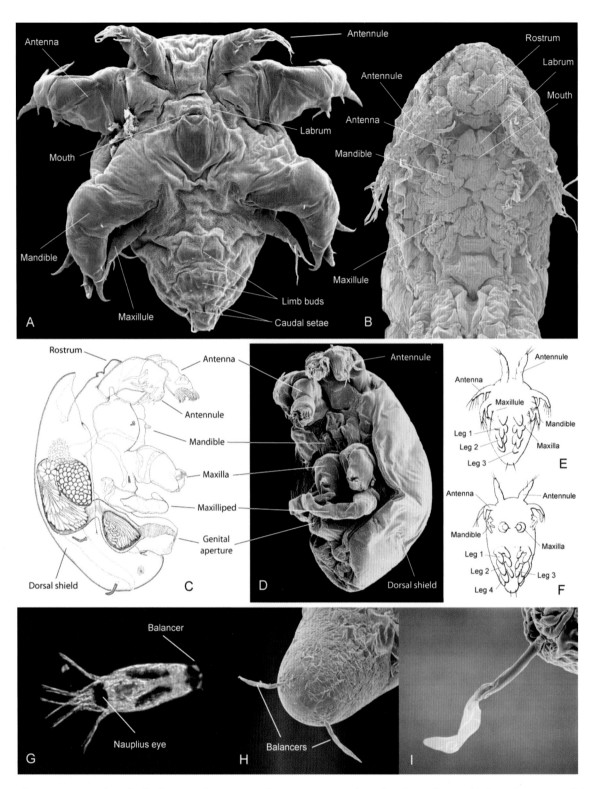

Fig. 27.3 Metanauplii and caligiform nauplii. A: intermediate-size metanauplius of *Caribeopsyllus amphiodiae* (Thaumatopsyllidae), with rudimentary limb buds, SEM, ventral view. B: recently molted copepodid I of *Thaumatopsyllus paradoxus* (Thaumatopsyllidae), showing the vestigial antennae, mandibles, and maxillules, SEM, ventral view. C and D: adult male of *Micrallecto fusii* (Micrallectidae), showing the metanaupliar organization C: drawing, lateral view. D: SEM, lateroventral view. E and F: drawings of *Chordeumium obesum* (Chordeumiidae). E: first metanauplius, showing leg buds 1–3, ventral view. F: second metanauplius, showing leg buds 1–4, ventral view. G: nauplius I of *Lepeophtheirus pectoralis* (Caligidae), light microscopy. H: hindbody of nauplius I of *Lepeophtheirus* sp., showing the balancers, SEM. I: closeup of the same. A courtesy of Masahiro Dojiri and Gordon Hendler; B original; C and D modified after Huys (2001); E and F modified after Jungersen (1914); G–I courtesy of Geoffrey Boxshall.

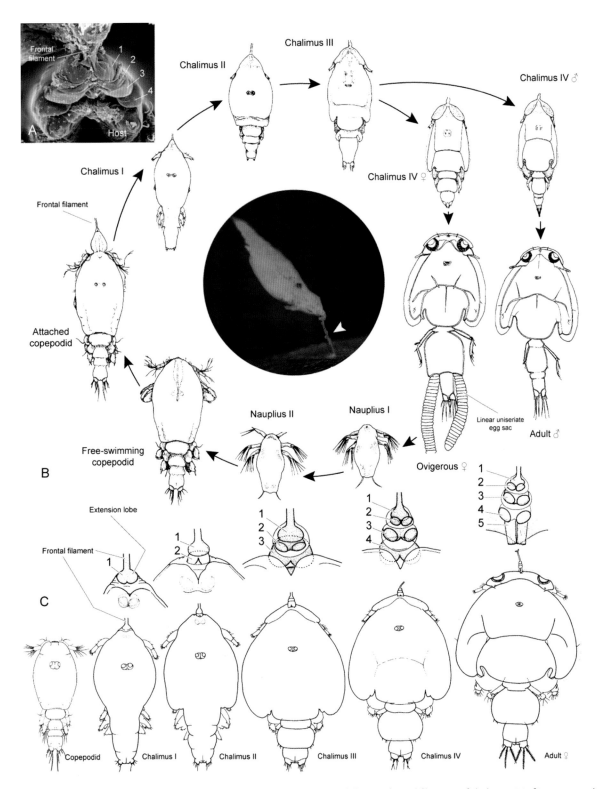

Fig. 27.4 Chalimus stages, showing the frontal filaments and extension lobes. A: frontal filament of chalimus IV of *Lernaeocera branchialis* (Pennellidae), with the numbers 1–4 representing the bulb-like extension lobes secreted around the base of the filament during each post-copepodid molt, SEM; adult males use their antennae to grasp early chalimus stages in the vicinity of their frontal attachment apparatus (which is not molted with the rest of the exoskeleton) when exhibiting mate guarding. B: drawing of the life cycle of *Caligus clemensi* (Caligidae); photograph (*center*) shows the attached chalimus of *C. elongatus*, with an arrow indicating the frontal filament. C: drawing of the post-naupliar development of *C. punctatus*; insets show the sequential addition of extension lobes at the origin of the filament during subsequent molts. A courtesy of Geoffrey Boxshall; B modified after Parker and Margolis (1964), Kabata (1972), and Raibaut (1985), with photograph courtesy of Øivind Øines; C modified after I.-H. Kim (1993).

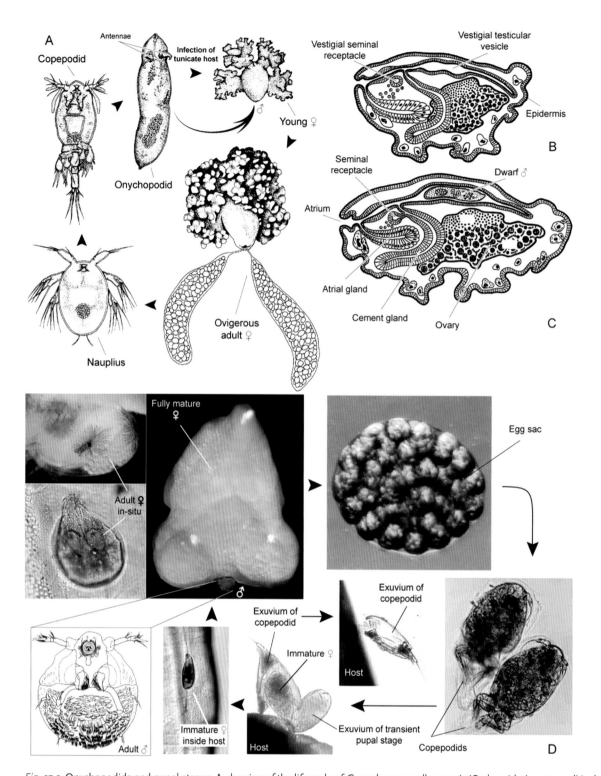

Fig. 27.5 Onychopodids and pupal stages. A: drawing of the life cycle of *Gonophysema gullmarensis* (Cyclopoida *incertae sedis*), dorsal view. B and C: schematic representations of the sagittal sections of *G. gullmarensis*, lateral view. B: young female. C: adult female; note the metamorphosed male in the testicular vesicle. D: life cycle of *Neomysidion rahotsu* (Nicothoidae), light microscopy (except for the drawing in the lower left). A modified after Bresciani and Lützen (1960) and Raibaut (1985); B and C modified after Bresciani and Lützen (1961); D modified after Ohtsuka et al. (2005, 2007).

Fig. 27.6 (opposite) Life cycles and post-mating metamorphosis. A: chart of types of life cycles of free-living and parasitic Copepoda (PM ♀ = post-metamorphic female); (a) Calanoida, Harpacticoida, free-living Cyclopoida; (b) Ergasilidae, in part (Urawa et al. 1980a, 1980b; Abdelhalim et al. 1991; Alston et al. 1996); (c) Notodelphyidae and Ascidicolidae (Dudley 1966); (d) Cancerillidae (Carton 1968), many poecilostome Cyclopoida associated with invertebrate hosts (e.g., Gibson and Grice 1978; Do et al. 1984; Costanzo and Calafiore 1985; Kuei and Björnberg 2002); (e) Philichthyidae (Izawa 1973); (f) Ergasilidae (in part) (Ben Hassine 1983); *(cont. on next page)*

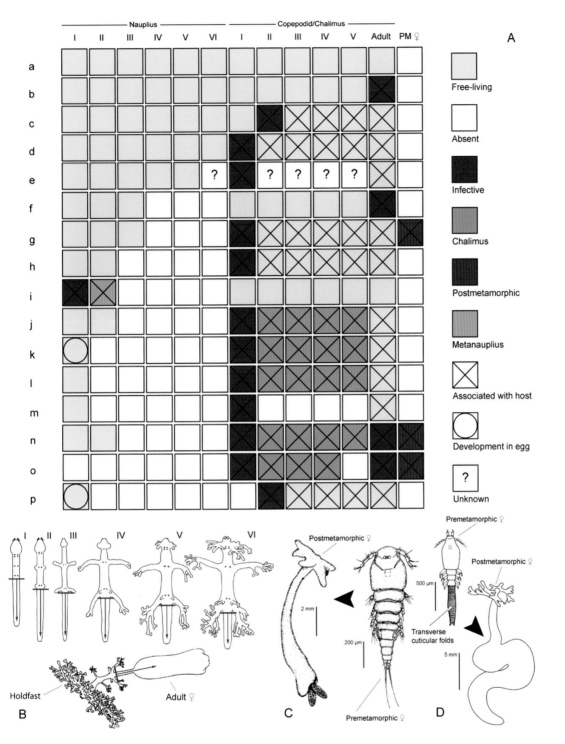

(g) Lernaeidae (Grabda 1963); (h) Lernanthropidae (Cabral et al. 1984); (i) Thaumatopsyllidae (Bresciani and Lützen 1962; Fosshagen 1970; Dojiri et al. 2008); (j) Caligidae (e.g., Kabata 1972; I.-H. Kim 1993; Ho and Lin 2004; Ohtsuka et al. 2009); (k) Lernaeopodidae, in part (e.g., *Salmincola californiensis*) (Kabata and Cousens 1973); (l) Lernaeopodidae, in part (e.g., *Alella macrotrachelus*) (Caillet 1979; Kawatow et al. 1980; Raibaut 1985); (m) Lernaeopodidae, in part (e.g., *Clavella adunca*) (Shotter 1971); (n) Pennellidae, in part (e.g., *Lernaeocera* spp., *Lernaeenicus sprattae*) (Scott and Scott 1913; Sproston 1942; Kabata 1958; Slinn 1970; T. Schram 1979); (o) Pennellidae, in part (e.g., *Cardiodectes medusaeus*) (P. Perkins 1983); (p) Chordeumiidae (*Parachordeumium amphiurae*) (Goudey-Perrière 1979). B: drawing of stages in the extensive metamorphosis of post-mated *Phrixocephalus cincinnatus* (Pennellidae), dorsal views; the first six stages are passed as the copepod traverses the eye of its flatfish host, and metamorphosis is completed when it penetrates the retina and becomes embedded in the choroid layer of the eye, where it develops an elaborate holdfast; the genital-abdominal region is marked off by a red transverse line and arrow. C: drawing of the metamorphosis of a post-mated female of *Lernaea cyprinacea* (Lernaeidae). D: drawing of the metamorphosis of a post-mated female of *Lernaeocera branchialis* (Pennellidae), dorsal views. A modified after Kabata (1981); B modified after Kabata (1969, 1979); C modified after Raibaut (1985); D modified after Boxshall (1992).

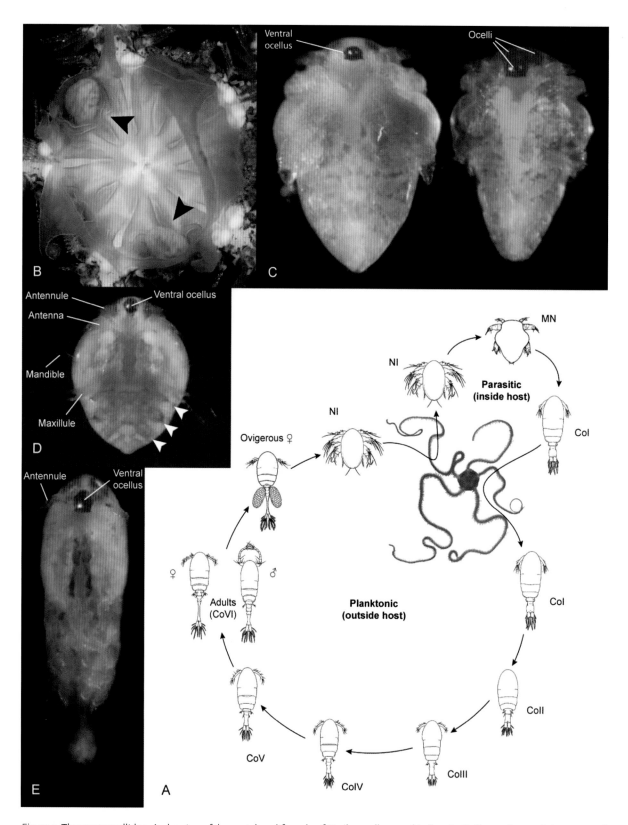

Fig. 27.7 Thaumatopsyllidae. A: drawing of the protelean life cycle of *Caribeopsyllus amphiodiae*. B: *Caribeopsyllus* sp. A (*sensu* Hendler and Kim 2010), with black arrows indicating two metanauplii in the stomach of *Ophiothrix angulata*, light microscopy, dorsal view. C: advanced metanauplii of *C. amphiodiae*, female (*left*) and male (*right*), showing the contrast between the sexes in the relative size of the nauplius eye's ventral ocellus, light microscopy, ventral view. D: metanauplius of *Caribeopsyllus* sp. A, with arrows showing the cephalic appendages and subcuticular primordia of the swimming legs, light microscopy, ventral view. E: copepodid I of *Caribeopsyllus* sp. A, light microscopy, ventral view. A modified after Dojiri et al. (2008); B, D, and E courtesy of Gordon Hendler; C courtesy of Gordon Hendler and Masahiro Dojiri.

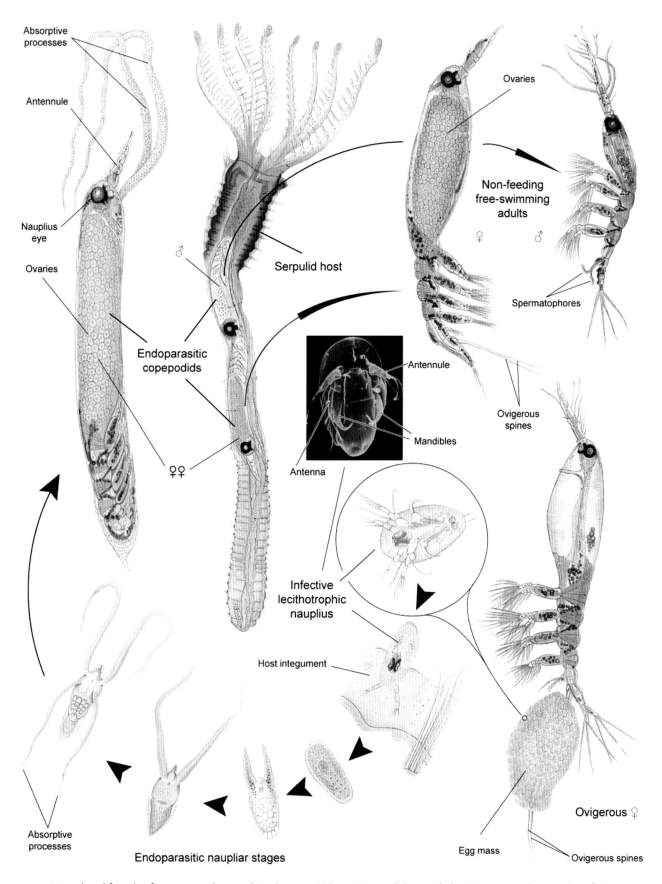

Absorptive processes

Antennule

Nauplius eye

Ovaries

♂

Endoparasitic copepodids

♀♀

Serpulid host

Ovaries

Non-feeding free-swimming adults

♀   ♂

Spermatophores

Ovigerous spines

Antennule

Mandibles

Antenna

Infective lecithotrophic nauplius

Host integument

Ovigerous ♀

Absorptive processes

Endoparasitic naupliar stages

Egg mass

Ovigerous spines

Fig. 27.8 Protelean life cycle of *Haemocera danae* (= ?*Cymbasoma rigidum*) (Monstrillidae), with the SEM (*center*) showing the infective nauplius of *Monstrilla hamatapex*. Drawing reconstructed from illustrations in Malaquin (1901); SEM modified after Grygier and Ohtsuka (1995).

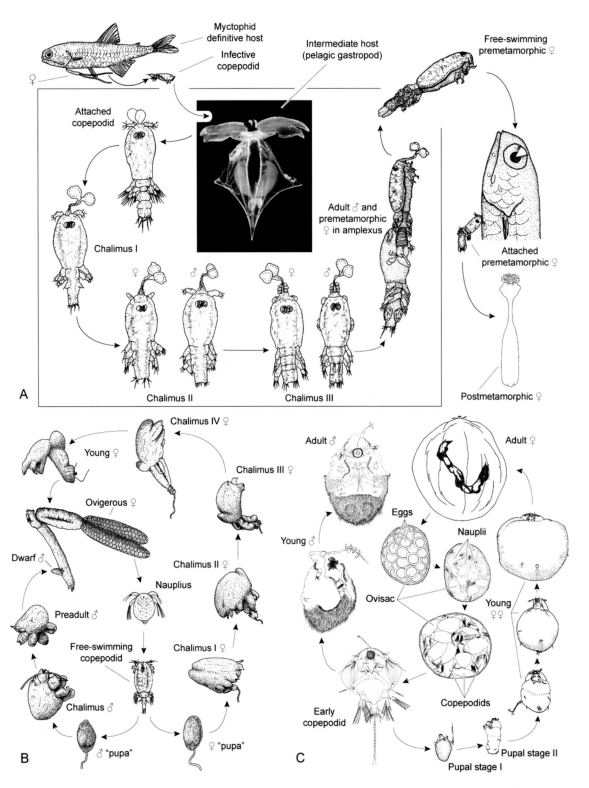

Fig. 27.9 Life cycles of parasitic copepods. A: drawing of the two-host life cycle of *Cardiodectes medusaeus* (Pennellidae), involving the thecosome gastropod intermediate host *Clio pyramidata* (photography, *center*) and the myctophid definitive host *Stenobrachius leucopsarus*. B: drawing of the life cycle of *Alella macrotrachelus* (Lernaeopodidae). C: drawing of the life cycle of *Hansenulus trebax* (Nicothoidae); a series of pupae and gradually advanced young females are illustrated at the same magnification as the adult female, showing the differences in body proportions caused by developing ovaries, which bring about the expansion of the trunk. A modified after P. Perkins (1983), with illustration of a post-metamorphic female modified after Boxshall and Halsey (2004) and photograph courtesy of Ron Gilmer and Richard Harbison; B modified after Caillet (1979), Kawatow et al. (1980), Raibaut (1985), and Benkirane (1987); C reconstructed from Heron and Damkaer (1986).

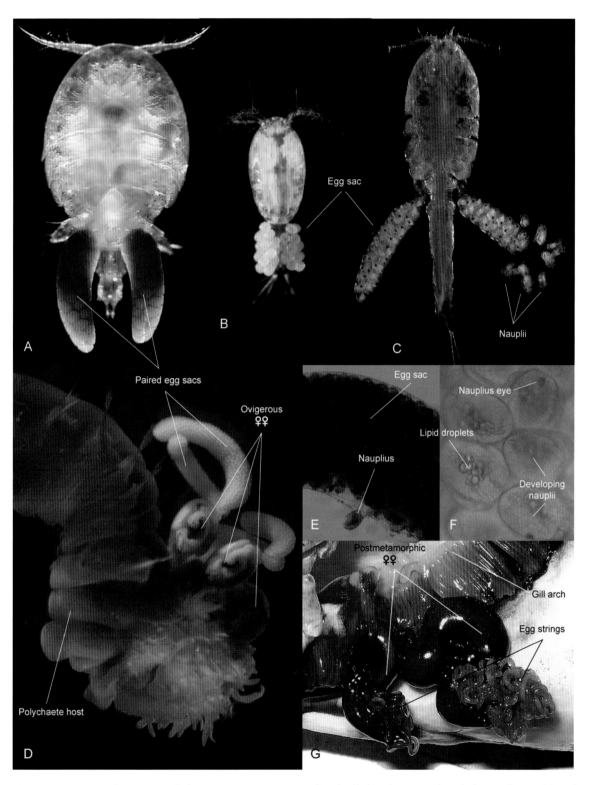

Fig. 27.10 Egg sacs and egg strings, light microscopy. A: ovigerous female of *Clausidium* sp. (Clausidiidae), with paired dorsolateral egg sacs, dorsal view. B: ovigerous female of *Cyclops* sp. (Cyclopidae), with paired laterodorsal egg sacs, dorsal view. C: ovigerous female of Lichomolgoidea sp., with paired lateral egg sacs, showing the eclosion of nauplii, dorsal view. D: three ovigerous females of *Melinnacheres steenstrupi* (Saccopsidae) attached to the gills of their terebellid host, *Terebellides stroemi*, lateral view. E: closeup of the egg sac of *M. steenstrupi*, showing the eclosion of the infective nauplius. F: closeup of eggs of *M. steenstrupi*, containing lecithotrophic nauplii at an advanced state of development. G: two post-metamorphic ovigerous females of *Lernaeocera branchialis* (Pennellidae), showing the spirally coiled uniseriate egg strings, attached to the gill arch of their gadid host, *Merlangius merlangus*, dorsal view. A and C courtesy of Arthur Anker; B courtesy of Jean-François Cart; D, E, and F original; G courtesy of Hans Hillewaert.

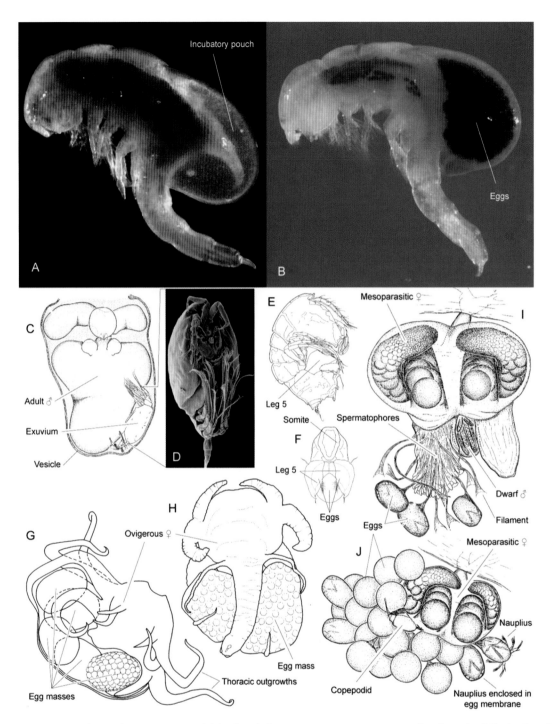

Fig. 27.11 A and B: *Pachypygus* sp. (Notodelphyidae), light microscopy. A: non-ovigerous female, with a fully developed incubatory pouch, lateral view. B: ovigerous female, with eggs retained inside the dorsal incubatory pouch, which is formed by the modification of the fourth pedigerous somite, lateral view. C: drawing of an adult male of *Nucellicola holmanae* (Chitonophilidae) enclosed in the membranous vesicle, together with the exuvium of the preceding copepodid, depicting a discrepancy in size as a result of hypermorphosis, ventral view. D: late copepodid (equivalent to CoIII) of *Lepetellicola brescianii* (Chitonophilidae), SEM, ventral view. E and F: drawings of *Parategastes sphaericus* (Tegastidae). E: adult female, lateral view. F: adult female, with foliaceous legs 5 forming a brood pouch shielding the eggs, ventral view. G: drawing of an ovigerous female of *Ophioika appendiculata* (Chordeumiidae), maintaining six egg masses in a subthoracic cage formed from the modified and ventrally downturned cephalic appendages and thoracic outgrowths, laterial view. H. drawing of an ovigerous female of *Parachordeumium amphiurae* (Chordeumiidae) holding a loose egg mass in the subthoracic brood cage, dorsal view. I: drawing of a mesoparasitic adult female of *L. brescianii* attached to a cocculiform host (Mollusca), showing an attached dwarf male, spermatophores, and eggs attached to the genital area via individual filaments, ventral view. J: ovigerous female of *L. brescianii* attached to a host, with offspring at different stages of development: eggs, a young nauplius enclosed in an egg membrane, a fully developed nauplius in the process of eclosion, and a copepodid I, ventral view. A and B courtesy of Arthur Anker; C modified after Huys et al. (2002); D, I, and J modified after Huys et al. (2002); E and F modified after Huys et al. (1996); G and H modified after Boxshall and Halsey (2004).

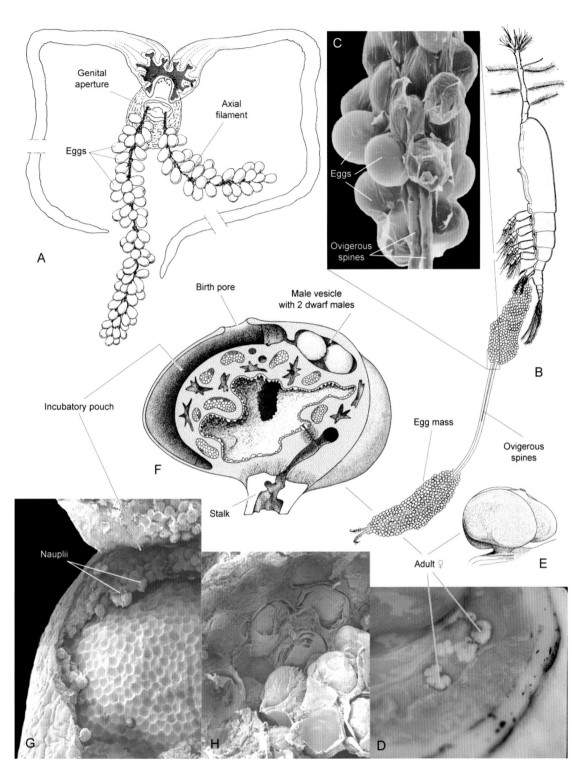

Fig. 27.12 A: drawing of an ovigerous female of *Phyllodicola petiti* (Phyllodicolidae) with eggs attached separately to the axial filament originating at the genital aperture, dorsal view. B: drawing of an ovigerous female of *Monstrilla longicornis* (Monstrillidae) with two egg masses,
lateral view. C: detail of an egg mass of *M. helgolandica* attached to ovigerous spines, SEM. D: females of *Pectenophilus ornatus* (Mytilicolidae) at various stages of development, attached to the gills of their bivalve host, *Patinopecten yessoensis*, light microscopy. E: drawing of
mature females of *Pectenophilus ornatus*, lateral view. F: drawing of the median section through an adult female *P. ornatus*, showing the
position of the incubatory pouch, birth pore, and dwarf males contained in the vesicle. G and H: *P. ornatus*, SEMs. G: adult female, with the
body wall partly removed to reveal the honeycomb-structured wall of the incubatory pouch. H: nauplii inside the brood pouch, at different
stages of eclosion. A modified after Laubier (1961); B and C modified after Huys and Boxshall (1991); D courtesy of Kazuya Nagasawa; E and
F modified after Nagasawa et al. (1988); G and H modified after Huys et al. (2006).

28

Robin J. Smith

Introduction to the Ostracoda

The Ostracoda are a diverse group of aquatic crustaceans, typically 0.3–5 mm long, although the predatory deep-sea *Gigantocypris* can reach lengths of more than 30 mm. The most distinctive feature of the Ostracoda is their calcitic carapace; it is a hard bivalved shell, hinged along the back, that can be closed to entirely cover and protect the non-mineralized body parts and appendages. The carapace lacks growth lines, as it is shed and regrown after each molt, and thus is distinguished from the bivalved carapaces of most other crustacean groups, such as the Cyclestherida and Spinicaudata.

The Ostracoda consist of 2 extant subclasses—the Myodocopa (with 2 orders: the Myodocopida and Halocyprida) and Podocopa (with 3 orders: the Platycopida, Podocopida, and Palaeocopida)—and extinct fossil taxa, such as the †Leperditicopida and †Metacopina. Embryonic and morphological studies have suggested that the 2 extant subclasses do not form a monophyletic group (e.g., Horne 2005; Wakayama 2007). Some molecular studies have supported this view, with, for example, the Podocopa showing closer affinities with the Branchiura than with the Myodocopa (Regier et al. 2005, 2008), although this was not the case in a later study (Regier et al. 2010).

Ostracods reproduce either parthenogenetically (in many freshwater species) or sexually. The Podocopa contain species with some of the longest spermatozoa known in the animal kingdom, reaching lengths up to 10 mm (for a review, see Wingstrand 1988), while some Myodocopa are famous for their bioluminescent displays at night, which are used to attract mates (Morin and Cohen 1991). Eggs are either brooded or released into the environment, and the eggs of some taxa can remain viable after desiccation.

Thanks to their hard mineralized carapaces that readily fossilize, ostracods are the most abundantly preserved arthropod in the fossil record; consequently, a large proportion of ostracodologists are engaged in studying fossil taxa. Traditionally, almost any fossil consisting of a bivalved hinged carapace lacking growth lines was considered an ostracod, but for the Cambrian †Bradoriida and †Phosphatocopida this is no longer the case; rare specimens with preserved appendages have revealed non-ostracod anatomies inside the carapaces (Hou et al. 1996; Siveter et al. 2003b; but also see Horne 2005). The Palaeozoic Palaeocopida are represented by thousands of extinct species and, reputedly, a handful of living species (e.g., Swanson 1991; Horne et al. 2002), although whether the extant species are really palaeocopids has been questioned (Siveter 2008). The extant Myodocopa have a very clear origin in the Silurian (Siveter et al. 2003a), and the extant Podocopa (excluding the Palaeocopida) have a fossil record extending back probably to the Early Ordovician (e.g., Siveter and Curry 1984). Fossilized ostracod larval carapaces and valves are common, but, as with the adults, fossil larval carapaces with preserved appendages inside are extremely rare (e.g., R. Smith 2000).

29 Ostracoda: Podocopa

ROBIN J. SMITH

GENERAL: The Podocopa is one of the 2 extant subclasses (the other being the Myodocopa) that form the Ostracoda, commonly called seed shrimps. The Podocopa is divided into 3 orders: the Platycopida, Podocopida, and Palaeocopida. Podocopans are small, typically 0.3–5 mm in length, and are characterized by a calcified bivalved carapace that entirely encloses the body. Many taxa have smooth carapaces, while others have ornamented carapaces with ridges, spines, pits, reticulations, and striations. The body typically sports eight pairs of appendages, plus the sexual organs, and has no clear division between the head, thorax, and abdomen. There are approximately 33,000 described species, 8,000 of which are living, and the 3 extant orders consist of 7 suborders and 10 superfamilies (Morin and Cohen 1991; Horne et al. 2002). They inhabit almost all aquatic environments, including the deep sea, coral reefs, brackish estuaries, beach deposits, rivers, lakes, springs, groundwater, and temporary puddles. Most podocopans are benthic, nektobenthic, or interstitial, but a small number of neustonic species, as well as species living semiterrestrially in the water films of wet leaf litter, are also known (Horne et al. 2004). The life cycle consists of seven or eight free-living larval/juvenile instars and the adult instar.

LARVAL TYPES

Instars (Including Nauplius): There are either eight or nine post-hatching stages, including the adult, that are known as instars (fig. 29.1), the first of which is considered to be the nauplius. All species show direct development, with all preadult instars consisting of a carapace that usually encloses the body, thus somewhat resembling the adult. No distinction is made between larvae and juveniles, and the stages are often referred to as the A–8 instar (i.e., the eighth stage before adult), A–7 instar, A–6 instar, and so forth.

MORPHOLOGY: The first instar typically consists of antennules, antennae, mandibles, and incipient caudal rami (the furcae of some authors), with the carapace totally enclosing the body (fig. 29.2A–D, G). The carapace is hinged along the dorsal margin in all instars of most groups, and it functions in a similar way to that of the adult, that is, opening along the anterior, ventral, and posterior margins to allow the limbs to be extended through the gap for locomotion and feeding (fig. 29.1E, H–J). The calcitic carapace is shed during each molt and regrown; because of this, it lacks growth lines. The caudal rami typically have an intermediate form in early instars, resembling a walking leg (figs. 29.1F; 29.2E, F). The fourth through seventh limbs are added in sequence, one pair at a time, during later molts. These limbs first appear as developmental buds and gain functionality during the subsequent molt. The sexual organs develop in the last two or three instars.

MORPHOLOGICAL DIVERSITY: While development is similar in many respects, there is some variation in the general morphology of instars among podocopan taxa. Podocopida and Platycopida species have a bivalved carapace in all instars, with the exception of the first instar of the podocopidan Darwinuloidea; the A-8 instars of the Darwinuloidea have a one-piece uncalcified flexible carapace covering the body (R. Smith and Kamiya 2008). The ontogenetic development of the Palaeocopida is incompletely studied, but the carapaces of the three smallest of the four known instars are remarkably different from those of the Podocopida and Platycopida, consisting of a shallow one-piece dorsal shield under which the body is suspended (fig. 29.2I) (Swanson 1989). The largest Palaeocopida larval instar that has been collected has a developed hinge along the dorsal margin, and the carapace (consisting of two valves) encloses the body, more closely resembling the carapaces of the other 2 orders.

The ontogeny of the appendages of the Podocopida is generally similar in all groups that have been studied, with differences restricted to the development of the fifth limbs, which change from walking legs into feeding appendages in some groups; the caudal rami, which appear late or become reduced

in some groups; and the maxillules of the commensal ento-
cytherids, which appear later, compared with other groups
(R. Smith and Martens 2000; R. Smith and Kamiya 2005, 2008).
Compared with the Podocopida, the Platycopida have a different ontogeny of the appendages, with the maxillules, fifth
limbs, and incipient sexual organs appearing earlier, and the
seventh limbs becoming reduced and eventually lost (Okada
et al. 2008). Without a complete range of Palaeocopida instars,
it is not clear how the development within this order differs
from that of the other 2 orders.

NATURAL HISTORY: Larval stages of ostracods are usually found in the same habitat as the adults. The first few
instars of nektobenthic species (many of the Cypridoidea) are
competent swimmers, despite the absence of the antennal
swimming setae, which appear later in the ontogeny. The
similar morphology of the feeding appendages of larval
stages with those of adults suggests similar feeding strategies, although for some species it has been noted that the
first and second stages are nourished by the remaining egg
yolk (Schreiber 1922; Kesling 1951). The Platycopida, Darwinuloidea, and some Cytheroidea brood eggs and the first few
instars in a posterior extension of the female carapace (e.g.,
Horne et al. 2002). Cytheroidean species are able to extrude
a sticky thread from the spinneret setae (exopodites) of the
antennae, and this has been observed in the first instar onward (fig. 29.2A). This is probably used to attach the larvae to
substrates, such as macrophytes in the marine intertidal zone,
to avoid being swept away by water movements (R. Smith
and Kamiya 2003).

PHYLOGENETIC SIGNIFICANCE: Within the Ostracoda, carapace features (notably hinges and pores) and appendages (especially antennules and antennae) of larvae have
been used to suggest phylogenetic relationships within genera,
families, and superfamilies (e.g., Tsukagoshi 1990; R. Smith
and Tsukagoshi 2005; R. Smith and Kamiya 2008). In a wider
context, ostracods, including podocopans, have long played an
important role in discussions of crustacean phylogeny, owing
in part to their extensive fossil record. At times they have been
placed among the "maxillopodan" groups, along with copepods, branchiurans, mystacocaridans, cirripedians, and others
(see the discussion in J. W. Martin and Davis 2001 and papers
cited therein), but more often they have been assigned to a
separate class (e.g., J. W. Martin and Davis 2001; Ahyong et al.
2011). Some evidence (both molecular and morphological)
suggests that the two major subclasses (the Podocopa and
Myodocopa) do not constitute a monophyletic assemblage

(Horne 2005; Wakayama 2007; Regier et al. 2005, 2008; but
also see Regier et al. 2010 for an opposing view). Because
of the unusual development of ostracods, with their unique
bivalved naupliar stages, larval characters have not featured
very prominently in these debates, other than to point out
the separate nature of ostracods from all other crustaceans.
Roessler (1998), however, interpreted an embryological stage
found in some podocopans as an additional naupliar stage (an
A-9 stage), which is completed within the egg. Unlike later
stages, it apparently lacks a bivalved carapace, and Roessler
(1998) suggested that it "makes the integration of podocopid
Ostracoda amongst other Crustacea easier."

HISTORICAL STUDIES: Historical studies documenting
the entire development of larvae of extant podocopans are
relatively few, and most have focused on the Cyprididae, such
as those by Claus (1868), G. Müller (1894), Schreiber (1922),
Kesling (1951), Fox (1964), Ghetti (1970), and Roessler (1983).
Scheerer-Ostermeyer (1940), Weygoldt (1960), and C. Hart
et al. (1985) studied the larval development of the Cytheroidea; Scheerer-Ostermeyer (1940) additionally studied that of
the Darwinuloidea. The only study on the Palaeocopida was
conducted by Swanson (1989). More recent work includes the
first studies on the larval development of the Platycopida and
Terrestricytheroidea. as well as more detailed descriptions
of the larvae of the Cytheroidea, Cypridoidea, and Darwinuloidea (R. Smith and Martens 2000; R. Smith and Kamiya
2002, 2003, 2005, 2008; Horne et al. 2004; Okada et al. 2008).

Selected References

Horne, D. J., R. J. Smith, J. E. Whittaker, and J. W. Murray. 2004.
    The first British record and a new species of the superfamily
    Terrestricytheroidea (Crustacea, Ostracoda): morphology,
    ontogeny, lifestyle and phylogeny. Zoological Journal of the
    Linnean Society 142: 253–288.
Okada, R., A. Tsukagoshi, R. J. Smith, and D. J. Horne. 2008. The
    ontogeny of the platycopid Keijcyoidea infralittoralis (Ostracoda:
    Podocopa). Zoological Journal of the Linnean Society 153:
    213–237.
Roessler, E. W. 1983. Estudios taxonómicos, ontogenéticos,
    ecológicos y etológicos sobre los ostrácodos de agua dulce
    en Colombia, 4: desarrollo postembrionario de Heterocypris
    bogotensis Roessler (Ostracoda, Podocopa, Cyprididae). Caldasia
    13: 755–776.
Smith, R. J., and T. Kamiya. 2008. The ontogeny of two species of
    Darwinuloidea (Ostracoda, Crustacea). Zoologischer Anzeiger
    247: 275–302.
Swanson, K. M. 1989. Manawa staceyi n. sp. (Punciidae, Ostracoda):
    soft anatomy and ontogeny. Courier Forschungsinstitut
    Senckenberg 113: 235–249.

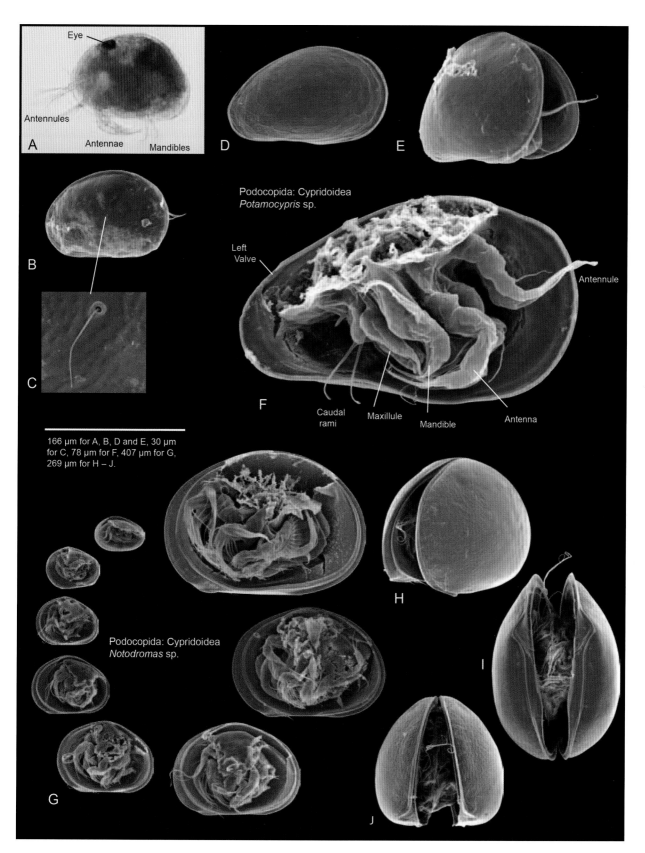

Eye

Antennules

A   Antennae   Mandibles

B

C

D

E

Podocopida: Cypridoidea
Potamocypris sp.

Left
Valve

Antennule

F   Caudal   Maxillule   Antenna
        rami   Mandible

166 μm for A, B, D and E, 30 μm
for C, 78 μm for F, 407 μm for G,
269 μm for H – J.

Podocopida: Cypridoidea
Notodromas sp.

H

G   I

J

Fig. 29.1 Cypridoidea (Podocopida), A: light microscopy, B–J: scanning electron microscopy. A–F: Instars of *Potamocypris*, undescribed species. A: A–8 instar, left lateral view. B: A–8 instar, right lateral view. C: detail of B. D: A–7 instar, right lateral view. E: A–6 instar, oblique frontal view. F: A–7 instar, with right valve removed. G–J: instars of *Notodromas*, undescribed species. G: instars A–8 through A–1, with left valve removed. H: A–3 instar, oblique frontal view. I: A–3 instar, ventral view. J: A–3 instar, frontal view. A–F original, of material from Kita Daito Jima, Japan; G–J original, of material from Kanazawa, Japan.

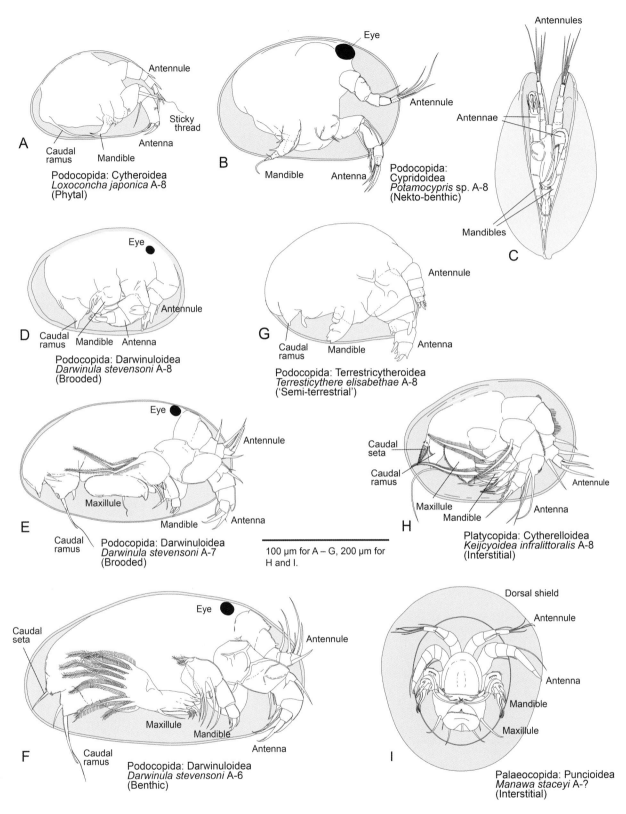

Fig. 29.2 Drawings of early larval stages of some podocopan ostracods. A, B, and D–H: right lateral views, with the right valve removed; for clarity, only one limb of each pair is shown; grey shading indicates the internal surface of the left valve. C and I: ventral views; the carapace is shaded grey. A modified after R. Smith and Kamiya (2003), reproduced with permission of Springer Science+Business Media; B and C original; D–F modified after R. Smith and Kamiya (2008), reproduced with permission of Elsevier; G modified after Horne et al. (2004), reproduced with permission of John Wiley and Sons; H reconstructed from drawings and SEM photographs in Okada et al. (2008); I reconstructed from SEM photographs in Swanson (1989).

30 | Ostracoda: Myodocopa

SHIN-ICHI HIRUTA
SHIMPEI HIRUTA

GENERAL: Myodocopan ostracods consist of 2 orders and 3 suborders—the Myodocopida, containing the Myodocopina; and the Halocyprida, containing the Halocypridina and Cladocopina (Horne et al. 2002)—with more than 600 extant species. Most Halocypridina species are holopelagic, and most Cladocopina species are benthic, with both groups living in shallow to abyssal waters. Most Myodocopida species also live from shallow to abyssal depths, mostly as benthos, although some are planktonic. Since information on the larval stages of the Myodocopina and Halocypridina is very limited (Tseng 1975; Kornicker and Sohn 1976; Kornicker and Iliffe 1989, 1995; Ikeda 1992), in this chapter we focus mostly on the post-embryonic development of the Myodocopina, a group that contains more than 400 extant species. The life cycle consists of four to seven larval stages, followed by the adult. All stages are free living.

LARVAL TYPES

Instars (Including Nauplius): Newly hatched myodocopid ostracods have a similar morphology to that of the adults, except for their smaller size, fewer appendages, and smaller number of bristles (figs. 30.1C1; 30.2A1, L1, R1, S1, T1). The free-living first larval stage, sometimes called the first instar and considered to be a nauplius, has a bivalved carapace that, when shut, encloses the entire soft body. The carapace is shed and regrown during every molt, whereas the body develops gradually during each molt (fig. 30.1A). Post-adult molting has not been confirmed (see Kornicker and Harrison-Nelson 1999). The total number of larval stages in the suborder Myodocopina is four to six: the Rutidermatidae and Sarsiellidae have four larval stages (figs. 30.2S, T); the Philomedidae has four or five (fig. 30.2L); the Cypridinidae has five or six (fig. 30.1C); and the Cylindroleberididae has four to six (figs. 30.2A, R). The suborder Halocypridina has six or seven larval stages (*Thaumatoconcha radiata* has six; *Conchoecia pseudodiscophora* and *C. elongata* have seven), and the Cladocopina (*Metapolycope duplex*) has six. Ostracod larval stages (instars) are designated

either as first instar, second instar, third instar, and so forth (in ascending order of size) or as A (= Adult), A–1 (= adult minus one), A-2, and so forth, in descending order.

MORPHOLOGY: Members of the Myodocopina hatch in almost the same developmental stage: the first instar has six developed pairs of appendages (antennules, antennae, mandibles, maxillules, fifth limbs, and furcae).

Instar Characters: In the early instars, members of the Myodocopina show almost the same development in the three characters listed below (fig. 30.1R). Although there are some exceptions, the process is remarkably consistent.

Occurrence of bristles on the fourth segment of the antennule: In the first larval stage, the fourth segment is without bristles. In the second stage, one dorsodistal bristle appears. In the next, one ventrodistal bristle is added. Although the third and fourth segments are fused in members of the Sarsiellidae and in *Rutiderma* species (Rutidermatidae), the bristles grow in the same way as described above (figs. 30.1E–G; 30.2B–D, U–W).

Sixth limb: In the first larval stage, the sixth limb is hirsute and without bristles. In the second stage, at least one bristle appears. Usually the limb has one anterior bristle, but some species have two anterior bristles, one anterior and one posterior bristle, or many anterior and ventral bristles. In the third stage, the limb develops into the adult form and is usually furnished with many bristles (figs. 30.1H–L; 30.2E–G, X–Z).

Seventh limb: In the first stage, the seventh limb either is not seen or is found as a very small bud-like process. In the second stage, all species exhibit this limb as a thumb-like process. In the third stage, the limb is elongate, but still bare. Bristles appear first at the fourth instar, so that a specimen with bristles on the limb is not younger than the fourth larval stage. In *Sarsiella* species, however, the seventh limb in males continues to show characteristics found in the first or second larval stages, even in the late larval and adult stages (figs. 30.1M–Q; 30.2H–K).

1 Fourth segment of the antennules without bristles; sixth limb without bristles ....Instar I
   Fourth segment of the antennules with bristle(s); sixth limb with bristle(s)..2
2 Fourth segment of the antennules with one dorsodistal bristle and without a ventrodistal bristle ....Instar II
   Fourth segment of the antennules with ventrodistal bristle(s) 3
3 Seventh limb without bristles* ....Instar III
Seventh limb with bristles ....Instar IV–Adult

*In species of *Sarsiella*, the limb in males continues to show the characteristics found in the first or second larval stages, even in the late larval and adult stages. For morphological differentiation of the appendages in the Halocypridina and Cladocopina, see Tseng (1975), Kornicker and Sohn (1976), Kornicker and Iliffe (1989, 1995), and Ikeda (1992).

Key: Above is a key to early myodocopid instars. A few species vary from the key, such as *Gigantocypris muelleri* (development of the antennule), *Paradoloria angulata* (antennule), *Amboleberis americana* (seventh limb), and *Cycloleberis christiei* (seventh limb) (Kornicker 1981; A. Cohen 1983; Hiruta 1983; Kornicker and Harrison-Nelson 1999), but it is nevertheless useful as a general guide.

It is sometimes difficult to distinguish between late larval stages and the adult, especially in females. The following characters may be useful for discriminating between them. (1) Carapace size. This is useful for distinguishing between late larvae and adults, and it is also useful in identifying instars (figs. 30.1C; 30.2A, L, R, S, T). The length of each instar is approximately 1.26 times the length of the preceding instar, although there may be overlap (Kornicker and Harrison-Nelson 1999). (2) Bristles of the seventh limb (both in number and in shape) (figs. 30.1P, Q, R; 30.2K). Instars later than instar III, as well as the adult, bear bristles on the seventh limb; earlier instars do not. The bristles in late larval stages are generally tapered. In the adult stage, the bristles have almost the same thickness along their entire length. (3) Endopodite of the antenna in the male (fig. 30.2M–P). In the male, which has a prehensile endopodite of the antenna, the morphology of the endopodite is often useful for the identification of stages if a sequence of instars is available for study. (4) Copulatory limb (male) and genitalia (female). The genitalia and copulatory limb appear in the late larval stages. They are not sclerotized, however, until the animal reaches maturity. (5) The presence of eggs in the brood pouch is useful for identifying adult females.

MORPHOLOGICAL DIVERSITY: The carapaces of the first through A–1 instars of both sexes in the Myodocopina resemble those of the adult females. The appendages, except for the sixth and seventh limbs, are well developed, even in the first larval stage, and are similar to those of the adult in general appearance. Accordingly, specimens in the first larval stage already possess the characters of each family to which they belong.

NATURAL HISTORY: Myodocopid ostracods reproduce sexually, although there are few observations of the mating process (A. Cohen and Morin 1990). Myodocopid larvae have a similar mode of life to that of the adults and are often found to-gether with adults. Some are planktonic, but most are benthic or epibenthic. Feeding styles include filter feeders, predators, scavengers, and detritus feeders. Females of the Myodocopina brood embryos within the posterodorsal carapace, releasing free-living instars. Most Halocyprida species release eggs directly into the sea. Halocypridina species are filter feeders, carnivores, and detritus feeders, whereas Cladocopina species are all carnivores.

PHYLOGENETIC SIGNIFICANCE: Because of the paucity of developmental studies on the Myodocopida, very little can be said concerning the phylogenetic significance of the larval stages. Some embryonic and morphological studies suggest that the 2 subclasses of extant ostracods are not a monophyletic group (e.g., Horne 2005; Wakayama 2007). This non-monophyletic view is also supported by molecular studies, where the Podocopa appears more closely allied to the Branchiura than to the Myodocopa (e.g., Regier et al. 2005, 2008), although this was not the case in a later study (Regier et al. 2010; also see chapters 28 and 29).

HISTORICAL STUDIES: Detailed studies of larvae of the Myodocopina were presented by Skogsberg (1920), Poulsen (1962, 1965), Kornicker (1969, 1981), and Hiruta (1977, 1978, 1979a, 1979b, 1980). Subsequent to those works, the total number of larval stages and the development of the carapace and appendages through ontogeny have been clarified by more recent investigations (A. Cohen 1983; Hiruta 1983; A. Cohen and Morin 1990; Kornicker 1992; Kornicker and Harrison-Nelson 1999, 2002).

Selected References

Cohen, A. C. 1983. Rearing and post-embryonic development of the Myodocopid ostracode *Skogsbergia lerneri* from coral reefs of Belize and the Bahamas. Journal of Crustacean Biology 3: 235–256.
Hiruta, S. 1983. Post-embryonic development of myodocopid Ostracoda. *In* R. F. Maddocks, ed. Applications of Ostracoda: Proceedings of the Eighth International Symposium on Ostracoda, Houston, Texas, July 26–29, 1982, 667–677. Houston: University of Houston–University Park, Department of Geosciences.
Horne, D. J., A. Cohen, and K. Martens. 2002. Taxonomy, morphology and biology of Quaternary and living Ostracoda. *In* J. A. Holmes and A. Chivas, eds. The Ostracoda: Applications

in Quaternary Research, 5–6. Geophysical Monograph No. 131. Washington, DC: American Geophysical Union.

Kornicker, L. S., and E. Harrison-Nelson. 1999. Eumeli Expeditions, Pt. 1: *Tetragonodon rex*, New Species, and General Reproductive Biology of the Myodocopina. Smithsonian Contributions to Zoology No. 602. Washington, DC: Smithsonian Institution Press.

Kornicker, L. S., and E. Harrison-Nelson. 2002. Ontogeny of *Rutiderma darbyi* (Crustacea: Ostracoda: Myodocopida: Rutidermatidae) and comparisons with other Myodocopina. Proceedings of the Biological Society of Washington 115: 426–471.

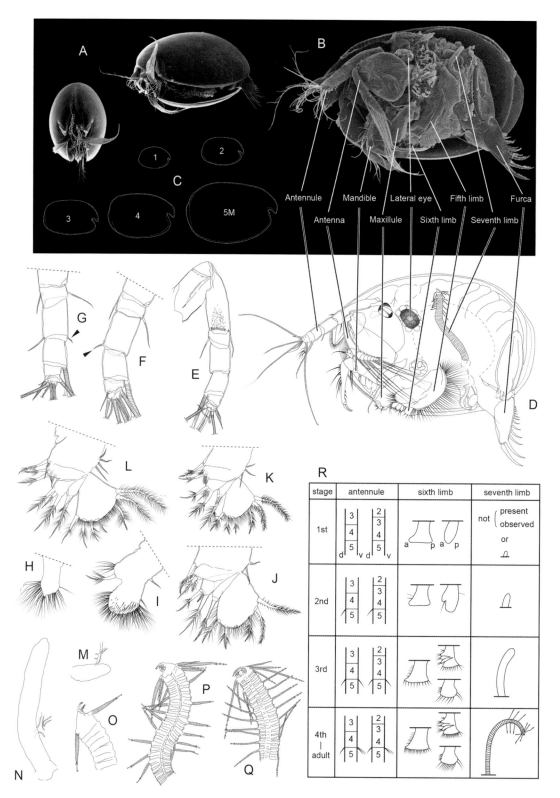

Fig. 30.1 Morphological differentiation of the carapace and appendages of *Vargula hilgendorfii* (Myodocopina: Cypridinidae), a famous bioluminescent species in Japan. A: fourth instar, anterior (*left*) and anteroventral (*right*) views, SEM. B: third instar, inside view (left valve removed), SEM. C: drawing of the first to fourth instars (numbered 1–4), sex undetermined, and fifth instar of a male (5M), lateral view of right side. D: drawing of the morphology of an adult male, lateral view. E–G: antennule of the first to third instars, with black arrows pointing to the bristles in the fourth articles in F and G. H–L: sixth limb of the first to fifth instars. M–P: seventh limb of the second to fifth instars. Q: seventh limb of an adult female. R: differentiation process of the antennule, sixth limb, and seventh limb in D (see further explanation in the text). Drawings of the appendages not to scale. A, B, and D original; C and E–Q modified after Hiruta (1980); R modified after Hiruta (1983).

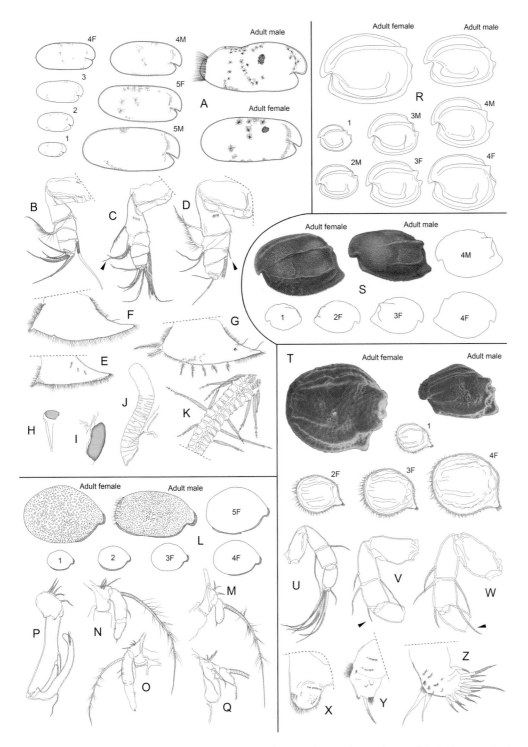

Fig. 30.2 Myodocopina (Cylindroleberididae, Philomedidae, Rutidermatidae, and Sarsiellidae), drawings (unless otherwise indicated). A–K: morphological differentiation of the carapace and appendages of *Bathyleberis yamadai* (Cylindroleberididae). A: first to fifth instars and adults; 1–3 = sex undetermined, 4F = female, 4M and 5M = male, lateral view. B–D: antennule of the first to third instars, with black arrows pointing to the bristles in C and D. E–G: sixth limb of the first to third instars. H–K: seventh limb of the first to fourth instars. L–Q: *Euphilomedes nipponica* (Philomedidae). L: first to fifth instars and adults, 1–2 = sex undetermined, 3F–5F = female, lateral view. M–Q: endopodite of the antenna. M–O: third to fifth instars. P: adult male. Q: adult female. R: *Asteropteron fuscum* (Cylindroleberididae), first to fourth instars and adults; 1 = sex undetermined, 2M–4M = male, 3F and 4F = female, lateral view. S: *Rutiderma darbyi* (Rutidermatidae), first to fourth instars and adults; 1 = sex undetermined, 2F–4F = female, 4M = male, SEMs of adults, lateral view. T–Z: *Sarsiella misakiensis* (Sarsiellidae). T: first to fourth instars and adults; 1 = sex undetermined, 2F–4F = female, SEMs of adults. U–W: antennule of the first to third instars, with black arrows pointing to the bristles in V and W. X–Z: sixth limb of the first to third instars. Drawings of the appendages not to scale. A–K and R modified after Hiruta (1979a); L–Q modified after Hiruta (1980); S with drawings modified after Kornicker and Harrison-Nelson (2002), and SEMs modified after Kornicker (1983); T–Z modified after Hiruta (1978).

31

JOEL W. MARTIN

Introduction to the Malacostraca

The crustacean class Malacostraca is an enormous assemblage that traditionally has been treated as consisting of 3 subclasses: the Phyllocarida (containing the relatively species-poor order Leptostraca, whose species are also known colloquially as sea fleas), the Hoplocarida (containing the entirely predatory stomatopods, or mantis shrimps, in the order Stomatopoda), and the Eumalacostraca, which contains the majority of all crustacean species within the megadiverse superorders Peracarida (with 9 extant orders) and Eucarida (containing the krill, decapods, and the monotypic Amphionidacea). Also included within the Eumalacostraca is the superorder Syncarida, containing the small orders Bathynellacea and Anaspidacea.

Malacostracans differ from other crustaceans in possessing 19 or 20 body somites, divided into a five-segmented cephalon, an eight-segmented thorax, and a six-segmented abdomen (pleon) (the Phyllocarida, with a seven-segmented abdomen, are the exception), all bearing distinctive appendages. The terminal somite of the pleon typically bears a telson, which, in combination with the uropods (the last pair of pleopods), forms a functional tail fan (also see Richter and Scholtz 2001).

Historically, most authors have agreed on the unity (mo-nophyly) of the Malacostraca, although some (e.g., F. Schram 1986) have excluded the Phyllocarida, and relationships within the Eumalacostraca have long been controversial. Some of the competing views of malacostracan phylogeny are found in the works of Siewing (1963), F. Schram (1981, 1984, 1986), Watling (1983, 1999), Mayrat and Saint Laurent (1996), F. Schram and Hof (1998), Wills (1998), Jarman et al. (2000), J. W. Martin and Davis (2001), Richter and Scholtz (2001), Spears et al. (2005), and Meland and Willassen (2007).

A few generalizations can be made for larval development within the Malacostraca (table 31.1). Phyllocarids are direct developers, as are many of the Eumalacostraca (including all of the Peracarida), whereas stomatopods and decapods have distinct and well-defined metamorphic larvae. Naupliar eclosion also occurs in some of the Eucarida (krill and dendrobranchiate shrimps), and these nauplii differ notably from those of most other crustaceans in being completely lecithotrophic (see Scholtz 2000). A few fossils of malacostracan larvae exist, which are almost exclusively fossils of stomatopod and decapod larvae from Jurassic and Cretaceous limestones, and these are described in chapter 32. In subsequent chapters, each of the above-mentioned groups is treated.

Table 31.1 Summary of larval development in the Malacostraca

Taxon	Direct	Metamorphic	Chapter(s)
Phyllocarida (leptostracans)	direct[1]		33
Hoplocarida (stomatopods)		antizoea, pseudozoea, erichthus, alima[2]	34
Eumalacostraca			
Syncarida (2 orders)	direct (Anaspidacea)	parazoea, bathynellid[3] (Bathynellacea)	35
Peracarida (9 orders, including amphipods, isopods, tanaids, cumaceans, and others)	direct[4]	manca, nauplioid, post-nauplioid, some others[4]	37–41
Eucarida			
Euphausiacea (krill)		nauplius, metanauplius, calyptopis, furcilia[5]	43
Amphionidacea	direct[6]		44
Decapoda (crabs, shrimps, lobsters, and their relatives)	some species	nauplius, zoea, decapodid[7]	46–54

[1] Embryonic and post-embryonic growth in phyllocarids, however, is often divided into two phases.

[2] Types and names of larvae vary according to their stomatopod lineage.

[3] Bathynellaceans are only slightly metamorphic at most; the parazoea has been called post-embryonic and the bathynellid phase is essentially a juvenile.

[4] Much variation exists in peracarid development, and a developmental stage called a manca occurs in many groups. Additionally, post-hatching stages are referred to as nauplioid and post-nauplioid larvae in the Mysida and Lophogastrica, and some developmental stages have received separate larval names in the Amphipoda (e.g., pantochelis, protopleon, physosoma) and in some parasitic isopod groups (e.g., praniza, zuphea).

[5] Development in krill proceeds along different pathways, depending on the type of egg brooding employed by the species in question.

[6] Although larval terminology has been applied in this group (e.g., the name amphion; see text), development is actually very gradual and without marked metamorphosis. See Kutschera et al. (2012).

[7] A nauplius phase occurs in the Decapoda only in the Dendrobranchiata (penaeoid and sergestoid shrimps). A variety of terms have been proposed for both the zoeal and the decapodid phases in different groups of decapods. Direct or abbreviated development also occurs in some decapods, such as crayfish and freshwater crabs, and some deep-sea species of other groups.

32

Fossil Malacostracan Larvae

Joachim T. Haug
Shane Ahyong
Carolin Haug

GENERAL: Mesozoic limestones—especially the Jurassic lithographic limestones of southern Germany (including the well-known Solnhofen deposits), but also the Cretaceous limestones of Lebanon and Brazil—are the only fossil lagerstätten that have yielded fossil malacostracan larvae to date. In the Solnhofen lithographic limestones, fossil palinurid larvae are extremely abundant and are known from thousands of specimens; stomatopod larvae are also known from the Solnhofen deposits.

LARVAL TYPES

Phyllosoma Larvae: Larvae of the Achelata (phyllosoma larvae; see chapter 51) are the most common among the fossil malacostracan larvae. This is quite astonishing, as these larvae appear to be very fragile. From the lithographic limestones of southern Germany, five different types of phyllosoma larvae can be distinguished. Type A (†*Phalangites priscus*; fig. 32.1E) is known from seven successive stages (Polz 1972), and type B (†*Palpipes cursor*; fig. 32.1D) is known from eight stages (Polz 1973). These two larval types are known from thousands of specimens and are most likely the early developmental stages of the 2 co-occurring species of †*Palinurina*: †*P. longipes*, and †*P. tenera* (Polz 1987; J. Haug et al. 2011b). Larva type C (†"*Dolichopus*" *tener*; fig. 32.1A–C) is much rarer, but here, too, several successive instars can be identified (Polz 1987). Type C could represent the larval stage of the supposed scyllarid †*Cancrinos claviger* (Polz 1996; J. Haug et al. 2009a, 2011b). Larva type D was described from a single specimen (Polz 1995), but comparable specimens are present in different collections (fig. 32.1F, G; pers. obs.). A description of a fifth type of phyllosoma larva was based on a single ill-preserved specimen and might be a representative of Scyllaridae *sensu stricto* (J. Haug et al. 2009a). Additionally, puerulus stages (the megalopa/decapodid of the Palinurida) are expected to be present in the Solnhofen lithographic limestones (Polz 1995), but they have not been identified as such with certainty.

The lithographic limestones from Lebanon are similar to those of southern Germany in many of their faunal components. Therefore it was not surprising that in 2009 the first phyllosoma larva from these deposits was reported (Pasini and Garassino 2009). In addition, these authors pointed out that more specimens are present in other collections, an observation that we can confirm (a second specimen was depicted by J. Haug et al. 2011b).

Isolated stalked compound eyes in a phosphatic Orsten-type preservation from limestone concretions from the Cretaceous Santana Formation of Brazil were described by G. Tanaka et al. (2009). These were also interpreted as fragments of phyllosoma larvae.

Zoea and Decapodid Larvae: Other fossil decapod larvae are extremely rare. Two specimens from the Santana Formation represent the only definitive fossil report of crab zoea larvae (Maisey and de Carvalho 1995). Fragmentary specimens from the Solnhofen lithographic limestones may also represent decapod zoea larvae (fig. 32.2K; J. Haug et al. 2011b), but their systematic affinities remain uncertain until better-preserved specimens are found. Possible megalopa/decapodid stages of the eryonid †*Cycleryon propinquus* may be represented by specimens that are usually identified as †"*Knebelia schuberti*" (Garassino and Schweigert 2006; J. Haug et al. 2011b).

Stomatopod Larvae: In addition to decapod larvae, larval stages of the Stomatopoda were found in the Solnhofen lithographic limestones. A large larva, possibly representing a new species, was reported by J. Haug et al. (2008) as the first definitive fossil stomatopod larva (fig. 32.2A–C). Two different larval stages of another stomatopod, †*Spinosculda ehrlichi*, are known (fig. 32.2D–I) (J. Haug et al. 2009b, 2010c). A small, rather poorly preserved specimen might be a larva of †*Sculda pennata*, the most common stomatopod in the Solnhofen deposits (fig. 32.2J; J. Haug et al. 2010c).

MORPHOLOGY: Only certain morphological aspects will be explained in this section, as the Mesozoic larvae are very similar to their extant relatives in their general morphology.

Phyllosoma Larvae: In the phyllosoma larvae, the antennules, antennae, and eyes are often preserved in their far anterior position (fig. 32.1A). The head shield, however, is usually not preserved, probably because most of these fossils are exuviae (but see the reconstruction in fig. 32.1C, based on a larva with a preserved shield). The phyllosoma larvae in general possess a very small pleon, except for larva D, which has a quite large pleon (fig. 32.1F). The fossil eyes of supposed phyllosoma larvae from the Santana Formation are preserved with squared facets (G. Tanaka et al. 2009). Yet the presence of squared facets draws the larval status of these specimens into question; at best they must represent very late larval stages, as earlier stages usually have hexagonal facets.

Zoea Larvae: The crab zoea larvae from the Santana Formation are known only from their eyes and head shield (carapace), which possesses a long rostrum (Maisey and de Carvalho 1995).

Stomatopod Larvae: One prominent character of fossil stomatopod larvae is their lanceolate uropodal exopods and endopods (fig. 32.2A, G, H; J. Haug et al. 2008, 2009b, 2010c). Another characteristic feature of extant stomatopod larvae, however—their long immobile rostrum—has been found in only one specimen to date, which probably represents a new species (fig. 32.2A, C; J. Haug et al. 2008). In other specimens the rostrum is usually not preserved. There are a few specimens that have lanceolate uropodal rami (a larval character), but also bear a short mobile rostrum (a post-larval character) (J. Haug et al. 2009b, 2011b). This combination of larval and post-larval characters may hint at the occurrence of an early megalopa *sensu* Villamar and Brusca (1988), which might also occur in the Solnhofen achelatans (J. Haug et al. 2009a).

MORPHOLOGICAL DIVERSITY: Fossil stomatopod larvae mainly differ from each other in the occurrence or lack of a long rostrum, but here preservational constraints probably had an impact. Species-specific characters, such as a pair of lateral spines on the sixth pleomere in †*Spinosculda ehrlichi* (J. Haug et al. 2009b), are already present in the larvae (fig. 32.2E, G, H); otherwise it probably would not be possible to assign these larval specimens to juveniles or adults of particular taxa. The single large stomatopod larva has, in addition to its long rostrum, posteriorly pointing spines on the shield, which are lacking in larvae of †*S. ehrlichi* (J. Haug et al. 2008).

Of the different types of fossil phyllosoma larvae, type D differs most from the other types. In general, it looks rather adult-like and is very large, but it still possesses prominent exopods at the thoracopods, which point to its larval status (fig. 32.1F, G) (Polz 1995). The phyllosoma larvae from Lebanon have a quite different morphology from those occurring in the Solnhofen deposits (pers. obs.), but more detailed investigations comparing larvae from these two deposits are needed.

NATURAL HISTORY: Based on the morphological similarities of phyllosoma and stomatopod larvae in the Mesozoic to extant larvae, their mode of life must have been very similar to that of larvae living today. The Mesozoic phyllosoma larvae were probably planktonic, with their large shield used for floating. The long rostrum and posterior spines on the shield in at least one fossil stomatopod larva also point to a planktonic mode of life (fig. 32.2A, C). Phyllosoma and stomatopod larvae were predators, which is clear from their raptorial appendages and claws (figs. 32.1B; 32.2A, C).

PHYLOGENETIC SIGNIFICANCE: The specimens described here are the only known Mesozoic fossils of malacostracan larvae. They are therefore important in establishing the presence of highly modified phyllosoma larvae (Decapoda) and stomatopod larvae in the Jurassic, and in allowing comparisons between extant and extinct malacostracan larvae.

HISTORICAL STUDIES: For a long time it was believed that larval stomatopods were present in Lebanon and in the Solnhofen lithographic limestones (Dames 1886; van Straelen 1938; H. Brooks 1969). It later became clear, however, that these supposed "stomatopods" were indeed members of the Thylacocephala, which are arthropods of uncertain affinities (e.g., Secretan 1985; Polz 1997; S. Lange et al. 2001).

The phyllosoma larvae were originally thought to be pantopod chelicerates (A. Müller 1962). The detailed studies of Polz (1971, 1972, 1973, 1984, 1987, 1995, 1996) corrected this view and identified the different types and stages, as well as the different kinds of embedding. Unfortunately, as some of these important works were published in German, these investigations have not been recognized by the scientific community in general.

Roger (1944) described a fossil from the Cretaceous of Lebanon, which he interpreted as an eryoneicus larva of a polychelid and described as a new species (†*Eryoneicus sahelalmae*). This interpretation was rejected by J. Haug et al. (2011b), and the specimen was reinterpreted as an immature achelate lobster (J. Haug et al. 2013a).

There are probably more fossil phyllosoma and megalopa/ decapodid types awaiting discovery and description. Apart from the possible zoea in poor preservation from the Solnhofen deposits (fig. 32.2K) (J. Haug et al. 2011b), the few larval findings from the Santana Formation and Lebanon suggest that more malacostracan larvae will be discovered there in the future (J. Haug and C. Haug 2013; J. Haug et al. 2013a).

Selected References

Haug, J. T., C. Haug, and M. Ehrlich. 2008. First fossil stomatopod larva (Arthropoda: Crustacea) and a new way of documenting Solnhofen fossils (Upper Jurassic, southern Germany). Palaeodiversity 1: 103–109.

Haug, J. T., C. Haug, D. Waloszek, and G. Schweigert. 2011b. The importance of lithographic limestones for revealing ontogenies in fossil crustaceans. Swiss Journal of Geosciences 104, Suppl. 1: S85–S98.

Polz, H. 1987. Zur Differenzierung der fossilen Phyllosomen (Crustacea, Decapoda) aus den Solnhofener Plattenkalken. Archaeopteryx 5: 23–32.

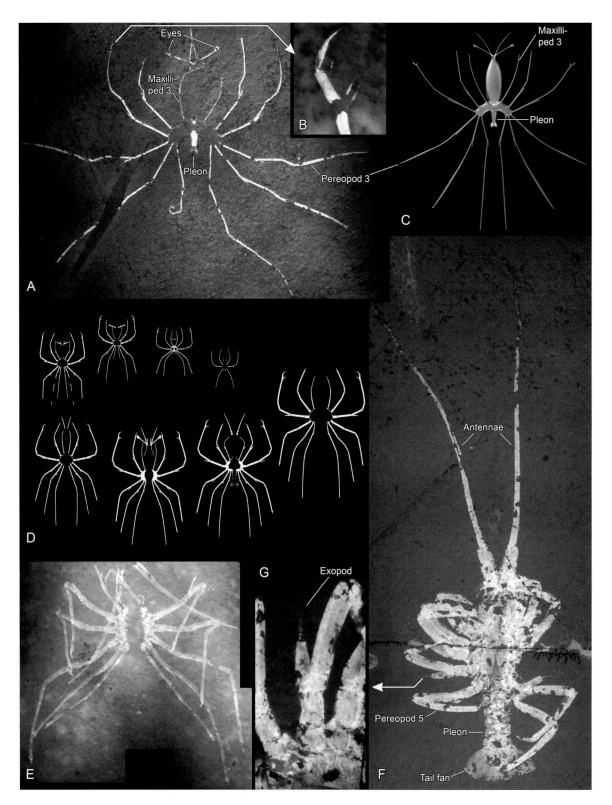

Fig. 32.1 Fossil phyllosomal larvae from the Solnhofen lithographic limestones. A–C: type C (†"*Dolichopus*" *tener*). A: photograph of an almost complete specimen, with the head shield missing; total length ca. 10 cm. B: closeup of the claw; note the preservation of the setae. C: 3D model, lateral view; the head shield is reconstructed from other specimens in which this structure is preserved. D: drawings of an ontogenetic series of type B (†*Palpipes cursor*); the smallest depicted stage is ca. 1.3 cm in total length, with largest depicted stage ca. 5.2. E: photograph of type A (†*Phalangites priscus*), with mainly the appendages preserved; total length is ca. 2.5 cm. F and G: photographs of type D. F: whole specimen, ventral view; total length is ca. 11 cm. G: detail of the thoracopod, with a multiannulated exopod (anterior to the left). A, B, and E–G specimens from the collection of Roger Frattigiani in Laichingen, Germany; C original; D modified after Polz (1973).

Fig. 32.2 Fossil stomatopod larvae from the Solnhofen lithographic limestones. A–C: giant yet-unnamed larva. A: photograph, lateral view; total length is ca. 18 mm; note the raptorial appendages, long immobile rostrum, and lanceolate uropodal rami (JME-SOS 8073). B: detail of the antennule; the single antennular elements are visible. C: 3D model of the specimen in A, lateral view. D–I: †*Spinosculda ehrlichi*. D: 3D model, lateral view. E: closeup of the spine, laterally on the sixth pleomere. F: closeup of the telson; note the lateral spines beginning to grow out (SMNS 67591). G: photograph of the holotype, dorsal view; the lanceolate uropodal rami and bifurcated telson indicate larval status; total length is ca. 9 mm (JME-SOS 8085). H: photograph of another specimen, dorsal view; total length is ca. 6.3 mm (SMNS 67634). I: telson of an early larval stage, with shorter posterior protrusions instead of spines, and probably lacking initial lateral spines. J: photograph of an assumed larva of †*Sculda pennata*; total length is ca. 5.1 mm. K: photograph of possible remains of a malacostracan larva; total length is ca. 4.5 mm (SMNS 67632); for similar, slightly better-preserved specimens, see J. Haug et al. (2011b). Abbreviations: JME = Jura-Museum Eichstätt; SMNS = Staatliches Museum für Naturkunde Stuttgart. A and B from JME; C and D original; E from the collection of Roger Frattigiani in Laichingen, Germany; F–H and K from SMNS; I from the collection of Marina and Matthias Wulf in Rödelsee, Germany (specimen 9203); J from the collection of Peter Rüdel in Gröbenzell, Germany (specimen 87133).

33

Leptostraca

Jørgen Olesen
Todd A. Haney
Joel W. Martin

GENERAL: The Leptostraca is a relatively small group (about 40 species) of marine malacostracan crustaceans, sometimes called sea fleas. Leptostracans have also been called hooded shrimps or thin-shelled shrimps (e.g., J. W. Martin and Haney 2009); *lepto* is Greek for "thin" or "delicate." Currently, the Leptostraca is divided among 3 families and 10 genera (Haney and Martin 2004), the most widespread and well-known genus being *Nebalia*. As in many other malacostracans, the body is divided into two limb-bearing regions, the thorax and the pleon. The head is exposed and has a unique movable rostrum covering a pair of stalked eyes. The carapace forms a laterally compressed shield that covers the thorax and is held together by an adductor muscle. Unique for leptostracans (among malacostracans), the thorax carries eight pairs of foliaceous serially similar limbs that are used for filtration and probably also for gas exchange (Cannon 1960). The seven-segmented pleon (telson excluded) is another special feature of the Leptostraca (maximally six segments in other malacostracans). Only the first six pleonites carry limbs (pleopods), of which only limbs 1–4 are biramous and relatively large. They are used for propelling the animal forward in a manner somewhat similar to that of harpacticoid copepods (noted by Sars 1896a and treated in more detail by Vannier et al. 1997). Most species live in shallow water and prefer muddy and/or organic substrates. They are reported to be detritivores or scavengers (e.g., J. W. Martin et al. 1996) on dead animal matter. A few species are known from the deep sea, and 2 species (*Nebaliopsis typica* and *Pseudonebaliopsis atlantica*) are pelagic (Walker-Smith and Poore 2001). One species (*Dahlella caldariensis*) is known from hydrothermal vents (Hessler 1984), and another species (*Speonebalia cannoni*) from marine caves (Bowman et al. 1985a). Most species are relatively small (5–15 mm in length), but *Nebaliopsis typica* is a giant (4–5 cm) (Brahm and Geiger 1966).

The life cycle consists of biphasic development within a brood chamber, followed by a hatching stage that is similar to the adult but lacking in setation and in the development of the fourth pair of pleopods.

LARVAL TYPES: Leptostracans do not have true larvae. The embryo, however, passes through several identifiable stages (separated by ecdyses) before leaving the brood chamber and becoming free swimming (Manton 1934, for *Nebalia bipes*), and these stages are sometimes called larvae. The stage that exits the mother's brood chamber looks like a small adult, but with less developed setation and only a rudiment of the fourth pair of pleopods (fig. 33.2). Therefore it is occasionally called a manca stage, but it is non-homologous to the manca of various peracarids. At this stage, the young can feed and move around freely in the water column, but the various appendages continue to develop their adult setation (Sars 1896a).

Phase 1: The development of *Nebalia* can be divided into two phases (based on Manton 1934). One phase, the embryonic phase, takes place within a vitelline membrane. It includes early germ-band formation and development of the caudal papilla, which is flexed ventrally, gradually becomes longer, and finally extends anteriorly (as far as the labrum) before the vitelline membrane disrupts (fig. 33.1A).

Phase 2: A second phase starts when the embryo has hatched from the vitelline membrane, and it lasts until the embryo is ready to leave the brood chamber. This phase is not embryonic in the strict sense, since development takes place outside the vitelline membrane (see Fritsch and Richter 2012 for definitions). In the second phase the caudal papilla is reflexed backward, the last body segments appear, and the appendages continue to develop. In *Nebalia bipes*, development passes through three stages, separated by ecdyses, before the embryo leaves the brood chamber and becomes free-swimming (Manton 1934). In *Nebaliopsis typica*, an aberrant large pelagic species, the presence of a ventral brood chamber was suggested by Linder (1943), based on preserved material. Cannon (1960), however, reported an embryonic-looking larva of *Nebaliopsis*, which he suggested was free-living. Later, Brahm and Geiger (1966) reported *Nebaliopsis* with eggs, developing under the carapace, that appeared to be contained in a "basket formed by the large and setose posterior pair of thoracic appendages that extend anteriad to the area of the mouth parts." They also reported that the embryos were shed when the specimens were placed in a fixative, so the "larva" reported by Cannon most likely was a released embryo.

MORPHOLOGY: The most detailed study of leptostracan development is that of Manton (1934) on *Nebalia bipes*, and much of the following summary is based on her work. Later supplements to Manton's work, using other techniques, include T. Williams (1998), Olesen and Walossek (2000), and Pabst and Scholtz (2009).

Phase 1: Phase 1 is the embryonic phase (development before hatching from the vitelline membrane). After the blastoderm cells have enclosed all the yolk, and the axes of the embryo have become apparent, the optic lobes and the three pairs of naupliar appendages make their appearance as early rudiments; later the labral rudiment appears as a transverse thickening between the antennae. Together, these structures are known as the germinal disc of the early embryo (see Manton 1934). The trunk region between the mandibles and the telson is formed by growth from teloblasts situated at the base of the Y-shaped germinal disc. All of the trunk segments are formed in series, in a backward direction (teloblastic growth). When the rudiments of the maxillules have appeared, a transverse furrow is formed behind them. As this furrow deepens, the part of the disc lying behind it folds backward and forms the caudal papilla. The caudal papilla elongates during development, forming more and more posterior segments. Finally, when the rudiments of all segments are differentiated internally, the caudal papilla extends anteriorly as far as the labrum (fig. 33.1A). At this stage the embryo is ready to hatch from the vitelline membrane.

Phase 2: Phase 2 involves development after hatching from the vitelline membrane. After leaving the vitelline membrane, the caudal papilla is reflexed backward and, at least in *Nebalia bipes*, development passes through three stages, separated by ecdyses, before the embryo leaves the brood chamber and becomes free-swimming (Manton 1934). Observations by Olesen and Walossek (2000) indicate some interspecific variation, so the number of free stages within the female's brood chamber is not necessarily the same in all species of *Nebalia* (or among the Leptostraca as a whole). During these stages, the yolk volume steadily decreases until it is all absorbed; the anterior globular part persists the longest.

What is presumably the first stage in *Nebalia longicornis*, after the vitelline membrane has burst, is shown in figure 33.1B–D. The caudal papilla is reflexed somewhat backward, but it is not yet entirely stretched out and still forms an angle with the anterior part of this embryo-like stage. The flexing zone is approximately in the region of the third pair of thoracopods (fig. 33.1B). All eight thorax segments are developed, but only two or three pleonites can be seen externally. A large telson, with a pair of undeveloped caudal rami, appears posteriorly. All cephalic appendages and all eight thoracopods are present as buds at different stages of development. The antennules, antennae, mandibles, and maxillules are all relatively long

uniramous laterally directed limb buds. A labral rudiment is present between the antennae. The maxillae and the thoracopods are small laterally directed biramous limbs, gradually becoming smaller posteriorly. The limb buds of pleopods are not yet present.

In a slightly later stage, the caudal papilla of *Nebalia longicornis* is almost entirely reflexed, and the pleon is even flexed slightly dorsally, foreshadowing the adult morphology (fig. 33.1E–G). Three pleonites are now clearly visible. The developmental stage of all limb buds is approximately the same as in the previously described stage. Despite the characteristic difference in morphology between adult thoracopods and pleopods, all limbs (except pleopods 4–6) are formed as laterally directed bifurcated limb buds that are quite similar in morphology (fig. 33.1H). A later stage of *Nebalia pugettensis* shows that the limb buds of pleopods 1–3 attain their vertical adult orientation before the thoracopods (fig. 33.1I). The same specimen shows the rudiment of a possible eighth pleon segment (fig. 33.1I).

A late stage of *Nebalia longicornis*, one that is just about ready to leave the brood chamber and become free-swimming, looks like a miniature version of the adult in many respects (fig. 33.2). A few of its morphological characteristics are highlighted here. Most yolk is absorbed, but some body parts still appear somewhat expanded (e.g., the thoracopods). The carapace is large, with lateral flanges covering most of the thoracopods laterally. The rostrum is large and hinged to the carapace. The eyes are large and cucumber-shaped. The antennules and antennae are large, and both have rudimentary adult-like segmentation. The maxillules have a long characteristic palp, extending posteriorly under the carapace. The maxillae have a two-segmented endopod, an unsegmented exopod, and endites, as in the adults. The thoracopods have the same components as in the adults (an endopod, exopod, and epipod), but setation along the inner margin of these limbs is rudimentary. There is a characteristic triangular protrusion between the thoracopods of each thoracomere, evolutionary aspects of which are discussed by Olesen and Walossek (2000). The pleon has seven segments. Pleomeres 1–3 carry well-developed biramous pleopods, with no sign of the protopod being subdivided into a coxa and basis. The fourth pair of pleopods (which, in adults, are similar in morphology to pleopods 1–3) are a pair of undeveloped buds. These are also retained after the embryo's release from the brood chamber, giving the first free-living stages a manca-like appearance. Pleopods 5 and 6 are small uniramous limbs, similar to those of the adults.

MORPHOLOGICAL DIVERSITY: Variation has been noted above.

NATURAL HISTORY: Leptostracans are dioecious. The larvae are brooded, and development takes place in a ventral brood chamber formed by the thoracopods and the carapace valves (but see *Nebaliopsis typica* below). The floor of the chamber is formed by the long setae of the thoracopodal endopods, which interlock with those of the neighboring and opposite thoracopods (Dahl 1985). The openings of the female gonopore are located within the brood chamber of the adult female, at the base of the eighth pair of thoracopods (Claus 1888), which is a typical position in malacostracans. In that work, Claus also reported that the male gonopores were in the "malacostracan position," which is in association with the sixth pair of thoracopods. Eggs change color during development, as noted for *Nebalia bipes* (Sars 1896a). There is much variation in the number of eggs carried, which seems, at least to some extent, to be species dependent, and to be generally lowest in tropical species. Clutches of 11 and 17 eggs were reported in a species of *Nebalia* (Lee and Morton 2005), of 2–16 eggs in *Paranebalia belizensis* (Modlin 1996), and of 27–88 eggs in *Nebalia hessleri* (Vetter 1996). Vetter also observed that in a laboratory culture of *Nebalia hessleri*, brooding females did not feed, probably because the presence of eggs interfered with the production of feeding currents. Manton (1934) reported peristaltic contractions of the embryonic gut both before and after hatching from the vitelline membrane, which is probably related to yolk absorption. Manton did not report any other types of movement (e.g., of limbs) while the embryos were situated in the brood chamber, neither before nor after hatching from the vitelline membrane. She also noted that hatching from this membrane seemed to be caused by a number of factors: (1) muscular movements of the body, (2) pressure caused by uneven growth of the embryonic tissue, and (3) pressure caused by the swelling of the yolk as it absorbs water.

PHYLOGENETIC SIGNIFICANCE: The Leptostraca has traditionally played an important role in discussions of crustacean phylogeny, especially with its status as either a primitive representative of the Malacostraca or as a "phyllopod" (branchiopod) being much debated (Milne-Edwards 1835; Claus 1872, 1888; Boas 1883; Sars 1885; Dahl 1984, 1987; F. Schram 1986; J. W. Martin and Christiansen 1996; J. W. Martin and Davis 2001). The current view is that the Leptostraca is the extant sister-group to the remaining malacostracans on both morphological and molecular grounds (Richter and Scholtz 2001; Jenner et al. 2009). Some aspects of leptostracan morphology, such as the presence of an extra seventh pleon segment and a pair of caudal rami on the telson, are often considered primitive. Leptostracans are grouped with a number of leptostracan-like fossils in the Phyllocarida, some of which may have been present already in the Cambrian, or at least in the Silurian (Rode and Lieberman 2002; Briggs et al. 2003). Developmental aspects of the Leptostraca have also traditionally played a major role in discussions of malacostracan and crustacean phylogeny. For example, many malacostracans have a stereotyped cell-division pattern in the post-naupliar germ band, including the presence of a ring of exactly 19 ecto-teloblasts in the caudal papilla, anterior to the telson (Scholtz 2000). Despite the absence of detailed investigations of the cell-division pattern of leptostracans, some evidence suggests that *Nebalia bipes* falls with this pattern (Manton 1934; Fischer and Scholtz 2010). It is well known that early development sometimes shows rudiments of structures that were present earlier in evolution but are entirely lost in adults. Olesen and Walossek (2000) reported an extra eighth pleomere (there are seven in adults) in a developing embryo of *Nebalia pugettensis*, which could represent an evolutionary rudiment. A similar example concerns the presence of possible rudimentary pleopodal epipods. Such epipods are absent in adults and late embryos of *Nebalia*, but Pabst and Scholtz (2009) located possible embryonic rudiments of epipods, indicating that they may have been present earlier in evolution. Much attention has been paid to the development of trunk limbs in *Nebalia*. As in other malacostracans, the leptostracan body is subdivided into two distinctly different body regions, a thorax and a pleon. Despite this division, Olesen and Walossek (2000) found that the early development of thoracopods 1–3 and pleopods 1–3 is similar, in the sense that they go through a phase as laterally directed bifurcate lobes (fig. 33.1H). This may indicate that the thorax and pleon in the Malacostraca have evolved from a homonomous limb series, with no tagmatization resembling that seen in many other crustaceans (e.g., branchiopods and cephalocarids), and that this series was only secondarily subdivided into a thorax and a pleon (Olesen and Walossek 2000; Pabst and Scholtz 2009).

HISTORICAL STUDIES: The habit in *Nebalia* of carrying its embryos or embryo-like stages in a brood chamber until they are adults was noted quite early (e.g., Krøyer 1847), but Metschnikoff (1868) was the first to treat the development of *Nebalia* in detail. He described all major phases of development and was, together with Claus (1872, 1888), among the first to argue for a malacostracan affinity for *Nebalia*. Other contributors include Robinson (1906), but the most comprehensive classical treatment of development in *Nebalia* remains that of Manton (1934).

Selected References

Manton, S. M. 1934. On the embryology of the crustacean *Nebalia bipes*. Philosophical Transactions of the Royal Society of London, B 223: 163–238.

Olesen, J., and D. Walossek. 2000. Limb ontogeny and trunk segmentation in *Nebalia* species. Zoomorphology 120: 47–64.

Pabst, T., and G. Scholtz. 2009. The development of phyllopodous limbs in Leptostraca and Branchiopoda. Journal of Crustacean Biology 29: 1–12.

Fig. 33.1 Embryos of *Nebalia* (Leptostraca), removed from the brood chamber of an ovigerous female, SEMs. A: embryo of *N. pugettensis*, immediately before hatching from the vitelline membrane. ventral view; note the well-developed cephalic limb buds and caudal papilla. B: embryo-like stage of *N. longicornis*, with a weakly reflexed caudal papilla, shortly after hatching from the vitelline membrane, lateral view. C: same as B, frontal view. D: same as B, ventral view. E: embryo-like stage of *N. longicornis*, with a fully reflexed caudal papilla, ventral view. F: same as E, lateral view. G: same as E, showing the limb buds, posterior view. H: caudal papilla of an embryonic stage of *N. pugettensis* (similar to the embryo in A), ventral view; note that the limb buds of the thoracopods and pleopods 1–3 are similar. I: later embryo-like stage of *N. pugettensis*, ventrolateral view; the limb buds of pleopod 1–3 have attained an upright position, different from that of the thoracopodal limb buds. A–I of material used in Olesen and Walossek (2000).

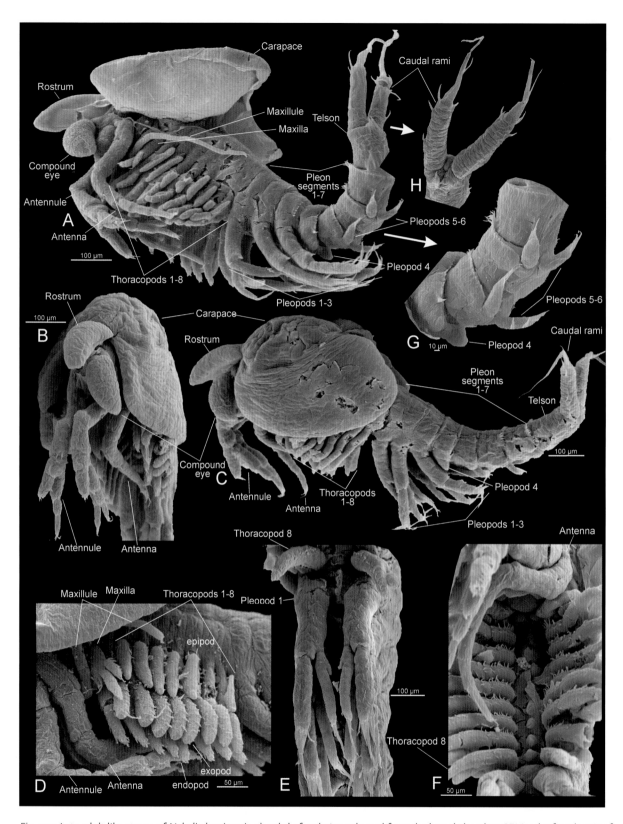

Fig. 33.2 Late adult-like stages of *Nebalia longicornis*, shortly before being released from the brood chamber, SEMs; the fourth pair of pleopods is not yet present. A: ventrolateral view, with most appendages exposed. B: rostrum, compound eyes, and antennules, frontal view. C: dorsolateral view, with the large carapace partly covering the thoracopods. D: thoracic region, ventrolateral view. E: first pairs of pleopods, anterior view. F: thoracic region, ventral view; note the rudimentary setation of the thoracopods and the sternitic triangular protrusions between the limbs. G: closeup of pleon segments 4–6, showing the undeveloped pleopod 4 and the small pleopods 5 and 6. H: closeup of the telson and the caudal rami, ventral view. A–H of material used in Olesen and Walossek (2000).

34 | Stomatopoda

SHANE T. AHYONG
JOACHIM T. HAUG
CAROLIN HAUG

GENERAL: Stomatopods (mantis shrimps) are well known as highly efficient predators. They are readily recognized by their triflagellate antennules, articulated rostrum, and subchelate maxillipeds, with the second developed as a massive raptorial claw. Almost 500 extant species are known, currently arrayed in 17 families and 7 superfamilies (F. Schram and Müller 2004; Ahyong et al. 2008, 2011). The Stomatopoda includes all extant representatives of the subclass Hoplocarida, with a fossil record that begins in the Palaeozoic (F. Schram 2007; J. Haug et al. 2010a). All known stomatopods are free-living, although some lysiosquilloids live in mated pairs. Most species live in shallow marine or estuarine habitats, but some occur on the outer continental shelf or slope.

The life cycle consists of a series of larval stages, all of which are free living, with development following one of three sequences: (1) antizoea, erichthus, post-larva, juvenile, adult (the Lysiosquilloidea); (2) pseudozoea, erichthus, post-larva, juvenile, adult (the Gonodactyloidea, Parasquilloidea, and Eurysquilloidea); and (3) pseudozoea, alima, post-larva, juvenile, adult (the Squilloidea).

LARVAL TYPES: Stomatopod larvae are among the most conspicuous crustacean larvae, not only because of their substantial size (up to 50 mm in length), but also because of their large functional raptorial maxilliped 2.

Antizoea, Pseudozoea: Among the two major hatching types, lysiosquilloids (and probably erythrosquilloids, owing to their shared maxilliped structure) hatch at an early stage of development, called an antizoea (fig. 34.1A, B), whereas larvae of other superfamilies (where known) hatch as pseudozoeae (figs. 34.1C; 34.2J).

Ericthus, Alima: Both antizoeae and pseudozoeae (except those of squilloids) develop into an erichthus (figs. 34.1F–H; 34.2B–F, K–Q). The squilloid pseudozoea (sometimes called a squillerichthus) develops into an alima (figs. 34.1D, E; 34.2G–I). The alima and erichthus proceed through post-larval and juvenile stages before becoming adults. For larval development in the superfamilies Bathysquilloidea and Erythrosquilloidea, only a single late pelagic bathysquilloid larva and erythrosquilloid post-larva have been identified (Manning 1991; Froglia 1992).

Differing larval types are defined for different stomatopod lineages, but the larvae of all lineages develop through several discrete ecological and functional stages. Early-stage larvae—typically the first two or three stages, termed propelagic stages (fig. 34.2J–L)—are lecithotrophic and live in the burrow of the female parent. Subsequent stages are pelagic (fig. 34.2M–Q) and capture prey using their raptorial claws. During the pelagic stages, thoracic (pereonal) and abdominal (pleonal) appendages develop progressively. Larvae having the full complement of thoracic appendages (five pairs of maxillipeds and three pairs of pereopods) and abdominal appendages (five pairs of pleopods and one pair of uropods) are at the synzoeal stage (e.g., fig. 34.1D–H).

Stomatopods typically have four to nine pelagic stages, although development in *Heterosquilla tricarinata* from New Zealand is abbreviated, with only one propelagic and two pelagic stages prior to the post-larva. As in natant decapods, the number of pelagic larval stages in at least some species of stomatopods is not constant, probably influenced by environmental conditions at the time of settlement (S. Morgan and Goy 1987). The molt from the last pelagic larva to the post-larva marks the transition from a planktonic to a benthic habit.

Stomatopod larvae are transparent, aside from the yellow-orange yolk of the propelagic stages, dark eye pigmentation at all stages, and minor pigmentation that may be present on maxillipeds 3–5 and mouthparts in the late pelagic stages. Significant body pigmentation is acquired prior to settlement at the post-larval stage.

MORPHOLOGY

Antizoeae: Antizoeae are the hatching larvae of lysiosquilloids, characterized by sessile eyes, uniflagellate antennules, five pairs of biramous thoracic appendages, the absence of abdominal appendages, and minimal segmentation. The carapace has a fixed spiniform rostrum, and slender posterolateral and posterodorsal spines; it is greatly enlarged and dorsally covers most of the abdomen, except for the posterior two or three somites and telson.

Pseudozoeae: Pseudozoeae are the hatching larvae of squilloids, gonodactyloids, parasquilloids, and probably eurysquilloids. Features include pedunculate eyes, biflagellate antennules, two pairs of uniramous thoracic appendages (maxillipeds 1 and 2), and four (squilloids) or five (other groups) pairs of natatory biramous pleopods. The carapace of pseudozoeae has similar spination to that in the antizoeae, but the former covers only the anterior thoracic somites, perhaps reflecting their later stage of development at hatching.

Erichthus: The erichthus has one or two (rarely three) intermediate denticles on the telson. The eyestalk is short, the antennal protopod stout, and the ventral surface of the antennular somite unarmed. The greatly enlarged maxilliped 2 is a functional raptorial claw, although additional teeth on the occlusal margin of the dactylus do not usually appear until the late pelagic or even first juvenile stages. Maxillipeds 3–5, pereopods, and uropods, all absent at hatching, first appear as buds and develop in size and form with progressive molts. Changes also occur in the larval eyes. The hemispherical and relatively unspecialized larval-type retina is gradually replaced during the late pelagic stages and in the post-larva by the specialized adult-type retina, which includes the specialized ommatidial midband containing the optical arrays and filter pigments of the adult (Cronin et al. 1995).

Alima: The alima is unique to squilloids and differs from the erichthus in having four or more intermediate denticles on the telson. The eyestalk is usually elongate, as is the antennal protopod. The antennular somite usually has a ventral spine. Development of the thoracic and abdominal appendages is similar to that of erichthus larvae. As in the erichthus, the adult-type retina develops adjacent to the larval retina in late pelagic larvae.

Post-Larva: Significant morphological changes occur between the last pelagic stage larva and the post-larva, which assumes a more or less adult form. Most significantly, the expansive larval carapace is reduced to a short shield-like form, with an articulated rather than a fixed rostrum. In squilloids, the larval-type retina is fully replaced by the adult-type retina at the molt to the post-larva. In other groups, remnants of the larval-type retina are usually still visible adjacent to the adult-type retina in the post-larva, but these remnants are completely lost by the first juvenile stage.

MORPHOLOGICAL DIVERSITY: Stomatopod larvae, especially in the late pelagic stages, may differ considerably among families and sometimes genera. The erichthus larvae of lysiosquilloids frequently have a proportionally higher and deeper carapace than that of other groups, as well as relatively wide abdominal somites. The pelagic larvae of other groups usually have a flatter carapace, and in squilloids the abdomen is typically slender. More specific morphological differences are in the form of the uropods, the telson, and the carapace. In particular, the spination of the carapace margins, the ventral spination of the rostrum and posterolateral spines, and the shape of the carapace can differ among species, genera, and families. Development of several structures, such as the mandibular palp and male penes, also varies among species; these features may be already evident as buds in the late pelagic larvae of some species, but in other species they may first appear only in post-larvae or juveniles. Likewise, differentiation between abdominal somite 6 and the telson varies among groups, as does differentiation of the distal and proximal uropodal exopod segments.

NATURAL HISTORY: Late-stage stomatopod larvae can be abundant in the plankton. At low latitudes, pelagic larvae are common year round, given the continuous reproductive cycle of tropical species. At high latitudes, larval densities may be seasonal, following annual reproductive cycles. *Heterosquilla tricarinata*, for instance, a temperate to subantarctic species from New Zealand, carries only a single brood per year, and its larvae appear in the plankton in spring or early summer, when planktonic food is most abundant (B. Williams et al. 1985).

Larval behavior has not been widely studied, but, in general, propelagic stages are positively thigmokinetic (clinging closely to the substrate) and negatively phototactic (avoiding light) (Dingle 1969). These behaviors may be adaptations that help retain the lecithotrophic propelagic stages within the parental burrow, during which time they are defended by the adult. At their transition to the feeding pelagic stage, however, larvae become positively phototactic. In common with many other zooplankters, stomatopod larvae vertically migrate at night to feed (Reaka 1986), when they can be collected in large numbers at light traps.

As with adults, larval stomatopods are aggressive predators. In captivity they readily prey on *Artemia* nauplii or other suitably sized zooplankton, and they have even been observed to attack and consume other stomatopod larvae (Alikunhi 1952). Suffice it to say that they use their raptorial second maxillipeds to great effect well before reaching adulthood.

The length of the larval life cycle is not known for most stomatopods, but it is probably shaped by both phylogenetic history and immediate environmental conditions, such as water temperature and available settling substrate. Larval duration is about 35 days in the tropical species *Neogonodactylus oerstedii* (Gonodactyloidea) (Provenzano and Manning 1978), about six weeks in the temperate *Squilla empusa* (Squilloidea) (S. Morgan and Provenzano 1979), one to two months in the temperate *Oratosquilla oratoria* (Squilloidea) (Hamano and Matsuura 1987), and almost nine months in the subantarctic *Pterygosquilla schizodontia* (Squilloidea) (Pyne 1972), each of which have full larval development. The abbreviated development of the temperate-subantarctic *Heterosquilla tricarinata* (Lysiosquilloidea) spans about 60–70 days (Greenwood and Williams 1984).

PHYLOGENETIC SIGNIFICANCE: Stomatopod larvae have been used often in attempts to elucidate their in-group phylogeny and to establish the relationship of stomatopods to other crustacean groups. The distinctive nature of the larvae, in conjunction with unique characters of the adults, confirm the monophyly of the group.

HISTORICAL STUDIES: Owing to the striking form and large size of stomatopod larvae, many past workers failed to recognize them as larval forms, instead naming them as separate genera and species, such as *Lysioerichthus*, *Pseuderichthus*, *Squillerichthus*, and *Smerdis*. To date, not all stomatopod larvae can be linked to known adults, so to minimize taxonomic confusion, the International Commission on Zoological Nomenclature suppressed most of these larval names as formal taxonomic names, although they are sometimes used nowadays as informal category names for various larval types. An exception is *Alima*, used as a formal genus name as well as the name for a squilloid larval type, the alima.

Several early workers recognized the fundamental larval types—the initial antizoea and pseudozoea, and later the erichthus and alima types (Claus 1871; W. Brooks 1886; Hansen 1895, 1926; Foxon 1932)—but it was Giesbrecht (1910) who formally recognized these larval differences as major taxonomic distinctions that now form the basis of the modern superfamilies Squilloidea, Gonodactyloidea, and Lysiosquilloidea (Manning 1980, 1995; Ahyong and Harling 2000; Ahyong 2001). W. Brooks (1886) attempted to infer phylogenetic relationships using larval developmental polarity. *Protosquilla* (Gonodactyloidea: Protosquillidae), having a fused abdominal somite 6 and telson in adults, was identified as primitive, because of the undifferentiated somite 6 and telson seen in many late pelagic larvae. Hansen (1895), however, showed that the fused abdominal somite 6 and telson of adult *Protosquilla* is not comparable with the undifferentiated condition of larvae. Far from being primitive, recent phylogenetic analyses of the Stomatopoda show that protosquillids are highly derived (Ahyong and Harling 2000; Ahyong and Jarman 2009; Porter et al. 2010). The early stage of development of antizoeae could be taken to imply that lysiosquilloids are basal stomatopods. Phylogenetic analyses, however, indicate that the plesiomor-

phic larval condition is closest to that of the gonodactyloid pseudozoea (Ahyong and Harling 2000).

Complete larval sequences are known for only a few species of stomatopods, including *Neogonodactylus oerstedii* (Manning and Provenzano 1963; Provenzano and Manning 1978), *Neogonodactylus wenneri* (S. Morgan and Goy 1987, as *Gonodactylus bredini*), *Heterosquilla tricarinata* (Greenwood and Williams 1984), *Oratosquilla oratoria* (Hamano and Matsuura 1987), and *Pterygosquilla schizodontia* (Pyne 1972, as *Squilla armata*). Partial larval series, however, are known for many species that have been hatched from eggs or traced through to post-larvae and juveniles from pelagic larvae (e.g., Gurney 1946; Alikunhi 1952; Townsley 1953; Alikunhi 1967; Michel 1969, 1970a; Michel and Manning 1972; S. Rodrigues and Manning 1992; Ahyong 2002; Feller et al. 2013). Most recently, fossil stomatopod larvae attributed to the extinct †Sculdidae have been identified from the Jurassic Solnhofen deposits (J. Haug et al. 2008, 2010c; C. Haug et al. 2009; also see chapter 32). Clearly, much remains to be learned about stomatopod development, past and present.

Selected References

Alikunhi, K. H. 1967. An account of the post-larval development, moulting and growth of the common stomatopods of the Madras coast. *In* Proceedings of the Symposium on Crustacea, Held at Ernakulam from January 12–15, 1965, pt. 2: 824–939. Mandapam Camp: Marine Biological Association of India.

Feller, K. D., T. W. Cronin, S. T. Ahyong, and M. L. Porter. 2013. Morphological and molecular description of the late-stage larvae of *Alima* Leach, 1817 (Crustacea: Stomatopoda) from Lizard Island, Australia. Zootaxa 3722: 22–32.

Greenwood, J. G., and B. G. Williams. 1984. Larval and early post-larval stages in the abbreviated development of *Heterosquilla tricarinata* (Claus, 1871) (Crustacea, Stomatopoda). Journal of Plankton Research 6: 615–635.

Hamano, T., and S. Matsuura. 1987. Egg size, duration of incubation, and larval development of the Japanese mantis shrimp in the laboratory. Nippon Suisan Gakkaishi 53: 23–39.

Morgan, S. G., and J. G. Goy 1987. Reproduction and larval development of the mantis shrimp *Gonodactylus bredini* (Crustacea: Stomatopoda) maintained in the laboratory. Journal of Crustacean Biology 7: 595–618.

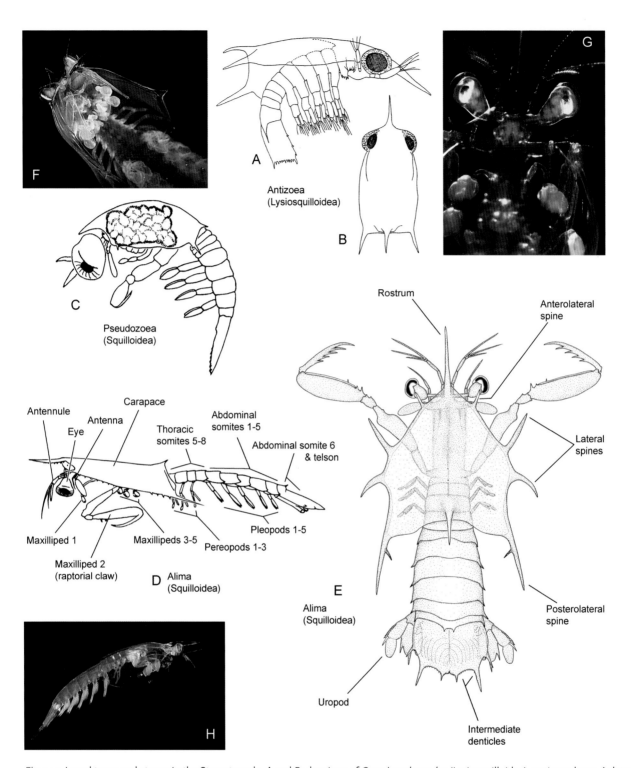

Fig. 34.1 Larval types and stages in the Stomatopoda. A and B: drawings of *Coronis scolopendra* (Lysiosquilloidea), antizoea larva. A: lateral view. B: carapace, dorsal view. C and D: drawings of *Oratosquilla oratoria* (Squilloidea). C: pseudozoea larva, lateral view. D: alima larva, synzoeal stage, lateral view. E: drawing of an alima larva, synzoeal stage, of *Anchisquilloidea mcneilli* (Squilloidea), dorsal view. F and G: erichthus larvae of the Lysiosquilloidea, synzoeal stage, light microscopy. H: erichthus larva of the Gonodactyloidea, synzoeal stage, lateral view, light microscopy. A and B modified after S. Rodrigues and Manning (1992); C and D modified after Hamano and Matsuura (1987); E original; F–H courtesy of Arthur Anker.

Fig. 34.2 (*opposite*) Drawings showing examples of larval diversity in the Stomatopoda. A: (*left*) late pelagic larva of *Bathysquilla crassispinosa* (Bathysquillidae: Bathysquilloidea), with pereopods and pleopods omitted, lateral view; (*right*) carapace, dorsal view. B: (*left*) erichthus larva of *Odontodactylus* sp. (Odontodactylidae: Gonodactyloidea), lateral view; (*right*) tail fan, dorsal view. C: (*right*) erichthus (*cont. on next page*)

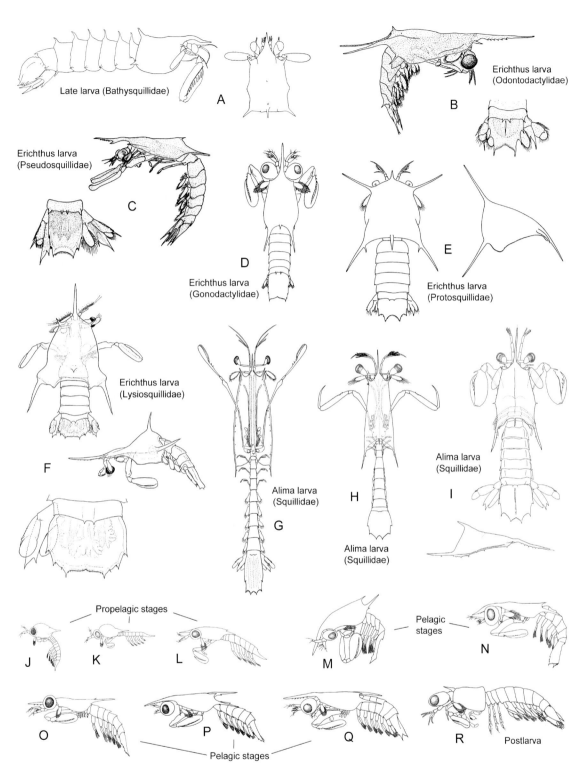

larva of *Pseudosquilla ciliata* (Pseudosquillidae: Gonodactyloidea), lateral view; (*left*) tail fan, dorsal view. D: erichthus larva of *Neogono-dactylus wenneri* (Gonodactylidae: Gonodactyloidea), dorsal view. E: (*left*) erichthus larva of *Chorisquilla tuberculata* (Protosquillidae), dorsal view; (*right*) carapace, lateral view. F: (*top*) erichthus larva of *Lysiosquilla* sp. (Lysiosquillidae: Lysiosquilloidea), dorsal view; (*center*) lateral view; and (*bottom*) telson. G: alima larva of *Alima neptuni* (Squillidae: Squilloidea), dorsal view. H: alima larva of *Oratosquilla oratoria* (Squil-lidae: Squilloidea), dorsal view. I: (*top*) alima larva of *Neoanchisquilla tuberculata* (Squillidae: Squilloidea), dorsal view; (*bottom*) carapace, lateral view. J–R: developmental stages of *Neogonodactylus wenneri* (Gonodactylidae: Gonodactyloidea), lateral views. A modified after Manning (1991), B and C modified after Townsley (1953); D modified after S. Morgan and Goy (1987); E modified after Michel and Manning (1972); F modified after Gamô (1979); G modified after Manning 1962); H modified after Komai and Tung (1929); I original; J–R modified after S. Morgan and Goy (1987).

35 Syncarida

HORST KURT SCHMINKE

GENERAL: Syncarids are small, presumably primitive crustaceans found mostly in subterranean waters. There are two orders of syncarids: the Anaspidacea and Bathynellacea. The Anaspidacea are confined to the Southern Hemisphere. In Australia they live in surface and subterranean waters, while in New Zealand and southern South America they are all subterranean. There are 5 families: the Anaspididae (5 species), Koonungidae (3 species), Psammaspididae (2 species), Patagonaspididae (1 species) and Stygocarididae (10 species). The Bathynellacea, which have a worldwide distribution and are found only in subterranean waters, contain 2 families: the Bathynellidae (90 species) and Parabathynellidae (186 species). Diversity in the Bathynellacea is very likely underappreciated (e.g., Camacho et al. 2012).

The life cycle differs between the 2 orders. In the Bathynellacea, it consists of a nauplioid, a parazoea, a bathynellid, and an adult. All except the nauplioid phase are free living, but the parazoeal phase can also be passed through in the egg. In the Anaspidacea, the hatchlings emerge from the egg as juveniles; other phases are passed through in the egg.

LARVAL TYPES

Nauplioid, Parazoea, and Bathynellid (Bathynellacea): According to Schminke (1981), the development of *Antrobathynella stammeri* passes through three phases (fig. 35.1A–C), termed a nauplioid, a parazoea (called a "post-embryonic phase" by Jakobi 1954) and a bathynellid (called a "pre-adult–phase" by Jakobi 1954). The nauplioid is a transitional embryonic phase passed through in the egg. There is no free-swimming nauplius. The parazoea is a larval phase, with its stages having particular larval characters. The bathynellid is a juvenile phase, with its stages not differing fundamentally in body structure from the adult, except for some partially developed or missing structures.

Direct Development (Anaspidacea): In the Anaspidacea, *Anaspides tasmaniae* has direct development (fig. 35.1D–F),

hatching from the egg as a young adult, which, however, has kept a few remnants of a larval organization (sessile eyes, the lack of a mandibular palp, and a notched telson). There are no free-living larval forms, but egg development passes through stages that appear to have counterparts in other crustacean groups.

MORPHOLOGY

Bathynellacea: The nauplioid (embryonic) phase is characterized by the lobe-like anlagen of the antennules, antennae, and mandibles. The parazoeal (larval) phase begins with hatching from the egg at a stage where only a single pair of thoracopods is functional. The subsequent two pairs of thoracopods appear in a stereotyped sequence, from front to rear. The main locomotory organs during this phase are the antennae, aided by the thoracopods as they develop. Segmentation of the thorax is complete from the beginning, while that of the pleon is only completed at the end of this phase. The parazoeal phase has three stages and ends when three pairs of thoracopods are functional. The parazoeal phase is a larval one, because the structure of its stages includes typical larval characters that are absent in the adult. The antennae are an example of one of these characters. After hatching from the egg, the exopods are longer than the unsegmented endopods. By their annulation and the many setae that they possess, the exopods give the impression of once having been composed of several segments. With the beginning of the bathynellid phase, the exopods start to degenerate by losing their annulations and several of their setae, thus becoming shorter than the endopods, which start to become segmented and elongate.

The stage with four thoracopods is the beginning of the bathynellid (juvenile) phase, because it resembles the adults in many respects. The body has the full complement of somites right from the beginning. Locomotion is accomplished solely by the thoracopods. Those still to be developed do not appear in a strictly stereotyped sequence, as in the preceding (parazoeal) phase, but instead appear in a variable sequence, allowing several possible combinations in one and the same species. It is therefore not known how many stages the bathynellid phase includes, but it seems that adulthood in females is reached after four stages, and in males after five.

There are reports (e.g., Miura and Morimoto 1953) of species hatching from the egg with four pairs of functional thoracopods (e.g., *Nihobathynella morimotoi*). These species have no free larval phase, because not only the nauplioid, but also the parazoea, are passed through in the egg. The same also applies to all species of Parabathynellidae studied thus far, all of which hatch from the egg with four or five pairs of functional thoracopods, thereby starting free life directly in the bathynellid phase.

Anaspidacea: What is known of the ontogeny of the Anaspidacea is due to V. Hickman's (1937) study of the embryology of *Anaspides tasmaniae*. Development begins with rudiments of three appendages that appear simultaneously as three pairs of thickenings. The area between the antennules also becomes thickened, forming the labrum. Simultaneously, an investing membrane (a first larval integument) is formed, which later plays an important role in hatching. V. Hickman used the term "egg nauplius" for this phase. In the following phase the two lobes of the antenna become more pronounced, the exopod being longer than the endopod. The mandible loses its two-lobed appearance and becomes pear-shaped. The caudal papilla of the egg nauplius starts to grow and split into two lobes, which later develop into forked branches. The first three pairs of thoracopods appear as anlagen, behind those of the maxillules and maxillae. A second larval integument is also formed. The transition to the next phase is not clearly defined and is more gradual. One characteristic of this phase is that both branches of the antenna elongate. The endopod finally greatly surpasses the exopod in length, and segmentation of the endopod begins. Another characteristic is the presence of the anlagen of the remaining thoracopods, except for the eighth thoracopod, which is delayed in development but still appears before the pleopods. All trunk somites and appendages are developed when *A. tasmaniae* hatches from the egg. This is different in 1 species of the Stygocarididae, however, as juveniles with only five functional thoracopods (including the maxillipeds) are known (Noodt 1970).

MORPHOLOGICAL DIVERSITY: Larvae of the Bathynellacea are fairly uniform and lack the diversity seen in other groups in the Crustacea. To date, there are no true larvae known in the Anaspidacea.

NATURAL HISTORY: Almost nothing is known about the natural history of developmental stages in the Bathynellacea, but a few observations were made by Jakobi (1954). When hatching from the egg, the parazoea of *Antrobathynella stammeri* has only one pair of thoracopods, and locomotion is clumsy and slow, so not infrequently it falls prey to swift cyclopoids. In case of danger it can passively lie on the bottom

and seem to be dead. After the appearance of the second pair of thoracopods, locomotion is skillful and quick. When the first pair of thoracopods is moved forward, the second pair is in contact with the ground, and the second pair follows as soon as the first has made contact with the ground. One seta of each furcal ramus is particularly long, and functionally these two setae take the place of the still-missing uropods in helping the larvae maintain balance. The larvae now make their first attempts to capture protozoans, which are treated with the mouthparts and antennae but normally still manage to escape. In *A. stammeri*, development from an egg to an adult on average takes 9 months.

In the Anaspidacea, V. Hickman (1937) made a few observations. Eggs are laid in the spring and summer, and development lasts from 32 to 35 weeks. The newly hatched *Anaspides tasmaniae* is 2.7 mm long. The main differences from the adults are that the eyes are sessile, the rostrum is absent, the telson is forked, and the pleopods lack endopods. Differences between the sexes appear about 20 weeks after hatching. The newly hatched shrimp is an active crawler. It also swims to the surface of the water from time to time, where it turns over, searching for food in calm water by moving about under the surface film in an inverted position. Adults can do the same, but they do so less frequently, because of their weight. Feeding starts a few days after hatching, when communication between the different parts of the gut (fore-, mid-, and hindgut) is established. When disturbed, newly hatched shrimps can lie with their back on the bottom for several seconds, as if dead. This death-feigning habit is unknown in adults.

PHYLOGENETIC SIGNIFICANCE: Schminke (1981) attempted to place the findings of *Antrobathynella stammeri* development in the context of the ontogeny of other malacostracans and revealed similarities with the development of, in particular, the Penaeoidea (Decapoda). The parazoea and bathynellid of the Bathynellacea show the same characteristics as the protozoea and zoea of the Penaeoidea. These similarities led Schminke (1981) to conclude that the Bathynellacea arose by progenesis. In the Penaeoidea, development leads from the zoeal to the post-larval phase through metamorphosis, while it breaks off prematurely in the Bathynellacea, with sexual maturity being reached before metamorphosis.

HISTORICAL STUDIES: Not much is known about the ontogeny of members of the Parabathynellidae, although an overview was given by Serban and Coineau (1990). The post-embryonic development of members of Bathynellidae is much better known, thanks in particular to a detailed study by Jakobi (1954) on *Antrobathynella stammeri*, but also in work by Serban (1985) on *Gallobathynella coiffaiti*. In his paper, Serban distinguished a few more stages in the post-embryonic development of *G. coiffaiti*, compared with *A. stammeri*. Post-embryonic development in *G. coiffaiti*, however, is also divided into two phases: a larval phase, lasting from hatching from the egg until the appearance of the fifth leg in functional form, and a juvenile phase. During the juvenile phase, sexual differences make their first appearance, and (probably with the onset of maturation) thoracopods 6–8 develop (in particular the complex copulatory male thoracopod 8) and reach their final form.

Schminke (1981) discussed *Antrobathynella stammeri* in the context of the ontogeny of other malacostracans. Serban's (1985) interpretation of the post-embryonic development of the Bathynellidae, however, is at variance with what is known of the post-embryonic development of other Crustacea. In this paper, Serban stressed the singularity of the post-embryonic development of the Bathynellidae, which he considered to be quite unlike anything known from other Crustacea.

Syncarids are said to have a direct development, yet their true development is blurred by the lack of metamorphosis (Schminke 1978). In the Bathynellacea, development breaks off before metamorphosis, while in the Anaspidacea their whole development before this event is passed through in the egg.

Selected References

Hickman, V. V. 1937. The embryology of the syncarid crustacean, *Anaspides tasmaniae*. Papers and Proceedings of the Royal Society of Tasmania for 1936: 1–35.

Jakobi, H. 1954. Biologie, Entwicklungsgeschichte und Systematik von *Bathynella natans* Vejdovsky. Zoologische Jahrbücher, Abteilung für Systematik, Ökologie und Geographie der Tiere 83(1/2): 1–62.

Schminke, H. K. 1981. Adaptation of Bathynellacea (Crustacea, Syncarida) to life in the interstitial ("Zoea Theory"). Internationale Revue der Gesamten Hydrobiologie 66: 575–637.

Serban, E. 1985. Le développement post-embryonnaire chez *Gallobathynella coiffaiti* (Delamare) (Gallobathynellinae, Bathynellidae, Bathynellacea). Travaux de l'Institut de Spéologie "Emile Racovitza" 24: 47–61.

Serban, E., and N. Coineau. 1990. Données concernant le développement post-embryonnaire dans la famille des Parabathynellidae Noodt (Bathynellacea, Podophallocarida, Malacostraca). Travaux de l'Institut de Spéologie "Emile Racovitza" 29: 3–24.

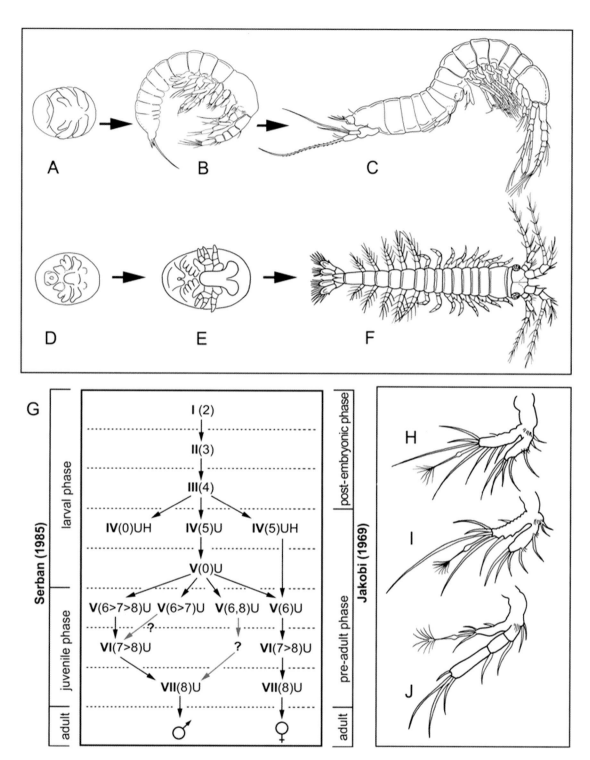

Fig. 35.1 Development in the Bathynellacea and Anaspidacea. A–C: drawings of *Antrobathynella stammeri* (Bathynellacea). A: nauplioid. B: parazoea. C: bathynellid, lateral view. D–F: drawings of *Anaspides tasmaniae* (Anaspidacea). D: egg nauplius. E: egg protozoea. F: post-larva, dorsal view. G–J: Bathynellacea. G: chart of phases and stages of development, according to Serban (1985) and Jakobi (1954, 1969); roman numerals = number of functional pairs of thoracopods; arabic numerals = number of pairs of thoracopods projecting from their somites as anlagen; H: hill-like (anlagen still inside somite but visible as a thickening on the outside); U = uropod present; for example, IV(5)U = stage with 4 pairs of functional thoracopods and a fifth projecting as a three-branched anlagen, with uropods present. H–J: drawings of *Antrobathynella stammeri* antennae. H: at stage I(2). I: at stage III(4). J: at stage IV(5)U. A and H–J modified after Jakobi (1954); B and C modified after Serban (1972); D–F modified after V. Hickman (1937); G modified after Schminke (1981).

36

Joel W. Martin

Introduction to the Peracarida

The Peracarida are an incredibly diverse group of mostly small crustaceans (with some notable exceptions). They are found in nearly all habitats and exhibit a wide range of lifestyles. Most are marine, but freshwater and terrestrial forms are common, as are parasites. Estimates of the number of described species are in the range of 20,000 to 25,000; there are undoubtedly thousands more still undiscovered and undescribed (e.g., see Ahyong et al. 2011; Poore and Bruce 2012). Peracarids appear to be united by the fact that development occurs within a female's brood pouch, formed by plates of the female's thoracic legs. Thus larvae per se are lacking. Traditionally, the group has been divided into 9 or 10 extant orders, including the highly diverse Amphipoda and Isopoda (together containing perhaps 300 extant families) and the much less diverse Tanaidacea, Cumacea, two groups of mysid shrimps (now treated as the Lophograstrida and Mysida), Spelaeogriphacea, Thermosbaenacea, and "Mictacea" (probably paraphyletic). Another order (the Bochusacea) has been proposed for one of the "mictacean" families and is accepted by some workers (Gutu and Iliffe 1998). These orders contain perhaps 350 families collectively (see J. W. Martin and Davis 2001; Ahyong et al. 2011).

Acceptance of the Peracarida as monophyletic is by no means settled. Analyses continue to question the group's monophyly and, if it is recognized as natural, debates would still remain over which taxa are contained within it. Some of the many studies devoted to this question include the works of Watling (1983), Pires (1987), H. Wagner (1994), Mayrat and Saint Laurent (1996), Gutu and Iliffe (1998), Hessler and Watling (1999), Watling (1999), Jarman et al. (2000), Richter and Scholz (2001), Jaume et al. (2006), and Meland and Willassen (2007) (much of which was summarized by Spears et al. 2005).

Peracarids do not have true larvae. Instead, in some species the eggs within the female's brood pouch each hatch as a miniature version of the adult, which is called a manca. The manca differs from the adult mostly in the fact that the final pair of thoracic legs is not yet present. Strictly speaking, then, we could be excused for excluding the peracarids from this atlas. Yet many display developmental patterns that can have a bearing on our understanding of relationships both within the Peracarida and of peracarids to other groups of Crustacea. Additionally, some peracarids exhibit unique morphological stages that have been called larvae in the past (e.g., the "Cystisoma larvae" of some hyperiid amphipods) or have juveniles that differ significantly from the adult (e.g., the praniza and microniscus forms of some parasitic isopods). For these reasons, we have included not only images of peracarid mancae and other post-hatching morphological stages, but also, for some groups, more developmental (late embryological) information than we have allowed for other crustacean taxa.

In the following chapters, all extant orders are treated, with the exception of the "Mictacea" (and thus the Bochusacea), for which no developmental studies exist (but see chapter 37 for some developmental information on "mictaceans"). Partly because of the paucity of information on some of these groups, we have combined into one chapter each what is known for the Thermosbaenacea, Spelaeogriphacea and "Mictacea" (chapter 37); Mysida and Lophograstrida (chapter 38); and Isopoda and Tanaidacea (chapter 40).

37

JØRGEN OLESEN
TOM BOESGAARD
THOMAS M. ILIFFE
LES WATLING

Thermosbaenacea, Spelaeogriphacea, and "Mictacea"

GENERAL: The Thermosbaenacea, Spelaeogriphacea, and "Mictacea" (which is probably paraphyletic) are three taxa of small, eyeless, unpigmented malacostracan crustaceans. They are represented by very few species from subterranean waters and, in the case of the "Mictacea," from the deep sea. The Spelaeogriphacea and "Mictacea" have peracarid affinities (a ventral marsupium). All three taxa, due to their hidden lifestyle, were discovered relatively late (Monod 1924; Gordon 1957; Bowman et al. 1985b). The Thermosbaenacea is the most diverse, with 34 known species; the Spelaeogriphacea has 4 species; and the "Mictacea" has 6 (Bowman et al. 1985b; H. Wagner 1994; Ohtsuka et al. 2002; Jaume et al. 2006; Jaume 2008). The Thermosbaenacea is primarily marine (but with only 5 euhaline species). Most of its species occur in anchialine environments associated with marine coastal areas. Most thermosbaenaceans have been found in caves, wells, and springs in the Caribbean and Mediterranean regions, but they also occur in Australia. Their distribution precisely matches that of the ancient Tethys Sea or its coastlines, so they are most probably relicts of a once widespread shallow-water marine Tethyan fauna, stranded in interstitial or crevicular groundwater during episodes of marine regression (Jaume 2008). The Spelaeogriphacea is less diverse, with 1 species in South Africa, 1 in Brazil, and 2 in Western Australia (Gordon 1957; Pires 1987; Poore and Humphreys 1998, 2003). The Spelaeogriphacea, like the Thermosbaenacea, are thought to be relicts of a more widespread shallow-water marine Tethyan fauna. The "Mictacea" currently has 6 species: 3 from caves or subterranean waters, and 3 from the deep sea (Bowman and Iliffe 1985; H. Sanders et al. 1985; Just and Poore 1988; Gutu and Iliffe 1998; Ohtsuka et al. 2002; Jaume et al. 2006).

No consensus has been reached regarding the classification of these 3 unique orders, and the Thermosbaenacea are sometimes considered non-peracarid, due to the presence of a dorsal brood pouch instead of a ventral marsupium, as in other peracarids (Siewing 1956; Richter and Scholtz 2001). The "Mictacea" and Thermosbaenacea were considered closely related

by Watling (1999), while Richter and Scholtz (2001) found the "Mictacea" and Spelaeogriphacea to be sister groups. Poore (2005) found evidence for a clade consisting of all three orders. Much attention has been paid to the status of the "Mictacea." Gutu and Iliffe (1998) and Gutu (1998) found evidence for a paraphyletic "Mictacea," with one of the "mictacean" families (the Mictocarididae) being close to Spelaeogriphacea, which they combined as a new order (the Cosinzeneacea) and, at the same time, created another order (the Bochusacea) to accommodate the Hirsutiidae, the remaining "mictacean" family. Jaume et al. (2006) also found reasons for splitting the "Mictacea" and suggested tanaidacean affinities for the Hirsutiidae. Hence, as suggested by Jaume et al. (2006), the relationships among the subgroups of the "Mictacea," the Spelaeogriphacea, and other peracarids are in need of reassessment.

The life cycle is very poorly known for these three groups. There are no known true larvae in any of the taxa. Thermosbaenaceans are unique in that developing embryos are carried in a dorsal brood pouch. In the Spelaeogriphacea and "Mictacea," development takes place in a ventral brood pouch.

LARVAL TYPES: No true larvae are known for any of these groups. Most is known about the development of the Thermosbaenacea. Very little is known for the Spelaeogriphacea; some data are presented here for the first time (fig. 37.1J–M). Almost no information exists for the "Mictacea" (but see the ovigerous female in fig. 37.1O).

Thermosbaenacea: Thermosbaenacean development is best known for *Thermosbaena mirabilis* (Siewing 1958; D. Barker 1962; Zilch 1972, 1974, 1975), but various aspects have also been reported for *Tethysbaena argentarii* (Stella 1959). Thermosbaenaceans are unique among peracarids in carrying their developing embryos in a dorsal brood chamber formed by a carapace. On average, 10 embryos are carried in each brood chamber (fig. 37.1B). The fertilized eggs / early embryos emerge from gonopores situated on the inner side of the sixth thoracopods. The eggs / embryos are first enclosed in a pair of pouches formed on each side of the body of the mother and partly covered by the carapace (fig. 37.1A) (Zilch 1972). These pouches protect the early embryos, are dissolved during early segmentation, and are lost when the embryos have developed a solid blastoderm. According to Zilch (1972, 1974), cleavage does not start until the embryo enters the brood pouch. Zilch (1974) divided the marsupial development of *Thermosbaena mirabilis* into 10 stages, 6 of which take place before the vitelline membrane bursts, and 4 afterward. A small ventrally curved caudal papilla appears in stage 4 (fig. 37.1C).

Spelaeogriphacea: Spelaeogriphacean development is largely unknown, but it takes place in a ventral marsupium, as in other peracarids. The marsupium is formed by oostegites on pereopods 2–5 (*Spelaeogriphus*) or on pereopods 1–5 (*Potiiocara* and *Mangkurtu*) (Gordon 1957, 1960b; Pires 1987; Poore and Humphreys 2003). Two different stages of *Spelaeogriphus lepidops* have been examined in relation to this chapter.

"Mictacea": "Mictacean" development is even less known than spelaeogriphacean development. Ovigerous females of *Mictocaris halope* (Mictocarididae) are known to have a typical peracaridan marsupium formed by medial non-setose oostegites of pereopods 1–5 (Bowman and Iliffe 1985). In hirsutiids, the oostegites are in an atypical position on the posterior surface of the coxa, rather than medially (H. Sanders et al. 1985; Just and Poore 1988; Gutu and Iliffe 1998; Ohtsuka et al. 2002; Jaume et al. 2006), something that has attracted considerable debate (see a summary in Boxshall and Jaume 2009) as either reflecting the presumed origin of oostegites as coxal epipods (Gutu and Iliffe 1998), or reflecting functional needs (H. Sanders et al. 1985; Jaume et al. 2006). Embryos in the marsupium are known for *Mictocaris halope* (Mictocarididae) (fig. 37.1O). In one hirsutiid (*Thetispelecaris yurikago*), one case of egg retention (one egg) by epipods (oostegites) has been reported, suggesting that these epipods, by definition, are indeed oostegites (Ohtsuka et al. 2002; Jaume et al. 2006).

MORPHOLOGY

Thermosbaenacea: Development is known in much detail for *Thermosbaena mirabilis*, due to the detailed studies by Zilch (1972, 1974); the following short summary is based on his work. During stages 1–3, which are all globular in shape and housed within the vitelline membrane, the early germ band is formed and starts to grow. In stage 3, the anlagen of the three pairs of naupliar appendages (the antennules, antennae, and mandibles) are present. In stage 4 (fig. 37.1C), which is still within the vitelline membrane, the anlagen for approximately six additional appendages (the maxillule, the maxilla, the maxilliped, and three thoracopods) have appeared as small undifferentiated buds, together with a small ventrally bent caudal papilla. In stage 5 (fig. 37.1D), the anterior appendages have become more differentiated, and a couple of additional thoracic limb buds have appeared; the caudal papilla has enlarged slightly. In stage 6 (fig. 37.1E), the limbs have become further differentiated and tubular limb buds of thoracopods 2–6 are now present. Distinct optical lobes are also present. The caudal papilla has enlarged significantly but is still ven-

trally curved, due to constraints presented by the presence of the vitelline membrane. A pair of rudimentary uropods is present posteriorly. The early carapace is represented by a transverse fold laterally, at the cephalon. Stage 7 (fig. 37.1F) is the first after the vitelline membrane has been shed, which results in a less curved body shape. All appendages have developed further, and the carapace margin is more pronounced. In stage 8 (fig. 37.1G), all appendages of the later adult are present, including the rudimentary pleopods 1 and 2. The optical lobes are large. In stage 9 (fig. 37.1H), the thoracopods have elongated significantly, and the posterior margin of the carapace is now clearly free. Stage 10 is the last stage within the marsupium; when the larva leaves, it is ready to feed.

Spelaeogriphacea: Two developmental stages (an intermediate and a late stage) of *Spelaeogriphus lepidops*—taken from marsupia of specimens collected in Bat Cave on Table Mountain, Cape Town, South Africa, in 1993 by Les Watling—are presented here and briefly described (fig. 37.1J–N). An intermediate stage (fig. 37.1J–L) is still very embryo-like, with a large yolk-inflated anterior part and a dorsally bent caudal papilla. The vitelline membrane has been shed. Anteriorly, in the germ band, there are a pair of large optical lobes, followed by relatively large antennular and antennal limb buds and smaller buds of the maxillules, maxillae, and maxillipeds. These are followed by five to six even smaller externally visible thoracopodal limb buds, which diminish in size posteriorly. A small lateral carapace lobe is present on each side, in a position corresponding roughly to the maxillipeds ventrally. An advanced late stage (fig. 37.1M, N), which still has much yolk anteriorly, has advanced buds of most limbs. Many adult structures, although in a primordial developmental stage, can be recognized. A labrum, a pair of paragnaths, and lateral carapace lobes (which are significantly larger than in the previously described stage) are also present.

No information exists about developmental stages of other spelaeogriphaceans, but Poore and Humphreys (2003) reported that embryos of *Mangkurtu kutjarra* flex dorsally, as do those of *Spelaeogriphus lepidops*. They also reported a manca stage of *M. kutjarra*, which they assumed is general for the Spelaeogriphacea.

"Mictacea": Basically nothing is known about its developmental stages, but Bowman and Iliffe (1985) illustrated an ovigerous female (also see fig. 37.1O), and Bowman et al. (1985b), in their diagnosis of the "Mictacea," stated that the embryos flex dorsally and the eggs hatch as a manca, lacking pereopod 7. It may well be, however, that the eggs hatch long before leaving the marsupium (as seems to be the case in spelaeogriphaceans), so this aspect, and "mictacean" development in general, needs more investigation.

MORPHOLOGICAL DIVERSITY: Little diversity is known, since information is available only for a few species.

NATURAL HISTORY: Among thermosbaenaceans, *Thermosbaena mirabilis* is the best-known species, thanks to the works by D. Barker (1962) and Zilch (1972, 1974). This species is known only from a remarkable habitat: public Roman baths fed by a hot spring (ca. 45°C) in an oasis in Tunisia (Monod 1924; Zilch 1972). *Thermosbaena mirabilis* is dioecious and fertilization is internal; mating was observed by Zilch (1972). According to D. Barker (1962), the breeding period extends from May to September, and a collection made by him in September yielded 50 breeding females out of 714 specimens. The peculiar thermosbaenacean habit of carrying their embryos on their backs (under the carapace) is unique among malacostracans, but it is seen in other crustaceans, such as cyclestheridan branchiopods (see chapter 10), some ostracods, and some ascothoracidan thecostracans (see Olesen 2013). Zilch (1972) noted that prior to the transfer of the embryos to the dorsal brood chamber, they are carried in a pair of dorsolateral egg pouches, which are attached to the basal parts of thoracopods 6 in a position from which the embryos emerge. These pouches degenerate, and the embryos are then carried freely under the carapace. It has been suggested that the ventilating epipods of *T. mirabilis*, which are directed into the brood chamber and generate a ventilation current, play a role in ventilating the embryos during development, and perhaps also in sucking the embryos into the brood chamber with the inhalant current (D. Barker 1962).

No natural history aspects are known regarding the development of the Spelaeogriphacea and "Mictacea," other than that some species in both taxa brood their offspring in a ventral marsupium, as in other peracarids.

PHYLOGENETIC SIGNIFICANCE: Relationships among these three small groups are unsettled, and the role of development figures prominently in the debate. In particular, questions of whether the Thermosbaenacea (with its dorsal brood pouch) belongs among the Peracarida, the status of the "Mictacea" and the proposed groups Cosinzeneacea and Bochusacea, and the relationships of these taxa to the Spelaeogriphacea remain unresolved. The unusual dorsal position of the brooded embryos caused Siewing (1956, 1958) to treat the Thermosbaenacea as a special taxon, the Pancarida, separate from the Peracarida. Certainly, based on brooding patterns alone, it is difficult to comprehend an in-group peracarid position for the Thermosbaenacea, which would imply the transformation from one advanced brooding system (in a ventral marsupium) to another (in a dorsal brood chamber). If such

a transformation has indeed taken place during evolution, it may have involved an intermediate phase, with the embryos carried in a pair of dorsolateral egg pouches attached to the basal part of the sixth thoracopods, as this is seen during the development of *T. mirabilis* embryos before they are transferred to a dorsal brood chamber. Another developmental aspect that suggests a position for the Thermosbaenacea independent from at least some peracarids is the ventrally flexed caudal papilla (fig. 37.1C–E), which is probably ancestral for the Malacostraca (Scholtz 2000; Richter and Scholtz 2001). In the Spelaeogriphacea and "Mictacea" (fig. 37.1K)—to which the Thermosbaenacea bear some similarities—and in the Cumacea, Isopoda, and Tanaidacea (see chapters 40 and 41), the caudal papillae are flexed dorsally. For further discussions, see works by Siewing (1956), Gutu (1998), Gutu and Iliffe (1998), Watling (1999), Richter and Scholtz (2001), Poore (2005), and Jaume et al. (2006).

HISTORICAL STUDIES: All three taxa have been discovered within the last 100 years, and since all live in subterranean habitats or in the deep sea, developmental information has been slow to appear. The Thermosbaenacea was described in 1924, but detailed studies on thermosbaenacean development only became available through works by D. Barker (1962) and Zilch (1972, 1974). The development of the Spelaeogriphacea and "Mictacea" still await thorough treatments.

Selected References

Barker, D. 1962. A study of *Thermosbaena mirabilis* (Malacostraca, Peracarida) and its reproduction. Quarterly Journal of Microscopical Science 103: 261–286.

Bowman, T. E., S. P. Garner, R. R. Hessler, T. M. Iliffe, and H. L. Sanders. 1985. Mictacea, a new order of Crustacea Peracarida. Journal of Crustacean Biology 5: 74–78.

Gordon, I. 1957. On *Spelaeogriphus*, a new cavernicolous crustacean from South Africa. Bulletin of the British Museum (Natural History), Zoology 5: 31–47.

Jaume, D., G. A. Boxshall, and R. N. Bamber. 2006. A new genus from the continental slope off Brazil and the discovery of the first males in the Hirsutiidae (Crustacea: Peracarida: Bochusacea). Zoological Journal of the Linnean Society 148: 169–208.

Zilch, R. 1974. Die Embryonalentwicklung von *Thermosbaena mirabilis* Monod (Crustacea, Malacostraca, Pancarida). Zoologischer Jahrbücher, Anatomie 93: 462–576.

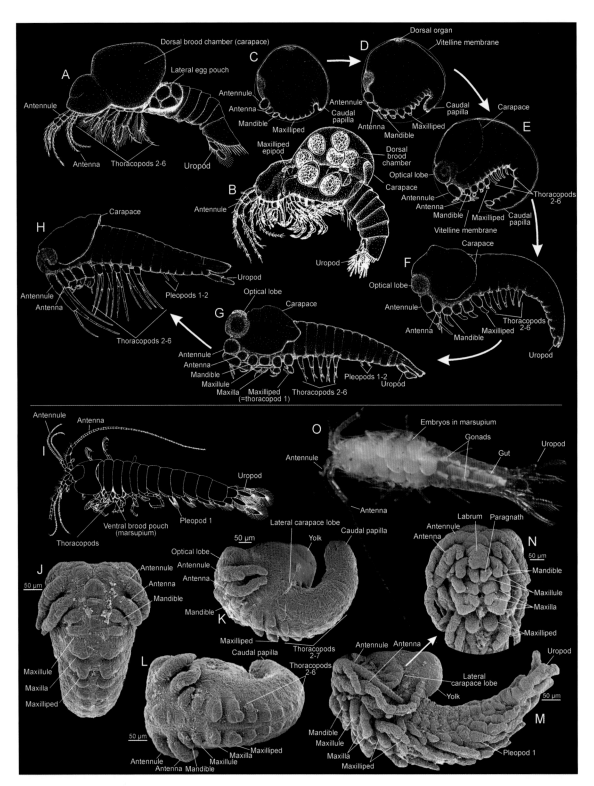

Fig. 37.1 Development of *Thermosbaena mirabilis* (Thermosbaenacea), *Spelaeogriphus lepidops* (Spelaeogriphacea), and *Mictocaris halope* ("Mictacea"). A–H: drawings of *Thermosbaena mirabilis*, lateral views. A: adult female, with early embryos still enclosed in the lateral egg pouches. B: adult female, with more advanced embryos now being enclosed freely within the dorsal brood chamber. C–H: developmental stages. C: embryonic stage 4. D: embryonic stage 5. E: embryonic stage 6. F: stage 7, the first stage after the disruption of the vitelline membrane. G: stage 8. H: stage 9. I–N: *Spelaeogriphus lepidops*. I: drawing of an adult female, with a ventral brood pouch (marsupium), lateral view. J–L: intermediate developmental stage, SEMs. J: ventral view. K: lateral view. L: lateroventral view. M and N: late developmental stage, SEMs. M: lateral view. N: anterior end, ventral view. O: adult female of *Mictocaris halope*, with a ventral brood pouch, light microscopy, dorsal view. A modified after Zilch (1972); B modified after D. Barker (1962); C–H modified after Zilch (1974); I modified after Gordon (1960b); J–N original, based on material collected in 1993 by Les Watling in Bat Cave on Table Mountain in Cape Town, South Africa; O original, based on material collected in 2012 by Tom Boesgaard, Thomas M. Iliffe, and Jørgen Olesen in Green Bay Cave in Bermuda.

38

Lophogastrida and Mysida

Carlos San Vicente
Guillermo Guerao
Jørgen Olesen

GENERAL: Lophogastrids and mysids are shrimp-like crustaceans that are known as opossum shrimps, due to the presence of a ventral brood pouch (marsupium) in mature females. They are readily distinguished from other shrimp-like crustaceans, such as euphausiids and carideans, by the presence of a statocyst in the proximal part of the endopod of the uropods. The statocyst is missing in the order Lophogastrida (the families Lophogastridae, Gnathophausiidae, and Eucopidae) and in the family Petalophthalmidae of the order Mysida. The marsupium is composed of either seven pairs of oostegites on the thoracopods (the Lophogastrida, Petalophthalmidae, and Boreomysinae) or two or three pairs (the Gastrosaccinae, Siriellinae, Ropalophthalminae, Mysidellinae, and Mysinae). The Lophogastrida and Mysida currently include 58 and 1,106 species, respectively (G. Anderson 2010a, 2010b). Lophogastrids are marine pelagic swimmers and have a cosmopolitan oceanic distribution. The majority of mysid species are marine (mainly coastal) and live in a suprabenthic (hyperbenthic) habitat. Although there are only a few freshwater species, they are widespread and often numerous; thus they are of ecological importance. The life cycle consists of an embryonic stage, a nauplioid stage, a post-nauplioid stage (intramarsupial stages), a juvenile, and an adult, all of which are free living. Developmental stages are all lecithotrophic.

LARVAL TYPES: Breeding females of both the Lophogastrida and Mysida carry their embryos and larvae in a marsupium, within which their entire embryonic and larval development takes place. Young individuals emerge from the marsupium as early juveniles; only in *Boreomysis arctica* have larval stages occasionally been found in the plankton (Jepsen 1965).

In non-ovigerous females, the mature ovary fills the posterior dorsolateral regions of the thorax, and the eggs present in the ovary can be easily observed through the carapace. Within the marsupium, the eggs emerge from the external genital openings of the oviducts, which are located on the body near the bases of the sixth pair of thoracopods. Eggs *sensu stricto* do not occur in the marsupium, since they are fertilized immediately once they are extruded from the oviducts (Mauchline 1980).

As described by Mauchline (1980), there are three main phases of marsupial development: the egg, an eyeless larva, and an eyed larva. Confusion has persisted in the literature, however, over the use of such terms as egg, embryo, and larva (Mauchline 1980; Cuzin-Roudy and Tchernigovteff 1985). The terminology used in this chapter follows Wittmann (1981a) in defining embryos as the developmental stage inside the egg membrane, and larvae (nauplioid and post-nauplioid) as the stages occurring after hatching.

Embryos: After fertilization, the development of the spherical embryo inside the egg membrane is termed the embryonic phase, which ends with the shedding of the egg membrane. The early embryos are spherical or subspherical, and their size is closely similar to that of ripe eggs within the oviducts. The term embryonic stage (Wittmann 1981a) refers to this egg-like embryo, which was called the "egg" by Mauchline (1972, 1973) or "early embryo" (stage 1 larva) by Mauchline (1980).

Nauplioid, Post-Nauplioid: Two successive main phases of post-embryonic development, differentiated by hatching (ecdysis) of the larvae, occur within the marsupium of lophogastrid and mysid females (K. B. Nair 1939; Matsudaira et al. 1952; Jepsen 1965; Davis 1966, 1968; Childress and Price 1978). These phases correspond to the stages described by Mauchline (1972, 1973) as eyeless larvae and eyed larvae (Mauchline 1980), and to the stages described as nauplioid and post-nauplioid stages by Wittmann (1981a). The emerging nauplioid form is immobile, yolk-filled, and encased in an unsegmented cuticle. It has the same appendages as a classic nauplius, but it lacks the nauplius eye and swimming setae. Later in this phase, all body segments and appendages of the adult are formed. The first larval molt results in the non-motile post-nauplioid form, which continues to rely on yolk reserves and has all its appendages, its carapace, and its eyes free. A second ecdysis, just before or after liberation from the marsupium, yields a free-living juvenile.

Depending on the species, the three intramarsupial phases (embryonic, nauplioid, and post-nauplioid) can be divided further into substages, based on successive morphological changes (Wittmann 1981a; Cuzin-Roudy and Tchernigovtzeff 1985; Greenwood et al. 1989; Wortham-Neal and Price 2002; Fockedey et al. 2006; Ghekiere et al. 2007). Development within a brood is usually synchronous, and, in some species, embryos and larvae are easily visible through the semitransparent and thin-walled marsupial oostegites (figs. 38.1A–C).

MORPHOLOGY: Intramarsupial development can be divided into three main phases (table 38.1).

Embryonic Phase: This phase begins when the eggs are fertilized within the brood pouch and ends when the egg membrane is completely shed. The early embryo, which is still within the egg membrane, is egg-like at first, but later has rudiments of the developing antennae and abdomen. The optical rudiments, antennules, and antennae, which are located along the anteroventral part of the embryo, and the abdominal rudiment, located at the posterior end, become more or less visible through the egg membrane. The embryo hatches as the abdominal rudiment straightens out, rupturing the egg membrane posteroventrally (fig. 38.1D).

Nauplioid Stage (Larval Stage 1): After hatching, the larva is dorsally concave (fig. 38.1D) and surrounded by a nauplioid cuticle that is formed under the egg membrane before hatching. Rudimentary antennules, antennae, and mandibles are present. The thoracic appendages develop during this stage, and the eyes become pigmented beneath the larval cuticle. This stage ends with the segmentation of the thorax and abdomen, as well as the appearance of thoracic chromatophores immediately before the formation of eye pigment, and it terminates in a molt (fig. 38.1D–I). The unsegmented abdominal rudiment resembles a thin tail and bears spines, setae, or cercopods at the posterior end (figs. 38.1F; 38.2F). These structures are transitory; at ecdysis between the nauplioid and post-nauplioid phases they disappear and are not replaced

(Wittmann 1981a). Yolk extends throughout the entire body anterodorsally, to the end of the abdominal rudiment. The cephalic appendages remain stub-like (figs. 38.1G; 38.2B, G, I). As the comma-shaped nauplioid larva develops in the marsupium, the yolk fits into the abdomen, the developing telson and uropods become visible under the larval cuticle, and the optic lobes become more bulbous (figs. 38.1H, I; 38.2J, K). The eight pairs of thoracic appendages elongate and become tube-like as development continues. As the optic lobes, thoracic appendages, and tail fan grow, the embryo consumes yolk, so that it retreats anteriorly as well as in the ventral abdominal area. The cephalic appendages have now elongated, and abdominal segmentation begins ventrally. In the late nauplioid stage, eye pigment begins to form. The naupliar cuticle is stretched, due to further development of the larva, with the cuticle being visible in the tail-fan region and optic lobes (figs. 38.1I; 38.2C, K). The optic lobes are fully bulbous and contain eye pigment; as this stage progresses, the eyestalks are formed, but because the cuticle encloses the eyes, they remain close to the body. The yolk, now enclosed within the gut region, continues to thin, causing the abdomen to become clearer. The posteriorly directed thoracic appendages remain relatively short and are pressed tightly together. The abdomen becomes completely segmented, and development of the telson and uropods is complete under the nauplioid cuticle. Pleopods are present but difficult to see, because of the presence of overlying thoracic appendages. The nauplioid phase ends when the naupliar cuticle is shed and the appendages and eyes become free to move.

Post-Nauplioid Stage (Larval Stage 2): This stage begins after the nauplioid cuticle is shed within the marsupium, and terminates when the yolk is absorbed inside the carapace/thoracic region (figs. 38.2D, E; 38.3A). Nauplioid exuviae are never found in the marsupium, since they are extremely thin and expelled as small pieces by water currents (Cuzin-Roudy and Tchernigovtzeff 1985). The molted larva has eyes on stalks; the eyestalks are free and bear facets distally. After molting, the post-nauplioid larva resembles the juvenile, except that the eyes remain directed dorsally rather than laterally.

Table 38.1 General features of lophogastrid and mysid intramarsupial stages

Larval stage	General features
embryonic phase	Egg-like, subspherical, enveloped by an egg membrane that is shed at the end of this stage. Yolk granules spread all over the embryo.
nauplioid stage (stage 1 = eyeless larva)	Early nauplioid: Larvae possess rudimentary antennules, antennae, and mandibles. Thoracic appendages develop, and segmentation of the thorax and abdomen progresses under the nauplioid cuticle.
	Late nauplioid: Segmentation of the thorax and abdomen is completed, and the eyes are pigmented under the nauplioid cuticle. Molting occurs at the end of this stage.
post-nauplioid stage (stage 2 = eyed larva)	Early post-nauplioid: Molted larvae have eyes on completed stalks. Eyestalks and thoracic appendages are mobile, elongated, and highly developed. The yolk forms an anterodorsal protuberance and is gradually absorbed into the digestive gland.
	Late post-nauplioid: As the yolk is gradually absorbed, the eyes rotate upward. The yolk becomes completely enclosed in the digestive glands and the digestive tract. The ventral chromatophores become functional.
	Larvae molt at the end of the post-nauplioid stage and are released from the marsupium as a juvenile form, with a developed statocyst containing a statolith.

Source: Modified after Mauchline 1980; Wittmann 1981a; Cuzin-Roudy and Tchernigovtzeff 1985; Wortham-Neal and Price 2002; Okumura 2003; and Fockedey et al. 2006.

Yolk is still present, and the dorsal region continues to be distinctly concave. A large yolk protrusion is located anterodorsally and projects between the eyes, but it is absorbed into the thorax as the stage progresses. The carapace is well developed in the posterior region of the thorax, but it is not completely formed in the anterior region. Within the carapace, the stomach becomes visible and the yolk protrusion is being absorbed. The larvae show no response to mechanical or chemical stimulation at this stage, though they are quite capable of various movements long before then (Berrill 1969). In the late post-nauplioid stage, the yolk is restricted to the thoracic region. All appendages of the adult stage are already visible; they are well developed and free. All chromatophores are fully functional, forming a variable network along the body, depending on the species. Cephalic appendages and telson setation are now more or less similar to adult forms (fig. 38.3B, C). The larvae usually respond immediately to mechanical stimulation by twitching violently, and this contraction usually initiates appendage activity (Berrill 1969).

The post-nauplioid stage ends with liberation from the marsupium and the second larval molt, which leads to a free-living juvenile. Young mysids are morphologically similar to the adults and develop the statolith within the vesicle of the statocyst, a feature absent in post-nauplioid larvae. Final emergence normally occurs at night, over a short period of time (a few minutes to an hour). Nocturnal emergence of the young is considered important for their survival, as it reduces their risk of predation (J. W. Green 1970). Liberated juveniles begin to swim and feed immediately after their release. After having released the juveniles, the female undergoes a sequence of molting, copulation, and subsequent new spawning.

MORPHOLOGICAL DIVERSITY: The morphological features of lophogastrid and mysid larvae have been documented by many authors. In all species studied to date, the different forms of the larval stages are similar (fig. 38.3D–N). Detailed larval morphology is insufficiently known in the Lophogastrida, and there are no documented data from the Petalophthalmidae (Mysida) or the subfamilies Heteromysinae and Mysidellinae.

Setation patterns on the posterior abdomen of the nauplioid phase of mysids are important in distinguishing species and, possibly, higher taxonomic categories (Wittmann 1981a). Telson setation patterns for the post-nauplioid phase and the juvenile stage have been documented for only a few mysid species. Comparisons of telson armature for post-nauplioid and juvenile *Americamysis bahia* indicate that these stages can be distinguished, based on this characteristic (Wortham-Neal and Price 2002).

NATURAL HISTORY: Mauchline (1980) and Wittmann (1984) presented extensive reviews of the literature dealing with seasonal, ecophysiological, and other aspects of size and breeding in lophogastrids and mysids.

In general, all embryos and larvae observed within the marsupium are at the same stage of development, and the spatial organization of individuals in the marsupium appears to be similar (Mauchline 1985). In all cases, nauplioid and post-nauplioid larvae are regularly oriented, with their anterior ends pointing posteriorly and their dorsal surfaces facing the ventral surface of the female. Wittmann (1978, 1981a) suggested that this orientation serves to package the greatest number of larvae in the least amount of space and prevents premature displacement of the young from the brood pouch.

The number of embryos and/or larvae contained in the marsupium ranges from a maximum of 350 in *Gnathophausia ingens* (Childress and Price 1978) to a minimum of 6 in *Mysidium colombiae* (Davis 1966, 1968). The number of marsupial individuals decreases with successive larval stages in almost all populations studied, as a consequence of intramarsupial mortality, which is attributed to premature loss of larvae from the brood pouch, cannibalism, and parasitism of larvae by epicaridean isopods (Mauchline 1973, 1980; Wittmann 1984). A small number of lost young may be captured by other breeding females and adopted into their brood pouches (Wittmann 1978; Mauchline 1980), but adoption appears to be of little importance for survival in natural populations (Wittmann 1984; Johnston and Ritz 2005). The intramarsupial mortality ratio in natural populations can be estimated by comparing the brood size at the beginning and the end of marsupial development. Natural populations of diverse mysid species have shown a mortality rate close to 10–25 percent during marsupial development (Mauchline 1973; Wittmann 1981b, 1984; Wooldridge 1981; San Vicente and Sorbe 1990, 1993, 1995, 2003). Fecundity (the number of young per brood) increases with increasing temperature (in relation to seasonal temperatures and/or latitudes).

The duration of larval development varies considerably among species, from a maximum of 530 days at 3.5°C reported in *Gnathophausia ingens* (Childress and Price 1978) to a minimum of 4 days in *Mesopodopsis orientalis* at 27°C (K. B. Nair 1939). As a general rule, the embryonic stage takes 25–40 percent of the total incubation period, the subsequent nauplioid stage lasts another 30–45 percent of this period, and the post-nauplioid stage accounts for the last 15–30 percent until release from the brood pouch (Wittmann 1984). The percentages of females carrying different types of young in natural populations may be used to estimate the duration of the different marsupial stages (Mauchline 1973; Wittmann 1981b, 1984). Meso- and bathypelagic species tend to produce eggs that, relative to their body size, are larger than epipelagic and coastal species (Mauchline 1980). The duration of larval development is also temperature dependent (Takahashi and Kawaguchi 2004). The yearly or seasonal reproduction pattern varies among different species and between populations of the same species at different latitudes (Mauchline 1980). Thus far, the bathypelagic lophogastrid *Gnathophausia ingens* is the only investigated species where the female produces a single brood (is semelparous) during her lifetime (Childress and Price 1978); other mysids are iteroparous, and the number of successive broods may be two, three, or more or less continuous throughout the year.

The second larval molt is associated with brood release

and takes place just before, during, or after emergence at night (Cuzin-Roudy and Tchernigovtzeff 1985). Although Amaratunga and Corey (1975) found that *Mysis stenolepis* larvae molted up to 3 hours after release, the molt of this non-motile post-nauplioid, a stage that would be subject to intense predation, including cannibalism by the female (Berrill 1969), probably would occur soon after liberation from the marsupium in the wild. Non-motile post-nauplioid larvae of *Americamysis bahia* were released from the marsupium at night, and within a short time molted to the free-swimming juvenile stage; this has been observed for several species of mysids (K. B. Nair 1939; Davis 1966; Berrill 1971; Clutter and Theilacker 1971; Wittmann 1978, 1981a; Cuzin-Roudy and Tchernigovtzeff 1985). Wittmann (1984), however, reported that *Leptomysis lingvura* began its release of young shortly before sunset. *Americamysis bahia* females aided the release of their broods by opening the oostegites, flexing the body upward, and manipulating the young with the adult's thoracic appendages. Similar behavior was reported by Wittmann (1978) for *L. lingvura* and *L. sardica*, but K. B. Nair (1939) and Amaratunga and Corey (1979) observed less involvement of the female during the larval liberation of *Mesopodopsis orientalis* and *Mysis stenolepis*, respectively.

PHYLOGENETIC SIGNIFICANCE: Historically, the mysids were treated as a single order; today they are treated as the separate orders Lophogastrida and Mysida. Studies addressing peracarid phylogeny in general, most of which include a detailed discussion of the lophogastrid/mysid problem, are cited in chapter 36. It is probable that, earlier in their evolution, mysids had planktonic larvae (Nilsson et al. 1986). The advent of the lecithotrophic phase and direct development may be one of the main reasons for their evolutionary success.

HISTORICAL STUDIES: According to Van Beneden (1861): "Il est presque impossible de ne pas faire de l'embryogénie quand des mysis tombent sous la main de l'observateur. On trouve nécessairement des embryons en voie de développement dans leur poche incubatrice, et il n'est guère possible de résister à la tentation d'étudier cette curieuse et instructive évolution." [It is nearly impossible to avoid studying embryology when mysids fall into the hands of the observer. One inevitably finds embryos in various stages of development in their incubation pouch, and it is hardly possible to resist the temptation to study this curious and instructive case of evolution]. Van Beneden compiled and discussed earlier works that focused primarily on the presence or absence of metamorphosis, and he was the first to describe the three main stages of mysid intramarsupial development. Among the earliest studies on mysid embryology are those of Nussbaum (1887) and Bergh (1893) on *Praunus flexuosus*, W. Wagner (1896) on *Neomysis integer*, W. Kane (1904) on *Mysis relicta*, and Manton (1928) on *Hemimysis lamornae lamornae*. Manton's study on *Hemimysis* is the most detailed and one of the most important works in the comparative embryology of the Crustacea

(Wittmann 1981a). The morphology of lophogastrid and mysid larvae has been documented by many authors. Most researchers have worked with preserved material or with live material that they observed only periodically during marsupial development (Manton 1928; K. B. Nair 1939; Matsudaira et al. 1952; Kinne 1955; Murano 1964; Jepsen 1965; Almeida Prado-Por 1966; Davis 1966, 1968; Berrill 1969; J. W. Green 1970; Brown and Talbot 1972; Quintero and de Roa 1973; Modlin 1979; Crescenti et al. 1994). Mauchline (1980) summarized the relevant information known to date on the morphology and biology in lophogastrids and mysids.

Attempts to culture larvae outside the marsupium of the female yielded little success until Wittmann (1981a) successfully reared *Leptomysis lingvura* larvae in vitro. Later works on development include those by Cuzin-Roudy and Tchernigovtzeff (1985) on *Siriella armata*; by Johnston et al. (1997) on *Anisomysis mixta australis*, *Tenagomysis tasmaniae*, and *Paramesopodopis rufa*; by Wortham-Neal and Price (2002) on *Americamysis bahia*; and by Vlasblom and Elgershuizen (1977), Fockedey et al. (2006), and Ghekiere et al. (2007) on *Neomysis integer*.

Selected References

Manton, S. M. 1928. On the embryology of a mysid crustacean, *Hemimysis lamornae*. Philosophical Transactions of the Royal Society of London, B 216: 363–463.

Mauchline, J. 1980. The Biology of Mysids and Euphausiids. Vol. 18 *In* J. H. S. Blaxter, F. S. Russell, and M. Younge, eds. Advances in Marine Biology. London: Academic Press.

Wittmann, K. J. 1981a. Comparative biology and morphology of marsupial development in *Leptomysis* and other Mediterranean Mysidacea (Crustacea). Journal of Experimental Marine Biology and Ecology 52: 243–270.

Wortham-Neal, J. L., and W. W. Price. 2002. Marsupial developmental stages in *Americamysis bahia* (Mysida: Mysidae). Journal of Crustacean Biology 22: 98–112.

Fig. 38.1 (opposite) Diverse females and larvae of *Hemimysis lamornae mediterranea*, light microscopy. A: brooding female, carrying embryos in the marsupium, lateral view. B: brooding female, carrying nauplioid larvae in the marsupium, lateral view. C: brooding female, carrying post-nauplioid larvae in the marsupium, dorsolateral view. D: nauplioid stages at hatching (*upper left* and *lower left*), with the egg membrane ruptured, and early nauplioid stages shortly after hatching (*the remainder*), lateral view. E–H: successive stages of nauplioid development, lateral views. I: detail of the cephalic appendages, early nauplioid stage. J: early nauplioid stage, posterior end, with the cercopods. A–J original, of material from the Ebro Delta (northwestern Mediterranean area).

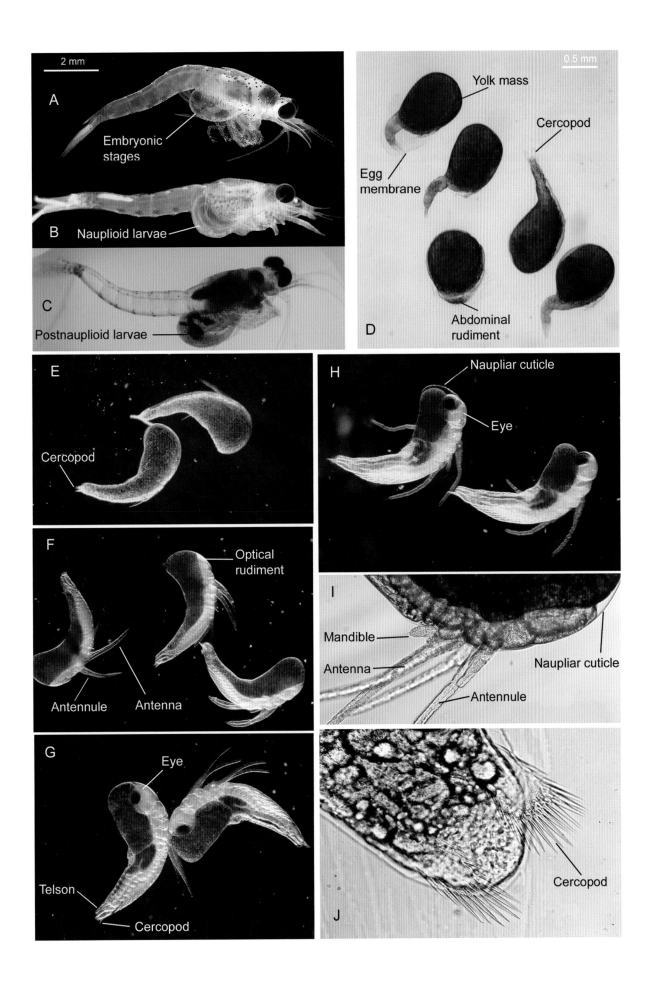

A

2 mm

Embryonic stages

B

Nauplioid larvae

C

Postnauplioid larvae

D

Yolk mass

Egg membrane

Cercopod

Abdominal rudiment

0.5 mm

E

Cercopod

F

Optical rudiment

Antennule

Antenna

G

Eye

Telson

Cercopod

H

Naupliar cuticle

Eye

I

Mandible

Antenna

Antennule

Naupliar cuticle

J

Cercopod

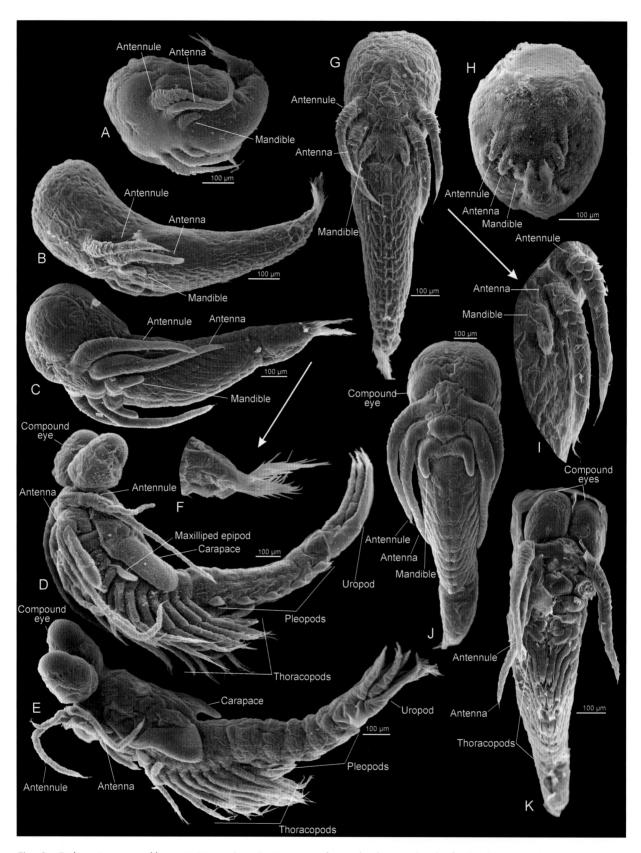

Fig. 38.2 Embryonic stage and larvae in *Praunus inermis*, SEMs. A: early nauplioid stage, shortly after hatching, lateral view. B and C: successive stages of nauplioid development, lateral view. D and E: post-nauplioid larvae, lateral views. F: early nauplioid stage, detail of the posterior end. G: nauplioid stage, ventral view. H: early embryonic stage, ventral view. I: nauplioid stage, detail of the cephalic appendages. J: nauplioid stage, ventral view. K: late nauplioid stage, ventral view. A–K original, of material from Limfjorden in Denmark.

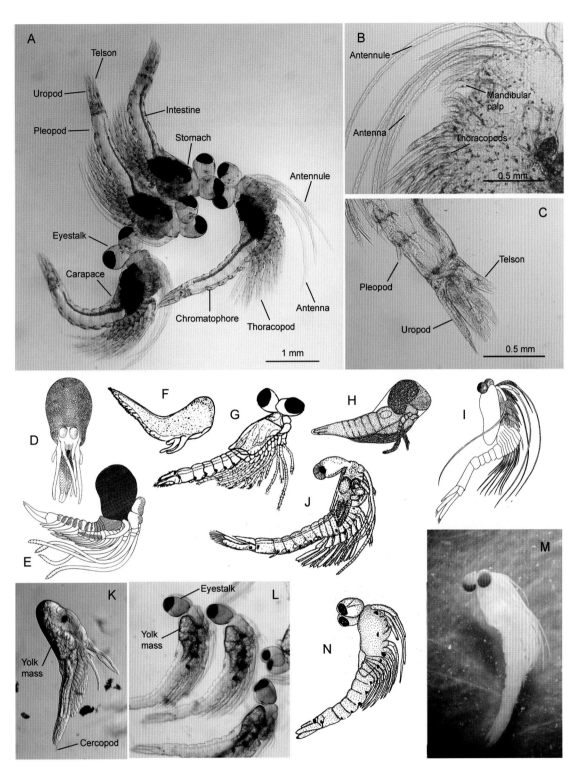

Fig. 38.3 A–C: Post-nauplioid larvae in *Hemimysis lamornae mediterranea*, light microscopy. A: post-nauplioid larvae, lateral view. B: detail of the cephalic appendages. C: detail of the posterior end of the body. D–M: selected examples of Lophohastrida and Mysida larval diversity, lateral views. D and E: drawings of *Eucopia sculpticauda*. D: nauplioid stage. E: post-nauplioid stage. F and G: drawings of *Leptomysis lingvura*. F: nauplioid stage. G: post-nauplioid stage. H: drawing of a nauplioid stage of *Mysidium columbiae*. I: drawing of a post-nauplioid stage of *Dactylamblyops latisquamosus*. J: drawing of a post-nauplioid stage of *Mesopodopsis orientalis*. K and L: *Siriella jaltensis*, light microscopy. K: nauplioid stage. L: post-nauplioid stage. M: post-nauplioid stage of *Antarctomysis maxima*, light microscopy. N: drawing of a post-nauplioid stage of *Neomysis intermedia*. A–C and K–M original, of material from the Ebro Delta (northwestern Mediterranean area); D, E, and I modified after Illig (1930); F and G modified after Wittmann (1981a); H modified after Davis (1966); J modified after K. B. Nair (1939); N modified after Murano (1964).

39

Amphipoda

CARSTEN WOLFF

GENERAL: The Amphipoda (beachhoppers and related groups) is one of the most species-rich groups within the Peracarida, with more than 170 extant families and perhaps half of the estimated 22,000 species of peracarids (J. W. Martin and Davis 2001; Ahyong et al. 2011). Amphipods are also one of the most ecologically diverse peracarid groups. They occur in marine, freshwater, and brackish environments, as well as in terrestrial ecosystems (e.g., Friend and Richardson 1986; J. Barnard and Karaman 1991; Gasca and Haddock 2004) and can be free-living, commensal, or even ectoparasitic. Although it is widely accepted that amphipods are divided into four subtaxa (the Caprellidea, Gammaridea, Hyperiidea, and Ingolfiellidea), the phylogenetic relationships of these groups are still imperfectly understood (F. Schram 1986; J. Barnard and Karaman 1991; Gruner 1993; J. W. Martin and Davis 2001). A typical (gammaridean) amphipod body is laterally compressed and divided into three major tagmata. The cephalothorax bears antennules, antennae, mandibles, maxillules, maxillae, and the first thoracopods (maxillipeds). The pereon bears seven pairs of uniramous thoracopods (pereopods), with the first two pairs usually modified as gnathopods. The pleon bears six biramous pleopods and has two functional units. The first three pairs of pleopods are longer and are used for swimming, while the last three pleopods (the uropods) are simpler in form and are located on the short urosome (the last three pleonic segments, together with the telson, at the terminal end). The urosome is reduced in caprellids and ingolfielids.

The life cycle consists of direct development for most amphipods (an embryo, a juvenile, and an adult); the life cycle of hyperiids involves a pantochelis stage (marsupial), protopleon stages 1–5 (post-marsupial), a juvenile, and an adult.

LARVAL TYPES: Amphipoda in general undergo direct development (fig. 39.1A–E) and therefore do not show any true larval forms. The hatched juveniles look like small versions of the adults (fig. 39.1E–G).

Embryonic Development: Because amphipods, like other peracarids, do not have true larvae, some details of their embryonic development are provided here for comparison with the development of other peracarids and with other crustaceans.

Post-Embryonic Development: Within the Hyperiidea—a group of parasitic/commensal amphipods usually hosted by gelatinous zooplankton, such as salps or scyphomedusae (Gasca and Haddock 2004)—some juveniles are unusual. Although hyperiid amphipods also show direct development (fig. 39.2A–D), some hatch into juvenile forms that show remarkable differences compared with adult forms. Although these are not true larvae in the traditional sense, these forms have been given larval names by many workers (such as "pantochelis larva" and "protopleon larva") (e.g., Laval 1963, 1980).

The first juvenile stage is characterized by completed segmentation of the pleon and by having a full set of differentiated pleopods (Dittrich 1987). Only the juvenile stages after the protopleon phase are able to swim and look for a new host (Dittrich 1987, 1988). The further differentiation of secondary sexual characters, such as oostegites, happens during subsequent molts (J. Kane 1963; Dittrich 1988). Some other post-embryonic stages within the Hyperiidea have received the distinct name "physosoma larva" (Woltereck 1904, 1905; Vinogradov 1999). Similar to their adults, physosoma larvae have a strongly swollen pereon and therefore are easy to identify within the water column (fig. 39.2G). Even these juvenile forms do not differ much at all from the adults, however, and the term larva is inappropriate here; such forms are mentioned in this chapter primarily for historical purposes.

MORPHOLOGY

Embryonic Development: Amphipoda generally undergo direct development and therefore pass through all embryonic stages within the egg membrane. As is typical for members of the Peracarida, embryonic development occurs in the female's ventral brood pouch. Amphipods show the same stereotyped cell-division pattern of the trunk ectoderm that is characteristic of malacostracan development (figs. 39.1A; 39.2B) (see Dohle et al. 2004). When the germ band elongates ventrally and segmentation starts, limbs form ventrally as small buds (fig. 39.1B). During ongoing development these buds elongate (fig. 39.2B), and they finally differentiate into distinct podomeres (fig. 39.1C, D) (Browne et al. 2005; Ungerer and Wolff 2005).

In contrast to isopods and tanaids, the developing embryo is curved toward the dorsal side, forming a caudal papilla (figs. 39.1C; 39.2D). Subsequently, the limb podomeres lose their round shape, and differentiation into tagmata (the cephalothorax, pereon, and pleon) takes place (fig. 39.1D) (Browne et al. 2005; Ungerer and Wolff 2005). Juveniles hatch within the marsupium (figs. 39.1E; 39.2E) and remain there for some time. They look like small adults, except that they lack secondary sexual characters. Post-marsupial development in most Amphipoda consists mainly of the gradual development of additional structures, such as oostegites in females and claspers in males (J. Kane 1963; Garcia-Schroeder and Araujo 2009).

Post-Embryonic Development: The hatchling stage resembles a small version of the adult in most groups. Some hyperiideans are an exception, having larval forms that differ from the adult; these have been given separate names historically. The hyperiid "pantochelis larva" is the first post-embryonic stage and remains inside the marsupium of the adult female (Laval 1963). Characteristically, the pleon is not completely differentiated and lacks appendages (fig. 39.2E, E′) (Laval 1980; Dittrich 1987). Ommatidia also are incompletely differentiated. This pantochelis stage is followed by "protopleon larvae" 1–4 (Laval 1980). At these stages the pleopods gradually develop, and the young leave the marsupium (usually at the first protopleon stage) and attach to the host (fig. 39.2F).

NATURAL HISTORY: Pantochelis larvae (Hyperiidea) stay inside the marsupium (Laval 1980). During the protopleon phase the young leave the marsupium and attach to specific hosts (the same individuals to which their mothers are attached). Larvae are unable to swim, but they sometimes possess modified grasping appendages adapted for movements on the host (Laval 1980). All larval forms are non-feeding. Only after larval development has been completed and the first juveniles have appeared does feeding on host tissue begin. At this time the juveniles are capable of swimming, and they start searching for a new host (Laval 1980; Dittrich 1987).

PHYLOGENETIC SIGNIFICANCE: See chapter 36 for a brief list of studies addressing overall peracarid systematics and phylogeny.

HISTORICAL STUDIES: Post-embryonic development of amphipods with larval forms was mentioned in the literature by the middle of nineteenth century. The first brief descriptions of the general biology of physosoma larvae were made with material from a German deep-sea expedition, the Valdivia Expedition (Woltereck 1904). More detailed studies on host-parasite interactions and the behavioral biology of post-embryonic stages of hyperiid amphipods were published later (e.g., Laval 1963; Dittrich 1987).

Selected References

Dittrich, B. 1987. Post-embryonic development of the parasitic amphipod *Hyperia galba*. Helgoländer Meeresuntersuchungen / Helgoland Marine Research 41: 217–232.
Garcia-Schroeder, D. L., and P. B. Araujo. 2009. Post-marsupial development of *Hyalella pleoacuta* (Crustacea: Amphipoda): stages 1–4. Zoologia 26: 391–406.
Laval, P. 1980. Hyperiid amphipods as crustacean parasitoids associated with gelatinous zooplankton. *In* M. Barnes, ed. Vol. 18, Oceanography and Marine Biology: An Annual Review, 11–56. Aberdeen, Scotland, UK: Aberdeen University Press.
Ungerer, P., and C. Wolff. 2005. External morphology of limb development in the amphipod *Orchestia cavimana* (Crustacea, Malacostraca, Peracarida). Zoomorphology 124: 89–99.
Woltereck, R. 1904. Zweite Mitteilung über die Hyperiden der Deutschen Tiefsee-Expedition: "Physosoma," ein neuer pelagischer Larventypus; nebst Bemerkungen zur Biologie von Thaumatops und Phronima. Zoologischer Anzeiger 27: 553–563.

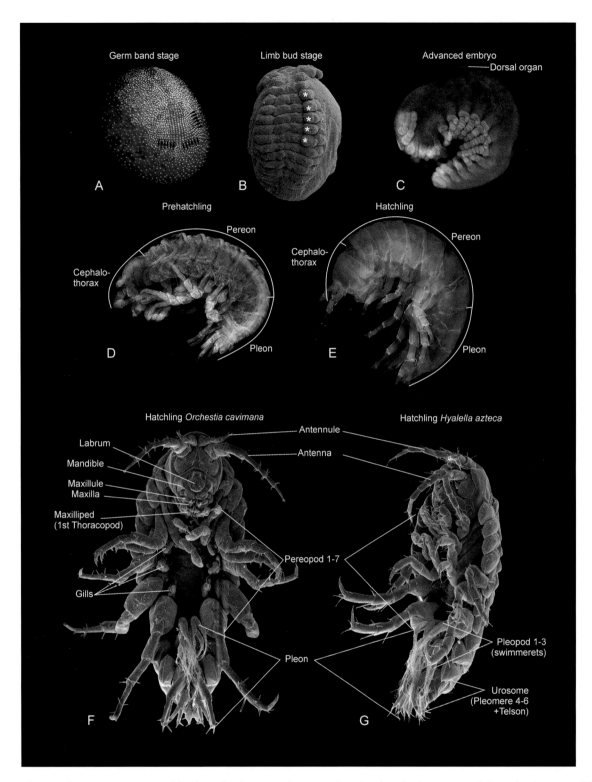

Fig. 39.1 Characteristic aspects of the direct development of gammaridean Amphipoda. A–F: stages of direct development of the talitrid amphipod *Orchestia cavimana* (Gammaridea). A: germ-disc stage, SEM, ventral view; cells of the prospective trunk ectoderm are arranged in a grid-like pattern (black arrows). B: limb-bud stage, SEM, ventral view; small limb buds (white stars) bulge out at the anterior thoracic segments; the developmental gradient from anterior (more developed) to posterior (less developed) is evident. C: advanced embryo, with a segmented limb anlagen, light microscopy, lateral view; the embryo shows a typical ventrally curved shape, with a dorsal organ on the dorsal side. D: embryo shortly before hatching (pre-hatchling), light microscopy, lateral view. E: hatchling, light microscopy, ventral view; a hatchling juvenile looks like a miniature adult, with the typical three tagmata (cephalothorax, pereon, and pleon). F: hatched juvenile of *Orchestia cavimana*, SEM, ventral view. G: hatched juvenile of the hyalellid amphipod *Hyalella azteca* (Gammaridea), SEM, ventrolateral view. A–G original, with A and C–E showing fluorescence staining: A, C, and D = Sytox green, which visualizes nuclei, and E = Congo red, which visualizes cuticular structures.

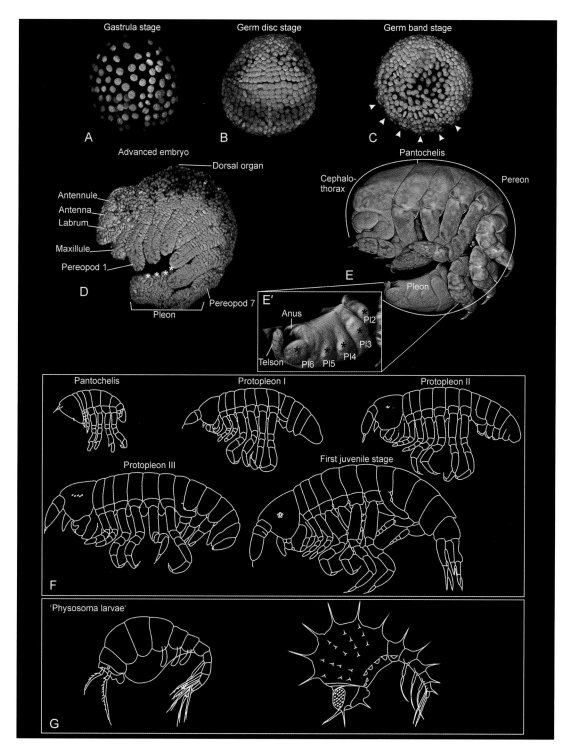

Fig. 39.2 Development of hyperiid Amphipoda. A–E: developmental stages of *Hyperia galba*, showing fluorescence staining, SEMs. A: gastrula stage. B: germ-disc stage, ventral view; ectodermal cells start to become arranged in a grid-like pattern. C: germ-band stage, with a bulging segmental anlagen (white arrowheads), lateral view. D: advanced embryo, with differentiating mouthparts and elongated pereopods, lateral view; the segmental anlagen of the pleon are evident (white stars). E: pantochelis larva hatched within the marsupium, lateral view. E': detail of the pleon, at pleonal segments 2–6 (the first segment is covered by the last pereopod); the anlagen of the limb buds (black stars) are evident. F: drawings of the post-embryonic (larval) development of the hyperiid amphipod *Vibilia armata* (Hyperiidea, Vibilioidea). G: drawings of post-embryonic stages of hyperiid amphipods referred to as physosoma larvae, lateral views; (*left*) physosoma of Lanceolidae; (*right*) physosoma of Cystisomatidae (the pereopods are cut off). A–E original, showing fluorescence staining: A–D = Sytox green, which visualizes nuclei, and E = Congo red, which visualizes cuticular structures; F modified after Laval (1963, 1980); G (*left*) modified after Woltereck (1905), (*right*) modified after Woltereck (1904).

40

CHRISTOPHER B. BOYKO
CARSTEN WOLFF

Isopoda and Tanaidacea

GENERAL: Isopods and tanaidaceans are often elongate, usually dorsoventrally depressed peracarid crustaceans. Combined, these 2 orders include an estimated 11,500 described species (Blazewicz-Paszkowycz et al. 2012; J. Williams and Boyko 2012). The majority of isopods and tanaidaceans exhibit direct development, passing from embryo to manca stages (the latter lacking the posterior pair of thoracopods, called pereopods in these taxa) to juveniles to adults. The oostegites of females are proximal plate-like limb branches of pereopods 2–5, and sometimes the maxilliped (thoracopod 1) (for a maximum of limbs 1–6; see Boxshall and Jaume 2009), which, together with the corresponding sternites, form an egg receptacle (the marsupium, but see Harrison 1984 for variations in brood-pouch formation). These, along with the gonopods (modified first pleopods) of males, are added in the molt(s) before the final molt to adulthood. Isopods, but not tanaids, molt biphasically, using body and/or limb movements to shed the anterior half of their exoskeleton prior to the posterior half (W. Johnson et al. 2001). Most of the parasitic isopods (e.g., the Cymothooidea) also show direct development, but members of the Gnathiidae and the epicarideans (the Bopyroidea and Cryptoniscoidea) have complex life cycles, with one or more juvenile stages differing radically from the adult forms. Before the life cycles were understood, some of these stages (e.g., praniza and microniscus) were once thought to represent taxa distinct from the adults. Additionally, males and females in these groups show the greatest amount of sexual dimorphism among isopods. Some species of terrestrial isopods (the Oniscoidea), and marine arcturid and antarcturid isopods, show parental care for their young after hatching.

The life cycle consists of the following stages in the groups treated in this chapter: manca I–III, juvenile, adult (most Isopoda); zuphea I, praniza I, zuphea II, praniza II, zuphea III, praniza III, non-feeding adult (the Gnathiidae); epicaridium, microniscus, cryptoniscus, bopyridium, adult (the Bopyroidea); epicaridium, microniscus, cryptoniscus, sac-like female

or cryptoniscoid-type male (the Cryptoniscoidea); and manca I–IV, juvenile, adult (the Tanaidacea).

LARVAL TYPES

Embryonic Development: Like other peracarids, isopods and tanaidaceans do not have true larvae. Instead, they hatch as miniature versions of the adult, with some exceptions (see below). Some details of their embryonic development are provided here for comparison with the development of other peracarids and with other crustaceans.

Manca: In most isopods, the embryo (fig. 40.1A–F) hatches into the first manca stage (fig. 40.1G) and then becomes a second, more fully developed manca (fig. 40.1H), although the latter stage still lacks the seventh pereopods (the last thoracopods). It is typically this second manca that emerges from the marsupium. Complete development of the posterior (seventh) pair of pereopods and differentiation of secondary sexual characteristics requires an additional third manca stage (Tomescu and Craciun 1987; Brum and Araujo 2007). Subsequent juvenile stages have seven pairs of fully formed pereopods and pleopods, except in those neotonous taxa lacking the seventh pereopods in adults (some anthuroids, and all gnathiids). Further differentiation of secondary sexual characters occurs during subsequent molts, and juveniles develop into either males or females, although protogyny and protandry are common (e.g., Araujo et al. 2004).

Praniza: The larval form in gnathiids is termed the praniza (figs. 40.2A; 40.3B–D, F–I), of which there are three postmarsupial instars. The praniza is a specialized stage for ectoparasitic feeding on fish hosts (fig. 40.3C).

Zuphea: An unfed praniza larva is sometimes called a zuphea larva (figs. 40.2B; 40.3A, E), with the term praniza reserved for instars swollen with blood from the host (Smit and Davies 2004). There are no juvenile stages between the three zuphea/praniza instars and the adult forms, which is unusual in isopods. Adult females are more similar to the praniza larvae but are readily distinguishable from them. Adult males are modified as non-feeding forms, with greatly enlarged mandibles used to defend their burrow openings and the resident female(s) (K. Tanaka 2007).

Epicaridium: Bopyroids (the Bopyridae, Dajidae, and Entoniscidae) and cryptoniscoids (7 families) exhibit the most complex isopod life cycle. In all species, the egg hatches into an epicaridium larva (fig. 40.2C, G) that moves into the water column to seek out its first host, a calanoid copepod (Boyko and Williams 2009).

Microniscus: Once it is attached to the host, the epicaridium larva molts into the microniscus (fig. 40.2D) form, which parasitizes the copepod host, sometimes with multiple larvae per copepod.

Cryptoniscus: After feeding on the hemolymph for an indeterminate amount of time (the habits of the microniscus are the most poorly known of the larval stages), the larva molts again to a third form, the cryptoniscus (fig. 40.2E, F). The cryptoniscus leaves its host copepod in search of its definitive host, which is a decapod (for bopyroids and some cryptoniscoids) or an ostracod, peracarid, nebaliacean, cirripede, or mysid.

Bopyridium: When a suitable host is found, the cryptoniscoid larva molts into a juvenile bopyridium that takes up residence in (only for entophiliine bopyrids and entoniscids) or on the host. The first larva to reach an uninfested host becomes a female, while the second and subsequent larvae become males. In at least 1 species, the male (through additional molts) may transform to become a female if the original female dies or is removed from the host (Reverberi and Pittoti 1943), but it is not known how widespread this phenomenon is within the epicarideans. In bopyroids, the male cryptoniscus becomes a male with a typical isopod form, but in cryptoniscoids the males retain their cryptoniscus appearance but possess reproductive capacity. Females in both groups become highly modified for reproduction, with some groups (entoniscids and cryptoniscoids) having females with vermiform or sac-like morphologies that make them unrecognizable as isopods. As is typical for parasites, the females produce enormous numbers of eggs, often tens of thousands per brood, compared with their free-living relatives (Boyko and Williams 2009).

In tanaidaceans, the embryo (fig. 40.2H) hatches into the first manca stage (or manca I hatchling), which remains inside the marsupium, and then becomes a second, more fully developed manca that emerges from the marsupium. There does not appear to be a molt between manca I and II forms. A third manca stage (fig. 40.2I) follows, with partially developed last thoracopods (sixth pereopods in tanaidaceans). The first and second pereonites are fused into the cephalothorax and bear the maxilliped and gnathopods, respectively. A fourth manca stage is probably present in all species, but as it is determined by the presence of budding pleopods, it cannot be readily identified in those species that lack pleopods as adults. Subsequent juvenile stage(s) have six pairs of fully formed pereopods (plus gnathopods) and pleopods (in those species that have pleo-

pods). Juveniles molt into either pre-copulatory preparatory males or females that are either gonochoristic, sequentially hermaphroditic, or simultaneously hermaphroditic (Holdich and Jones 1983; Larsen 2003).

MORPHOLOGY

Embryonic Development: The vast majority of the Isopoda and Tanaidacea show a superficial intralecithal cleavage type. Only some parasitic isopods (the epicarideans) display either a total cleavage or an intermediate mode (Strömberg 1971). After a few cleavage cycles, nuclei start moving to the egg surface and form a blastoderm (fig. 40.1A). Shortly after gastrulation, a transverse row of ectoteloblasts (black arrows in fig. 40.1B) differentiates from scattered blastoderm cells and, as is typical for malacostracan development, generates the trunk ectoderm by an invariant cell-division pattern (e.g., Dohle et al. 2004). In isopods and tanaidaceans, the germ band expands around the egg periphery, so that the developing embryo is curved toward the dorsal side. The degree of segmental differentiation shows a more or less continual decrease along the ventral germ band in an anteroposterior direction. Limbs form ventrally as small buds (fig. 40.1C), which elongate (fig. 40.1D) and then differentiate into podomeres (fig. 40.1E) (Wolff 2009). All pereopods and pleopods are present, except those of the seventh pereonite, which are lacking at this stage.

Manca: The embryonic development of an isopod is essentially complete by the hatching of a manca within the mother's marsupium (fig. 40.1F, F′). The body of the manca is typically divided into three major tagmata (the cephalothorax, pereon, and pleon). The cephalothorax bears two pairs of antennae (also called antennules and antennae), a pair of mandibles, two pairs of maxillae (also called maxillules and maxillae), and the first pair of thoracopods (called the maxillipeds). The pereon bears six (instead of seven) pairs of uniramous thoracopods. Note that isopod and tanaidacean workers differ in their use of pereopod, with thoracopods 3–8 considered pereopods 2–7 in isopods, but pereopods 1–6 in tanaidaceans. The pleon bears six biramous limbs, some of which are specialized as respiratory structures (e.g., F. Schram 1986; Gruner 1993). The last (sixth) pleonite is fused to the telson, forming a pleotelson, and bears the uropods. In general, both isopods and tanaidaceans require two post-marsupial molts (fig. 40.1G, H), through manca stages, for the complete development of the last pair of pereopods and the differentiation of secondary sexual characteristics (Tomescu and Craciun 1987; Brum and Araujo 2007). After this, they are classified as juveniles (Araujo et al. 2004).

Praniza, Zuphea: The praniza, known only in gnathiid isopods, is a manca-like stage (i.e., it lacks the last pereopod) and is full of the host's blood. It is sometimes called a zuphea when unfed. The zuphea differs only in size; it is less swollen because the gut is empty (fig. 40.2 A, B).

Epicaridium: The epicaridium is distinct from all larval stages in free-living isopods (fig. 40.2G). It vaguely resembles a sphaeromatid isopod but has six pairs of pereopods, the last being longest and, in some taxa (e.g., entoniscids), highly modified. The cephalon is rounded and extends ventrally, nearly covering the mouthparts. The antennae are long, while the antennules are very short. Eyespots are often present. There are seven pereomeres, and six pleomeres bearing slender pleopods. An anal tube is present, although the length of this tube can vary. The uropods are biramous.

Microniscus: The microniscus is more elongate than the epicaridium, with the cephalon directed forward instead of ventrally. The pereopods have only rudimentary dactyli, and buds of the developing seventh pereopods are present. The sixth pair of pereopods, while still somewhat longer than the others, is much more similar to all other pairs in this stage. The pleopods are plate-like, the uropods are weakly biramous, and the anal tube has usually disappeared.

Cryptoniscus: The cryptoniscus is elongate, rounded anteriorly and tapering posteriorly. There are seven pairs of pereopods. All articles of the first two pairs usually are relatively short and bear short stout dactyli, while the other pairs have more elongate articles and bear long thin dactyli (the sixth pair is now very similar to, though often slightly longer than, the fifth pair). The telson often has a serrate posterior edge and bears six pairs of plate-like pleopods, as well as biramous uropods with unequal branch lengths.

MORPHOLOGICAL DIVERSITY: The manca larvae essentially look like smaller, less developed versions of the adults in both isopods and tanaidaceans, and they exhibit at least some of the characters of their respective species. In contrast, the zuphea and praniza larvae (as well as the females) of different gnathiid species all look extremely similar to each other. Thus they have generally not been used in traditional morphological taxonomic distinctions between species in this group, but there are color-pattern differences (fig. 40.3B, D–I), as well as morphological ones (see G. Wilson et al. 2011). Cryptoniscoid larvae show characters that allow for placement in their higher taxon group (the Bopyridae, Entoniscidae, Dajidae, or Cryptoniscoidea), but, for the majority of species, few larvae have been reared and/or matched with other parts of the life cycle. Even less is known about the epicaridium larvae, although those of the entoniscids are distinct from all others in the possession of an extremely elongate and modified sixth pereopod (fig. 40.1G). Nothing is known about morphological diversity in the microniscus stage, as these are infrequently described, though often collected.

NATURAL HISTORY: Most isopods are marine, although both terrestrial (ca. 35% of species, Schmalfuss 2003) and freshwater (ca. 9%, G. Wilson 2008) representatives are known. The majority of isopods, and all tanaids, have limited dispersal capabilities, as the manca larvae are released in the local environment of the parent and cannot travel far from their place of birth. The larvae of gnathiids are able to use fish hosts as dispersal aids, and the microniscus larvae of epicarideans can use their intermediate copepod hosts for the same purpose; both types of larvae can also swim. It is unclear if the copepod hosts are more important in dispersal, or less so, than the definitive crustacean hosts, particularly if the definitive hosts are pelagic and wide ranging, as is the case with many carideans and penaeids. Tanaids and isopods are both known from hadal depths (Jamieson et al. 2010), although the deepest bopyrids are known from approximately 5,000 m (Birstein and Zarenkov 1970) and the deepest gnathiids from approximately 4,000 m (Camp 1988), which may reflect the distributional limitations of intermediate or definitive hosts.

PHYLOGENETIC SIGNIFICANCE: See chapter 36 for a brief list of studies addressing overall peracarid systematics and phylogeny.

HISTORICAL STUDIES: The earliest studies associated the epicaridean and cryptoniscus stages with adults, as female isopods often contained unreleased epicaridean larvae, and immature females on hosts were often accompanied by cryptonisci. The microniscus stage was originally described by F. Müller (1871) as a new genus of epicarideans parasitic on copepods and, later (Giard and Bonnier 1887), was considered to represent a distinct family, but Sars (1896–1899a) correctly considered them to be truly larval in nature. Despite this, authors as late as 1948 continued to describe species in this larval "genus" (e.g., Gnanamuthu and Krishnaswamy 1948). Similarly, both the zuphea and praniza larvae of gnathiids were described as genera by Risso (1816) and Latreille (1817), respectively, and the gnathiid males also had their own genus, *Anceus* (Risso 1816). The relationships between *Praniza*, *Anceus*, and adult gnathiids were shown by Hesse (1855, 1858, 1864) but these were not readily accepted by all researchers at the time (e.g., Bate 1858).

Selected References

Boyko, C. B., and J. D. Williams. 2009. Crustacean parasites as phylogenetic indicators in decapod evolution. *In* J. W. Martin, K. A. Crandall, and D. L. Felder, eds. Decapod Crustacean Phylogenetics, 197–220. Crustacean Issues No. 18. Boca Raton, FL: CRC Press.

Larsen, K. 2003. Proposed new standardized anatomical terminology for the Tanaidacea (Peracarida). Journal of Crustacean Biology 23: 644–661.

Smit, N. J., and A. J. Davies. 2004. The curious life-style of the parasitic stages of gnathiid isopods. Advances in Parasitology 58: 289–391.

Tomescu, N., and C. Craciun. 1987. Post-embryonic development in *Porcellio scaber* (Crustacea: Isopoda). Pedobiologia 30: 345–350.

Wolff, C. 2009. The embryonic development of the malacostracan crustacean *Porcellio scaber* (Isopoda, Oniscidea). Development Genes and Evolution 219: 545–564.

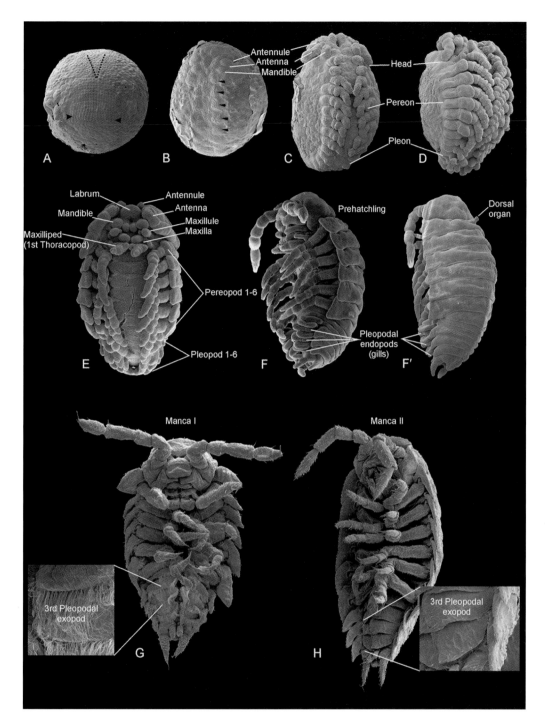

Fig. 40.1 Direct development of the oniscidean woodlouse *Porcellio scaber*, SEMs. A–F: some events in direct embryonic development (stages according to Wolff 2009). A: germ-band stage, ventral view; a transverse row of ectoteloblasts (black arrowheads) generates a grid pattern of cells (through a strict cell division) that represents the post-naupliar ectoderm (trunk primordium); in front of the forming grid pattern, the naupliar (head) region shows a characteristic bilobed appearance (black dotted line). B: limb-bud stage, with the embryo gradually lengthening along the ventral curvature, ventral view; the segments of antenna 1, antenna 2, the mandible, and the first thoracic segments (black arrows) bear limb primordia as small ventrally pointed protrusions. C: fused labrum stage, ventrolateral view. D: maxilliped differentiation stage, ventral view; due to the further differentiation of the appendages, the later tagmata (head, pereon, and pleon) are evident. E: dorsal-closure pleon stage, ventral view; the first thoracopods differentiate into medially fused maxillipeds that form the posterior border of the cephalothorax, the pereon bears six uniramous pereopods, and the pleon has six biramous pleopods. F: pre-hatchling stage, lateral view. F′: dorsolateral view, with parts of the dorsal organ evident between the cephalothorax and the pereon, lateral view; the embryo's shape changes from a ventrally to a dorsally flexed curvature, and the endopods of pleopods 2–3 differentiate into respiratory organs (gills). G: manca 1 stage, ventral view; hair tufts on the posterior rim of the exopods of pleopod 2–5 are a typical feature for this stage. H: manca 2 stage, ventrolateral view; the molt into the manca 2 stage happens within the mother's marsupium. A–F original, G and H from Wolff (2009), reproduced with permission from Springer Science+Business Media.

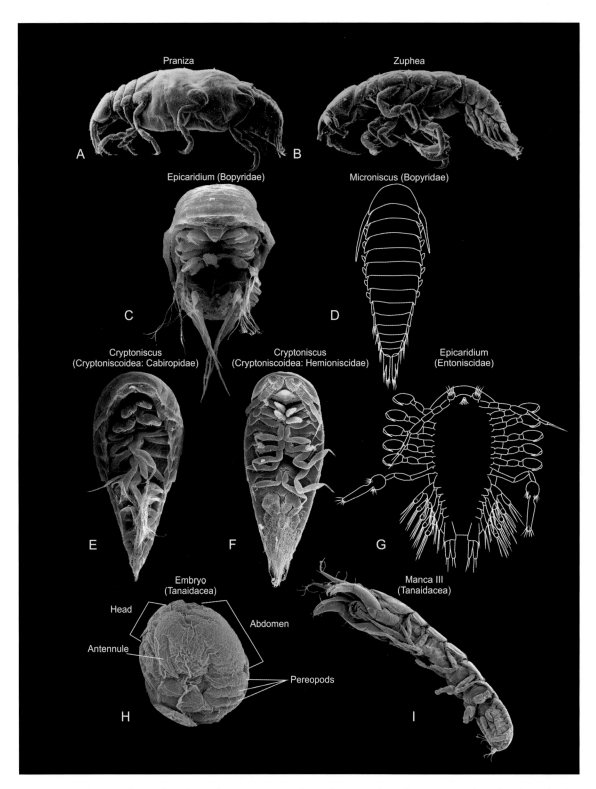

Fig. 40.2 Larval stages of isopods and tanaidaceans, SEMs (unless otherwise indicated). A–G: isopods, with indirect development. A: praniza (satiated) larva of *Elaphognathia cornigera* (Gnathiidae), lateral view. B: zuphea (unsatiated) larva of *E. cornigera* (Gnathiidae), lateral view. C: epicaridium larva of *Orthione griffenis* (Bopyridae), frontal view. D: drawing of a microniscus larva of a bopyrid isopod, dorsal view. E: cryptoniscus larva of *Cabirops bombyliophila* (Cryptoniscoidea: Cabiropidae), ventral view. F: cryptoniscus larva of *Hemioniscus pagurophilus* (Cryptoniscoidea: Hemioniscidae), ventral view. G: drawing of the epicaridium of *Achelion occidentalis* (Entoniscidae), ventral view. H and I: Tanaidacea. H: embryo of *Leptochelia* sp., ventral view I: manca III of *Leptochelia* sp., ventrolateral view. A and B courtesy of Katsuhiko Tanaka (GODAC, Japan); C, E, and F courtesy of Jason Williams (Hofstra University, New York); D modified after Sars (1896–1899); G modified after Hartnoll (1966); H and I courtesy of Kim Larsen (CIMAR/CIIMAR, Portugal).

Fig. 40.3 Larval stages of gnathiid isopods, light microscopy. A–C: *Elaphognathia cornigera*. A: zuphea (unsatiated) larva instar I, dorsal view. B: praniza (satiated) larva instar III, dorsal view. C: praniza (satiated) larva instar III feeding, lateral view. D: praniza (satiated) larvae instar III of *Gnathia trimaculata*, dorsal view. E: undescribed gnathiid zuphea, dorsal view. F: undescribed gnathiid praniza, possibly conspecific with the zuphea in E, dorsal view. G: praniza of *G. aureomaculosa*, dorsal view. H: praniza of *G. grutterae*, dorsal view. I: praniza of *G. grandilaris*, dorsal view. E–I not to scale. Material in E and F from the Great Barrier Reef in Australia. A–C courtesy of Katsuhiko Tanaka (GODAC, Japan); D–I courtesy of Nico Smit (North West University, Potchefstroom, South Africa).

41

SARAH GERKEN
JOEL W. MARTIN

Cumacea

GENERAL: Cumaceans are small benthic peracarids with a distinctive shape. The carapace is large and inflated, and the thorax and abdomen are relatively slender. Currently, 8 families and approximately 100 genera are recognized. There are about 1,500 described species, most of which live in the first few centimeters of sand or mud substrates in marine or brackish waters, from tidal to abyssal depths (Gerken 2005; Watling 2005). Sexual dimorphism is often pronounced. Cumaceans are typically very small, from 1 to 10 mm long, but some species in colder waters can be 30 mm long. All known species have direct development, with the egg hatching in the marsupium and developing into the first manca stage, which is released from the marsupium looking similar to the adults, but lacking the last pair of pereopods. Thus the life cycle consists of an egg, manca I, manca II, a subadult, and an adult, all of which are free living.

LARVAL TYPES: Cumaceans have several distinct life stages, but, as in all peracarids, they have no true larval stages.

Manca: The cumacean embryo goes through several stages in the marsupium (fig. 41.1A), hatching from the egg and then developing through 5 or 6 stages (fig. 41.1B–E) (Bishop 1982; Akiyama and Yamamoto 2004a) into the first manca stage (fig. 41.1F), which is defined as being entirely without the fifth (last) pair of pereopods. Sars (1900a) observed the development from an egg to a manca stage for several different genera, but he described the process only for the embryos of *Diastylis lucifera*, while stating that development was "essentially alike" in the other genera (*Leucon, Lamprops,* and *Pseudocuma*). The embryos are released from the brood pouch as the first manca stage. Manca I then molts into the second manca stage, where the final pair of pereopods are indicated by a ventral expansion on pereonite 5 (fig. 41.1G). Sars (1900a) suggested that *Diastylis lucifera* released its young from the brood pouch at the manca II stage, with the limb buds evident when the young were released. While this may be true for *D. lucifera*, it is not the rule, as it is generally reported that young are released

from the brood pouch at the manca I stage, and in field collections it is common to encounter manca I individuals with food in the gut.

MORPHOLOGY

Manca: The manca is defined by its lack of a fifth (last) pair of pereopods (fig. 41.1F, G). Compared with adults, mancae can be difficult to identify to species, as they tend to have fewer setae, less developed carapace ornamentation, and different proportions. The telson and uropods are more alike in proportion to each other than they are in the adult, and the appendages generally have articles that are more similar in length. The pereopods may be shorter than in the adult, and the first manca lacks the last pair of pereopods. Relative to the carapace, the abdomen is generally shorter than in the adult.

MORPHOLOGICAL DIVERSITY: The ceratocumatid genus *Ceratocuma* entirely lacks the fifth pair of pereopods in adults, making it easy to confuse adults with manca stages. The lampropid genus *Stenotyphlops* can also be confusing, as the fifth pair of pereopods is represented only by a tiny two-article filament; thus manca I and manca II stages are very difficult to identify, except by their smaller size relative to the adults, which is not helpful if there are no other individuals at a different stage of development present for size comparison. Depending on the species, sexual dimorphism may or may not begin to be evident in the manca stage. Juvenile or subadult males and females are quite similar in appearance, with the only differences being the development of additional pereopod exopods, pleopods, and antennae on the males, and the development of brood plates in the females. In males, the developing exopods, pleopods, and antennae typically are formed over several molts, increasing in size and complexity with each molt until attaining their final form in a terminal molt. In the terminal molt, long setae are added to the pereopod exopods and pleopods, the antennal peduncle becomes setose, and the antennal flagellum is present in its final form, with distinct annulations and setae. Also, the male typically acquires a carapace in the terminal molt that may vary quite remarkably from that of the female, with reductions in ornamentation and texture, and with an elongate shape. The changes in the adult male carapace, relative to the female and subadult forms, vary with the species.

NATURAL HISTORY: All cumaceans are members of the benthic infauna. A wide variety of life-history strategies are exhibited, even though relatively few studies exist (e.g., Corey 1969, 1976; Bishop 1982; Bishop and Shalla 1994; Cartes and Sorbe 1996; Akiyama and Yamamoto 2004a, 2004b). These strategies range from continuous twice- or thrice-yearly semelparous reproduction in intertidal and shallow subtidal species (Corey 1969, 1976; Bishop and Shalla 1994; Corbera et al. 2000) to iteroparous deep-sea species with discrete seasonal reproduction, life spans in excess of three years, and brooding periods in excess of one year (see Gerken 2005).

PHYLOGENETIC SIGNIFICANCE: See chapter 36 for a brief list of studies addressing overall peracarid systematics and phylogeny.

HISTORICAL STUDIES: The first cumacean was described in 1780 (and mentioned in the literature even earlier; see Holthuis 1964), and a second was described in 1804 (Gerken 2005). Studies on their development, however, were delayed, in part because of initial arguments over whether cumaceans were distinct organisms or the larvae of other crustaceans. Krøyer (1841) discovered a cumacean brooding eggs, and the argument was finally resolved. Essentially all of our current knowledge of cumacean development comes from the meticulous study by Sars (1900a), which remains the standard reference.

Selected References

Gerken, S. 2005. Cumaceans: Life History of the Cumacea. Online at http://afsag.uaa.alaska.edu/?page=history/.

Sars, G. O. 1900a. An Account of the Crustacea of Norway, with Short Descriptions and Figures of All the Species. Vol. 3, Cumacea. Bergen, Norway: Bergen Museum.

Watling, L. 2005. Cumacea World Database. Online at www.marinespecies.org/cumacea/.

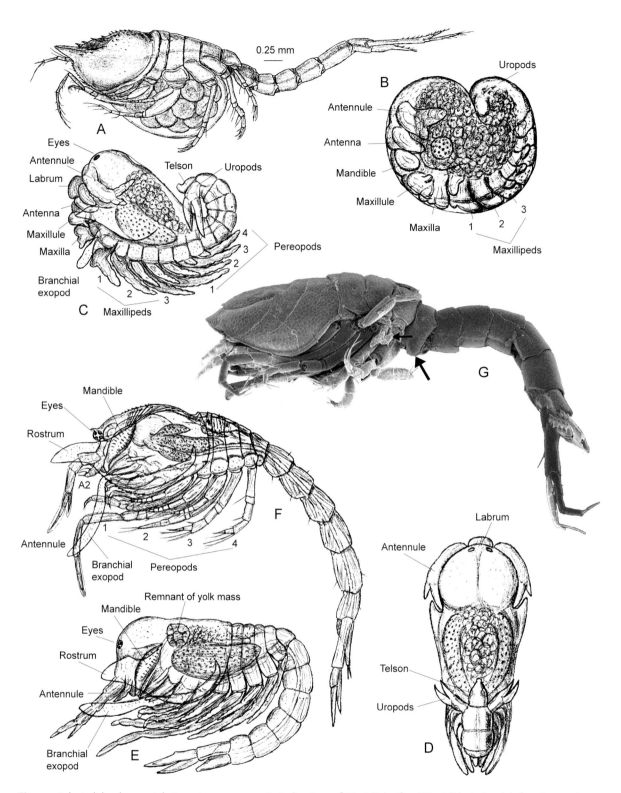

Fig. 41.1 Selected developmental stages in cumaceans. A–F: drawings of *Diastylis lucifera* (Diastylidae). A: adult female, carrying eggs within a ventral brood pouch, lateral view. B: embryo, freed from the chorion but with the thoracic limbs still within the larval skin, viewed from the left side. C: later embryo, now with all limbs exposed, lateral view. D: same as C, dorsal view. E: embryonic stage immediately following C, lateral view; note the abrupt change in the body flexure, going from dorsally curved to ventrally curved. F: embryo in "one of the last stages" (quoting Sars 1900a), which appears to be a manca I stage, lateral view. G: manca II stage of a species of *Lamprops* (Lampropidae), showing an enlarged ventral bulge (larger black arrow) on the last pereonite, indicating where the final (fifth) pereopod will appear, SEM, lateral view; this specimen is a male, indicated by the presence of an exopod on the fourth pereopod (smaller black arrow) and by the carapace ridges. A–F modified after Sars (1900a); G original.

42 | Introduction to the Eucarida

Joel W. Martin

The Eucarida is usually treated as a superorder within the subclass Malacostraca. It contains only three orders: the Euphausiacea (krill), Amphionidacea (monotypic, *Amphionides reynaudii*) and Decapoda (crabs, shrimps, lobsters, and their many relatives). Eucarids are distinguished from other malacostracans by having the carapace fused dorsally to all of the thoracic somites, and by the possession of compound eyes borne on stalks.

The concept of the Eucarida has not been seriously challenged; that is, nearly all authors agree that krill and the decapods are closely related. Richter and Scholtz (2001), however, posited that the Eucarida is paraphyletic, arising from within the Peracarida, and other authors have argued for an alignment of krill with either the Mysida (e.g., Jarman et al. 2000) or the Haplopoda (Stomatopoda) (e.g., Meland and Willassen 2007). *Amphionides* is a different story, and much less is known about it (see chapter 44). Salient works discussing the monophyly (or lack thereof) of the Eucarida include those previously cited for the phylogeny and classification of the Malacostraca.

Eucarids tend to have a more primitive mode of larval development, in that both krill and dendrobranchiate shrimps hatch as nauplius larvae. Furthermore, free-living nauplii of both the Euphausiacea and Dendrobranchiata are all lecithotrophic, whereas in most other groups of crustaceans the nauplius is a feeding stage. This functional difference has been suggested to be of phylogenetic significance (Scholtz 2000; also see the discussion in Koenemann et al. 2007a). Other decapods apparently bypass the nauplius stage during development, with eclosion occurring at a later stage (usually called the zoea), which is characterized by thoracic limb locomotion, followed in turn by a decapodid stage (variously referred to as a megalopa, glaucothoe, puerulus, or other name). In the monotypic Amphionidacea, the earliest stages of *Amphion* remain unknown, though they are unlikely to be naupliar (see chapter 44).

43

JOEL W. MARTIN
JAIME GÓMEZ-GUTIÉRREZ

Euphausiacea

GENERAL: Euphausiaceans (commonly called krill, or euphausiids) are elongate shrimp-like marine holoplanktonic and micronektonic crustaceans, often found in huge swarms (schools) that can attain enormous biomass (Everson 2000). Juvenile and adult phases have well-developed thoracic gills that are not covered by the carapace, large dark compound eyes, and no chelae on any of their five thoracic legs (pereopods), although species of *Stylocheiron* and *Nematoscelis* have elongate second or third legs. Additionally, most euphausiid species have photophores in the cephalothorax and/or the trunk (abdomen), with a few exceptions, like *Thysanopoda myniops* (Brinton 1987). There are 86 extant species, in 2 families: the Bentheuphausiidae (containing the single species *Bentheuphausia amblyops*) and Euphausiidae (all other species, divided among 10 genera) (Baker et al. 1990; Brinton et al. 1999; J. W. Martin and Davis 2001). Together, these species constitute one of the most ecologically and economically important groups of the pelagic Crustacea. Krill are a major component of many marine food webs. They are well known as the primary food of baleen whales, and they form part of the diet of seals, seabirds, fishes, celphalopods, carnivorous zooplankton, and even humans (as "okiami") and their domesticated animals (Endo and Nicol 1997; Everson 2000). Krill also are of great interest to marine biologists and oceanographers, in that they can be used as zoogeographic indicators of oceanic water masses, and as an aid to understanding speciation in the pelagic ecosystem (Brinton 1962, 1975; Brinton et al. 1999). The life cycle consists of nauplii (I and II), a pseudometanauplius (only in some sac-spawning genera), a metanauplius, a calyptopis, a furcilia, a juvenile, and an adult (figs. 43.1; 43.2).

LARVAL TYPES: Of the known species, 61 are broadcast spawners, shedding their eggs directly into the water (fig. 43.1A), and the other 25 are sac spawners, protecting the embryos in a membranous ovigerous sac attached to the last pair of thoracic legs (fig. 43.1B) (Mauchline 1971; Gómez-Gutiérrez 2002). Euphausiids use at least five distinct hatching mecha-

nisms, with significant variability between broadcast- and sac-spawning species, and even within the same species. Thus euphausiids may hatch as a nauplius, pseudometanauplius, metanauplius, or calyptopis I stage (Gómez-Gutiérrez 2002, 2003; Gómez-Gutiérrez and Robinson 2005; Gómez-Gutiérrez 2006) (fig. 43.1A, B).

Nauplius: Eggs most often hatch as a nauplius stage, which molts into a second nauplius (Marshall and Hirche 1984).

Pseudometanauplius, Metanauplius: In broadcast spawners, the more advanced second nauplius sometimes is analogous to the pseudometanauplius of sac-spawning species. The pseudometanauplius then develops into a metanauplius (fig. 43.2B).

Calyptopis, Furcilia: The metanauplius, in turn, goes through a calyptopis phase, with 3 stages, numbered I–III (fig. 43.2C–E); a furcilia phase, with a variable number of stages (fig. 43.2F–L); and, finally, the juvenile and adult phases (Mauchline 1971; Brinton et al. 1999). Furcilia stages vary in the number of terminal telson spines (fig. 43.2O–U), the development of the antennae, and the degree of development of the trunk appendages. The term cyrtopia, coined by Dana (1852) as a new genus and later used to refer to late furcilia larvae (e.g., by Sars 1885), has now been abandoned (Silas and Mathew 1977; Brinton et al. 1999). The calyptopis and furcilia phases are unique to the Euphausiacea (Mauchline and Fisher 1969; Silas and Mathew 1977; Brinton et al. 1999).

MORPHOLOGY

Nauplius: The euphausiid nauplius is typical of the nauplii of many crustacean groups (see chapter 2), except that it is totally lecithotrophic. The body is oval and unsegmented, with no compound eyes and only three pairs of limbs. The antennules are simple (unbranched), whereas both the antennae and mandibles are biramous and natatory (figs. 43.2A; 43.3A).

Pseudometanauplius: This stage is present only in some genera with a sac-spawning strategy. The overall shape resembles the nauplius, but the swimming function of the mandibles has been lost. The mandibles and maxillae are present as small buds.

Metanauplius: The overall shape resembles the nauplius, but the swimming function of the mandibles has been lost. The antennules and antennae are still natatory in function. The mandibles, maxillae, and maxillipeds are present as small buds (figs. 43.2B; 43.3B, C).

Calyptopis: The body is now composed of two parts. The carapace is distinct and forms an anterior hood-like expansion. The trunk is unsegmented in calyptopis I, and becomes segmented in calyptopis II. Calyptopis I has a functional mouth and anus, and starts external feeding in this stage. The compound eyes are imperfectly developed and immobile, and they are covered by the carapace (often called the head shield). The mandibles, maxillae, and maxillipeds are distinct, but the thoracic legs and pleopods have not yet appeared. The uropods are visible as buds in calyptopis II and completely developed in calyptopis III (figs. 43.2C–E; 43.3D–J).

Furcilia: The compound eyes are now fully developed, mobile, and project beyond the sides of the carapace. The antennae are still natatory in function. The anterior pairs of thoracic legs and pleopods are developing. In later furcilia larvae, the antennular flagella become elongate and distinctly articulated, the antennae are transformed and are no longer natatory, and the trunk legs and gills appear in succession. The spines of the telson decrease in number in successive furcilia developmental stages; the number is variable among and within species (figs. 43.2F–U; 43.3K, L).

Juvenile: All legs are now developed. The telson has only one terminal spine plus two pairs of posterolateral spines, thus assuming its final form (Mauchline 1971). The juvenile essentially has an adult form but lacks gonads. For each species, the animal's size at first maturity is typically half the maximum total length for that species (Ross and Quetin 2000) (fig. 43.1A, B).

MORPHOLOGICAL DIVERSITY: In general the nauplius, pseudometanauplius (in sac-spawning species), metanauplius, and calyptopis phases exhibit very regular ontogenetic and development patterns. They are difficult to identify to species level, but often they can be identified (particularly in stages older than the metanauplius) using subtle morphological differences in the serration and spination of the carapace, and in size and/or body shape (Brinton 1975; Knight 1975, 1976, 1978, 1980; Brinton et al. 1999). The furcilia phase includes three or more (up to six) stages, with the number of stages varying among and within species. In most species living in relatively stable environments (e.g., *Nematoscelis dificilis, Euphausia sanzoi, E. gibboides, E. eximia,* and *E. fallax*), furcilia development follows regular or fixed sequences of decreasing terminal spines on the telson and the appearance of trunk legs, whereas coastal species of *Nyctiphanes, Thysanoessa, Meganyctiphanes,* and *Euphausia* do not follow this regular pattern (Lebour 1926; R. Macdonald 1927; Einarsson 1945; Boden 1950; Gopalakrishnan 1973; Le Roux 1973, 1974; Knight 1975, 1976, 1978, 1980, 1984; Silas and Mathew 1977; Gómez-Gutiérrez 1996; Antezana and Melo 2008).

NATURAL HISTORY: The euphausiid life cycle is relatively well known (Gurney 1947; Mauchline and Fisher 1969; Brinton et al. 1999) (fig. 43.1A, B). The eggs of euphausiid broadcast spawners sink to distinct depths, except for a few species, such as *Euphausia crystallorophias,* which spawn neutrally buoyant eggs (Quetin and Ross 1984; Ikeda 1986; Harrington and Thomas 1987). Many tropical and subtropical species have continuous reproduction throughout the year (Gómez-Gutiérrez 1995). Temperate and polar species typically reproduce and hatch in the spring, with larval stages developing and molting as they swim upward. Larval stages are nourished by yolk reserves through the metanauplius stage (which is a non-feeding stage). A mouth and digestive tract appear by the calyptopis I stage, allowing these larvae to feed on marine snow, phytoplankton, and microzooplankton. This ontogenetic development is concurrent with a vertical migration to the photic zone, where phytoplankton is more abundant. Yolk reserves are now depleted, and the calyptopis I is completely dependent on external food intake. Swimming (via trunk appendages) begins in the furcilia stage, with a variable number of trunk legs (pleopods) becoming functional at each subsequent molt, along with a decrease in the number of telson spines (fig. 43.2F–U). The number of pleopods added at a given furcilia stage can vary, even within a single species, and environmental conditions can also play a role (Knight 1984). Calyptopes and early furcilia larvae typically make shorter vertical diel migrations than juveniles and adults. The juveniles and adults are commonly found in the upper 20–30 meters at night, but they may inhabit an epipelagic (<200 m in depth), mesopelagic (200–600 m), or bathypelagic (>600 m) habitat (Brinton 1962; Mauchline and Fisher 1969), often with significant downward migrations during daylight hours.

PHYLOGENETIC SIGNIFICANCE: Euphausiaceans play an important phylogenetic role, both as a basal group among the eucarids and as a proposed sister-group to the Decapoda (see chapter 42). In some ways krill are morphologically similar to dendrobranchiate shrimps, a group considered by many to be basal to the Decapoda, and these two taxa are the only eucarid groups to exhibit naupliar eclosion from the eggs. Additionally, their nauplii are different from those of most other crustacean nauplii, being lecithotrophic rather than actively feeding. This difference has been suggested to be of phylogenetic significance (Scholtz 2000; Koenemann et al. 2007a).

HISTORICAL STUDIES: In part because of their tremendous ecological and economic importance, larval development in euphausiids has been extensively studied. The odd names calyptopis and furcilia that are assigned to the unique larval phases of euphausiid development are attributable to Sars (1885). Sars followed the generic assignments of Dana (1852), who assumed that these forms were adults. Important historical treatments of larval morphology include a monograph on *Meganyctiphanes norvegica* embryology and early larval development by Sars (1885), and works by Fraser

(1936) and Silas and Mathews (1977). In their monumental krill monographs, Mauchline and Fisher (1969) and Mauchline (1980) summarized most of the knowledge about larval biology and ecology that was then known. Studies in the early morphology and taxonomy of euphausiids improved our ability to identify larvae of several krill genera and species (e.g., Boden 1950; Gopalakrishnan 1973; Knight 1975). Brinton (1975) described the early larval stages of 14 euphausiid species, including 7 relatively rare mesopelagic species of the genus *Thysanopoda*. Information on egg biometry was recently updated by Gómez-Gutiérrez et al. (2010). There are regional reviews of the biology, taxonomy, and ecology of the euphausiids in the Antarctic (Marr 1962), South Atlantic (Gibbons et al. 1999), and North Pacific (Brodeur and Yamamura 2005), but the most comprehensive and updated review of the taxonomy of the larvae, juveniles, and adults is found in the interactive multimedia compact disc by Brinton et al. (1999), from which selected material is currently available for free online in the Marine Species Identification Portal (http://species-identification.org/index.php/). We dedicate this chapter to Margaret Knight (1925–2011), Scripps Institution of Oceanography, for her inspiration and prominent work in the taxonomy of larval stages of the Euphausiacea.

Selected References

Brinton, E. 1975. Euphausiids of Southeast Asian Waters: Scientific Results of Marine Investigations of the South China Sea and the Gulf of Thailand, 1959–1961. NAGA Report No. 4. La Jolla: University of California, Scripps Institution of Oceanography.

Brinton, E., M. D. Ohman, A. W. Townsend, M. Knight, and A. Bridgeman. 1999. Euphausiids of the World Ocean. World Biodiversity Database CD-ROM Series, Windows Version 1.0. Amsterdam: Expert Center for Taxonomic Identification.

Mauchline, J. 1971. Euphausiacea Larvae. Zooplankton Sheet No. 135/137. [Copenhagen:] Conseil International pour l'Exploration de la Mer.

Mauchline, J., and L. R. Fisher. 1969. The Biology of Euphausiids. Vol. 7 *in* F. S. Russell and C. M. Yonge, eds. Advances in Marine Biology. London: Academic Press.

Silas, E. G., and K. J. Mathew. 1977. A critique to the study of larval development in Euphausiacea. Proceedings of the Symposium on Warm Water Zooplankton, Held in Goa, 14–19 October 1976, 571–582. Special Publication, National Institute of Oceanography. Goa, India: National Institute of Oceanography. Online at http://eprints.cmfri.org.in/6024/.

Life cycle phase, eggs freely spawned

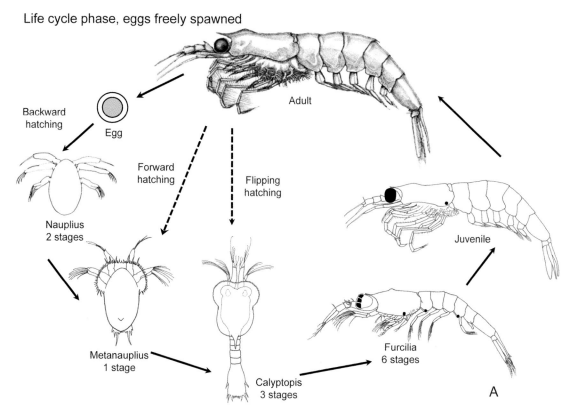

Backward
hatching

Egg

Forward
hatching

Flipping
hatching

Adult

Nauplius
2 stages

Juvenile

Metanauplius
1 stage

Calyptopis
3 stages

Furcilia
6 stages

A

Life cycle phase, eggmasses brooded

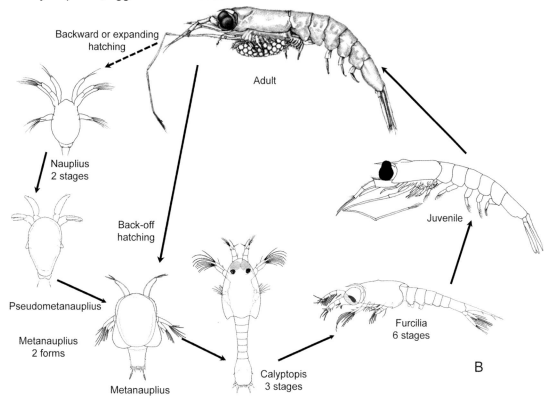

Backward or expanding
hatching

Adult

Nauplius
2 stages

Back-off
hatching

Juvenile

Pseudometanauplius

Metanauplius
2 forms

Metanauplius

Calyptopis
3 stages

Furcilia
6 stages

B

Fig. 43.1 Drawings of the life cycles of euphausiids, showing the five known hatching mechanisms for the Euphausiidae. A: broadcast-spawning euphausiids. B: sac-spawning euphausiids. A modified after Brinton et al. (1999); B from Gómez-Gutiérrez (2003), reproduced with permission of Oxford University Press.

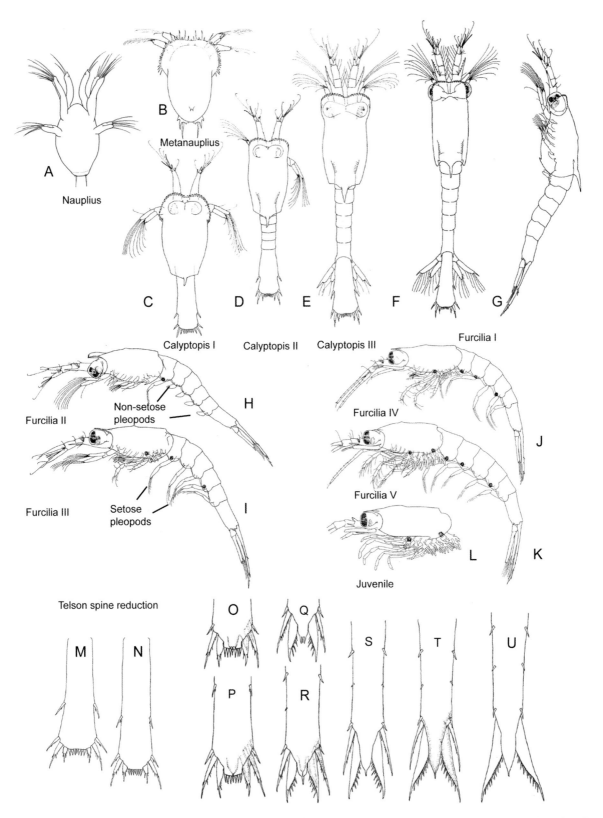

Fig. 43.2 Euphausiid larval phases. A: nauplius of *Euphausia pacifica*, dorsal view. B–U: *E. eximia* larvae. B: metanauplius, dorsal view. C: calyptopis I, dorsal view. D: calyptopis II, dorsal view. E: calyptopis III, dorsal view. F and G: furcilia I. F: dorsal view. G: lateral view. H–K: furcilia II–V, respectively, lateral views. L: anterior section of a juvenile, lateral view. M–T: telson spine reduction during furcilia-phase ontogeny, dorsal views. M: furcilia I. N: furcilia II. O: furcilia III, pre-molt to 3 terminal spines. P: furcilia III, pre-molt to 1 terminal spine. Q: furcilia IV, with 3 terminal spines. R: furcilia IV, with 1 terminal spine. S: furcilia V. T: furcilia VI. U: juvenile. A from Gómez-Gutiérrez (2003), reproduced with permission of Oxford University Press; B–U modified after Knight (1980).

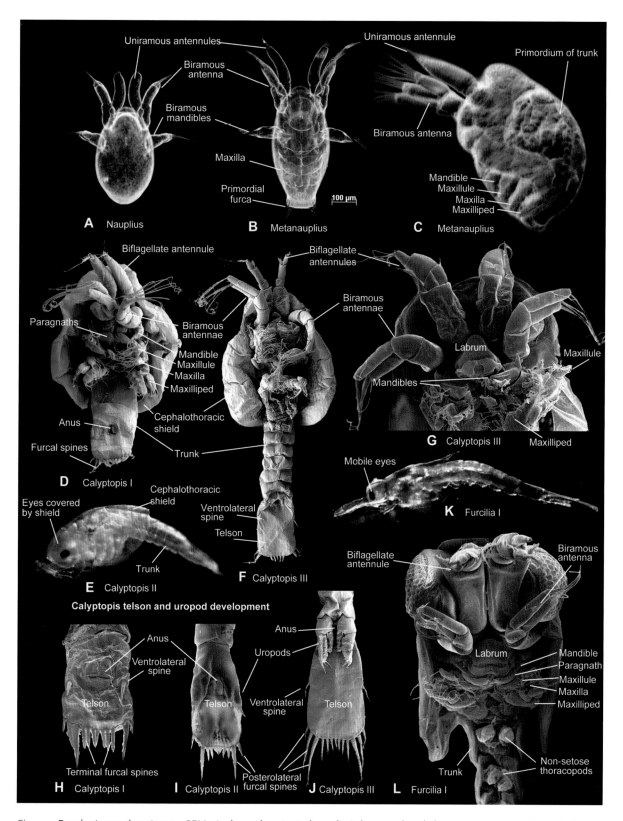

Fig. 43.3 Developing euphausiacean, SEMs (unless otherwise indicated). A: live nauplius, light microscopy, ventral view. B: live meta-nauplius of a broadcast-spawning species, light microscopy, ventral view. C: metanauplius of a sac-spawning species, left lateral view. D: calyptopis I, ventral view. E: live calyptopis III, light microscopy, lateral view. F: calyptopis III, ventral view. G: calyptopis III cephalothorax, ventral view. H and I: telson morphology, ventral views. H: calyptopis I. I: calyptopis II. J: calyptopis III, ventral view. K: live furcilia I, light microscopy, lateral view. L: furcilia I cephalothorax, ventral view. A and B courtesy of Israel Ambríz-Arreola (University of Guadalajara, Mexico); C, E, G–I, K, and L original; D, F, and J courtesy of Andreas Maas and Dieter Waloszek (University of Ulm, Germany).

44 Amphionidacea

JOEL W. MARTIN
VERENA KUTSCHERA

GENERAL: The sole reported species of the order Amphionidacea and family Amphionidae is *Amphionides reynaudii*, a small (to 30 mm long) pelagic shrimp-like crustacean found in all oceans between latitudes 36° N and 36° S. The species, although morphologically quite variable, is characterized by having a very thin, almost membranous carapace that is somewhat inflated and fringed with setae; reduced or vestigial mandibles and maxillules in adult females; biramous thoracopods, the first of which (and no others) is modified into a maxilliped; and large first pleopods in females, which extend forward to form a brood pouch with the carapace (Fransen 2010). Males are rare; nearly all records are of females or larvae. Larvae are also uncommon, although D. Williamson (1973) examined 251 larvae from the International Indian Ocean Expedition. According to Fransen (2010), "no author has seen an intact adult male or female and the figure provided by [D.] Williamson (1973) [reproduced here as fig. 44.1H] is a composite drawing based on several specimens." Relatively little is known about adults or larvae.

The life cycle includes eggs (assumed to be fertilized and then carried loose in the female's brood pouch) that probably hatch as an amphion (zoeal) larval stage (number unknown, but probably 20), followed by post-larvae (metamorphosed larvae), which are essentially small adults (Heegaard 1969; Kutschera et al. 2012).

LARVAL TYPES

Amphion Larva: It is not known if a free naupliar stage exists, but this is considered unlikely, due to the retention of eggs by the female and the advanced nature of the zoeal stages (Fransen 2010). *Amphionides reynaudii* passes through several (number unknown, but possibly as many as 20) zoeal stages (Kutschera et al. 2012), somewhat similar to what is seen in the Caridea, with which they were once grouped (Heegaard 1969). Although called a zoea by D. Williamson (1973) (and a mysis by Heegaard 1969), this larva differs from the true zoea of the Decapoda in that only the first pair of thoracopods is

developed as a maxilliped (fig. 44.2A–C, F, G). For this reason it has been called an amphion larva by some workers (e.g., D. Williamson 1973, F. Schram 1986). Nearly all we know of larval development of *A. reynaudii* comes from the work of Heegaard (1969), who unfortunately employed the terms promysis, mysis, and post-larva. The recent work of Kutschera et al. (2012), displaying the first photographs of *Amphionides* larvae, applies a more neutral terminology and refers to stages. Morphologically, amphion larvae progress more or less gradually toward the adult form.

MORPHOLOGY: Larval morphology was described by Heegaard (1969). Gurney (1942) reported a very early larval form that lacked a rostrum; Heegaard (1969) could not confirm this, but referred to the possibility of this promysis stage. The first zoeal stage (fig. 44.1A) has antennules, antennae, three mouthparts (all well developed), and three thoracopods. Only the first thoracopod develops into a true maxilliped; thoracopods 2 and 3 are natatory (fig. 44.2A–C, E–H). The fourth thoracopod appears as a bud in the second zoeal stage (fig. 44.1B), and uropod buds also begin to develop (fig. 44.1C). Other appendages appear more or less gradually in subsequent stages (figs. 44.1D–F; 44.2A–C), until the adult form is reached. Thoracopods bear large spines (fig. 44.1G) that form a feeding basket in most stages (fig. 44.2E). Also, the carapace bears a post-rostral spine and a vestigial dorsal organ in most stages (Heegaard 1969). Sexual differentiation begins in zoeal stage 9. Later stages can be differentiated into two distinct forms, based on the thickness of the antennular peduncle and the lengths of the antennular flagella (see D. Williamson 1973; Fransen 2010). According to Fransen (2010), "specimens in the last two zoeal stages with a robust antennular peduncle always have a uniramous eighth thoracopod, a small pleurobranch on the eighth somite, and the first pleopod short and biramous, and in the last zoeal stage a paired testis and paired genital openings in the coxae of the eighth thoracopods are present [in males]." In contrast, "those specimens in which the last two zoeal stages have a slender antennular peduncle possess neither an appendage on the eighth somite, nor a pleurobranch, and have a uniramous first pleopod. Ovaries develop and the genital pores on the coxae of the sixth pair of thoracopods become visible in the last zoeal stage [in females]." The female eighth thoracopod never develops in the larvae, and it is also absent in the adult.

MORPHOLOGICAL DIVERSITY: There is considerable morphological variability in *Amphionides*, which has resulted in descriptions of numerous "species" and erroneous reports on the number of larval stages. Gurney (1936b, 1942) recognized 9 zoeal stages, whereas Heegaard (1969) believed there were 13. D. Williamson (1973) showed that the stage 12 and stage 13 zoeae described by Heegaard (1969) were simply the last zoeal stage in males and females, respectively. Kutschera et al. (2012) assumed that 20 zoeal stages occur.

NATURAL HISTORY: The adult is a tropical to subtropical species, known from all oceans between 36° N and 36° S latitudes (Heegaard 1969; Fransen 2010). Most records of adult females are from depths of 2,000 m or more, but most of the known larvae of *Amphionides reynaudii* have come from depths of less than 100 m, and metamorphosed larvae are known mostly from depths of more than 700 m (Heegaard 1969; D. Williamson 1973; Fransen 2010). The scarcity of males in collections might be because of their better swimming ability (and thus their greater ability to avoid nets), or because their life span is short (D. Williamson 1973; Fransen 2010). It is assumed that hatching takes place as the eggs float upward after their release from the female's brood pouch (Fransen 2010). Because different larval stages are found throughout the year, spawning probably takes place year round. Development is very slow and gradual (Heegaard 1969; Kutschera et al. 2012). Almost nothing else is known about this species.

PHYLOGENETIC SIGNIFICANCE: *Amphionides* is an important taxon, in that it has been proposed as the sister-group to the Decapoda. Larval evidence has been important in discussions of decapod and eucarid phylogeny (e.g., D. Williamson 1973; Fransen 2010; Kutschera et al. 2012). The amphion larva is essentially a zoeal larva, but an important difference is that it has only the first thoracic limb modified into a feeding appendage (the maxilliped).

HISTORICAL STUDIES: The genus *Amphion* was established for a larval stage by H. Milne-Edwards (1832). Zimmer (1904) first used the name *Amphionides*, not recognizing the adult as the parent of the larval genus (see D. Williamson 1973; Fransen 2010). The order Amphionidacea was established by D. Williamson (1973) for the sole known species; earlier accounts of multiple species have largely been discounted as variations (D. Williamson 1973; Fransen 2010). Suggestions of similarity between the larvae of *Amphionides* and the phyllosoma larvae of spiny lobsters have been made by several authors, but these similarities are now thought to be the result of convergence (Fransen 2010). For discussions of systematic relationships that include evidence from the larval stages, see D. Williamson (1973), Fransen (2010), and Kutschera et al. (2012). The paucity of material is probably partly to blame for the general lack of knowledge about this species and its development.

Selected References

Fransen, C. H. J. M. 2010. Order Amphionidacea Williamson, 1973. *In* J. Forest and J. C. von Vaupel Klein, eds., F. R. Schram and M. Charmantier-Daures, advisory eds. The Crustacea: Complementary to the Volumes Translated from the French of the Traité de Zoologie (Founded by P.-P. Grassé). Vol. 9, pt. A, Eucarida: Euphausiacea, Amphionidacea, and Decapoda, 83–95. Leiden, Netherlands: Brill.

Heegaard, P. 1969. Larvae of Decapod Crustacea: The Amphionidae. Dana Report No. 77. Copenhagen: Høst.

Kutschera, V., A. Maas, D. Waloszek, C. Haug, and J. T. Haug. 2012. Re-study of larval stages of *Amphionides reynaudii* (Eucarida, Malacostraca, Crustacea *s. l.*) with modern imaging techniques. Journal of Crustacean Biology 32(6): 916–930.

Williamson, D. I. 1973. *Amphionides reynaudii*, representative of a proposed new order of eucaridan Malacostraca. Crustaceana 25: 35–50.

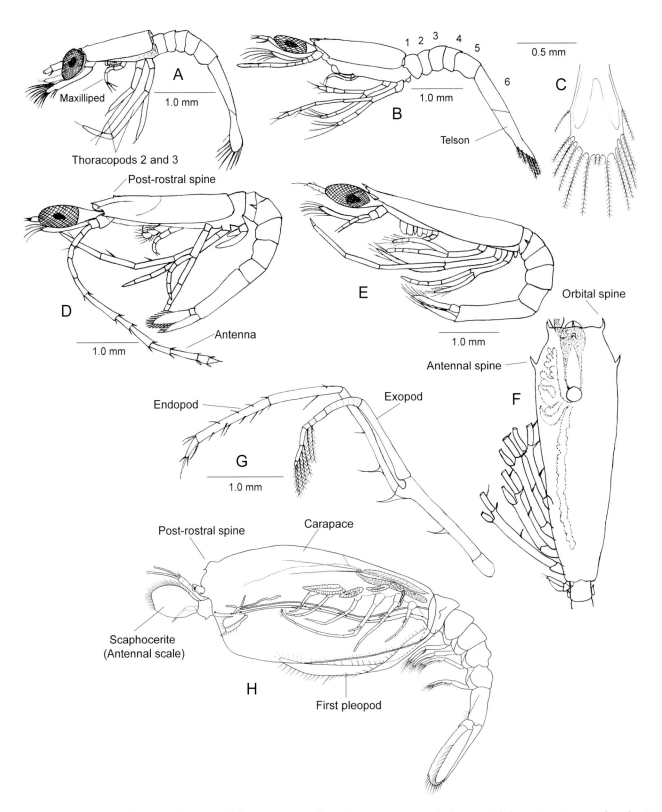

Fig. 44.1 Drawings of selected amphion (zoeal) larvae. A: stage 1, lateral view. B: stage 2, with pleomere (abdomen) somites numbered 1–6, lateral view. C: telson of stage 2, showing the uropod buds developing beneath the cuticle, ventral view. D: stage 4, lateral view. E: stage 5, lateral view. F: stage 8, carapace and proximal parts of the left-side appendages, dorsal view. G: third maxilliped (third thoracic appendage), showing the thorn-like spines of the feeding basket. H: reconstruction of an adult female (= post-larva of Heegaard 1969), lateral view. A–G modified after Heegaard (1969); H modified after D. Williamson (1973).

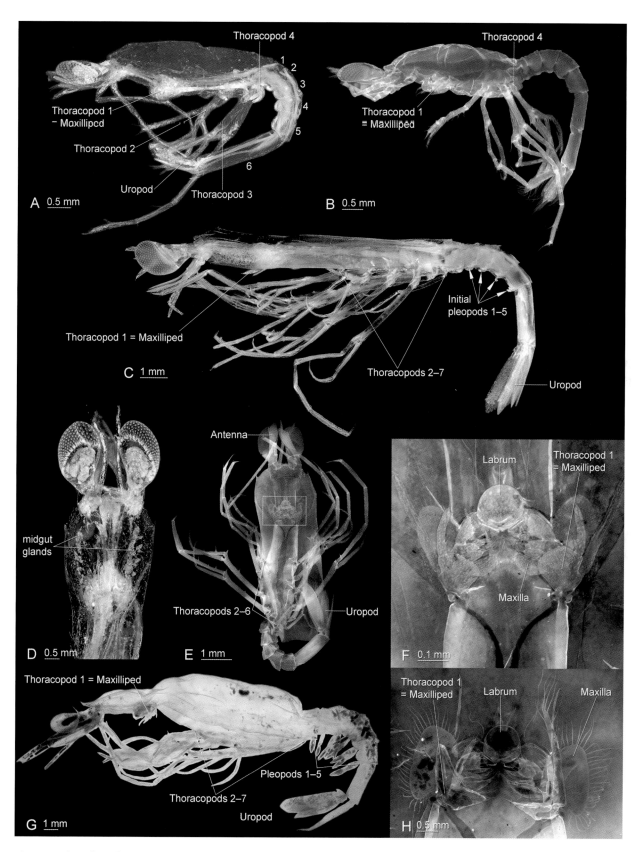

Fig. 44.2 Selected amphion (zoeal) larvae, light microscopy. A: stage 4, with the pleomere somites numbered 1–6, lateral view. B: stage 6, lateral view. C: stage 10, lateral view. D: stage 8, displaying the midgut glands, dorsal view. E: stage 9, emphasizing the feeding basket built by the thoracopods, ventral view; the rectangle is magnified in F. F: stage 9 thoracopod 1, functioning as a maxilliped. G: stage 14, lateral view. H: stage 14 thoracopod 1, functioning as a maxilliped. A–H modified after Kutschera et al. (2012).

45

Joel W. Martin

Introduction to the Decapoda

The Decapoda is a diverse and morphologically disparate group of malacostracan crustaceans. Most decapods are familiar to everyone, and many of them are consumed by humans all over the world. Included within this order are the familiar crabs, lobsters, shrimps, and hermit crabs, as well as many other lesser-known groups. The Decapoda have been referred to as the "pinnacle of crustacean evolution" (e.g., Crandall et al. 2009), because of their great diversity and successful exploitation of so many habitats. They differ from other eucarids in having three pairs of maxillipeds and five pairs of pereopods, the first of which is often (though not always) a heavy chelate appendage.

The Decapoda is usually divided into two large suborders: the Dendrobranchiata (containing the commercially harvested penaeoid shrimps), distinguished by having eggs that are not borne on the abdominal pleopods and that hatch as nauplii, and the Pleocyemata (all other groups), in which females carry the eggs on the pleopods and where eclosion is always post-naupliar. Gill morphology is another important distinguishing feature (see J. W. Martin et al. 2007). The Pleocyemata include the colorful stenopodid cleaner shrimps, the true shrimps of the Caridea, the diverse and successful freshwater (and marine) crayfish and lobster groups (Astacidea), the familiar hermit crabs and their relatives (in the Anomura), the true crabs (Brachyura), and several other large and speciose assemblages: 10 infraorders, 233 families, and 17,635 species all told (e.g., De Grave et al. 2009; Ahyong et al. 2011; Shen et al 2013).

Recent compilations have provided useful analyses of the overall diversity of living decapods. We now have relatively current lists of all living species of marine lobsters (T. Chan 2010); caridean, dendrobranchiate, procarididean, and stenopodidean shrimps (De Grave and Fransen 2011); anomurans (e.g., Baba et al. 2008; Boyko and McLaughlin 2010; McLaughlin et al. 2010; Osawa and McLaughlin 2010); and brachyuran (true) crabs (Ng et al. 2008). Even fossil families and genera have been compiled recently (De Grave et al. 2009; Schweitzer et al. 2010), such that our overall knowledge of extinct and ex-

tant decapods is better than at any point in history. Yet perhaps the only thing agreed upon in the classification and phylogeny of the Decapoda is that the group is monophyletic. Beyond that, there have always been many competing hypotheses about their internal relationships, what R. Brusca and Brusca (2003) referred to as a "popular carcinological pastime." Even the lower limit of the Brachyura—the question of what is, and what is not, a crab—has proved surprisingly elusive to resolve.

Concerning their larval development, it is fair to say that more is known about decapods than for all other groups of crustaceans combined. Nearly all decapods hatch at a post-naupliar stage and have both zoeal (planktonic) and decapodid larval forms. The latter are morphologically and ecologically transitional larvae that are usually referred to as a megalopa (e.g., in the true crabs and carideans), a puerulus (in achelatan lobsters), or a glaucothoe (in anomurans) (table 45.1). Only the dendrobranchiates (penaeoid shrimps) hatch as a nauplius larva, a condition assumed to be ancestral to the Decapoda in general. Subsequent development in dendrobranchs—although they may go through orthonaupliar, metanaupliar, zoeal (including protozoeal and mysis), and several decapodid stages—is usually called regular anamorphic (e.g., Anger 2001), referring to the gradual changes in morphology during ontogeny, rather than truly metamorphic, as in other decapod groups. Because of the great morphological diversity of decapod larvae, many early workers mistakenly assumed that some of these forms were adults. As a result, a large number of genus-level names were applied, and many of these names are still encountered in the literature on decapod larvae. The result is a vast body of literature that can be very confusing. We have attempted to clarify the use of all previously employed decapod larval names in table 45.2. The basic morphological progression of the appendages of decapod larvae is summarized in table 45.3.

Of the 9 infraorders of decapods currently recognized in the Pleocyemata, larvae remain unknown only for the Procaridoida and Glypheidea. All major groups save these two are treated in the chapters that follow.

Table 45.1 Summary of commonly employed terminology for larval phases in the Decapoda

Taxon	Hatch as nauplius (planktonic)	Zoea (planktonic)	Decapodid (usually benthic or demersal)	Chapter
Dendrobranchiata	yes	protozoea, mysis	decapodid	46
Pleocyemata				
Stenopodidea	no	zoea	decapodid	47
Caridea	no	zoea	decapodid	48
Astacidea	no	zoea	decapodid	49
Gebiidea, Axiidea (Thalassinidea)	no	zoea	megalopa	50
Achelata	no	naupliosoma, phyllosoma	puerulus, nisto	51
Polychelida	no	eryoneicus zoea	eryoneicus megalopa	52
Anomura	no	zoea	glaucothoe, megalopa	53
Brachyura	no	zoea	megalopa	54

Note: For information on the pre-zoeal stage, see individual chapters.

Table 45.2 Historically proposed names for various larval forms in the Decapoda and the current status of those names

Name	Original usage	Current taxon	Status
Penaeoidea			
Cerataspides[1]	aristeid larva	Aristeidae	no longer used[2]
Cerataspis[1]	aristeid larva	Aristeidae, *Plesiopenaeus*	no longer used[2]
Cryptopus	= *Cerataspis*	Aristeidae	no longer used
Euphema	*Gennadas* mysis	Benthesicymidae	no longer used
Hoplites	*Gennadas* mysis	Benthesicymidae	no longer used
Loxopis	penaeoid larva	Penaeiodea	no longer used
Metazoea[1]	third penaeid protozoea	Penaeidae	used occasionally[3]
Mysis[1]	dendrobranchiate larva	Dendrobranchiata	used occasionally[3]
Opisthocaris	*Solenocera* larva	Solenoceridae	no longer used
Opthalmeryon	= *Cerataspis*	Aristeidae	no longer used
Peteinura	= *Cerataspis*	Aristeidae	no longer used
Platysacus	*Solenocera* larva	Solenoceridae	no longer used
Protozoea[1]	dendrobranchiate larva	Dendrobranchiata	used occasionally[3]
Rachitia	*Penaeopsis* protozoea	Penaeidae	no longer used
Sergestoidea			
Acanthosoma[1]	*Sergestes* mysis	Sergestidae	no longer used[2]
Elaphocaris[1]	*Sergestes* protozoea	Sergestidae	no longer used
Erichthina	*Lucifer* larva	Sergestidae	no longer used
Mastigopus[1]	sergestid decapodid	Sergestidae	no longer used
Podopsis	sergestid larva	Sergestidae	no longer used
Sceletina	sergestid larva	Sergestidae	no longer used
Sciacaris	*Sergestes* larva	Sergestidae	no longer used
Xylaphocaris	*Sergestes* protozoea	Sergestidae	no longer used
Caridea			
Ampliplectus	caridean larva	Caridea	no longer used
Anebocaris	alpheid larva	Alpheidae	no longer used
Anisocaris	*Discias?* larva	Pasiphaeidae	no longer used
Atlantocaris	*Thalassocaris?* larva	Pandalidae	no longer used
Bentheocaris	*Acanthephyra* larva	Oplophoridae	no longer used
Boreocaris	*Pandalus* larva	Pandalidae	no longer used
Camptocaris	caridean larva	Caridea	no longer used
Caricyphus	oplophorid larva	Oplophoridae	no longer used
Caulurus	caridean larva	Caridea	no longer used
Copiocaris	pandalid larva	Pandalidae	no longer used
Coronocaris	palaemonid larva	Palaemonidae	no longer used
Diaphoropus	alpheid larva	Alpheidae	no longer used
Dymas	*Pandalus borealis* larva	Pandalidae	no longer used
Eretmocaris	*Lysmata* larva	Lysmatidae	used occasionally[3]
Falcicaris	pasiphaeid larva?	Pasiphaeidae	no longer used
Hippocaricyphus	hippolytid larva	Hippolytidae	no longer used
Hoplocaricyphus	*Acanthephyra* larva	Oplophoridae	no longer used
Icotopus	pandalid larva	Pandalidae	no longer used
Kyptocaris	pandalid larva	Pandalidae	no longer used
Mesocaris	palaemonid larva	Palaemonidae	used occasionally[3]
Miersia	*Lysmata seticaudata* larva	Lysmatidae	no longer used
Odontolophus	palaemonid larva	Palaemonidae	no longer used
Oligocaris	pandalid larva	Pandalidae	no longer used
Pandacaricyphus	pandalid larva	Pandalidae	no longer used
Parathanas	alpheid larva	Alpheidae	no longer used
Parva	*Acanthephyra purpurea* decapodid	Oplophoridae	no longer used
Posydon	caridean larva	Caridea	no longer used
Procletes	*Heterocarpus ensifera* larva	Pandalidae	no longer used
Retrocaris	palaemonid larva	Palaemonidae	used occasionally[3]
Rhomaleocaris	palaemonid larva?	Palaemonidae	no longer used
Polychelida			
Amphion	*Polycheles* larva	Amphionidacea	no longer used[4]
Eryoneicus[1]	*Polycheles* larva	Polychelidae	in use[5]

Table 45.2 Continued

Name	Original usage	Current taxon	Status
Achelata			
Naupliosoma[1]	achelate pre-zoea	Palinuridae, Scyllaridae	in use[5]
Nisto[1]	scyllarid decapodid	Scyllaridae	in use[5]
Phyllosoma[1]	achelate larva	Palinuridae, Scyllaridae	in use[5]
Puerulus[1]	palinurid decapodid	Palinuridae	in use[5]
Gebiidea and Axiidea (former Thalassinidea)			
Anomalocaris	callianassinid larva	Callianassidae	no longer used[6]
Oodeopus	callianassinid larva	Callianassidae	no longer used
Sesterius	axiid larva	Axiidea	no longer used
Urozoea	axiid larva	Axiidea	no longer used
Stenopodidea			
Embryocaris	*Stenopus* larva	Stenopodidae	no longer used
Anomura			
Euacanthus	*Porcellana* larva	Porcellanidae	no longer used
Glaucothoe[1]	anomuran decapodid	Anomura	in use[5]
Grimothea	*Munida* decapodid	Munididae	no longer used
Problemacaris	anomuran larva	Anomura	no longer used
Prophylax	pagurid larva	Paguridae	no longer used
Zoeides	*Hippa* larva	Hippidae	no longer used
Zoontocaris	galatheid larva	Galatheidae	no longer used
Brachyura			
Acanthocaris	raninid larva	Raninidae	no longer used
Acanthotribola	brachyuran larva	Brachyura	no longer used
Cyllene	portunid larva	Portunidae	no longer used
Cyllenula	brachyuran larva	Brachyura	no longer used
Dohrnia	*Xantho* larva	Xanthidae	no longer used
Hemisphaerium	brachyuran larva	Brachyura	no longer used
Hyadella	brachyuran larva	Brachyura	no longer used
Lonchophorus	= *Pluteocaris*, dorippid larva?	Dorippidae	no longer used
Marestia	grapsid decapodid	Grapsidae	no longer used
Megalopa[1]	brachyuran decapodid	Brachyura	in use[5]
Monolepis	brachyuran larva	Brachyura	no longer used
Paradesmarestia	*Pachygrapsus* larva?	Grapsidae	no longer used
Paramonolepis	brachyuran larva	Brachyura	no longer used
Pluteocaris	= *Lonchophorus*, dorippid larva?	Dorippidae	no longer used
Protodesmarestia	grapsid larva?	Grapsidae	no longer used
Protomonolepis	*Pilumnus* larva?	Pilumnidae	no longer used
Pseudomonolepis	brachyuran larva	Brachyura	no longer used
Pterocaris	brachyuran larva	Brachyura	no longer used
Quadribola	brachyuran larva	Brachyura	no longer used
Spinaria	brachyuran decapodid	Brachyura	no longer used
Tribola	*Plagusia* decapodid	Plagusiidae	no longer used
Tricuspidella	brachyuran larva	Brachyura	no longer used
Zoea[1]	brachyuran larva	Brachyura	in use[5]

Source: Compiled by J. W. Goy and J. W. Martin.

[1] See glossary for definition.

[2] No longer used as a valid genus name but sometimes still employed as a general descriptive term for these distinctive larvae, in the same sense that the terms megalopa and zoea are used for distinct phases in the life cycle of brachyurans. See chapter 46 and Bracken-Grissom et al. (2012).

[3] The use of these names for general stages of development (i.e., not as valid genus names) is often dependent on the author. See the corresponding chapters.

[4] A few early workers thought that the *Amphion* larva, known only from plankton, might be the larval stage of the bizarre lobster genus *Polycheles*. More often the word amphion is used to refer to the larval stages of the genus *Amphionides* in the Amphionidacea, but even that usage has been questioned. See chapter 44 and Kutschera et al. (2012).

[5] In general use as a descriptive term for a developmental phase but not as a valid genus name.

[6] Not to be confused with the large Cambrian predator of the genus *Anomalocaris*, a name that is still valid and used frequently in the paleontological literature.

Table 45.3 Synopsis of appendage development and function in decapod larvae

Development		Larval phase			Post-larval phases	
Ecological role		Pelagic/planktonic		Transitional	Benthic/demersal[1]	
Somite	Appendage	Nauplius	Zoea	Decapodid	Juvenile	Adult
cephalon	eyes	simple	cmpd	cmpd	cmpd	cmpd
	antennule	loco	sensory	sensory	sensory	sensory
	antenna	loco	loco/se[2]	sensory	sensory	sensory
	mandible	loco	feeding	feeding	feeding	feeding
	maxillule	nf	feeding	feeding	feeding	feeding
	maxilla	nf	feeding	feeding	feeding	feeding
thorax	maxilliped 1	nf	loco	feeding	feeding	feeding
	maxilliped 2	nf	loco	feeding	feeding	feeding
	maxilliped 3	absent	nf/loco	feeding	feeding	feeding
	pereopod 1	absent	nf/loco	loco	loco	loco
	pereopod 2	absent	nf/loco	loco	loco	loco
	pereopod 3	absent	nf/loco	loco	loco	loco
	pereopod 4	absent	nf/loco	loco	loco	loco
	pereopod 5	absent	nf/loco	loco	loco	loco
	maxilliped 1	nf	loco	feeding	feeding	feeding
—	gills	absent	nf	present	present	present
abdomen[4]	pleopod 1	absent	nf	loco	em/loco[3]	end/loco[3]
	pleopod 2	absent	nf	loco	em/loco[3]	end/loco[3]
	pleopod 3	absent	nf	loco	em/loco[3]	end/loco[3]
	pleopod 4	absent	nf	loco	em/loco[3]	end/loco[3]
	pleopod 5	absent	nf	loco	em/loco[3]	end/loco[3]
—	sexual organs	absent	absent	absent[5]	nf	present

Source: Modified from work by Paul F. Clark, presented as part of his doctoral thesis; used by permission.

Abbreviations: Absent = appendage absent; cmpd = paired compound eyes; em = endopods metamorphosing, becoming adult-like in form and function but not yet fully functional; end = endopods functional (primarily as egg-carrying appendages in females and as reproductive structures in males); feeding = appendage present and functional for feeding; loco = appendage present and functioning for locomotion; nf = appendage developing but not yet functional; present = present and functional, but not used in feeding or locomotion; simple = simple (not compound).

[1] With the exception of the Sergestoidea.

[2] In some protozoea larvae, the antenna remains locomotory.

[3] In both the Dendrobranchiata and Caridea, the pleopods retain a locomotory function, in addition to their function in reproduction.

[4] Also referred to as a pleon by some researchers.

[5] With some exceptions (e.g., the well-developed male pleopods found in the eryoneicus megalopa stage of polychelid lobsters). See chapter 52.

46

Dendrobranchiata

JOEL W. MARTIN
MARIA M. CRIALES
ANTONINA DOS SANTOS

GENERAL: The Dendrobranchiata contains the economically important penaeoid and sergestoid shrimps. Among other characters, members of this suborder are recognized by their caridoid facies (elongate shrimp-like shape), dendrobranchiate (highly branching) gills, typically chelate first three pairs of legs, a second abdominal somite with pleura that do not overlap those of the first somite, and eggs that are shed directly into the water and hatch at a naupliar stage (see Pérez-Farfante and Kensley 1997; Tavares et al. 2009; Tavares and Martin 2010). There are approximately 500 extant species, currently partitioned among 7 families in 2 superfamilies. Nearly all are marine, with species found from shallow tropical waters to depths of 1,000 m or more on the continental slopes. Almost half of the known species are members of the Penaeidae, which inhabit shallow and inshore tropical and subtropical waters; species in this group form the basis of massive seafood industries. The Aristeidae, Benthesicymidae, and Sergestidae are predominantly deep-water families and contain species that are either deep benthic dwellers or are members of the meso- and bathypelagic fauna; some species of Sergestidae, however, are found in fresh water. Most Solenoceridae occur in deeper offshore waters. Species of the Sicyoniidae inhabit coastal waters up to 200 m in depth. Species of the Luciferidae are all planktonic (Pérez-Farfante and Kensley 1997; Tavares and Martin 2010). The life cycle consists of a nauplius, a zoea (including protozoea and mysis stages), a decapodid, a juvenile, and an adult. All stages are free living. Larval development is regular anamorphic (Anger 2001), meaning that the appearance of characters and shape changes are gradual.

LARVAL TYPES: Nearly all species shed their eggs directly into the water. Luciferids, which brood their eggs for a short time on the posterior thoracopods, are the exception.

Nauplius: Dendrobranchiates are unique among decapods in having their eggs hatch as free-swimming nauplii; thus they are also the only decapods with three larval phases (a nauplius, a zoea, and a decapodid).

Zoea (Protozoea, Mysis), Decapodid: Neither the transition from naupliar to zoeal stages, nor the transition from decapodid to early juvenile stages, is truly metamorphic. Nevertheless, larvae can be partitioned into a nauplius, a zoeal phase (subdivided into protozoea and mysis stages), and (usually) several decapodid stages (in this group, also called post-larvae). For most decapod groups, the term post-larva refers to any developmental stage beyond the decapodid, and especially to juveniles. In the Dendrobranchiata, however, the term post-larva has been used for the transitional phase between pelagic larvae and benthic juveniles, covering several decapodid stages. There are different substages within all of these stages; the number varies among and within the families. The Penaeidae usually have 5–6 naupliar, 6 zoeal (3 protozoea + 3 mysis), and several (1–9) decapodid stages before settlement (Dall et al. 1990). All described solenocerids have 5–6 nauplii and 5 zoeae (3 protozoeae + 2 mysis) (Calazans 2000). In the Benthesicymidae, the genus *Gennadas* exhibits 4 mysis stages (Rivera and Guzmán 2002). Although the pattern of appendage development in the 2 superfamilies is essentially identical, the gross morphology of the larvae in the two groups can be quite distinctive (Tavares and Martin 2010). Sergestoid larvae tend to be more spinous, to the extent that in the past, distinctive names have been applied to them, such as elaphocaris (for the protozoea) and acanthosoma (for the zoea or mysis stage). The bizarre "larval genera" *Cerataspides* and *Cerataspis* are widely thought to belong to the penaeiods; *Cerataspis monstrosa* recently was shown to be the larval stage of the aristeid shrimp *Plesiopenaeus armatus* (Bracken-Grissom et al. 2012), whereas *Cerataspides* has never been attributed with certainly to an adult genus (Heegaard 1966; S. Morgan et al. 1985; Bracken-Grissom et al. 2012).

MORPHOLOGY

Nauplius (fig. 46.1): Early nauplii are characterized by having three pairs of propulsive limbs (antennules, antennae, and mandibles) and a naupliar eye (not illustrated here; see chapter 2); these larvae are often called orthonauplii (D. Williamson 1982a). The antennules are always uniramous and lack a flagellum. The antennae are typically biramous. They may be unsegmented in early nauplii, but the peduncle always consists of two segments by the end of this stage. The endopod is always unsegmented, and the exopod is segmented throughout its length. The endopod and exopod of the mandibles are unsegmented in penaeids. The mandibular coxa (including the gnathobase) becomes the body of the mandible in the next stage; the rest of the appendage is retained as a palp (D. Williamson 1982a). Later naupliar stages (fig. 46.1)—characterized by buds of the maxillules, maxillae, and the first and second maxillipeds, as well as by masticatory swellings (gnathobases) on the mandibles—are often referred to as metanauplii (e.g., Dall et al. 1990). The end of the naupliar stage is reached when the first two maxillipeds are developed, the telson is bifurcate, and the carapace bud is developed. This carapace, once acquired, is never lost, but it may undergo one or more metamorphoses. A simple carapace is present in the two naupliar stages of *Lucifer*, but it is absent in those of *Sergestes* (D. Williamson 1982a). The number of naupliar stages in dendrobranchiates varies from five to eight.

Protozoea: Zoeal development in dendrobranchiates is divided into two stages: the protozoea (fig. 46.2) and the mysis (fig. 46.3). The protozoea stage exhibits three distinct substages. In the first substage, all five pairs of head appendages are functional. The antennules and antennae maintain their ancestral natatory function, a function shared with the first two pairs of thoracopods. The mandibles are now feeding appendages; the endopod and exopod are lost, and the masticatory surface is divided into an incisor process and a molar process (Dall et al. 1990). Between the two processes are a variable number of long movable serrate teeth, one of which is a *lacinia mobilis* (*sensu* Moore and McCormick 1969). All thoracic somites are discernible; the more posterior thoracic appendages are absent or rudimentary (D. Williamson 1982a). A carapace, attached at the somite of the maxilla, covers only part of the thorax, to about the fourth segment (Dall et al. 1990). The carapace of the protozoea in the Penaeidae and Sergestidae usually lacks a rostrum. The carapace is unarmed in penaeids, but solenocerids possess many spines in all stages (e.g., fig. 46.2E, H). For example, protozoea I of *Pleoticus muelleri* has a pair of frontal spines on the anterior portion of the carapace, as does *Mesopenaeus tropicalis* (Calazans 2000). Sergestids usually bear dorsal and lateral spines or processes on the carapace, which are often elaborately branched (fig. 46.2F). Unstalked compound eyes, as well as the nauplial eye, are present beneath the carapace (Dall et al. 1990). In penaeid and sergestid protozoeae, the eyes are covered in stage I only (D. Williamson 1982a). Frontal organs are also present only in this substage. Swimming and feeding are now virtually continuous (Dall et al. 1990). The pleon is still unsegmented and ends in a large bilobed telson, which seems to have either a sensory or a cleaning function. Dall et al. (1990) confirmed the use of the telson for cleaning the antennae and mouthparts, and they also noted that the telson can work as a hydroplane

to change direction. Forward propulsion is provided by the antennae.

In the second protozoea, the compound eyes become stalked; the frontal organ disappears; and a rostrum, supraorbital spines, and most leg rudiments appear (Dall et al. 1990). The carapace in penaeids has a rostral spine and a pair of supraorbital spines in protozoeae II and III (D. Williamson 1982a). The rostral spines in solenocerids can range from smooth in *Pleoticus muelleri* to spinulate in *Solenocera necopina* and *Mesopenaeus tropicalis* (Calazans 2000). The pleon is divided into six segments, and the telson is not separated from the sixth somite (Dall et al. 1990).

The third protozoea is often referred to as a metazoea, because the remaining thoracopods are prominent and biramous, although still not functional (D. Williamson 1982a). In this stage the biramous and setose uropods appear, and the telson separates from the sixth pleomere. In larval sergestoids, the uropods are non-functional in the protozoea stage and do not become functional until the mysis stage.

Mysis (fig. 46.3): With the molt to the first mysis stage, the larvae undergo major changes in their appearance, and their bodies begin to look shrimp-like. The most significant change is the development of pereopods, with large exopods; these are now functional locomotory appendages. In some Penaeoidea, however, the posterior exopods do not become functional until mysis II (D. Williamson 1982a). The larva now swims backward. The body is vertical, with the telson up, and slowly spins on the vertical axis; this motion is augmented by rapid backward thrusts from the flexion of the pleon (Dall et al. 1990). The carapace conforms more closely to and covers the thoracic somites. During this stage in the penaeids, the rostral, hepatic, and/or pterygostomial spines appear for the first time. In the antennules of the Penaeoidea and Sergestoidea, the peduncle divides into three segments and the inner ramus appears (D. Williamson 1982a). The antennae, which have lost their locomotory function, also change in appearance: the exopods of the antennae become flattened antennal scales, and a statocyst appears near the base of the antennule (Dall et al. 1990). The mandibles again become biramous when the bud of the mandibular palp appears. The maxillae are largely unchanged; only the outer lobe grows a proximal extension and develops many marginal setae (D. Williamson 1982a). The setose epipod of the maxillule disappears, while the epipod on the maxilla enlarges with each molt to form the scaphognathite in the juvenile. All three maxillipeds are now functional, rudimentary chelae appear on the first three pereopods, and gill rudiments appear on the thoracopods in the later substages (Dall et al. 1990).

The number of mysis stages recorded in the literature is quite variable. Dall et al. (1990) defined three mysis stages, on the basis of pleopod development: the first mysis stage has no (or barely perceptible) pleopod buds on the first five pleomeres; the second mysis has prominent non-articulated pleopod buds; and the third mysis has small, two-segmented, lightly setose but non-functional pleopods on the first five pleomeres.

Decapodid (fig. 46.4): There are no dramatic changes in morphology with the molt to the decapodid phase. In this stage all exopods are reduced or lost, and the uniramous pleopods become large, setose, and functional, taking over as the sole locomotory appendages. The larva again swims forward, with a rhythmic beating of the pleopods (Dall et al. 1990). All of the non-sensory cephalic appendages and the three maxillipeds assume new functions, as mouthparts. The chelae on the first three pereopods are now functional, with small teeth and short bristles terminally. The supraorbital spines disappear. The antennae change slightly, with both distal rami of the first antennae becoming segmented. The mandibular palp, present since the midmysis stage, is usually segmented, setose, and probably functional; the *lacinia mobilis* is often reduced or absent (Dall et al. 1990). The endopods of both maxillae become unsegmented, degenerate, and palp-like. Both the endopod and exopod of the first maxilliped are vestigial. The endopod of the second maxilliped becomes recurved, its setation changes dramatically, and the exopod degenerates. Gills on the thoracopods are still only rudimentary. The telson continues to narrow distally. It is only faintly cleft in penaeids, but in some solenocerids (e.g., *Pleoticus muelleri* at this stage) the telson is triangular in shape, with a pointed tip (Calazans 2000). Subsequently, through a gradual series of molts, the eventual adult condition is reached.

MORPHOLOGICAL DIVERSITY: Despite having relatively few species (ca. 500), there is considerable diversity within each of the larval types in the Dendrobranchiata, particularly between penaeoids and sergestoids. The carapace of sergestoids usually has a pair of frontal processes, a pair of long lateral processes, and one posterior dorsal process, which can, in turn, bear spines or setae and secondary branching. The carapace of penaeoids is typically smooth, except in the Solenoceridae. They possess spines that are often robust and short, lacking secondary branching, and the margin of the carapace is usually serrated. Mysis stages of the Solenoceridae possess fewer spines, and the serrate margin of the carapace disappears gradually as they develop further. In older stages of sergestoids, the carapace processes become smaller, extra pairs of processes can appear on the carapace, and the telson becomes smaller than the uropods. In the protozoeal stages, the few known luciferid larvae possess a smooth carapace, a well-developed rostrum, and three simple spines on the posterior margin. In the mysis stages the carapace has a small rostrum, and only a pair of anterolateral spines and a pair of hepatic spines. The bulky shape of the spectacular carapace of *Cerataspides* and *Cerataspis* (fig. 46.3F)—with its various tubercles, horns, spines, and large oil droplets (contained in dorsal carapace tubercles)—most likely provides buoyancy to support their pelagic life.

NATURAL HISTORY: Most adult penaeids are benthic dwellers that inhabit shallow waters, whereas the aristeids, benthesicymids, and sergestids are deep benthic dwellers or form part of the meso- and bathypelagic fauna. Larval development of penaeids usually occurs in offshore waters, where females lay demersal or pelagic eggs. After a few hours, the eggs develop into free-swimming nauplii. During development, planktonic stages (nauplius, protozoea, mysis, and decapodid) pass through a series of feeding and behavioral changes. Naupliar phases do not feed; protozoeal stages feed mainly on phytoplankton; and mysis and decapodid stages ingest microalgae and zooplankton, such as copepods (e.g., Ewald 1965). Protozoeal and mysis stages of many penaeids perform diel vertical migrations in the water column. Later in their development, decapodids change their behavior and migrate vertically, with a tidal-cycle rhythm to enhance their horizontal advection to coastal areas (e.g., Rothlisberg et al. 1995; Criales et al. 2010). Larval stages develop rapidly, needing only 10–15 days to reach the first decapodid stage. Decapodids remain in planktonic stages for an additional 10–15 days, until they become juveniles ready to settle. There is little information about the larval ecology of deep benthic or pelagic species, but from the few available studies we can infer that the sequence of development, the duration of the larval stages, and feeding strategies are similar to those of the coastal penaeids (e.g., Calazans 2000).

PHYLOGENETIC SIGNIFICANCE: Dendrobranchiates are widely thought to be the least derived of the extant Decapoda. The fact that they, alone among decapod groups, shed their eggs directly into the sea, as well as exhibit naupliar eclosion and essentially anamorphic (for decapods) development, are major reasons for this assumption. Additionally, their nauplii are lecithotrophic, a characteristic that they share with krill (the Euphausiacea)—most other crustacean nauplii are planktotrophs—a fact that some workers (e.g., Scholtz 2000) have argued is of phylogenetic significance. Larval characters have been used to support phylogenetic studies of dendrobranchs compared with other decapods, and systematic studies within the group (e.g., see dos Santos and Lindley 2001; Tavares and Martin 2010).

HISTORICAL STUDIES: Because of the great commercial importance of penaeoid shrimps in fisheries and aquaculture, larval development is known for several species, especially for the genera *Penaeus* and *Metapenaeus*. For many other penaeid genera (e.g., *Tanypenaeus* and *Heteropenaeus*), however, no descriptions of any species exist. Furthermore, for deep-water or pelagic species, descriptions of larval stages are limited to a few species per family, and to larvae caught in plankton. Regional keys exist for identifying penaeoid larvae and postlarvae using laboratory-reared and field-caught larvae. Cook (1966) developed a comprehensive and useful key for penaeid larvae from the Gulf of Mexico, using the setal formulae of the protopod and endopod of the antenna for the protozoeal phase, and the spine patterns of the body for mysis and decapodid stages. These features, in addition to the shape and number of spines of the telson and rostrum, have been the main criteria used for the development of other generic identification keys. Some of the most valuable keys for penaeid larvae in other regions are those of Boschi and Scelzo (1969) for 3 penaeid genera of the Argentinean coast, Hassan (1974) for penaeid larvae of Pakistan, Paulinose (1982) and C. Jackson et al. (1989) for the larvae of Indo-Pacific penaeid genera, and Calazans (1993) for 15 penaeid genera of southern Brazilian waters. Other valuable generic identification keys have been developed for sergestid larvae (Gurney and Lebour 1940; Knight and Omori 1982; Hashizume 1999). Dall et al. (1990), based on some of the above regional keys and descriptions, developed an identification key for the larvae of 5 dendrobranchiate families and 10 genera of the Penaeidae. Dos Santos and Lindley (2001) produced a more recent key to genera and species of dendrobranchiate larvae for the species occurring in the northeastern Atlantic Ocean.

Different terms have been proposed for the various larval stages. In the Penaeidae, larval morphology and behavior were first fully described by F. Müller (1864), who used the terms nauplius, protozoea, mysis, and post-larva (megalopa), as did D. Williamson (1982a) and Dall et al. (1990). More recently, Anger (2001) used a division similar to that of D. Williamson (1982a), with three different phases (nauplius, zoea, and decapodid), avoiding the term megalopa because its use is most commonly restricted to true crabs (Brachyura), although it

has also been used for the decapodid stage in other taxa, such as anomurans and thalassinideans. Larvae also have played an important role in the classification of adult dendrobranchiates (see Tavares and Martin 2010).

Selected References

Dall, W., B. J. Hill, P. C. Rothlisberg, and D. J. Sharples. 1990. The biology of the Penaeidae. Advances in Marine Biology 27: 1–489.

dos Santos, A., and J. A. Lindley. 2001. Crustacea Decapoda, Larvae, 2: Dendrobranchiata (Aristeidae, Benthesicymidae, Penaeidae, Solenoceridae, Sicyonidae, Sergestidae, and Luciferidae). International Council for the Exploration of the Sea Identification Leaflets for Plankton / Fiches d'Identification du Plancton, Leaflet No. 186. Copenhagen: International Council for the Exploration of the Sea.

Gurney, R. 1943. The larval development of two penaeid prawns from Bermuda of the genera *Sicyonia* and *Penaeopsis*. Proceedings of the Zoological Society of London B113: 1–16.

Heegaard, P. 1966. Larvae of Decapod Crustacea: The Oceanic Penaeids *Solenocera-Cerataspis-Cerataspides*. Dana Report No. 67. Copenhagen: Høst.

Williamson, D. I. 1982a. Larval morphology and diversity. *In* L. G. Abele, ed. The Biology of Crustacea. Vol. 2, Embryology, Morphology, and Genetics, 43–110. New York: Academic Press.

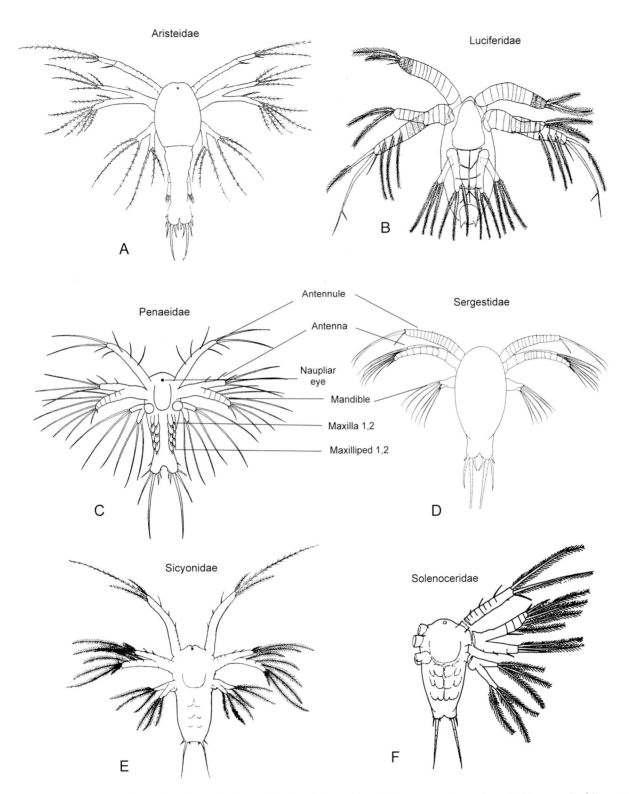

Figure 46.1 Drawings of examples of nauplius larvae of the Dendrobranchiata. A: last-stage metanauplius of *Aristaeomorpha foliacea* (Aristeidae), dorsal view. B: nauplius stage III of *Lucifer typus* (Luciferidae), ventral view. C: nauplius stage V of *Rimapenaeus constrictus* (Penaeidae), ventral view. D: nauplius stage IV of *Sergestes similis* (Sergestidae), dorsal view. E: nauplius stage III of *Sicyonia carinata* (Sicyonidae), ventral view. F: nauplius stage IV of *Pleoticus muelleri* (Solenoceridae), showing the right-side antennule, antenna, and mandible, ventral view. A modified after Heldt (1955); B modified after Hashizume (1999); C modified after Pearson (1939, as *Trachypenaeus constrictus*); D modified after Knight and Omori (1982); E modified after Heldt (1938); F modified after Iorio et al. (1990).

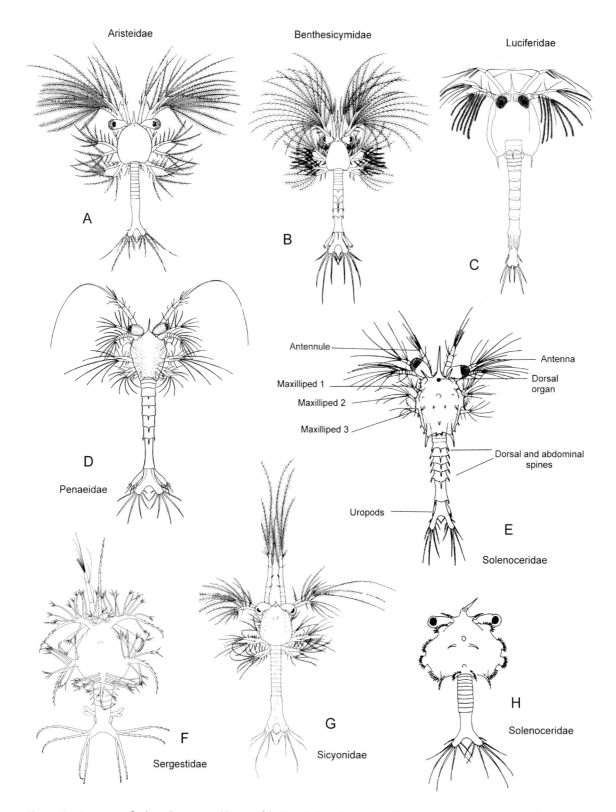

Figure 46.2 Drawings of selected protozoeal larvae of the Dendrobranchiata, dorsal views. A: protozoea stage II of *Aristeus antennatus* (Aristeidae). B: protozoea stage III of *Gennadas elegans* (Benthesicymidae). C: protozoea stage III of *Lucifer typus* (Luciferidae). D: protozoea stage III of *Rimapenaeus constrictus* (Penaeidae). E: protozoea stage III of *Pleoticus muelleri* (Solenoceridae). F: protozoea stage III of *Sergestes similis* (Sergestidae). G: protozoea stage II of *Sicyonia carinata* (Sicyonidae). H: protozoea stage III of *Solenocera necopina* (Solenoceridae), appendages not shown. A modified after Heldt (1955); B modified after Heldt (1938); C modified after Hashizume (1999); D modified after Pearson (1939, as *Trachypenaeus constrictus*); E modified after Iorio et al. (1990); F modified after Knight and Omori (1982); G modified after Heldt (1938); H modified after Calazans 2000.

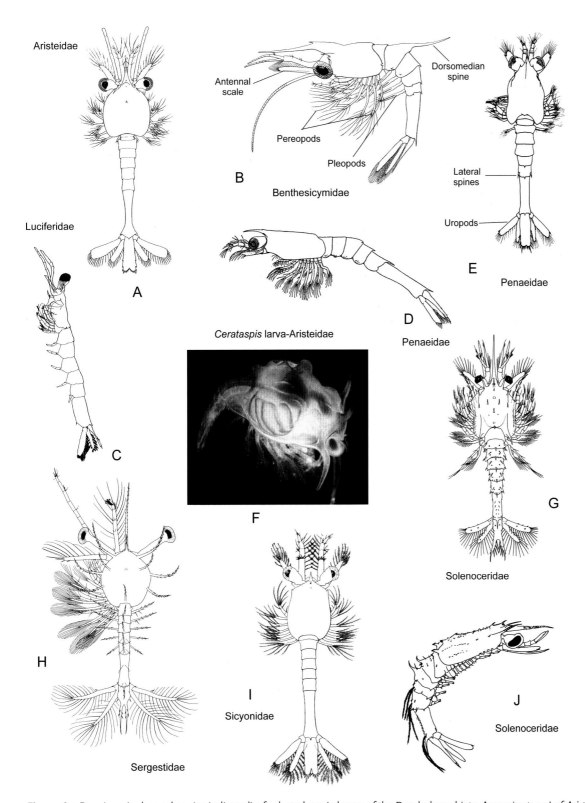

Figure 46.3 Drawings (unless otherwise indicated) of selected mysis larvae of the Dendrobranchiata. A: mysis stage I of *Aristaeomorpha foliacea* (Aristeidae), dorsal view. B: mysis stage IV of *Gennadas elegans* (Benthesicymidae), lateral view. C: mysis stage II of *Lucifer intermedius* (Luciferidae), lateral view. D: mysis stage I of *Rimapenaeus constrictus* (Penaeidae), lateral view. E: mysis stage I of *Metapenaeopsis palmensis* (Penaeidae), dorsal view. F: mysis stage (stage number undetermined) of *Cerataspis* sp. (Aristeidae), light microscopy, frontal view. G: mysis stage II of *Pleoticus muelleri* (Solenoceridae), dorsal view. H: mysis stage I of *Sergestes similis* (Sergestidae), with left-side appendages, dorsal view. I: mysis stage I of *Sicyonia carinata* (Sicyonidae), dorsal view. J: mysis stage II of *Solenocera necopina* (Solenoceridae), lateral view. A modified after Heldt (1955); B and I modified after Heldt (1938); C modified after Hashizume (1999); D modified after Pearson (1939, as *Trachypenaeus constrictus*); E modified after C. Jackson et al. 1989; F courtesy of Maria Criales; G modified after Iorio et al. (1990); H modified after Knight and Omori (1982); J modified after Calazans (2000).

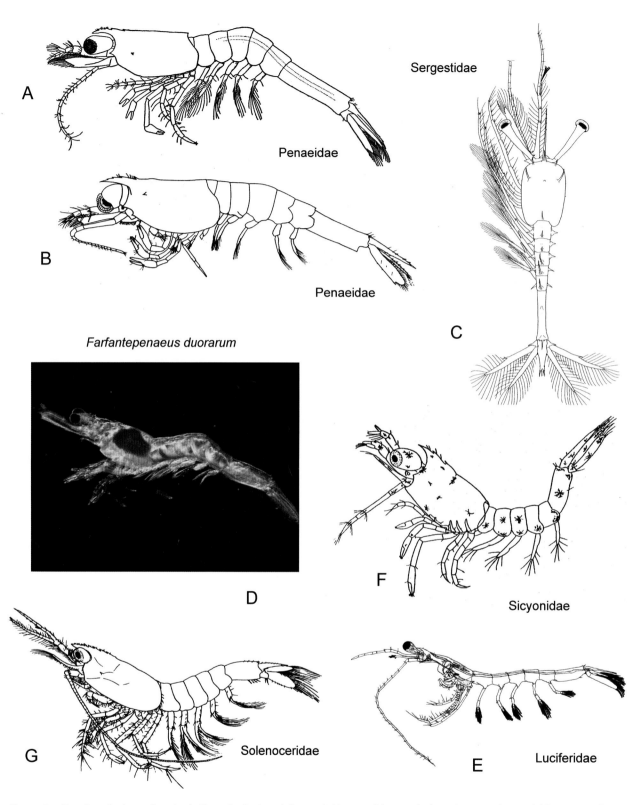

Figure 46.4 Drawings (unless otherwise indicated) of selected decapodid larvae of the Dendrobranchiata. A: decapodid first stage of *Rima-penaeus constrictus* (Penaeidae), lateral view. B: decapodid first stage of *Metapenaeopsis palmensis* (Penaeidae), lateral view. C: decapodid first stage of *Sergestes similis* (Sergestidae), with left-side appendages, dorsal view. D: advanced decapodid stage of *Farfantepenaeus duorarum* (Penaeidae), light microscopy, lateral view. E: decapodid first stage of *Lucifer intermedius* (Luciferidae), lateral view. F: decapodid first stage of *Sicyonia wheeleri* (Sicyonidae), lateral view. G: decapodid first stage of *Pleoticus muelleri* (Solenoceridae), lateral view. A modified after Pearson (1939, as *Trachypenaeus constrictus*); B modified after C. Jackson et al. (1989); C modified after Knight and Omori (1982); D courtesy of Maria Criales; E modified after Hashizume (1999); F modified after Gurney (1943); G modified after Iorio et al. (1990).

47 | Stenopodidea

JOSEPH W. GOY

GENERAL: Stenopodideans are a unique group of very colorful shrimp-like crustaceans with a particular blend of natant (swimming) and reptant (crawling) characteristics that make them difficult to place in the decapod family tree. They are one of the smallest infraorders of decapods, with only about 70 described species in 3 families (De Grave et al. 2009; Goy 2010). The Stenopodidea can be divided into three distinct marine ecological groups: littoral/sublittoral forms frequenting rocky bottoms or coral reefs; deep-water forms on muddy substrates or mostly as commensals in hexactinellid glass sponges; and stygobitic species in anchialine caves on islands. One species, *Stenopus hispidus*, has become a widely recognized symbol of tropical marine biodiversity. Stenopodideans are rather homogenous as adults, but they have several peculiar types of larvae that suggest some relationship to the Thalassinidea (= Glypheidea and Axiidea) and Nephropsidea. The life cycle consists of a pre-zoea, a zoea, a decapodid, a juvenile, and an adult. All stages are free living. Some species that are commensal in hexactinellid sponges hatch as a decapodid and then molt to the juvenile and adult.

LARVAL TYPES: The Stenopodidea have a range of developmental patterns, from direct to abbreviated to extended larval development (Gurney 1942; Felder et al. 1985; Gore 1985; Seridji 1990).

Pre-Zoea: A pre-zoea has been found in species of *Spongicola* (Bate 1888; Saito and Konishi 1999), *Stenopus* (W. Brooks and Herrick 1891; Goy 1990), and *Microprosthema* (J. W. Martin and Goy 2004), in which the pre-zoeal cuticle is shed and, very quickly thereafter, the setae and spines of the zoeal appendages and telson are extended. This stage can only swim by flexing and extending its body, as the appendages are not capable of independent movement.

Zoea: Except for species having direct development, stenopodideans have a series of planktonic larval stages, each of which is called a zoea. The number of zoeal stages can be few (4–5 in *Microprosthema*) or as many as 9–11 (in *Stenopus*).

Abbreviated development, where hatchlings are in advanced zoeal stages, has been documented in *Richardina spinicincta* (Kemp 1910a), and this may also occur in other deep-water species.

Decapodid: The zoeae are followed by a decapodid, which is a transitional form, taking on a more benthic lifestyle. Direct development has been confirmed for *Spongicoloides koehleri* (Kemp 1910b), *Spongiocaris hexactinellicola* (Berggren 1993), and *Spongicola japonicus* (Saito and Konishi 1999). Development beyond the decapodid for *Microprosthema* sp. (Raje and Ranade 1975) and *M. validum* (Ghory et al. 2005) shows a gradual change in size and morphological characteristics, such as an increase in spines on the rostrum, and in setae on the appendages and telson.

MORPHOLOGY

Pre-Zoea: When a pre-zoea is present, it is covered by the embryonic cuticle; the rostrum is turned under the carapace, but the antennae are fully extended. The appendages and telson are developed, but their setae are not fully extended. The pre-zoeae are feeble swimmers, moving by using their antennae and flexing their bodies. The pre-zoeal cuticle is shed very rapidly (within 6 hours), and the rostrum and the setae on the appendages and telson are then extended.

Zoea: The zoeal stages follow the typical development pattern of the Eucarida, with the first stage bearing sessile eyes, the second stage with stalked eyes, and the third stage with developed uropods. Most of the descriptions of stenopodidean zoeae are based on specimens from plankton collections, with the authors hypothesizing species assignments based on adult zoogeographical distributions (Cano 1892b; Gurney 1924, 1936a; Lebour 1941; Kurian 1956; Bourdillon-Casanova 1960; D. Williamson 1970, 1976b; Seridji 1985, 1990). Stenopodideans are unique among the decapods in having four pairs of natatory limbs at hatching: three pairs of maxillipeds, and one pair of pereopods. Other diagnostic characters common to larvae of the Stenopodidea were summarized by Gurney

Table 47.1 Comparison of larval stages of stenopodidean shrimps

Species	No. of larval stages		Time from hatching to decapodid	Reference(s)
	Zoea	Decapodid		
Stenopus hispidus	9–11	1	10–253 days	Castro and Jory 1983; Goy 1990
S. scutellatus	9–11	1	43–77 days	Goy 1990; Fletcher et al. 1995
Richardina spinicincta	0	1	a few hours	Kemp 1910a
Microprosthema sp.	4	1	14–15 days	Raje and Ranade 1975
M. validum	5	1	12–17 days	Ghory et al. 2005
Spongicola japonicus	0	1	a few hours	Saito and Konishi 1999
Spongicoloides koehleri	0	1	a few hours	Kemp 1910b

(1924, 1942) and Lebour (1941). The pleon in later stages is bent at the third somite, almost at a right angle; the pleura are sometimes pointed; and pleomere 5 often has a median ventral spine. The endopods of pereopods 1–3 arise from below the basis. The rostrum is very long and sometimes serrated, with prominent supraorbital spines. The mandibles are very large, fitting into a deep notch in the carapace. The maxillules have vestigial endopods. The hind margin of the telson is deeply indented in the first and second stages, bearing seven setae in the first stage and eight setae in the second stage, with the second seta reduced to a thalassinid hair. In later stages the posterior margin of the telson is almost straight, with a short spine at each angle, and six or four long setae in between. Finally, the appearance of the first pleopod is delayed until the end of larval life. Variations in the number of zoeal stages for stenopodidean shrimps are summarized in table 47.1.

Abbreviated development is known in *Richardina spinicincta* (Kemp 1910a), where the late embryo has all of the appendages except the uropods, and the telson is deeply cleft. Lebour (1941) described a first-stage zoea of this species (as Stenopid A) with the endopods of the maxillules fully developed and the exopods bearing three setae, whereas these structures are vestigial in all other known first-stage stenopodidean larvae. The endopods of the maxilla and first maxilliped are more developed in *R. spinicincta*, compared with other first-stage larvae in the Stenopodidea. Gurney (1942) thought that the features of this zoea represented primitive characteristics, since these features normally appear later in development.

Decapodid: This stage immediately follows the zoea. It is a transitional phase, with concomitant changes in locomotion, feeding, and habitat. The decapodid assumes a more benthic lifestyle, closer to that of the adult. Chelipeds are developed, and setose natatory pleopods shift the swimming responsibilities from the zoeal thoracopods. Some stenopodideans still have the pleon flexed at right angles (at the third pleomere) in this stage. Development beyond the decapodid shows a gradual change in size and morphological characters toward the adult.

MORPHOLOGICAL DIVERSITY: The second telson process of all larvae of the Stenopodidea is reduced and hair-like, and is called a thalassinid hair (figs. 47.1D; 47.3C). In addition to stenopodideans, within the Decapoda this process is found

only in the larvae of the Thalassinidea (both the Gebiidea and Axiidea), Enopolometopoidea, Anomura, and Dromiacea. Stenopodidean zoeae have conspicuously large mandibles, and exopods develop on all but the last two thoracic appendages. Larvae of the Stenopodidae have long rostrums, spines on the abdominal pleura, usually a dorsal spine on the third abdominal somite, and vestigial endopods on the maxillules (figs. 47.1; 47.2). Later stages have prominent supraorbital spines. Larval stages of the Spongicolidae have much shorter rostrums, generally smooth abdominal pleura, and supraorbital spines that are either relatively short or absent (fig. 47.3).

Larvae of *Stenopus* spp. exhibit marked-time molting (Gore 1985; Anger 2001), where the first five molts in the developmental pattern show considerable morphological change (fig. 47.2); subsequent molts show just an increase in size, with little morphological change (Goy 1990). Lebour (1941), however, captured a *Stenopus hispidus* zoea measuring 21.5 mm in total length that molted into a decapodid measuring only 10 mm in total length. Larvae of *Stenopus hisipdus* measuring more than 31 mm in total length, with fully developed pleopods, have been collected in plankton (Gurney 1936a; Lebour 1941; D. Williamson 1976b). Yet D. Williamson (1976b) collected a specimen measuring 19 mm that had no trace of developing pleopods. These teleplanic larvae have been collected in the open ocean over deep water, as well as obtained from laboratory rearing (Goy 1990; Fletcher et al. 1995; D. Zhang et al. 1997). It is clear that they are able to delay metamorphosis, continuing to molt and grow until they reach coastal waters (D. Williamson 1976b).

Since the last comprehensive study of planktonic stenopodidean zoeae (D. Williamson 1976b), numerous new species have been described in the Stenopodidea, and a few species have been cultured in the laboratory (Raje and Ranade 1975; Castro and Jory 1983; Goy 1990; Fletcher et al. 1995; D. Zhang et al. 1997; J. W. Martin and Goy 2004; Ghory et al. 2005). Thus it is now possible to better relate the unidentified planktonic larvae to described species. *Stenopus* spp. (Gurney 1924) represent larvae of *Stenopus scutellatus* (fig. 47.1G) and *S. spinosus*. *Stenopus* II of Gurney (1936a) (fig. 47.2A) and Stenopid D of Lebour (1941) represent larvae of *Stenopus scutellatus*, and *Stenopus* III of Gurney (1936a) (fig. 47.2D) is a zoea of *S. spinosus*. Stenopid I of Gurney (1936a) and Stenopid B of Lebour (1941) are larvae of *Microprosthema semilaeve*, while Stenopids C, E,

F, and possibly A are larvae of *Microprosthema* species. Species Ind 1 (fig. 47.2C), Species Ind 3 (fig. 47.2E), and Species Ind 5 of D. Williamson (1976b) are zoeae of *Stenopus pyrsonotus*, *Juxtastenopus spinulatus*, and *Microprosthema validum*, respectively. Seridji (1990) described some planktonic larval stages of *Stenopus spinosus* off the Algerian coast that differed considerably from other descriptions (Cano 1892b; Kurian 1956; Bourdillon-Casanova 1960; Pretus 1991); these probably were zoeae of a species of *Odontozona*.

Cano (1892b) described and illustrated the decapodid of *Stenopus spinosus*. The adult characters were present in this stage, but the pleon was still flexed at right angles at the third pleomere. The decapodid of *S. hispidus* was attained in the laboratory by Lebour (1941) from a molted last stage zoea (fig. 47.2F, G). It had the adult color pattern and the flexed third abdominal pleomere. The adult spination was gradually acquired with successive molts, and the remnants of the pereopodal exopods were lost. Lebour (1941) described two additional decapodids, which she labeled Stenopid B (fig. 47.3D) and Stenopid C; her Stenopid B is *Microprosthema semilaeve* (Felder et al. 1985; J. W. Martin and Goy 2004; Goy 2010). Larvae and post-larval forms of *M. semilaeve* described by Raje and Ranade (1975) represent a new species of *Microprosthema* from the Indian Ocean (Felder et al. 1985; J. W. Martin and Goy 2004; Goy 2010). Development beyond the decapodid for *Microprosthema* sp. (Raje and Ranade 1975) and *M. validum* (fig. 47.3F) (Ghory et al. 2005) shows a gradual change in their sizes and morphological characters, including an increase in the number of rostral spines and the number of setae on the appendages and telson. Kemp (1910b) described the newly hatched decapodid of *Spongicoloides koehleri*, which exhibits direct development. This decapodid closely resembles the adult, except for its pereopodal exopods and sessile uropods. Saito and Konishi (1999) described the direct development of *Spongicola japonicus*, which hatches as a decapodid resembling the adult. The uropods were free on this decapodid, but there were rudimentary exopods on the first three pereopods, a rudimentary mandibular palp, and a rudimentary first pleopod.

NATURAL HISTORY: Most of the information on stenopodidean larvae is based on material collected from the plankton. Larvae of the Stenopodidea are not very common, but they usually have a unique appearance, and some can attain considerable sizes. Gurney (1924, 1936a) and Lebour (1941) described 13 species of planktonic larvae, mostly from the Atlantic; D. Williamson (1976b) added 5 more species from the Pacific and attempted to place them in adult genera. The only works describing stenopodidean zoeae from eggs hatched in the laboratory are by W. Brooks and Herrick (1891) on *Stenopus hispidus*, Raje and Ranade (1975) on *Microprosthema* sp., J. W. Martin and Goy (2004) on *M. semilaeve*, and Ghory et al. (2005) on *M. validum*. *Stenopus hispidus* and *S. scutellatus*, however, pass through 9–11 instars in 120–253 days and 43–77 days, respectively (Castro and Jory 1983; Goy 1990; Fletcher et al. 1995; D. Zhang et al. 1997). This contrasts with the shorter pattern seen in *Microprosthema* sp. (Raje and Ranade

1975), which has four zoeal stages metamorphosing to the decapodid in 14–15 days, and *M. validum* (Ghory et al. 2005), which passes through five zoeal stages before reaching the decapodid stage in 12–17 days. Live larvae are very transparent, with bright red pigment in patches or spots on the carapace and the middle of the abdomen, as well as sometimes on the antennae, rostrum, and telson. In late-stage zoeae, grooves on the thorax and abdomen stand out in red.

As in most decapods, the decapodid is a transitional stage from the plankton to the more benthic habitats of the adult, with these larvae undergoing considerable morphological and ecological changes. Decapodids have been collected in the plankton (Cano 1892b), but they are much weaker swimmers than the zoeae. In terms of coloration, there is more red in the decapodid than in the zoea, giving this stage a more diffuse reddish appearance. Decapodids in the genus *Stenopus* take on the color patterns of their respective adult species.

PHYLOGENETIC SIGNIFICANCE: Stenopodideans have long been a source of interest and curiosity among decapod workers, as their relationship to other decapods is unclear. Their current position as a separate infraorder of the Pleocyemata (e.g., De Grave et al. 2009; Ahyong et al. 2011) is supported by their unique larval development: among the decapods, only stenopodideans possess four pairs of natatory limbs at hatching, a pleon sharply bent at the third somite, enormous mandibles, and several other characters that rather easily set them apart from other decapod larvae (see Gurney 1924, 1942; Lebour 1941). Because their overall morphology is more shrimp- or lobster-like, various workers have postulated their close affinity to caridean shrimps; but the Stenopodidea also share a distinctive thalassinid seta with larvae of the Thalassinidea (the Gebiidea and Axiidea), Enoplometopoidea, Anomura, and some (primitive dromiacean) Brachyura.

HISTORICAL STUDIES: Bate's (1888) description of the larva from an egg of *Spongicola venustus* was the first description of a stenopodidean pre-zoea. W. Brooks and Herrick (1892) studied the embryology of *Stenopus hispidus* from specimens collected in the Bahamas, as well as the hatching pre-zoea and first-stage zoea of this species. They confused plankton-caught larvae of *Callianassa*, *Lysmata*, and *Sergestes* with those belonging to *Stenopus*, however, one of the first evidences of the difficulty of defining the developmental sequence of decapod larvae from plankton material. Cano (1892b) described the larvae and decapodid of *Stenopus spinosus*, but Ortmann (1893) gave a new name (*Embryocaris stylicauda*) to a late larva of *S. spinosus*. Kemp (1910a, 1910b) described probable abbreviated development in *Richardina spinicincta* from young in the egg, as well as describing a newly hatched decapodid of *Spongicolides koehleri*. A series of researchers then described stenopodidean larvae from the plankton, with a few trying to relate them to adults (Gurney 1924, 1936a; Lebour 1941; Kurian 1956; Bourdillon-Casanova 1960; D. Williamson 1970, 1976b; Seridji 1985, 1990; Pretus 1991; Fernandes et al. 2007, 2010). Only a few publications have discussed laboratory culture of stenopodidean larvae (Raje and Ranade 1975; Saito and Konishi 1999; J. W. Martin and Goy 2004; Ghory et al. 2005). With our present state of knowledge of stenopodidean larval development, the group as a whole may be closely related to the reptant decapods (Gurney 1942; D. Williamson 1983, 1988a). Species of *Stenopus* are much sought after in the marine aquarium trade (Calado et al. 2003), but their long larval duration and low survival rates impair the commercial rearing of these highly priced ornamental shrimps.

Selected References

Ghory, F. S., F. A. Siddiqui, and Q. B. Kazmi. 2005. The complete larval development including juvenile stage of *Microprosthema validum* Stimpson, 1860 (Crustacea: Decapoda: Spongicolidae), reared under laboratory conditions. Pakistan Journal of Marine Science 14: 33–64.

Gurney, R. 1936a. Larvae of decapod Crustacea, 1: Stenopodidea. Discovery Reports 12: 379–392.

Gurney, R. 1942. Larvae of Decapod Crustacea. London: Ray Society.

Lebour, M. V. 1941. The stenopid larvae of Bermuda. Pp. 161–181 *in* R. Gurney and M. V. Lebour. On the Larvae of Certain Crustacea Macrura, Mainly from Bermuda. Journal of the Linnean Society London, Zoology 41: 89–181.

Williamson, D. I. 1976b. Larvae of Stenopodidae (Crustacea: Decapoda) from the Indian Ocean. Journal of Natural History 10: 497–509.

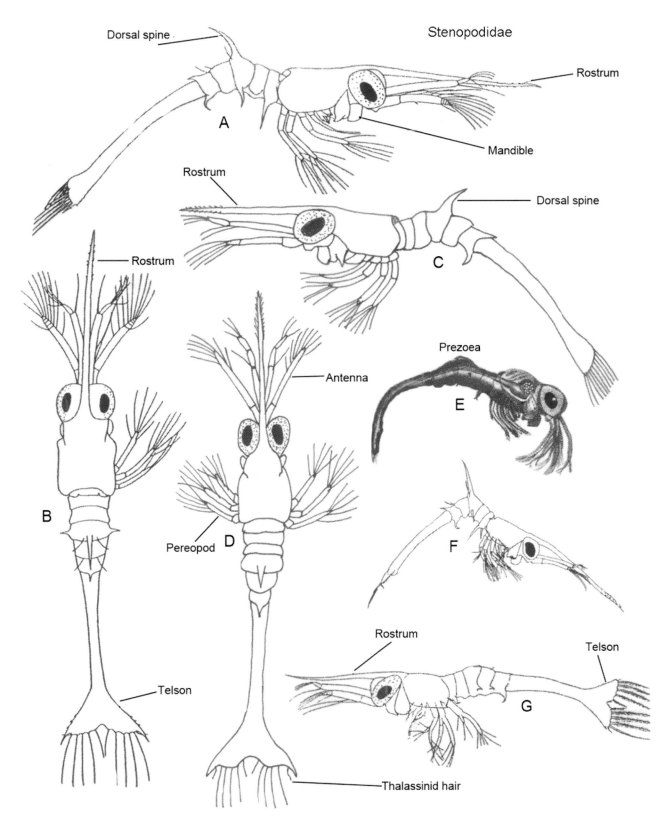

Fig. 47.1 Drawings of pre-zoeal and first zoeal stages of the Stenopodidae. A and B: zoea I of *Stenopus hispidus*. A: lateral view. B: dorsal view, with the right-hand pereopods shown. C and D: zoea I of *S. cyanoscelis*. C: lateral view. D: dorsal view. E: pre-zoea of *S. hispidus*, lateral view. F: zoea I of *S. spinosus*, lateral view. G: zoea I of *S. scutellatus*, lateral view. A and B modified after Gurney (1936a); E modified after W. Brooks and Herrick (1891); F modified after Pretus (1991); G modified after Gurney (1924).

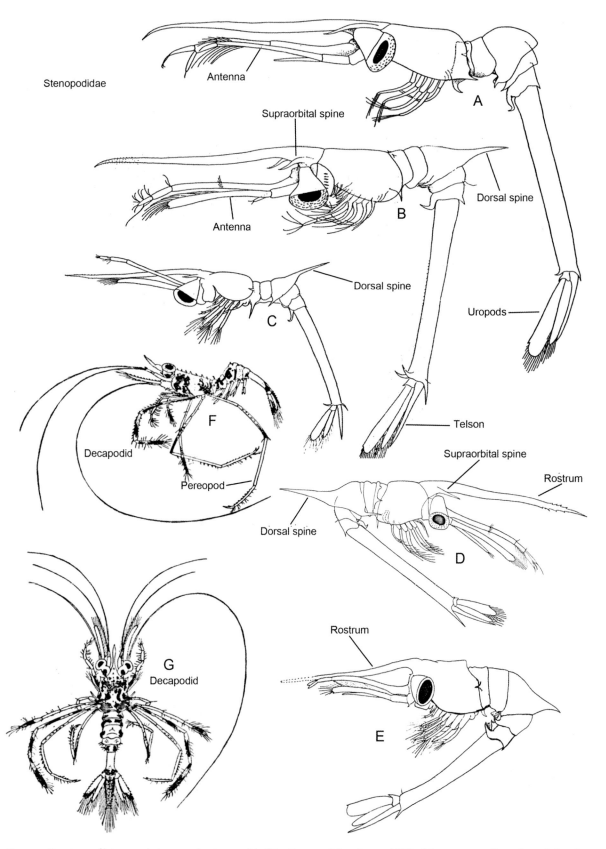

Fig. 47.2 Drawings of later zoeal stages and a decapodid of the Stenopodidae. A: zoea VIII? of *Stenopus scutellatus*, lateral view. B: zoea VI? of *S. hispidus*, lateral view. C: zoea IV? of *S. pyrsonotus*, lateral view. D: zoea VI? of *S. spinosus*, lateral view. E: late zoea, *Juxtastenopus spinulatus*, lateral view. F and G: decapodid of *S. hispidus*. F: lateral view. G: dorsal view. A and D modified after (Gurney 1936a); B modified after Gurney (1924): C and E modified after D. Williamson (1976b); F and G modified after Lebour (1941).

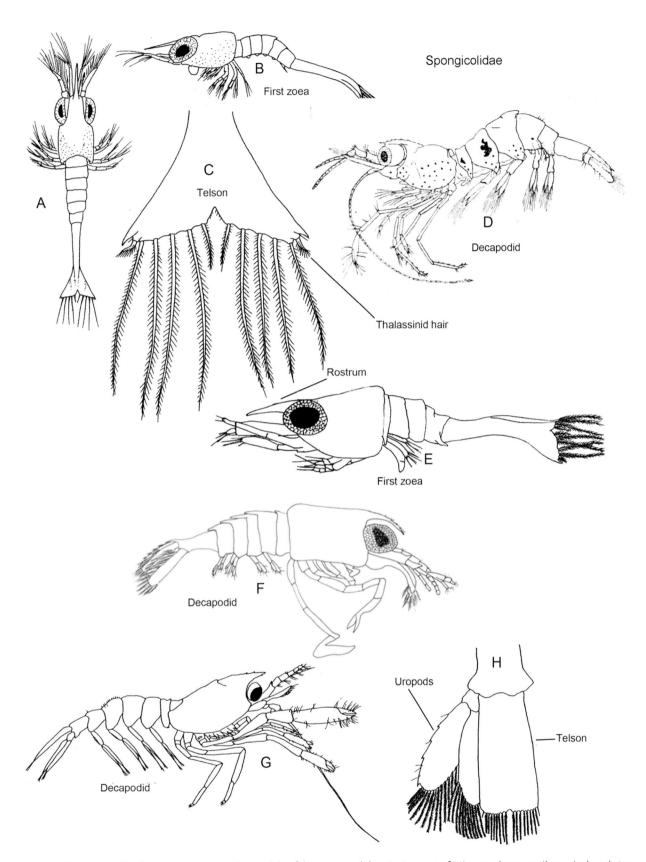

Spongicolidae

Fig. 47.3 Drawings of the first zoeal stage and decapodids of the Spongicolidae. A–C: zoea I of *Microprosthema semilaeve*. A: dorsal view.
B: lateral view. C: telson, dorsal view. D: decapodid of *M. semilaeve*, lateral view. E–G: *M. validum*. E: zoea I, lateral view. F: decapodid, lateral view. G: decapodid, lateral view. H: telson of *Spongicola japonicus*, dorsal view. A–C modified after J. W. Martin and Goy (2004); D modified after Lebour (1941); E–G modified after Ghory et al. (2005); H modified after Saito and Konishi (1999).

48 Caridea

Guillermo Guerao
José A. Cuesta

GENERAL: Carideans are commonly called shrimps or prawns (as are dendrobranchiates). They are the largest group of the shrimp-like Decapoda, with an estimated 3,438 extant species, partitioned among 389 genera, 36 families, and 16 superfamilies (Holthuis 1993; Bracken et al. 2009a; De Grave et al. 2009; Fransen and De Grave 2009; De Grave and Fransen 2011). Carideans differ from dendrobranchiate shrimps in various characters: the pleuron of the second abdominal somite overlaps those of the first and the third somites; the first pereopods are usually chelate, the second pereopods are always chelate, and the third pereopods are never chelate; the gill structure is lamellar (phyllobranch); and the fertilized eggs are incubated by the female and remain attached to the pleopods. Caridean shrimps occur in all aquatic habitats on the planet, from the deep sea (including hydrothermal vents) to shallow tropical and subtropical marine waters and freshwater systems. Species of the genus *Merguia* (Hippolytidae) are even considered semiterrestrial (Abele 1970; S. Gilchrist et al. 1983). Freshwater caridean shrimps account for approximately one quarter of all described Caridea (De Grave et al. 2008). Among freshwater families, the two most speciose are the nearly exclusively freshwater Atyidae and the Palaemonidae, which also have brackish water and marine representatives (De Grave et al. 2008). The majority of caridean shrimps are free living, but they commonly establish associations with other organisms (Bauer 2004). The life cycle consists of a zoea, a decapodid (not always clearly distinguishable), a juvenile, and an adult phase. Larval stages are free living, except for the zoeae and decapodid of the genus *Dugastella* (Cuesta et al. 2006; Huguet et al. 2011).

LARVAL TYPES

Zoea: Caridean shrimps are characterized by a series of planktonic shrimp-like larval stages, each of which is called a zoea. The number of zoeal stages can be as few as 1 or 2 (e.g., many freshwater Palaemonidae and Atyidae) or as many as 11 (e.g., the Rhynchocinetidae) or 13 (e.g., the Pandalidae).

These stages are often designated ZI, ZII, ZIII, and so forth. The number of zoeal stages can also vary within a species, due to environmental characteristics.

Decapodid: The post-zoeal stage (usually called a decapodid) is very similar to the juvenile stages, with setose natatory pleopods on the abdominal somites. Decapodids show a general resemblance to the adult, although some larval characters are retained. The transition between the zoeal and decapodid phase is a gradual morphological change, and thus is not truly metamorphic.

Post-Larva: The post-larva is a decapodid in which the zoeal pereopodal exopods are reduced. In the juvenile, they are lost completely (Anger 2001).

MORPHOLOGY

Zoea: While there is no concise definition that would apply to the whole group, several characters are generally applicable (Gurney 1942; D. Williamson 1982a; Haynes 1985). The carapace has a rostrum or rostral spine that is directed forward (although it may be absent), is styliform, and usually does not have teeth or spines in ZI. The presence or absence of certain spines on the carapace is useful for distinguishing among families and for identifying one or more stages (fig. 48.1C). All zoeae have a pair of compound eyes, which are almost always sessile in the first stage and stalked in subsequent stages (with the peduncle having a variable relative length); a nauplius eye is never seen. In early zoeal stages, the antennules have an unsegmented peduncle that later becomes three-segmented. The antennae are biramous, the antennal exopods (scaphocerites) are usually segmented in ZI, and the unsegmented endopods bear one long apical seta (fig. 48.1G). This seta may be fused with the endopod to form a long spine; there may also be a small spine or another seta. The mandibles are usually asymmetrical and possess an incisor, a molar region, and a *lacinia mobilis*. The maxillules always have coxal and basial endites and an endopod; reduced exopods are present in some species (fig. 48.1I). The maxillae always have coxal and basial endites, an endopod, and an exopod (the scaphognathite) (fig. 48.1 J). Three pairs of maxillipeds are functional upon hatching. The maxillipeds are biramous and bear natatory setae on the exopods. Maxillipeds 2 and 3 have an endopod that is longer than that of maxilliped 1 (fig. 48.1K–M); the number of natatory setae on the exopod of each maxilliped is helpful for distinguishing zoea I of different shrimp species at the family level. The pereopods appear in succession, except when development is abbreviated; but pereopod 5 may develop before pereopods 3 and 4 in the Alpheidae. The exopods bear long natatory setae on some or all pereopods. The presence of exopods on certain pereopodal pairs is an important morphological character for identifying shrimp larvae. The abdomen (pleon) is long and segmented (fig. 48.1A–D). The abdomen of a ZI larva typically consists of five segments plus a terminal telson (pleonite 6 is not separated from the telson); the sixth pleonite usually appears in ZIII. The presence or absence of posterolateral spines on the pleon is often an important character for placing larvae into a known caridean family. The shape of the telson is useful in determining the stage of development. For caridean larvae with extended development, the typical number of telsonic setae is 7 + 7 in ZI and 8 + 8 in later stages. Pleopods are usually absent or rudimentary in early zoeal stages, and biramous in later stages. The pleopods are setose in post-zoeal stages (decapodid or juvenile). Uropods are typically absent in early zoeae (ZI and ZII) and develop in ZIII.

Decapodid: In the decapodid stage, the abdomen now bears setose natatory pleopods. Some larval characters are retained,

including well-developed exopods with natatory setae on the maxillipeds and/or pereopods. The decapodid stage shows a general resemblance to the adult, and it is often indistinguishable from the juvenile stage.

Post-Larva: In the post-larva the pereopodal exopods are reduced; otherwise, in all respects it is a decapodid stage. In the juvenile, these exopods are lost completely (Anger 2001). The term post-larva, however, is highly ambiguous. It implies a non-larval nature, but it has been applied to both last larval and early juvenile stages in the Caridea.

MORPHOLOGICAL DIVERSITY: Despite the high number of caridean species, and the diversity of adult forms, the general morphology of their larvae (zoea and decapodid) does not show the same degree of diversity. In contrast to other groups of decapods with strong differences between larval and adult forms (e.g., brachyuran crabs), caridean larvae commonly resemble the adult shrimps. The differences between the zoeal and decapodid larval phases are also minimal. Variability is restricted mainly to the carapace, abdominal denticulation and spinulation, appendage setation, and the different degree of development for some appendages (see figs. 48.2; 48.3; table 48.1). Although morphological variability is not high, it is present at all taxonomic levels, even intrageneric ones. For this reason it is not always possible to define a set of larval features that would allow for the characterization of most caridean families, or for differentiation among them. Another problem in attempting to characterize their larval

Table 48.1 General features of caridean larval stages for extended and abbreviated development

Caridean larval stage	Extended development[1]	Abbreviated development[2]
zoea I	Eyes sessile. Antenna with segmented exopod (scaphocerite) and unsegmented endopod. First 3 maxillipeds natatory. Pereopods absent or present as a bud. Pleopods and uropods absent. Telson not separated from the sixth abdominal somite (pleonite), with 14 setae. Usually planktotrophic.	Eyes sessile. Antenna with unsegmented exopod and flagellated (segmented) endopod. First 3 maxillipeds natatory. Pereopods present. Pleopods usually present but not yet functional. Uropods absent. Telson not separated from the sixth pleonite, with 14 or more setae. Lecithotrophic.
zoea II	Eyes stalked. May have additional spines on the carapace. Scaphocerite segmented. Developing pereopod 1 usually visible. Pleopods and uropods absent. Telson not separated from the sixth pleonite. Usually planktotrophic.	Eyes stalked. May have additional spines on the carapace. Scaphocerite unsegmented. Pleopods present but not yet functional. Uropods absent. Telson not separated from the sixth pleonite. Lecithotrophic.
zoea III	Antennal endopod segmented. May have additional spines on the carapace. Pereopods further developed but not yet functional. Pleopods absent. Uropods present. Telson separated from the sixth pleonite. Usually planktotrophic.	Antennal endopod segmented. May have additional spines on the carapace. Pereopods functional, very similar to juveniles and adults. Pleopods present, biramous and segmented. Uropods present. Telson separated from the sixth pleonite. Lecithotrophic. Last zoeal stage.
zoea IV and higher	May have additional spines on the carapace. Scaphocerite unsegmented. Pereopods and pleopods further developed but not yet functional. Usually planktotrophic.	
decapodid	Mouthparts fully functional. Pereopods developed and functional. Pleopods present, biramous, segmented, and with natatory setae. In most cases, there are no clear differences from the first juvenile stage. Mainly benthotrophic.	Mouthparts fully functional. Pereopods developed and functional. Pleopods present, biramous, segmented, and with natatory setae. In most cases, there are no clear differences from the first juvenile stage. Lecithotrophic (obligate or facultative) or benthotrophic.

[1] Many exceptions exist within almost all families.

[2] Different types of abbreviated development are known (see Magalhães 1988; Guerao 1993; Jalihal et al. 1993; Lai and Shy 2009; Rodríguez and Cuesta 2011).

morphology at the family level is the lack of accurate descriptions for a high percentage of species. Any conclusion at this point is premature, although some attempts have been made. Morphological differences were established among larvae of several genera of the Hippolytidae (D. Williamson 1957) and Crangonidae (D. Williamson 1960), but most morphological studies have focused on differentiation at the intrageneric level (Geiselbrecht and Melzer 2009; Landeira et al. 2009; Terossi et al. 2010). A single feature can rarely be used to identify a group, but this does apply in *Lysmata* zoeae, all of which are characterized by an unusually long fifth pereopod, with a greatly dilated propodus (fig. 48.3E). The larvae sit on the bottom, using this pereopod, along with the telson, like a tripod. The most striking features shown by some caridean species are the long spines on some of the pleonites (see fig. 48.3 D, J, L, M). These spines are not present in the adult form. Caridean larvae are mainly translucent, but they may have chromatophores with different pigments that can give a colored appearance.

NATURAL HISTORY: Zoeae and decapodids of caridean shrimps are found in marine, brackish, estuarine, and freshwater plankton, where they play an important role in food webs. Zoeae of many species swim upside down; at the decapodid stage they turn to swim upright. Although caridean larvae are generally planktotrophs, feeding on phyto- and zooplankton, there are cases of species or of stages (decapodid) that are benthotrophic or that feed on detritus. There are also lecithotrophic larvae in species with abbreviated development. The number of larval stages shows great inter- and intraspecific variability, from 1 to 3 zoeal stages (in abbreviated development) to 9 to 13 zoeal stages (in extended development), and from 1 to 5 decapodid stages. There are no clear patterns between the number of larval stages and taxonomic groups, as occurs in some other taxa (e.g., brachyuran crabs). In several cases this plasticity has been associated with the effects of physical or chemical and nutritional changes (see Anger 2001). Special attention has been paid to the larval requirements of species of commercial interest, including species used for human consumption (mainly *Macrobrachium* spp.) and in the ornamental/aquarium trade (*Lysmata* and *Periclimenes*). More information on these topics is found in Valenti et al. (2009) and Calado (2008), respectively. While the larvae of some species remain with the parental population (especially in species with abbreviated development), in other freshwater species the larvae migrate downstream and develop in estuaries and offshore waters, and later migrate upstream at juvenile stages. For more information on caridean ecology and larval life history, see Bauer (2004).

PHYLOGENETIC SIGNIFICANCE: Although there is a large body of literature on caridean larvae, there is little with respect to their use in phylogenetic studies, except in

comparing the Caridea to other decapod groups (see the discussion in J. W. Martin and Davis 2001). Carideans differ from dendrobranchiate shrimps by carrying their eggs and developing embryos on their pleopods, and by the fact that eclosion is zoeal rather than naupliar (along with other adult characters). Carideans share these characters with all pleocyematan decapods (D. Williamson 1988a; Richter and Scholtz 2001). Unfortunately, larval characters have not yet been used to explore the phylogeny of caridean shrimps, even at intrafamilial level, although some authors have suggested it (e.g., Anker et al. 2006 for the Alpheidae). All studies on caridean phylogeny have been based on adult morphology (Christoffersen 1990; Holthuis 1993; Fransen and De Grave 2009) and, more recently, on molecular analyses (Baeza et al. 2009; Bracken et al. 2009a; Li et al. 2011).

HISTORICAL STUDIES: According to Gurney (1942), the Société Hollandaise des Sciences de Haarlem offered a prize in 1840 to anyone who could determine whether metamorphosis existed in decapods, as had been proposed by J. Thompson (1828), but the prize was apparently not claimed. It was Joly (1843) who took up the question, using *Atyaephyra desmarestii*, from which he obtained the larval stages. Joly concluded that there was metamorphosis in this species, as well as in *Palaemonetes* and *Crangon* (DuCane 1839). Thus the first real evidence for confirmation of metamorphosis in decapods came from caridean species. Following these studies, many important publications on caridean larvae appeared, especially from plankton samples. Notable among these were the works of Sars (1900b, 1906, 1912), H. Williamson (1901, 1915), Gurney (1923, 1927, 1942) and Lebour (1931, 1932a, 1932b), among others. Lebour made an important contribution to our knowledge of British caridean larval morphology, with detailed descriptions that included, in some cases, coloration patterns, something exceptional even today. The first complete larval development from a reared caridean (*Palaemonetes vulgaris*) was carried out by Faxon (1879). That study, along with one of *Homarus* by S. Smith (1873), were also the first complete larval series known for a decapod species.

Selected References

Gurney, R. 1942. Larvae of Decapod Crustacea. London: Ray Society.

Lebour, M. V. 1931. The larvae of the Plymouth Caridea, 1 and 2: 1, the larvae of the Crangonidae; 2, the larvae of the Hippolytidae. Proceedings of the Zoological Society of London 101: 1–9.

Williamson, D. I. 1960. Crustacea, Decapoda: Larvae, 7; Caridea, Family Crangonidae, Stenopodidea. Fiches d'Identification du Plancton, Leaflet No. 90. Copenhagen: International Council for the Exploration of the Sea.

Williamson, D. I. 1982a. Larval morphology and diversity. In L. G. Abele, ed. The Biology of Crustacea. Vol. 2, Embryology, Morphology, and Genetics, 43–110. New York: Academic Press.

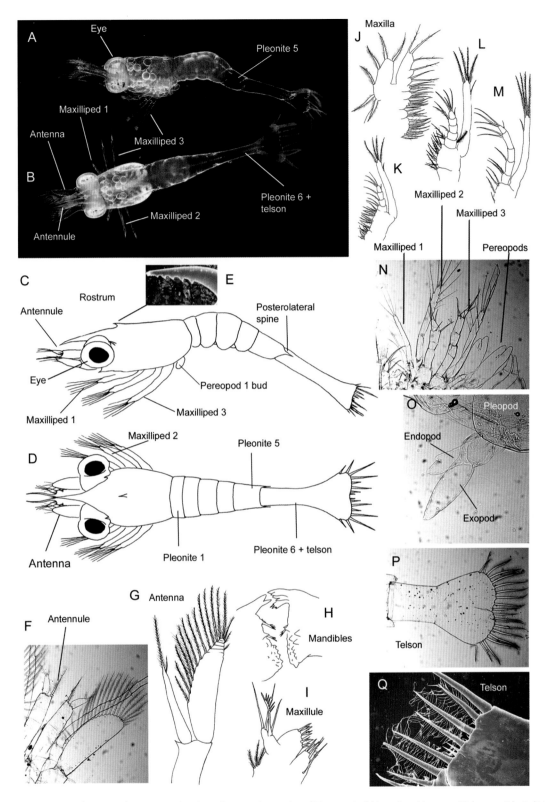

Fig. 48.1 Caridean zoeal stages, with selected appendages. A and B: zoea I of *Macrobrachium* sp. (Palaemonidae), light microscopy. A: lateral view. B: dorsal view. C and D: schematic drawings of caridean zoea (based on a palaemonid zoea II). C: lateral view. D: dorsal view. E: carapace dorsal spine of *Palaemon elegans*, SEM. F: antennule and antenna of zoea I of *Palaemonetes zariquieyi*, light microscopy. G: drawing of an antenna of a palemonid zoea I; H–M: drawings of Atyidae. H: mandibles. I: maxillule. J: maxilla. K–M: drawings of maxillipeds. K: maxilliped 1. L: maxilliped 2. M: maxilliped 3. N–P: *P. zariquieyi*, light microscopy. N: zoea II, maxillipeds and pereopods. O: zoea II, pleopod, endopod, and exopod. P: zoea I, telson. Q: zoea IV, posterior margin of the telson of *Palaemon elegans*, SEM. A–Q original.

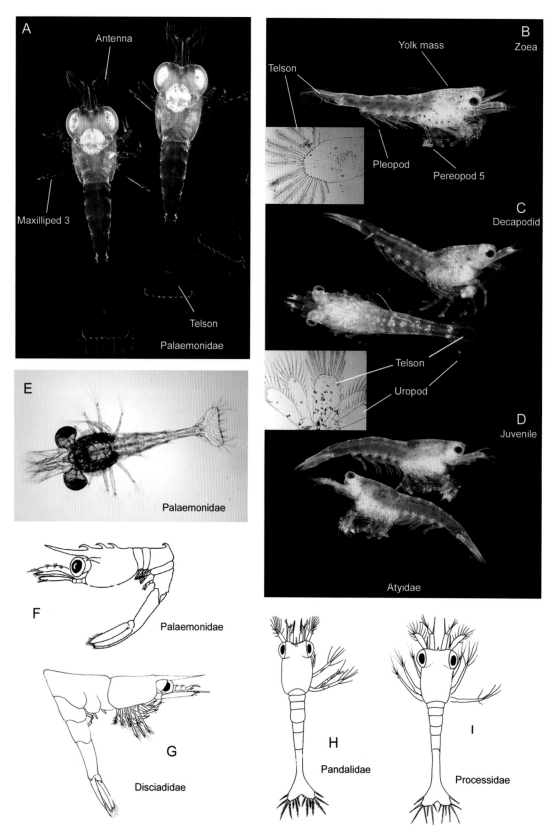

Fig. 48.2 Examples of larval diversity in the Caridea. A: zoea I of *Palaemon elegans* (Palaemonidae), light microscopy, dorsal view. B–D: *Neocaridina denticulata* (Atyidae), showing the extremely abbreviated larval development, light microscopy. B: zoea, lateral view. C: decapodids, lateral and dorsal views. D: juveniles, lateral views. E: zoea II of *Macrobrachium ohione* (Palaemonidae), light microscopy, ventral view. F: drawing of zoea V of *Palaemon macrodactylus* (Palaemonidae), lateral view. G: drawing of a last zoea of *Discias atlanticus* (Disciadidae), lateral view. H: drawing of zoea I of the Pandalidae, with right-side appendages, dorsal view. I: drawing of zoea I of *Processa* sp. (Processidae), dorsal view. A–D original; E courtesy of Ray Bauer; F modified after González-Ortegón and Cuesta (2006). G–I modified after Gurney (1942).

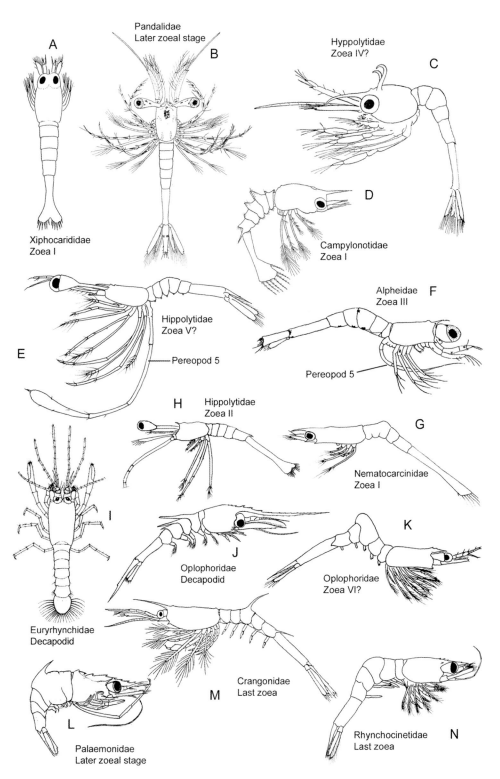

Fig. 48.3 Drawings of examples of larval diversity in the Caridea (*continued*). A: zoea I of *Xiphocaris* sp. (Xiphocarididae), dorsal view. B: later zoeal stage of *Parapandalus richardi* (Pandalidae), dorsal view. C: zoea IV? of *Caridion steveni* (Hippolytidae), lateral view. D: zoea I of *Campylotus* sp. (Campylonotidae), lateral view. E: zoea V? of *Lysmata* sp. (Hippolytidae), lateral view. F: zoea III of *Athanas nitescens* (Alpheidae), lateral view. G: zoea I of *Nematocarcinus* sp. (Nematocarcinidae), lateral view. H: zoea II of *Merguia rhizophorae* (Hippolytidae), lateral view. I: decapodid of *Euryrhynchus burchelli* (Euryrhynchidae), dorsal view. J: decapodid of *Janicella spinicauda* (Oplophoridae), lateral view. K: zoea VI? of *Acanthephyra* sp. (Oplophoridae), lateral view. L: later zoeal stage of the Palaemonidae (*Palaemon*?), lateral view. M: last zoea of *Pontophilus norvegicus* (Crangonidae), lateral view. N: last zoea of *Rhynchocinetes* sp. (Rhynchocinetidae), lateral view. A original; B modified after Lebour (1940); C modified after Lebour (1930); D and G modified after Thatje et al. (2001); E, F, K, and N modified after Gurney (1942); H modified after S. Gilchrist et al. (1983); I from Magalhāes (1988), reproduced with permission of Brill; J from Fernandes et al. (2007), used by permission; L modified after Gurney (1924); M modified after D. Williamson (1960).

49

JOSEPH W. GOY

Astacidea

GENERAL: The Infraorder Astacidea contains the marine clawed lobsters and freshwater crayfishes. The group consists of five superfamilies: 1 extinct (the Palaeopalaemonoidea), 2 for the marine clawed lobsters (the Enoplometopoidea and Nephropoidea), and 2 for freshwater crayfishes (the Astacoidea and Parastacoidea), with about 650 living species in the 5 extant families (De Grave et al. 2009). Many of the marine clawed lobsters (*Homarus*, *Nephrops*, and *Metanephrops*) and freshwater crayfishes are commercial species that have a long research history on all aspects of their biology. The American lobster (*Homarus americanus*), European lobster (*H. gammarus*), and Norway lobster (*Nephrops norvegicus*) have been the subjects of research since the last half of the 1800s. Most species in the infraorder have abbreviated development (Gore 1985; Rabalais and Gore 1985; Anger 2001). The life cycle varies according to the superfamily: a pre-zoea, a zoea, a decapodid, a juvenile, and an adult for the Enoplometopoidea and Nephropoidea; and a decapodid (or juvenile) and an adult for the Astacoidea and Parastacoidea.

LARVAL TYPES: All species of the Nephropoidea, Astacoidea, and Parastacoidea studied to date have an abbreviated development (Gurney 1942; Rabalais and Gore 1985; Anger 2001). The 2 species of the Enoplometopoidea studied thus far have extended development (Gore 1985; Anger 2001).

Pre-Zoea: In many nephropoideans and enoplometopoideans there is a brief pre-zoeal stage (F. Berry 1969; Uchida and Dotsu 1973; Wear 1976; Charmantier et al. 1991; Iwata et al. 1991, 1992; Abrunhosa et al. 2007; Okamoto 2008). The pre-zoeal stage is followed by one to three zoeal stages in the Nephropoidea and at least eight zoeal stages in the Enoplometopoidea, and these zoeal stages are subsequently followed by the molt to the decapodid.

Zoea: The marine clawed lobsters (Nephropoidea) have an advanced abbreviated development (Gore 1985) in which the young hatch as zoeae, but in a more developed state than that seen in other decapods, so that, by comparison, the larval de-

velopment of nephropoideans is shorter in both duration and the number of molts. The reef lobsters (Enoplometopoidea) have an extended development, with at least eight zoeal stages before molting to the decapodid (Gore 1985; Anger 2001).

Decapodid: In the Nephropoidea, after a reduced number of morphologically advanced instars, the zoea undergoes a metamorphic molt to a decapodid. In the Enoplometopoidea, there are at least eight large-sized zoeal stages before the molt to a decapodid, although this molt from zoea to decapodid is not as metamorphic as it is in the Nephropoidea.

Direct development is seen in one nephropoidean (Okamoto 2008), and this is the norm seen in the Astacoidea and Parastacoidea. In this type of development, the hatchlings are decapodids or juveniles, with all the appendages of the adult except the first pleopods and uropods. Gurney (1942) felt that since the uropods do not appear until the second molt, the first three stages correspond to the first three stages of decapods with normal free-living larvae.

MORPHOLOGY

Pre-Zoea: This stage of marine clawed lobsters is enveloped by a thin membrane, which bends the rostrum toward the mouthparts and bends any abdominal spines toward the dorsal surface of the abdomen. After the first molt these processes become erect, and they remain so throughout the following zoeal stages.

Zoea: Nephropoidean lobster zoeae are large. They bear functional pereopods, with fully segmented endopods and natatory exopods; usually a large rostrum; and spines on the abdominal somites. The telson is crescent-shaped, with a large median spine. The antennules are unsegmented, and the mandibles have an unsegmented palp. The first and second maxillipeds have exopods without setae. The second zoea has pleopod buds. In the third zoea, these limbs become biramous and segmented, but they are still not yet functional. The uropods appear in the third zoea. In the molt to decapodid, the pleopods become functional, but the first pleopod is still absent. In advanced development, the legs all appear at the same time, and development proceeds directly to the adult form. Table 49.1 summarizes the variation in the number of zoeal stages for nephropid lobsters. A remarkable stage 2 zoea from the Great Barrier Reef was tentatively referred to the genus *Nephropsis* by Gurney (1938); this zoea had a fully developed first pereopod, with a large chela, while the antennae and telson had their normal forms for this stage (fig. 49.2G, H). In this paper, however, Gurney made no mention of the second telsonal process, which resembles the thalassinid (anomuran) seta (the thalassinid, or anomuran, hair of some authors) found in all stenopodidean, thalassinidean, and anomuran larvae. Iwata et al. (1991) felt this was the stage 2 zoea of an *Enoplometopus*, since morphologically it closely resembled the second zoea of *E. occidentalis*. Both the Pacific *E. occidentalis* and the Atlantic *E. antillensis* have been reared in the laboratory through eight zoeal stages, in 55 days and 64–96 days, respectively (Iwata et al. 1991; Abrunhosa et al. 2007). At the eighth zoeal stage in both species, the pleopods were still not present, so the molt to the decapodid must occur some time later in their development. Comparing the larval development in these 2 species shows that *E. antillensis* lacks pereopods in the pre-zoea, but they appear in the third zoeal stage. In *E. antillensis*, the first pereopod arises as a subchelate structure in the second zoeal stage, but it becomes entirely formed only in the fourth zoeal stage (Abrunhosa et al. 2007); while in *E. occidentalis* the first pereopod is completely formed by the first zoeal stage (Iwata et al. 1991).

Decapodid: In nephropid lobsters, this stage resembles the adult, yet the decapodid continues to swim for several days before becoming more benthic in habit. At metamorphosis from zoeal stages, the swimming function shifts from the thorax to the abdomen, with the first pleopod still absent. By the next molt, exopods are completely absent from the pereopods, and the first pleopod appears. The pereopod chela of the decapodid of *Enoplometopus* is very large, but the second to fifth pereopods are still in larval form (Gurney 1938). Freshwater crayfishes regularly hatch bearing all adult appendages except the first pleopods and uropods (E. Andrews 1907; Gurney 1942; Suter 1977), and the young remain attached to the parent until fully formed. Although crayfish young hatch with all of the appendages of the adult, they still have some of the characters seen in nephropid first-stage zoeae, such as

sessile eyes and no uropods. While no appendages posterior to the maxillipeds are functional, the maxillipeds possess more of a larval form. The uropods do not appear until the second molt, as in all normal larval series, and they are similar to the three free stages seen in the ontogeny of *Homarus*. The first pleopod appears in stage four in crayfishes.

MORPHOLOGICAL DIVERSITY: Since most species of the Astacidea have abbreviated development, there is much morphological conformity in the nephropoidean marine clawed lobsters, on one hand, and the freshwater crayfishes on the other. *Nephrops norvegicus* (fig. 49.2A–F), with a prezoea and three zoeal stages, shows the typical lobster pattern. The pre-zoea (fig. 49.2E) has the rostrum and abdominal spines folded under, close to the body, but the long arms of the crescent-shaped telson are free. The first zoea has a large rostrum, but no supraorbital spines. The eyes are sessile. Abdominal somites 3–5 each have a dorsal spine; those on somites 4 and 5 are very long and bent forward. Somite 6 has a pair of long dorsal spines. The telson is a narrow crescent, each arm of which is very long, with the posterior margin bearing a large median spine and numerous setae. The antennules are unsegmented, and the mandibles have an unsegmented palp. The first and second maxillipeds have non-setose exopods, but all legs are present, each with setose exopods, and legs 1–3 are chelate. The pleopods and uropods are absent. The second zoea is not much different, although the eyes are stalked, and there are large rudimentary pleopods on abdominal somites 2–5. In the third zoea, the uropods are present, the pleopods are setose, and the arms of the telson are bent forward. The decapodid assumes the general characters of the adult, but the uropodal exopods remain undivided. There are no pleopods on the first abdominal somite, but throughout successive juvenile stages these and other external sexual characters gradually develop. The larvae of *Nephrops* and *Homarus* (fig. 49.1) resemble each other closely, and the latter also has only three zoeal stages. *Homarus*, however, lacks the very long dorsal spines on the abdomen, the telson is not drawn out into a sickle shape, the second telsonal spine is slightly reduced, and the exopod of the first maxilliped is setose. The chelipeds are similar in morphology in the early juvenile stages of *Homarus*, but internal changes started in the juveniles will produce the differentiation into cutter and crusher claws (W. Costello and Lang 1979).

Table 49.1 Comparison of larval stages of nephropsid lobsters

Species	No. of larval stages		Time from hatching to decapodid	Reference(s)
	Zoea	Decapodid		
Homarus americanus	3	1	10–20 days	S. Smith 1873
H. gammarus	3	1	15–35 days	H. Williamson 1905
Nephrops norvegicus	3	1	14–20 days	H. Williamson 1905; Jorgensen 1925
Metanephrops thomsoni	2	1	4–9 days	Uchida and Dotsu 1973
M. sagamiensis	2	1	6–8 days	Iwata et al. 1992
M. challenger	1	1	3–4 days	Wear 1976
M. japonicas	0	1	a few hours	Okamoto 2008

NATURAL HISTORY: *Homarus* spp. and *Nephrops norvegicus* are of great commercial value, and there has been much research on these species, but mainly from the perspective of their potential for aquaculture. Behavioral responses of the larvae of these marine clawed lobsters, as related to their swimming ability, have also been a topic of interest (Ennis 1973; Neil et al. 1976; D. Jackson and Macmillan 2000). All freshwater crayfishes have a unique direct development, in that it is accompanied by a filament (telson thread) found in no other decapod group. Hatchlings of astacoideans remain attached to maternal pleopods by this filament for a short time (Gurney 1935; Thomas 1973; Felder et al. 1985; Scholtz 1995). This telson thread is the molted embryonic cuticle, which attaches the telson to the empty egg case. When this breaks, the hatchlings cling to the adult's pleopods by means of specially hooked first chelipeds (fig. 49.4F, G) (E. Andrews 1907; Price and Payne 1984; Yamanaka et al. 1997; Scholtz and Kawai 2002; Vogt and Tolley 2004). In the Parastacoidea, the hatchling also bears a telson thread, but, once broken, the hatchling grasps the adult's pleopodal setae with small recurved hooks on its fourth and fifth pereopods (fig. 49.3D, H) (Wood-Mason 1876; Suter 1977; Hamr 1992; Scholtz 1995; Rudolph and Rojas 2003; Burton et al. 2007).

Post-embryonic development in the Astacoidea and Parastacoidea varies among species, but it generally consists of several stages that are attached to the mother: an initial embryonic stage, followed by as many as three lecithotrophic stages. Subsequent juvenile stages are not lecithotrophic, instead bearing adult appendages and having decreasing dependence on the mother (Suter 1977; Price and Payne 1984; Felder et al. 1985; Hamr 1992; Scholtz and Kawai 2002). Freshwater crayfishes are born immobile and helpless, with unusable limbs. Extended maternal care allows the young crayfishes to survive in strong water flows and provides protection from predation (Scholtz and Kawai 2002; Burton et al. 2007). Vogt and Tolley (2004) conducted an extensive study of the hatchlings of the parthenogenetic form of *Procambarus fallax* (P. Martin et al. 2010), commonly called Marmorkrebs (marbled crayfish). Freshly hatched juveniles have a spherical body, resembling that of the embryos, and a large yolk supply. This stage is safely attached to the empty egg case by a telson thread. Shortly after hatching, this stage starts flipping its abdomen, so the hooked first pereopods can attach to the adult pleopods or to an empty egg case.

PHYLOGENETIC SIGNIFICANCE: Because most astacideans undergo advanced or direct development, larval (zoeal) stages (and thus their characters) are not as available for phylogenetic purposes as they are in other decapod groups. Even though most of the Astacidea exhibit abbreviated development, their developmental pattern follows the typical pattern seen in the zoeal development of all the Eucarida, where the first stage has sessile eyes, the second stage has stalked eyes, and the uropods develop in the third stage. Although adults are somewhat similar to other lobster-like taxa (e.g., adult stenopodideans, gebiideans, and axiideans), most zoeae of the Astacidea lack the thalassinid seta of the telson shared by larvae of these other groups. The Enoplometopoidea are the exception in the infraorder, sharing this character, as well as many other larval morphological characters, with the thalassinidean larvae of the Axiidea (see chapter 50), whose zoeal stages are so similar to those of the Astacidea that they are often referred to as the homarine line of thalassinidean larvae. Whether this resemblance to astacidean larvae is convergent or not remains unknown. Based on larval morphology, Iwata et al. (1991) suggested that the genus *Enoplometopus* (and thus the entire Enoplometopoidea) may belong to the Axiidea rather than to the nephropsids (see chapter 50).

HISTORICAL STUDIES: The first thorough description of the larval development of *Homarus americanus* was by S. Smith (1873), and for *Nephrops norvegicus*, by Jorgensen (1925). Other early works on nephropidean larvae are summarized in Gurney (1942). The direct development of freshwater crayfishes and their attachment to maternal pleopods was summarized by Wood-Mason (1876) and Gurney (1935) for the Parastacoidea, and by E. Andrews (1907) and Gurney (1935) for the Astacoidea.

Selected References

Charmantier, G., M. Charmantier-Daures, and D. E. Aiken. 1991. Metamorphosis in the lobster *Homarus* (Decapoda): a review. Journal of Crustacean Biology 11: 481–495.

Jorgensen, O. M. 1925. The early stages of *Nephrops norvegicus* from the Northumberland plankton, together with a note on the post-larval development of *Homarus vulgaris*. Journal of the Marine Biological Association of the United Kingdom, n.s., 8: 870–876.

Price, J. O., and J. F. Payne. 1984. Post-embryonic to adult growth in the crayfish *Orconectes neglectus chaenodactylus* Williams, 1952 (Decapoda, Astacidea). Crustaceana 46: 176–194.

Suter, P. J. 1977. The biology of two species of *Engaeus* (Decapoda: Parastacidae) in Tasmania, 2: life history and larval development, with particular reference to *E. cisternarius*. Australian Journal of Marine and Freshwater Research 28: 85–93.

Wear, R. G. 1976. Studies on the larval development of *Metanephrops challengeri* (Balss, 1914) (Decapoda, Nephropidae). Crustaceana 30: 113–122.

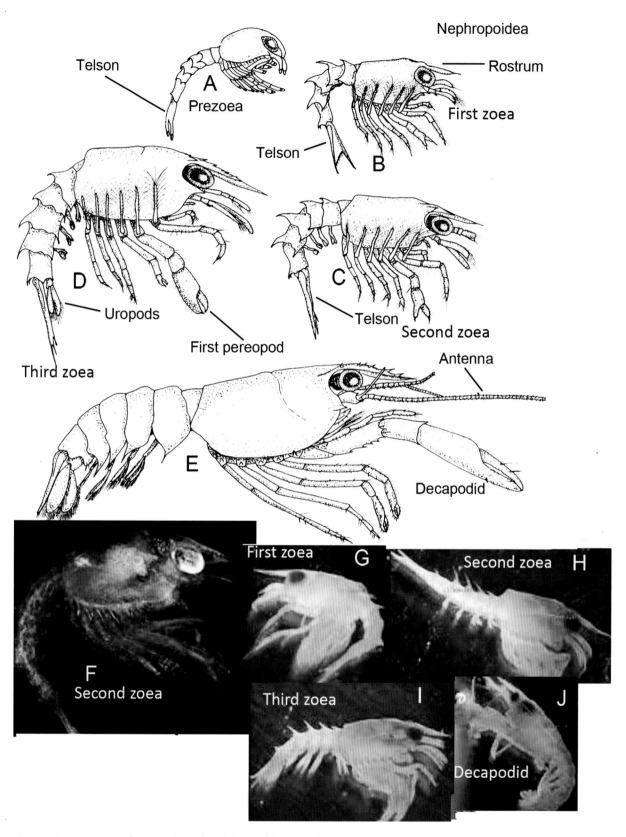

Fig. 49.1 A–E: Drawings of pre-zoeal, zoeal, and decapodid stages of the lobster *Homarus americanus* (Nephropoidea), lateral views. A: pre-zoea. B: zoea I. C: zoea II. D: zoea III. E: decapodid. F: zoea of *Homarus gammarus*, light microscopy, lateral view. G–J: zoeal and decapodid stages of *Homarus americanus*, photographs, lateral views. G: zoea I. H: zoea II. I: zoea III. J: decapodid. A–E and G–J modified after Charmantier et al. (1991); F original.

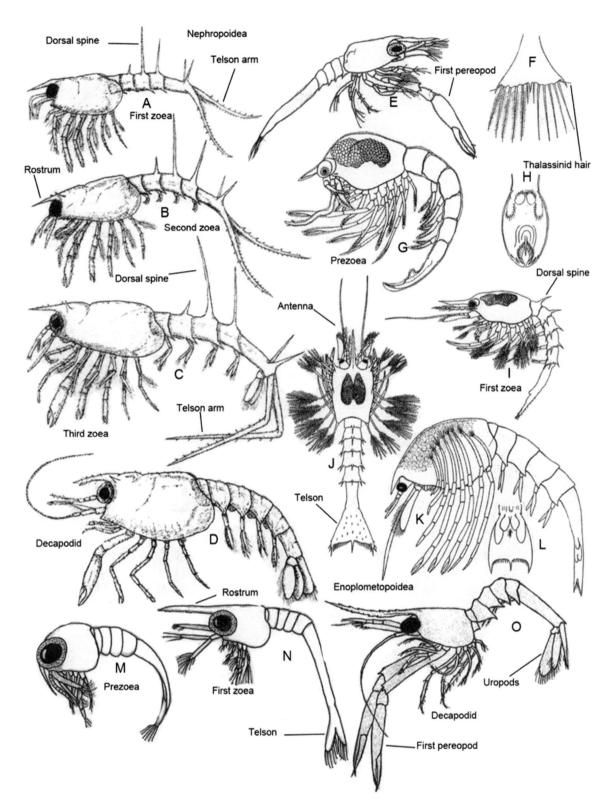

Fig. 49.2 Drawings of the Nephropoidea and Enoplometopoidea. A–D: zoeal and decapodid stages of the lobster *Nephrops norvegicus* (Nephropoidea). A: zoea I, lateral view. B: zoea II, lateral view. C: zoea III, lateral view. D: decapodid, lateral view. E and F: *Nephropsis* sp.?. E: zoea II, lateral view. F: telson, dorsal view. G–J: pre-zoea and zoea of *Metanephrops challengeri* (Nephropoidea). G: pre-zoea, lateral view. H: pre-zoeal telson, dorsal view. I: zoea I, lateral view. J: zoea I, dorsal view. K and L: pre-zoea of *M. andamanicus* (Nephropoidea). K: pre-zoea, lateral view. L: pre-zoeal telson, dorsal view. M and N: *Enoplometopus antillensis* (Enoplometopoidea). M: pre-zoea, lateral view. N: zoea I, lateral view. O: decapodid of *Enoplometopus* sp., lateral view. A–D modified after Jorgensen (1925); E, F, and O modified after Gurney (1938a); G–J modified after Wear (1976); K and L modified after F. Berry (1969), M and N modified after Abrunhosa et al. (2007).

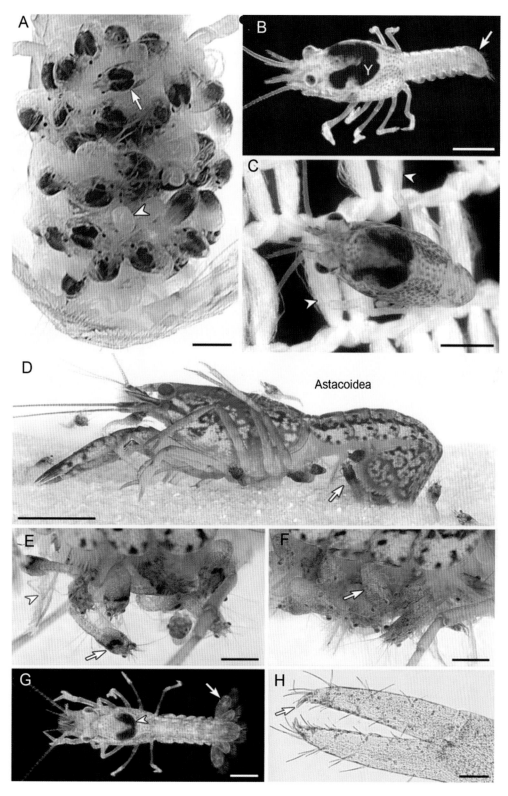

Fig. 49.3 Juvenile stages of the freshwater crayfish Marmorkrebs (*Procambarus fallax*), light microscopy. A: juvenile stage II (white arrows), under the maternal abdomen. B: free-moving juvenile stage II, removed from the maternal pleopod, dorsal view; the white arrow points to the telson, without uropods. C: juvenile stage II in a typical attachment posture, clinging to a synthetic net with the first pereopods (white arrows), dorsal view. D: juvenile stage III, mounted on the mother from all sides, lateral view. E and F: juvenile stage III (white arrows), under the mother's abdomen. G: free-moving juvenile stage III, with a typical tail fan composed of the telson and uropods (white arrow). H: first pereopod of juvenile stage III, with short and slightly curved terminal spikes (white arrow). Scale bars = 1 cm. A–H from Vogt and Tolley (2004), reproduced with permission of John Wiley and Sons.

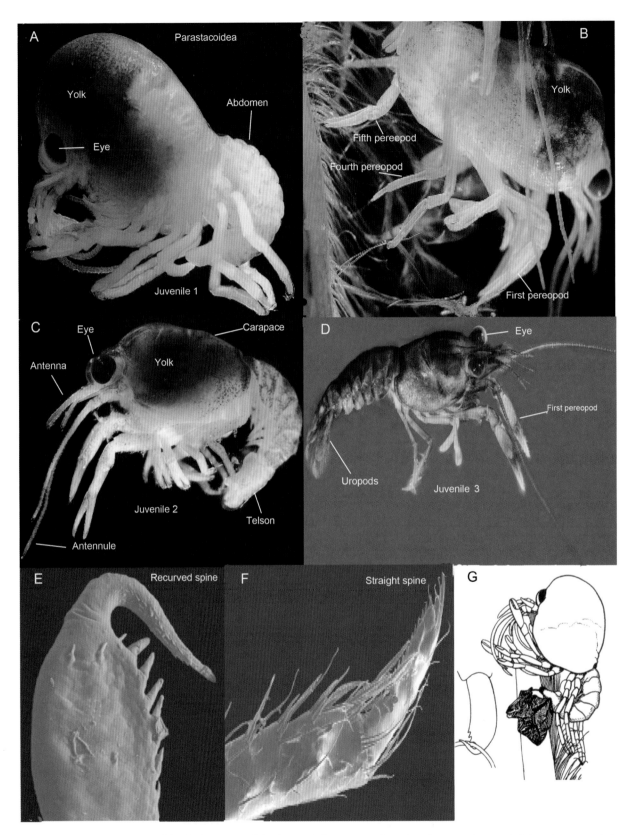

Fig. 49.4 A–F: juvenile stages of the freshwater crayfish *Cherax cainii* (Parastacoidea). A and B: juvenile stage I, light microscopy. A: lateral view. B: dorsolateral view. C: juvenile stage II, light microscopy, lateral view. D: juvenile stage III, light microscopy, dorsolateral view. E: recurved spine on the dactylus of the fifth pereopod of juvenile stage II, SEM. F: straight spine on the dactylus of the fifth pereopod, SEM. G: drawing of juvenile stage I (*right*) and the dactylus of the fourth pereopod (*left*) of *Cherax preissii* (Parastacoidea). A–F from Burton et al. (2007), reproduced with permission of *American Midland Naturalist*; G modified after Gurney (1942).

50

GERHARD POHLE
WILLIAM SANTANA

Gebiidea and Axiidea
(= Thalassinidea)

GENERAL: Gebiids and Axiids are widely referred to as ghost shrimps (Callianassidae), mud shrimps (Upogebiidae), mud lobsters (Thalassinidae), or lobster shrimps (Axiidae) (McLaughlin et al. 2005). The two groups presently include 13 families, containing 615 extant species that represent about 4.2 percent of all decapods (De Grave et al. 2009). Recent molecular studies have recognized thalassinideans as a paraphyletic group, now separated into the infraorders Axiidea and Gebiidea (Bracken et al. 2009b; Robles et al. 2009). But because of common usage, in this chapter we still refer to them collectively as thalassinideans. Thalassinideans appear shrimp-like, with elongate and flexible subcylindrical bodies and variable sclerotization of the exoskeleton. The group has a near-global distribution—from 71° N to 55° S latitudes (Robles et al. 2009)—within estuarine and marine habitats. Here they inhabit soft sediments, mainly within intertidal and shallow subtidal waters, although they are also found to depths exceeding 2,500 m (Heard et al. 2007). Many are specialized to construct and live in elaborate burrows in sand or mud, where they constitute an important bioturbation component that affects sediment properties and processes (Griffis and Suchanek 1991). The life cycle consists of a pre-zoea, a zoea, a decapodid (megalopa), a juvenile, and an adult, all of which are free living.

LARVAL TYPES: The larvae of thalassinideans can be separated into two distinct groups, those resembling anomuran larvae, now treated as the Gebiidea, and those resembling the larvae of nephropid lobsters, which form the "homarine group," now treated as the Axiidea (Gurney 1938, 1942). Collectively, larval information is very limited, available only for approximately 13 percent of the known species and 25 percent of the genera. Complete larval descriptions are available for only about one third of the larval accounts, and the larvae of at least 4 families (the Callianideidae, Ctenochelidae, Michelidae, and Thomasiniidae) presently are unknown (Pohle et al. 2011). Also, larval descriptions of some species are based on plankton-collected material, and thus are of uncertain parentage. Consequently, assignment of these larvae to genera/species should be viewed with caution.

Pre-Zoea: As in other decapods, a pre-zoea can be the hatching stage in some thalassinidean species (fig. 50.2A–C). This stage is non-feeding, of very short duration, and thought to have very poor mobility (Vaugelas et al. 1986; Strasser and Felder 2000).

Zoea: The subsequent zoeal phase (fig. 50.1A, K) consists of two to seven stages. While usually characteristic for a species, the number of zoeal stages can vary intraspecifically, which appears to be more common in callianassids (Strasser and Felder 1999, 2000).

Megalopa: The zoea is followed by a decapodid (megalopa), which resembles the adults more than the previous larval stages (fig. 50.1Q, R). The decapodid has also been referred to as the first post-larva. Subsequent stages change more gradually in morphology than, for example, in the brachyurans. Consequently, they are at times referred to as second and third post-larvae and the like (e.g., Shenoy 1967), but they effectively represent juvenile stages (Felder et al. 1985). An apparent exception to this developmental sequence is *Upogebia savignyi*, a sponge commensal that lacks a free larva and has hatchlings that are basically in adult form (Gurney 1937).

MORPHOLOGY

Pre-Zoea: Few larval accounts present information on this short-lived stage (e.g., Pohle et al. 2011). The cephalothorax is rounded and generally without spines, other than an incipient folded rostrum (fig. 50.2A–C). The carapace and appendages are covered by a thin cuticle, with the setae and spines of the first zoea still invaginated.

Zoea: The zoeae are elongated shrimp-like stages that are characterized by having thoracic appendages, with natatory setae, that are used for locomotion (fig. 50.1A, K). Chromatophores with distinct patterns, colors, and locations are usually present (fig. 50.5) but often go unreported, because they fade in preserved specimens. The first zoeal stage has its eyes fused to the carapace and bears five abdominal somites plus a telson (fig. 50.2E). The spinal and/or setal armament of the telson changes during zoeal development, as may its shape in some taxa (e.g., compare the telson of fig. 50.2E with 50.2G). Subsequent stages have movable eyes, and a sixth abdominal somite is added no later than the last zoeal stage (fig. 50.2G). Typically, and in common with anomurans, the second telsonic outgrowth ("thalassinidean seta" or "thalassinidean hair") from the disto-lateral margin is much reduced (sometimes minute, missed, or not illustrated) compared to the other spines or setae in that general location (fig. 50.3H, I). This feature, found in gebiidean and axiidean zoeae, is known to be shared with anomurans but may also be present in astacidean lobsters, such as *Homarus americanus* (personal observation). The rostrum and other carapace spines are variously developed, depending on the taxon. Progressing posteriorly, the appendages consist of pairs of antennules, antennae, mandibles, maxillules, maxillae, and maxillipeds 1–3 (fig. 50.1B–I). The pereopods (fig. 50.1L–P) appear in different stages of development at various zoeal stages, ranging from undeveloped in species with many zoeal stages to fully differentiated in species with few zoeal stages. The range of development of the abdominal pleopods (fig. 50.1A) among taxa is similar to that of the pereopods. The uropods (fig. 50.1J, K) appear in later zoeal stages in some species.

Megalopa: This stage, characterized by a switch to abdominal appendages for swimming, morphologically is distinctly different from the zoeae. The megalopa is closer to the adult form, including the carapace and fully developed uropods (fig. 50.1Q, R). Appendages (e.g., antennae) usually differ markedly from those of the zoeae. This includes the mouthparts (e.g., the mandibles, which now have fully developed palps). Comparatively, the maxillipeds have reduced exopods but more fully developed endopods and epipods. The pleopods, of which there may be three or four pairs (depending on the group), are now marginally covered with full arrays of plumose setae and are used for swimming. The exopods of the pereopods are lost or greatly reduced (e.g., fig. 50.4 D, H), while the endopods are enlarged and adorned with many different setae. Depending on the group, either the first (fig. 50.4B) or the first two pereopods (fig. 50.4H) are modified as chelipeds.

MORPHOLOGICAL DIVERSITY: The recent division of thalassinideans into the Gebiidea and Axiidea, based on molecular evidence, is consistent with groupings proposed much earlier by Gurney (1938), based on larval morphology. He distinguished between broadly equivalent anomuran and homarine lines, consisting, respectively, of the Laomediidae and Upogebiidae, and the Axiidae and Callianassidae. Larvae described since Gurney's time generally fit this grouping, including those of the Thalassinidae and Axianassidae within the gebiidean/anomuran line.

Gebiidea ("Anomuran Group" of Gurney 1938): In this group the zoeal carapace never has anterolateral spinules.

The rostrum is usually round and unarmed, relatively short, and typically less than one quarter of the carapace length (fig. 50.1A). There are no dorsal spines on the abdomen (figs. 50.1A; 50.2D–H, J–L; 50.3A, C, E, F). The median telson spine is absent or small, and it is always absent in ZI (figs. 50.1J; 50.2 E, G, L; 50.3C, F). Pereopod 5 is always uniramous (fig. 50.2K) and never has an exopod. The maxilliped 3 endopod, if present, is inserted near the base of the basipodite (fig. 50.1I) and is often rudimentary, having no or few segments and setae.

Among gebiideans, larvae of the Laomediidae stand out, with zoeae that are relatively long and narrow in shape, with a distinct single-spined carapace that substantially elongates in the more flattened neck area behind the eyes during successive zoeal stages (figs. 50.2D–G; 50.5B). The asymmetrical mandibles, with their drawn-out sickle-shaped component on one side, are another distinctive trait (fig. 50.2I), as are the procurved ventral hooks on the abdomen (figs. 50.2D, F, H, J–L; 50.5A). The curvature of the rostrum and the relative position of the mandibles on the carapace are features that separate genera (fig. 50.2M–O). The megalopa is known for only 3 species among 14 laomediid larval accounts (Pohle et al. 2011), insufficient to define clear family characteristics or morphological diversity. The megalopa of *Naushonia*, however, has a prominent first leg, bearing a sickle-shaped last segment (fig. 50.4A) that is not present in *Jaxea*, which instead bears laterally armed abdominal somites (fig. 50.4B) not present in the former.

Species of *Axiannassa* also have zoeae with laomediid-type mandibles (fig. 50.3A, B), but they lack the distinct ventral abdominal hooks of various laomediids, instead bearing a pair of lateral spines on the last somite (fig. 50.3A). This, along with their distinct mandible location and lack of rostrum curvature (fig. 50.2P), supports the recent separation of the genus from the Laomediidae and its transfer to the resurrected Axianassidae (Tudge and Cunningham 2002). Some features in the megalopa of *Axiannassa* species that support this separation include the absence of a *linea thalassinica* (longitudinal groove) on the carapace and the absence of a transverse suture on the uropods (fig. 50.4C), both of which are present in *Jaxea* and *Naushonia* (fig. 50.4A, B).

Zoeae of *Thalassina anomala* (fig. 50.3C, D), the only known larval stages of the Thalassinidae (Sankolli 1967), share the sickle-shaped mandible of the Laomediidae and Axianassidae, but they lack the abdominal hooks of the former and the distinct abdominal spines of the latter.

Larvae of the Upogebiidae, known from more than a dozen species (Pohle et al. 2011), are also poorly armed, with such features generally restricted to a rostral spine (fig. 50.3E) and, in some species, a lateral pair of spines on abdominal somite 5 (fig. 50.3F). The latter thus resemble *Axianassa* (fig. 50.3A), but lack its distinct sickle-shaped mandibles (fig. 50.3B). Upogebiid megalopae typically are also poorly armed, lacking distinct spines on the abdomen (fig. 50.4D, E), and the carapace often has a rounded rostrum (fig. 50.4F). In few species (e.g., *U. darwinii*), the megalopa is armed with lateral spinules, but these have a blunt end that does not extend beyond the eyes (fig. 50.4G). In common with all known gebiidean megalopae, only the first pereopod bears a chela or subchela (fig. 50. 4D, E).

Axiidea ("Homarine Group" of Gurney 1938): In this group the zoeal carapace often bears anterolateral spinules (fig. 50.3M, O). The rostrum is usually flat (but it may be round in the first zoea), armed with serrations/spinules (fig. 50.1K), and relatively long, typically one half or more of the carapace length (fig. 50.3G, I–P). There is at least one dorsal abdominal spine. With few exceptions, all stages have a median telson abdominal spine (fig. 50.3H, I, N), often larger than in the Gebiidea. Pereopod 5 is biramous in some species (fig. 50.3K). The maxilliped 3 endopodite is terminally inserted (fig. 50.3L), with the endopod often being segmented and bearing setae.

Within the Axiidea, most known larvae are either axiids or callianassids. Among the Axiidae, those of *Axius* are the most heavily armed, best exemplified by zoeae of *A. serratus* (fig. 50.3G), which have a long rostrum between the eyes and antennal spines below them, an abdomen with somites bearing as many as four pairs of spines, and three large terminal spines on a triangular telson (fig. 50.3H). More than other axiids, the zoeae of *Axius* most closely resemble the zoeae of lobsters (the Homaridae), which are not thalassinids. A number of similar wild-caught larvae, well described by Gurney (1924, 1938), remain to be assigned to species. Laboratory-reared first-stage zoeae of *Axiopsis serratifrons* (fig. 50.3J) are distinct in lacking a median telsonic process. Presumed zoeae of *Calocarides coronatus* (fig. 50.3K) and *Calocaris macandrae* (fig. 50.3 L, M) are the most poorly armed axiid zoeae, but inadequate descriptions, and variances in the armature of wild-caught specimens (e.g., fig. 50.3L, M) (Gurney 1942 vs. Bourdillon-Casanova 1960), indicate the need for more in-depth work. The megalopa of axiids typically loses much of its abdominal armature but develops a strongly armed rostrum (fig. 50.4H, I).

The Callianassidae, a group presently in a state of taxonomic flux, has the largest assemblage of species, with known larvae for 10 genera. This list includes species now removed from *Callianassa*, and some still assigned to that genus but whose status remains contentious and may change (Pohle et al. 2011). Consequently, these larvae currently cannot be clearly differentiated by genus. With some variations, two basic and distinct types of zoeae are apparent, however, as first mentioned by Gurney (1942). Both share a carapace with a toothed rostrum and an anterolateral margin; a distinct oversized dorsal spine on the second abdominal somite; and a lobster-like telson, with a median posterior spine flanked by lateral spines (figs. 50.3N, O; 50.5E). These zoeae differ in the armature of abdominal somites 3–5: one type having only a small median posterodorsal spine (fig. 50.3N), and the other instead bearing a distinct comb-like denticulated ridge (figs. 50.1K; 50.3O). The typical callianassid megalopa (fig. 50.4J) differs from that of axiids in having an unarmed rostrum (fig. 50.4K) while, as axiideans, both families possess a chela on the first two pereopods (fig. 50.4H, J). The latter is a major

difference from known gebiidean megalopae (fig. 50.4A–E). The first zoea of *Neaxius vivesi* (fig. 50.3P), presently the sole known larva of the Strahlaxiidae, does not conform to axiidean characteristics.

NATURAL HISTORY: The duration of egg incubation and the subsequent larval period is dependent on environmental factors, primarily temperature and salinity, as well as developmental constraints. The duration of the zoeal phase has been estimated from as little as 2–3 days to as long as 5–6 months (Tamaki et al. 1996). The short-lived pre-zoeal hatching stage lacks setae on the appendages and swims poorly, if at all. The zoeal stages that follow are truly planktonic, initially with maxillipeds, and later also pereopods, as natatory appendages that bear exopods with long swimming setae. A striking exception are the larvae of *Callichirus kraussi*, which lie on their backs in the parent's burrow water until metamorphosis takes place, after 3–5 days (Forbes 1973). Generally, the zoeae are predators that start feeding soon after hatching. Some species with relatively few larval stages, however, such as the callianassids *Lepidophthalmus siriboia*, *L. louisianensis*, and *Pestarella tyrrhena*, have larvae with a sufficient yolk reserve to complete their larval development with little or no external food supply (Thessalou-Legaki et al. 1999; Nates and McKenney 2000; Abrunhosa et al. 2008). Such species show relatively poorly developed feeding appendages and foregut (Abrunhosa et al. 2006).

The megalopa represents the settlement stage. Functionally it is a transitional phase, settling out of the plankton to a benthic existence. Correspondingly, the maxillipeds have metamorphosed from natatory appendages in zoeal stages to functional mouthparts, resembling those of juveniles and adults. Similarly, the pereopods are now fully developed, allowing for ground locomotion and the grasping of objects and prey.

PHYLOGENETIC SIGNIFICANCE: The relatively recent division of thalassinideans into the Gebiidea and Axiidea, based on molecular evidence (Bracken et al. 2009b; Robles et al. 2009), is consistent with the groupings (the anomuran group and the homarine group, respectively) proposed much earlier by Gurney (1938), which were based on larval morphology. Salient characters of phylogenetic significance in each lineage are noted above under the individual taxonomic headings. For example, the recognition of *Axiannassa* as a genus of the Axianassidae rather than the Laomediidae is based largely on larval characters (Tudge and Cunningham 2002), and zoeae in both of these families bear a resemblance to larvae of the Thalassinidae. The phylogenetic status of the tropical reef lobsters (all species of *Enoplometopus*) remains unclear. While the genus is currently considered a member of a separate superfamily (Saint Laurent 1988) within the infraorder Astacidea (De Grave et al. 2009), other workers, based on adult (Holthuis 1974, 1983) and larval (Iwata et al. 1991) morphological evidence, placed *Enoplometopus* within the Axiidae, among the infraorder Axiidea. Based on larvae, Gurney (1938) did not consider this genus to be an axiid.

HISTORICAL STUDIES: A standout is a classic series of studies by Gurney (1924, 1938, 1942) on decapod larvae, with extensive treatments devoted to thalassinideans. These publications included meticulous descriptions and illustrations that set the standard in terms of detailed larval descriptions. They also established the diphyletic nature of larvae, based on a comprehensive morphological analysis that is still applicable today in the genetics-based infraordinal grouping. Also, Pohle et al.'s (2011) summary of the copious body of literature on larval descriptions of thalassinideans (with updated taxonomy) should be of useful service in future studies.

Selected References

Gurney, R. 1924. Crustacea, pt. 9: Decapod Larvae. Vol. 8, Natural History Reports: British Antarctic ("Terra Nova") Expedition, 1910, Zoology, 37–202. London: British Museum (Natural History).

Gurney, R. 1938. Larvae of decapod Crustacea, 5: Nephropsidea and Thalassinidea. Discovery Reports 17: 291–344.

Gurney, R. 1942. Larvae of Decapod Crustacea. London: Ray Society.

Pohle, G., W. Santana, G. Jansen, and M. Greenlaw. 2011. Plankton-caught zoeal stages and megalopa of the lobster shrimp *Axius serratus* (Decapoda: Axiidae) from the Bay of Fundy, Canada, with a summary of axiidean and gebiidean literature on larval descriptions. Journal of Crustacean Biology 31: 82–99.

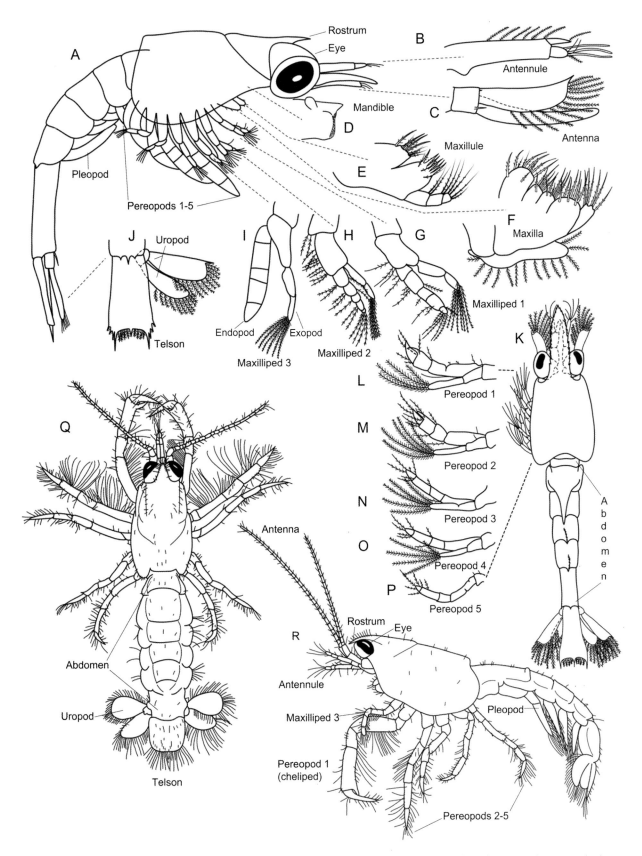

Fig. 50.1 Drawings of zoeal and megalopal stages (with selected appendages) of the Gebiidea and Axiidea. A: zoeal stage (Gebiidea: Upogebiidae), lateral view. B–I: appendages (as labeled). J: zoeal telson and uropods, dorsal view. K: zoeal stage (Axiidea: Callianassidae), dorsal view. L–P: appendages (as labeled). Q and R: megalopa of *Upogebia affinis* (Gebiidea: Upogebiidae). Q: dorsal view. R: lateral view. A–J modified after Ngoc-Ho (1981); K–P modified after Dolgopolskaia (1969); Q and R modified after Andryszak (1986).

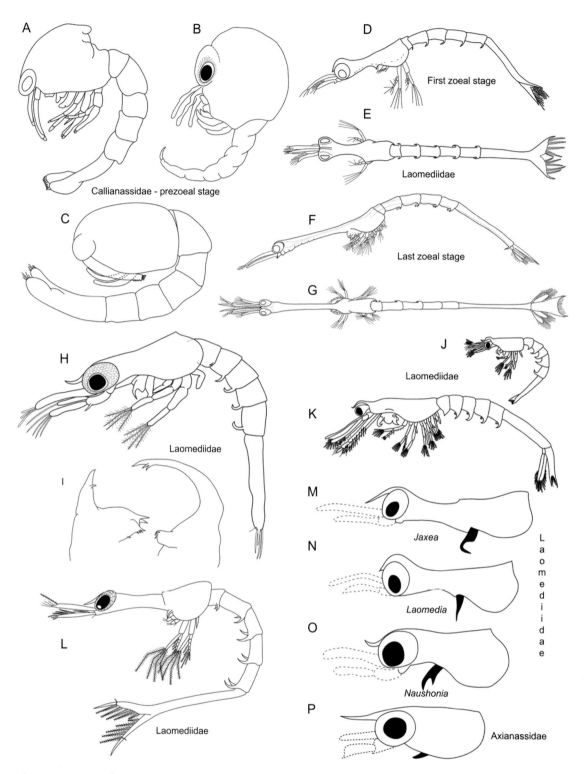

Fig. 50.2 Drawings of pre-zoeae and zoeal stages of the Gebiidea and Axiidea. A–C: pre-zoea of the Callianassidae, lateral views. A: *Callichirus seilacheri*. B: *Glypturus armatus*. C: *Lepidophthalmus sinuensis*. D–G: *Laomedia astacina* (Gebiidea: Laomediidae). D and E: zoea I. D: lateral view. E: dorsal view. F and G: zoea V. F: lateral view. G: dorsal view. H and I: *Naushonia* sp. (Gebiidea: Laomediidae). H: zoea I, lateral view. I: mandibles. J and K: *Naushonia crangonoides* (Gebiidea: Laomediidae). J: zoea I, lateral view. K: zoea VI, lateral view. L: zoea I of *Jaxea novaezealandiae* (Gebiidea: Laomediidae), lateral view. M–P: zoeal cephalothorax and relative mandible position (solid black), various genera. M: *Jaxea* (Gebiidea: Laomediidae). N: *Laomedia* (Gebiidea: Laomediidae). O: *Naushonia* (Gebiidea: Laomediidae). P: *Axianassa* (Gebiidea: Axianassidae). A modified after Aste and Retamal (1983, as *Callianassa garthi*); B modified after Vaugelas et al. (1986, as *Callichirus armatus*); C modified after Nates et al. (1997); D–G modified after Fukuda (1982); H, I, and M–P modified after Konishi (2001); J and K modified after Goy and Provenzano (1978); L modified after Wear and Yaldwyn (1966).

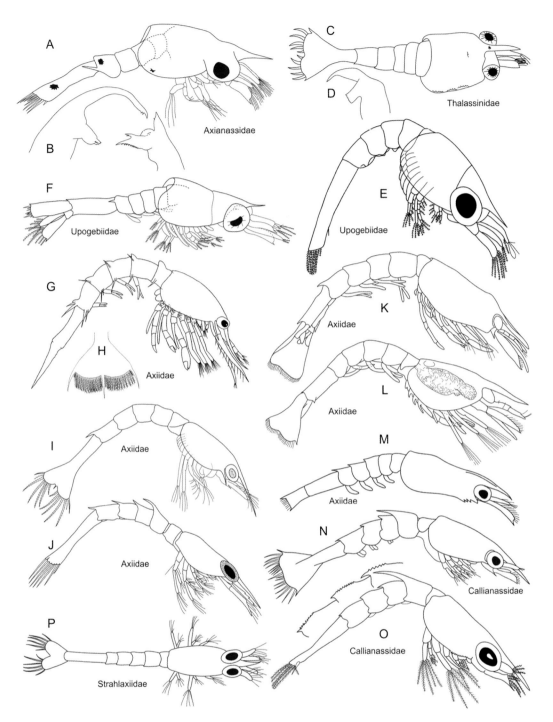

Fig. 50.3 Drawings of details of zoeal larvae in the Gebiidea and Axiidea. A and B: zoea I of *Axianassa australis* (Gebiidea: Axianassidae).
A: lateral view. B: mandibles. C and D: zoea I of *Thalassina* anomala (Gebiidea: Thalassinidae). C: dorsolateral view, appendages not shown.
D: mandible. E: zoea I of *Upogebia darwinii* (Gebiidea: Upogebiidae), lateral view. F: zoea V of *U. paraffinis*, lateral view. G and H: zoea I
of *Axius serratus* (Axiidea: Axiidae). G: lateral view. H: telson, dorsal view. I: zoea I of *Eiconaxius* sp. (Axiidea: Axiidae), lateral view. J: zoea I
of *Axiopsis serratifrons* (Axiidea: Axiidae), lateral view. K: zoea II of *Calocarides coronatus* (Axiidea: Axiidae), lateral view. L and M: zoea II of
Calocaris macandrae (Axiidea: Axiidae). L: lateral view. M: lateral view, appendages not shown. N: zoea I of *Sergio mirim* (Axiidea: Callianassi-
dae), lateral view. O: zoea I of *Nihonotrypaea petalura* (Axiidea: Callianassidae), with an enlarged view of abdominal somites 3–5, lateral view.
P: zoea I of *Neaxius vivesi* (Axiidea, Strahlaxiidae), dorsal view. A and B modified after Strasser and Felder (2005); C and D modified after
Sankolli (1967); E modified after Ngoc-Ho (1977); F modified after Melo and Brossi-Garcia (2000) G and H modified after Pohle et al. (2011);
I modified after Gurney (1924); J modified after S. Rodrigues (1994); K modified after Elofsson (1959); L modified after Gurney (1942);
M modified after Bourdillon-Casanova (1960); N modified after S. Rodrigues (1984, as *Callichirus mirim*); O modified after Konishi et al.
(1990, as *Callianassa petalura*); P modified after Berril (1975).

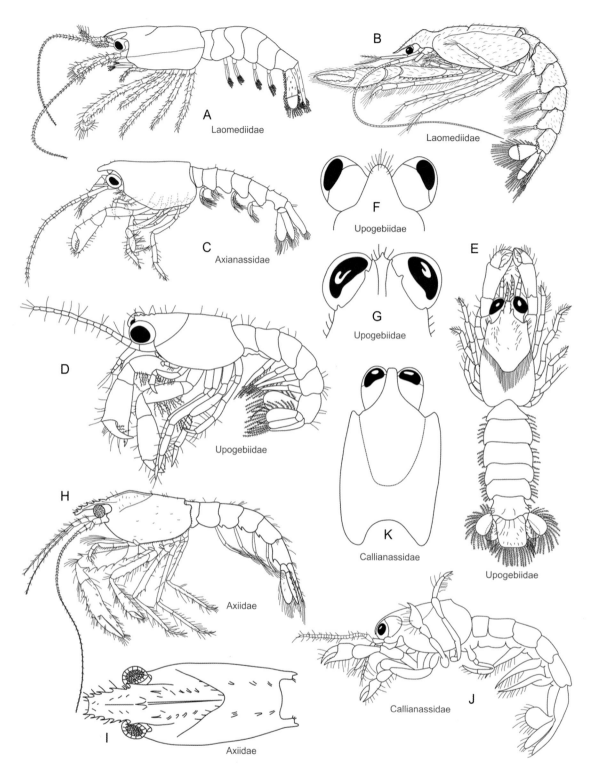

Fig. 50.4 Drawings of megalopal stages of the Gebiidea and Axiidea. A: megalopa of *Naushonia crangonoides* (Gebiidea: Laomediidae), lateral view. B: megalopa of *Jaxea novaezealandiae* (Gebiidea: Laomediidae), lateral view. C: megalopa of *Axianassa australis* (Gebiidea: Axianassidae), lateral view. D: megalopa of *Upogebia darwinii* (Gebiidea: Upogebiidae), lateral view; E: megalopa of *U. quddusiae*, dorsal view. F: cephalothorax rostrum and eyes of *U. edulis*, dorsal view. G: cephalothorax rostrum and eyes of *U. kempi*, dorsal view. H and I: megalopa of *Axius* sp. (Axiidea: Axiidae). H: lateral view. I: cephalothorax, dorsal view. J and K: megalopa of *Nihonotrypaea ?harmandi* (Axiidea, Callianassidae). J: lateral view. K: cephalothorax, dorsal view. A modified after Goy and Provenzano (1978); B modified after Wear and Yaldwyn (1966); C modified after Strasser and Felder (2005); D modified after Ngoc-Ho (1977); E modified after Siddiqui and Tirmizi (1995); F modified after Shy and Chan (1996); G modified after Shenoy (1967); H and I modified after Kurata (1965); J and K modified after Konishi et al. (1999, as *Callianassa* sp.).

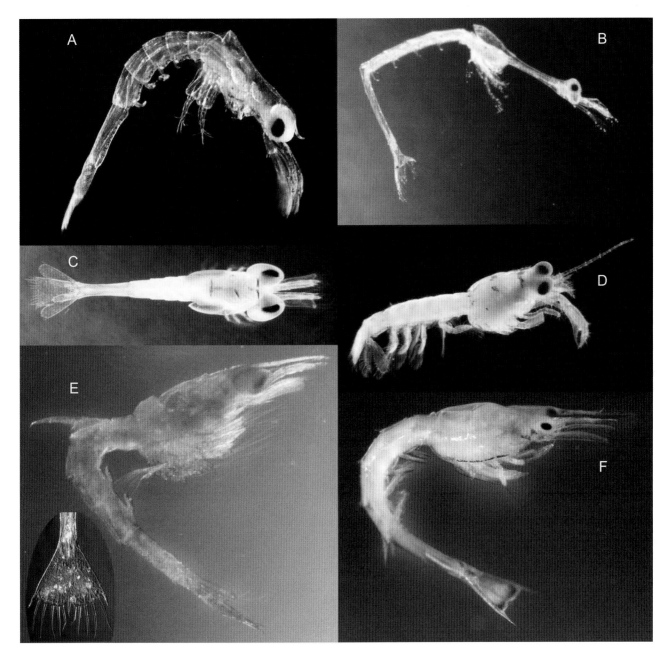

Fig. 50.5 Larvae of the Gebiidea and Axiidea, light microscopy. A: zoea I of *Naushonia* sp., lateral view; B: zoea II of *Jaxea nocturna*, lateral view; C: zoea III of *Upogebia* sp., dorsal view; D: megalopa of *U. deltaura*, dorsolateral view; E: zoea V of *Callianassa subterranea*, lateral view; the inset shows the telson of zoea I, dorsal view; F: zoea I of *Axius stirhynchus*, dorsolateral view. A modified after Konishi (2001); B–F modified after J. Martin (2001).

51

Achelata

FERRAN PALERO
PAUL F. CLARK
GUILLERMO GUERAO

GENERAL: The Achelata (formerly called the Palinura) includes the spiny and slipper lobsters. They are decapod crustaceans characterized by the lack of chelae on all pereopods as adults (except for the small grooming chela of the fifth pereopod in females) and by the phyllosoma, a transparent, flat-bodied larval phase adapted for long-distance dispersal (Scholtz and Richter 1995). Adult habitats range from the intertidal zone through the deep sea, down to 1,000 m in depth. Most of the Achelata are omnivorous scavengers, and some spiny lobsters have been implicated as key predators in a variety of benthic habitats (Edgar 1990). Recent evidence confirms that 2 families can be defined within the Achelata: the Palinuridae and Scyllaridae (Davie 1990; Palero et al. 2009a). Palinurids are found in practically all temperate and tropical seas, between 65° N and 60° S latitudes, with most of the species and highest abundances reported from the tropics (Holthuis 1991). Approximately 60 species of spiny and coral lobsters have been described and assigned to 12 extant genera, which can be grouped into the stridentes (spiny lobsters characterized by a unique sound-emitting stridulatory apparatus) and the silentes (R. George and Main 1967). Several palinurid genera are of economic value. Scyllarids are also widespread in tropical and temperate waters. Known as slipper lobsters, scyllarids are characterized by their short paddle-like antennae and granular flattened cephalothorax (Holthuis 1991; Scholtz and Richter 1995). Within the Scyllaridae, 4 subfamilies, with approximately 80 described species, are recognized (Holthuis 1985, 2002). The Scyllarinae include half of the known species and are the most diverse group. They are known from tropical lagoons and reefs, the continental shelf and slopes, and deep-sea ridges and seamounts. Few scyllarids are of significant interest to commercial fisheries. The life cycle consists of a naupliosoma (Jasus, Ibacus, Scyllarides), a phyllosoma, a puerulus (the Palinuridae) or nisto (the Scyllaridae), a juvenile, and an adult, all of which are free living.

LARVAL TYPES

Naupliosoma: The naupliosoma is considered a brief non-feeding pre-zoea. It has been reported in some palinurid and scyllarid genera (Jasus in J. Gilchrist 1913; Ibacus in Harada 1958; Scyllarides in Crosnier 1972). Whenever the naupliosomal stage is present, it lasts from only a few minutes to hours and quickly leads to the first-stage phyllosoma (see Crosnier 1972). The validity of this pre-zoeal stage is a matter of opinion, however, and it may be the result of a premature rupture of the egg.

Phyllosoma: The phyllosoma is a transparent flat-bodied zoeal phase, with natatory thoracopodal exopods (D. Williamson 1969). There are usually from 7 to 13 phyllosoma stages, although M. Johnson and Knight (1966) reported over 25 ecdyses in Panulirus. The duration of the phyllosoma stage differs greatly among achelates, ranging from a few months to almost 2 years (Booth et al. 2005; Bradford et al. 2005). Advanced stages in which gills have not yet appeared are often referred to in the literature as "subfinal" stages; as soon as gill buds are visible (on the dorsal margin of the thorax), the phyllosoma is said to be in its "final stage" (fig. 51.4A). The final-stage phyllosoma metamorphoses into the decapodid phase (Kaestner 1970; Felder et al. 1985), where the swimming function is taken over by the well-developed pleopods of the abdomen (pleon).

Puerulus, Nisto: The decapodid phase is called a puerulus in the Palinuridae and a nisto in the Scyllaridae (Gordon 1953; Palero et al. 2009b). Both the puerulus and the nisto are transitional stages between a planktonic and a benthic existence and, after settlement, metamorphose into the adult-like juvenile.

MORPHOLOGY

Naupliosoma: If recognized, this phase is characterized by large biramous antennae (figs. 51.2A; 51.4C) provided with several plumose setae that are adapted for swimming, while the maxillipeds are coiled up and functionless (Crosnier 1972). Naupliosomae have been documented only in Achelata genera in which the stage I phyllosoma possesses a biramous antenna.

Phyllosoma: The word phyllosoma is derived from the Greek *phyllos* ("leaf") and *soma* ("body"). According to its functional morphology, this aberrant larval form belongs to the zoea type (D. Williamson 1969). The Achelata are among the only crustaceans to hatch with stalked eyes (fig. 51.1A). The eyestalk is usually made up of a narrow proximal section that becomes wider distally and ends at the cornea (fig. 51.3B). The proximal part is always flexible and can be longer than one third of the total body length (e.g., *Arctides*). A median eye is also present, at least in the early zoeal stages. The antennal exopod is a simple spine in most phyllosoma larvae (figs. 51.1; 51.2). A carapace (cephalon) is present, providing a dorsal and lateral covering for the head and at least part of the thorax (figs. 51.1C; 51.2). It is always fused with the dorsal surface of the head, and its attachment is restricted to this region in all phyllosomae. No lateral carapace projections have been described, but some palinurid phyllosomae have a rostrum (e.g., *Phyllamphion*; fig. 51.2F). The maxillule and maxilla are always present. Thoracopod 1 (maxilliped 1) is greatly reduced, and it is absent in early stages of the Scyllaridae (fig. 51.3B) and some species of *Panulirus*. The endopods of thoracopods 2–6 are segmented. The first phyllosoma stage of the Scyllaridae and some palinurid genera (*Jasus*, fig. 51.2D) only shows exopods in pereopods 1 and 2, but all phyllosomae of stridentes genera also have an expopod on maxilliped 3 (fig. 51.2). The late phyllosoma stages have setose exopods either on all pereopods or on all but the last. The abdomen remains comparatively small and unsegmented until the late stages (figs. 51.2B–F; 51.4D–G). The telson is rounded in all stages, and the uropods develop much more gradually than in other decapods. The Achelata are exceptional in that, although the uropods appear before the pleopods, they develop and acquire setae gradually. No intermediate forms are known connecting phyllosomae with more conventional zoeae.

Puerulus (Palinuridae): Externally, the puerulus resembles an adult lobster in form, although it is somewhat dorsoventrally flattened and has proportionally large natatory pleopods (Gordon 1953). These two features are found in all palinurid pueruli known to date and indicate that the puerulus is specialized for swimming. The characteristics of the puerulus have been described in the stridentes genera *Palinurus*, *Palinustus*, and *Panulirus*, and in the silentes genera *Jasus*, *Palinurellus*, and *Projasus* (see Palero et al. 2008a and the references therein).

Nisto (Scyllaridae): The nisto is a transparent, relatively short-lived phase that shows an almost adult-like morphology. The carapace length can reach up to 20 mm in some species (fig. 51.4L, M). Little is known about the biology of the nisto, compared with the puerulus phase; they are not commonly caught using standard larval sampling methods, which are aimed at the study of settlement and/or recruitment of commercial lobster species (Booth et al. 2005). Only a few Scyllarinae nistos have been described to date, and, of these, only two have been reared to an identifiable lobster, either from laboratory-reared larvae or from late-stage phyllosoma larvae caught in the plankton (Ito and Lucas 1990).

MORPHOLOGICAL DIVERSITY: Differences among phyllosomae from the Palinuridae (figs. 51.1; 51.2) and Scyllaridae (figs. 51.3; 51.4) lie mostly in the morphology of the peduncular segments of the antennae (flattened and with a broad base in the Scyllaridae) and in the antennal flagellum (much longer in the Palinuridae) (Gurney 1936c; M. Johnson 1971a). The absence of an exopod on the third maxilliped is not confined to the Scyllaridae; it is also absent in *Jasus*.

Palinuridae: Given that the antennal flagella are not fully developed in early stages, distinguishing phyllosoma larvae of *Jasus* from scyllarid larvae can be difficult. The most useful diagnostic features of the *Jasus* first phyllosoma appear to be an antennal endopod that is slightly longer than the exopod, an antennule with a fairly prominent spine or seta near the lateral margin, and a dorsolateral (subexopodal) spine on the basis of legs 1–3. After the first few stages, *Jasus* phyllosomae can be distinguished from those of the Scyllaridae by the length of the antennae (fig. 51.2D). Among the most striking phyllosoma larvae are those discovered by Reinhardt (1849), for which he proposed a new generic name, *Phyllamphion*. Two types of phyllamphion are known to date: *Phyllamphion elegans*, with a rectangular shield (fig. 51.2F), and *Phyllamphion santuccii*, with a circular shield (fig. 51.2C). These larvae probably belong to *Palinurellus* and *Puerulus* (Sims 1966; Michel 1970b). *Phyllamphion* larvae deviate from other palinurid phyllosomae in having an expanded cephalon, a heavy spine-like process on the first peduncular segment of the antennae, eyes on relatively short stalks, and a well-defined palp on the maxillae. Other types of palinurid phyllosomae worth mentioning are those with chelate pereopods. M. Johnson and Robertson (1970) reviewed the main traits of *Justitia* phyllosomae and highlighted the fact that, after the early stages, pereopods 1 and 3 terminate in a chela and subchela, respectively. *Palinustus* phyllosomae also possess chelate pereopods 1, 3, and 4 and a subchelate pereopod 2 (fig. 51.1L) (Palero et al. 2010a). The phyllosoma of *Palinustus* can be further distinguished by its antennules (extending beyond the end of the antennal peduncle) and by the proximal segment of the antennular peduncles (which is much longer than other segments). *Panulirus* phyllosomae closely resemble those of *Palinurus*, but they have a pear-shaped cephalon (fig. 51.2E) (see McWilliam 1995 for a categorization of *Panulirus* species into four morphological types).

Scyllaridae: *Parribacus*, *Scyllarides*, and *Arctides* phyllosomae are distinguished from other genera of the Scyllaridae by their lack of subexopodal spines and their rounded or pear-shaped cephalic shield (fig. 51.4A, F, G). Although some morphometric ratios can be used to separate species (Coutures 2001), there is no direct evidence showing clear-cut generic differences in the early stages. *Parribacus* phyllosomae are remarkable for their large final size (a total length of up to 80 mm), without much development of the antennae, mouthparts, and abdomen (fig. 51.4A, B) (M. Johnson 1971a). *Ibacus* and *Thenus* phyllosomae display abbreviated development, and their newly hatched larvae are larger in size (>3 mm in total length) than those of any other Achelata. Harada (1958) provided a detailed morphological description of these larvae. *Thenus* phyllosomae have a cephalic shield that is broader than long (much broader than the thorax), and in the late stages they possess anteriorly projecting shoulders, similar to those of *Ibacus* phyllosomae (fig. 51.4D). The phyllosomae of the Scyllarinae are the most diverse, with a cephalic shield that is wider than the thorax and subcircular, subquadratic, or trapezoidal in outline (fig. 51.3). In later stages the abdomen is broad at the base, forming a direct tapering continuation of the thorax (see Robertson 1968). The antennae in early stages are uniramous, much shorter than the antennules, and in later stages are broad, with a lateral process (fig. 51.3D, E). Phyllosoma length in the final stage varies among species, ranging from about 9 to 28 mm. Scyllarinae phyllosomae are difficult to separate into species, due to their similarity, especially in the early stages (Booth et al. 2005).

NATURAL HISTORY: The ecology and behavior of phyllosomae are not well understood. A relationship among phyllosomae and gelatinous zooplankton has been observed in several studies, even though it is unclear if this association is for food, protection, transport, or some combination of these. Some palinurid phyllosomae are known to prey on ctenophores, hydromedusans, chaetognaths, and larval fish (D. Dexter 1972), and some scyllarid larvae are ectoparasitic on scyphomedusans (Shojima 1963). Gelatinous zooplankters are more common inside the continental shelf break than beyond it, and the association of phyllosomae with them may play a role in controlling larval dispersal in some species. Long-distance transport of larvae has been documented for some palinurid lobsters, for which the phyllosoma phase may last for up to two years (Bradford et al. 2005; R. George 2005). Metamorphosis is probably dictated by the nutritional state of the phyllosoma, rather than by environmental cues (McWilliam and Phillips 1997). Pueruli apparently navigate to the shallow-water juvenile nurseries by means of a complex receptor system, formed by the antennae, and a pinnate setal system, which enables them to orient to cues associated with coastlines (Phillips and Macmillan 1987). Compared with spiny lobsters, scyllarids have been the subject of fewer studies, particularly those following these animals into recruitment. Much of the early life of most scyllarid lobsters is spent in waters away from parental grounds, but, like the duration of larval life, the extent of dispersal varies widely among species (Booth et al. 2005). For example, species belonging to *Parribacus*, *Scyllarides*, and *Arctides* have larval distributions generally much farther from shore than those of *Evibacus* and many other scyllarinid species. Recent work has shown that some scyllarinids complete their larval development within the lagoons formed by coral-island barrier reefs (Booth et al. 2005). The ecology of the post-larval (decapodid) stage of scyllarids is not well known, because nistos are less commonly captured than phyllosomae. Nistos collected off eastern Australia were active at night but buried themselves in sandy substrates by day (Barnett et al. 1986), and nistos of *Scyllarus* have been found buried in coarse granitic sand as well as in fine to medium sands and mud (Palero et al. 2009b).

PHYLOGENETIC SIGNIFICANCE: The highly modified and unusual nature of their phyllosomae caused early confusion and debate as to which group of decapods the achelatans might belong, and it has even led to more recent speculation that achelatans might not be decapods at all, or that they might be the result of hybridization events combining genomes of unrelated organisms (the "larval transfer" hypothesis of D. Williamson 1988a, 1988b, 1992; D. Williamson and Vickers 2007). Larval characters also have been useful for consistently separating the Palinuridae from the Scyllaridae, and for establishing relationships among genera in both groups.

HISTORICAL STUDIES: An obscure publication (Forster 1782) briefly describing a *Phyllamphion* specimen (as *Cancer cassideus*) is the earliest-known account of an Achelata larva. Because of their specialized body form, much doubt and debate existed in earlier years regarding the systematic position of phyllosoma larvae. The controversy started with the establishment of *Phyllosoma* by Leach (1818). Milne-Edwards (1830b) regarded phyllosomae as a separate group of the Stomatopoda, dividing them into "phyllosomes ordinaires" (with a distinct well-developed abdomen that was narrower than the thorax), "phyllosomes brevicaudes" (with a rudimentary abdomen situated in an emargination of the thorax), and "phyllosomes laticaudes" (with a broad abdomen continuous with the thorax). Later, Dohrn (1870) hatched eggs from *Scyllarus* and showed that the larvae were identical to the phyllosomae obtained from plankton by Claus (1863). Richters (1873) distinguished between the phyllosomae attributed to the Palinuridae and Scyllaridae on the basis of the morphology of the antennae and showed that "phyllosomes ordinaires" were in fact the larval stages of the Palinuridae. John Gilchrist (1913) was the first to report the existence of the naupliosoma, and Gurney (1936c) described a large collection of phyllosomae from the *Discovery* Expedition (1901–1904), establishing the identity of 9 genera. Martin Johnson made a remarkable contribution by describing phyllosomae from the South China Sea (M. Johnson 1971a) and the Pacific Ocean (M. Johnson 1971b). More recent records of phyllosomae are those by McWilliam et al. (1995) from the Indo-Pacific, Inoue et al. (2004) from Japanese waters, and Webber and Booth (2001) from New Zealand. More recently, the use of morphological descriptions, combined with DNA markers, has allowed phyllosomae caught in the wild to be identified to particular species (Palero et al. 2008b, 2010b).

Selected References

Gurney, R. 1936c. Larvae of decapod Crustacea, 3: Phyllosoma. Discovery Reports 12: 400–440.

Johnson, M. W. 1971a. On palinurid and scyllarid lobster larvae and their distribution in the South China Sea (Decapoda, Palinuridea). Crustaceana 21: 247–282.

McWilliam, P. S. 1995. Evolution in the phyllosoma and puerulus phases of the spiny lobster genus *Panulirus* White. Journal of Crustacean Biology 15: 542–557.

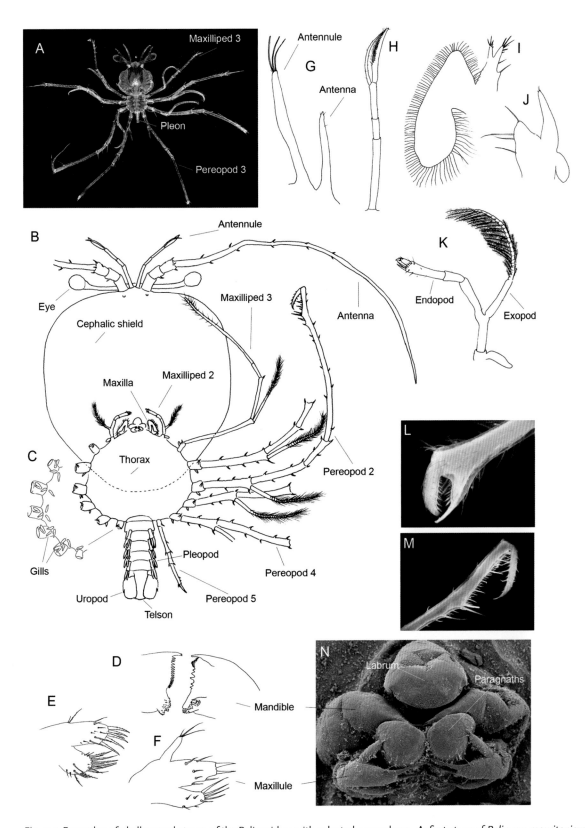

Fig. 51.1 Examples of phyllosomal stages of the Palinuridae, with selected appendages. A: first stage of *Palinurus mauritanicus*, light microscopy. B–F: *Palinustus* sp. B: final stage, ventral view. C: left side of the thorax, dorsal view. D: mandibles. E: maxillule. F: maxillule with palp. G: *Palinurus* sp., antennules and antenna of the first stage. H–M: *Palinustus* sp. H: antennule of the final stage. I: maxilla. J: maxilliped 1. K: maxilliped 2. L: protopod and dactyl of pereopod 1, light microscopy. M: protopod and dactyl of pereopod 2, light microscopy. N: mouthparts of a first-stage phyllosoma of *Palinurus mauritanicus*, SEM. A and D–M original; B from Palero et al. (2010a), reproduced with permission of *Zootaxa* 2403: 42–58; N modified after Palero and Abelló (2007).

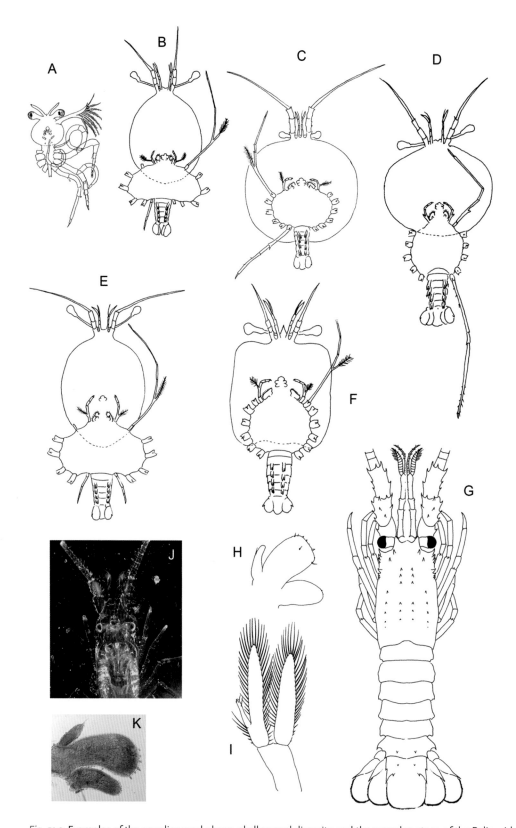

Fig. 51.2 Examples of the naupliosomal phase, phyllosomal diversity, and the puerulus stage of the Palinuridae (with some appendages shown). A: drawing of the naupliosoma of *Jasus*, dorsal view. B– F: drawings of phyllosomae, ventral views. B: *Justitia* sp. C: *Phyllamphion santuccii* (= *Puerulus* sp.). D: *Jasus* sp. E: *Panulirus* sp. F: *Phyllamphion elegans* (= *Palinurellus* sp.). G–I: drawings of a puerulus of *Palinurus mauritanicus* G: dorsal view. H: maxillule. I: pleopod. J: puerulus of *Palinurus mauritanicus*, light microscopy. K: maxillule of a puerulus, light microscopy. A modified after Archey (1915); B modified after Michel (1971); C modified after Sekiguchi et al. (1996); D modified after D. Miller (1986); E modified after Goldstein et al. (2008); F modified after Sims (1966); G modified after Guerao et al. (2006b); H–K original.

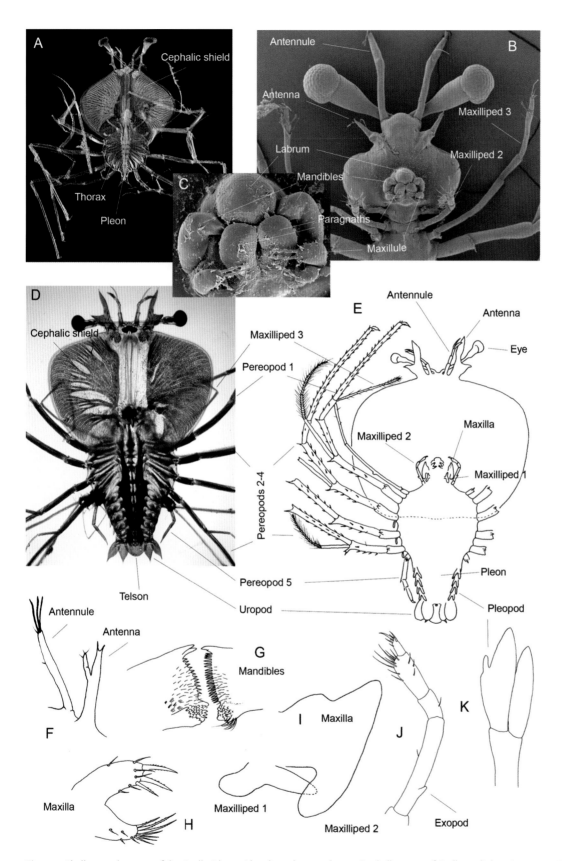

Fig. 51.3 Phyllosomal stages of the Scyllaridae, with selected appendages. A: phyllosoma of *Scyllarus*, light microscopy. B: stage 1 of *Scyllarus* sp., SEM, ventral view. C: mouthparts, SEM. D and E: final stage of *Scyllarus* sp. D: light microscopy, ventral view. E: drawing, ventral view. F: antennules and antenna of stage 1. G–K: drawings of the final stage. G: mandibles. H: maxillule. I–K: drawings of the final stage. I: maxilla and maxilliped 1. J: maxilliped 2. K: pleopod. A courtesy of the Museo de Ciencias Naturales de Tenerife; B–K originals.

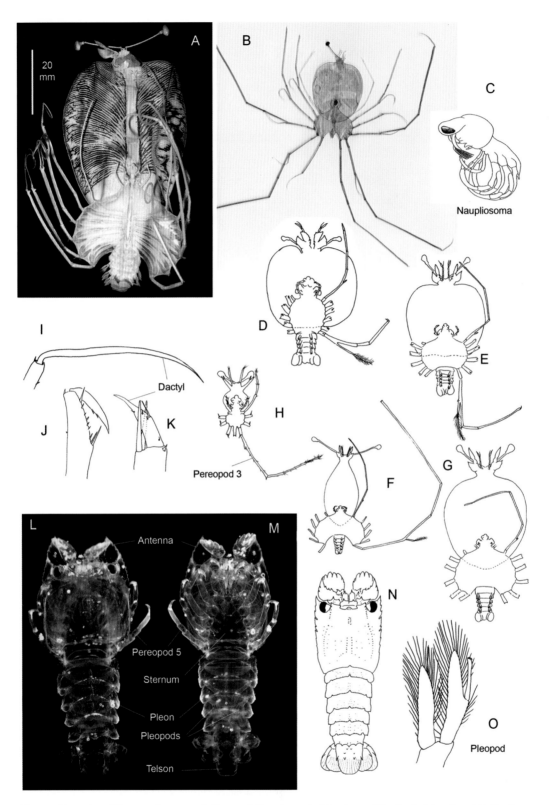

Fig. 51.4 Examples of the naupliosoma, phyllosomal diversity, and nisto stage of the Scyllaridae. A and B: *Parribacus* sp. A: final stage, light microscopy, ventral view. B: subfinal stage, light microscopy, ventral view. C: drawing of a naupliosoma of *Scyllarides* sp., lateral view. D–G: drawings of phyllosomes, ventral views. D: *Ibacus* sp. E: *Evibacus princeps*. F: *Arctides* sp. G: *Scyllarides* sp. H: drawing of stage 1 of *Scyllarus* sp., dorsal view. I: drawing of a pereopod 3 dactyl. J: drawing of pereopod 1, protopod, and dactyl of *Parribacus* sp. K: drawing of pereopod 1, protopod, and dactyl of *Scyllarus* sp. L and M: nisto of *Scyllarus*, light microscopy. L: dorsal view. M: ventral view. N: drawing of a nisto, dorsal view. O: drawing of the pleopod of a nisto. A, B, and F–M original; C modified after Crosnier (1972); D modified after Marinovic et al. (1994); E modified after Robertson (1969); N modified after Palero et al. (2009b).

52 Polychelida

JOEL W. MARTIN

GENERAL: Polychelids are a relatively species-poor group of deep-sea lobsters with an unusual flattened carapace, greatly reduced eyes, chelae on the first four (and sometimes five) pereopods, and delicate elongate first pereopods (chelipeds). This infraorder, the last remnant of a mostly extinct lineage, is represented by only 1 currently recognized extant family, the Polychelidae, with 6 extant genera and approximately 40 species, most of which belong to the genera *Stereomastis* and *Polycheles* (Firth and Pequegnat 1971; Wenner 1979; Galil 2000; Ahyong 2009; De Grave et al. 2009). The life cycle consists of a pre-zoea, an eryoneicus zoea, an eryoneicus megalopa, a juvenile, and an adult, all of which are free living.

LARVAL TYPES: Polychelid larvae are among the most bizarre larvae of all the Decapoda, if not the entire Crustacea. For years their enormous size obscured their recognition as larval forms, and, as with many other decapod groups, the larvae were assigned to genera (usually *Eryoneicus*), as though they were adults. Today the name eryoneicus is the accepted term for the large globular larval stage common to this group.

Pre-Zoea: The first hatching stage was not described with certainty (from a captured ovigerous female) until 1996 (Guerao and Abelló 1996a). It is a pre-zoeal stage, with poorly developed pleopods and uropods, appendages sheathed in a membranous covering, and natatory setae that have not yet emerged, so the larva cannot swim. D. Williamson (1983) also mentioned a pre-zoeal stage from plankton samples.

Eryoneicus Zoea: Following the pre-zoeal stage, by definition the next few stages (the exact number is unknown) are zoeal, with swimming accomplished by the thoracopodal expopods. Later stages, by definition, are megalopal (decapodid of Kaestner 1970; Felder et al. 1985), as the swimming function is taken over by the well-developed abdominal pleopods, despite the fact that in polychelids the megalopae are planktonic rather than benthic. The number of zoeal and megalopal stages is not known with certainty for any species. The stage that would be the ecological equivalent of the true

megalopa (a transitional stage from planktonic to adult habitat) has not been identified, but most likely it is merely a juvenile rather than a discrete larval stage. The eryoneicus zoea is here defined as an early larval stage of a polychelid lobster in which swimming is accomplished by the natatory epipods of the thoracopods (the maxillipeds and first two pereopods).

Eryoneicus Megalopa: The eryoneicus megalopa is a decapodid stage, but it is unusual in its shape and tremendous size. It is defined here as a later polychelid larval stage, where swimming is accomplished by well-developed abdominal pleopods.

MORPHOLOGY: The eryoneicus larva is immediately recognizable by its large size and by the distinctive nearly spherical shape of the carapace.

Pre-Zoea: Paradoxically, the hatching stage (in at least 1 species) is a tiny (the carapace is ca. 1.0 mm in diameter) pre-zoea (fig. 52.1A), with poorly defined limbs ensheathed by the cuticle, and with no emerged natatory setae on the exopods (Guerao and Abelló 1996a). The first two pereopods are chelate on hatching.

Eryoneicus Zoea: Zoeal stages are characterized by setose natatory exopods on the maxillipeds and at least the first two legs, and by a rostrum that projects beyond the length of the antennules (fig. 52.1B).

Eryoneicus Megalopa: Later (megalopal) stages are extremely large and inflated (figs. 52.1C, D; 52.2). The inflated carapace bears many spines, and the rostrum is reduced, not projecting beyond the antennules. Two stout modified spines, with a few thorns on the blunt apex—termed "piliers" ("pillar" or "column") by earlier workers (e.g., Bernard 1953)—are seen on most eryoneicus larvae, one at the front and one toward the posterior of the carapace, along the midline (figs. 52.1D, E; 52.2B–G); differences in the terminal morphology of these spines could be species-specific (Bernard 1953). These structures, which are lost in the adults, might be homologous to dorsal organs found in similar places in other decapod larvae (Gurney 1942; J. W. Martin and Laverack 1992; Laverack et al. 1996). In late megalopal stages, all pereopods (1–4 or 5) are well developed and chelate. The fingers of the pereopodal chelae are finely pectinate. Individual teeth overlap and bear minute striations, with these teeth becoming smaller and more sparse in more posterior legs (fig. 52.1F–K). The pereopods bear well-developed gills (pleuro-, arthro-, and podobranchs; Faxon 1895). The abdomen is always shorter than the carapace, and its appendages are well developed. The uropods and telson form a functional tail fan, and the telson is terminally acute, considered by some to be a primitive (shrimp-like) character. Well-developed male reproductive structures (first pleopods), terminating in a spatulate blade (similar to what is seen in adults), appear in late megalopal stages. Some specimens exceed 6 cm in carapace length and 8 cm in total length. This size, coupled with their spherical fat-filled carapace, makes them by far the most massive of all known crustacean larvae.

MORPHOLOGICAL DIVERSITY: Morphological variation is relatively minor among the 40 or so described known larval forms. Differences among species (fig. 52.2) lie mostly in the morphology of the frontal margin, the number of carapace spines, the occurrence of these spines in well-defined rows, the spination of the rostral region, and the overall shape of the carapace (D. Williamson 1983; Boyko 2006). The carapace of some specimens appears weakly calcified and membranous, whereas others have a firm rigid carapace; this does not appear to be size-related. Some specimens appear to have flocculent debris on the front half of the carapace.

NATURAL HISTORY: Almost nothing is known about the natural history of eryoneicus larvae, apart from the fact that they can occur at depths as great as 3,700 m (Wicksten 1981). They are usually considered bathypelagic. In the only laboratory-rearing study, newly hatched larvae did not swim, "remained on the bottom of the container," and did not eat Artemia nauplii offered to them (Guerao and Abelló 1996a). The obvious disparity in size between the tiny hatching stage (ca. 1.0 mm) and the enormous megalopal stages (more than 6.0 cm), a 60-fold increase, might be related to, and indicative of, a long larval life. Gurney (1942) suggested that the size and shape of the carapace, and the overall spination of the carapace and abdomen, may aid flotation. The adult genus Cardus, more squat and oval as an adult than other genera, may have arisen from the eryoneicus larval stage via neoteny (Ahyong 2009). Neoteny has also been posited as the explanation for the precocious appearance of well-developed first pleopods in larger male larvae. According to D. Williamson (1982a), the duration of the zoeal stage has been shortened, but "a relatively long pelagic life has been retained or reacquired by lengthening the following phase [the megalopa]" as a "secondary extension of larval life." Gordon (1960a) reported an eryoneicus larva found in the stomach of a deep-sea scyphozoan jellyfish; another report (Wicksten 1981) noted the remains of a bony fish in a larva's mouthfield. Possible fossil (Upper Jurassic) records of eryoneicus larvae were mentioned

by Roger (1946), but his interpretation was rejected by J. Haug et al. (2009a; also see J. Haug et al. 2008, 2013a). Coloration, where known, is usually translucent white or orange to red or purplish-red; the antennae, tips of the chelipeds, and other appendages can be darker red (Faxon 1895, his plate B, his fig. 2; Boyko 2006). Yellowish fat deposits are sometimes visible through the carapace. Fig. 52.1C is the only known photograph of a live eryoneicus larva.

PHYLOGENETIC SIGNIFICANCE: The bizarre morphology of the eryoneicus stages hindered correct attribution of these larvae to adults species for many years, and as recently as 1979, Saint Laurent argued that eryoneicus larvae were adults. We now know that they are larvae of the ancient (Upper Triassic) lobster family Polychelidae, a group that is related to the extinct eryonids and is unquestionably monophyletic, although relationships of polychelids to other groups of lobsters are still unclear. Grouped with the Achelata (to constitute the Palinura) in many previous studies, polychelids are now thought to represent a separate lineage, one that might be the sister-group to all extant reptant (crawling) decapod taxa (see Ahyong 2009).

HISTORICAL STUDIES: Early authors (e.g., Bate 1882; Faxon 1895; Bouvier 1905; Selbie 1914; Gurney 1942; Bernard 1953) used the name Eryoneicus (sometimes as "Eryonicus") as a genus name, as though the larvae were adults of some bathypelagic lobster species. Most workers (especially Sund 1920; Gurney 1942; Bernard 1953), however, realized or suspected that these were larval forms. The genus name Eryoneicus was suppressed in 1964 by the International Commission on Zoological Nomenclature (ICZN 1964). As recently as 1979, Saint Laurent continued to argue that the larvae were adults that had been secondarily adapted to a bathypelagic existence. In part this confusion stemmed from the presence of well-developed and apparently functional pleopods in larger males, which no other decapod crustacean larvae possess. To date, only one larval stage has been positively correlated with a known adult, the result of a laboratory-rearing study of the larvae of Polycheles typhlops (by Guerao and Abelló 1996a). Other larvae, however, have been associated (with varying degrees of confidence) with known adults, based on the spination and ornamentation of the carapace and abdomen (e.g., Bernard 1953; Boyko 2006).

Selected References

Bernard, F. 1953. Decapoda Eryonidae (Eryoneicus et Willemoesia). Dana Report No. 37. Copenhagen: Høst.

Guerao, G., and P. Abelló. 1996a. Description of the first larval stage of Polycheles typhlops (Decapoda: Eryonidea: Polychelidae). Journal of Natural History 30: 1179–1184.

Harvey, A. W., J. W. Martin, and R. Wetzer. 2001. Phylum Arthropoda: Crustacea. In C. Young, M. Sewell, and M. Rice, eds. Atlas of Marine Invertebrate Larvae, 337–369. London: Academic Press.

Williamson, D. I. 1983. Crustacea, Decapoda: Larvae, 8; Nephropidea, Palinuridea, and Eryonidea. Fiches d'Identification du Zooplancton No. 167/168. Copenhagen: Conseil International pour l'Exploration de la Mer.

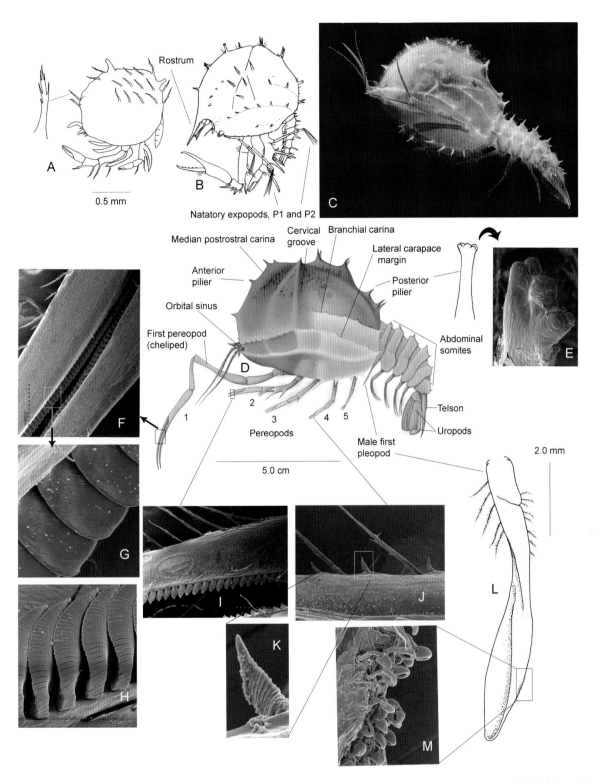

Fig. 52.1 Eryoneicus larvae of the Polychelida. A: drawing of the first hatching stage, a non-swimming pre-zoea of *Polycheles typhlops*. B: drawing of an early eryoneicus zoea (stage unknown), with a well-developed rostrum and natatory (swimming) exopods on pereopods 1 and 2. C: live eryoneicus megalopa (probably *Polycheles pacificus*), ventrolateral view; note the cocked position of the chela of pereopod 1. D: painting of "*Eryonicus caecus*" (probably *Polycheles pacificus*). E: tip of a pilier on the carapace, SEM. F–K: pereopodal chelae and teeth, SEMs. F: fingers of pereopod 1. G: overlapping striated teeth, magnified from F. H: cheliped teeth of pereopod 1 (from a different specimen), oblique view. I: fingers of the chela of pereopod 2, with more widely separated teeth. J: fixed finger (propodus) of the chela of pereopod 4, with widely separated striated teeth. K: one tooth in J, magnified. L: pleopod 1 of a male eryoneicus megalopa, drawn from an SEM specimen. M: cincinnuli (modified hook-like setae), for attaching to a matching pleopod on the opposite side, SEM. Material in C from the Pacific Ocean, mesopelagic zone. A modified after Guerao and Abelló (1996a); B modified after Gurney (1942) and Selbie (1914); C courtesy of David Wrobel (used by permission); D from Faxon (1895, his plate B, his fig. 2), with the terminology mostly following Galil (2000, his fig. 1); E–M original.

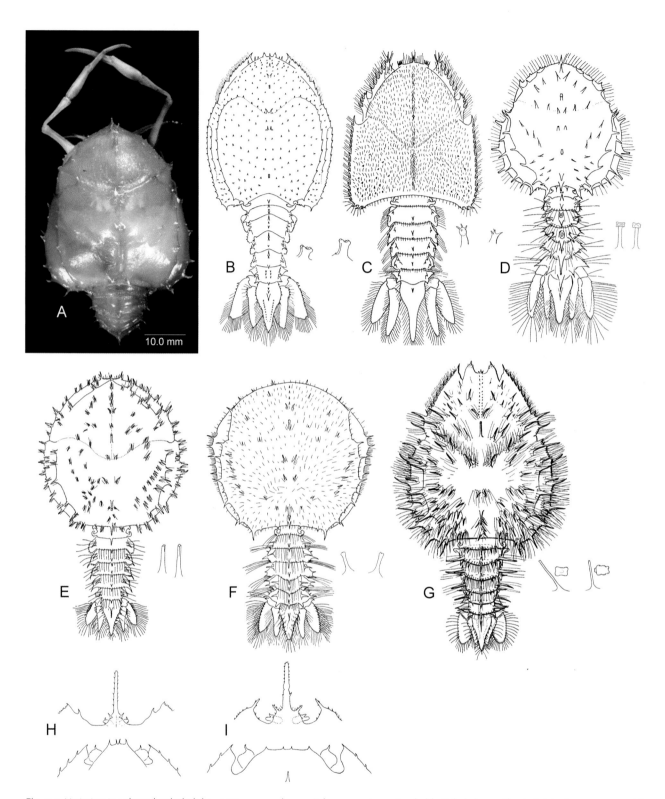

Fig. 52.2 Variation in selected polychelid eryoneicus megalopae. A: large specimen (probably *Polycheles pacificus*), light microscopy, used for the SEM images in fig. 52.1. B–G: drawings of a variety of eryoneicus megalopae, dorsal views, with anterior and posterior piliers drawn to the lower right of each specimen (as identified by Bernard 1953): B: "*Eryoneicus taningi.*" C: "*E. inermis*" (= *Cardus crucifer*). D: "*E. interme-dius.*" E: "*E. australis.*" F: "*E. gurneyi.*" G: "*E. transiens.*" H and I: drawings of the frontal region, showing the reduction in the rostrum from the zoeal (upper) to megalopal (lower) stages (as identified by Bernard 1953), dorsal views. H: "*E. spinoculatus.*" I: "*E. atlanticus.*" Images not to scale. A modified after Harvey et al. (2001); B–I modified after Bernard (1953).

53 Anomura

Alan Harvey
Christopher B. Boyko
Patsy McLaughlin
Joel W. Martin

GENERAL: The Anomura contains a remarkably diverse array of decapod crustaceans. Adults are common benthic inhabitants from the intertidal zone to abyssal depths, and from the equator to the poles. In addition, they are conspicuous components of atypical decapod habitats, such as hydrothermal vents, bodies of fresh water, and even terrestrial environs. The group includes pelagic swimmers, burrowers, and portable-shelter carriers. Despite this diversity, adult anomurans share certain attributes, such as the reduced posterior pair of pereopods, that justify their grouping in a single infraorder. The current classification of this notoriously intractable taxon includes 7 superfamilies: the Aegloidea, Chirostyloidea, Galatheoidea, Hippoidea, Lithodoidea, Lomisoidea, and Paguroidea (after McLaughlin et al. 2007, 2010; Ahyong et al. 2010; Schnabel and Ahyong 2010; Schnabel et al. 2011). Anomuran larvae also exhibit considerable diversity; consequently, larval development is less easily generalized than in its sister clade, the Brachyura (true crabs; see chapter 54). Larval development is best known for the superfamily Paguroidea, which has the largest number of extant genera and species (in excess of 120 and 1,100, respectively). In comparison with the knowledge of available brachyuran developmental sequences, however, information even for paguroids is woefully meager. The life cycle consists of a pre-zoea, a zoea (0–11 stages), a decapodid (usually called a megalopa or glaucothoe), a juvenile, and an adult. None are parasitic, although some porcellanids are obligatory commensals of cnidarians and echinoderms.

LARVAL STAGES: Direct development (without free-swimming larval stages on hatching) has been documented only in members of the exclusively freshwater superfamily Aegloidea and for a few paguroids in the family Diogenidae (e.g., Dechancé 1963; W. Rodrigues and Hebling 1978; G. Morgan 1987; López-Greco et al. 2004).

Pre-Zoea: A non-feeding pre-zoea, preceding the typical first larval stage, has been recorded in at least a few species in all anomuran superfamilies, except the monotypic Lomisoidea (Gore 1970).

Zoea: The vast majority of anomuran embryos hatch as a planktonic zoeal larva that swims with its thoracic appendages. Anomuran zoeae have a generally shrimp-like body form, a segmented pleon, and a moderately broad triangular telson. These larvae pass through one to numerous zoeal stages before the metamorphic molt to the next stage, the decapodid, which in this group is usually called a megalopa or glaucothoe. Zoeal stages vary in number, not only among families, but also within and among genera of individual families. Examples of abbreviated and advanced development are found in all superfamilies except the Hippoidea.

Megalopa (or Glaucothoe): The megalopa is a morphological and behavioral transitional stage from the pelagic free-swimming zoea to the benthic adult. The megalopa can swim using its pleopods, but it can also use its pereopods to walk. Morphologically, megalopae are much more similar to adults than they are to zoeae. In all known cases (except for the direct-developing Aegloidea), there is a single megalopal stage, although Wilkens et al. (1990) stated that the advanced zoea II of *Munidopsis polymorpha* molts directly to a first-stage crab.

MORPHOLOGY

Pre-Zoea: The pre-zoeal cuticle is a translucent, extremely thin membrane made up of three layers (Quintana and Konishi 1988). This cuticle uniformly covers the pre-zoeal body and protrudes as long extensions on the antennules, antennae, and telson. The duration of the pre-zoeal stage varies from a few minutes (J. MacDonald et al. 1957; Quintana and Konishi 1986) to a few hours (Shenoy and Sankolli 1967).

Zoea (figs. 53.1–53.3): Most anomuran zoeae have an ovoid carapace that covers the cephalothorax (fig. 53.1B, C). The eyes are sessile in zoea I (fig. 53.2C), but stalked in zoea II and beyond (fig. 53.2A, B). The antennules are initially uniramous buds. The exopod articulates from the protopod (fig. 53.2G) in zoea III; if the endopodal lobe articulates in the zoeal phase, it does so in zoea IV. The antennae are biramous, with one prominent spine, and often a second smaller spine, on the protopod distally, at the base of the endopod (fig. 53.2H). The scaphocerite (exopod) typically is blade-like, not fused to the protopod, and provided with a terminal spine and a row of marginal plumose setae. When functional in the zoeal stages, the mandibles are usually asymmetrical, with a distinct incisor and a prominently toothed molar process; a palp may develop in the ultimate stage (fig. 53.2I). The endopods of the maxillules may be unsegmented or divided into two or three segments, each provided with one to three setae (fig. 53.2J). The maxillae (fig. 53.2K) typically each consist of bilobed coxal and basial endites provided with few marginal setae, an unsegmented (but often bilobed) endopod, and a scaphognathite that varies widely within the infraorder, particularly in the early zoeal stages. The first and second maxillipeds (fig. 53.2L, M), representing the first two of the eight developing thoracic appendages, are used for propulsion by the zoea. Each maxilliped consists of a protopod divided into a coxa and a basis, a partially two-segmented exopod with terminal natatory setae (four in zoea I, increasing with each stage), and a segmented endopod (five segments in the first maxilliped, four in the second). The third maxilliped initially is a naked uniramous rudiment; in zoea II the exopod acquires natatory setae, and an endopod bud appears. This bud enlarges in subsequent stages but remains unsegmented (fig. 53.2N). Pereopod buds are probably often underreported, especially in the older literature, but they typically appear by zoea III.

By zoea III, the sixth pleomere of the pleon usually is distinctly separated from the fifth (fig. 53.2A). In the majority of anomuran zoea I, the telson is somewhat fan-shaped and slightly to distinctly bilobed (fig. 53.3G), although the median cleft is reduced or lost in subsequent stages (fig. 53.2F). Each terminal margin is provided with seven processes on either side of a median cleft. The first (outermost) of these seven processes is a fused or articulated spine. The second is represented by a fine seta, usually referred to as the anomuran hair, although this is something of a misnomer, as it is lacking in most hippoids but is present in stenopodids, thalassinoids (the Axiidea and Gebiidea), and primitive brachyurans (Burkenroad 1981; also see chapter 54). Telsonal processes 3–7 are articulated and plumodenticulate. In zoea II, an additional medial pair of processes, similar to but smaller than processes 3–7, is added. Uniramous (sometimes biramous) pleopod buds appear on somites 2–5 in the ultimate zoeal stage (fig. 53.2B). The uropods usually make their appearance as biramous but unarticulated buds during zoea III; both the exopod and the endopod articulate from the protopod in zoea IV, when the exopod also usually develops an anterodistal spine (fig. 53.2F).

Megalopa (figs. 53.4–53.7): Metamorphosis to a megalopa (often called a glaucothoe in anomurans) brings major morphological changes. Indeed, megalopae and adults primarily differ by degree. Megalopae have paired biramous swimming pleopods on pleomeres 2–5 (figs. 53.4B, L; 53.5B, L) and disproportionately large eyes. The armature and setation of the carapace, abdomen, and appendages are reduced and proportionately smaller in megalopae than in adults, and the antennular and antennal flagella have fewer segments than do those of adults. Megalopae lack the primary and secondary sexual characteristics of the adults.

MORPHOLOGICAL DIVERSITY: The zoeal phase of anomuran development exhibits a staggering range of morphological expression, both across and within superfamilies. In addition, the rate of development of individual traits can differ considerably within species, with such traits having either abbreviated or advanced development from those of related species with typical development. Zoea I in these species often have traits characteristic of later stages in species with regular development, such as stalked eyes; mandibular palps; a distinct sixth pleomere; and well-developed antennules, third maxillipeds, pleopods, and pereopods. In porcellanids, these latter three characters show pronounced intrastage growth. Another source of morphological variation is primary lecithotrophy. Common traits in species with non-feeding zoeae are symmetrical, unarmed, and often reduced mandibles, as well as maxillules and maxillae with unsegmented endopods and reduced endites that lack marginal setae.

Galatheoidea (figs. 53.1G–M; 53.2C–E; 53.3G, H; 53.4; 53.6D–J): Most galatheoid species have four or five zoeal stages, which are often variable within species. Advanced development, either directly observed or inferred from the large size and small number of eggs per clutch, is apparently widespread in *Munidopsis*, which has two or three zoeal stages in the 2 species reared in the laboratory. All porcellanids pass through two zoeal stages, except for *Petrocheles*, which has five zoeal stages (a recent molecular study by Hiller et al. 2006 suggests that this genus may not be a member of the Porcellanidae).

Pre-zoeae, typically lasting no more than a few hours, have only been reported for porcelain crabs. Galatheoid zoeae have a rostral spine that is usually armed with lateral spinules and sometimes expanded at the base. Mostly notably, it is greatly elongated in the Porcellanidae, up to eight times as long as the carapace (figs. 53.1M; 53.2C; 53.3H). The posterolateral corners of the carapace vary from being unarmed in the Munidopsidae (fig. 53.1H) and some Munididae, to having spines that can be more than four times as long as the carapace (in some porcellanids). In *Munidopsis*, the eyes are rudimentary and partially obscured by the broad overhanging rostrum (fig. 53.1G, H). The antennal protopod is unarmed in porcellanids, but it has two ventrodistal spines in *Munidopsis*. The antennal exopod is elongate in the Porcellanidae and Munididae; in the former it is a slender spine, with few to no plumose setae, whereas in the Munididae its shape is quite variable, with plumose setae present along the margin of the proximal half. In the Porcellanidae and Munidopsidae, the first maxilliped usually has a four-segmented endopod. The posterodorsal margins of the abdominal somites usually have minute denticles and posterolateral spines on somites 4 and 5. Porcellanid zoeae have a unique trapezoidal telson, as long as to much longer than wide (fig. 53.2E). In the Munididae, the telson is deeply bifurcate, and the furcae of the telson are usually armed (fig. 53.1K). The fourth paired processes become robust fused spines in zoea III of galatheids and munidids.

Many features of galatheoid megalopae anticipate the distinctiveness of the adults. For example, the antennal peduncles consist of only four segments, and they lack an acicle and a supernumerary segment (fig. 53.4D). The carapace is typically somewhat longer than wide (fig. 53.6D–J). The telson often has faint delineations of what will become the adult telsonal plates (fig. 53.6H). Megalopae are unknown in *Munidopsis serricornis*, and reportedly absent in *M. polymorpha*.

Chirostyloidea (figs. 53.1N–Q; 53.3I; 53.6K): Larvae are unknown for the Kiwaidae. In the Chirostylidae, complete larval development is known only for *Chirostylus stellaris* (Fujita and Clark 2010), which undergoes abbreviated development. Its two lecithotrophic zoeal stages each last only one day, with little morphological differences between the two. Other first-stage zoeae show similar evidence of advanced lecithotrophic development, except for *Eumunida*, whose zoeae are, in addition, more similar to paguroid than to galatheoid zoeae. In the first stage zoea of *Eumunida*, the posterior margins of the carapace are rounded and lack either spinules or posterolateral spines (fig. 53.1N). The rostrum is slightly shorter than the carapace. Abdominal somites 3–5 have posterolateral spines. In species with advanced zoeae, the carapace is globular, spinose, and somewhat sculptured; the rostrum is shorter than the carapace, and may be armed (figs. 53.1P, Q; 53.3I).

Species with advanced development differ considerably from *Eumunida*. The maxillipeds are unique. The exopods of the first and second maxillipeds are greatly reduced in size, unsegmented, and arise anteriorly from the protopods; the exopods may have four normal natatory setae, or else four or eight rudimentary setae. There are six abdominal somites. Somites 3–6 bear large posterolateral spines (fig. 53.1P), and in *Chirostylus*, somites 2–5 also have large posterodorsal spines (fig. 53.3I). The telson shape varies widely, but it consistently has anterolateral setae and at least ten pairs of posterior marginal setae, and it lacks the anomuran hairs.

Megalopae, known only for *Chirostylus stellaris* (Fujita and Clark 2010), show few noteworthy features. In general they resemble adults, but with reduced spination.

Paguroidea (figs. 53.1T–BB; 53.2A, B, F–N; 53.3A–F; 53.5; 53.7A–F): With 6 families and over 1,100 species, the Paguroidea is easily the largest and most diverse group of anomurans. Our current understanding of their larval development varies widely across families. No larval information is available for the monotypic Pylojacquesidae; only a few zoeae (probably prematurely hatched) and plankton-caught megalopae are known for the Pylochelidae; and no complete larval series are available for the Parapaguridae. A few diogenids have direct development, with eggs hatching into megalopae. Advanced development is present in the Pylochelidae, but otherwise is uncommon. Similarly, primary lecithotrophy is probable in pylochelids but rare in other families; secondary lecithotrophy is known in a few diogenids, but appears to be widespread in the Paguridae. Four zoeal stages is the norm for the Paguri-

dae, whereas the number varies from 2 to 8 in diogenids and coenobitids.

Pre-zoeae have been reported for a few pagurids. Many pagurid zoeae have posterolateral carapace spines. The endopods of the antennules never articulate in the Paguridae (fig. 53.2G). A mandibular palp is present in the ultimate stage in most species (fig. 53.2I), although lacking in many pagurids and present in zoea I in pylochelids. The scaphognathite usually does not develop an epipod until the ultimate stage (fig. 53.2K). The articulation of pleomeres 5 and 6 can be delayed or incomplete in some pagurids. The armature of pleomeres 2–5 varies widely (fig. 53.1T, V). Endopod buds of the uropods do not articulate in pagurids. Parapagurids lose the first telson process by zoea IV, whereas some diogenids lose the fourth process by zoea III.

Paguroid megalopae vary in size by an order of magnitude (e.g., cephalothorax length = 0.4 mm in *Pagurites perspicax* vs. >4 mm in *Dardanus* and some parapagurids). The shield is delineated from the posterior carapace (figs. 53.5A; 53.7A–F). The presence and development of ocular acicles varies greatly within families. The right cheliped is larger than the left in pagurids, parapagurids, and the diogenid *Petrochirus*; the left is larger than the right in *Diogenes*. The fourth pereopods are reduced and usually subchelate. The uropods and the dactyl and propodus of the fourth and fifth pereopods usually have at least a few corneous scales (fig. 53.5A, B). The uropods are asymmetrical (the left is larger than the right) in pagurids and some diogenids.

Lithodoidea (figs. 53.1CC–FF; 53.7G–I): Most lithodoids have four zoeal stages, although *Paralomis* and some *Lithodes* have only two stages. A brief pre-zoeal stage appears to be common. The zoeae are advanced, even in those species with four instars, and many structures are more developed in lithodoid zoeae than in most other anomurans of the same stage. There is considerable variation among and within genera in the degree of advancement of these characters, although lithodids tend to be somewhat more advanced than hapalogasterids. In the antennae of zoea I, the endopods are elongate and terminate in a simple or bifid spine. The uropods, when present, are uniramous and often reduced.

In the megalopae, the armature of the carapace and appendages is reduced, compared with that of the adults. The mandibles are reduced in at least some species with lecithotrophic megalopae. The right cheliped is usually noticeably larger than the left. The pleon is not bent under the thorax in the megalopae (fig. 53.7G–I). As in the zoeae, the uropods are either uniramous or lacking.

Lomisoidea (fig. 53.1R, S): Only first-stage zoeae are known from *Lomis hirta*, the sole member of this superfamily (Cormie 1993; McLaughlin et al. 2004). These zoeae show some evidence of advanced development, and at least facultative lecithotrophy. The carapace is unarmed, with a moderately long rostrum. The blade-like scaphocerite has three or four prominent spines on the outer margin. The endopods of the first maxilliped have four segments. The pleon has the standard five free segments (not four, as stated by Cormie, according to McLaughlin et al. 2004). Pleomeres 3–5 have posterolateral spines, and pleomeres 2–5 have one medial plus two pairs of posterodorsal spines.

Hippoidea (figs. 53.1A–F; 53.6A–C): Hippoids typically pass through 3–6 zoeal stages. There is evidence for marked-time molting (in the absence of sand for megalopal burrowing) in the Hippidae (Harvey 1993) and perhaps the Albuneidae (Sanchez and Aguilar 1975), which may explain both the large number (up to 11) of zoeal stages (Knight 1967; Siddiqi and Ghory 2006) and very large size (carapace length of at least 6.5 mm; J. W. Martin and Ormsby 1991) recorded in this superfamily.

The carapace is globular. It has a long recurved rostral spine, which increases disproportionately with each stage in the Hippidae. Posterolateral margins are unarmed in the Blepharipodidae, but the margins have a long spine in the other 2 families (fig. 53.1A–F). The eyes are apparently stalked in zoea I in some Hippidae. The antennular endopods appear as buds in the ultimate stage, except in the Hippidae, where they are lacking. Whether, and when, the exopods articulate varies among the families. The antennal exopods are reduced or absent in zoea I, and remain absent in *Hippa*. Hippids lack a mandibular palp. The maxillae of the Hippidae are remarkable for their complete lack (or fusion) of endopods and endites. The scaphognathite of all families has numerous distal setae and a well-developed, initially naked epipod. The first maxilliped of blepharipodids has four segments; third maxilliped buds do not appear until at least zoea IV and remain uniramous in the Hippidae. Pereopod buds appear in the first two zoeal stages in albuneids and blepharipodids. The pleomere armature varies across families; pleomere 6 remains fused to the telson in the Hippidae. Blepharipodids do not develop pleopods on pleomere 5 (and sometimes 4). The endopods of the uropods appear as small buds in stages 3 or 4, and they articulate in the ultimate stage (except in the Blepharipodidae). The telson of blepharipodids is typically anomuran, but its shape and setation are highly unusual in the other 2 families (fig. 53.1A, D, E).

In the megalopa, the upper ramus of the antennule has an exceptional number of segments (18–44) in the Blepharipodidae and Albuneidae, reflecting the unusual specialization of the antennules as breathing tubes in these burrowers. Similarly, in *Emerita*, in which adults filter feed by using their greatly elongated plumose antennae, megalopa have exceptionally well-developed antennae. Paired pleopods are lacking on pleomere 5 (and sometimes 4) in the megalopae of the Blepharipodidae, as in their zoeae. The telson of hippoids has numerous marginal setae (short in hippids), but the shape varies across families (fig. 53.6A–C).

Aegloidea: The freshwater Aeglidae undergo direct development, bypassing not only the zoeal but also the megalopal phase, hatching as a benthic juvenile (fig. 53.6L).

NATURAL HISTORY: In most species, zoeae are pelagic and planktotrophic. The number of zoeal stages is typically phylogenetically constrained, whereas the duration of the zoeal phase is largely controlled by external factors, primarily temperature and food availability (Anger 2001). Settlement out of the plankton and into the benthos occurs during the megalopal stage (but see Harvey 1993). Many anomuran taxa have been reared in the laboratory up until the megalopal stage, only to suffer heavy or complete mortality without molting to the first juvenile crab stage (Provenzano 1962; Christiansen and Anger 1990; Harvey 1992; Fujita and Osawa 2005). Although some authors suggest that this indicates that the metamorphosis to the post-larval stage is particularly difficult (Knight 1970; Gore 1972), for several species this problem has been solved by providing an appropriate settlement cue, such as shells for marine hermit crabs or land for terrestrial hermit crabs (Harvey 1992, 1996; Brodie 1999), and some species clearly delay metamorphosis in the absence of these cues (Harvey 1996). Newly settled megalopae often initiate behaviors characteristic of subsequent post-larval stages. This is the first stage that can recognize and use empty snail shells in hermit crabs (Agassiz 1875), can burrow in hippoids (Knight 1968, 1970; Harvey 1993), and can filter feed in porcellanids (Gonor and Gonor 1972). Lecithotrophy is widespread in the Anomura. It may occur in some to all zoeal stages (Rice and Provenzano 1966; Wilkens et al. 1990; Harvey 1992; Konishi and Taishaku 1994; Clark and Ng 2008), in megalopae (P. Miller and Coffin 1961; Dawirs 1981; Harvey and Colasurdo 1993), or in both (Anger 1996; Harvey 1996).

PHYLOGENETIC SIGNIFICANCE: The Anomura is a diverse group with a notoriously unstable taxonomic history. Every element of its classification—including its name ("Anomura" vs. "Anomala"), its composition, interfamilial relationships, and monophyly of component superfamilies and families—has been a source of confusion and controversy. Larval morphology has played a limited but important role in this discussion. Based primarily on larval features, J. MacDonald et al. (1957) proposed that paguroid hermit crabs were more closely related to lithodids and galatheoids, while coenobitoidean hermit crabs were more closely related to thalassinoids. Although using a small data set (by modern standards), this paper strongly influenced most classifications published over the next quarter century. Van Dover et al. (1982) surveyed published descriptions of 340 species of anomuran and brachyuran larvae and found a strong phylogenetic signal in the scaphognathite of the maxilla. McLaughlin et al. (2004) used megalopal and early juvenile stages in their critical assessment of the lithodid-pagurid relationship, but decided against using the zoeal phase in their phylogenetic analysis, due to complications introduced by certain aspects of zoeal biology, namely lecithotrophy and cross-taxon variability in developmental rates.

HISTORICAL STUDIES: Surprisingly, a controversial issue among carcinologists up until the mid-1800s was whether decapods underwent metamorphic development. Meanwhile, a number of genera were erected for larval anomurans. Perhaps the best known of these is *Glaucothoe*, a genus of pelagic crustaceans that turned out to be the exceptionally large megalopae of parapagurid hermit crabs. Until recently, hermit crab megalopae (itself derived from *Megalopa*, a generic name given to brachyuran larval forms) were known as glaucothoe. Other generic names given to larval anomurans include *Grimothea* (= *Munida* megalopae), *Prophylax* (= paguroid megalopae), *Zoeides* (= hippid zoeae), and *Zoontocaris* (= galatheoid zoeae). Rathke (1840, 1842) provided the first description of the larval stages of an anomuran (*Pagurus bernhardus*) from material collected in the plankton. Agassiz (1875) first reported that in hermit crabs, the megalopa was the initial stage to use shells; M. Thompson (1903) enhanced and expanded Agassiz's findings, as well as providing the first detailed description of the internal anatomy of anomuran larvae. Until the mid-1950s, when researchers started using *Artemia* nauplii to feed decapod zoeae, most descriptions of anomuran larvae were limited to first-stage zoeae hatched from ovigerous females or from material collected from the plankton; Robert Gurney and Marie Lebour were particularly successful at identifying and describing anomuran and other decapod larvae during this phase. Modern studies on anomuran larvae no longer rely on plankton-collected material, which makes these larvae more likely to be correctly identified and their developmental sequences more likely to be complete. Reviews of anomuran larval studies are lacking. R. H. Gore, P. A. McLaughlin, A. J. Provenzano, and their students have published a formidable body of larval descriptions from the North American coasts, as have K. Konishi, M. Osawa, and S. Shokita for Asia.

Although rearing methods have led to more (and more complete) larval developmental sequences, there are still many species that have not been reared past the first zoeal stage (e.g., Provenzano 1971; Saito and Konishi 2002; Guerao et al. 2006a; Clark and Ng 2008), or past the megalopal phase (Provenzano 1962; Christiansen and Anger 1990; Harvey 1992; Fujita and Osawa 2005). Many species in the first category are deeper-water species, which may suggest that larvae of these species have additional requirements not met in the standard culture environment, whereas the inability to metamorphose to the juvenile stage often seems to come from a metamorphic cue missing from the culture environment, such as conspecifics, a suitable substrate, and empty snail shells (G. Jensen 1989; Harvey and Colasurdo 1993; Harvey 1996).

Selected References

Anger, K. 2001. The Biology of Decapod Crustacean Larvae. Crustacean Issues No. 14. Lisse, Netherlands: A. A. Balkema.

Baba, K., Y. Fujita, I. S. Wehrtmann, and G. Scholtz. 2011. Developmental biology of squat lobsters. *In* G. C. B. Poore, S. T. Ahyong, and J. Taylor (eds.), The Biology of Squat Lobsters. 105–148. Boca Raton, FL: CRC Press; Collingwood, Victoria, Australia: CSIRO Publishing.

Gurney, R. 1942. Larvae of Decapod Crustacea. London: Ray Society.

MacDonald, J. D., R. B. Pike, and D. I. Williamson. 1957. Larvae of the British species of *Diogenes*, *Pagurus*, *Anapagurus*, and *Lithodes* (Crustacea, Decapoda). Proceedings of the Zoological Society of London 128: 209–257.

Williamson, D. I., and K. G. Von Levetzow. 1967. Larvae of *Parapagurus diogenes* (Whitelegge) and some related species (Decapoda, Anomura). Crustaceana 12: 179–192.

Fig. 53.1 (*following page*) Drawings of examples of zoeal diversity (first zoeal stage). A–F: Hippoidea. A and B: *Paraleucolepidopa myops* (Albuneidae), zoea IV. A: dorsal view. B: lateral view. C and D: *Emerita holthuisi* (Hippidae). C: zoea I, dorsal view. D: zoea III, dorsal view. E and F: *Blepharipoda occidentalis* (Blepharipodidae), zoea I. E: dorsal view. F: lateral view. G–M: Galatheoidea. G and H: *Munidopsis tridentata* (Munidopsidae), zoea II. G: dorsal view. H: lateral view. I and J: *Galathea intermedia* (Galatheidae), zoea II. I: dorsal view. J: lateral view. K and L: *Pleuroncodes planipes* (Munididae), zoea V. K: dorsal view. L: lateral view. M: *Allopetrolisthes angulosus* (Porcellanidae), zoea II, lateral view. N–Q: Chirostyloidea. N and O: *Eumunida annulosa* (Eumunididae), zoea I. N: dorsal view. O: lateral view. P and Q: *Uroptychus* cf. *politus* (Chirostylidae), zoea I. P: dorsal view. Q: lateral view. R and S: *Lomis hirta* (Lomisidae), zoea I. R: dorsal view. S: lateral view. T–BB: Paguroidea. T and U: *Coenobita compressus* (Coenobitidae), zoea IV. T: dorsal view. U: lateral view. V and W: *Areopaguristes nigroapiculus* (Diogenidae), zoea I. V: dorsal view. W: lateral view. X and Y: *Phimochirus holthuisi* (Parapaguridae), zoea I. X: dorsal view. Y: lateral view. Z and AA: *Parapagurus diogenes* (Paguridae), zoea IV. Z: dorsal view. AA: lateral view. BB: *Pylocheles mortensenii* (Pylochelidae), lateral view. CC–FF: Lithodoidea. CC and DD: *Cryptolithodes typicus* (Hapalogastridae), zoea IV. CC: dorsal view. DD: lateral view. EE and FF: *Acantholithodes hispidus* (Lithodidae), zoea III. EE: dorsal view. FF: lateral view. A–F modified after Knight (1970); C and D modified after Siddiqi and Ghory (2006); E and F from M. Johnson and Lewis (1942), reproduced with permission of the Marine Biological Laboratory, Woods Hole, MA; G and H from Samuelsen (1972), reproduced with permission of Marine Biology Editorial Office, Institute of Marine Research, Bergen, Norway; I and J modified after Christiansen and Anger (1990); K and L from Boyd (1960), reproduced with permission of the Marine Biological Laboratory, Woods Hole, MA; M modified after Wehrtmann et al. (1996). N and O modified after Guerao et al. (2006a); P and Q modified after Pike and Wear (1969); R and S modified after Cormie (1993); T and U modified after Brodie and Harvey (2001); V and W modified after Kornienko and Korn (2011); X and Y modified after D. Williamson and Von Levetzow (1967); Z and AA modified after Gore and Scotto (1983); BB modified after Saito and Konishi (2002); CC and DD modified after J. Hart (1965); EE and FF modified after Hong et al. (2005).

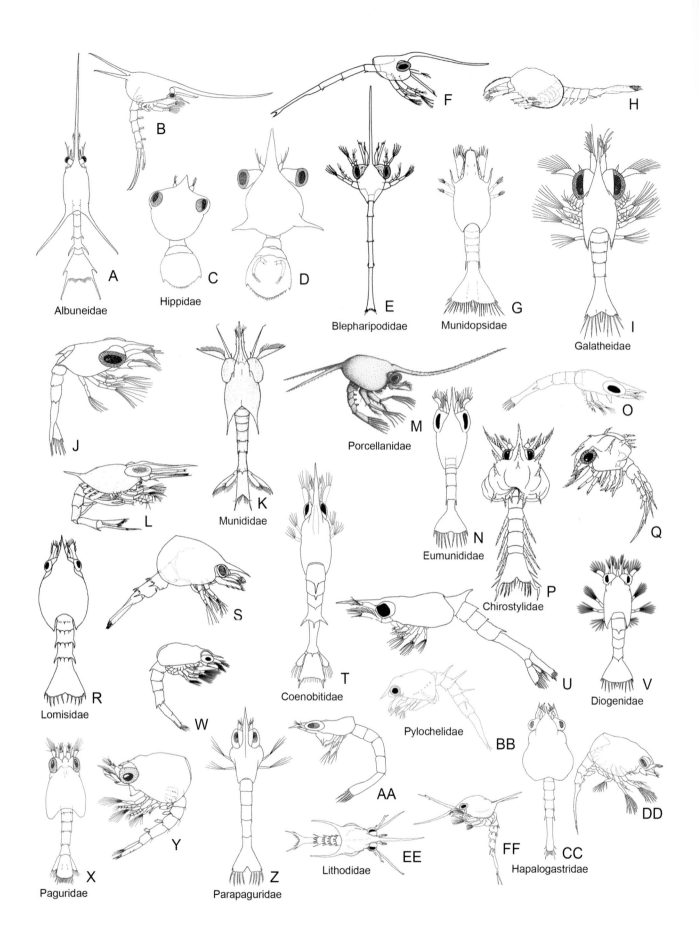

Albuneidae

Hippidae

Blepharipodidae

Munidopsidae

Galatheidae

Porcellanidae

Munididae

Eumunididae

Chirostylidae

Lomisidae

Coenobitidae

Diogenidae

Pylochelidae

Paguridae

Parapaguridae

Lithodidae

Hapalogastridae

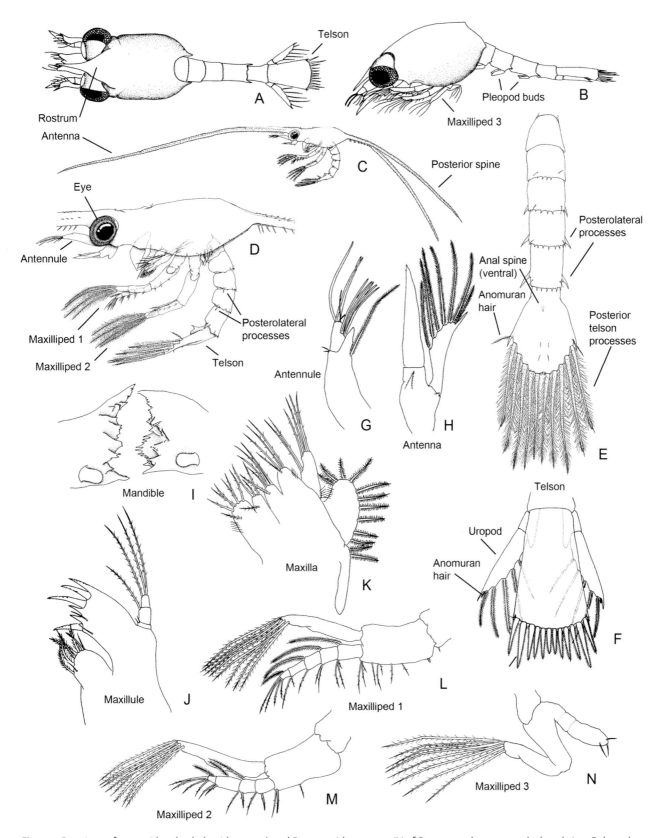

Fig. 53.2 Drawings of paguroid and galatheoid zoeae. A and B: paguroid-type zoea IV of *Parapagurodes constans*. A: dorsal view. B: lateral view. C–E: galatheoid-type (Porcellanidae) zoea I of *Novorostrum decorocrus*. C: lateral view. D: detail, lateral view. E: telson. F–N: selected appendages of paguroid a-type zoea IV of *Parapagurodes constans*. F: telson. G: antennule. H: antenna. I: mandible. J: maxillule. K: maxilla. L: maxilliped 1. M: maxilliped 2. N: maxilliped 3. A, B, and F–N modified after Hong and Kim (2002); C–E modified after Fujita and Osawa (2005).

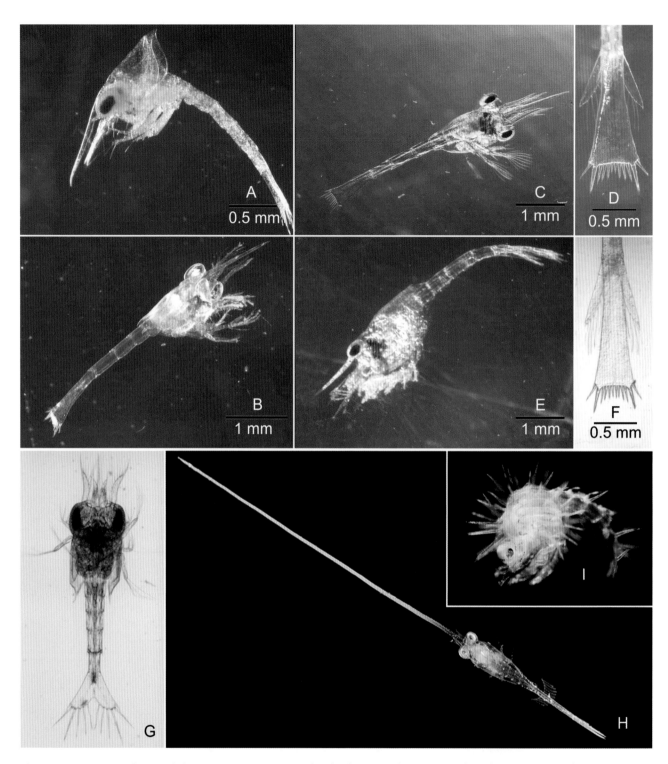

Fig. 53.3 Anomuran zoeal stages, light microscopy. A–F: *Pagurus bernhardus* (Paguridae). A: zoea I, lateral view. B: zoea II, dorsal view. C: zoea III, dorsal view. D: zoea III, telson. E: zoea IV, lateral view. F: zoea IV, telson. G: Galatheidae, unspecified zoea, dorsal view. H: Porcellanidae, unspecified zoea, dorsal view. I: *Chirostylus ortmanni* (Chirostylidae), lateral view. A–F courtesy of Klaus Anger; G courtesy of the Alfred Wegener Institute for Polar and Marine Research; H courtesy of Arthur Anker; I courtesy of Peter Ng.

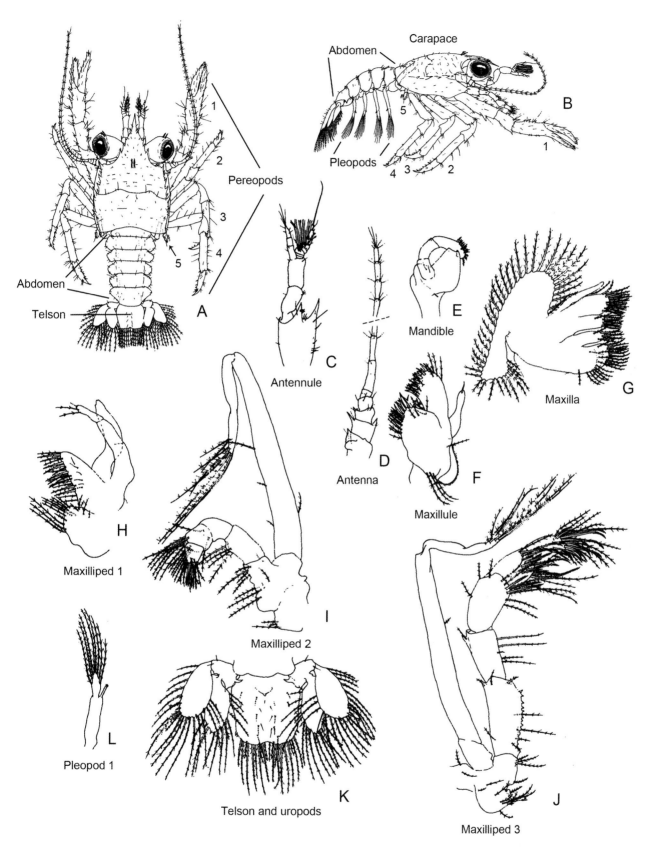

Fig. 53.4 Drawings of *Galathea rostrata*. A and B: galatheoid-type megalopa. A: dorsal view. B: lateral view. C–L: selected appendages of megalopa. C: antennule. D: antenna. E: mandible. F: maxillule. G: maxilla. H: maxilliped 1. I: maxilliped 2. J: maxilliped 3. K: telson and uropods. L: pleopod 1. A–L modified after Gore (1979).

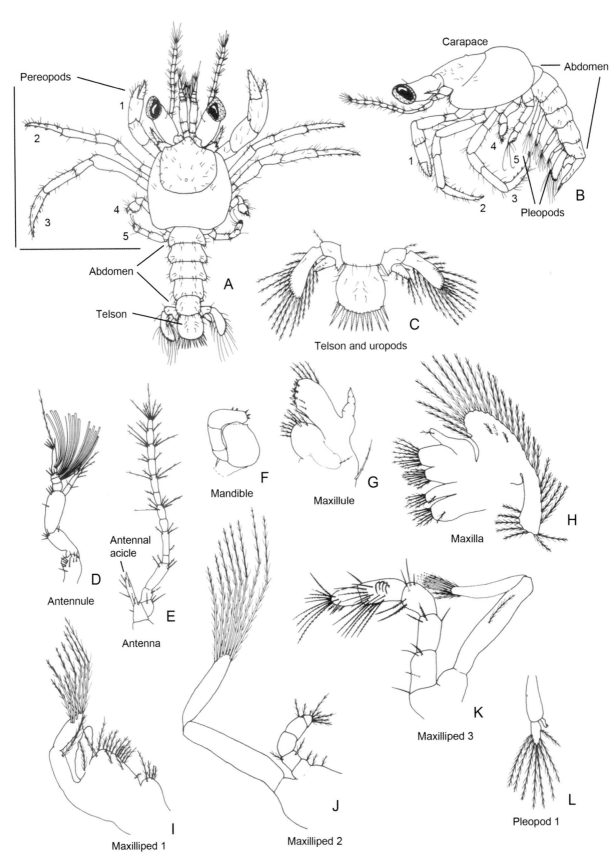

Fig. 53.5 Drawings of *Phimochirus holthuisi*. A and B: paguroid-type megalopa. A: dorsal view. B: lateral view. C–L: selected appendages of megalopa. C: telson and uropods. D: antennule. E: antenna. F: mandible. G: maxillule. H: maxilla. I: maxilliped 1. J: maxilliped 2. K: maxilliped 3. L: pleopod 1. A–L modified after Gore and Scotto (1983).

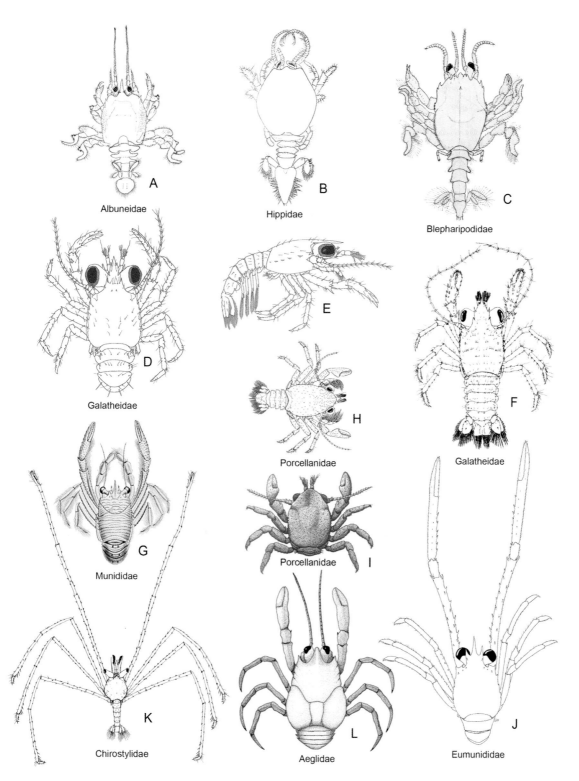

Fig. 53.6 Drawings of examples of anomuran megalopal diversity. A–C: Hippoidea. A: *Lepidopa benedicti* (Albuneidae), dorsal view. B: *Emerita holthuisi* (Hippidae), dorsal view. C: *Blepharipoda doelloi* (Blepharipodidae), dorsal view. D–I: Galatheoidea. D and E: *Galathea intermedia* (Galatheidae). D: dorsal view. E: lateral view. F: *Sadayoshia edwardsii* (Galatheidae), dorsal view. G: *Munida gregaria* (Munididae), dorsal view. H: *Novorostrum decorocrus* (Porcellanidae), dorsal view. I: *Allopetrolisthes angulosus* (Porcellanidae), dorsal view. J and K: Chirostyloidea. J: *Eumunida sternomaculata* (Eumunididae), dorsal view. K: *Chirostylus stellaris* (Chirostylidae), dorsal view. L: *Aegla schmitti* (Aeglidae), dorsal view. A modified after Stuck and Truesdale (1986); B modified after Siddiqi and Ghory (2006); C modified after Boschi et al. (1968); D and E modified after Christiansen and Anger (1990); F modified after Fujita and Shokita (2005); G modified after Dana (1855); H modified after Fujita and Osawa (2005); I modified after Wehrtmann et al. (1996); J modified after Saint Laurent and Macpherson (1990); K from Fujita and Clark (2010), reproduced with permission of *Crustacean Research* (formerly *Research on Crustacea*); L from Teodósio and Masunari (2007), reproduced with permission of *Nauplius*.

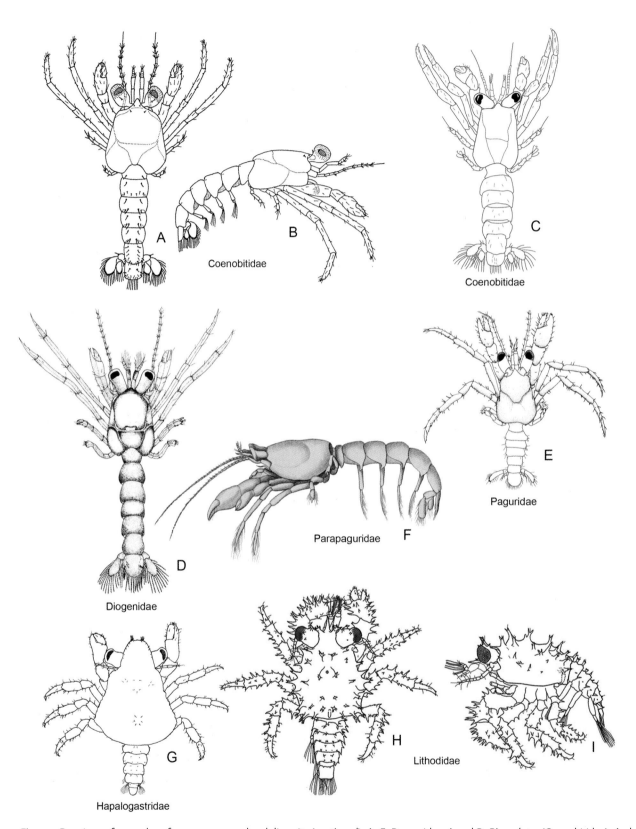

Fig. 53.7 Drawings of examples of anomuran megalopal diversity (*continued*). A–F: Paguroidea. A and B: *Birgus latro* (Coenobitidae). A: dorsal view. B: lateral view. C: *Coenobita compressus* (Coenobitidae), dorsal view. D: *Clibanarius sclopetarius* (Diogenidae), dorsal view. E: *Parapagurodes constans* (Paguridae), dorsal view. F: *Glaucothoe peronii* (Parapaguridae), lateral view. G–I: Lithodoidea. G: *Cryptolithodes typicus* (Hapalogastridae), dorsal view. H and I: *Lopholithodes foraminatus* (Lithodidae). H: dorsal view. I: lateral view. A and B modified after Reese and Kinzie (1968); C modified after Brodie and Harvey (2001); D modified after Brossi-Garcia (1987); E modified after Hong and Kim (2002); F modified after Milne-Edwards (1830a); G modified after J. Hart (1965); H and I modified after Duguid and Page (2009).

54 Brachyura

JOEL W. MARTIN

GENERAL: Brachyurans (true crabs) are easily recognized by their typically wide, dorsoventrally compressed carapace, reduced abdomen tucked beneath the cephalothorax, and first legs modified as relatively large chelipeds (claws). They are the most speciose group of decapods, with nearly 7,000 extant species in 93 families and 38 superfamilies (Ng et al. 2008; De Grave et al. 2009). Morphological variation in both adults and larvae can be extreme. Adults inhabit a wide variety of marine, freshwater, and terrestrial habitats, from the deep sea to cloud forests and phytotelmata. Most are marine, but nearly 1,300 species are known from freshwater systems (Cumberlidge and Ng 2009). The vast majority are free living, but some are commensal, and a few are considered parasitic (pinnotherids), because of tissue damage to the host. The life cycle typically consists of a pre-zoea (not known in all species, and possibly an artifact), a zoea, a megalopa, a juvenile, and an adult.

LARVAL TYPES

Pre-Zoea: A brief non-feeding pre-zoeal hatching stage (fig. 54.1A–E) (also see earlier decapod groups, such as the Penaeioidea, Caridea, Anomura, and Polychelida), considered by some authors to be late embryonic rather than larval, has been reported for several crab families (see review by Hong 1988a), such as cancrids (Quintana 1984a; Quintana and Saelzer 1986), majids (Webber and Wear 1981), epialtids (Guerao and Abelló 1996b), oregoniids (Haynes 1973), panopeids (Montú et al. 1988), portunids (Campbell and Fielder 1987), and atelecyclids (Sasaki and Mihara 1993). Because of its very short duration (often a matter of minutes), this stage probably occurs in nature more widely than is currently recognized (Quintana 1984a). The pre-zoea swims poorly and is not seen in the plankton. Not all laboratory hatchings, even of the same species, exhibit this stage, and some workers feel that its presence is an artifact and indicates suboptimal conditions at hatching (Campbell and Fielder 1987; but see Hong 1988a, who argues that it is a normal part of development).

Zoea: Except where larval stages are bypassed during development (e.g., direct development, which occurs in most freshwater species; also see Wear 1967 for a description of embryonic zoeal stages in a marine species), brachyuran crabs are characterized by having a series of planktonic shrimp-like larval stages, each of which is called a zoea (figs. 54.1–54.5). The number of zoeal stages can be as few as one or two (e.g., most majids) or as many as 8 (some portunids) or 12 (e.g., *Plagusia*), and it is usually consistent within a family (see Rice 1980). These stages are often designated ZI, ZII, ZIII, and so forth. Abbreviated development (loss of some zoeal stages) is also common (Rabalais and Gore 1985; Clark 2000).

Megalopa: The zoea is followed by a functionally and ecologically transitional form called a megalopa (figs. 54.6–54.10)—also referred to as a megalops or megalop (the decapodid of Kaestner 1970; Felder et al. 1985)—a stage known to occur only in the Malacostraca. The megalopa is sometimes unfortunately referred to as a post-larval stage, mostly in the older literature. Rarely, there can be more than one megalopa, or no megalopal stage (e.g., in hymenosomatids). The megalopa is followed by a benthic juvenile first-crab stage that more closely resembles a miniature version of the adult.

MORPHOLOGY

Pre-Zoea: The pre-zoeal stage (fig. 54.1A–E), if present, is ensheathed by a cuticular covering and is therefore nonfeeding. Pre-zoeae swim poorly, with a jerking motion via abdominal flexing (Campbell and Fielder 1987; Hong 1988a); more often they sink to the bottom. Appendages are identifiable mostly as poorly developed limb buds. Zoeal setae, if visible, are not evaginated, and the spines are folded under the cuticle. The pre-zoeal sheath itself is lightly plumose, especially where it covers the antennae and telson. The pre-zoeal cuticle does not merely reflect the underlying zoeal cuticle (fig. 54.1C–E), and it may represent "a remnant of an ancestral free swimming stage" (see Hong 1988a; Ingle 1992).

Zoea: All zoeae possess a carapace that covers the head and the anterior part of the thorax and its limbs (figs. 54.1F, G; 54.2). The carapace has free (unattached) ventral and posterolateral borders, and it typically has a prominent dorsal spine, a rostral spine (the rostrum), and a pair of lateral spines. Other combinations exist, however, and dorsal and lateral spines may be lacking in some species. The typical spination imparts an overall triangular shape to the zoea in higher crab families; the zoea is more shrimp-like (longer than wide) in dromiaceans and other presumed primitive crab groups, such as dromiids, homolids, and latreillids (e.g., fig. 54.3A–M). The carapace often bears sparse setae, usually in pairs. The posteroventral border bears a row of setae, the first of which often differs from the rest and is referred to as the anterior seta (e.g., Clark et al. 1998; Clark 2000) (fig. 54.1F). In many groups, a dorsal organ (fig. 54.1F, G, H) of unknown function is found between the eyes and just anterior to the dorsal spine (J. W. Martin and Laverack 1992; Laverack et al. 1996); there may also be a posterior dorsal organ, which can be on tubercles (e.g., D. Williamson 1982b).

All zoeae have a pair of large compound eyes (figs. 54.1F, G; 54.2A, B), which are almost always sessile in the first stage and stalked in subsequent stages. A nauplius eye is not seen in brachyuran larvae. In early-stage zoeae, the antennules are simple unsegmented subcylindrical exopods that bear sensory setae (aesthetascs). Later zoeal stages may subdivide the peduncles and add a dorsal flagellum, a ventral flagellum, or both. The antennae are biramous; the antennal exopods are typically rod-like (dromioids and homoloids are exceptions). The mandibles are usually at least slightly asymmetrical, usually lack a palp in early stages (exceptions are seen in some species with abbreviated development), and possess incisor and molar regions that are discernible but much less developed than in adults. The maxillules and maxillae are always present, though their degree of development varies among taxa (D. Williamson 1982a). The first two pairs of maxillipeds are biramous and bear natatory setae on the expods. Of the eight pairs of thoracic appendages found in adults (the first three correspond to maxillipeds, the last five to pereopods, or legs), stage I zoeae may have as few as only the first two pairs, or as many as all eight pairs may be visible, although this latter situation is usually seen only in species with abbreviated development. In most taxa, all eight pairs appear by the final zoeal stage.

The abdomen (fig. 54.1F, G, I) is relatively long and segmented. The abdomen of a first-stage zoea typically consists of five segments plus a terminal telson; the last (fifth) segment subdivides in the next molt or two. In a few cases, first-stage zoeae have all six segments, and in a few brachyurans, the last abdominal segment is fused to the telson (e.g., Lucas 1971). The abdomen usually bears a pair of dorsolateral knobs (sometimes referred to as dorsolateral processes) on somites 2 and 3, and sometimes has acute posterolateral processes on somites 3–5. The shape of the telson (fig. 54.1G, I, J) varies widely, but it is typically strongly forked. The terminal margin of the telson is concave (except in pinnotherids and leucosiids) and bears several pairs of plumodenticulate setae, usually 3 pairs in the first zoea (dromiids again are an exception), and increasing in number in later stages. The pleopods appear as small buds in late-stage zoeae. The uropods are typically absent in early brachyuran zoeae, but in some taxa they develop in late zoeal stages. Chromatophores (faded to obscurity in preserved specimens) occur in consistent places on the carapace, abdomen, and some appendages; their location and colors have been used as taxonomic and systematic aids (e.g., Aikawa 1929; Gurney 1942).

Nearly all features of zoeae larvae (see Ingle 1992 and Clark et al. 1998 for terminology) have been used as taxonomic or phylogenetic characters, and there is a large body of literature on zoeal characters in crab phylogeny (e.g., D. Williamson 1976a; Rice 1980, 1981a, 1983; J. W. Martin 1984; Marques and Pohl 1995; McLay et al. 2001; Clark 2009). Zoeal stages in a series typically increase in size by a factor of between 1.0 and 1.5, although there are many exceptions (Rice 1968; Gore 1985). Table 54.1 summarizes general changes in zoeal stages.

Megalopa: The megalopa (figs. 54.6; 54.7) is characterized by a dorsoventrally compressed crab-like carapace, well-developed chelipeds, and functional setose natatory pleopods on abdominal somites 2–5. The last abdominal segment bears uropods, and these, along with the telson, form a tail fan (fig. 54.6A, L), although the uropods are also frequently reduced in size. These changes indicate a functional shift from swimming with thoracopods in the zoeal stages to swimming with abdominal pleopods in the megalopa (Felder et al. 1985; J. W. Martin 1988; D. Williamson 1982a). All mouthparts are now employed in feeding, and all appendages of the juvenile crab are now recognizable. Megalopae are morphologically intermediate between the zoeal and juvenile crab stages, but they are far more similar to juveniles, with some adult characters—such as a paddle-like fifth leg in some portunids, chelate fourth and fifth legs in dorippids, and a widened carapace in some pinnotherids—already apparent. Megalopae, however, often possess features of both zoeae and juveniles, in addition to some characters unique to the megalopal stage (Rice 1981b, 1988; J. W. Martin 1988). With this variable mix of characters, the megalopal stage shows even greater diversity than does the zoeal phase. Generally, megalopae have proportionally larger eyes (hence the name) and smaller pereopods than do adults, and the carapace is usually narrower and smoother. The carapace can be unarmed or bear a large dorsal spine or spines; a rostrum, extended out from in front of the crab, is present in some species. A dorsal organ has been reported, but it is less conspicuous than in zoeal larvae (Barrientos and Laverack 1986; J. W. Martin and Laverack 1992). In some taxa (notably some portunids), large spines extend posteriorly from the posterolateral corners of the fourth sternal plate. The abdomen invariably projects posteriorly, and its appendages are setose pleopods that are now used for swimming. The abdomen, if it is flexed beneath the cephalothorax, is only loosely bent. Megalopae typically exhibit bilateral symmetry, even in those taxa with pronounced asymmetries as adults (e.g., chela asymmetry in *Uca* and *Calappa*), although some

Table 54.1 General features of brachyuran larval stages

Brachyuran larval stage	General features[1]
pre-zoea	Non-swimming, often considered late embryonic. Entire body and appendages ensheathed in the cuticle; appendages lack setae. Very brief duration, often only a few minutes. Not known in all groups.
zoea I	Eyes sessile. Carapace with 1 pair of posterodorsal setae. Antennule and antenna usually without endopod buds. First 2 maxillipeds natatory; exopods segmented, with 4 long distal setae. No visible third maxilliped. No uropods. Telson not separated from the sixth abdominal somite. Telson forked, typically with 3 pairs of setae on the posterior border.
zoea II	Eyes stalked. Antennule and antenna endopods appear as buds. First 2 maxilliped exopods now with 6 natatory setae. Developing third maxilliped may be visible. No uropods. Telson not separated from the sixth abdominal somite (except in some majoids). Telson may have 4 pairs of posterior setae.
zoea III	Eyes stalked. Mandibular palp may be present as a bud. First 2 maxilliped exopods now with 8 natatory setae. Developing third maxilliped usually visible. Developing pereopods (including the bilobed cheliped) and pleopods usually present as buds. Uropod buds sometimes present. Telson separated from the sixth abdominal somite. Telson may have additional pairs of setae.
zoea IV and higher[2]	Eyes stalked. Mandibular palp bud present. Maxillipeds now with 10 or more natatory setae on the exopods. Pereopod and pleopods further developed but not yet functional. Telson separated from the sixth abdominal somite. Telson may have additional pairs of setae.
megalopa (decapodid)	Eyes stalked. Mandible with setose palp. Maxillipeds not natatory, instead functioning as mouthparts. Pereopods developed and fully functional, the first always chelate, the fifth often with specialized curved serrate setae and spines. Pleopods biramous, segmented, and natatory. Uropods setose. Telson usually rounded or truncate.

[1] Many exceptions exist within almost all families. This is especially true for dromiaceans, homolids, raninids, and majoids (modified after Pohle et al. 1999; also see Ingle 1992).

[2] Not all crabs have more than 4 zoeal stages, and many have fewer. In general, the more zoeal stages there are, the more delayed the morphological development.

chela asymmetry is seen in the megalopae of some species. Paired pleopods may be held together by interlocking toothed spines (cincinnuli) to facilitate their functioning as a single unit (fig. 54.7L, M). In many species, the dactyls of the fifth pereopods bear 2–5 long, curved, specialized serrate sensory setae, unique to this stage of development (fig. 54.6E). The ventral surfaces of the dactylus of pereopods 2–5 often bear stout serrate spines not seen in later crab stages (fig. 54.6D). Gill morphology and ontogeny in zoeal and megalopal stages was described by Hong (1988b).

MORPHOLOGICAL DIVERSITY: Diversity within the constraints of the above descriptions, in terms of both size and shape, is extremely high. Within the broad term zoea lies a bewildering diversity of shapes, sizes, and ornamentation (figs. 54.2–54.5). The examples in the figures used in this chapter are extremely limited and selective; not all families, or even superfamilies, are represented. Zoeae are known with multiple carapace spines (fig. 54.2D) and with no spines (fig. 54.4N), with spinose or smooth abdomens, with armed or unarmed telsons, and with brightly colored chromatophores or with nearly transparent bodies. Even within a given family, zoeal morphology may be strikingly different among species, such as the zoeae of the Hymenosomatidae (fig. 54.4T, V) and Leucosiidae (fig. 54.4N, P). The distinctly anomuran features of the zoea larvae of dromiacean crabs—the flattened antennal exopod, and a telson with anomuran seta (the "hair-like second telson process")—have led many workers to exclude them from the Brachyura on this basis alone (e.g., Gurney 1942; Rice 1980). These features have also given rise to hypotheses of horizontal gene transfer and hybridization between widely separated lineages (e.g., D. Williamson 1988a, 1988b, 1992). Megalopal stages are also highly diverse across the

Brachyura (figs. 54.7–54.10): some bear large posterior spines off the sternal plate (e.g., some portunoids; fig. 54.9T), some bear a strong ischial hook on the cheliped (e.g., some portunoids and xanthoids; fig. 54.7K) (see Atkins 1954; J. W. Martin 1988), some possess an acute rostrum (fig. 54.9K), some have dorsal carapace spines (e.g., homolids and latreillids; fig. 54.8E–G), and some lack the specialized sensory setae on the fifth pereopod (e.g., bythograeids and at least some dorippids; fig. 54.8M, S). Zoeae are typically nearly transparent, or at least more so than the megalopae; the latter can be fragile and transparent or very robust and fully pigmented. More than 30 distinct setal types have been described from zoeae and megalopae of a single species (Pohle and Telford 1981).

NATURAL HISTORY: Zoeal stages are common components of marine plankton and many food chains. Zoeae are usually weak swimmers that depend mostly on currents, and on vertical migration into and out of currents, for transportation. An extensive body of literature exists on their swimming ability, depth regulation, recruitment and dispersal, vertical migration, feeding, larval release and hatching rhythms, responses to various pressures (such as barokinesis), predator avoidance, development, physiology, and bioenergetics. For entry into this literature, interested readers should consult the publications from the labs of K. Anger, C. E. Epifanio, R. B. Forward, A. Hines, S. G. Morgan, J. Paula, S. D. Sulkin, and their colleagues (e.g., Sulkin 1984; Morgan 1985, 2001; Hines 1986; Paula 1989; Anger 2001; Epifanio 2007; J. Cohen and Forward 2009). Zoeal swimming forward is accomplished via the biramous maxillipeds, and rearward by abdominal thrusting/flexing. Zoeae are predators and will eat a variety of phyto- and zooplankton; in the laboratory the larvae of many species have been reared on *Artemia* nauplii and/or rotifers. Feeding is accomplished by trapping prey items between the flexed abdomen and the mouthparts. Lecithotrophy in zoeae is rare, but it has been documented for some deep-water species (e.g., Taishaku and Konishi 2001). The duration of zoeal stages in the plankton depends on environmental factors, as well as on phylogenetic constraints (e.g., see Hines 1986); the number of stages tends to be conservative within a given family.

The megalopa is a morphological and ecological transitional stage from the plankton to the (usually) benthic habitat of the adult. Swimming is accomplished via the setose pleopods of the extended abdomen and by abdominal flexure. Although megalopae are usually illustrated with legs outstretched, during swimming their legs may be draped over the carapace (which often has distinct grooves or depressions to accommodate them), or even hooked over the eyes (Atkins 1954), presumably to increase hydrodynamic efficiency (fig. 54.10I–M). Many adult structures, such as the internal gastric mill, are already fully developed in the megalopa. Megalopae tend to be opportunistic feeders, consuming live plankters (including other crustaceans and small fishes) or scavenging, and they can be aggressive feeders, holding prey items with their chelipeds. Several reports exist of mass strandings of huge numbers of megalopae along beaches; the cause is not known.

PHYLOGENETIC SIGNIFICANCE: Crab larvae have been used extensively in phylogenetic studies to determine relationships within families, within the Brachyura as a whole, and of the Brachyura to other groups of decapods (especially anomurans). A brief selection of such studies would include the works of D. Williamson (1976a), Rice (1980, 1981a, 1983), J. W. Martin (1984, 1988; J. W. Martin et al. 1985), McLay et al. (2001), and Clark (2009) on xanthoids, and Marques and Pohle (1995) on pinnotheroids, to name only a few from an enormous body of literature. So widespread is the use of crab larvae in phylogeny that practically all modern descriptions of crab larvae now include comments on their systematic and phylogenetic implications. Particularly noteworthy in this regard are the zoeae of dromiacean crabs and their similarity to larvae of the Anomura, causing some workers to question their inclusion among the true crabs, as well as giving rise to odd hypotheses, such as hybridization between distantly related groups (the "larval transfer" hypothesis of D. Williamson 1988a, 1988b, 1992; D. Williamson and Vickers 2007).

HISTORICAL STUDIES: An enormous body of literature exists on the larval stages of crabs. Classic studies on morphology include those of Gurney (1942), Lebour (1928, 1944), Bourdillon-Casanova (1960), D. Williamson (1976a, 1982a), Rice (1980, 1981a), Wear (1965, 1967, 1968), J. W. Martin (1984), Wear and Fielder (1985), Ingle (1992), Clark et al. (1998), and Pohle et al. (1999), among many others. Most treatments tend to be regional (e.g., Wear 1965; Ingle 1992; Kornienko and Korn 2010). Although tremendous progress has been made in the ensuing years, Gurney (1942) remains the classic and indispensible reference.

The genus *Zoea* was created for what turned out to be the planktonic larval stages of several species of brachyuran crabs. As biologists began to better understand the life-history patterns of crustaceans, the term zoea eventually came to refer to any crustacean larva with functional thoracic appendages, replacing numerous taxon-specific names for the same stage (see the glossary; also see Gurney 1942; D. Williamson 1969, 1982a; Harvey et al. 2001). Zoeal stages are described in the literature more often than are megalopal stages, a result of rearing studies that do not come to completion. Mortality in rearing and in nature is typically very high. Often larvae have been described from plankton-collected samples, and in those cases their identity with adults often cannot be ascertained. Zoeae are predators and particulate feeders, and they can be raised in the laboratory on brine shrimp (*Artemia*) nauplii, rotifers, algae, yeast, or some combination of these foods. Rearing techniques are now relatively well documented (e.g., Rice and Williamson 1970). At least one report exists of brachyuran zoeae in the fossil record (the Lower Cretaceous of Brazil), from the gut contents of a fossil planktivorous fish (Maisey and de Carvalho 1995).

The terminology of the megalopal stage also has been quite confusing. Referred to occasionally in the past as a postlarva (e.g., Gurney 1942), this term has fallen into disfavor, because megalopae are clearly distinct from subsequent juvenile stages. Like zoea larvae, many of these transitional forms were originally described as new species (or even genera) before their developmental status was recognized. Once these links were established, the now-defunct generic name often became the name of the transitional stage for that group. The earliest of these, *Megalopa*, is the transitional form of a brachyuran crab, and the term megalopa is often used (not without controversy; see Felder et al. 1985) as the general term across the Decapoda for the transitional larval stages where swimming is achieved through pleopods (D. Williamson 1982a). Megalopal-centered morphological studies include those of Rice (1981b, 1988), Quintana (1986a), and J. W. Martin (1988). Most modern studies are now based on laboratory-reared larvae from an identified egg-bearing adult.

Selected References

Ingle, R. W. 1992. Larval Stages of Northeastern Atlantic Crabs: An Illustrated Key. London: Chapman & Hall.

Pohle, G., F. L. Mantelatto, M. L. Negreiros-Fransozo, and A. Fransozo. 1999. Larval Decapoda (Brachyura). *In* D. Boltovskoy, ed. South Atlantic Zooplankton, vol. 2, 1281–1351. Leiden, Netherlands: Backhuys.

Rice, A. L. 1980. Crab zoeal morphology and its bearing on the classification of the Brachyura. Transactions of the Zoological Society of London 35: 271–424.

Wear, R. G., and D. R. Fielder. 1985. The Marine Fauna of New Zealand: Larvae of the Brachyura (Crustacea, Decapoda). New Zealand Oceanographic Institute Memoir No. 92. Wellington: New Zealand Department of Scientific and Industrial Research.

Williamson, D. I. 1982a. Larval morphology and diversity. *In* L. G. Abele, ed. The Biology of Crustacea. Vol. 2, Embryology, Morphology, and Genetics, 43–110. New York: Academic Press.

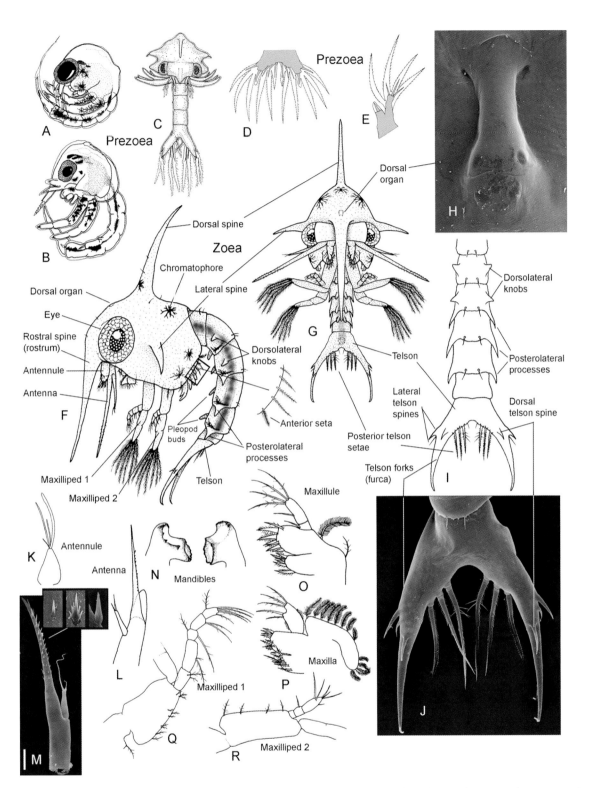

Fig. 54.1 Drawings (unless otherwise indicated) of pre-zoeal and zoeal stages, with selected appendages. A and B: *Erimacrus isenbeckii* (Atelecyclidae). A: early (just after hatching) pre-zoeal stage, lateral view. B: slightly later pre-zoeal stage, lateral view. C: pre-zoea of *Cancer coronatus*, dorsal view. D and E: pre-zoea of *Anamathia rissoana* (Epialtidae), showing the ensheathing cuticle differing from (not reflecting) the underlying morphology. D: telson, dorsal view. E: antenna. F and G: brachyuran zoea (based on a xanthid), F: lateral view. G: frontal view. H: dorsal organ of zoea I of *Xantho poressa* (Xanthidae), SEM. I: xanthid zoeal abdomen and telson, dorsal view. J: telson of zoea I of *X. poressa*, SEM, dorsal view. K: antennule. L: antenna. M: antenna of zoea I of *Portunus acuminatus* (Portunidae), with insets of selected setules and denticles from the spinous process, SEM. N–P: mouthparts, based on zoea III of *Sesarma erythrodactyla* (Sesarmidae). N: mandibles. O: maxillule. P: maxilla. Q and R: maxillipeds, showing only the endopods (the natatory setae of the exopods not shown). A and B modified after Sasaki and Mihara (1993); C modified after Quintana (1984a); D and E modified after Guerao and Abelló (1996b); F, G, K, and L original; H, J, and M courtesy of Roland Meyer, used by permission; I modified after Meyer et al. (2004), N–R modified after C. Kim and Ko (1985).

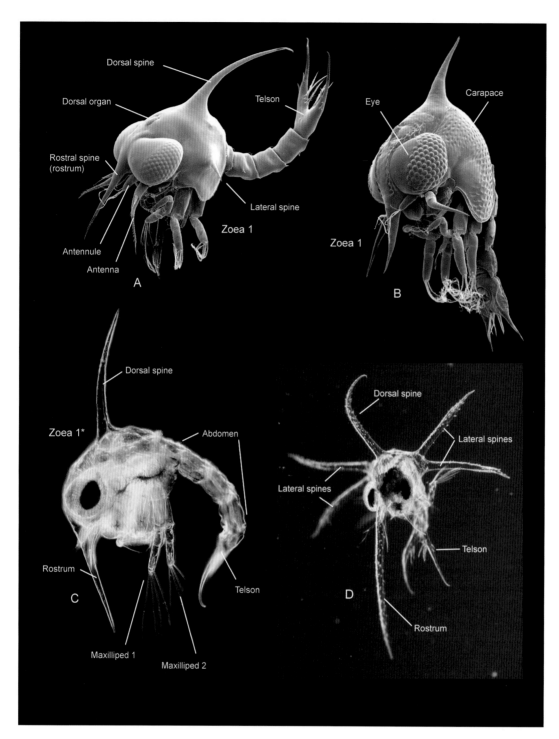

Fig. 54.2 Zoeal stages. A: zoea I of *Portunus acuminatus* (Portunidae), SEM. B: zoea I of *Goniopsis pulchrata* (Grapsidae), SEM. C: *Carcinus maenas* (zoeal stage undetermined, but probably zoea I), light microscopy. D: unidentified multispined zoea (probably of the Tetralii-dae; compare with fig. 54.5D), light microscopy. Material in D is from the Great Barrier Reef in Australia. A modified after Meyer et al. (2006), used by permission; B courtesy of Roland Meyer, used by permission; C courtesy of Visuals Unlimited, used by permission; D courtesy of Tim Hellier / Norbert Wu Productions, www.norbertwu.com (2005, 2010), used by permission.

Fig. 54.3 (*opposite*) Drawings of examples of zoeal diversity. A–E: Dromiidae. A and B: zoea I of *Petalomera wilsoni*. A: dorsal view. B: lateral view. C–E: *Dromidia antillensis*. C: zoea VII, dorsal view (appendages not shown). D: zoea II, antenna. E: zoea II, telson (the black arrow indicates the anomuran hair-like seta), dorsal view. F–H: zoea IV of *Homola* sp. (Homolidae). F: lateral view. G: carapace, dorsal view. H: tel-son, dorsal view. I: zoea I of *Homola barbata*, lateral view. J–M: zoea II of Latreillidae (*Latreillia ?elegans* [designated "ASM 11" in the original publication]). J: lateral view; K: carapace, dorsal view; L: abdomen and telson, dorsal view. M: antenna. N and O: zoea I (*cont. on next page*)

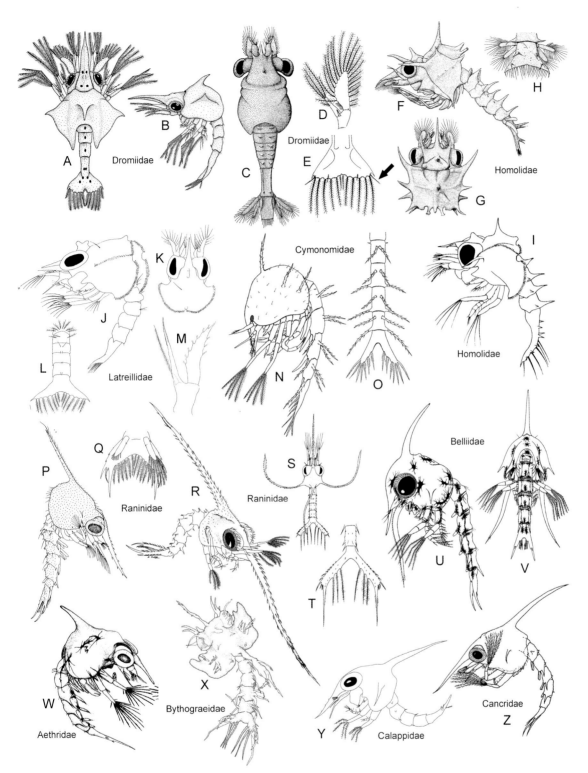

of *Cymonomus bathamae* (Cymonomidae). N: lateral view. O: abdomen and telson, dorsal view. P–T: Raninidae. P and Q: zoea III of *Raninoides benedicti*. P: lateral view. Q: telson, ventral view. R: zoea I of *Ranina ranina*, dorsolateral view. S and T: zoea I of *Lyreidus tridentatus*. S: dorsal view. T: telson, dorsal view. U and V: zoea I of *Heterozius rotundifrons* (Belliidae). U: lateral view. V: posterior view. W: zoea IV of *Hepatus epheliticus* (Aethridae), lateral view. X: zoea I of *Bythograea thermydron* (Bythograeidae), dorsal view. Y: early zoea of *Calappa japonica* (Calappidae), lateral view. Z: zoea IV of *Cancer setosus* (Cancridae), lateral view. A and B modified after Wear (1970); C–E modified after Rice and Provenzano (1966); F–H modified after Rice and von Levetzow (1967); I modified after Rice and Provenzano (1970), as reproduced in Wear and Fielder (1985); J–M modified after Rice and Williamson (1977); N and O modified after Wear and Batham (1975); P and Q modified after Knight (1968); R modified after Sakai (1971); S and T modified after Wear and Fielder (1985); U and V modified after Wear (1968), as reproduced in Wear and Fielder (1985); W modified after Costlow and Bookhout (1962); X modified after Van Dover et al. (1984); Y modified after Taishaku and Konishi (1995); Z modified after Quintana and Saelzer (1986).

Fig. 54.4 (*opposite*) Drawings of examples of zoeal diversity (*continued*). A and B: zoea I of *Trichopeltarion fantasticum* (Atelecyclidae). A: lateral view. B: posterior view. C and D: zoea I of *Carpilius corallinus* (Carpiliidae). C: lateral view. D: frontal view. E: zoea III of *Corystes casivelaunus* (Corystidae), lateral view. F: zoea I of *Ethusa microphthalma* (Ethusidae), lateral view. G and H: Dorippidae. G: early zoea of *Dorippe* sp. (from plankton), lateral view. H: telson of zoea IV of *D. frascone*, dorsal view. I and J: zoea III of *Menippe nodifrons* (Menippidae). I: lateral view. J: telson, dorsal view. K and L: Oziidae. K: zoea I of *Ozius truncatus*, lateral view. L: zoea I of *Epixanthus dentatus*, lateral view. M: zoea IV of *Goneplax rhomboides* (Goneplacidae), lateral view. N–Q: Leucosiidae. N and O: "last zoea" of *Philyra platycheira*. N: lateral view. O: telson, dorsal view. P and Q: zoea III of *Leucosia craniolaris*. P. frontal view. Q: telson, dorsal view. R: zoea II of *Spiroplax spiralis* (Hexapodidae), lateral view. S: zoea II of *Doclea muricata* (Epialtidae), dorsolateral view. T–V: Hymenosomatidae. T and U: "final stage" of *Halicarcinus* sp. T: lateral view. U: telson, dorsal view. V: zoea I of *Elamenopsis demeloi*, lateral view. W: zoea I of *Orithyia sinica* (Orithyidae), lateral view. X and Y: zoea I of *Dorhynchus thompsoni*. X: lateral view. Y: posterior view. Z and AA: zoea II of *Mithrax pleuracanthus* (Majidae). Z: lateral view. AA: telson, dorsal view. BB and CC: zoea I of *Chionoecetes bairdi* (Oregoniidae). BB: lateral view. CC: frontal view. DD and EE: zoea IV of *Parthenope serrata* (Parthenopidae). DD: lateral view. EE: telson, dorsal view. FF and GG: zoea III of *Pilumnus dasypodus* (Pilumnidae). FF: lateral view. GG: frontal view. HH and II: zoea IV of *Thia scutellata* (Thiidae). HH: lateral view. II: telson, dorsal view. A and B modified after Wear and Fielder (1985); C and D modified after Laughlin et al. (1983); E modified after Ingle and Rice (1971); F modified after J. W. Martin and Truesdale (1989); G modified after Cano (1892a); H modified after Quintana (1987); I and J modified after Scotto (1979); K modified after Wear (1968); L modified after Saba et al. (1978); M modified after Ingle and Clark (1983); N and O modified after Quintana (1986a); N and O modified after Quintana (1986b); R modified after Pereyra Lago (1988); S modified after Krishnan and Kannupandi (1987); T and U modified after Wear (1965); V modified after Kakati (1988); W modified after Hong (1976); X and Y modified after D. Williamson (1982b); Z and AA modified after Goy et al. (1981); BB and CC modified after Haynes (1973); DD and EE modified after Yang (1971); FF and GG modified after Bookhout and Costlow (1979); HH and II modified after Ingle (1984).

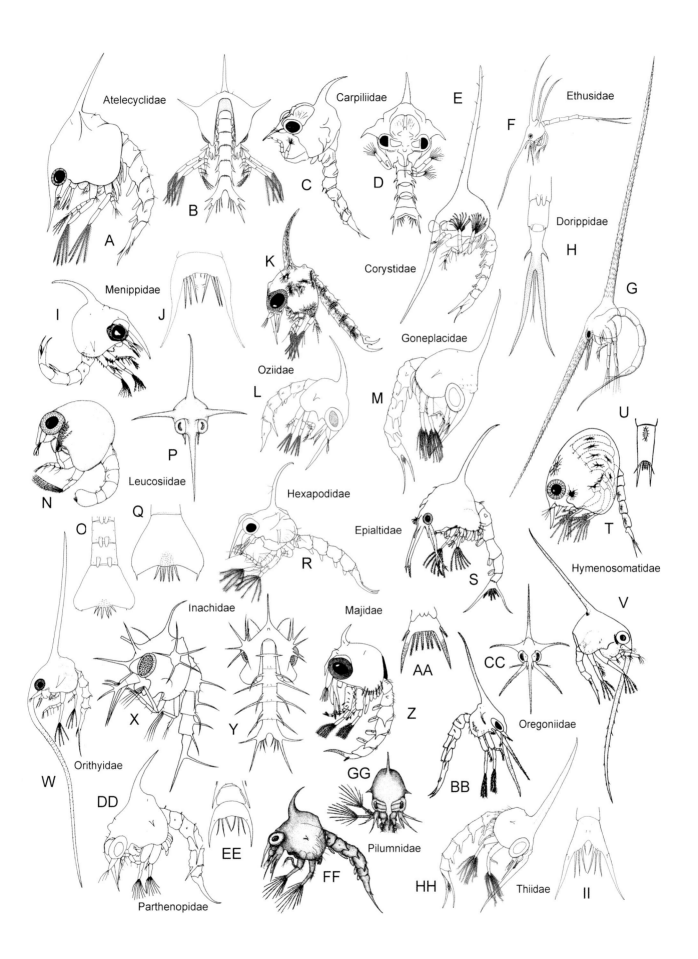

Atelecyclidae

Carpiliidae

E

F Ethusidae

A

B

C

D

Corystidae

Dorippidae

H

G

K

Menippidae

J

I

Goneplacidae

Oziidae

L

M

U

Leucosiidae

P

N

Hexapodidae

Hymenosomatidae

O

Q

R

Epialtidae

T

S

V

Inachidae

Majidae

AA

CC

X

Y

Z

Oregoniidae

W

Orithyidae

GG

BB

DD

EE

FF

Pilumnidae

HH

Thiidae

II

Parthenopidae

Fig. 54.5 (opposite) Drawings of examples of zoeal diversity *(continued)*. A and B: zoea V of *Callinectes similis* (Portunidae). A: lateral view. B: frontal view. C: zoea IV of *Geryon quinquedens* (Geryonidae), lateral view. D and E: Tetraliidae. D: assumed zoea III of *Tetralia* (from plankton), dorsal view. E: zoea I of *Tetralia rubridactyla*, frontal view. F: zoea I of *Quadrella maculosa* (Trapeziidae), frontal view, and abdomen, ventral view. G: zoea III of *Panopeus austrobesus* (Panopeidae), lateral view. H: zoea III of *Xantho poressa* (Xanthidae), lateral view. I and J: zoea III of *Troglocarcinus corallicola* (Cryptochiridae). I: lateral view. J: abdomen, dorsal view. K: zoea III of *Cardisoma guanhumi* (Gecarcinidae), lateral view. L–O: Plagusiidae. L and M: zoea II of *Percnon gibbesi*. L: lateral view. M: frontal view. N: plankton-caught presumed plagusiid, dorsolateral view. O: zoea IV of *Plagusia depressa*, frontal view. P: zoea IV of *Metasesarma rubripes* (Sesarmidae), lateral view. Q and R: zoea IV of *Cyclograpsus intermedius* (Varunidae). Q: lateral view. R: frontal view. S and T: zoea III of *Scopimera inflata* (Dotillidae). S: lateral view. T: telson, dorsal view. U: zoea II of *Uca thayeri* (Ocypodidae), lateral view. V: zoea VI of *Ucides cordatus* (Ucididae), lateral view. W–Z: Pinnotheridae. W and X: zoea III of *Dissodactylus mellitae*. W: lateral view. X: frontal view. Y: telson of zoea (stage not given) of *Pinnotheres novaezealandiae*, dorsal view. Z: abdomen of zoea IV of *Pinnixa longipes*, dorsal view. A and B modified after Bookhout and Costlow (1977); C modified after H. Perkins (1973); D modified after M. George and John (1975); E and F modified after Clark and Ng (2006); G modified after Montú et al. (1988); H modified after Rodríguez and Martin (1997); I and J modified after Scotto and Gore (1981); K modified after Costlow and Bookhout (1968); L–N modified after Paula and Hartnoll (1989); O modified after K. Wilson and Gore (1980); P modified after H. Diaz and Ewald (1968); Q and R modified after C. Kim and Jang (1986); S and T modified after Fielder and Greenwood (1985); U modified after Anger et al. (1990); V modified after Domingues Rodrigues and Hebling (1989); W and X modified after Marques and Pohl (1996); Y modified after Wear (1965); Z modified after Bousquette (1980).

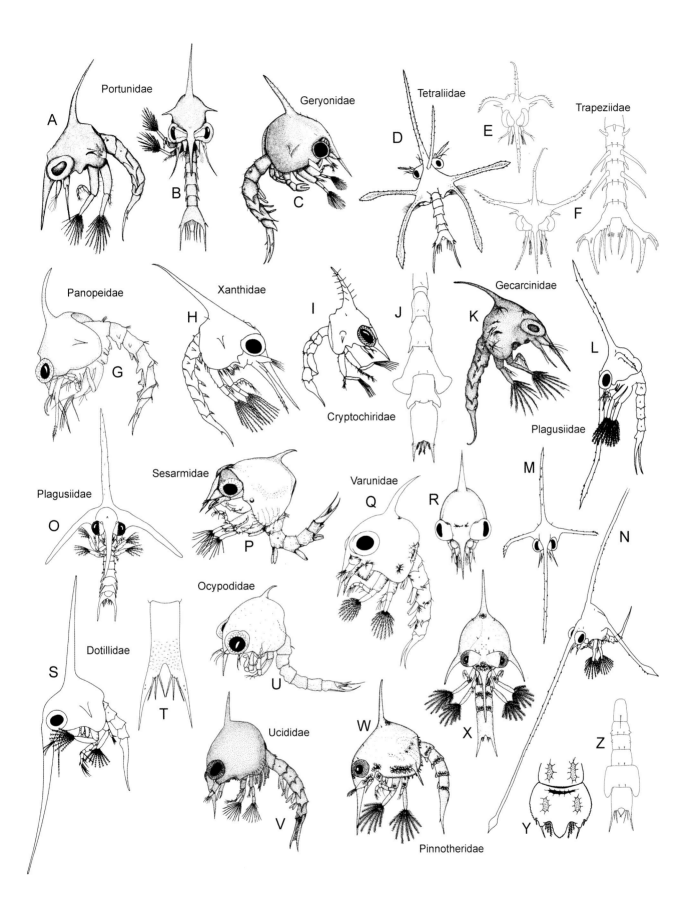

Portunidae

A

B

Geryonidae

C

Tetraliidae

D

E

Trapeziidae

F

Panopeidae

G

Xanthidae

H

I

Cryptochiridae

J

Gecarcinidae

K

Plagusiidae

L

Plagusiidae

O

Sesarmidae

P

Varunidae

Q

R

M

N

S

Dotillidae

T

Ocypodidae

U

Ucididae

V

W

X

Y

Z

Pinnotheridae

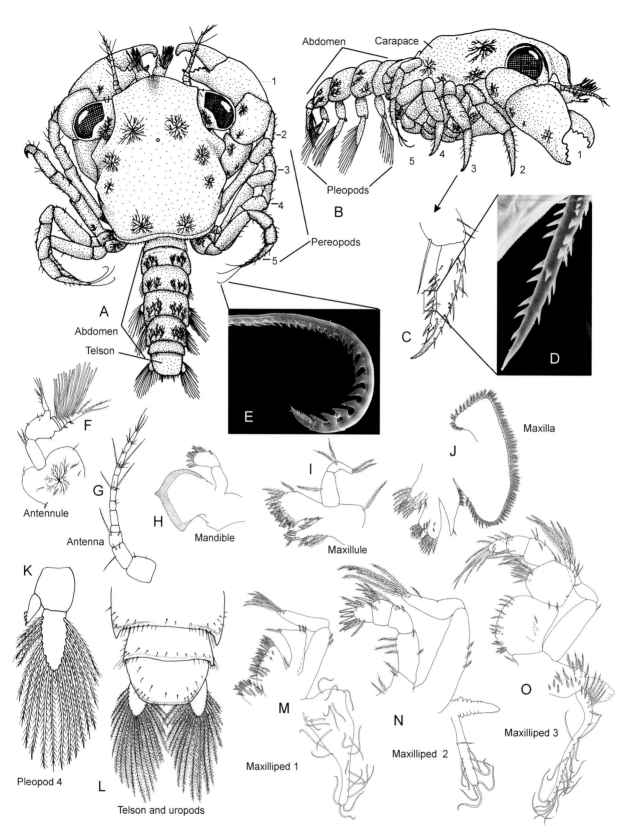

Fig. 54.6 Drawings (unless otherwise indicated) of megalopa and selected appendages. A and B: megalopa of *Menippe adina* (Menippidae). A: dorsal view. B: lateral view. C: dactylus of pereopod 3. D and E: from *M. mercenaria*. D: single dactylar spine (box in C), SEM. E: sensory seta of dactylus of pereopod 5, SEM. F: antennule. G: antenna. H: mandible. I: maxillule. J: maxilla. K: pleopod 4, with a cincinnuli-bearing endopod to the left, frontal view. L: telson and uropods of the vent crab *Bythograea thermydron*. M: maxilliped 1. N: maxilliped 2. O: maxilliped 3. A–E modified after J. W. Martin et al. (1988); F–K and M–O original; L modified after J. W. Martin and Dittell (2007).

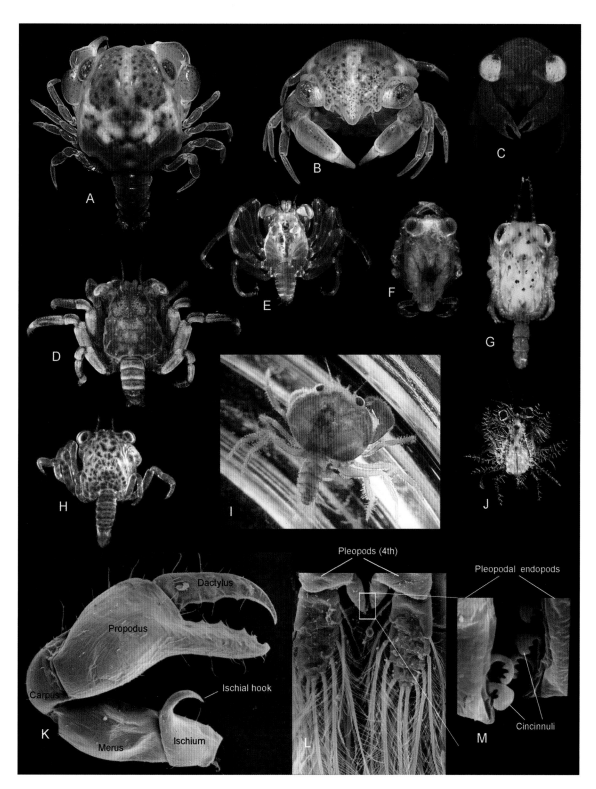

Fig. 54.7 Various megalopae. A–J: light microscopy. A and B: large colorful megalopa (possibly *Carpilius*). A: dorsal view. B: frontal view. C: undetermined species, frontal view. D: grapsoid megalopa (probably *Euchirograpsus*). E–H: undetermined megalopae, dorsal views. I: megalopa of the hydrothermal vent crab *Bythograea thermydron*, dorsal view. J: undetermined spider (majoid) crab megalopa, dorsal view. K–M: SEMs. K: cheliped of a megalopa of *Panopeus* sp. (Panopeidae), with a large recurved ischial hook. L: paired fourth pleopods of a *Panopeus* megalopa, showing natatory setae and endopods. M: closeup of the hook-like cincinnuli setae of the pleopodal endopods (box in L). Images not to scale. Material in A and B from the western Pacific; material in D from lobster traps off the French Frigate Shoals in Hawaii; material in E–H from Moorea. A–C courtesy of Arthur Anker; D courtesy of Leslie Harris; E–H courtesy of Hsiu-Lin Chin; I modified after J. W. Martin and Dittell (2007); J courtesy of Cheryl Clarke Hopcroft; K–M modified after J. W. Martin (1988).

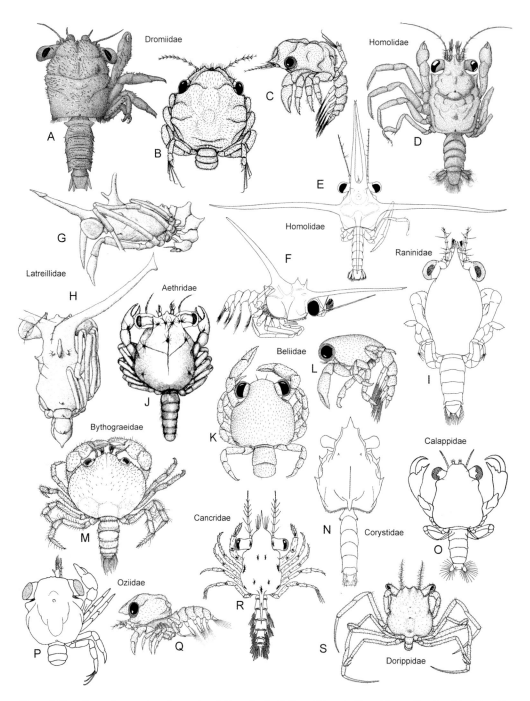

Fig. 54.8 Drawings of examples of megalopal diversity. A–C: Dromiidae. A: *Dromidia antillensis*, with right-side appendages, dorsal view. B and C: *Petalomera wilsoni*. B: dorsal view. C: lateral view. D–F: Homolidae. D: *Homola barbata*, with some appendages, dorsal view. E and F: *Paramola petterdi*. E: dorsal view. F: lateral view. G and H: *Latriellia elegans* (Latreillidae). G: lateral view. H: dorsal view. I: *Raninoides benedicti* (Raninidae), dorsal view. J: *Hepatus epheliticus* (Aethridae), dorsal view. K and L: *Heterozius rotundifrons* (Bellidae). K: dorsal view. L: lateral view. M: *Bythograea thermydron* (Bythograeidae), dorsal view. N: *Corystes cassivelaunus* (Corystidae), dorsal view. O: *Calappa flammea* (Calappidae), dorsal view. P and Q: Oziidae. P: *Epixanthus dentatus*, with right-side appendages, dorsal view. Q: *Ozius truncatus*, lateral view. R: *Cancer novaezealandiae* (Cancridae), dorsal view. S: *Dorippe frascone* (Dorippidae), dorsal view. A modified after Rice and Provenzano (1966); B, C, E, and F modified after Wear and Fielder (1985); D modified after Rice (1964); G and H modified after Rice (1982); I modified after Knight (1968); J modified after Costlow and Bookhout (1962); K–L and Q modified after Wear (1968); M modified after J. W. Martin and Dittel (2007); N modified after Ingle and Rice (1971); O modified after Lebour (1944); P modified after Saba et al. (1978); R modified after Wear (1965); S modified after Quintana (1987).

Fig. 54.9 (*opposite*) Drawings of examples of megalopal diversity (*continued*). A–D: Leucosiidae. A: *Nucia laminata*, with some appendages, dorsal view. B and C: *Leucosia craniolaris*. B: lateral view. C: dorsal view. D: *L. craniolaris*, frontal view. E: *Spiroplax spiralis* (Hexapodidae), dorsal view. F: *Goneplax rhomboides* (Goneplacidae), with right-side appendages, dorsal view. G: *Doclea muricata* (Epialtidae), (*cont. on next page*)

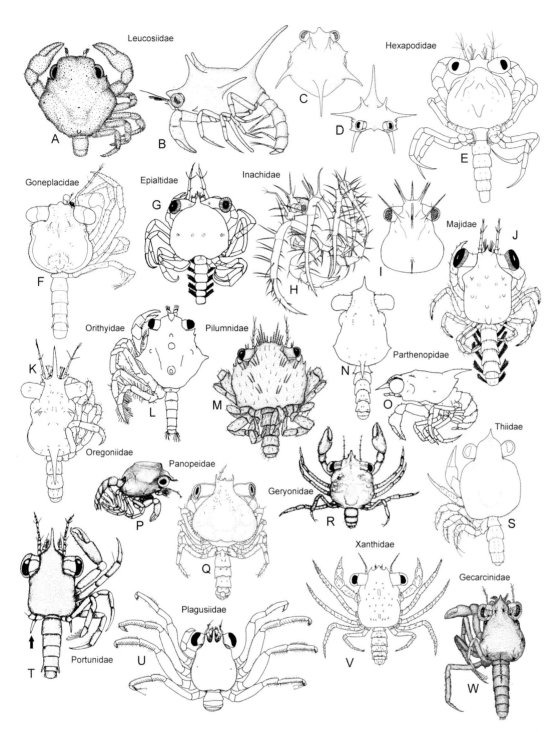

dorsal view. H and I: *Dorhynchus thompsoni* (Inachidae). H: ventral view. I: carapace, dorsal view. J: *Mithrax pleuracanthus* (Majidae), dorsal view. K: *Hyas araneus* (Oregoniidae), with right-side appendages, dorsal view. L: *Orithyia sinica* (Orithyidae), with left-side appendages, dorsal view. M: *Pilumnus novaezealandiae* (Pilumnidae), dorsal view. N and O: *Parthenope serrata* (Parthenopidae). N: dorsal view. O: lateral view. P and Q: Panopeidae. P: *Panopeus herbstii*, lateral view. Q: *P. austrobesus*, dorsal view. R: *Geryon quinquedens* (Geryonidae), dorsal view. S: *Thia scutellata* (Thiidae), with left-side appendages, dorsal view. T: *Callinectes similis* (Portunidae), with some appendages (the black arrow indicates the spine coming off the fourth sternal plate), dorsal view. U: *Percnon gibbesi* (Plagusiidae), dorsal view. V: *Xantho poressa* (Xanthidae), dorsal view. W: *Cardisoma guanhumi* (Gecarcinidae), with right-side appendages, dorsal view. A–C modified after Quintana (1986b); D modified after Quintana (1986a); E modified after Pereyra Lago (1988); F modified after Ingle and Clark (1983); G modified after Krishnan and Kannupandi (1987); H and I modified after D. Williamson (1982b); J modified after Goy et al. (1981); K modified after Christiansen (1973); L modified after Y.-H. Kim and Chung (1990); M modified after Wear (1967); N and O modified after Yang (1971); P modified after Costlow and Bookhout (1961); Q modified after Montú et al. (1988); R modified after H. Perkins (1973); S modified after Ingle (1984); T modified after Bookhout and Costlow (1977); U modified after Paula and Hartnoll (1989); V modified after Rodríguez and Martin (1997); W modified after Costlow and Bookhout (1968).

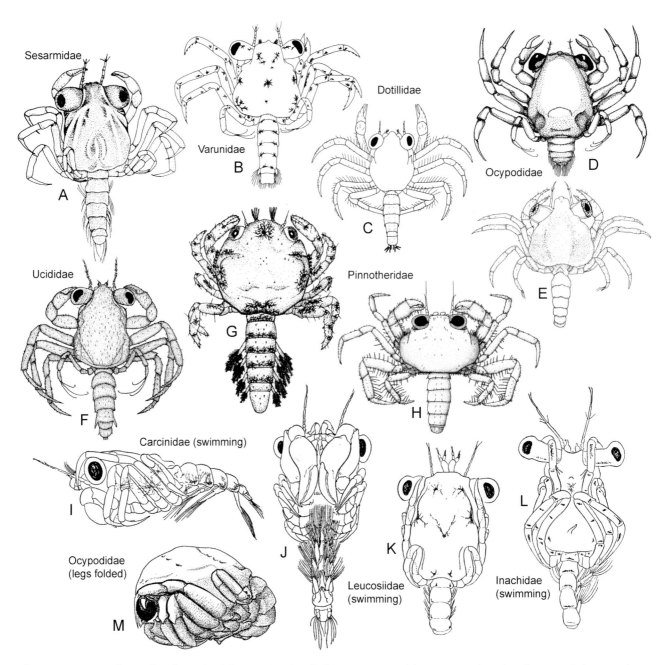

Fig. 54.10 Drawings of examples of megalopal diversity (*continued*). A: *Metasesarma rubripes* (Sesarmidae), dorsal view. B: *Cyclograpsus intermedius* (Varunidae), dorsal view. C: *Scopimera inflata* (Dotillidae), dorsal view. D and E: Ocypodidae, dorsal views. D: *Ocypode occidentalis*. E: *Uca thayeri*. F: *Ucides cordatus* (Ucididae), dorsal view. G and H: Pinnotheridae, dorsal views. G: *Dissodactylus mellitae*. H: *Pinnixa longipes*. I–L: megalopae in a swimming position. I: *Carcinus maenas* (Carcinidae), lateral view. J and K: *Ebalia tuberosa* (Leucosiidae). J: ventral view. K: dorsal view. L: *Macropodia* sp. (Inachidae), dorsal view. M: *Ocypode albicans* (Ocypodidae), with legs folded against the body, lateral view. A modified after H. Diaz and Ewald (1968); B modified after C. Kim and Jang (1986); C modified after Fielder and Greenwood (1985); D and M modified after Crane (1940); E modified after Anger et al. (1990); F modified after Domingues Rodriguez and Hebling (1989); G modified after Marques and Pohl (1996); H modified after Bousquette (1980); I–L modified after Atkins (1954).

55

JOEL W. MARTIN
JØRGEN OLESEN
JENS T. HØEG

Summary and Synopsis

GENERAL TRENDS

This atlas, in presenting and comparing the majority of what is known about the larval morphology of all major crustacean groups—extant and extinct, marine and freshwater—is the first of its kind. Therefore, it is logical for us to take this opportunity to ask if there are obvious trends or insights afforded by this comparison. Can we make any general comments concerning larval development in the Crustacea? What do we see, looking at the group as a whole, that might not be obvious by studying development within each taxon?

Certainly there are some commonalities that would apply to all groups, and these have been noted by nearly all previous workers. For example, it seems clear that a naupliar larval phase must have been present ancestrally, as there is at least a vestige of this phase in all major lineages. Indeed, a growing body of fossil evidence points to the fact that an earlier head larva, a larval type slightly longer than the nauplius, was part of the ground plan of the Euarthropoda, and that the nauplius of true crustaceans is a slightly shortened version of this larval form (see chapter 3).

It is also clear that many groups have post-naupliar larval phases that, at least in some cases, are the primary dispersal stages. The nauplius phase is often the primary (or only) dispersive phase, but in many groups the nauplius is followed by a second, sometimes more dispersive phase that is morphologically distinct from the nauplius; a common example is the malacostracan zoea. This second phase, in turn, can be followed by yet another phase that is typically morphologically and ecologically different, one that facilitates the transition from a planktonic larval life to the demersal or benthic life of juveniles and adults (e.g., the cyprid phase of cirripedes, or the decapodid phase of decapods). Table 55.1, although necessarily brief and general (refer to individual chapters for more details), summarizes the hatching, dispersive, and transitional phases in the major groups of crustaceans.

As significant as dispersal may be, it is also important to recognize that dispersal is not the only advantage in having larvae. Another benefit is in growth, facilitated by life in an environment that differs from that of the adult, thereby avoiding competition for the same resources (see Strathmann 1993; Wary 1995; C. Hickman 1999 *in* B. Hall and Wake 1999). Having morphologically and ecologically distinct larvae very likely provides more than one advantage for the organism under study, given the wide variety of taxa in which larvae occur, and it is not always clear which benefit is the primary driving mechanism for their evolution.

ORIGIN OF CRUSTACEAN LARVAE

Partly because of the exquisite preservation of and the equally exquisite research that has been conducted on the Orsten fauna (see chapter 3), we know without any doubt that nauplius-like larvae existed as long ago as the Cambrian. The oldest unquestionably crustacean larval stage is a metanauplius from the early Cambrian of China, about 525 million years ago (X. Zhang et al. 2010), a larval stage that in some ways resembles the larvae of extant cirripedes (see chapter 22). This in itself is remarkable, as it speaks not only of the age of the Crustacea as a whole, but also of the age of groups within the Crustacea, groups that were already modified at least slightly from a presumed ancestral and more general larval series. It is clear that the nauplius has been around for well over 500 million years, an amount of time that has allowed endless evolutionary experimentation. Therefore it is not surprising that we see a large diversity of forms within the basic naupliar constraints. By the Upper Cambrian, an assortment of nauplii and nauplius-like larvae (including head larvae) can be found, most of which, however, cannot be assigned to an adult (see chapter 3). Yet some major taxa are already recognizable from these larval forms. For example, the fossil †*Rehbachiella kinnekullensis* shows branchiopod affinities and has been suggested to be the sister taxon to all extant Branchiopoda.

Was the crustacean nauplius originally a feeding larva,

Table 55.1 Brief and generalized overview of hatching and primary modes of development in the major extant lineages of Crustacea

Taxon	Nauplius phase(s) present	Hatching phase[1]	Primary dispersal phase(s)[1]	Ecol./morph. transitional phase(s)	Type of development[1]	Chapter(s)
Anostraca	yes	nauplius	nauplius/adult	none	anamorphic	5
Notostraca	yes	nauplius	nauplius/adult	none	anamorphic	7
Laevicaudata	yes	nauplius[2]	nauplius[2]/adult	none	partly anamorphic[2]	8
Spinicaudata	yes	nauplius	nauplius/adult	none	partly anamorphic	9
Cyclestherida	no	embryo-juvenile	juvenile/adult	none	direct	10
Anomopoda	no	embryo-juvenile	juvenile/adult	none	direct	11
Ctenopoda	no	embryo-juvenile	juvenile/adult	none	direct	12
Haplopoda	yes[3]	nauplius	nauplius, juvenile, adult	pre-juvenile	partly anamorphic	13
Onychopoda	no	embryo-juvenile	juvenile/adult	none	direct	14
Remipedia	yes	nauplius	nauplius?	none	anamorphic	15
Cephalocarida	yes	nauplius	nauplius?	none	anamorphic	16
Facetotecta	yes	nauplius	nauplius	y-cyprid	incompletely known	18
Ascothoracida	yes	nauplius[4]	nauplius	a-cyprid	metamorphic	19
Acrothoracica	yes	nauplius[5]	nauplius	cyprid	metamorphic	20
Rhizocephala	yes	nauplius[6]	nauplius	cyprid	metamorphic	21
Thoracica	yes	nauplius[7]	nauplius	cyprid	metamorphic	22
Tantulocarida	possibly	nauplius?	tantulus	tantulus	metamorphic	23
Branchiura	yes	metanauplius-like stage[8]	metanauplius and later stages	none	partly anamorphic	24
Pentastomida	no	primary larva	primary larva	nymph	weakly metamorphic	25
Mystacocarida	yes	nauplius	nauplius, post-naupliar stages	none	anamorphic	26
Copepoda	yes	nauplius	nauplius or copepodid	varies with taxon; often copepodid	metamorphic	27
Podocopa	yes	nauplius	post-naupliar instars	none	partly anamorphic	29
Myodocopa	yes	nauplius	post-naupliar instars	none	partly anamorphic	30
Leptostraca	no	embryo-juvenile	juvenile	none	direct	33
Stomatopoda	no	antizoea, pseudo-zoea	antizoea, pseudozoea, erichthus, alima	post-larva	metamorphic	34
Syncarida	no	parazoea[9]	parazoea	bathynellid	weakly metamorphic	35
Peracarida	no	manca	manca, juvenile	none	direct	36–41
Euphausiacea	yes	nauplius	nauplius, calyptopis, furcilia	furcilia	metamorphic	43
Amphionidacea	no	amphion[10]	amphion	none	weakly metamorphic	44
Dendrobranchiata	yes	nauplius	zoea, mysis	decapodid	metamorphic	46
Stenopodidea	no	zoea	zoea	decapodid	metamorphic	47
Caridea	no	zoea	zoea	decapodid	metamorphic	48

or did planktotrophy arise later from larvae that were originally lecithotrophic? Here, too, we have some clues from the Cambrian. The early Cambrian metanauplius described from China was already well adapted for swimming and feeding, with an overall morphology that "closely mirrors that of corresponding developmental stages of living barnacles and copepods" (X. Zhang et al. 2010). The limbs appear to have served the combined functions of locomotion and feeding. Indeed, in a wide diversity of marine invertebrate taxa—such as crustaceans, brachiopods, and the Ambulacraria (hemichordates plus echinoderms)—the ancestral form appears to have been a feeding larva, suggesting that, at least in these taxa, planktotrophy came first (Strathmann 1993, 2007, 2012). Yet other authors (e.g., Scholtz 2000; Peterson 2005; Servais et al. 2008) have argued that the original larval form of crustaceans (and of most marine groups) must have been lecithotrophic, with planktotrophy arising later (probably multiple times) from non-feeding larvae. These arguments, and their eventual resolution, will have a bearing not only on debates over the

origin of crustacean larvae in general, but also on such issues as the separate origin of the naupliar larvae of the euphausiaceans (krill) and dendrobranchiate shrimps, which have no feeding ability whatsoever (e.g., Scholtz 2000).

By the Devonian, we have fossils of another branchiopod (†*Lepidocaris rhyniensis*, from the Rhynie chert, approximately 400 million years ago), confirming that branchiopods are an ancient lineage (see chapter 6).

By the Mesozoic, it is reasonable to assume that the major extant larval forms for all crustacean groups had evolved and were in most ways similar to how we find them today. This is based on the fact that the highly modified larvae of achelate lobsters (the odd, flattened phyllosoma stage), despite the fact that they appear to be highly derived (compared with most other decapod zoeal larvae), were present in the Mesozoic, as were the larvae of the Stomatopoda (see chapter 32). Fossils of true zoea larvae—from brachyuran crabs, based on two specimens from the Cretaceous (Maisey and de Carvalho 1995)—also are known, albeit not well preserved or identifi-

Table 55.1 Continued

Taxon	Nauplius phase(s) present	Hatching phase[1]	Primary dispersal phase(s)[1]	Ecol./morph. transitional phase(s)	Type of development[1]	Chapter(s)
Astacidea	no	zoea	zoea	decapodid	metamorphic	49
Gebiidea, Axiidea	no	zoea	zoea	decapodid	metamorphic	50
Achelata	no	zoea[11]	zoea	decapodid[11]	metamorphic	51
Polychelida	no	zoea[12]	zoea	decapodid[12]	metamorphic	52
Anomura	no	zoea	zoea	decapodid[13]	metamorphic	53
Brachyura	no	zoea	zoea	decapodid[14]	metamorphic	54

Note: For each group, only the presumed ground-pattern (ancestral) type of development is listed; for details and the many exceptions in each group, see the relevant chapter.

[1] Hatching phase: The term nauplius here encompasses orthonauplius, metanauplius, and any other form of nauplius larvae. Primary dispersal phase(s): This category takes into account the fact that in some taxa, such as many of the branchiopods, the adults swim faster than the larvae and are probably more responsible for dispersal than are the larvae. Similarly, dispersal across wide geographic regions is often accomplished through the transfer of eggs (cysts) rather than by traditional larval dispersal. A question mark (?) indicates uncertainty. Type of development: This category is somewhat arbitrary, in that different components of a given life cycle might (or might not) exhibit metamorphosis.

[2] Laevicaudata: Nauplii are highly modified in this group. The last larval stage has been called a heilophore larva in the past. Type of development: Although listed as partly anamorphic, the molt to the juvenile stage is highly metamorphic.

[3] Haplopoda (*Leptodora*): There are 2 distinct types of development in this group: one that hatches as a nauplius, and one with direct development.

[4] Ascothoracida: Some members of this group hatch as a-cyprids.

[5] Acrothoracica: Most members of this group hatch as cyprids.

[6] Rhizocephala: Most members of this group hatch as cyprids.

[7] Thoracica: Some members of this group hatch as cyprids.

[8] Branchiura: Sometimes referred to as metanaupliar, despite the fact that these larvae are well developed and therefore more like juveniles. We refer to them as metanauplius-like.

[9] Syncarida: This applies only to the Bathynellacea; the Anaspidacea undergo direct development.

[10] Amphionidacea: The amphion is referred to as a zoea by some workers; it differs from a true zoea in the number of thoracopods that are modified for feeding/swimming.

[11] Achelata: In the Achelata the zoeal stage is more commonly referred to as a phyllosoma, and the decapodid stage is more commonly referred to as a puerulus (in the Palinuridae) or a nisto (in the Scyllaridae).

[12] Polychelidae: It is suggested here that these distinctive zoeae be referred to as eryoneicus zoeae, to preserve the traditional name eryoneicus for the very distinctive and globose zoeal larvae. The name eryoneicus megalopa is similarly suggested here to recognize the distinctive nature of the decapodid larvae.

[13] Anomura: The decapodid stage is commonly referred to as a megalopa (or glaucothoe) in this group.

[14] Brachyura: The decapodid stage of brachyurans is almost universally referred to as a megalopa.

able to genera or species. It is therefore reasonable to assume that crustacean larvae have been in their present form for a very long time, a fact that is also evident from independent phylogenetic analyses (those not dependent on fossils) of various crustacean lineages.

RELATIONSHIPS BASED ON LARVAE

It is reassuring to note that some groups, long thought to be monophyletic on the basis of their adult morphology, are indeed characterized by distinctive larval phases or larval characters found in no other group, strongly supporting their phylogenetic unity.

In the Branchiopoda, for example, the naupliar stages are distinguished by reduced and unsegmented antennules, a specific morphology for the antennal naupliar process, and mandibles that are uniramous. These characters are present even in the bizarre larvae of the Laevicaudata (see chapter 8). This larval-based monophyly has been confirmed by nearly all morphological and molecular analyses of branchiopod phylogeny, lending great credibility to the use of crustacean larval features as sound phyletic indicators.

Another example is in the diverse Cirripedia, where all known naupliar larvae are immediately identifiable by their unique frontolateral horns and related pores and glands, characters that are seen in no other group of crustaceans. Indeed, this feature has often been used as a uniting character for this large and diverse lineage.

In the Malacostraca, the only two taxa with naupliar eclosion (the Euphausiacea and Dendrobranchiata) have nauplii that are lecithotrophic, rather than planktotrophic. Naupliar eclosion is widely regarded as a primitive feature; thus the presence of nauplii in these two groupings has long been used to argue for the basal position of dendrobranchiate shrimps within the Decapoda, and for the sister-group position of the euphausiids. The fact that both groups have lecithotrophic nauplii prompted Scholtz (2000) to suggest that these larvae are derived independently from those of all other crustaceans, whose nauplii are almost (but not quite) exclusively plankto-trophic.

Does the number of naupliar stages, as opposed to (or in addition to) the morphology of those stages, have phylogenetic significance? It's a possibility, and a recent paper by Oakley et al. (2013) proposed the taxon name Hexanauplia to encompass those taxa with exactly six naupliar stages (the Thecostraca and Copepoda), a grouping based primarily on new phylogenomic data. It is well known that in the decapod groups, certain families are characterized by the number of zoeal stages (e.g., Rice 1980).

We might also ask what, if any, insights into higher-level phylogeny are available from the information accumulated in this book. For example, is there also larval support for the recent and well-supported molecular phylogeny of all arthropods (Regier et al. 2010) shown in chapter 1? Or for the recognition of the Ecdysozoa? Or for other hypotheses of relationships among the Crustacea, or between the crustaceans and other arthropodous groups?

Perhaps not surprisingly, the concept of the Ecdysozoa is not much altered, being neither supported nor weakened by our compilation of information on crustacean larvae, probably because of the vast diversity contained within this proposed clade, which includes the arthropodous Hexapoda, Crustacea, Myriapoda, and Chelicerata, and the more vermiform Nematoda, Nematomorpha, Priapulida, Kinorhyncha, and Loricifera (see Telford et al. 2008). The Tardigrada and Onychophora are also sometimes included in discussions of the Ecdysozoa (see J. W. Martin 2013). With this incredible range of taxa, it would be surprising indeed if larval features existed that were shared by all of them. Indeed, many of these taxa have no larvae at all, and the character "lack of a primary larva" has even been used to unite them (Telford et al. 2008), although that odd concept does not apply to the most diverse groups in the clade.

Within the Arthropoda, there is a similar paucity of larval data for comparative purposes. The phylum is so diverse that comparing larval characters between, for example, insects and crustaceans is bound to be fruitless.

Within the Crustacea, some obvious affinities, based on larval features, have been alluded to already, such as the obvious monophyly of branchiopods and thecostracans. Are there other insights? Yes, but mostly in a negative vein; that is, most of the larval evidence is not supportive of recently proposed phylogenetic schemes. For example, there is no obvious larval evidence for combining the Cephalocarida with the Remipe-

dia in a grouping called the Xenocarida, as proposed by Regier et al. (2010). It is true that naupliar stages of both classes are blind (they lack a naupliar eye), but this loss is not unique to these two taxa and can easily be explained away as an adaptation to a lightless environment.

Is there larval evidence to support the inclusion of the Pentastomida within the Crustacea, which is strongly suggested by molecular evidence (e.g., Regier et al. 2010), yet thrown into doubt by the discovery of Cambrian fossils (see chapter 25)? Here, the answer is no. That is, there are no obvious features of pentastomid development that would suggest that they are crustaceans. In the Branchiura, the class that molecular evidence suggests is closely related to the pentastomids, in the metanauplius we do see a pre-oral spine for penetration of the host integument. But this penetration device has no connection to the mouth, so it is not homologous to the stylet of the pentastomid nymphs, which is actually on the mouth itself and is not pre-oral. Thus any similarities of penetration spines between the two groups are only superficial and should simply be seen as convergent adaptations to parasitism.

A stellar, and positive, example of the usefulness of larvae for phylogenetic and systematic purposes is furnished by the Facetotecta. Facetotectans are known only as larvae (y-nauplii followed by y-cyprids), yet these larvae share some similarities with the Cirripedia and Ascothoracica, such as the presence of lattice organs. For this reason, facetotectans have long been suggested as phylogenetically close to cirripedes and ascothoracidans. Recent molecular evidence has verified this beyond doubt, and thus has confirmed the monophyly of the Thecostraca. It is remarkable, and a testament to the phylogenetic significance of larval morphology, that the Facetotecta is the only major crustacean taxon known exclusively from larvae, and yet there is no doubt about the phylogenetic affiliations of this subclass.

In general, tremendous progress has been made in the field of larval morphology, and more answers are coming soon. The field is bright and full of promise, with new methodologies, new discoveries, and a new generation of scientists offering fresh insights. Yet in summary, we are forced to conclude that the overall comprehensive look at development in all crustaceans, marine and freshwater, extant and extinct, in this atlas leaves us in an all-too-familiar place, with more questions than answers, as is so often the case in the study of the Crustacea. We hope that this compilation stirs the imagination and leads to further increases in our understanding of this diverse and fascinating group.

GLOSSARY

Abbreviated development—Development in which the developmental sequence is of fewer stages or shorter duration than is normal (for a given taxon) and/or the hatching stage is at a later stage of development than for most species of the taxon (D. Williamson 1982a; Gore 1985; Rabalais and Gore 1985). It has also been used in a more ecological sense, as an intraspecific term when development is shortened because of limited food availability.

Acanthosoma—An older term (no longer used) referring to the Sergestidae (Dendrobranchiata: Sergestoidea) larvae of the mysis stage.

A-cyprid—The a-cyprid is a cypridoid larva that is found exclusively in the Ascothoracida, characterized by having a bivalved shield and a distal claw and claw guard on the antennules for mechanical attachment.

Adhesive gland—See **cement gland**.

Aesthetasc (aesthetasc seta)—A thin-walled chemosensory seta, typically tubular and with parallel (not tapering) sides. It is found on the antennules of decapods and their larvae, and elsewhere in other crustaceans (e.g., in copepods and all larvae of the Thecostraca).

Alima—A larval stage of squilloid stomatopods, characterized by the presence of four or more intermediate denticles on the telson and a small ventral spine on the antennular somite. Unlike many other larval names, *Alima* is also a formal, valid genus name in the Stomatopoda.

Amphion larva—A name given to the zoea-like or mysid (following Heegaard 1969) larval stages of the monotypic order Amphionidacea (*Amphionides reynaudii*), characterized by having only the first thoracopod modified into a maxilliped. Recent studies (e.g., Kutschera et al. 2012) have abandoned this usage in favor of numbering the stages as 1, 2, 3, and so forth.

Anal somite—The terminal segment of the body in many crustacean groups, bearing the anus either terminally or dorsally; it represents the telson or a somatic complex incorporating abdominal somite(s) and the telson. In copepods, the anal somite usually bears paired appendages (caudal rami), and the anus is usually protected by an anal operculum (a small median process on the dorsal surface of the anal somite).

Anal spine—A spine found in, and characteristic of, the sixth somite of some caridean shrimp larvae.

Anamorphosis (anamorphic development)—A type of indirect development in which a hatchling or larva with few segments develops into the adult gradually, via a series of small stepwise changes in body morphology, with progressively more segments and appendages added during morphogenesis (e.g., the gradual development of the nauplius into the adult in the Anostraca and Notostraca).

Androdioecious—A rare reproductive situation in which a species is composed of both male populations and hermaphroditic populations. Among the Crustacea, it occurs in some branchiopods and cirripedes (Weeks et al. 2006).

Anomuran seta (anomuran hair)—Reduced hair-like second telson process (seta), typically pappose or plumose, common to—and indicative of—anomuran and some (notably dromiacean) brachyuran zoeae; sometimes termed an anomuran hair. It is distinct from all other types of telson setae and occurs on or near the outermost part of the telson's posterior margin. It is also present in larvae of the Stenopodidea and both thalassinid groups. See **thalassinid seta**.

Antenna—The second pair of cephalic appendages, typically used for swimming and feeding in early larval forms (and in some adults, especially in the Branchiopoda), but usually used for sensory perception in later larvae and adults. Sometimes referred to as second antenna or antenna 2, and abbreviated A2.

Antennule—The first pair of cephalic appendages, usually natatory in early (e.g., naupliar) larvae but mostly chemosensory in later (zoeal, megalopal, etc.) larvae and adults. Sometimes referred to as first antenna or antenna 1, and abbreviated A1. Cirripede cypris larvae are unique in having antennules that are used for bipedal walking over the substratum.

Anterior seta—Distinctive short, densely plumose seta, the first in a row of setae along the posteroventral margin of the brachyuran zoeal carapace (the "soire antérieure" of Bourdillon-Casanova 1960). In older literature, it is sometimes referred to as the "majid seta" (or "majid spine"), because the term was incorrectly thought to be restricted to that taxon (Clark et al. 1998).

Antizoea (plural = antizoeae)—The hatching larval type of lysiosquilloid stomatopods, characterized by sessile eyes, uniramous antennules, five pairs of biramous thoracic appendages, the absence of abdominal appendages, and minimal segmentation.

Anus—The posterior opening of the digestive tract, typically located ventrally, at the posterior margin of the last abdominal somite.

Apophysis—A spinous process on a swimming-leg segment, typically sexually dimorphic (present in males only). It may originate as a produced segmental margin, or it may be derived by modification of a setal element present in the female.

Appendage—A paired extension (limb) of a somite, usually composed of serially repeated elements (the antennae, mandibles, thoracopods, etc.).

Architecture—A general term sometimes used to refer to the morphological organization of the crustacean body.

Armature—A collective term for the setae and spines carried on the limbs and caudal rami.

Arthrite—A movable, articulating, sclerotized extension of a basal segment (a basis or a fused coxa and basis) of an appendage that is moved by muscles, as in the precoxal endite of the maxillule in copepods. See **naupliar process**.

Arthritic process—See **naupliar process**.

Arthrodial membrane—A flexible, unsclerotized membrane, part of the exoskeleton between the sclerotized parts of two somites or between two segments of an appendage.

Atrium oris—A pre-oral cavity in some crustaceans that is delimited ventrally by the posteriorly directed labrum, dorsally by the ventral surface of the cephalon just

behind the mouth, and laterally by the paragnaths (when present) and mandibles.

Balancers—A pair of typically flaccid and swollen setal elements representing the caudal rami in nauplii (NI–NII) of siphonostomatoid fish parasites (the Caligidae, Dissonidae, Pandaridae, and Kroyeriidae). They are not sensory in function, but appear to be correlated with the presence of linear, uniseriate egg sacs. They represent the positional homologue of caudal ramus seta 4 in other podoplean copepods (Huys et al. 2007).

Basial endite—A lobe produced from the inner margin of the limb basis (sometimes written as "basal endite").

Basipod—The most proximal limb portion of all the post-antennular appendages of arthropods. The basipod bears two rami (the endopod and exopod) and is retained in all major arthropod lineages (the Trilobita, Chelicerata, Crustacea, Hexapoda, and Myriapoda).

Bathynellid—The juvenile phase of species of the Bathynellacea.

Benthic—Of or pertaining to the benthos (the bottom of a lake, sea, or ocean); bottom-dwelling.

Benthotrophic larvae—A benthonic larval stage that derives its nourishment by feeding in the benthos (commonly on detritus).

Biramous—Referring to an appendage (limb) having two branches (rami) coming off of the limb basis: the outer branch, or exopodite (exopod), and the inner branch, or endopodite (endopod).

Blastoderm—The early developmental layer of cells forming the walls of the blastula.

Bopyridium—The juvenile (post-larval) stage of parasitic isopods in the Bopyroidea (epicarideans), which develops into a male or female adult.

Brood pouch—Any enclosure containing developing embryos or young. This may be in the form of modified limbs or outgrowths of limbs (e.g., in notostracans and anostracans), or plates of limbs (in peracarids), or carapace valves (in some predatory cladocerans). See **marsupium**.

Bud—A general term usually referring to the earliest stage of a developing limb (e.g., for copepods, see Ferrari and Dahms 2007). A limb bud does not articulate, is often considered functionless in immature stages, and—in copepods—bears at least the crown group of terminal setae.

Calyptopis (plural = calyptope)—A phase of larval development exclusive to the euphausiids, characterized by compound eyes beneath the carapace and the absence of pleopods on the abdomen. The calyptopis phase consists of three stages and is followed by the furcilia phase. The abdomen becomes segmented in stage 2. Uropods develop in stage 3, along with the separation of the telson from the sixth abdominal segments. There are no pleopods on the abdomen.

Caudal papillae—Paired outgrowths at the terminal end of certain crustacean larvae.

Caudal ramus (plural = rami)—Paired setiferous appendages on the posterior margin of the anal somite in copepods.

Cement gland—A term used for two separate and non-homologous structures. (1) A multicellular gland in cirripede cyprids and adults. In cyprids it terminates on the antennule and secretes the adhesive substance used for attachment. In adults it terminates underneath the body or peduncle and serves to securely attach the body during growth. (2) In anostracans, the term refers to a pair of glands that lie dorsal and proximal to the eggs in the ovisac and produce the outer coat of the eggs.

Cephalic shield—Cuticular shield covering the dorsal surface of the head region (and sometimes of more posterior somites, such as in copepods, branchiurans, and notostracans) that is formed by the fusion of the dorsal surfaces (tergites) of the cephalon, and sometimes ventrolaterally and/or caudally extended to cover trunk somites. Synonymous with head shield.

Cephalic stylet—A conspicuous intromittent organ used by the tantulus larva to puncture the integument of its crustacean host, and operated by paired protractor and retractor muscles (Huys 1991). After attachment these muscles atrophy rapidly, and the stylet often assumes an asymmetrical position inside the cephalon.

Cephalosome—The anterior tagma of the body in copepods, covered by the dorsal cephalic shield. It consists of five cephalic somites and the first thoracic (maxilliped-bearing) somite.

Cephalothoracic suction cup—An attachment device formed by the ventral surface of the dorsoventrally flattened prosome in bomolochiform fish parasites (poecilostome Cyclopoida) (Dojiri and Cressey 1987; Huys et al. 2012). The integrity of the posterior wall of the sucker is typically safeguarded by the modified lamelliform first pair of swimming legs. The ventral concavity of the cephalothorax can sometimes be sealed around most of its perimeter by a membranous extension applied to the surface of the host, offering optimal suction efficiency.

Cephalothorax—A tagma consisting of the fused cephalic somites (five) plus at least one thoracic somite. The appendage of the latter somite is always specialized as a feeding appendage (maxilliped).

Cerataspis, Cerataspides—Two "larval genus" names assigned to several remarkable large, elongate, and presumed penaeoidean shrimp larvae that are most often found in the stomach contents of neustonic fish predators, such as tuna, dolphin, and skipjack. The adult of *Cerataspis monstrosa* is now known (Bracken-Grissom et al. 2012; also see chapter 45).

Chalimus—A post-copepodid developmental stage, or one of the copepodid stages, of some parasitic copepods (e.g., many siphonostomatoid fish parasites, and some caligid-like parasitic copepods), characterized by possession of a "frontal filament" (not to be confused with other uses of frontal filament; see below) for attachment to the host.

A maximum of four chalimus stages (chalimi) can occur in the life cycle, being homologous to copepodids II–V of free-living copepods.

Cincinnulus (plural = cincinnuli)—A spine or modified seta that is short and hooked. It serves to connect the endopods of paired abdominal pleopods, so that they function in unison.

Compound eye—An eye that consists of several to numerous, almost identical components, the ommatidia. The sensory input of the compound eye is processed by at last two retinoptic neuropils that are connected to the syncerebrum (Richter et al. 2010). Compound eyes have been present beginning either at the level of the Arthropoda *sensu stricto* or earlier (*sensu* Maas et al. 2004).

Copepodid (copepodite)—A post-naupliar stage found exclusively in the Copepoda (though not in all copepods), marked by the absence of a naupliar enditic (coxal) process on the antenna, the presence (usually) of articulating thoracic somites and more than three functional limbs, and the presence (often) of articulating abdominal somites. There are typically five copepodid stages in the life cycle prior to the adult, designated copepodid I through V (CoI–CoV, or CI–CV).

Coxa—A term for the proximal portion of the limb in various arthropods.

Coxal spine—A spine projecting from the coxa of (typically) a thoracic appendage.

Coxal endite—A lobe produced from the inner margin of the coxa.

Cryptogonochorism—An extreme form of sexual dimorphism found in some copepods (the Xenocoelomatidae and the unrelated genus *Gonophysema*) and rhizocephalans. Dwarf males penetrate the female and move into a special invaginated receptacle, where they undergo a metamorphic reduction so that little tissue except the gonads remains. The male effectively becomes a functional testis, resulting in a pseudohermaphroditic condition (Bresciani and Lützen 1961, 1974; Bocquet et al. 1970).

Cryptoniscus—Third larval stage of parasitic isopods in the Bopyroidea and Cryptoniscoidea (epicarideans). This stage seeks a definitive crustacean host. In the Cryptoniscoidea, the adult male retains the cryptoniscus appearance.

Cyprid—Another term for the cypris larva of cirripedes (sometimes called the c-cyprid). See **cypris larva**.

Cypridoid larva—The terminal pelagic instar in thecostracan development, being covered by a carapace (head shield) with lattice organs, and having frontal filaments, prehensile antennules for attachment, and six pairs of natatory thoracopods. See **a-cyprid**, **cypris larva**, and **y-cyprid**.

Cypris larva—The cypris larva (sometimes called the c-cypris or c-cyprid) is a type of cypridoid larva, unique to cirripedes, which follows the naupliar phase. It is highly specialized and is characterized by a bivalved shield, internal compound eyes, four segmented prehensile antennules with a specialized attachment organ, and six pairs of natatory thoracic appendages. It attaches by secretion from a cement gland. See **cypridoid larva**.

Cyrtopia—An older name (no longer used) referring to later furcilia larvae of the euphausiids. These stages show the antennular flagella becoming elongate and distinctly articulate, the antennae becoming transformed and no longer being natatory, and the posterior legs and gills appearing in succession.

Decapodid—A generic term introduced by Kaestner (1970) (synonymous with D. Williamson's 1969 use of "megalopa") for the last larval phase in malacostracans, especially when the larval series has two distinctly different larval phases (e.g., the anomuran glaucothoe and the brachyuran megalopa, distinct from their zoeal larvae). It is a larval phase that represents the transition between the planktonic zoeal phase and the (usually benthic) juvenile phase in decapods. Functionally, it is the stage in which the pleopods become the appendages of locomotion.

Diapause—A significant period of quiescence during development.

Direct development—A form of abbreviated development in which there is an absence (or suppression) of larval stages (free-swimming or otherwise), such that the hatching stage is a juvenile. It thus differs very little or not at all from adult morphology, apart from (typically) size and reproductive maturation.

Dorsal organ—This term has been applied to a variety of structures that are unlikely to be homologous. In the Crustacea it has been used to describe (1) an embryonic organ that probably regulates the osmotic pressure of the body during development in some Peracarida (e.g., fig. 39.1C) and (2) a nearly circular or oval morphological structure, in the center of which is often a pit surrounded by four bumps or pores, that is located on the midline of the head, usually dorsal to and slightly posterior to the eyes, recognizable in many larval and adult branchiopods (e.g., fig. 7.1B, C, F) and larval malacostracans. It is sometimes called the "sensory dorsal organ" (after Laverack et al. 1996). Its function is not known in all groups (J. W. Martin and Laverack 1992).

Dorsocaudal spine—A slightly upturned spine, projecting posterodorsally from the supra-anal flap above the anus in copepods, cirripedes, and several of the Orsten taxa.

Dorsolateral knobs—See **dorsolateral processes**.

Dorsolateral processes—Knobs or protrusions (sometimes referred to as dorsolateral knobs) on the dorsolateral surface of abdominal somites 2 and 3 in brachyuran zoeal stages.

Egg mass—A mass of eggs attached to the body of a female copepod but not enclosed within a common outer membrane; also used for other groups of crustaceans bearing large numbers of eggs.

Egg nauplius—Embryonic stage with the naupliar appendages more distinctly developed than the more

posterior limbs. It is found in many cladocerans and in malacostracans that exhibit direct development, such as decapods and some peracarids (Scholtz 2000).

Egg sac—A batch of one or more eggs contained within a sac-like membrane that emerges from and remains attached to the genital opening(s); typically paired.

Elaphocaris—An older term (no longer used) given to Sergestidae (Dendrobranchiata) larvae of the protozoea stage.

Embryonized larvae—Larvae that exhibit at least some naupliar characters while still contained within the egg membrane; used synonymously with "egg larvae" or "egg nauplius" by some authors. See **nauplioid**.

Endites—Medial setae- or spine-bearing lobes of the basis, coxa, and proximal segments of a limb, typically on the maxillules, maxillae, and maxillipeds, but also on the thoracopods of non-malacostracans.

Ephippium—A modified portion of the shed carapace valves that functions as a pouch containing resting eggs in some Cladocera.

Epicaridium—The first larval stage of parasitic isopods in the Bopyroidea and Cryptoniscoidea (epicarideans). This stage seeks the copepod intermediate host.

Erichthus—A larval stage of non-squilloid stomatopods, characterized by the presence of one or two intermediate denticles on the telson and a ventrally unarmed antennular somite.

Eryoneicus—The large, globose larvae of deep-sea lobsters of the family Polychelidae. In early stages, where swimming is accomplished by the natatory exopods of the thoracopods (the maxillipeds and the first two pereopods), it is an "eryoneicus zoea"; in late stages, where swimming is accomplished by well-developed abdominal pleopods, it is called an "eryoneicus megalopa" (both terms used here for the first time).

Eryoneicus megalopa—A late (post-zoeal) polychelid larval stage, where swimming is accomplished by well-developed abdominal pleopods.

Eucrustacea—The monophyletic group containing all extant crustacean species as well as some fossil species, characterized by a number of autapomorphies (Waloszek 2003a, 2003b), including the orthonauplius as a feeding and locomotory hatching stage.

Extension lobe—A bulb of material secreted around the base of the frontal filament in fish-parasitic siphonostomatoid copepods, appearing at each of the four molts occurring between the attached infective copepodid stage and chalimus IV. Extension lobes accumulate around the origin of the frontal filament during successive molts.

Externa—The external reproductive sac of rhizocephalan parasites, which (in the virginal state) is invaded by a male trichogon.

Flagellum—An extension of the endopod (usually of the antennules and antennae) that is long and multiarticulate.

Frontal filaments—A term with at least two definitions in the Crustacea. (1) Paired pre-antennular filaments, innervated by the lateral protocerebrum, of unknown but presumably sensory function, found on the antero-ventral head region of the Remipedia and Branchiopoda, and in the mantle cavity of thecostracan cyprioid larvae, containing branching ciliary extensions (Fritsch and Richter 2010). (2) The (non-homologous) unpaired filament on the anterior midline of the head, used as a permanent attachment device by the chalimus stages of the large, fish-parasitic clade of siphonostomatoid copepods (the Caligidae, Cecropidae, Hatschekiidae, Lernaeopodidae, Pandaridae, and Pennellidae) and by various stages (copepodid, pupal stages, and adult male) in the life cycle of the Nicothoidae.

Frontal horns—See **frontolateral horns**.

Frontal organ—A term that has been applied widely and inconsistently in crustacean literature. It can refer to sensory cells on the anterior surface of the cephalon of malacostracans, and to the haft organ (frontal eye) in non-malacostracans (e.g., the Bellonci organ of ostracods).

Frontolateral horn glands—Glands (also called frontal horn glands) exiting on the frontal horns of cirripede nauplii; also found exiting with pores anteroventrally on the cypris carapace.

Frontolateral horns—A pair of horns (sometimes called frontal horns) unique to, and projecting anterolaterally from, the head shield of cirripede nauplii. These horns bear a specialized gland (the frontolateral gland, or frontal horn gland) exiting from the tip. They are unique to cirripede nauplii.

Frontolateral pores—Pores found on either side of a cypris carapace in cirripedes. They are the exit points for the glands carried over from the frontal horns of the preceding nauplius phase.

Furca—A pair of articulated caudal outgrowths, usually bearing marginal setation. This term is often used interchangeably with caudal furca or caudal rami.

Furcilia—A phase of larval development, exclusive to the euphausiids (krill), that follows the calyptopis phase and is characterized by eyes that are no longer under the carapace; progressive development of the thoracic legs, pleopods, and photophores; and progressive reduction in the number of telson spines.

Genital complex—A compound body region in adult female copepods, formed by the fusion of the fifth pedigerous and genital double-somites.

Genital double-somite—The double-somite resulting from the fusion of the genital and first abdominal somites in adult female copepods. The fusion typically occurs at the final molt from copepodid V to adult.

Germ band (germband, germ layer)—Part of the developing embryo that gives rise to the visibly segmented part of the animal, usually consisting of the gnathal, thoracic, and abdominal segments and including the mesoderm, ventral ectoderm, and dorsal epidermis.

Germinal disc—The small, often round, single-cell layer of cells from which the embryo develops; synonymous with blastodisc. See **germ band**.

Glaucothoe (glaucothöe)—The decapodid larval stage of anomurans, sometimes used synonymously with megalopa, although the latter term is more often associated with the Brachyura.

Gnathobase—A projecting medial extension of the coxal segment of an appendage, particularly of the antenna (in naupliar stages) or mandible (the proximal endite), often bearing a toothed cutting edge distally. See **naupliar process**.

Gonochoric—Having separate sexes in different individuals.

Gymnoplean—The body plan in which the prosome-urosome boundary lies between the fifth pedigerous somite and the genital somite.

Gymnopleans—Copepods in which the major body articulation of the adult is between the sixth and seventh thoracic somites. There is also a significant difference in size between these two somites, exhibited by members of the orders Platycopioida and Calanoida. See **podopleans**.

Head larva—The hatching stage in the ground pattern of Crustacea *sensu lato* (*sensu* some authors, e.g., Stein et al. 2008), thought by some (e.g., Walossek 1999) to have been retained from the ground pattern of the Euarthropoda and to consist of an ocular segment, four limb-bearing segments, and an undifferentiated hindbody.

Head shield—See **cephalic shield**.

Heilophore—A larval stage characteristic of clam shrimps; the last nauplius stage before the bivalved first juvenile. The term means "carrier of a lip"; it was introduced by Botnariuc (1947) and used by Petrov (1992) and Eder (2002).

Hermaphroditic—Possessing both male and female reproductive organs within a single individual.

Heterochrony—Evolutionary changes caused by variation in the relative time of appearance and the rate of development of features.

Heterogonic (heterogony)—Alternating between parthenogenetic and sexual-reproduction life cycles.

Hypermorphosis—Excessive growth of some part or appendage of the body.

Hypostome—Sclerotized area anterior to the mouth opening, found in the ground pattern of the Euarthropoda and retained in crustaceans (Maas et al. 2004). See **labrum**.

Instar—A discrete stage in a growth series, delimited by molts; often used synonymously with stage.

Intercoxal sclerite—A ventral, flat, exoskeletal plate uniting the contralateral pair of thoracic limbs in copepods. It typically connects the coxae of a pair of swimming legs (legs 1–5 and occasionally the maxillipeds) and ensures that they beat in unison; also called a "coupler" or "interpodal bar."

Interna—The internal part (a ramified rootlet system) of rhizocephalan cirripedes that develops from an injected vermigon.

Interstitial—Living within the labyrinth of pore spaces (interstices) among sand grains or other sediment particles.

Juvenile—A young organism that has completed its larval stages but differs in size and sexual maturity from the adult. Early juveniles are often referred to as post-larvae.

Kentrogon—The infective larval stage of kentrogonid rhizocephalans (the Cirripedia). It develops from a settled female cyprid and injects the vermigon larva into the host through a stylet.

Labrophora—A clade recognized by some workers to contain the extinct Phosphatocopida and the extinct and extant crown-group crustaceans (i.e., equal to the Mandibulata minus the ateloceratans; see chapter 3).

Labrum—A lobe-like, muscular outgrowth of the posterior surface of the hypostome that extends over the mouth opening, usually as a small flap between the bases of the antennules. Together with the hypostome, it forms a hypostome-labrum complex in some taxa. Evolutionarily, it is probably an outgrowth of the mouth membrane.

Lacinia mobilis—An enlarged, nearly articulated spine of the mandible that is adjacent to the incisor process in some peracarids.

Larva—A broad term referring to any immature, post-embryonic form of an animal that differs morphologically from the adult and develops into the adult either gradually (via anamorphosis) or by more abrupt changes in morphology (metamorphosis).

Laterocaudal spines—Small, caudolaterally directed spines on either side of the caudal margin of the telson in various crustacean larvae.

Lattice organs—Fine, cuticular chemosensory structures, usually in five pairs, found in (and unique to) the head shield of thecostracan cypridoid larvae. Ontogenetically, they develop from setae on the naupliar head shield.

Lecithotrophic—Pertaining to developmental stages (usually nauplii) in which the yolk of an egg provides all of the nourishment for the animal. Such nauplii typically are non-feeding and lack a functional mouth and digestive tract.

Lecithotrophic larva—A larva that lives off yolk supplied via the egg.

Linea anomurica—A longitudinal lateral groove along the dorsolateral part of the carapace that typically defines the upper margin of the branchiostegite in anomurans. Compare with *linea thalassinica*, which is probably homologous.

Linea thalassinica—A longitudinal lateral groove along the dorsolateral part of carapace, extending from the anterior margin near the eye to the posterior margin in most thalassinideans (the Gebiidea and Axiidea).

Manca—A larval stage of direct-developing peracarids, including most isopods and all tanaidaceans and cumaceans. The first manca stage is passed within the marsupium of the mother and is typically recognized by the fact that it lacks the final pereopod seen in adults.

Mandible—The third pair of cephalic appendages in crustaceans, posterior to the antennae and anterior to the maxillules (when present), used by adults in the mastication of food.

Marked-time molting—A type of development in which subsequent molts show an increase in size (sometimes a significant increase) but little or no morphological change. It is possibly a mechanism for delaying metamorphosis to the next larval phase (Gore 1985; Goy 1990; Anger 2001).

Marsupium—A female pouch (brood pouch) in which developing eggs or embryos are brooded. In the Crustacea, it can be formed by a variety of methods (Harrison 1984); most often it is a ventral, basket-like pouch formed by the side branches of the pereopods (oostegites) and the corresponding sternites (e.g., in most of the Peracarida).

Masticatory spine—See **naupliar process**.

Mastigopus—An older term (no longer used) given to Sergestidae (Dendrobranchiata) larvae of the decapodid stage.

Maxilla—The second post-mandibular limb of crustaceans, specialized as a mouthpart in nearly all extant crustaceans (cephalocarids are an exception).

Maxillule—The first post-mandibular limb of crustaceans, specialized as a mouthpart.

Median eye—An eye connected to a paired or unpaired median anterior neuropil of the syncerebrum by one or several median-eye nerves (Richter et al. 2010). It is a simple, light-sensitive organ found in many arthropod groups. See **nauplius eye**.

Megalopa—A functionally and ecologically transitional larval stage between the planktonic zoeal phase and the benthic juvenile stage in crabs (the Brachyura). It is the brachyuran decapodid stage, characterized by having fully functional mouthparts and by swimming via propulsion of the pleopods. The term is also sometimes used for other decapodid stages within the Decapoda (e.g., anomurans). See **decapodid**.

Mesoparasitic—Referring to parasitic copepods that live partly embedded in their host, usually with the anterior end forming an anchor process (a holdfast, rootlet system, endosoma, etc.).

Metamorphosis (metamorphic development)—Although originally referring to any change in form, in biological applications the term now more commonly refers to a profound change in morphology (typically accompanied by behavioral and functional changes) during the life cycle of an organism.

Metanauplius—A naupliar larva characterized by having additional appendages (at least as a bud or seta) on the thorax, and usually with some additional indications of segmentation (or incipient segmentation) beyond the three head-appendage somites. It follows the orthonauplius (if present). In euphausiids, the mandibles at this stage are reduced and are no longer used in swimming.

Metazoea—The third protozoea (particularly in dendro-branchiate shrimps) is often referred to as a metazoea when the remaining thoracopods, although still not functional, are prominent and biramous.

Microniscus—The second larval stage of parasitic isopods in the Bopyroidea and Cryptoniscoidea (epicarideans). This stage feeds on a copepod intermediate host.

Mixotrophic—Pertaining to naupliar development (in copepods), consisting of a combination of early naupliar stages, typically rich in yolk and non-feeding, and later planktotrophic naupliar stages with a fully developed antennary coxal endite, functional mouth, and alimentary tract.

Multiseriate—Referring to egg sacs (in copepods) containing two or more rows of eggs.

Mysis—The larval stage of the Dendrobranchiata in which the carapace covers all of the thoracic somites, and the thoracic appendages (the maxillipeds) are used for locomotion. *Mysis* is also a valid genus name in the mysidacean family Mysidae.

Nährboden—A placenta-like organ within the closed brood chamber of some cladocerans (the Onychopoda and Monidae, and *Penilia*) that provides nutrition for the development of the embryos. This term was first introduced by Weismann (1876–1879), but also see Patt (1947) and Egloff et al. (1997).

Naupliar eye—An eye consisting of a cluster of three or four median eyes that form a single structural unit but are separated from one another by pigment layers (Richter et al. 2010). It is median and unpaired, although early anlagen in some groups appear paired. The naupliar eye is characteristic of the nauplius larva and, in some taxa, persists into the adult stages.

Naupliar process—A process extending from the coxal segment of the antenna or (less frequently) the mandible of some crustacean naupliar larvae; also referred to in the literature as a masticatory spine, gnathobase, arthritic process, coxal endite, or arthrite. It is not uniform in form or function across all crustacean taxa.

Naupliar shield—The fused dorsal cuticle of the three anterior cephalic appendage-bearing somites of the nauplius. It may also extend laterally and posteriorly to cover parts of the naupliar body.

Nauplioid—A form of embryonized larva. The naupliar appendages (the antennules, antennae, and mandibles) are more distinctly developed than the more posterior limbs. Nauplioids are found in many malacostracans that exhibit direct development, such as decapods and some peracarids. The term also refers to the non-swimming developmental stage of a nauplius found in the brood sac of some mysids (Wittmann 1981a; Scholtz 2000; also see chapter 38).

Naupliosoma—A pre-zoeal larval phase found in some Achelata genera. It is characterized by large, biramous antennae provided with several plumose setae that are adapted for swimming, and by coiled-up and functionless maxillipeds (Crosnier 1972).

Nauplius (plural = nauplii)—The earliest crustacean larval phase of extant Crustacea (the Eucrustacea, *sensu* Walossek 1999), consisting of a rounded or oval body, three pairs of functional appendages (the uniramous antennules, biramous antennae, and mandibles), and (typically) a nauplius eye. The antenna often bears a setiferous enditic process (naupliar process) on its coxa, but this may not be present on nauplii that do not feed. Different ontological forms have been referred to as orthonauplius, metanauplius, and the like. See **metanauplius** and **orthonauplius**.

Nauplius eye—See **naupliar eye**.

Neonata—The term means "new born" and is used to describe newly released cladoceran embryos. Neonata are not separate instars, since they cannot be distinguished from later embryos in the brood chamber and no molt separates them.

Nisto—The decapodid phase of scyllarid lobsters (the Achelata; usually restricted to the genus *Scyllarus*). It is a transparent, relatively short-lived phase that shows an adult-like morphology but is dorsoventrally flattened and bears large natatory pleopods.

Nymph—A later, worm-like larval stage of the Pentastomida that is infective to the definitive host. It is distinguished from a primary larva by lacking a penetration apparatus, being more elongate, and having annulations.

Oligoarthra—A collective name for all harpacticoid copepods (excluding the Polyarthra).

Onychopodid—A modified larval stage in the life cycle of *Gonophysema gullmarensis* (Copepoda *incertae sedis*), characterized by a simple, elongate, sac-like body provided with paired grasping antennae, which are used to attach to the skin of its ascidian host prior to penetration. Male onychopodids enter the female via the atrial pore and undergo a drastic metamorphic reduction, so that little tissue except the testes remains. See **cryptogonochorism**.

Oostegites—Thin lamellar lobes (side branches) extending medially from the coxa of adult female pereopods (thoracopods 2–8) that interlock with each other and/or overlap to form the ventral part of the marsupium (brood pouch) in the Peracarida (except the Thermosbaenacea). They are an autapomorphy of the peracarids.

Oral disc—The adhesive disc located anteriorly on the ventral surface of the cephalon of a tantulus larva, often concealed by a surrounding sheathing membrane (apron). This term is used for the permanent attachment of the larval head to its crustacean host during the subsequent development of both sexual adults and the parthenogenetic female.

Ornamentation—A general collective term for cuticular spinulation, denticulation, or markings on the surface of the body or appendages. It is used to refer to surface structures that do not penetrate the integument.

Orsten—A special type of preservation of organisms in calcium phosphate, resulting in uncompressed, three-dimensionally preserved fossils (primarily crustaceans and their larvae) from the Cambrian, approximately 525 to 488 million years ago (Maas et al. 2006).

Orthonauplius—A simple, early-stage nauplius larva with no external signs of segmentation, an undifferentiated hindbody, and no appendage buds other than the antennules, antennae, and mandibles. It is considered autapomorphic for the Eucrustacea (*sensu* Walossek 1999).

Ovigerous spines—The long, paired spinous processes on the ventral surface of the genital double-somite of female monstrillid copepods, used for carrying eggs.

Pantochelis—The first post-embryonic (larval) stage of hyperiid amphipods, in which the undifferentiated pleon lacks pleopods. This stage is passed within the marsupium of the adult female.

Parazoea—The early larval phase of the Bathynellacea.

Pedigerous somite—A thoracic somite bearing a pair of swimming legs.

Peltatulus—A larval stage characteristic of spinicaudatan clam shrimps. The term is applied to a stage approximately in the middle of the larval series, in which the carapace has become visible as a pair of dorsolateral swellings but is still not free of the trunk. This term (used by Kapler 1960; Petrov 1992; and Eder 2002) is derived from *peltastes*, a Greek word for an ancient type of leather shield with a shape roughly like the early carapace (shield) anlagen in spinicaudatans.

Penetration apparatus—A complex of structures at the anterior end of the primary larva of the Pentastomida that presumably facilitates burrowing into the visceral tissues of the first (intermediate) host.

Penetration spine—A short, stiff, chitinous rod that is part of the penetration apparatus in pentastomid primary larvae.

Phase—A morphologically discrete period of larval life; it is often subdivided into a number of stages or instars. Different phases are typically marked by dramatic changes in morphology and in the function of the appendages (e.g., the zoeal and megalopal phases of crabs, or the naupliar and copepodid phases of copepods).

Phyllamphion—A palinurid phyllosoma larva characterized by an expanded cephalon that covers the thorax. These larvae probably belong to *Palinurellus* and *Puerulus* (Sims 1966; Michel 1970b).

Phyllobranch—A type of decapod gill bearing a double series of plate-like branches from the axis.

Phyllosoma—The dorsoventrally flattened, transparent (or translucent), leaf-like planktonic zoeal stage of Achelata lobsters, usually with a very large head shield and a small pleon.

Physosoma—A term used for some post-embryonic stages within the Hyperiidea (Amphipoda). The pereonic region of a physosoma larva is characteristically strongly enlarged, for passive floating in the water column.

Pilier—A stout, blunt spine tipped with minute bumps (thorns)—one anterior and one posterior—found on the dorsal midline of the carapace of eryoneicus larvae. It is possibly homologous to (an extended form of) the dorsal organs of other decapod larvae (Gurney 1942; Bernard 1953).

Planktotrophic, planktotrophic larva—Referring to a planktonic-dispersing larva that derives its nourishment from active feeding in plankton, rather than depending on yolk.

Pleon—The abdomen in malacostracans (and sometimes in other crustaceans); a body tagma composed primitively of six somites bearing pleopods, uropods, and a telson. Many workers employ the term pleon (rather than abdomen) when discussing malacostracans, because it is not thought to be homologous with the abdomen of other crustacean groups.

Pleonite (pleomere)—Somite of the abdomen. See **pleon**.

Podopleans—Copepods in which the major body articulation of the adult is between the fifth and sixth thoracic somites. Often there is a significant difference in size between these two somites, exhibited by members of the orders Misophrioida, Harpacticoida, Cyclopoida, Geleyelloida, and Siphonostomatoida. Podoplean refers to the body plan in which the prosome-urosome boundary lies between the fourth and fifth pedigerous somites. See **gymnopleans**.

Polyarthra—Copepods of the families Canuellidae and Longipediidae, traditionally classified as the most primitive Harpacticoida, and considered by some authors (e.g., Tiemann 1984; Dahms 2004b) as a separate basal clade in the Copepoda.

Posterolateral processes—The often acute, posterior projections of the posterolateral borders of abdominal (pleomeric) somites 2–5 in brachyuran zoeae; these are also seen in other decapod groups.

Post-larva—A broadly used and ill-defined term in larval literature that may include any stage after the last true larval form, especially in groups with a gradual transformation in molts after the final larval phase. For example, in brachyuran crabs this would be the stage following the megalopal (decapodid) phase, and it would also include subsequent juveniles (e.g., the second and third post-larva). Historically, it has often been used to refer to the decapodid larval stage itself; it is still used in that sense by workers in the field of crustacean aquaculture. In this atlas, post-larva is considered synonymous with "early juvenile" and the latter is thus an unnecessary term.

Post-nauplioid—A nauplioid, non-swimming larva within the brood chamber of mysids that bears additional appendages beyond the first three pairs. See **nauplioid**.

Praniza—A larval form of gnathiid isopods specialized for ectoparasitic feeding on fish hosts. The term praniza is used when the stage is satiated (swollen with blood); zuphea is sometimes used for unsatiated (unfed) larva (Smit and Davies 2004).

Pre-oral spine—A stylet-like spine found in some branchiuran genera, presumed to be used to puncture and penetrate the host's integument and promote hemorrhaging (see chapter 24).

Pre-zoea—Often considered an embryonic stage, the pre-zoea is a short-duration hatching stage in decapods (see chapters 53 and 54). The pre-zoea is typically ensheathed in a thin cuticle, with poorly developed pleopods and uropods and with natatory setae that are invaginated (have not yet emerged). It cannot swim (or swims jerkily) and is non-feeding. Some workers consider the pre-zoea to be an artifact of rearing.

Primary larva—The hatching stage of the Pentastomida (sometimes referred to only as "the larva"), distinguished from subsequent (nymphal) stages by having paired short, claw-tipped limbs; a penetration apparatus; and the absence of annulations on the cuticle.

Proctodeum—The posterior ectodermal region of the alimentary canal; the hindgut.

Propelagic stage—The initial lecithotrophic larval stages of stomatopods (typically the first two or three) that remain in the burrow of the female parent.

Prosome—The anterior tagma of the body in copepods (up to the major body articulation), consisting of the cephalosome and either four (podoplean tagmosis) or five (gymnoplean tagmosis) free pedigerous somites.

Protelean—An organism that is parasitic when larval or immature, but free-living as an adult. Examples in the Crustacea include all species in the Tantulocarida and some families in the Copepoda (the Monstrillidae and Thaumatopsyllidae).

Protopleon—A larval stage of hyperiid amphipods in which the pleon has only gradually developed pleopods. This stage is unable to swim.

Protozoea—In the Dendrobranchiata, this refers to a larva in which the carapace does not cover all of the thoracic somites. It is a transitional stage between the nauplius phase and the mysis stage, as the larva still uses its antennules and antennae as locomotory appendages. The mandibles are already feeding appendages. Additionally, this larva possesses a pair of compound eyes and two pairs of functional maxillipeds.

Proximal endite—A movable setiferous endite, located in the membrane medially underneath the basipod. The proximal endite is an autapomorphic structure of Crustacea *sensu lato*, first appearing only on the third cephalic limb in late larval stages.

Pseudibacus—The decapodid phase of palinurid lobsters (the Achelata), usually restricted to the genus *Scyllarides*.

Pseudometanauplius—A larval form that follows the nauplius and precedes the metanauplius, with an elongated abdomen but without the development of any limbs beyond the first three, commonly seen in the development of euphausiids (krill). It typically has an elongate, pear-like shape and is slightly more slender than a larva in the metanauplius stage.

Pseudozoea—The hatching stomatopod larval type of squilloids, gonodactyloids, parasquilloids, and probably eurysquilloids. These larvae have pedunculate eyes, biramous antennules, two pairs of uniramous thoracic appendages (maxillipeds 1 and 2), and four or five pairs of natatory biramous pleopods.

Puerulus—The decapodid phase of palinurid lobsters (the Achelata). This phase resembles the adult lobster in form, although it is somewhat dorsoventrally flattened and has proportionally large natatory pleopods.

Pupal stage—In the Crustacea, it refers to one or more transient stages (pupae) in the post-naupliar phase of development in nicothoid and lernaeopodid parasitic copepods (the Siphonostomatoida), usually attached to the host by a frontal filament.

Regular development—The numerically predominant type of ontogenetic growth and developmental staging that consistently recurs in a given taxon (Gore 1985; Rabalais and Gore 1985).

Scaphocerite—The scale-shaped, usually rigid outer branch (exopod) of the antenna in select malacostracans. In decapod zoeae this is usually flattened and blade-like, with a row of marginal plumose setae.

Scaphognathite—The outer (lateral) lamellar expansion of the maxilla, consisting of the exopod and often the epipod, usually with a marginal row of plumose setae.

Second nauplius—A term restricted to the euphausiid literature, used to describe a naupliar larva bearing a pair of spines at the posterior part of the trunk.

Somite—A segment or division of the body; a composite group of elements (usually exoskeletal, muscular, and nervous) which makes up a serially repeated component of the body (Ferrari and Dahms 2007).

Source segment—A segment from which a limb is patterned by the formation of new segment elements. It is homologous to the formative zone.

Spermatophore—A cuticular capsule containing spermatozoa, transferred from the male to the female during mating.

Stadium—See **instar** and **stage**.

Stage—A period of development (in a given larval phase) between two molts, during which time the exoskeleton does not change. A synonym of stadium and instar.

Stomodeum—The anterior ectodermal portion of the alimentary canal; the foregut.

Stylet—The hollow, cuticular structure of the kentrogon larva, used for injecting a motile, multicellular, vermiform larval stage (the vermigon) into the host. The term is also (rarely) used to refer to one of the articulated outgrowths of the terminal somite that is paired to form the caudal furca in some taxa.

Suprabenthic—Referring to organisms that live near the bottom of the sea.

Synzoea—The pelagic larval stage of stomatopods having a full complement of thoracic appendages (five pairs of maxillipeds and three pairs of pleopods) and abdominal appendages (five pairs of pleopods and paired uropods).

Tagma (plural = tagmata)—A functional and morphologically distinct section of the body of an arthropod, formed by the fusion of two or more somites and defined by a common function, such as in the head, thorax, and abdomen of insects or the cephalothorax of decapod crustaceans.

Tagmosis—The division of the body into functional regions (tagma).

Tantulus larva—The infective, non-feeding stage in the life cycle of the Tantulocarida.

Teleplanic larvae—Larvae capable of (and characterized by) long-distance dispersal, which often also includes delayed metamorphosis, with continued molting and growth increases, but with no morphological change (marked-time molting). These larvae are known primarily in the Decapoda (see chapter 47).

Tessmann's larvae—An older term referring to the a-cyprid larvae of the Laurida (Ascothoracida) found in plankton.

Teloblastic growth—Growth in which segments develop one at a time from a growth zone just ahead of (anterior to) the telson, so that the youngest segment is just anterior to the growth zone.

Telson—The posterior, non-somitic (lacking ganglia, limbs, excretory organs, etc.) portion of the body of a crustacean.

Telson thread—A thread (filament) formed from the embryonic cuticle. It connects the telson of early juvenile stages of both Astacoidea and Parastacoidea crayfishes to the maternal pleopods.

Tergopleura—Broad lateral extensions of each tergite of certain body segments in arthropods, covering the basal parts of limbs.

Thalassinid seta (thalassinid hair)—The second (counting from the outside margin) hair-like, plumose seta on the telson of zoeae in the Stenopodidea, Thalassinidea, Anomura, and Dromiacea. See **anomuran seta**.

Thoracopod—A thoracic limb; one of the paired appendages of each somite of the thorax.

Trichogon—A minute instar formed after a male rhizocephalan cyprid settles on a virginal female parasite. The trichogon then becomes implanted as a cryptic dwarf male (R. Brusca and Brusca 2003). It is probably homologous to the vermigon instar in females.

Trunk—The unsegmented, post-cephalic part of a larva (and, more broadly, of any bilaterian metazoan). It is also sometimes used to refer to adults when there is no clear distinction between the thorax and the abdomen (e.g., in adult notostracans).

Umbilical cord—In crustaceans, a tubular tissue connection found in metamorphosed tantulus larvae and used for transporting nutrients (host body fluids or pre-digested tissues) via the larval head to the developing sexual female or male. It typically degenerates when development is complete.

Uniseriate—Referring to egg sacs containing only a single row of eggs, which often are disc-like in shape.

Urosome—The posterior tagma of the body in copepods, which is located posterior to the major articulation and consists of the fifth pedigerous somite (in podopleans, but not in gymnopleans) and the abdominal somites.

Vermigon—A motile, multicellular, vermiform (worm- or slug-shaped) larval stage that is injected into the host via the stylet of a rhizocephalan kentrogon larva.

Vitelline membrane—A membrane directly adjacent to the outer surface of the plasma membrane of an egg.

Y-cyprid—The cypridoid larva of the Facetotecta. It has a univalved carapace, and a large segmented abdomen and telson.

Y-larvae—The enigmatic larval stages of the Facetotecta, typically including a series of five y-nauplii and one y-cypris stage.

Y-nauplius—The nauplius larva of the Facetotecta, with a distinctly faceted, shield-like dorsal cuticle (naupliar shield).

Yolk—The part of the egg that serves as the food source for the developing embryo and/or subsequent larval stages (in lecithotrophic larvae).

Y-organism—The as-yet-unidentified parental form of the y-larvae (Glenner et al. 2008).

Ypsigon—The unsegmented, slug-like parasitic stage that succeeds the y-larva in at least some facetotectans (Glenner et al. 2008). See **vermigon**.

Zoea (plural = zoeae)—An early, typically planktonic larval phase in the Decapoda. It swims by way of natatory setae on the exopods of the (usually first two) maxillipeds, has an abdomen (pleon) without (or with rudimentary) uropods, and possesses paired compound eyes. In the Dendrobranchiata, the zoea is not an early stage, but instead follows the metanaupliar stage and is synonymous with the mysis stage. Some publications use "zoeas" for the plural.

Zuphea—A name sometimes used to refer to the unsatiated (unfed) praniza larva of gnathiid isopods (Smit and Davies 2004). See **praniza**.

REFERENCES

Abdelhalim, A. I., J. W. Lewis, and G. A. Boxshall. 1991. The life-cycle of *Ergasilus sieboldi* Nordmann (Copepoda: Poecilostomatoida), parasitic on British freshwater fish. Journal of Natural History 25: 559–582.

Abele, L. G. 1970. Semi-terrestrial shrimp (*Merguia rhizophorae*). Nature 226: 661–662.

Abele, L. G., W. Kim, and B. E. Felgenhauer. 1989. Molecular evidence for inclusion of the phylum Pentastomida in the Crustacea. Molecular Biology and Evolution 6: 685–691.

Abrunhosa, F. A., M. A. Pires, J. F. Lima, and P. A. Coelho-Filho. 2006. Developmental morphology of mouthparts and foregut of the larvae and post-larvae of *Lepidophthalmus siriboia* Felder and Rodrigues, 1993 (Decapoda: Callianassidae). Acta Amazonica 36: 335–342.

Abrunhosa, F. A., M. W. P. Santana, and M. A. B. Pires. 2007. The early larval development of the tropical reef lobster *Enoplometopus antillensis* Lütken (Astacidea, Enoplometopidae) reared in the laboratory. Revista Brasileira de Zoologia 24: 382–396.

Abrunhosa, F. A., D. J. B. Simith, C. A. M. Palmeira, and D. C. B. Arruda. 2008. Lecithotrophic behaviour in zoea and megalopa larvae of the ghost shrimp *Lepidophthalmus siriboia* Felder and Rodrigues, 1993 (Decapoda: Callianassidae). Annals of the Brazilian Academy of Sciences 80: 639–646.

Adams, D. K., D. J. McGillicuddy Jr., L. Zamudio, A. M. Thurnherr, X. Liang, O. Rouxel, C. R. German, and L. S. Mullinea. 2011. Surface-generated mesoscale eddies transport deep-sea products from hydrothermal vents. Science 332: 580–583.

Addis, A., F. Biagi, A. Floris, E. Puddu, and M. Carcupino. 2007. Larval development of *Lightiella magdalenina* (Crustacea, Cephalocarida). Marine Biology (Berlin) 152: 1432–1793.

Agar, W. E. 1908. Note on the early development of a cladoceran (*Holopedium gibberum*). Zoologischer Anzeiger 33: 420–427.

Agassiz, A. 1875. Instinct? in hermit crabs. American Journal of Science and Arts, Ser. 3, 10: 290–291.

Ahyong, S. T. 2001. Revision of the Australian Stomatopod Crustacea. Records of the Australian Museum, Suppl. 26: 1–326.

Ahyong, S. T. 2002. Stomatopod Crustacea from the Marquesas Islands: results of MUSORSTOM 9. Zoosystema 24: 347–372.

Ahyong, S. T. 2009. The polychelidan lobsters: phylogeny and systematics (Polychelida: Polychelidae). *In* J. W. Martin, K. A. Crandall, and D. L. Felder, eds. Decapod Crustacean Phylogenetics, 369–398. Crustacean Issues No. 18. Boca Raton, FL: CRC Press.

Ahyong, S. T., and C. Harling. 2000. The phylogeny of the stomatopod Crustacea. Australian Journal of Zoology 48: 607–642.

Ahyong, S. T., and S. N. Jarman. 2009. Stomatopod interrelationships: preliminary results based on analysis of three molecular loci. Arthropod Systematics & Phylogeny 67: 91–98.

Ahyong, S. T., T. Y. Chan, and Y. C. Liao. 2008. A Catalog of the Mantis Shrimps (Stomatopoda) of Taiwan. Taipei, Taiwan, R.O.C.: National Science Council.

Ahyong, S. T., J. K. Lowry, M. Alonso, R. N. Bamber, G. A. Boxshall, P. Castro, S. Gerken, G. S. Karaman, J. W. Goy, D. S. Jones, K. Meland, D. C. Rogers, and J. Svavarsson. 2011. Subphylum Crustacea Brünnich, 1772. *In* Z.-Q. Zhang, ed. Animal biodiversity: an outline of higher-level classification and survey of taxonomic richness. Special Issue, Zootaxa 3148: 165–191.

Aikawa, H. 1929. On larval forms of some Brachyura. Records of Oceanographic Works in Japan 2: 17–55.

Akiyama, T., and M. Yamamoto. 2004a. Life history of *Nippoleucon hinumensis* (Crustacea: Cumacea: Leuconidae) in Seto Inland Sea of Japan, 1: summer diapauses and molt cycle. Marine Ecology Progress Series 284: 211–225.

Akiyama, T., and M. Yamamoto. 2004b. Life history of *Nippoleucon hinumensis* (Crustacea: Cumacea: Leuconidae) in Seto Inland Sea of Japan, 2: non-diapausing subpopulation. Marine Ecology Progress Series 284: 227–235.

Aldred, N., and A. S. Clare. 2008. The adhesive strategies of cyprids and development of barnacle-resistant marine coatings. Biofouling 24: 351–363.

Alekseev, V. R. 2002. Copepoda. *In* C. H. Fernando, ed. A Guide to Tropical Freshwater Zooplankton: Identification, Ecology and Impact on Fisheries, 123–188. Leiden, Netherlands: Backhuys.

Alikunhi, K. H. 1952. An account of the stomatopod larvae of the Madras plankton. Records of the Indian Museum 49: 239–319.

Alikunhi, K. H. 1967. An account of the post-larval development, moulting and growth of the common stomatopods of the Madras coast. *In* Proceedings of the Symposium on Crustacea, Held at Ernakulam from January 12–15, 1965, pt. 2: 824–939. Mandapam Camp: Marine Biological Association of India.

Almeida, W. O., and M. L. Christoffersen. 1999. A cladistic approach to relationships in Pentastomida. Journal of Parasitology 85: 695–704.

Almeida, W. O., A. T. Silva-Souza, and D. L. Sales. 2010. Parasitism of *Phalloceros harpagos* (Cyprinodontiformes: Poeciliidae) by *Sebekia oxycephala* (Pentastomida: Sebekidae) in the headwaters of the Cambé River, Paraná State, Brazil. Brazilian Journal of Biology 70: 457–458.

Almeida Prado-Por, M. S. 1966. Notes on the development stages of *Schistomysis spiritus* (Norman, 1860). Anais da Academia Brasileira de Ciências 38: 349–353.

Alston, S., G. A. Boxshall, and J. W. Lewis. 1996. The life-cycle of *Ergasilus briani* Markewitsch, 1933 (Copepoda: Poecilostomatoida). Systematic Parasitology 35: 79–110.

Amaratunga, T., and S. Corey. 1975. Life history of *Mysis stenolepis* Smith (Crustacea, Mysidacea). Canadian Journal of Zoology 53: 942–952.

Anderson, D. T. 1967. Larval development and segment formation in the branchiopod crustaceans *Limnadia stanleyana* King (Conchostraca) and *Artemia salina* (L.) (Anostraca). Australian Journal of Zoology 15: 47–91.

Anderson, D. T. 1973. Embryology and Phylogeny in Annelids and Arthropods. New York: Pergamon Press.

Anderson, D. T. 1982. Embryology. *In* L. G. Abele, ed. The Biology of Crustacea. Vol. 2, Embryology, Morphology, and Genetics, 1–41. New York: Academic Press.

Anderson, D. T. 1986. The circumtropical barnacle *Tetraclita divisa* (Nilsson-Cantell) (Balanomorpha, Tetraclitidae): cirral activity and larval development. Proceedings of the Linnaean Society of New South Wales 109: 107–116.

Anderson, D. T. 1994. Barnacles: Structure, Function, Development and Evolution. London: Chapman & Hall.

Anderson, G. 2010a. Lophogastrida Classification, January, 2010. Online at http://peracarida.usm.edu/LophogastridaTaxa.pdf.

Anderson, G. 2010b. Mysida Classification, January, 2010. Online at http://peracarida.usm.edu/MysidaTaxa.pdf.

Anderson, L. I., and N. H. Trewin. 2003. An Early Devonian arthropod fauna from the Windyfield cherts, Aberdeenshire, Scotland. Palaeontology 46: 467–509.

Anderson, L. I., W. R. B. Crighton, and H. Hass. 2004. A new univalve crustacean from the Early Devonian Rhynie chert hot-spring complex. Transactions of the Royal Society of Edinburgh, Earth Sciences 94: 355–369.

Andrews, E. A. 1907. The young of the crayfishes *Astacus* and *Cambarus*. Smithsonian Contributions to Knowledge No. 35. Washington, DC: Smithsonian Institution.

Andrews, T. F. 1948. The parthenogenetic reproductive cycle of the cladoceran *Leptodora kindtii*. Transactions of the American Microscopic Society 67: 54–60.

Andryszak, B. L. 1986. *Upogebia affinis* (Say): its post-larval stage described from Louisiana plankton, with a comparison to post-larvae of other species within the genus and notes on its distribution. Journal of Crustacean Biology 6: 214–226.

Anger, K. 1996. Physiological and biochemical changes during lecithotrophic larval development and early juvenile growth in the northern stone crab, *Lithodes maja* (Decapoda: Anomura). Marine Biology (Berlin) 126: 283–296.

Anger, K. 2001. The Biology of Decapod Crustacean Larvae. Crustacean Issues No. 14. Lisse, Netherlands: A. A. Balkema.

Anger, K., M. Montú, C. D. Bakker, and L. Loureiro Fernandes. 1990. Larval development of *Uca thayeri* Rathbun, 1900 (Decapoda: Ocypodidae) reared in the laboratory. Meeresforschung 32: 276–294.

Anker, A., S. T. Ahyong, P. Y. Noel, and A. R. Palmer. 2006. Morphological phylogeny of alpheid shrimps: parallel preadaptation and the origin of a key morphological innovation, the snapping claw. Evolution 60: 2507–2528.

Antezana, T., and C. Melo. 2008. Larval development of Humboldt Current krill, *Euphausia mucronata* G. O. Sars 1883 (Malacostraca: Euphausiacea). Crustaceana 81: 305–28.

Araujo, P. B., A. F. Quadros, M. M. Augusto, and G. Bond-Buckup. 2004. Postmarsupial development of *Atlantoscia floridana* (van Name, 1940) (Crustacea, Isopoda, Oniscidea): sexual differentiation and size at onset of sexual maturity. Invertebrate Reproduction and Development 45: 221–230.

Aste, A., and M. A. Retamal. 1984. Desarollo larval de *Callianassa uncinata* H. Milne-Edwards, 1837 (Decapoda, Callianassidae) bajo condiciones de laboratorio. Gayana Zoología 48: 41–56.

Atienza, D., E. Saiz, A. Skovgaard, I. Trepat, and A. Calbet. 2008. Life history and population dynamics of the marine cladoceran *Penilia avirostris* (Branchiopoda: Cladocera) in the Catalan Sea (NW Mediterranean). Journal of Plankton Research 30: 345–357.

Atkins, D. 1954. Leg disposition in the brachyuran megalopa when swimming. Journal of the Marine Biological Association of the United Kingdom 33: 627–636.

Avenant, A., J. G. van As, and G. C. Loots. 1989. On the hatching and morphology of *Dolops ranarum* larvae (Crustacea: Branchiura). Journal of Zoology (London) 217: 511–519.

Avenant-Oldewage, A., and L. Everts. 2010. *Argulus japonicus*: sperm transfer by means of a spermatophore on *Carassius auratus* (L.). Experimental Parasitology 126: 232–238.

Avenant-Oldewage, A., and E. Knight. 1994. A diagnostic species compendium of the genus *Chonopeltis* Thiele, 1900 (Crustacea: Branchiura) with notes in its geographical distribution. Koedoe 37: 41–56.

Baba, K., E. Macpherson, G. C. B. Poore, S. T. Ahyong, A. Bermudez, P. Cabezas, C.-W. Lin, M. Nizinski, C. Rodrigues, and K. E. Schnabel. 2008. Catalogue of squat lobsters of the world (Crustacea: Decapoda: Anomura—families Chirostylidae, Galatheidae and Kiwaidae). Zootaxa No. 1905. Auckland, NZ: Magnolia Press.

Baba, K., Y. Fujita, I. S. Wehrtmann, and G. Scholtz. 2011. Developmental biology of squat lobsters. *In* G. C. B. Poore, S. T. Ahyong, and J. Taylor (eds.), The Biology of Squat Lobsters. 105–148. Boca Raton, FL: CRC Press; Collingwood, Victoria, Australia: CSIRO Publishing.

Baeza, J. A., C. D. Schubart, P. Zillner, S. Fuentes, and R. T. Bauer. 2009. Molecular phylogeny of shrimps from the genus *Lysmata* (Caridea: Hippolytidae): the evolutionary origins of protandric simultaneous hermaphroditism and social monogamy. Biological Journal of the Linnean Society 96: 415–424.

Baker, A. de, B. P. Boden, and E. Brinton. 1990. A Practical Guide to the Euphausiids of the World. London: Natural History Museum.

Baldass, F. 1937. Entwicklung von *Holopedium gibberum*. Zoologische Jarhbücher, Abteilung für Anatomie und Ontogenie der Tiere 63: 399–454.

Baldass, F. 1941. Entwicklung von *Daphnia pulex*. Zoologische Jarhbücher, Abteilung für Anatomie und Ontogenie der Tiere 67: 1–60.

Banaja, A. A., J. L. James, and J. Riley. 1975. An experimental investigation of a direct life-cycle in *Reighardia sterna* (Diesing, 1864), a pentastomid parasite of the herring gull (*Larus argentatus*). Parasitology 71: 493–503.

Bannister, J. L., and J. R. Grindley. 1966. Notes on *Balaenophilus unisetus* P. O. C. Aurivillius, 1879, and its occurrence in the southern hemisphere (Copepoda, Harpacticoida). Crustaceana 10: 296–302.

Baqai, I. U. 1963. Studies on the post-embryonic development on the fairy shrimp *Streptocephalus seali* Ryder. Tulane Studies in Zoology 10: 91–120.

Barker, D. 1962. A study of *Thermosbaena mirabilis* (Malacostraca, Peracarida) and its reproduction. Quarterly Journal of Microscopical Science 103: 261–286.

Barker, M. F. 1976. Culture and morphology of some New Zealand barnacles (Crustacea: Cirripedia). New Zealand Journal of Marine and Freshwater Research 10: 139–158.

Barlow, D. I., and M. A. Sleigh. 1980. The propulsion and use of water currents for swimming and feeding in larval and adult *Artemia*. In G. Persoone, P. Sorgeloos, O. Roels, and E. Jaspers, eds. The Brine Shrimp *Artemia*. Vol. 1, Morphology, Genetics, Radiobiology, Toxicology, 61–73. Wetteren, Belgium: Universa Press.

Barnard, J. L., and G. S. Karaman. 1991. The families and genera of marine gammaridean Amphipoda (except marine gammaroids), pt. 2. Records of the Australian Museum, Suppl. 13: 419–866.

Barnard, K. H. 1955. South African parasitic Copepoda. Annals of the South African Museum, Cape Town 41: 223–314.

Barnett, B. M., Hartwick, R. F., and Milward, N. E. 1986. Descriptions of the nisto stage of *Scyllarus demani* Holthuis, two unidentified *Scyllarus* species, and the juvenile of *Scyllarus martensii* Pfeffer (Crustacea: Decapoda: Scyllaridae), reared in the laboratory; and behavioural observations of the nistos of *S. demani*, *S. martensii*, and *Thenus orientalis* (Lund). Australian Journal of Marine and Freshwater Research 37: 595–608.

Barrientos, Y., and M. S. Laverack. 1986. The larval crustacean dorsal organ and its relationship to the trilobite median tubercle. Lethaia 19: 309–313.

Barton, D. P. 2007. Pentastomid parasites of the introduced Asian house gecko, *Hemidactylus frenatus* (Gekkonidae), in Australia. Comparative Parasitology 74: 254–259.

Bartsch, I. 1996. Parasites of the Antarctic brittle star *Ophiacantha disjuncta* (Ophiacanthidae, Ophiuroidea): redescription of the copepod *Lernaeosaccus ophiacanthae* Heegaard, 1952. Mitteilungen aus dem Hamburgischen Zoologischen Museum und Institut 93: 63–72.

Bate, C. S. 1858. On *Praniza* and *Anceus*, and their affinity to each other. Annals and Magazine of Natural History, ser. 3, 2: 165–172.

Bate, C. S. 1882. *Eryoneicus*, a new genus allied to *Willemoesia*. Annals and Magazine of Natural History, ser. 5, 10: 456–458.

Bate, C. S. 1888. Report on the Crustacea Macrura collected by H.M.S. *Challenger* during the years 1873–76. Vol. 24, *in* C. W. Thomson and J. Murray, eds. Reports on the Scientific Results of the Voyage of H.M.S. *Challenger* during the Years 1873–76, Zoology. London: Eyre & Spottiswoode.

Bauer, R. T. 2004. Remarkable shrimps: adaptations and natural history of the Carideans. Norman: University of Oklahoma Press.

Belk, D. 1984. Antennal appendages and reproductive success in the Anostraca. Journal of Crustacean Biology 4: 66–71.

Belk, D., and F. R. Schram. 2001. A new species of anostracan from the Miocene of California. Journal of Crustacean Biology 21: 49–55.

Belmonte, G. 2005. Y-nauplii (Crustacea, Thecostraca, Facetotecta) from coastal waters of the Salento Peninsula (southeastern Italy, Mediterranean Sea) with descriptions of four new species. Marine Biology Research 1: 254–266.

Benesch, R. 1969. Zur Ontogenie und Morphologie von *Artemia salina* L. Zoologische Jahrbücher, Abteilung für Anatomie und Ontogenie der Tiere 86: 307–458.

Ben Hassine, O. K. 1983. Les copépodes parasites des poissons Mugilidae en Méditerranée occidentale (côtes françaises et tunisiennes): morphologie, bio-écologie, cycles évolutifs. Thèse d'état, Université de Montpellier II.

Benkirane, O. 1987. Recherches sur l'organe de fixation des Lernaeopodidae (Copepoda, Siphonostomatoida). Thèse 3e cycle, Université de Montpellier II.

Bennike, O. 1998. Fossil egg sacs of *Diaptomus* (Crustacea: Copepoda) in Late Quaternary lake sediments. Journal of Paleolimnology 19: 77–79.

Berggren, M. 1993. *Spongiocaris hexactinellicola*, a new species of stenopodidean shrimp (Decapoda: Stenopodidae) associated with hexactinellid sponges from Tartar Bank, Bahamas. Journal of Crustacean Biology. 13: 784–792.

Bergh, R. S. 1893. Beiträge zur Embryologie der Crustaceen, 1: zur Bildungsgeschichte des Keimstreifens von Mysis. Zoologische Jahrbücher, Abteilung für Anatomie und Ontogenie der Tiere 6: 491–526.

Bernard, F. 1953. Decapoda Eryonidae (*Eryoneicus* et *Willemoesia*). Dana Report No. 37. Copenhagen: Høst.

Berrill, M. 1969. The embryonic behavior of the mysid shrimp, *Mysis relicta*. Canadian Journal of Zoology 47: 1217–1221.

Berrill, M. 1971. The embryonic development of the avoidance reflex of *Neomysis americana* and *Praunus flexuosus* (Crustacea: Mysidacea). Animal Behavior 19: 707–713.

Berrill, M. 1975. The burrowing, aggressive and early larval behaviour of *Neaxius vivesi* (Bouvier) (Decapoda, Thalassinidea). Crustaceana 29: 92–98.

Berry, P. F. 1969. The Biology of *Nephrops andamanicus* Wood-Mason (Decapoda, Reptantia). Oceanic Research Institute Investigational Report No. 22. Durban, South Africa: Oceanographic Research Institute.

Berry, W. 1926. Description and notes on the life history of a new species of *Eulimnadia*. American Journal of Science 65: 429–433.

Bielecki, J., B. K. K. Chan, J. T. Høeg, and A. Sari. 2009. Antennular sensory organs in cyprids of balanomorphan cirripedes: standardizing terminology using *Megabalanus rosa*. Biofouling 25: 203–214.

Birstein, Y. A., and N. A. Zarenkov. 1970. Bottom decapods

(Crustacea Decapoda) of the Kurile-Kamchatka Trench area. Akademiya Nauk SSSR, Trudy Instituta Okeanologii Imeni P. P. Shirshova 86: 420–426.

Bishop, J. D. D. 1982. The growth, development and reproduction of a deep sea cumacean (Crustacea: Peracarida). Zoological Journal of the Linnean Society 74: 359–380.

Bishop, J. D. D., and S. Shalla. 1994. Discrete seasonal reproduction in an abyssal peracarid crustacean. Deep Sea Research, pt. 1, 41: 1789–1900.

Björnberg, T. K. S. 1965. Observations on the development and the biology of the Miracidae Dana (Copepoda: Crustacea). Bulletin of Marine Science 15: 512–520.

Björnberg, T. K. S. 1972. Developmental Stages of Some Tropical and Subtropical Planktonic Marine Copepods. Studies on the Fauna of Curaçao and Other Caribbean Islands 40, No. 136. The Hague: M. Nijhoff.

Blazewicz-Paszkowycz, M., R. Bamber, and G. Anderson. 2012. Diversity of Tanaidacea (Crustacea: Peracarida) in the world's oceans: how far have we come? PLoS One 7: e33068.

Boas, J. E. V. 1883. Studien über die Verwandtschaftsbeziehungen der Malakostraken. Morphologisches Jahrbuch 8: 485–579.

Bocquet, C. 1953. Sur un copépode harpacticoïde mineur, Diarthrodes feldmanni, n. sp. Bulletin de la Société Zoologique de France 78: 101–105.

Bocquet, C., J. Bocquet-Védrine, and J. P. L'Hardy. 1970. Contribution à l'étude du développement des organes génitaux chez Xenocoeloma alleni (Brumpt), copépode parasite de Polycirrus caliendrum Claparède. Cahiers de Biologie Marine 11: 195–208.

Boden, B. P. 1950. The post-naupliar stages of the crustacean Euphausia pacifica. Transactions of the American Microscopy Society 69: 373–381.

Boikova, O. S. 2005. Postembryonic development in Diaphanosoma brachyurum (Lievin, 1848) (Crustacea: Ctenopoda: Sididae). Hydrobiologia 537: 7–14.

Boikova, O. S. 2008. Comparative investigation of the late embryogenesis of Leptodora kindtii (Focke, 1844) (Crustacea: Branchiopoda), with notes on types of embryonic development and larvae in Cladocera. Journal of Natural History 42: 2389–2416.

Bonnier, J. 1903. Sur deux types nouveaux d'épicarides parasites d'un cumacé et d'un schizopode. Comptes Rendus Hebdomadaires des Séances de l'Académie des Sciences 136: 102–103.

Bookhout, C. G., and J. D. Costlow. 1977. Larval development of Callinectes similis reared in the laboratory. Bulletin of Marine Science 27: 704–728.

Bookhout, C. G., and J. D. Costlow. 1979. Larval development of Pilumnus dasypodus and Pilumnus sayi reared in the laboratory (Decapoda, Brachyura, Xanthidae). Crustaceana, Suppl. 5: 1–16.

Booth, J. D., W. R. Webber, H. Sekiguchi, and E. Coutures. 2005. Diverse larval recruitment strategies within the Scyllaridae. New Zealand Journal of Marine and Freshwater Research 39: 581–592.

Borgstrøm, R., and P. Larsson. 1974. The first three instars of Lepidurus arcticus (Pallas), (Crustacea: Notostraca). Norwegian Journal of Zoology 22: 45–52.

Borutzky, E. V. 1929. Zur Frage über den Ruhezustand bei Copepoda–Harpacticoida: Dauereier bei Canthocamptus arcticus Lilljeborg. Zoologischer Anzeiger 83: 225–233.

Boschi, E. E., and M. A. Scelzo. 1969. Nuevas Campañas Exploratorias Camaroneras en el Litoral Argentino, 1967–1968: Con Referencias al Plancton de la Región. Proyecto de Desarrollo Pesquero, Servicio Información Técnica 16. Mar del Plata, Argentina: Proyecto de Desarrollo Pesquero.

Boschi, E. E., B. Goldstein, and M. A. Scelzo. 1968. Metamorfosis del crustáceo Blepharipoda doelloi Schmitt de las aguas de la provincia de Buenos Aires (Decapoda, Anomura, Albuneidae). Physis (Buenos Aires) 27: 291–311.

Botnariuc, N. 1947. Contributions à la connaissance des phyllopodes conchostracés de Roumanie. Notationes Biologicae 5: 68–158.

Botnariuc, N., and T. Orghidan. 1953. Phyllopoda. Vol. 4, fasc. 2, Fauna Republicii Populare Române: Crustacea. Bucharest: Editura Academiei Republicii Populare Române.

Botnariuc, N., and N. Viña Bayés. 1977. Contribution à la connaissance de la biologie de Cyclestheria hislopi (Baird), (Conchostraca: Crustacea) de Cuba. In T. Orghidan, ed. Résultats des Expéditions Biospéléogiques Cubano-Roumaines à Cuba, 257–262. Bucharest: Editura Academiei Republicii Socialiste Romania.

Bouchet, G C. 1985. Redescription of Argulus varians Bere, 1936 (Branchiura, Argulidae) including a description of its early development and first larval stage. Crustaceana 49: 30–35.

Bouligand, Y. 1960. Notes sur la famille des Lamippidae, pt. 1. Crustaceana 1: 258–278.

Bourdillon-Casanova, L. 1960. Le méroplancton du Golfe de Marseille: les larves de crustacés décapodes. Recueil des Travaux de Station Marine d'Endoume 30: 1–286.

Bousquette, G. D. 1980. The larval development of Pinnixa longipes (Lockington, 1877) (Brachyura: Pinnotheridae), reared in the laboratory. Biological Bulletin 159: 592–605.

Bouvier, E.-L. 1905. Palinurides et eryonides recueillis dans l'Atlantique oriental pendant les campagnes de l'Hirondelle et de la Princesse-Alice. Bulletin du Musée Océanographique de Monaco 28: 1–7.

Bowman, T. E., and T. M. Iliffe. 1985. Mictocaris halope, a new unusual peracaridan crustacean from marine caves on Bermuda. Journal of Crustacean Biology 5: 58–73.

Bowman, T. E., J. Yager, and T. M. Iliffe. 1985a. Speonebalia cannoni, n. gen., n. sp., from the Caicos Islands, the first hypogean leptostracan (Nebaliacea: Nebaliidae). Proceedings of the Biological Society of Washington 98: 439–446.

Bowman, T. E., S. P. Garner, R. R. Hessler, T. M. Iliffe, and H. L. Sanders. 1985b. Mictacea, a new order of Crustacea Peracarida. Journal of Crustacean Biology 5: 74–78.

Boxshall, G. A. 1974. The population dynamics of Lepeophtheirus pectoralis (Müller): seasonal variation in abundance and age structure. Parasitology 69: 361–371.

Boxshall, G. A. 1990. Precopulatory mate guarding in copepods. Bijdragen tot de Dierkunde 60: 209–213.

Boxshall, G. A. 1991. A review of the biology and phylogenetic relationships of the Tantulocarida, a subclass of Crustacea recognized in 1983. Verhandlungen der Deutschen Zoologischen Gesellschaft 84: 271–279.

Boxshall, G. A. 1992. Copepoda. In F. W. Harrison and A. G. Humes, eds. Microscopic Anatomy of Invertebrates. Vol. 9, Crustacea, chapter 7. New York: Wiley-Liss.

Boxshall, G. A., and D. Defaye. 1996. Classe des Mystacocarides (Mystacocarida Pennak et Zinn, 1943). In J. Forest, ed. Traité de Zoologie: Anatomie, Systématique, Biologie. Vol. 7, fasc. 2, Crustacés: Généralités (suite) et Systématique (Céphalocarides à Syncarides), 409–424. Paris: Masson.

Boxshall, G. A., and S. H. Halsey. 2004. An Introduction to Copepod Diversity, 2 vols. Andover, UK: Ray Society.

Boxshall, G. A., and R. Huys. 1989. New tantulocarid, *Stygotantulus stocki*, parasitic on harpacticoid copepods, with an analysis of the phylogenetic relationships within the Maxillopoda. Journal of Crustacean Biology 9: 126–140.

Boxshall, G. A., and R. Huys. 1998. The ontogeny and phylogeny of copepod antennules. Philosophical Transactions of the Royal Society of London, B 353: 765–786.

Boxshall, G. A., and D. Jaume. 2009. Exopodites, epipodites and gills in crustaceans. Arthropod Systematics & Phylogeny 67: 229–254.

Boxshall, G. A., and R. J. Lincoln. 1983. Tantulocarida, a new class of Crustacea ectoparasitic on other crustaceans. Journal of Crustacean Biology 3: 1–16.

Boxshall, G. A., and R. J. Lincoln. 1987. The life cycle of the Tantulocarida (Crustacea). Philosophical Transactions of the Royal Society of London, B 315: 267–303.

Boxshall, G. A., and W. Vader. 1993. A new genus of Tantulocarida (Crustacea) parasitic on an amphipod host from the North Sea. Journal of Natural History 27: 977–988.

Boxshall, G. A., R. Huys, and R. J. Lincoln. 1989. A new species of *Microdajus* (Crustacea: Tantulocarida) parasitic on a tanaid in the northeastern Atlantic, with observations on *M. langi* Greve. Systematic Parasitology 14: 17–30.

Boyd, C. M. 1960. The larval stages of *Pleuroncodes planipes* Stimpson (Crustacea, Decapoda, Galatheidae). Biological Bulletin 118, 17–30.

Boyko, C. B. 2006. New and historical records of polychelid lobsters (Crustacea: Decapoda: Polychelidae) from the Yale Peabody Museum collections. Bulletin of the Peabody Museum of Natural History 47: 37–48.

Boyko, C. B., and P. A. McLaughlin. 2010. Annotated checklist of anomuran decapod crustaceans of the world (exclusive of the Kiwaoidea and families Chirostylidae and Galatheidae of the Galatheoidea), 4: Hippoidea. Raffles Bulletin of Zoology, Suppl. 23: 139–151.

Boyko, C. B., and J. D. Williams. 2009. Crustacean parasites as phylogenetic indicators in decapod evolution. *In* J. W. Martin, K. A. Crandall, and D. L. Felder, eds. Decapod Crustacean Phylogenetics, 197–220. Crustacean Issues No. 18. Boca Raton, FL: CRC Press.

Braband, A., S. Richter, R. Hiesel, and G. Scholtz. 2002. Phylogenetic relationships within the Phyllopoda (Crustacea, Branchiopoda) based on mitochondrial and nuclear markers. Molecular Phylogenetics and Evolution 25: 229–244.

Bracken, H. D., S. De Grave, and D. L. Felder. 2009a. Phylogeny of the infraorder Caridea based on mitochondrial and nuclear genes (Crustacea: Decapoda). Advances in Decapod Phylogenetics. *In* J. W. Martin, K. A. Crandall, and D. L. Felder, eds. Decapod Crustacean Phylogenetics, 281–308. Crustacean Issues No. 18. Boca Raton, FL: CRC Press.

Bracken, H. D., A. Toon, D. L. Felder, J. W. Martin, M. Finley, J. Rasmussen, F. Palero, and K. A. Crandall. 2009b. The decapod tree of life: compiling the data and moving toward a consensus of decapod evolution. Arthropod Systematics & Phylogeny 67: 99–116.

Bracken-Grissom, H. D., D. L. Felder, N. L. Vollmer, J. W. Martin, and K. A. Crandall. 2012. Phylogenetics links monster larva to deep-sea shrimp. Ecology and Evolution 2012. doi:10.1002 /ece3.347.

Bradford, R. W., B. D. Bruce, S. M. Chiswell, J. D. Booth, A. Jeffs, and S. Wotherspoon. 2005. Vertical distribution and diurnal migration patterns of *Jasus edwardsii* phyllosomas off the east coast of the North Island, New Zealand. New Zealand Journal of Marine and Freshwater Research 39: 593–604.

Brady, G. S. 1894. On *Fucitrogus Rhodymeniæ*, a gall-producing copepod. Journal of the Royal Microscopical Society 1894: 168–170.

Brahm, C., and S. R. Geiger. 1966. On the biology of the pelagic crustacean *Nebaliopsis typica* G. O. Sars. Bulletin of the Southern California Academy of Sciences 65: 41–46.

Branch, G. M. 1974. *Scutellidium patellarum*, n. sp., a harpacticoid copepod associated with *Patella* spp. in South Africa, and a description of its larval development. Crustaceana 26: 179–200.

Bratcik, R. J. 1980. Morfologicheskiye osobennosti postembrionalnogo razvitiya *Caenestheria* sp. (Conchostraca, Cyzicidae). Trudy Instituta Biologii Vnutrennikh vod Akademii Nauk SSSR 44/47: 66–71.

Brattström, H. 1948. On the Larval Development of the Ascothoracid *Ulophysema öresundense* Brattström. Vol. 2, Studies on *Ulophysema öresundense*. Undersökningar över Öresund No. 33. Kungliga Fysiografiska Sällskapets i Lund Förhandlingar, n.s., 59, No. 5. Lund, Sweden: C. W. K. Gleerup.

Brauer, F. 1874. Vorlaufige Mitteilungen über die Entwicklung und Lebensweise des *Lepidurus productus*. Sitzungsberichte der Kaiserlichen Akademie der Wissenschaften, Mathematisch-Naturwissenschaftliche Classe 69: 130–141.

Brendonck, L., D. C. Rogers, J. Olesen, S. Weeks, and W. R. Hoeh. 2008. Global diversity of large branchiopods (Crustacea: Branchiopoda) in freshwater. Hydrobiologia 595: 167–176.

Bresciani, J. 1961. Some features of the larval development of *Stenhelia* (*Delavalia*) *palustris* Brady, 1868 (Copepoda Harpacticoida). Videnskabelige Meddelelser fra Dansk Naturhistorisk Forening i Kjøbenhavn 123: 237–247.

Bresciani, J. 1986. The fine structure of the integument of free-living and parasitic copepods: a review. Acta Zoologica (Stockholm) 67: 125–145.

Bresciani, J., and J. Lützen. 1960. *Gonophysema gullmarensis* (Copepoda Parasitica): an anatomical and biological study of an endoparasite living in the ascidian *Ascidiella aspersa*, 1: anatomy. Cahiers de Biologie Marine 1: 157–184.

Bresciani, J., and J. Lützen. 1961. *Gonophysema gullmarensis* (Copepoda Parasitica): an anatomical and biological study of an endoparasite living in the ascidian *Ascidiella aspersa*, 2: biology and development. Cahiers de Biologie Marine 2: 347–371.

Bresciani, J., and J. Lützen. 1962. Parasitic copepods from the west coast of Sweden including some new or little known species. Videnskabelige Meddelelser fra Dansk Naturhistorisk Forening 124: 367–408.

Bresciani, J., and J. Lützen. 1974. On the biology and development of *Aphanodomus* Wilson (Xenocoelomidae), a parasitic copepod of the polychaete *Thelepus cincinnatus*. Videnskabelige Meddelelser fra Dansk Naturhistorisk Forening 137: 25–63.

Brian, A. 1919. Sviluppo larvale della *Psamathe longicauda* Ph. e dell'*Harpacticus uniremis* Kröy. (Copepodi Harpacticoidi) (descrizione della serie copepodiforme). Atti della Società Italiana di Scienze Naturali e del Museo Civico di Storia Naturale 58: 29–58.

Brickner, I., and J. T. Høeg. 2010. Antennular specialization in cyprids of coral associated barnacles. Journal of Experimental Marine Biology and Ecology 392: 115–124.

Briggs, D. E. G., M. D. Sutton, D. J. Siveter, and D. J. Siveter. 2003. A new phyllocarid (Crustacea: Malacostraca) from the Silurian Fossil-Lagerstätte of Hertfordshire, UK. Proceedings of the Royal Society of London, B 271: 131–138.

Brinton, E. 1962. The distribution of Pacific euphausiids. Bulletin of the Scripps Institution of Oceanography 8: 51–270.

Brinton, E. 1975. Euphausiids of Southeast Asian Waters: Scientific Results of Marine Investigations of the South China Sea and the Gulf of Thailand, 1959–1961. NAGA Report No. 4. La Jolla: University of California, Scripps Institution of Oceanography.

Brinton, E. 1987. A new abyssal euphausiid *Thysanopoda minyops*, with comparisons of the eye size, photophores, and associated structures among deep-living species. Journal of Crustacean Biology 7: 636–666.

Brinton, E., M. D. Ohman, A. W. Townsend, M. Knight, and A. Bridgeman. 1999. Euphausiids of the World Ocean. World Biodiversity Database CD-ROM Series, Windows Version 1.0. Amsterdam: Expert Center for Taxonomic Identification.

Brodeur, R., and O. Yamamura, eds. 2005. Micronekton of the North Pacific. PICES Science Report No. 30. Sidney, BC: North Pacific Marine Science Organization.

Brodie, R. J. 1999. Ontogeny of shell-related behaviors and transition to land in the terrestrial hermit crab, *Coenobita compressus* H. Milne-Edwards. Journal of Experimental Marine Biology and Ecology 241: 67–80.

Brodie, R. [J.], and A. W. Harvey. 2001. Larval development of the land hermit crab *Coenobita compressus* H. Milne-Edwards reared in the laboratory. Journal of Crustacean Biology 21: 715–732.

Brooks, H. K. 1969. Palaeostomatopoda. *In* R. C. Moore, ed. Treatise on Invertebrate Paleontology, pt. R, Arthropoda 4, vol. 2 Decapoda, R533–R535. Lawrence: University of Kansas Press and Geological Society of America.

Brooks, W. K. 1886. Report on the Stomatopoda collected by H.M.S. *Challenger* during the years 1873–76. Vol. 16, *in* C. W. Thomson and J. Murray, eds. Reports on the Scientific Results of the Voyage of H.M.S. *Challenger* during the Years 1873–76, Zoology. London.

Brooks, W. K., and F. H. Herrick. 1891. The embryology and metamorphosis of the Macrura. Memoirs of the National Academy of Science 5: 321–576.

Brossi-Garcia, A. L. 1987. Morphology of the larval stages of *Clibanarius sclopetarius* (Herbst, 1796) (Decapoda, Diogenidae) reared in the laboratory. Crustaceana 52, 251–275.

Browman, H. I., S. Kruse, and J. O'Brien. 1989. Foraging behavior of the predaceous cladoceran, *Leptodora kindtii*, and escape responses of their prey. Journal of Plankton Research 11: 1075–1088.

Brown, A. C., and M. S. Talbot. 1972. The ecology of the sandy beaches of the Cape Peninsula, South Africa, 3: a study of *Gastrosaccus psammodytes* Tattersall (Crustacea: Mysidacea). Transactions of the Royal Society of South Africa 40: 309–333.

Browne, W. E., A. L. Price, M. Gerberding, and N. H. Patel. 2005. Stages of embryonic development in the amphipod crustacean, *Parhyale hawaiensis*. Genesis 42: 124–149.

Brum, P. E. D., and P. B. Araujo. 2007. The manca stages of *Porcellio dilatatus* Brandt (Crustacea, Isopoda, Oniscidea). Revista Brasileira de Zoologia 24: 493–502.

Brusca, G. J. 1975. General Patterns of Invertebrate Development. Eureka, CA: Mad River Press.

Brusca, R. C., and G. J. Brusca. 2003. Invertebrates, 2nd ed. Sunderland, MA: Sinauer Associates.

Burkenroad, M. D. 1981. The higher taxonomy and evolution of Decapoda (Crustacea). Transactions of the San Diego Society of Natural History 19: 251–268.

Burmeister, H. 1835. Beschreibung einiger neuen oder weniger bekannten Schmarotzerkrebse, nebst allgemeinen Betrachtungen ueber die Gruppe, welcher sie angehoeren. Nova Acta Physico-Medica Academiae Caesareae Leopoldino-Carolinae Naturae Curiosorum (Acta der Kaiserlichen Leopoldinisch-Carolinischen Deutschen Akademie der Naturforscher) 17: 269–336.

Burton, T., B. Knott, D. Judge, P. Vercoe, and A. Brearley. 2007. Embryonic and juvenile attachments structures in *Cherax cainii* (Decapoda: Parastacidae): implications for maternal care. American Midland Naturalist 157: 127–136.

Butorina, L. G. 1998. Resting eggs and hatching of young *Polyphemus pediculus* (Crustacea, Onychopoda). *In* L. Brendonck, L. de Meester, and N. Hairston, eds. Proceedings of the Symposium "Diapause in the Crustaceae—with Invited Contributions on Non-Crustacean Taxa," held in Ghent, August 24–29, 1997. Special Issue, Advances in Limnology 52: 521–534.

Butorina, L. G. 2000. A review of the reproductive behavior of *Polyphemus pediculus* (L.) Müller (Crustacea: Branchiopoda). Hydrobiologia 427: 13–26.

Cabral, P., F. Coste, and A. Raibaut. 1984. Cycle évolutif de *Lernanthropus kroyeri* van Beneden, 1851, copépode branchial hématophage du loup *Dicentrarchus labrax* (Linné, 1758) dans des populations naturelles et en élevage. Annales de Parasitologie Humaine et Comparée 59: 189–207.

Caillet, C. 1979. Biologie comparée de *Caligus minimus* Otto, 1848 et de *Clavellodes macrotrachelus* (Brian, 1906), copépodes parasites de poissons marins. Thèse 3e cycle, Université de Montpellier II.

Calado, R. 2008. Marine Ornamental Shrimp: Biology, Aquaculture and Conservation. Oxford: Wiley-Blackwell.

Calado, R., J. Lin, A. L. Ryne, R. Araújo, and L. Narcisco. 2003. Marine ornamental decapods: popular, pricey, and poorly studied. Journal of Crustacean Biology. 23: 963–973.

Calado, R., C. Bartilotti, L. Narciso, and A. dos Santos. 2004. Redescription of the larval stages of *Lysmata seticaudata* (Risso, 1816) (Crustacea, Decapoda, Hippolytidae) reared under laboratory conditions. Journal of Plankton Research 26: 737–752.

Calazans, D. 1993. Key to the larvae and decapodids of genera of the infraorder Penaeidea from the southern Brazilian coast. Nauplius 1: 45–62.

Calazans, D. 2000. Taxonomy of solenocerid larvae and distribution of larval phases of *Pleoticus muelleri* (Bate, 1888) (Decapoda: Solenoceridae) on the southern Brazilian coast. *In* J. C. von Vaupel Klein and F. R. Schram, eds. The Biodiversity Crisis and Crustacea, 565–576. Crustacean Issues No. 12. Rotterdam: A. A. Balkema.

Calman, W. T. 1909. Crustacea. *In* E. R. Lankester, ed. A Treatise on Zoology, pt. 7, fasc. 3. London: Adam & Charles Black.

Cals, P., and J. Cals-Usciati. 1982. Développment postembryonnaire des crustacés Mystacocarides: le problème des différences présumées entre les deux espèces voisines *Derocheilocaris remanei* Delamare-Deboutteville et Chappuis et *Derocheilocaris typicus* Pennak et Zinn. Comptes Rendus Hebdomadaires des Séances de l'Académie des Sciences, ser. 3, 294: 505–510.

Camacho, A. I., B. A. Dorda, and I. Rey. 2012. Undisclosed taxonomic diversity of Bathynellacea (Malacostraca: Syncarida) in the Iberian Peninsula revealed by molecular data. Journal of Crustacean Biology 32: 816–826.

Camp, D. K. 1988. *Bythognathia yucatanensis*, new genus, new species from abyssal depths in the Caribbean Sea, with a list of gnathiid species described since 1926 (Isopoda: Gnathiidea). Journal of Crustacean Biology 8: 668–678.

Campbell, G. R., and D. R. Fielder. 1987. Occurrence of a prezoea in two species of commercially exploited portunid crabs (Decapoda, Brachyura). Crustaceana 52: 202–206.

Cannon, H. G. 1921. The early development of the summer egg of a Cladoceran (*Simocephalus vetulus*). Quarterly Journal of Microscopical Science 65: 627–642.

Cannon, H. G. 1926. On the post-embryonic development of the fairy-shrimp (*Chirocephalus diaphanus*). Journal of the Linnean Society London, Zoology 36: 401–416.

Cannon, H. G. 1928. On the feeding mechanism of a fairy shrimp, *Chirocephalus diaphanus* Prévost. Transactions of the Royal Society of Edinburgh 55: 807–822.

Cannon, H. G. 1960. Leptostraca. *In* H-E. Gruner, ed. Dr. H. G. Bronns Klassen und Ordnungen des Tierreichs, vol. 5, sect. 1, bk. 4, pt. 1. Leipzig: Akademische Verlagsgesellschaft Geest & Portig.

Cannon, H. G., and Leak, F. M. C. 1933. On the feeding mechanism of the Branchiopoda. Philosophical Transactions of the Royal Society of London, B 222: 267–339.

Cano, G. 1892a. Sviluppo post-embrionale dei Dorippidei, Leucosiadi, Corystoidei e Grapsidi. Reale Accademia delle Scienze Fisiche e Matematiche, Napoli, della Società Italiana delle Scienze 8, ser. 3, 4: 1–14.

Cano, G. 1892b. Sviluppo post-embrionale dello *Stenopus spinosus* Risso: studio morfologico. Bollettino della Società Natural Napoli 1(5): 134–137.

Carcupino, M., A. Floris, A. Addis, A. Castelli, and M. Curini-Galletti. 2006. A new species of the genus *Lightiella*: the first record of Cephalocarida (Crustacea) in Europe. Zoological Journal of the Linnean Society 148: 209–220.

Carpenter, J. H. 1999. Behavior and ecology of *Speleonectes epilimnius* (Remipedia, Speleonectidae) from surface water of an anchialine cave on San Salvador Island, Bahamas. Crustaceana, Jan H. Stock Memorial Issue 72: 779–991.

Cartes, J. E., and J. C. Sorbe. 1996. Temporal population structure of deep-water cumaceans from the western Mediterranean slope. Deep Sea Research, pt. 1, 43(9): 1423–1438.

Carton, Y. 1968. Développement de *Cancerilla tubulata* Dalyell parasite de l'ophiure *Amphipholis squamata* Della Chiaje. Crustaceana, Suppl. 1: 11–28.

Castellani, C., A. Maas, D. Waloszek, and J. T. Haug. 2011. New pentastomids from the Late Cambrian of Sweden: deeper insight of the ontogeny of fossil tongue worms. Palaeontographica: Beiträge zur Naturgeschichte der Vorzeit, A, Paleozoology-Stratigraphy 293: 95–145.

Castro, A. de, and D. E. Jory. 1983. Preliminary experiments on the culture of the banded coral shrimp, *Stenopus hispidus*. Journal of Aquatic Science 3: 84–89.

Chan, B. K. K. 2003. Studies on *Tetraclita squamosa* and *Tetraclita japonica* (Cirripedia: Thoracica), 2: larval morphology and development. Journal of Crustacean Biology 23: 522–547.

Chan, T.-Y. 2010. Annotated checklist of the world's marine lobsters (Crustacea: Decapoda: Astacidea, Glypheidea, Achelata, Polychelida). Raffles Bulletin of Zoology, Suppl. 23: 153–181.

Chandler, G. T., and J. W. Fleeger. 1984. Tube-building by a marine meiobenthic harpacticoid copepod. Marine Biology (Berlin) 82: 15–19.

Channing, A., and D. Edwards. 2009. Yellowstone hot spring environments and the palaeo-ecophysiology of Rhynie chert plants: towards a synthesis. Plant Ecology and Diversity 2: 111–143.

Chappuis, P. A. 1916. *Viguierella coeca* Maupas: ein Beitrag zur Entwicklungsgeschichte der Crustaceen (unter Benützung eines Manuskriptes von E. Maupas). Revue Suisse de Zoologie 24: 521–564.

Charmantier, G., M. Charmantier-Daures, and D. E. Aiken. 1991. Metamorphosis in the lobster *Homarus* (Decapoda): a review. Journal of Crustacean Biology 11: 481–495.

Chen, H.-N., J. T. Høeg, and B. K. K. Chan. 2013. Morphometric and molecular identification of individual barnacle cyprids from wild plankton: an approach to detecting fouling and invasive barnacle species. Biofouling: 29: 133–145.

Childress, J. J., and M. H. Price. 1978. Growth rate of the bathypelagic crustacean *Gnathophausia ingens* (Mysidacea: Lophogastridae), 1: dimensional growth and population structure. Marine Biology (Berlin) 50: 47–62.

Christiansen, M. E. 1973. The complete larval development of *Hyas araneus* (Linnaeus) and *Hyas coarctatus* Leach (Decapoda, Brachyura, Majidae) in the laboratory. Norwegian Journal of Zoology 21: 63–89.

Christiansen, M. E., and K. Anger. 1990. Complete larval development of *Galathea intermedia* Lilljeborg reared in laboratory culture (Anomura: Galatheidae). Journal of Crustacean Biology 10, 87–111.

Christoffersen, M. L. 1990. A new superfamily classification of the Caridea (Crustacea: Pleocyemata) based on phylogenetic pattern. Journal of Zoological Systematics and Evolutionary Research 28: 94–106.

Christoffersen, M. L., and J. E. De Assis. 2013. A systematic monograph of the Recent Pentastomida, with a compilation of their hosts. Zoologische Mededeelingen 87: 1–206.

Cisne, J. L. 1982. Origin of the Crustacea. *In* L. G. Abele, ed. The Biology of Crustacea. Vol. 1, Systematics, the Fossil Record, and Biogeography, 64–92. New York: Academic Press.

Clare, A. S. 1995. Chemical signals in barnacles: old problems, new approaches. *In* F. R. Schram and J. T. Høeg, eds. New Frontiers In Barnacle Evolution, 49–67. Crustacean Issues No. 10. Rotterdam: A. A. Balkema.

Clare, A. S. 2011. Toward a characterization of the chemical cue to barnacle gregariousness. *In* T. Breithaupt and M. Thiel, eds. Chemical Communication in Crustaceans, 431–450. New York: Springer.

Clark, P. F. 2000. Interpreting patterns in chaetotaxy and segmentation associated with abbreviated brachyuran zoeal development. Invertebrate Reproduction and Development 38: 171–181.

Clark, P. F. 2009. The bearing of larval morphology on brachyuran phylogeny. *In* J. W. Martin, K. A. Crandall, and D. L. Felder, eds. Decapod Crustacean Phylogenetics, 221–241. Crustacean Issues No. 18. Boca Raton, FL: CRC Press.

Clark, P. F., and P. K. L. Ng. 2006. First stage zoeas of *Quadrella* Dana, 1851 (Crustacea: Decapoda: Brachyura: Xanthoidea: Trapeziidae) and their affinities with those of *Tetralia* Dana, 1851, and *Trapezia* Latreille, 1828. Hydrobiologia 560: 267–294.

Clark, P. F., and P. K. L. Ng. 2008. The lecithotrophic zoea of *Chirostylus ortmanni* Miyake & Baba, 1968 (Crustacea: Anomura: Galatheoidea: Chirostylidae) described from laboratory hatched material. Raffles Bulletin of Zoology 56: 85–94.

Clark, P. F., D. K. Calazans, and G. W. Pohle. 1998. Accuracy and standardisation of brachyuran larval descriptions. Invertebrate Reproduction and Development 33: 127–144.

Claus, C. 1863. Ueber einige Schizopoden und niedere Malakostraken Messinas. Zeitschrift für Wissenschaftliche Zoologie 13: 422–453.

Claus, C. 1868. Beitrage zur Kenntnis der Ostracoden, 1: Entwicklungsgeschichte von Cypris. Schriften der Gesellschaft zur Beförderung der Gesammten Naturwissenschaften zu Marburg 9: 151–166.

Claus, C., 1871. Die Metamorphose der Squilliden. Abhandlungen der Königlichen Gesellschaft der Wissenschaften zu Göttingen 16: 111–163.

Claus, C. 1872. Ueber den Bau und die systematische stellung von Nebalia, nebst Bemerkungen über das seither unbekannte Männchen dieser Gattung. Zeitschrift für Wissenschaftliche Zoologie 22: 323–330.

Claus, C. 1873. Zur Kenntniss des Baues und der Entwicklung von Branchipus stagnalis und Apus cancriformis. Abhandlungen der Königlichen Gesellschaft der Wissenschaften in Göttingen 18: 93–140.

Claus, C. 1875. Über die entwicklung, organisation und systematische Stellung der Arguliden. Zeitschrift für Wissenschaftliche Zoologie 25: 1–68.

Claus, C. 1877. Zur Kenntnis des Baues und der Organisation der Polyphemiden. Denkschrift der Kaiserlichen Academie der Wissenschaften in Wien 37: 137–160.

Claus, C. 1886. Untersuchungen über die Organisation und Entwicklung von Branchipus und Artemia nebst vergleichenden Bemerkungen über andere Phyllopoden. Arbeiten aus dem Wiener Zoologischen Institut 6: 1–140.

Claus, C. 1888. Über den Organismus der Nebaliiden und die systematische Stellung der Leptostraken. Arbeiten aus dem Zoologischen Institut der Universität Wien 8: 1–149.

Clutter, R. I., and G. H. Theilacker. 1971. Ecological efficiency of a pelagic mysid shrimp: estimates from growth, energy budget, and mortality studies. U.S. Fishery Bulletin 69: 93–115.

Cohen, A. C. 1983. Rearing and post-embryonic development of the Myodocopid ostracode Skogsbergia lerneri from coral reefs of Belize and the Bahamas. Journal of Crustacean Biology 3: 235–256.

Cohen, A. C., and J. G. Morin. 1990. Patterns of reproduction in ostracodes: a review. Journal of Crustacean Biology 10: 184–211.

Cohen, J., and R. B. Forward Jr. 2009. Zooplankton diel vertical migration—a review of proximate control. In R. N. Gibson, R. J. A. Atkinson, and J. D. M. Gordon, eds. Oceanography and Marine Biology, an Annual Review, vol. 47, 77–110. Boca Raton, FL: CRC Press.

Cohen, R. G., S. G. R. Gil, and C. G. Vélez. 1998. The post-embryonic development of Artemia persimilis Piccinelli & Prosdocimi. Hydrobiologia 391: 63–80.

Collis, S., and G. Walker. 1994. The morphology of the nauplius stages of (Crustacea: Cirripedia: Rhizocephala). Acta Zoologica (Stockholm) 75: 297–303.

Cook, H. L. 1966. A generic key to the protozoean, mysis and post-larval stages of the littoral Penaeidae of the northwestern Gulf of Mexico. U.S. Fishery Bulletin 65: 437–447.

Corbera, J., C. San Vicente, and J. C. Sorbe. 2000. Small-scale distribution, life cycle and secondary production of Cumopsis goodsiri in Creixell Beach (western Mediterranean). Journal of the Marine Biological Association of the United Kingdom 80: 271–282.

Corey, S. 1969. The comparative life histories of three Cumacea (Crustacea): Cumopsis goodsiri (Van Beneden), Iphinoe trispinosa (Goodsir), and Pseudocuma longicornis (Bate). Canadian Journal of Zoology 47(4): 695–704.

Corey, S. 1976. The life history of Diastylis sculpta Sars, 1871 (Crustacea, Cumacea). Canadian Journal of Zoology 54: 615–619.

Cormie, A. K. 1993. The morphology of the first zoea stage of Lomis hirta (Lamarck, 1818) (Decapoda, Lomisidae). Crustaceana 64: 249–255.

Costanza, R., R. d'Arge, R. de Groot, S. Farber, M. Grasso, B. Hannon, K. Limburg, S. Naeem, R. V. O'Neill, J. Paruelo, R. G. Raskin, P. Sutton, and M. van den Belt. 1997. The value of the world's ecosystem services and natural capital. Nature 387: 253–260.

Costanzo, G., and N. Calafiore. 1985. Larval development of Herrmannella rostrata Canu, 1891 (Copepoda, Poecilostomatoida, Sabelliphilidae) of the Lake Faro (Messina) reared in the laboratory. Memorie di Biologia Marina e di Oceanografia 15: 141–154.

Costello, M. J. 2009. The global economic cost of sea lice to the salmonid farming industry. Journal of Fish Diseases 32: 115–118.

Costello, W. J., and F. Lang. 1979. Development of the dimorphic claw closer muscles of the lobster Homarus americanus, 4: changes in functional morphology during growth. Biological Bulletin 156: 179–195.

Costlow, J. D., and C. G. Bookhout. 1961. The larval stages of Panopeus herbstii Milne-Edwards reared in the laboratory. Journal of the Elisha Mitchell Scientific Society 77: 33–42.

Costlow, J. D., and C. G. Bookhout. 1962. The larval development of Hepatus epheliticus (L.) under laboratory conditions. Journal of the Elisha Mitchell Scientific Society 78: 113–125.

Costlow, J. D., and C. G. Bookhout. 1968. The complete larval development of the land-crab, Cardisoma guanhumi Latreille in the laboratory (Brachyura, Gecarcinidae). Crustaceana, Suppl. 2: 259–270.

Coull, B. C., and B. W. Dudley. 1976. Delayed naupliar development of meiobenthic copepods. Biological Bulletin, Woods Hole Marine Biological Laboratory 150: 38–46.

Coutures, E. 2001. On the first phyllosoma stage of Parribacus caledonicus Holthuis, 1960, Scyllarides squammosus (H. Milne-Edwards, 1837) and Arctides regalis Holthuis, 1963 (Crustacea, Decapoda, Scyllaridae) from New Caledonia. Journal of Plankton Research 23: 745–751.

Crandall, K., M. L. Porter, and M. Pérez-Losada. 2009. Crabs, shrimps, and lobsters (Decapoda). In S. B. Hedges and S. Kumar, eds. The Timetree of Life, 293–297. New York: Oxford University Press.

Crane, J. 1940. Eastern Pacific expeditions of the New York Zoological Society, 18: on the post-embryonic development of brachyuran crabs of the genus Ocypode. Zoologica (Scientific Contributions of the New York Zoological Society) 25: 65–82.

Crescenti, N., G. Costanzo, and L. Guglielmo. 1994. Developmental stages of Antarctomysis ohlinii Hansen, 1908 (Mysidacea) in Terra Nova Bay, Ross Sea, Antarctica. Journal of Crustacean Biology 14: 383–395.

Criales, M. M, M. Robblee, J. A. Browder, H. Cardenas, and T. Jackson. 2010. Nearshore concentration of pink shrimp Farfantepenaeus duorarum post-larvae in northern Florida Bay in relation to the nocturnal flood tide. Bulletin of Marine Science 86: 51–72.

Cristescu, M. E. A., and P. D. N. Hebert. 2002. Phylogeny and adaptive radiation in the Onychopoda (Crustacea, Cladocera): evidence from multiple gene sequences. Journal of Evolutionary Biology 15: 838–849.

Cronin, T. W., N. J. Marshall, R. L. Caldwell, and D. Pales. 1995. Compound eyes and ocular pigments of crustacean larvae (Stomatopoda and Decapoda, Brachyura). Marine and Freshwater Behaviour and Physiology 26: 219–231.

Crosnier, A. 1972. Naupliosoma, phyllosomes et pseudibacus de *Scyllarides herklotsi* (Herklots) (Crustacea, Decapoda, Scyllaridae) récoltés par l'Ombango dans le sud du Golfe de Guinée. Cahiers ORSTOM, Océanographie 10: 139–149.

Cuesta, J. A., E. Palacios-Theil, P. Drake, and A. Rodríguez. 2006. A new rare case of parental care in decapods. Crustaceana 79: 1401–1405.

Cumberlidge, N., and P. K. L. Ng. 2009. Systematics, evolution, and biogeography of freshwater crabs. *In* J. W. Martin, K. A. Crandall, and D. L. Felder, eds. Decapod Crustacean Phylogenetics, 491–508. Crustacean Issues No. 18. Boca Raton, FL: CRC Press.

Cuzin-Roudy, J., and C. Tchernigovtzeff. 1985. Chronology of the female molt cycle in *Siriella armata* M.-Edw. (Crustacea: Mysidacea) based on marsupial development. Journal of Crustacean Biology 5: 1–14.

Dahl, E. 1956. Some crustacean relationships. *In* K. G. Wingstrand, ed. Bertil Hanström: Zoological Papers in Honour of His Sixty-Fifth Birthday, Nov. 20, 1956, 138–147. Lund, Sweden: Lund Zoological Institute.

Dahl, E. 1984. The subclass Phyllocarida (Crustacea) and the status of some early fossils: a neontologist's view. Videnskabelige Meddelelser fra Dansk Naturhistorisk Forening 145: 61–76.

Dahl, E. 1985. Crustacea Leptostraca, principles of taxonomy and a revision of European shelf species. Sarsia 70: 135–165.

Dahl, E. 1987. Malacostraca maltreated: the case of the Phyllocarida. Journal of Crustacean Biology 7: 721–726.

Dahms, H.-U. 1989. First record of a lecithotrophic nauplius in Harpacticoida (Crustacea, Copepoda) collected from the Weddell Sea (Antarctica). Polar Biology 10: 221–224.

Dahms, H.-U. 1990a. Naupliar development of *Paraleptastacus brevicaudatus* Wilson, 1932 (Copepoda: Harpacticoida: Cylindropsyllidae). Journal of Crustacean Biology 10: 330–339.

Dahms, H.-U. 1990b. The first nauplius and the copepodite stages of *Thalestris longimana* Claus, 1863 (Copepoda, Harpacticoida, Thalestridae) and their bearing on the reconstruction of phylogenetic relationships. Hydrobiologia 202: 33–60.

Dahms, H.-U. 1990c. Naupliar development of Harpacticoida (Crustacea, Copepoda) and its significance for phylogenetic systematics. Microfauna Marina 6: 169–272.

Dahms, H.-U. 1992. Metamorphosis between naupliar and copepodid phases in the Harpacticoida. Philosophical Transactions of the Royal Society of London, B 333: 221–236.

Dahms, H.-U. 1993a. Naupliar development of *Scutellidium hippolytes* (Copepoda, Harpacticoida) and a comparison of nauplii within Tisbidae. Hydrobiologia 250: 1–14.

Dahms, H.-U. 1993b. Comparative copepodid development in Tisbidimorpha *sensu* Lang, 1948 (Copepoda, Harpacticoida) and its bearing on phylogenetic considerations. Hydrobiologia 250: 15–37.

Dahms, H.-U. 1993c. Copepodid development in Harpacticoida (Crustacea, Copepoda). Microfauna Marina 8: 195–245.

Dahms, H-U. 2000. Phylogenetic implications of the crustacean nauplius. Hydrobiologia 417: 91–99.

Dahms, H.-U. 2004a. Post-embryonic apomorphies proving the monophyletic status of the Copepoda. *In* J.-S. Hwang, J.-S. Ho, and C.-T. Shih, eds. Contemporary Studies on Copepoda. Special Issue, Zoological Studies 43(2): 446–453.

Dahms, H.-U. 2004b. Exclusion of the Polyarthra from Harpacticoida and its reallocation as an underived branch of Copepoda (Arthropoda, Crustacea). Invertebrate Zoology 1: 29–51.

Dahms, H.-U., and J. Bresciani 1993. Naupliar development of *Stenhelia (D.) palustris* from the North Sea. Ophelia 37: 101–116.

Dahms, H.-U., and G. R. F. Hicks. 1996. Naupliar development of *Parastenhelia megarostrum* (Copepoda: Harpacticoida), and its bearing on phylogenetic relationships. Journal of Natural History 30: 11–22.

Dahms, H-U., and P. Y. Qian. 2004. Life histories of the Harpacticoida (Copepoda, Crustacea): a comparison with meiofauna and macrofauna. Journal of Natural History 38: 1725–1734.

Dahms, H.-U., M. Bergmans, and H. K. Schminke. 1990. Distribution and adaptations of sea ice inhabiting Harpacticoida (Crustacea, Copepoda) of the Weddell Sea (Antarctica). P.S.Z.N. [Pubblicazioni della Stazione Zoologica di Napoli] 1, Marine Ecology 11: 207–226.

Dahms, H.-U., S. Lorenzen, and H. K. Schminke. 1991. Phylogenetic relationships within the taxon Tisbe (Copepoda, Harpactocoida) as evidenced by naupliar characters. Zeitschrift für Zoologische Systematik und Evolutionsforschung 29: 450–465.

Dahms, H.-U., N. V. Schizas, and T. C. Shirley. 2005. Naupliar evolutionary novelties of *Stenhelia peniculata* (Copepoda, Harpacticoida) from Alaska affirming taxa belonging to different categorial rank. Invertebrate Zoology 2: 1–14.

Dahms, H-U., J. A. Fornshell, and B. J. Fornshell. 2006. Key for the identification of crustacean nauplii. Organisms, Diversity & Evolution 6: 47–56.

Dahms, H.-U., S. Chullasorn, N. V. Schizas, P. Kangtia, W. Anansatitporn, and W.-X. Yang. 2009. Naupliar development among the Tisbidae (Copepoda: Harpacticidae) with a phylogenetic analysis and naupliar description of *Tisbe thailandensis* from Thailand. Zoological Studies 48: 780–796.

Dall, W., B. J. Hill, P. C. Rothlisberg, and D. J. Sharples. 1990. The biology of the Penaeidae. Advances in Marine Biology 27: 1–489.

Dames, W. 1886. Ueber einige Crustaceen aus den Kreideablagerungen des Libanon. Zeitschrift der Deutschen Geologischen Gesellschaft 38: 551–576.

Damkaer, D. M. 2002. The copepodologist's cabinet: a biographical and bibliographical history. Memoirs of the American Philosophical Society 240: 1–300.

Dana, J. D. 1852. Crustacea, pt. 1. *In* United States Exploring Expedition, during the Years 1838, 1839, 1840, 1841, 1842, under the Command of Charles Wilkes, U.S.N. Philadelphia: C. Sherman.

Dana, J. D. 1855. Crustacea: Atlas. *In* United States Exploring Expedition, during the Years 1838, 1839, 1840, 1841, 1842, under the Command of Charles Wilkes, U.S.N. Philadelphia: C. Sherman.

Dana, J. D., and E. C. Herrick. 1837. Description of the *Argulus catostomi*, a new parasitic crustaceous animal. American Journal of Science 31: 297–308.

Darwin, C. 1859. On the Origin of Species by Means of Natural Selection; or, The Preservation of Favoured Races in the Struggle for Life. London: John Murray.

Davie, P. J. F. 1990. A new genus and species of marine crayfish, *Palibythus magnificus*, and new records of *Palinurellus* (Decapoda: Palinuridae) from the Pacific Ocean. Invertebrate Taxonomy 4: 685–695.

Davis, C. C. 1966. A study of the hatching process in aquatic invertebrates, 22: multiple membrane shedding in *Mysidium columbiae* (Zimmer) (Crustacea: Mysidacea). Bulletin of Marine Science 16: 124–131.

Davis, C. C. 1968. Mechanisms of hatching in aquatic invertebrate eggs. Oceanography and Marine Biology 6: 325–376.

Dawirs, R. R. 1981. Elemental composition (C, N, H) and energy in the development of *Pagurus bernhardus* (Decapoda: Paguridae) megalopa. Marine Biology (Berlin) 64: 117–123.

Debaisieux, P. 1953. Histologie et histogénèse chez *Argulus foliaceus* L. (Crustacé, Branchiure). Cellule 55: 245–290.

de Beer, G. R. 1958. Embryos and Ancestors. New York: Oxford University Press.

Dechancé, M. 1964. Sur une collection de crustacés pagurides de Madagascar et des Comores. Cahiers ORSTOM, Océanographie 2: 27–46.

De Grave, S., and C. H. J. M. Fransen. 2011. Carideorum catalogus: the Recent species of the dendrobranchiate, stenopodidean, procarididean and caridean shrimps (Crustacea: Decapoda). Zoologische Mededeelingen Leiden 85: 195–589.

De Grave, S., Y. Cai, and A. Anker. 2008. Global diversity of shrimps (Crustacea: Decapoda: Caridea) in freshwater. Hydrobiologia 595: 287–293.

De Grave, S., N. D. Pentcheff , S. T. Ahyong, T.-Y. Chan, K. A. Crandall, P. C. Dworschak, D. L. Felder, R. M. Feldmann, C. H. J. M. Fransen, L. Y. D. Goulding, R. Lemaitre, M. E. Y. Low, J. W. Martin, P. K. L. Ng, C. E. Schweitzer, S. H. Tan, D. Tshudy, and R. Wetzer. 2009. A classification of living and fossil genera of decapod crustaceans. Raffles Bulletin of Zoology, Suppl. 21: 1–109.

Dejdar, E. 1930. Die Korrelatione zwischen Kiemensäckchen un Nackenschild bei Phyllopoden. Zeitschrift für Wissenschaftliche Zoologie 136: 423–452.

Delage, Y. 1884. Évolution de la *Sacculina* (Thomps.) crustacé endoparasite de l'ordre nouveau des kentrogonides. Archives de Zoologie Expérimentale et Générale, ser. 2, 2: 417–736.

Delamare-Deboutteville, C. 1953. Recherches sur l'écologie et la répartition du mystacocaride *Derocheilocaris remanei* Delamare et Chappuis en Méditérranée. Vie et Milieu 4: 459–469.

Delamare-Deboutteville, C. 1954. Recherches sur les crustaces souterrains, 3: le développement postembryonnaire des Mystacocarides. Archives de Zoologie Expérimentale et Générale 91: 25–34.

Della Croce, N., and S. Bettanin. 1965. Sviluppo embrionale della forma partenogenetica di *Penilia avirostris* Dana. Cahiers de Biologie Marine 6: 269–275.

deWaard, J. R., V. Sacharova, M. E. A. Cristescu, E. A. Remigio, T. J. Crease, and P. D. N. Hebert. 2006. Probing the relationships of the branchiopod crustaceans. Molecular Phylogenetics and Evolution 39: 491–502.

Dexter, B. L. 1981. Setogenesis and molting in planktonic crustaceans. Journal of Plankton Research 3: 1–13.

Dexter, D. M. 1972. Molting and growth in laboratory reared phyllosomes of the California spiny lobster, *Panulirus interruptus*. California Fish and Game 58: 107–115.

Diaz, H., and J. J. Ewald. 1968. A comparison of the larval development of *Metasesarma rubripes* (Rathbun) and *Sesarma ricordi* H. Milne-Edwards (Brachyura, Grapsidae) reared under similar laboratory conditions. Crustaceana, Suppl. 2: 225–248.

Diaz, W., and F. Evans. 1983. The reproduction and development of *Microsetella norvegica* (Boeck) (Copepoda, Harpacticoida) in Northumberland coastal waters. Crustaceana 45: 113–130.

Dick, T. A., B. Dixon, and A. Choudhury. 1991. *Diphyllobothrium*, *Anisakis*, and other fish borne parasitic zoonoses. Southeast Asian Journal of Tropical Medicine and Public Health 22: 150–152.

Dingle, H. 1969. Ontogenetic changes in phototaxis and thigmokinesis in stomatopod larvae. Crustaceana 16: 108–110.

Dittrich, B. 1987. Post-embryonic development of the parasitic amphipod *Hyperia galba*. Helgoländer Meeresuntersuchungen / Helgoland Marine Research 41: 217–232.

Dittrich, B. 1988. Studies on the life cycle and reproduction of the parasitic amphipod *Hyperia galba* in the North Sea. Helgoländer Meeresuntersuchungen / Helgoland Marine Research 42: 79–98.

Do, T. T., T. Kajihara, and J.-S. Ho. 1984. The Life History of *Pseudomyicola spinosus* (Raffaele & Monticelli, 1885) from the Blue Mussel, *Mytilus edulis galloprovincialis* in Tokyo Bay, with Notes on the Production of Atypical Male. Bulletin of the Ocean Research Institute, University of Tokyo No. 17. Tokyo: Ocean Research Institute, University of Tokyo.

Dodds, G. S. 1926. Entomostraca from the Panama Canal zone with description of one new species. Occasional Papers of the Museum of Zoology, University of Michigan 174: 1–27.

Dohle, W., M. Gerberding, A. Hejnol, and G. Scholtz. 2004. Cell lineage, segment differentiation, and gene expression in crustaceans. *In* G. Scholtz, ed. Evolutionary Developmental Biology of Crustacea, 95–134. Crustacean Issues No. 15. Lisse, Netherlands: A. A. Balkema.

Dohrn, A. 1870. Untersuchungen über Bau und Entwicklung de Arthropoden, 6: zur Entwicklungsgechichte der Panzerkrebse (Decapoda Loricata). Zeitschrift für Wissenschaftliche Zoologie 20: 249–271.

Dojiri, M., and R. F. Cressey. 1987. Revision of the Taeniacanthidae (Copepoda: Poecilostomatoida) Parasitic on Fishes and Sea Urchins. Smithsonian Contributions to Zoology No. 447. Washington, DC: Smithsonian Institution Press.

Dojiri, M., G. Hendler, and I.-H. Kim. 2008. Larval development of *Caribeopsyllus amphiodiae* (Thaumatopsyllidae: Copepoda), an enterozoic parasite of the brittle star *Amphiodia urtica*. Journal of Crustacean Biology 28: 281–305.

Dolgopolskaia, M. A. 1969. Larvae of Decapoda Macrura and Anomura. *In* F. D. Mordukhai-Boltovskoi, ed. Vol. 2, Keys to the Fauna of Black and Azov Seas, 307–362. Kiev, Russia: Naukova Dumka [in Russian].

Domingues Rodrigues, M., and N. J. Hebling. 1989. *Ucides cordatus cordatus* (Linnaeus, 1763) (Crustacea, Decapoda), complete larval development under laboratory conditions and its systematic position. Revista Brasileira de Zoologia 6: 147–166.

dos Santos, A., and J. A. Lindley. 2001. Crustacea Decapoda, Larvae, 2: Dendrobranchiata (Aristeidae, Benthesicymidae, Penaeidae, Solenoceridae, Sicyonidae, Sergestidae, and Luciferidae). International Council for the Exploration of the Sea Identification Leaflets for Plankton / Fiches d'Identification du Plancton, Leaflet No. 186. Copenhagen: International Council for the Exploration of the Sea.

DuCane, C. 1839. 20—Letter from Captain DuCane to the Rev. Leonard Jenyns on the subject of the metamorphosis of Crustacea. Annals and Magazine of Natural History 2: 178–181.

Dudley, P. L. 1964. Some gastrodelphyid copepods from the Pacific coast of North America. American Museum Novitates 2194: 1–51.

Dudley, P. L. 1966. Development and Systematics of Some Pacific

Marine Symbiotic Copepods: A Study of the Biology of the Notodelphyidae, Associates of Ascidians. University of Washington Publications in Biology No. 21. Seattle: University of Washington Press.

Duguid, W. D. P. and L. R. Page. 2009. Larval and early post-larval morphology, growth, and behaviour of laboratory reared *Lopholithodes foraminatus* (brown box crab). Journal of the Marine Biological Association of the United Kingdom 89: 1607–1626.

Dumont, H. J., and W. Hollwedel. 2009. *Leptodora kindtii* (Focke, 1844) from Bremen, Germany: discovered, forgotten, and rediscovered. Crustaceana 82: 1457–1461.

Dumont, H. J., and S. V. Negrea 2002. Introduction to the Class Branchiopoda. Guides to the Identification of the Microinvertebrates of the Continental Waters of the World No. 19. Leiden, Netherlands: Backhuys.

Eder, E. 2002. SEM investigations of the larval development of *Imnadia yeyetta* and *Leptestheria dahalacensis* (Crustacea: Branchiopoda: Spinicaudata). Hydrobiologia 486: 39–47.

Edgar, G. J. 1990. Predator-prey interactions in seagrass beds, 3: Impacts of the western rock lobster *Panulirus cygnus* George on epifaunal gastropod populations. Journal of Experimental Marine Biology and Ecology 139: 33–42.

Egborge, A. B. M., and N. Ozoro. 1989. *Cyclestheria hislopi* (Baird, 1895) (Branchiopoda: Conchostraca) in Yewa River, Nigeria. Archive für Hydrobiologie 115: 137–148.

Egloff, D. A., P. W. Fofonoff, and T. Onbé. 1997. Reproductive biology of marine cladocerans. Advances in Marine Biology 31: 79–167.

Einarsson, H. 1945. Euphausiacea. Dana Report No. 27. Copenhagen: Reitzel.

Elfimov, A. S. 1995. Comparative morphology of the thoracican cyprid larvae: studies of the carapace. *In* F. R. Schram and J. T. Høeg, eds. New Frontiers in Barnacle Evolution, 137–152. Rotterdam: A. A. Balkema.

Elofsson, R. 1959. A new decapod larva referred to *Calocarides coronatus* (Trybom). Publications from the Biological Station, Espegrend 7: 1–10.

Elofsson, R. 1963. The nauplius eye and frontal organs in Decapoda (Crustacea). Sarsia 12: 1–68.

Elofsson, R. 1965. The nauplius eye and frontal organs in Malacostraca (Crustacea). Sarsia 19: 1–54.

Elofsson, R. 1966. The nauplius eye and frontal organs of the non-Malacostraca (Crustacea). Sarsia 25: 1–128.

Elofsson, R. 1971. Some observations on the internal morphology of Hansen's nauplius y (Crustacea). Sarsia 46: 23–40.

Endo, Y., and S. Nicol. 1997. Krill Fisheries of the world. FAO Fishery Technical Report No. 367. Rome: Food and Agriculture Organization of the United Nations.

Ennis, G. P. 1973. Behavioral responses to changes in hydrostatic pressure and light during larval development of the lobster *Homarus gammarus*. Journal of the Fisheries Research Board of Canada 30: 1349–1360.

Epifanio, C. E. 2007. Biology of larvae. *In* V. S. Kennedy and L. E. Cronin, eds. The Blue Crab: *Callinectes sapidus*, 513–533. College Park: Maryland Sea Grant College, University of Maryland.

Eriksson, S. 1934. Studien über die Fangapparate der Branchiopoden nebst einigen phylogenetischen Bemerkungen. Zoologiske Bidrag från Uppsala 15: 23–287.

Esslinger, J. H. 1962. Morphology of the egg and larva of *Porocephalus crotali* (Pentastomida). Journal of Parasitology 48: 457–462.

Esslinger, J. H. 1968. Morphology of the egg and larva of

Raillietiella furcocerca (Pentastomida) from a Colombian snake (*Clelia clelia*). Journal of Parasitology 54: 411–416.

Everson, I., ed. 2000. Krill Biology, Ecology and Fisheries. Fish and Aquatic Resources Series No. 6. Oxford: Blackwell Science.

Ewald, J. J. 1965. The laboratory rearing of pink shrimp, *Penaeus duorarum*, Burkenroad. Bulletin of Marine Science 15: 436–449.

Fahrenbach, W. H. 1962. The biology of a harpacticoid copepod. Cellule 62: 303–376.

Fain, A. 1961. Les pentastomides de l'Afrique centrale. Annales de Musée Royale de l'Afrique Centrale, ser. 8, 92: 1–115.

Fain, A. 1964. Observations sur le cycle évolutif du genre *Raillietiella* (Pentastomida). Bulletin de la Classe des Sciences de l'Académie Royale de Belgique 50: 1036–1060.

Falk-Petersen, S., S. Timofeev, V. Pavlov, and J. R. Sargent. 2007. Climate variability and possible effects on arctic food chains: the role of Calanus. *In* J. B. Ørbæk, R. Kallenborn, I. Tombre, E. Nøst Hegseth, S. Falk-Petersen, and A. H. Hoel, eds. Arctic-Alpine Ecosystems and People in a Changing Environment, 147–166. Berlin: Springer Verlag.

Fanenbruck, M., and S. Harzsch. 2005. A brain atlas of *Godzilliognomus frondosus* Yager, 1989 (Remipedia, Godzilliidae) and comparison with the brain of *Speleonectes tulumensis* Yager, 1987 (Remipedia, Speleonectidae): implications for arthropod relationships. Arthropod Structure and Development 34: 343–378.

Fanenbruck, M., S. Harzsch, and J. W. Wägele. 2004. The brain of the Remipedia (Crustacea) and an alternative hypothesis on their phylogenetic relationships. Proceedings of the National Academy of Sciences (USA) 101: 3868–3873.

Faxon, W. 1879. On the development of *Palaemonetes vulgaris*. Bulletin of the Museum of Comparative Zoology 5: 303–330.

Faxon, W. 1895. Reports on an Exploration Off the West Coasts of Mexico, Central and South America, and Off the Galapagos Islands, in Charge of Alexander Agassiz, by the U.S. Fish Commission Steamer *Albatross*, during 1891, Lieut.-Commander Z. L. Tanner, U.S.N. Commanding. Vol. 15, The Stalk-Eyed Crustacea. Memoirs of the Museum of Comparative Zoology at Harvard College No. 18.

Fayers, S. R. 2003. The biota and palaeoenvironments of the Windyfield chert, Early Devonian, Rhynie, Scotland. PhD diss., University of Aberdeen.

Fayers, S. R., and N. H. Trewin. 2003. A new crustacean from the Early Devonian Rhynie chert, Aberdeenshire, Scotland. Transactions of the Royal Society of Edinburgh, Earth and Environmental Science 93: 355–382.

Fayers, S. R., and N. H. Trewin. 2004. A review of the palaeo-environments and biota of the Windyfield chert. Transactions of the Royal Society of Edinburgh, Earth Sciences 94: 325–339.

Felder, D. L., J. W. Martin, and J. W. Goy. 1985. Patterns in early post-larval development of decapods. *In* A. M. Wenner, ed. Larval growth, 163–225. Crustacean Issues No. 2. Rotterdam: A. A. Balkema.

Feller, K. D., T. W. Cronin, S. T. Ahyong, and M. L. Porter. 2013. Morphological and molecular description of the late-stage larvae of *Alima* Leach, 1817 (Crustacea: Stomatopoda) from Lizard Island, Australia. Zootaxa 3722: 22–32.

Fernandes, L. D. A., M. F. De Souza, and S. L. C. Bonecker. 2007. Morphology of Oplophorid and Bresiliid larvae (Crustacea, Decapoda) of Southwestern Atlantic plankton, Brazil. Pan-American Journal of Aquatic Sciences 2: 199–230.

Fernandes, L. D. A., B. J. F. S. Peixoto, E. V. De Almeida, and S. L. C. Bonecker. 2010. Larvae of the family Stenopodidae

(Crustacea: Stenopodidea) from South Atlantic Ocean. Journal of the Marine Biological Association of the United Kingdom 90: 735–748.

Ferrari, F. D. 1988. Developmental patterns in numbers of ramal segments of copepod post-maxillipedal legs. Crustaceana 54: 256–293.

Ferrari, F. D. 1991. Using patterns of appendage development to group taxa of *Labidocera*, Diaptomidae and Cyclopidae (Copepoda). *In* S.-i. Uye, S. Nishida, and J.-S. Ho, eds. Proceedings of the Fourth International Conference on Copepoda, Karuizawa, Japan, 16–20 September 1990. Special Volume, Bulletin of the Plankton Society of Japan: 115–127.

Ferrari, F. D. 1998. Setal developmental patterns of the thoracopods of cyclopid copepods (Cyclopoida) and their use in phylogenetic inference. Journal of Crustacean Biology 18: 471–489.

Ferrari, F. D., and H.-U. Dahms. 2007. Post-Embryonic Development of the Copepoda. Crustaceana Monographs No. 8. Leiden, Netherlands: Brill.

Ferrari, F. D., and M. J. Grygier. 2003. Comparative morphology among trunk limbs of *Caenestheriella gifuensis* and *Leptestheria kawachiensis* (Crustacea: Branchiopoda: Spinicaudata). Zoological Journal of the Linnean Society 139: 547–564.

Ferrari, F. D., and V. N. Ivanenko. 2005. Copepodid stages of *Euryte longicauda* (Cyclopoida, Cyclopidae, Euryteinae) from the White Sea associated with the bryozoan *Flustra foliacea*. Journal of Crustacean Biology 25: 353–374.

Ferrari, F. D., and H. Ueda. 2005. Development of the fifth leg of copepods belonging to the calanoid superfamily Centropagoidea (Crustacea). Journal of Crustacean Biology 25: 333–352.

Ferrari, F. D., V. N. Ivanenko, and H.-U. Dahms. 2010. Body architecture and relationships among basal copepods. Journal of Crustacean Biology 30: 465–477.

Ficker, G. 1876. Zur Kenntniss der Entwicklung von *Estheria ticinensis* Bals.-Criv. Arbeiten aus dem Zoologisch-Vergleichend-Anatomischen Institute der Wiener Universität 74: 407–421.

Fielder, D. R., and J. G. Greenwood. 1985. The complete larval development of the sand bubbler crab, *Scopimera inflata* H. Milne-Edwards, 1837 (Decapoda, Ocypodidae), reared in the laboratory. Crustaceana 48: 133–146.

Fiers, F. 1998. Female leg 4 development in Laophontidae (Harpacticoida): a juvenile adaptation to precopulatory behaviour. *In* H.-U. Dahms, T. Glatzel, H. J. Hirche, S. Schiel, and H. K. Schminke, eds. Proceedings of the Sixth International Conference on Copepoda, held in Oldenburg/Bremerhaven on July 29–August 3, 1996. Special Volume, Journal of Marine Systems 15: 41–51.

Firth, R. W., Jr., and W. E. Pequegnat. 1971. Deep-sea lobsters of the families Polychelidae and Nephropidae (Crustacea, Decapoda) in the Gulf of Mexico and Caribbean Sea. Texas A&M Research Foundation Reference 71-11T. College Station: Texas A&M University, Department of Oceanography.

Fischer, A., and G. Scholtz. 2010. Axogenesis in the stomatopod crustacean *Gonodactylaceus falcatus* (Malacostraca). Invertebrate Biology 129: 59–76.

Fletcher, D. J., I. Kötter, M. Wunsch and I. Yasir. 1995. Preliminary observations on the reproductive biology of ornamental cleaner prawns. International Zoological Yearbook 34: 73–77.

Fockedey, N., A. Ghekiere, A., S. Bruwiere, C. R. Janssen, and M. Vincx. 2006. Effect of salinity and temperature on the intra-marsupial development of the brackish water mysid *Neomysis*

integer (Crustacea: Mysidacea). Marine Biology (Berlin) 148: 1339–1356.

Forbes, A. T. 1973. An unusual abbreviated larval life history in the estuarine burrowing prawn *Callianassa kraussi* (Crustacea: Decapoda: Thalassinidea). Marine Biology (Berlin) 22: 361–365.

Fornshell, J. A. 2012. Key to marine arthropod larvae. Arthropods (IAEES) 1: 1–12.

Forró, L., N. M. Korovchinsky, A. A. Kotov, and A. Petrusek. 2008. Global diversity of cladocerans (Cladocera: Crustacea) in freshwater. Hydrobiologia 595: 177–184.

Forster, J. R. 1782. Nachricht von einem neuen Insekte. Naturforscher 17: 206–213.

Fosshagen, A. 1970. *Thespesiopsyllus paradoxus* (Sars) (Copepoda, Cyclopoida) from western Norway. Sarsia 42: 33–40.

Fox, H. M. 1964. On the larval stages of cyprids and on *Siphlocandona* (Crustacea, Ostracoda). Proceedings of the Zoological Society of London 142: 165–176.

Foxon, G. E. H. 1932. Report on stomatopod larvae, Cumacea and Cladocera. *In* Vol. 4, no. 11, Great Barrier Reef Expedition 1928–29, Scientific Reports, 375–398. London: British Museum of Natural History.

Frangoulis, C., E. D. Christou, and J. H. Hecq. 2005. Comparison of marine copepod outfluxes: nature, rate, fate and role in the carbon and nitrogen cycles. Advances in Marine Biology 47: 253–309.

Fransen, C. H. J. M. 2010. Order Amphionidacea Williamson, 1973. *In* J. Forest and J. C. von Vaupel Klein, eds., F. R. Schram and M. Charmantier-Daures, advisory eds. The Crustacea: Complementary to the Volumes Translated from the French of the Traité de Zoologie (Founded by P.-P. Grassé). Vol. 9, pt. A, Eucarida: Euphausiacea, Amphionidacea, and Decapoda, 83–95. Leiden, Netherlands: Brill.

Fransen, C. H. J. M., and S. De Grave. 2009. Evolution and radiation of shrimp-like decapods: an overview. *In* J. W. Martin, K. A. Crandall, and D. L. Felder, eds. Decapod Crustacean Phylogenetics, 245–260. Crustacean Issues No. 18. Boca Raton, FL: CRC Press.

Fraser, F. C. 1936. On the development and distribution of the young stages of krill (*Euphausia superba*). Discovery Reports 14: 3–192.

Frey, D. 1980. The non-swimming chydorid Cladocera of wet forests, with descriptions of a new genus and two new species. Internationale Revue der Gesamten Hydrobiologie 65: 613–641.

Friend, J. A., and A. M. M. Richardson. 1986. Biology of terrestrial amphipods. Annual Review of Entomology 31: 25–48.

Fritsch, M., and S. Richter. 2010. The formation of the nervous system during larval development in *Triops cancriformis* (Bosc) (Crustacea, Branchiopoda): an immunohistochemical survey. Journal of Morphology 271: 1457–1481.

Fritsch, M., and S. Richter. 2012. Nervous system development in Spinicaudata and Cyclestherida (Crustacea, Branchiopoda)—comparing two different modes of indirect development by using an event pairing approach. Journal of Morphology 273: 672–695.

Fritsch, M., O. R. P. Bininda-Emonds, and S. Richter. 2013. Unraveling the origin of Cladocera by identifying heterochrony in the developmental sequences of Branchiopoda. Frontiers in Zoology 10: 35.

Fritsch, M., T. Kaji, J. Olesen, and S. Richter. 2013. The development of the nervous system in Laevicaudata (Crustacea, Branchiopoda): insights into the evolution and homologies of branchiopod limbs and "frontal organs." Zoomorphology 132: 163–181.

Froglia, C. 1992. Stomatopod Crustacea of the Ligurian Sea. Doriana 6(275): 1–10.

Fryer, G. 1956. A report on the parasitic Copepoda and Branchiura of the fishes of Lake Nyasa. Proceedings of the Zoological Society of London 127: 293–344.

Fryer, G. 1959. A report on the parasitic Copepoda and Branchiura of the fishes of the Lake Bangweulu (Northern Rhodesia). Proceedings of the Zoological Society of London 132: 517–550.

Fryer, G. 1961. Larval development in the genus Chonopeltis (Crustacea: Branchiura). Proceedings of the Zoological Society of London 137: 61–69.

Fryer, G. 1964. Further studies on the parasitic Crustacea of the African freshwater fishes. Proceedings of the Zoological Society of London 143: 79–102.

Fryer, G. 1966. Branchinecta gigas Lynch, a non-filter-feeding raptatory anostracan, with notes on the feeding habits of certain other anostracans. Proceedings of the Linnean Society of London 177: 19–34.

Fryer, G. 1968. Evolution and adaptive radiation in the Chydoridae (Crustacea, Cladocera): a study in comparative functional morphology and ecology. Philosophical Transactions of the Royal Society of London, B 254: 221–385.

Fryer, G. 1974. Evolution and adaptive radiation in the Macrothricidae (Crustacea: Cladocera): a study in comparative functional morphology and ecology. Philosophical Transactions of the Royal of Society of London, B 269: 137–274.

Fryer, G. 1977. On some species of Chonopeltis (Crustacea: Branchiura) from the rivers of the extreme South West Cape region of Africa. Journal of Zoology (London) 182: 441–455.

Fryer, G. 1983. Functional ontogenetic changes in Branchinecta ferox (Milne-Edwards) (Crustacea, Anostraca). Philosophical Transactions of the Royal Society of London, B 303: 229–343.

Fryer, G. 1987a. A new classification of the branchiopod Crustacea. Zoological Journal of the Linnean Society 91: 357–383.

Fryer, G. 1987b. Quantitative and qualitative: numbers and reality in the study of living organisms. Freshwater Biology 17: 177–189.

Fryer, G. 1988. Studies on the functional morphology and biology of the Notostraca (Crustacea: Branchiopoda). Philosophical Transactions of the Royal Society of London, B 321: 27–124.

Fryer, G. 1991. Functional morphology and the adaptive radiation of the Daphniidae (Branchiopoda: Anomopoda). Philosophical Transactions of the Royal Society of London, B 331: 1–99.

Fryer, G. 1995. Phylogeny and adaptive radiation within the Anomopoda: a preliminary exploration. Hydrobiologia 307: 57–68.

Fryer, G. 1998. The role of copepods in freshwater ecosystems. Special Volume, Journal of Marine Systems 15: 71–73.

Fujita, Y., and P. F. Clark. 2010. The larval development of Chirostylus stellaris Osawa, 2007 (Crustacea: Anomura: Chirostylidae) described from laboratory-reared material. Crustacean Research 39: 55–66.

Fujita, Y., and M. Osawa. 2005. Complete larval development of the rare porcellanid crab, Novorostrum decorocrus Osawa, 1998 (Crustacea: Decapoda: Anomura: Porcellanidae), reared under laboratory conditions. Journal of Natural History 39: 763–778.

Fujita, Y., and S. Shokita. 2005. The complete larval development of Sadayoshia edwardsii (37) (Decapoda: Anomura: Galatheidae) described from laboratory-reared material. Journal of Natural History 39: 865–886.

Fukuda, Y. 1982. Zoeal stages of the burrowing mud shrimp Laomedia astacina de Haan (Decapoda: Thalassinidea: Laomediidae) reared in the laboratory. Proceedings of the Japanese Society of Systematic Zoology 24: 19–31.

Galil, B. S. 2000. Crustacea, Decapoda: review of the genera and species of the family Polychelidae Wood-Mason, 1874. In A. Crosnier, ed. Vol. 21, Résultats des Campagnes MUSORSTOM, 285–387. Mémoires du Muséum National d'Histoire Naturelle No. 184. Paris: Éditions du Muséum.

Galil, B. S., P. F. Clark, and J. T. Carlton, eds. 2011. In the Wrong Place—Alien Marine Crustaceans: Distribution, Biology and Impacts. Invading Nature No. 6. New York: Springer.

Gamô, S. 1969a. Notes on three species of harpacticoid Copepoda, Porcellidium sp., Peltidium ovale Thompson & A. Scott, and Dactylopusia (?) platysoma Thompson & A. Scott, from Tanabe Bay. Publications of the Seto Marine Biological Laboratory 16: 345–361.

Gamô, S. 1969b. Further notes on Paramenophia platysoma (Thompson & A. Scott) (= Dactylopusia platysoma Thompson & A. Scott), harpacticoid Copepoda from Tanabe Bay, Kii Peninsula. Proceedings of the Japanese Society of Systematic Zoology 5: 19–22.

Gamô, S. 1979. Notes on a giant stomatopod larva taken south-east of Mindanao, Philippines (Crustacea). Science Reports of the Yokohama National University, sect. 2, Biological and Geological Sciences 26: 11–18.

Garassino, A., and Schweigert, G. 2006. The Upper Jurassic Solnhofen decapod crustacean fauna: review of the types from old descriptions (infraorders Astacidea, Thalassinidea, and Palinura). Memorie della Società Italiana di Scienze Naturali e del Museo Civico di Storia Naturale in Milano 34(1): 1–64.

Garcia-Schroeder, D. L., and P. B. Araujo. 2009. Post-marsupial development of Hyalella pleoacuta (Crustacea: Amphipoda): stages 1–4. Zoologia 26: 391–406.

Gasca, R., and S. H. D. Haddock. 2004. Associations between gelatinous zooplankton and hyperiid amphipods (Crustacea: Peracarida) in the Gulf of California. Hydrobiologia 530/531: 529–535.

Gauld, D. T. 1959. Swimming and feeding in crustacean larvae: the nauplius larva. Proceedings of the Zoological Society of London 132: 31–50.

Geiselbrecht, H., and R. R. Melzer. 2009. Morphology of the first zoeal stage of the partner shrimp Periclimenes amethysteus Risso, 1827 (Decapoda: Caridea: Palaemonidae: Pontoniinae) studied with the scanning EM. Zootaxa 2140: 45–55.

George, M. J., and S. John. 1975. On a rare brachyuran crab zoea from the south-west coast of India. Current Science 44: 162–163.

George, R. W. 2005. Evolution of life cycles, including migration, in spiny lobsters (Palinuridae). New Zealand Journal of Marine and Freshwater Research 39: 503–514.

George, R. W., and A. Main. 1967. The evolution of spiny lobsters (Palinuridae): a study of evolution in the marine environment. Evolution 21: 803–820.

Gerken, S. 2005. Cumaceans: Life History of the Cumacea. Online at http://afsag.uaa.alaska.edu/?page=history/.

Gerschler, M. W. 1911. Monographie der Leptodora kindtii (Focke). Archive für Hydrobiologie 6: 415–466.

Ghekiere, A., N. Fockedey, T. Verslycke, M. Vincx, and C. R. Janssen. 2007. Marsupial development in the mysid Neomysis integer

(Crustacea: Mysidacea) to evaluate the effects of endocrine-disrupting chemicals. Ecotoxicology and Environmental Safety 66: 9–15.

Ghetti, P. F. 1970. The taxonomic significance of ostracod larval stages, with examples from the Burundi ricefields. Bollettino di Zoologia 37: 103–120.

Ghory, F. S., F. A. Siddiqui, and Q. B. Kazmi. 2005. The complete larval development including juvenile stage of *Microprosthema validum* Stimpson, 1860 (Crustacea: Decapoda: Spongicolidae), reared under laboratory conditions. Pakistan Journal of Marine Science 14: 33–64.

Giard, A., and J. Bonnier. 1887. Contributions à l'étude des bopyriens. Travaux de l'Institut Zoologique de Lille et du Laboratoire de Zoologie Maritime de Wimereux (Pas-de-Calais) No. 5. Lille: Imprimerie L. Danel.

Gibbons, M. J., V. Spiridinov, and G. Tarling. 1999. Euphausiids. *In* D. Boltovskoy, ed. South Atlantic Zooplankton, 1241–1279. Leiden, Netherlands: Backhuys.

Gibson, V. R., and G. D. Grice. 1978. The developmental stages of a species of *Corycaeus* (Copepoda: Cyclopoida) from Saanich Inlet, British Columbia. Canadian Journal of Zoology 56: 66–74.

Giesbrecht, W. 1882. Beitrage zur Kenntniss einiger Notodelphyiden. Mitteilungen aus der Zoologischen Station zu Neapel 3: 293–372.

Giesbrecht, W. 1910. Stomatopoden, pt. 1. Fauna und Flora des Golfes von Neapel und der Angrenzenden Meeres-Abschnitte, Monographie No. 33. Berlin: R. Friedländer & Sohn.

Giesbrecht, W. 1913. Crustacea. *In* A. Lang, ed. Handbuch der Morphologie der Wirbellosen Tiere. Vol. 4, Arthropoda, 9–252. Jena, Germany: G. Fischer.

Gieskes, W. W. C. 1979. The Cladocera of the North Atlantic and the North Sea: biological and ecological studies. PhD diss., McGill University.

Gilchrist, J. D. F. 1913. A free swimming nauplioid stage in *Palinurus*. Journal of the Linnean Society of London 32: 225–231.

Gilchrist, S. L., L. E. Scotto, and R. H. Gore. 1983. Early zoeal stages of the semiterrestrial shrimp *Merguia rhizophorae* (Rathbun, 1900) cultured under laboratory conditions (Decapoda, Natantia, Hippolytidae) with a discussion of characters in the larval genus *Eretmocaris*. Crustaceana 45: 238–259.

Glenner, H. 2001. Cypris metamorphosis, injection and earliest internal development of the kentrogonid rhizocephalan *Loxothylacus panopaei* (Gissler), Crustacea: Cirripedia: Rhizocephala: Sacculinidae. Journal of Morphology 249: 43–75.

Glenner, H., and J. T. Høeg. 1994. Metamorphosis in the Cirripedia Rhizocephala and the homology of the kentrogon and trichogon. Zoologica Scripta 23: 161–173.

Glenner, H., and J. T. Høeg. 1995a. A new motile, multicellular stage involved in host invasion of parasitic barnacles (Rhizocephala). Nature 377: 147–150.

Glenner, H., and J. T. Høeg. 1995b. Scanning electron microscopy of cyprid larvae in *Balanus amphitrite amphitrite* (Crustacea: Cirripedia: Thoracica: Balanomorpha). Journal of Crustacean Biology 15: 523–536.

Glenner, H., J. T. Høeg, A. Klysner, and B. Brodin Larsen. 1989. Cypris ultra-structure, metamorphosis and sex in seven families of parasitic barnacles (Crustacea: Cirripedia: Rhizocephala). Acta Zoological (Stockholm) 70: 229–242.

Glenner, H., M. J. Grygier, J. T. Høeg, P. G. Jensen, and F. R.

Schram. 1995. Cladistic analysis of the Cirripedia Thoracica. Zoological Journal of the Linnean Society 114: 365–404.

Glenner, H., J. T. Høeg, J. J. O'Brien, and T. D. Sherman. 2000. The invasive vermigon stage in the parasitic barnacles *Loxothylacus texanus* and *L. panopaei* (Sacculinidae): closing of the rhizocephalan life cycle. Marine Biology (Berlin) 136: 249–257.

Glenner, H., J. T. Høeg, M. J. Grygier, and Y. Fujita. 2008. Induced metamorphosis in crustacean y-larvae: towards a solution to a 100-year-old riddle. BMC Biology 6: 21.

Glenner, H., J. T. Høeg, J. Stenderup, and A. V. Rybakov. 2010. The monophyletic origin of a remarkable sexual system in akentrogonid rhizocephalan parasites: a molecular and larval structural study. Experimental Parasitology 125: 3–12.

Gnanamuthu, C. P., and S. Krishnaswamy. 1948. Isopod parasites of free-living copepods of Madras. Proceedings of the Indian Academy of Sciences, B 27: 119–126.

Goldstein, J. S., H. Matsuda, and T. Takenouchi. 2008. A description of the complete development of larval Caribbean spiny lobster, *Panulirus argus* (Latreille, 1804) in culture. Journal of Crustacean Biology 28: 306–327.

Gomes, A. L. S., and J. C. O. Malta. 2002. Postura, desenvolvimento e eclosão dos ovos de *Dolops carvalhoi* Lemos de Castro (Crustácea, Branchiura) em laboratório, parasita de peixes da Amazônia central. Revista Brasileira de Zoologia 19, Suppl. 2: 141–149.

Gómez-Gutiérrez, J. 1995. Distribution patterns, abundance and population dynamics of the euphausiids *Nyctiphanes simplex* and *Euphausia eximia* in the west coast of Baja California, México. Marine Ecology Progress Series 119: 63–76.

Gómez-Gutiérrez, J. 1996. Ecology of early larval development of *Nyctiphanes simplex* Hansen (Euphausiacea) off the southwest coast of Baja California, México. Bulletin of Marine Science 58: 131–146.

Gómez-Gutiérrez, J. 2002. Hatching mechanism and delayed hatching of the eggs of three broadcast euphausiid species under laboratory conditions. Journal of Plankton Research 24: 1265–1276.

Gómez-Gutiérrez, J. 2003. Hatching mechanism and accelerated hatching of the eggs of a sac-spawning euphausiid *Nematoscelis difficilis*. Journal of Plankton Research 25: 1397–1411.

Gómez-Gutiérrez, J. 2006. Hatching mechanisms and death of euphausiid embryos during hatching: evidences for evolutionary reversal of the free-living nauplius? Océanides 21: 63–79.

Gómez-Gutiérrez, J., and C. J. Robinson. 2005. Embryonic, early larval development time, hatching mechanism and interbrood period of the sac-spawning euphausiid *Nyctiphanes simplex* Hansen. Journal of Plankton Research 27: 279–295.

Gómez-Gutiérrez, J., W. T. Peterson, and C. B. Miller. 2010. Embryo biometry of three broadcast spawning euphausiid species applied to identify cross-shelf and seasonal spawning patterns along the Oregon coast. Journal of Plankton Research 32: 739–760.

Gonor, S. L., and J. J. Gonor. 1972. Feeding, cleaning, and swimming behavior in larval stages of porcellanid crabs (Crustacea: Anomura). U.S. Fishery Bulletin 71: 225–234.

González-Ortegón, E., and J. A. Cuesta. 2006. An illustrated key to species of *Palaemon* and *Palaemonetes* (Crustacea: Decapoda: Caridea) from European waters, including the alien species *Palaemon macrodactylus*. Journal of the Marine Biological Association of the United Kingdom 86: 93–102.

Gooding, R. U. 1957. On some Copepoda from Plymouth, mainly associated with invertebrates, including three new species. Journal of the Marine Biological Association of the United Kingdom 36: 195–221.

Gopalakrishnan, K. 1973. Development and growth studies of the euphausiid *Nematoscelis difficilis* (Crustacea) based on rearing. Bulletin of the Scripps Institution of Oceanography 20. Berkeley: University of California Press.

Gordon, I. 1953. On the puerulus stage of some spiny lobsters (Palinuridae). Bulletin of the British Museum (Natural History), Zoology 2: 17–42.

Gordon, I. 1957. On *Spelaeogriphus*, a new cavernicolous crustacean from South Africa. Bulletin of the British Museum (Natural History), Zoology 5: 31–47.

Gordon, I. 1960a. Note on the decapod crustacean form [*sic*] *Stygiomedusa fabulosa* Russell. Crustaceana 1: 169–172.

Gordon, I. 1960b. On a *Stygiomysis* from the West Indies, with a note on *Spelaeogriphus* (Crustacea, Peracarida). Bulletin of the British Museum (Natural History), Zoology 6: 285–324.

Gore, R. H. 1970. *Petrolisthes armatus*: a redescription of larval development under laboratory conditions (Decapoda, Porcellanidae). Crustaceana: 76–89.

Gore, R. H. 1972. *Petrolisthes platymerus*: the development of larvae in laboratory culture (Crustacea: Decapoda: Porcellanidae). Bulletin of Marine Science 22: 336–354.

Gore, R. H. 1979. Larval development of *Galathea rostrata* under laboratory conditions, with a discussion of larval development in the Galatheidae (Crustacea Anomura). U.S. Fishery Bulletin 76: 781–806.

Gore, R. H. 1985. Molting and growth in decapod larvae. *In* A. M. Wenner, ed. Larval Growth, 1–65. Crustacean Issues No. 2. Rotterdam: A. A. Balkema.

Gore, R. H., and L. E. Scotto. 1983. Studies on decapod Crustacea from the Indian River region of Florida, 27: *Phimochirus holthuisi* (Provenzano, 1961) (Anomura: Paguridae); the complete larval development under laboratory conditions, and the systematic relationships of its larvae. Journal of Crustacean Biology 3: 93–116.

Goudey-Perrière, F. 1979. *Amphiurophilus amphiurae* (Hérouard), crustacé copépode parasite des bourses génitales de l'ophiure *Amphipholis squamata* Della Chiaje, échinoderme: morphologie des adultes et étude des stades juvéniles. Cahiers de Biologie Marine 20: 201–230.

Goy, J. W. 1990. Components of reproductive effort and delay of larval metamorphosis in tropical marine shrimp (Crustacea: Decapoda: Caridea and Stenopodidea). PhD diss., Texas A&M University.

Goy, J. W. 2010. Infraorder Stenopodidea Claus 1872. *In* J. Forest and J. C. von Vaupel Klein, eds., F. R. Schram and M. Charmantier-Daures, advisory eds. The Crustacea: Complementary to the Volumes Translated from the French of the Traité de Zoologie (Founded by P.-P. Grassé). Vol. 9, pt. A, Eucarida: Euphausiacea, Amphionidacea, and Decapoda, 215–265. Leiden, Netherlands: Brill.

Goy, J. W., and A. J. Provenzano Jr. 1978. Larval development of the rare burrowing mud shrimp *Naushonia crangonoides* Kingsley (Decapoda: Thalassinidea: Laomediidae). Biological Bulletin 154: 241–261.

Goy, J. W., C. G. Bookhout, and J. D. Costlow Jr. 1981. Larval development of the spider crab *Mithrax pleuracanthus* Stimpson reared in the laboratory (Decapoda: Brachyura: Majidae). Journal of Crustacean Biology 1: 51–62.

Grabda, J. 1963. Life cycle and morphogenesis of *Lernaea cyprinacea* L. Acta Parasitologica Polonica 11: 169–198.

Grabda, J. 1974. Contribution to the knowledge of biology of *Cecrops latreillii* Leach, 1816 (Caligoida: Cecropidae), the parasite of the ocean sunfish *Mola mola* (L.). Acta Ichthyologica et Piscatoria 3: 61–74.

Green, E. P., R. P. Harris, and A. Duncan. 1992. The production and ingestion of faecal pellets by nauplii of marine calanoid copepods. Journal of Plankton Research 14: 1631–1643.

Green, J. 1958. *Dactylopusioides macrolabris* (Claus) (Copepoda: Harpacticoida) and its frond mining nauplius. Proceedings of the Zoological Society of London 131: 49–54.

Green, J. W. 1970. Observations on the behavior and larval development of *Acanthomysis sculpta* (Tattersall), (Mysidacea). Canadian Journal of Zoology 48: 289–292.

Greenwood, J. G., and B. G. Williams. 1984. Larval and early postlarval stages in the abbreviated development of *Heterosquilla tricarinata* (Claus, 1871) (Crustacea, Stomatopoda). Journal of Plankton Research 6: 615–635.

Greenwood, J. G., M. B. Jones, and J. Greenwood. 1989. Salinity effects on brood maturation of the mysid crustacean *Mesopodopsis slabberi*. Journal of the Marine Biological Association of the United Kingdom 69: 683–694.

Gresty, K. A., G. A. Boxshall, and K. Nagasawa. 1993. The fine structure and function of the cephalic appendages of the branchiuran parasite, *Argulus japonicus* Thiele. Philosophical Transactions of the Royal Society of London, B 339: 119–135.

Greve, L. 1965. A new epicaridean from western Norway, parasite on Tanaidacea. Sarsia 20: 15–19.

Griffis, R. B., and T. H. Suchanek. 1991. A model of burrow architecture and trophic modes in thalassinidean shrimp (Decapoda: Thalassinidea). Marine Ecology Progress Series 79: 171–183.

Grobben, C. 1879. Die Entwicklungsgeschichte der *Moina rectirostris*: zugleich ein Beitrag zur Kentniss der Anatomie der Phyllopoden. Arbeiten aus dem Zoologischen Institute der Universität Wien und der Zoologischen Station in Triest 2: 203–268.

Grube, A. E. 1853. Bemerkungen über die Phyllopoden nebst einer Uebersicht ihrer Gattungen und Arten. Archiv für Naturgeschichte 19: 71–172.

Gruner, H.-E. 1993. Crustacea. *In* H.-E. Gruner, A. Kaestner, K. G. Grell. G. Hartwich, and H. H. Dathe, eds. Lehrbuch der Speziellen Zoologie. Vol. 1, Klasse Crustacea, Wirbellose Tiere 4, Arthropoda (ohne Insecta), 448–1030. Jena, Germany: Gustav Fischer Verlag.

Grygier, M. J. 1987. New records, external and internal anatomy, and systematic position of Hansen's y-larvae (Crustacea: Maxillopoda: Facetotecta). Sarsia 72: 261–278.

Grygier, M. J. 1991. Towards a diagnosis of the Facetotecta (Crustacea: Maxillopoda: Thecostraca). Zoological Science 8: 1196.

Grygier, M. J. 1992. Laboratory rearing of ascothoracidan nauplii (Crustacea: Maxillopoda) from plankton at Okinawa, Japan. Publications of the Seto Marine Biological Laboratory 35: 235–251.

Grygier, M. J. 1996. Classe des thécostracés (Thecostraca Gruvel, 1905): sous-classe des Facetotecta (Facetotecta Grygier, 1985). *In* J. Forest, ed. Traité de Zoologie: Anatomie, Systématique,

Biologie. Vol. 7, fasc. 2, Crustacés: Généralités (suite) et Systématique (Céphalocarides à Syncarides), 425–432. Paris: Masson.

Grygier, M. J., and J. T. Høeg. 2005. Ascothoracida (Ascothoracids). In K. Rohde, ed. Marine Parasites, 149–154. Wallingford, UK: CABI; Collingwood, Victoria, Australia: CSIRO.

Grygier, M. J., and S. Ohtsuka. 1995. SEM observation of the nauplius of Monstrilla hamatapex, new species, from Japan and an example of upgraded descriptive standards for monstrilloid copepods. Journal of Crustacean Biology 15: 703–719.

Grygier, M. J., and S. Ohtsuka. 2008. A new genus of monstrilloid copepods (Crustacea) with anteriorly pointing ovigerous spines and related adaptations for subthoracic brooding. Zoological Journal of the Linnean Society 152: 459–506.

Guerao, G. 1993. The larval development of a fresh-water prawn Palaemonetes zariquieyi Sollaud, 1939 (Decapoda, Palaemonidae) reared in the laboratory. Crustaceana 64(2): 226–241.

Guerao, G., and P. Abelló. 1996a. Description of the first larval stage of Polycheles typhlops (Decapoda: Eryonidea: Polychelidae). Journal of Natural History 30: 1179–1184.

Guerao, G., and P. Abelló. 1996b. Morphology of the prezoea and the first zoea of the deep-sea spider crab Anamathia rissoana (Brachyura, Majidae, Pisinae). Scientia Marina 60: 245–251.

Guerao, G., E. Macpherson, S. Samadi, B. Richer de Forges, and M.-C. Boisselier. 2006a. First stage zoeal descriptions of five Galatheoidea species from western Pacific (Crustacea: Decapoda: Anomura). Zootaxa 1227: 1–29.

Guerao, G., D. Díaz, and P. Abelló. 2006b. Morphology of puerulus and post-puerulus stages of the spiny lobster Palinurus mauritanicus (Decapoda: Palinuridae). Journal of Crustacean Biology 26: 480–494.

Günzl, H. 1978. Der Ankerapparat von Sida crystallina (Crustacea, Cladocera), 1: Bau und Funktion. Zoomorphologie 90: 197–204.

Gurney, R. 1923. The larval stages of Processa canaliculata, Leach. Journal of the Marine Biological Association of United Kingdom 13: 245–265.

Gurney, R. 1924. Crustacea, pt. 9: Decapod Larvae. Vol. 8, Natural History Reports: British Antarctic ("Terra Nova") Expedition, 1910, Zoology, 37–202. London: British Museum (Natural History).

Gurney, R. 1926. The nauplius larva of Limnetis gouldi. Revue der Gesamten Hydrobiologie und Hydrogeographie 16: 114–117.

Gurney, R. 1927. Report on the larvae of the Crustacea Decapoda. In Zoological Results of the Cambridge Expedition to the Suez Canal, 1924. Transactions of the Zoological Society of London 22: 231–286.

Gurney, R. 1932. British Fresh-Water Copepoda. Vol. 2, Harpacticoida. London: Ray Society.

Gurney, R. 1933a. British Fresh-Water Copepoda. Vol. 3, The Classification of the Cyclopoida and the Parasitic Forms Derived from Them. London: Ray Society.

Gurney, R. 1933b. Notes on some Copepoda from Plymouth. Journal of the Marine Biological Association of the United Kingdom 19: 299–304.

Gurney, R. 1934. The development of Rhincalanus. Discovery Reports 9: 207–214.

Gurney, R. 1935. The mode of attachment of the young in the crayfishes of the families Astacidae and Parastacidae. Annals and Magazine of Natural History, ser. 10, 16: 553–555.

Gurney, R. 1936a. Larvae of decapod Crustacea, 1: Stenopodidea. Discovery Reports 12: 379–392.

Gurney, R. 1936b. Larvae of decapod Crustacea, 2: Amphionidae. Discovery Reports 12: 392–399.

Gurney, R. 1936c. Larvae of decapod Crustacea, 3: Phyllosoma. Discovery Reports 12: 400–440.

Gurney, R. 1937. Notes on some decapod Crustacea from the Red Sea, 2: the larvae of Upogebia savignyi (Strahl). Proceedings of the Zoological Society of London B107: 85–101.

Gurney, R. 1938. Larvae of decapod Crustacea, 5: Nephropsidea and Thalassinidea. Discovery Reports 17: 291–344.

Gurney, R. 1942. Larvae of Decapod Crustacea. London: Ray Society.

Gurney, R. 1943. The larval development of two penaeid prawns from Bermuda of the genera Sicyonia and Penaeopsis. Proceedings of the Zoological Society of London B113: 1–16.

Gurney, R. 1946. Notes on stomatopod larvae. Proceedings of the Zoological Society of London 116: 133–175.

Gurney, R. 1947. Some notes on the development of the Euphausiacea. Proceedings of the Zoological Society of London 117: 49–64.

Gurney, R., and M. V. Lebour. 1940. Larvae of Crustacea Decapoda, 6: the genus Sergestes. Discovery Reports 20: 1–68.

Gutu, M. 1998. Spelaeogriphacea and Mictacea (partim) suborders of a new order, Cosinzeneacea (Crustacea, Peracarida). Travaux du Museum d'Histoire Naturelle "Grigore Antipa" 40: 121–129.

Gutu, M., and T. M. Iliffe. 1998. Description of a new hirsutiid (n. g., n. sp.) and reassignment of the family Hirsutiidae from order Mictacea to the new order, Bochusacea (Crustacea, Peracarida). Travaux du Museum d'Histoire Naturelle "Grigore Antipa" 40: 93–120.

Gutzmann, E., P. Martínez Arbizu, A. Rose, and G. Veit-Köhler. 2004. Meiofauna communities along an abyssal depth gradient in the Drake Passage. Deep-Sea Research, pt. 2: Topical Studies in Oceanography 51: 1617–1628.

Hadfield, M. G. 2011. Biofilms and marine invertebrate larvae: what bacteria produce that larvae use to choose settlement sites. Annual Review of Marine Science 3: 453–470.

Hadfield, M. G. 2012. Why larvae settle where they do: the importance of bacteria. Tenth International Larval Biology Symposium, Berkeley, California: 11 (abstract).

Hairston, N. G., Jr, R. A. Van Brunt, C. M. Kearns, and D. R. Engstrom. 1995. Age and survivorship of diapausing eggs in a sediment egg bank. Ecology 76: 1706–1711.

Hakalahti, T., and E. T. Valtonen. 2003. Population structure and recruitment of the ectoparasite Argulus coregoni Thorell (Crustacea: Branchiura) on a fish farm. Parasitology 127: 79–85.

Hakalahti, T., A. F. Pasternak, and E. T. Valtonen. 2003. Seasonal dynamics of egg laying and egg-laying strategy of the ectoparasite Argulus coregoni (Crustacea: Branchiura). Parasitology 128: 655–660.

Hakalahti, T., H. Häkkinen, and E. T. Valtonen. 2004. Ectoparasitic Argulus coregoni (Crustacea: Branchiura) hedge their bets: studies on egg hatching dynamics. Oikos 107: 295–302.

Hall, B. K., and M. H. Wake, eds. 1999. The Origin and Evolution of Larval Forms. San Diego: Academic Press.

Hall, J. R. 1972. Aspects of the biology of Derocheilocaris typica, 2: distribution. Marine Biology (Berlin) 12: 42–52.

Hall, J. R., and R. R. Hessler. 1971. Aspects in the population dynamics of Derocheilocaris typica (Mystacocarida, Crustacea). Vie et Milieu 22: 305–326.

Hallberg, E., R. Elofsson, and M. J. Grygier. 1985. An ascothoracid compound eye. Sarsia 70: 167–171.

Hamano, T., and S. Matsuura. 1987. Egg size, duration of incubation, and larval development of the Japanese mantis shrimp in the laboratory. Nippon Suisan Gakkaishi 53: 23–39.

Hamr, P. 1992. Embryonic and post-embryonic development in the Tasmanian freshwater crayfishes *Astacopsis gouldi*, *Astacopsis franklini*, and *Parastacoids tasmanicus tasmanicus* (Decapoda: Parastacidae). Australian Journal of Marine and Freshwater Research 43: 861–878.

Haney, T. A., and J. W. Martin. 2004. A new genus and species of leptostracan (Crustacea: Malacostraca: Phyllocarida) from Guana Island, British Virgin Islands, and a review of leptostracan genera. Journal of Natural History 38: 447–469.

Hansen, H. J. 1895. Isopoden, Cumaceen und Stomatopoden der Plankton Expedition. Vol. 2 (G, c), Ergebnisse der Plankton Expedition der Humboldt Stiftung. Kiel, Germany: Lipsius & Tischer.

Hansen, H. J. 1897. The Choniostomatidæ: a family of Copepoda, parasites on Crustacea Malacostraca. Copenhagen: Andr. Fred. Høst & Son.

Hansen, H. J. 1899. Die Cladoceren und Cirripedien der Plankton-Expedition. Vol. 2 (G, d), Ergebnisse der Plankton-Expedition der Humboldt-Stiftung. Kiel, Germany: Lipsius & Tischer.

Hansen, H. J. 1926. The Stomatopoda of the Siboga Expedition. Siboga Expeditie 1899–1900, No. 35. Leiden, Netherlands: E. J. Brill.

Haq, S. M. 1965. Development of the copepod *Euterpina acutifrons* with special reference to dimorphism in the male. Proceedings of the Zoological Society of London 144: 175–201.

Harada, E. 1958. Notes on the naupliosoma and newly hatched phyllosoma of *Ibacus ciliatus* (von Siebold). Publications of the Seto Marine Biological Laboratory 7: 173–180.

Harding, J. P. 1954. The copepod *Thalestris rhodymeniae* (Brady) and its nauplius, parasitic in the seaweed *Rhodymenia palmata* (L.) Grev. Proceedings of the Zoological Society of London 124: 153–161.

Hardy, A. 1970. The Open Sea: Its Natural History. Vol. 1, The World of Plankton. London: Collins.

Harrington, S., and P. G. Thomas. 1987. Observations on spawning by *Euphausia crystallorophias* from waters adjacent Enderby Land (East Antarctica) and speculations on the early ontogenetic ecology of neritic euphausiids. Polar Biology 7: 93–95.

Harris, R. P., P. H. Wiebe, J. Lenz, H. R. Skjoldal, and M. Huntley, eds. 2000. ICES Zooplankton Methodology Manual. San Diego: Academic Press.

Harrison, K. 1984. The morphology of the sphaeromatid brood pouch (Crustacea: Isopoda: Sphaeromatidae). Zoological Journal of the Linnean Society 82: 363–407.

Hart, C. W., Jr., L. C. Hayek, J. Clark, and W. H. Clark. 1985. The Life History and Ecology of the Entocytherid Ostracod *Uncinocythere occidentalis* (Kozloff and Whitmann) in Idaho. Smithsonian Contributions to Zoology No. 419. Washington, DC: Smithsonian Institution Press.

Hart, J. F. L. 1965. Life history and larval development of *Cryptolithodes typicus* Brandt (Decapoda, Anomura) from British Columbia. Crustaceana 8: 255–277.

Hartnoll, R. G. 1966. A new entoniscid from Jamaica (Isopoda, Epicaridea). Crustaceana 11: 45–52.

Harvey, A. W. 1992. Abbreviated larval development in the Australian terrestrial hermit crab *Coenobita variabilis* McCulloch (Anomura: Coenobitidae). Journal of Crustacean Biology 12: 196–209.

Harvey, A. W. 1993. Larval settlement and metamorphosis in the sand crab *Emerita talpoida* (Crustacea: Decapoda: Anomura). Marine Biology (Berlin) 117: 575–581.

Harvey, A. W. 1996. Delayed metamorphosis in Florida hermit crabs: multiple cues and constraints (Crustacea: Decapoda: Paguridae and Diogenidae). Marine Ecology Progress Series 141: 27–36.

Harvey, A. W., and E. A. Colasurdo. 1993. Effects of shell and food availability on metamorphosis in the hermit crabs *Pagurus hirsutiusculus* (Dana) and *Pagurus granosimanus* (Stimpson). Journal of Experimental Marine Biology and Ecology 165: 237–249.

Harvey, A. W., J. W. Martin, and R. Wetzer. 2001. Phylum Arthropoda: Crustacea. *In* C. Young, M. Sewell, and M. Rice, eds. Atlas of Marine Invertebrate Larvae, 337–369. London: Academic Press.

Hashizume, K. 1999. Larval development of seven species of *Lucifer* (Dendrobranchiata, Sergestoidea), with a key for the identification of their larval forms. *In* F. R. Schram and J. C. von Vaupel Klein, eds. Vol. 1, Crustaceans and the Biodiversity Crisis: Proceedings of the Fourth International Crustacean Congress, Amsterdam, The Netherlands, July 20–24, 1998, 753–779. Leiden, Netherlands: Brill.

Hassan, H. 1974. A generic key to the Penaeid larvae of Pakistan. Agriculture Pakistan 25: 227–236.

Haug, J. T. and C. Haug. 2013. An unusual fossil larva, the ontogeny of achelatan lobsters, and the evolution of metamorphosis. Bulletin of Geosciences 88: 195–206.

Haug, C., J. T. Haug, and D. Waloszek. 2009. Morphology and ontogeny of the Upper Jurassic mantis shrimp *Spinosculda ehrlichi*, n. gen., n. sp. from southern Germany. Palaeodiversity 2: 111–118.

Haug, C., J. T. Haug, S. R. Fayers, N. H. Trewin, C. Castellani, D. Waloszek, and A. Maas. 2012. Exceptionally preserved nauplius larvae from the Devonian Windyfield chert, Rhynie, Aberdeenshire, Scotland. Paleontologia Electronica 15: 2–24A.

Haug, J. T., C. Haug, and M. Ehrlich. 2008. First fossil stomatopod larva (Arthropoda: Crustacea) and a new way of documenting Solnhofen fossils (Upper Jurassic, southern Germany). Palaeodiversity 1: 103–109.

Haug, J. T., C. Haug, D. Waloszek, A. Maas, M. Wulf, and G. Schweigert. 2009a. Development in Mesozoic scyllarids and implications for the evolution of Achelata (Reptantia, Decapoda, Crustacea). Paleodiversity 2: 97–110.

Haug, J. T., A. Maas, and D. Waloszek. 2009b. Ontogeny of two Cambrian stem crustaceans, †*Goticaris longispinosa* and †*Cambropachycope clarksoni*. Palaeontographica, A 289: 1–43.

Haug, J. T., D. Waloszek, C. Haug, and A. Maas. 2010a. High-level phylogenetic analysis using developmental sequences: the Cambrian †*Martinssonia elongata*, †*Musacaris gerdgeyeri* gen. et sp. nov. and their position in early crustacean evolution. Arthropod Structure and Development 39: 154–173.

Haug, J. T., A. Maas, and D. Waloszek. 2010b. †*Henningsmoenicaris scutula*, †*Sandtorpia vestrogothiensis* gen. et sp. nov. and heterochronic events in early crustacean evolution. Transactions of the Royal Society of Edinburgh, Earth and Environmental Science 100: 311–350.

Haug, J. T., C. Haug, A. Maas, V. Kutschera, and D. Waloszek. 2010c. Evolution of mantis shrimps (Stomatopoda, Malacostraca) in the light of new Mesozoic fossils. BMC Evolutionary Biology 10: 290.

Haug, J. T., J. Olesen, A. Maas, and D. Waloszek. 2011a. External morphology and post-embryonic development of *Derocheilo-caris remanei* (Mystacocarida) revisited, with a comparison to the Cambrian taxon *Skara*. Journal of Crustacean Biology 32: 668–692.

Haug, J. T., C. Haug, D. Waloszek, and G. Schweigert. 2011b. The importance of lithographic limestones for revealing ontogenies in fossil crustaceans. Swiss Journal of Geosciences 104, Suppl. 1: S85–S98.

Haug, J. T., D. Audo, S. Charbonnier, and C. Haug. 2013a. Diversity of developmental patterns in achelate lobsters—today and in the Mesozoic. Development Genes and Evolution 223: 363–373.

Haug, J. T., A. Maas, C. Haug, and D. Waloszek. 2013b. Evolution of crustacean appendages. *In* M. Thiel and L. Watling, eds. Functional Morphology and Diversity. Vol. 1, Natural History of the Crustacea, 34–73. New York: Oxford University Press.

Haynes, E. [B.] 1973. Description of prezoeae and stage I zoeae of *Chionoecetes bairdi* and *C. opilio* (Oxyrhyncha, Oregoniidae). U.S. Fishery Bulletin 71: 769–774.

Haynes, E. B. 1985. Morphological development, identification, and biology of larvae of Pandalidae, Hippolytidae, and Crangonidae (Crustacea, Decapoda) of the northern North Pacific ocean. U.S. Fishery Bulletin 83(3): 253–288.

Heard, R. W., R. A. King, D. M. Knott, B. P. Thoma, and S. Thornton-DeVictor. 2007. A guide to the Thalassinidea (Crustacea: Malacostraca: Decapoda) of the South Atlantic Bight. NOAA Professional Paper NMFS 8. Seattle: U.S. Dept. of Commerce, National Oceanic and Atmospheric Administration, National Marine Fisheries Service, Scientific Publications Office.

Heath, H. 1924. The external development of certain phyllopods. Journal of Morphology 38: 453–483.

Heegaard, P. E. 1947. Contribution to the phylogeny of the arthropods. Spolia Zoological Musei Hauniensis 8: 1–236.

Heegaard, P. E. 1951. Antarctic parasitic copepods and an ascothoracid cirriped from brittle-stars. Videnskabelige Meddelelser fra Dansk Naturhistorisk Forening i Kjøbenhavn 113: 171–190.

Heegaard, P. 1966. Larvae of Decapod Crustacea: The Oceanic Penaeids *Solenocera–Cerataspis–Cerataspides*. Dana Report No. 67. Copenhagen: Høst.

Heegaard, P. 1969. Larvae of Decapod Crustacea: The Amphionidae. Dana Report No. 77. Copenhagen: Høst.

Heldt, J. H. 1938. La reproduction chez les crustacés décapodes de la famille des pénéides. Annales de l'Institut Océanographique du Monaco 18: 21–306.

Heldt, J. H. 1955. Contribution à l'étude de la biologie des crevettes pénéides *Aristeomorpha foliacea* (Risso) et *Aristeus antennatus* (Risso) (formes larvaires). Bulletin de la Société des Sciences Naturelles de Tunisie 8(1–2): 1–29.

Hendler, G., and M. Dojiri. 2009. The contrariwise life of a parasitic, pedomorphic copepod with a non-feeding adult: ontogenesis, ecology, and evolution. Invertebrate Biology 128: 65–82.

Hendler, G., and I.-H. Kim. 2010. Larval biology of thaumatopsyllid copepods endoparasitic in Caribbean ophiuroids. Journal of Crustacean Biology 30: 206–224.

Hentschel, E. 1967. Die postembryonalen Entwicklungsstadien von *Artemia salina* Leach bei verschiedenen Temperaturen (Anostraca, Crustacea). Zoologischer Anzeiger 180: 372–384.

Heron, G., and D. Damkaer. 1986. A new nicothoid copepod

parasitic on mysids from northwestern North America. Journal of Crustacean Biology 6: 652–665.

Herring, P. J. 1988. Copepod luminescence. *In* G. A. Boxshall and H. K. Schminke, eds. Biology of Copepods: Proceedings of the Third International Conference on Copepoda, held at the British Museum (Natural History), London, August 10–August 14, 1987. Hydrobiologia 167/168: 183–195.

Herzig, A., and B. Auer. 1990. The feeding behaviour of *Leptodora kindtii* and its impact on the zooplankton community of Neusiedler See (Austria). Hydrobiologia 198: 107–117.

Hesse, C. E. 1855. Recherches sur les crustacés désignés par les noms d'ancée et pranize. Comptes Rendus Hebdomadaires des Séances de l'Académie des Sciences 41: 970.

Hesse, C. E. 1858. Mémoire sur la transformation des pranizes en ancées, sur les moeurs et les habitudes de ces crustacés. Comptes Rendus Hebdomadaires des Séances de l'Académie des Sciences 46: 568.

Hesse, C. E. 1864. Mémoire sur les pranizes et les ancées. Mémoires des Savants Étrangers de l'Académie des Sciences 18: 231–302.

Hessler, R. R. 1969. Mystacocarida. *In* R. C. Moore, ed. Treatise of Invertebrate Paleontology, pt. R, Arthropoda 4, vol. 2 Decapoda, R120–R128. Lawrence: University of Kansas Press and Geological Society of America.

Hessler, R. R. 1984. *Dahlella caldariensis*, new genus, new species: a leptostracan (Crustacea, Malacostraca) from deep-sea hydrothermal vents. Journal of Crustacean Biology 4: 655–664.

Hessler, R. R. 1992a. Reflections on the phylogenetic position of the Cephalocarida. Acta Zoologica (Stockholm) 73: 315–316.

Hessler, R. R. 1992b. Mystacocarida. *In* R. P. Higgins and H. Thiel, eds. Introduction to the Study of Meiofauna, rev. ed., 377–379. Washington, DC: Smithsonian Institution Press.

Hessler, R. R., and R. Elofsson 1992. Cephalocarida. *In* F. W. Harrison and A. G. Humes, eds. Microscopic Anatomy of Invertebrates. Vol. 9, Crustacea, chapter 2. New York: Wiley-Liss.

Hessler, R. R., and W. A. Newman. 1975. A trilobitomorph origin for the Crustacea. Fossils and Strata 4: 437–59.

Hessler, R. R., and H. L. Sanders. 1966. *Derocheilocaris typicus* Pennak & Zinn (Mystacocarida) revisited. Crustaceana 11: 141–155.

Hessler, R. R., and L. Watling. 1999. Les péracarides: un groupe controversé. *In* J. Forest, ed. Traité de Zoologie: Anatomie, Systématique, Biologie. Vol. 7, Crustacés, fasc. 3A: Péracarides, 1–10. Mémoires de l'Institute Océanographique, Fondation Albert I, Prince de Monaco No. 19. Paris: Masson.

Hessler, R. R., R. Elofsson, and A. Y. Hessler. 1995. Reproductive system of *Hutchinsoniella macracantha* (Cephalocarida). Journal of Crustacean Biology 15: 493–522.

Heymons, R. 1926. Pentastomida. *In* H. Balss, ed. Vol. 3(1), Handbuch der Zoologie, eine Naturgeschichte der Stämme des Tierreiches 69–131. Berlin: W. de Gruyter.

Heymons, R. 1935. Pentastomida. Vol. 5, sect. 4, bk. 1 *in* H. G. Bronns, ed. Klassen und Ordnungen des Tierreichs. Leipzig: Akademische Verlagsgesellschaft.Hickman, C. S. 1999. Larvae in invertebrate development and evolution. *In* B. K. Hall and M. H. Wake, eds. The Origin and Evolution of Larval Forms, 21–59. San Diego: Academic Press.

Hickman, V. V. 1937. The embryology of the syncarid crustacean, *Anaspides tasmaniae*. Papers and Proceedings of the Royal Society of Tasmania for 1936: 1–35.

Hill, L. L., and R. E. Coker. 1930. Observations on mating habits of *Cyclops*. Journal of the Elisha Mitchell Scientific Society 45: 206–220.

Hiller, A., H. Kraus, M. Almon, and B. Werding. 2006. The *Petrolisthes galathinus* complex: species boundaries based on color pattern, morphology and molecules, and evolutionary interrelationships between this complex and other Porcellanidae (Crustacea: Decapoda: Anomura). Molecular Phylogenetics and Evolution 40: 547–569.

Hines, A. H. 1986. Larval patterns in the life histories of brachyuran crabs (Crustacea, Decapoda, Brachyura). Bulletin of Marine Science 39: 444–466.

Hipeau-Jacquotte, R. 1978. Relation entre âge de l'hôte et type de développement chez un copépode ascidicole Notodelphyidae. Comptes Rendus Hebdomadaires des Séances de l'Académie des Sciences, D 287: 1207–1210.

Hipeau-Jacquotte, R. 1987. Ultrastructure and presumed function of the pleural dermal glands in the atypical male of the parasitic copepod *Pachypygus gibber* (Crustacea: Notodelphyidae). Journal of Crustacean Biology 7: 60–70.

Hiruta, S. 1977. A new species of the genus *Sarsiella* Norman from Hokkaido, with reference to the larval stages (Ostracoda: Myodocopina). Journal of the Faculty of Science, Hokkaido University, ser. 6, Zoology 21: 44–60.

Hiruta, S. 1978. Redescription of *Sarsiella misakiensis* Kajiyama from Hokkaido, with reference to the larval stages (Ostracoda: Myodocopina). Journal of the Faculty of Science, Hokkaido University, ser. 6, Zoology 21: 262–278.

Hiruta, S. 1979a. A new species of the genus *Bathyleberis* Kornicker from Hokkaido, with reference to the larval stages (Ostracoda: Myodocopina). Journal of the Faculty of Science, Hokkaido University, ser. 6, Zoology 22: 99–121.

Hiruta, S. 1979b. Redescription of *Asteropteron fuscum* (G. W. Müller) from Amakusa, Kyushu, with reference to the larval stages (Ostracoda: Myodocopina). Proceedings of the Japanese Society of Systematic Zoology 17: 15–30.

Hiruta, S. 1980. Morphology of the larval stages of *Vargula hilgendorfii* (G. W. Müller) and *Euphilomedes nipponica* Hiruta from Japan (Ostracoda: Myodocopina). Journal of Hokkaido University of Education, sect. 2B, Biology 30: 145–167.

Hiruta, S. 1983. Post-embryonic development of myodocopid Ostracoda. *In* R. F. Maddocks, ed. Applications of Ostracoda: Proceedings of the Eighth International Symposium on Ostracoda, Houston, Texas, July 26–29, 1982, 667–677. Houston: University of Houston–University Park, Department of Geosciences.

Ho, J.-S. 1966. Larval stages of *Cardiodectes* sp. (Caligoida: Lernaeoceriformes), a copepod parasitic on fishes. Bulletin of Marine Science 16: 159–199.

Ho, J.-S., and J.-S. Hong. 1988. Harpacticoid copepods (Thalestridae) infesting the cultivated wakame (brown alga, *Undaria pinnatifida*) in Korea. Journal of Natural History 22: 1623–1637.

Ho, J.-S., and C.-l. Lin. 2004. Sea lice of Taiwan (Copepoda: Siphonostomatoida: Caligidae). Keelung, Taiwan, R.O.C.: Sueichan Press.

Høeg, J. T. 1984. Size and settling behaviour in male and female cypris larvae of the parasitic barnacle *Sacculina carcini* Thompson (Crustacea: Cirripedia: Rhizocephala). Journal of Experimental Marine Biology and Ecology 76: 145–156.

Høeg, J. T. 1985a. Cypris settlement, kentrogon formation and host invasion in the parasitic barnacle *Lernaeodiscus porcellanae*

(Müller) (Crustacea: Cirripedia: Rhizocephala). Acta Zoologica (Stockholm) 66: 1–45.

Høeg, J. T. 1985b. Male cypris settlement in *Clistosaccus paguri* Lilljeborg (Crustacea: Cirripedia: Rhizocephala). Journal of Experimental Marine Biology and Ecology 89: 221–235.

Høeg, J. T. 1987a. Male cypris metamorphosis and a new male larval form, the trichogon, in the parasitic barnacle *Sacculina carcini* (Crustacea: Cirripedia: Rhizocephala). Philosophical Transactions of the Royal Society of London, B 317: 47–63.

Høeg, J. T. 1987b. The relation between cypris ultrastructure and metamorphosis in male and female *Sacculina carcini* (Crustacea, Cirripedia). Zoomorphology 107: 299–311.

Høeg, J. T. 1990. "Akentrogonid" host invasion and an entirely new type of life cycle in the rhizocephalan parasite *Clistosaccus paguri* (Thecostraca: Cirripedia). Journal of Crustacean Biology 10: 37–52.

Høeg, J. T. 1992. The phylogenetic position of the Rhizocephala: are they truly barnacles? Acta Zoologica (Stockholm) 73: 323–326.

Høeg, J. T. 1995. The biology and life cycle of the Cirripedia Rhizocephala. Journal of the Marine Biological Association of the United Kingdom 75: 517–550.

Høeg, J. T., and J. Lützen. 1993. Comparative morphology and phylogeny of the family Thompsoniidae (Cirripedia, Rhizocephala, Akentrogonida), with descriptions of three new genera and seven new species. Zoologica Scripta 22: 363–386.

Høeg, J. T., and J. Lützen. 1995. Life cycle and reproduction in the Cirripedia Rhizocephala. *In* A. D. Ansell, R. N. Gibson, and M. Barnes, eds. Vol. 33, Oceanography and Marine Biology: An Annual Review, 427–485. London: UCL Press.

Høeg, J. T., and J. Lützen. 1996. Rhizocephala. *In* J. Forest, ed. Traité de Zoologie: Anatomie, Systématique, Biologie. Vol. 7, fasc. 2, Crustacés: Généralités (suite) et Systématique (Céphalocarides à Syncarides), 541–568. Paris: Masson.

Høeg J. T., and O. S. Møller. 2006. When similar beginnings leads to different ends: constraints and diversity in cirripede larval development. Invertebrate Reproduction and Development 49: 125–142.

Høeg, J. T., and A. V. Rybakov. 1992. Revision of the Rhizocephala Akentrogonida (Crustacea: Cirripedia), with a list of all the species and a key to the identification of families. Journal of Crustacean Biology 12: 600–609.

Høeg, J. T., B. Hosfeld, and P. G. Jensen. 1998. TEM studies of lattice organs of cirripede cypris larvae (Crustacea, Thecostraca, Cirripedia). Zoomorphology 118: 195–205.

Høeg, J. T., N. C. Lagersson, and H. Glenner. 2004. The complete cypris larva and its significance in thecostracan phylogeny. *In* G. Scholtz, ed. Evolutionary and Developmental Biology of Crustacea, 197–215. Crustacean Issues No. 15. Lisse, Netherlands: A. A. Balkema.

Høeg, J. T., H. Glenner, and J. Shields. 2005. Cirripedia Thoracica and Rhizocephala (barnacles). *In* K. Rohde, ed. Marine Parasites, 154–165. Wallingford, UK: CABI; Collingwood, Victoria, Australia: CSIRO.

Høeg J. T., M. Pérez-Losada, H. Glenner, G. A. Kolbasov, and K. A. Crandall. 2009a. Evolution of morphology, ontogeny and life cycles within the Crustacea Thecostraca. Arthropod Systematics & Phylogeny 67: 179–217.

Høeg, J. T., Y. Achituv, B. Chan, C. I. Chan, M. Pérez-Losada, and P. G. Jensen. 2009b. Cypris morphology in the barnacles

Ibla and *Paralepas* (Cirripedia: Thoracica): implications for understanding cirripede evolution. Journal of Morphology 270: 241–255.

Høeg, J. T., D. Maruzzo, K. Okano, H. Glenner, and B. K. K. Chan. 2012. Metamorphosis in balanomorphan, pedunculated and parasitic barnacles: a video based analysis. Integrative and Comparative Biology 52: 337–347.

Hoeh, W. R., N. D. Smallwood, D. M. Senyo, E. G. Chapman, and S. C. Weeks. 2006. Evaluating the monophyly of *Eulimnadia* and the Limnadiinae (Branchiopoda: Spinicaudata) using DNA sequences. Journal of Crustacean Biology 26: 182–192.

Holdich, D. M., and J. A. Jones. 1983. Tanaids. Cambridge: Cambridge University Press.

Holthuis, L. B. 1964. The earliest published record of a cumacean. Crustaceana 7: 317–318.

Holthuis, L. B. 1974. Biological results of the University of Miami deep-sea expeditions, 106: the lobsters of the superfamily Nephropidea of the Atlantic Ocean (Crustacea: Decapoda). Bulletin of Marine Science (University of Miami) 24: 723–884.

Holthuis, L. B. 1983. Notes on the genus *Enoplometopus,* with descriptions of a new subgenus and two new species (Crustacea, Decapoda, Axiidae). Zoologische Mededeelingen Leiden 56: 281–298.

Holthuis, L. B. 1985. A revision of the family Scyllaridae (Crustacea: Decapoda: Macrura), 1: subfamily Ibacinae. Zoologische Verhandelingen 218: 1–130.

Holthuis, L. B. 1991. Marine Lobsters of the World: An Annotated and Illustrated Catalogue of Species of Interest to Fisheries Known to Date. FAO Fisheries Synopsis No. 125, FAO Species Catalog vol. 13. Rome: Food and Agriculture Organization of the United Nations.

Holthuis, L. B. 1993. The Recent Genera of the Caridean and Stenopodidean Shrimps (Crustacea, Decapoda), with an Appendix on the Order Amphionidacea, rev. ed. Ed. C. H. J. M. Fransen and C. van Achterberg. Leiden, Netherlands: Nationaal Natuurhistorisch Museum.

Holthuis, L. B. 2002. The Indo-Pacific scyllarine lobsters (Crustacea, Decapoda, Scyllaridae. Zoosystema 24: 499–683.

Hong, S. Y. 1976. Zoeal stages of *Orithyia sinica* (Linnaeus) (Decapoda, Calappidae) reared in the laboratory. Publication of the Institute of Marine Sciences, National Fisheries University of Busan 9: 17–23.

Hong, S. Y. 1988a. The prezoeal stage in various decapod crustaceans. Journal of Natural History 22: 1041–1075.

Hong, S. Y. 1988b. Development of epipods and gills in some pagurids and brachyurans. Journal of Natural History 22: 1005–1040.

Hong, S. Y., and M. H. Kim. 2002. Larval development of *Parapagurodes constans* (Decapoda: Anomura: Paguridae) reared in the laboratory. Journal of Crustacean Biology 22: 882–893.

Hong, S. Y., R. I. Perry, J. A. Boutillier, and M. H. Kim. 2005. Larval development of *Acantholithodes hispidus* (Stimpson) (Decapoda: Anomura: Lithodidae) reared in the laboratory. Invertebrate Reproduction and Development 47: 101–110.

Horne, D. J. 2005. Homology and homoeomorphy in ostracod limbs. Hydrobiologia 538: 55–80.

Horne, D. J., A. Cohen, and K. Martens. 2002. Taxonomy, morphology and biology of Quaternary and living Ostracoda. *In* J. A. Holmes and A. Chivas, eds. The Ostracoda: Applications in Quaternary Research, 5–6. Geophysical Monograph No. 131. Washington, DC: American Geophysical Union.

Horne, D. J., R. J. Smith, J. E. Whittaker, and J. W. Murray. 2004. The first British record and a new species of the superfamily Terrestricytheroidea (Crustacea, Ostracoda): morphology, ontogeny, lifestyle and phylogeny. Zoological Journal of the Linnean Society 142: 253–288.

Hou, X.-G., D. J. Siveter, M. Williams, D. Walossek, and J. Bergström. 1996. Appendages of the arthropod *Kunmingella* from the Early Cambrian of China: its bearing on the systematic position of the Bradoriida and the fossil record of the Ostracoda. Philosophical Transactions of the Royal Society of London, B 351: 1131–1145.

Hsü, F. 1933. Studies on the anatomy and development of a fresh water phyllopod, *Chirocephalus nankinensis* (Shen). Contributions of the Biological Laboratory of the Scientific Society of China, Zoological Series 9: 119–163.

Huang, Y., C. Callahan, and M. G. Hadfield. 2012. Recruitment in the sea: bacterial genes required for inducing larval settlement in a marine worm. Nature: Scientific Reports 228. doi:10.1038/srep00228.

Huguet, D., J. E. García Muñoz, J. E. García Raso, and J. A. Cuesta. 2011. Extended parental care in the freshwater shrimp genus *Dugastella* Bouvier (Decapoda, Atyidae, Paratyinae). Crustaceana 84(2): 251–255.

Hulsemann, K. 1991. Tracing homologies in appendages during ontogenetic development of calanoid copepods. *In* S-i. Uye, S. Nishida, and J.-S. Ho, eds. Proceedings of the Fourth International Conference on Copepoda, Karuizawa, Japan, 16–20 September 1990. Special Volume, Bulletin of the Plankton Society of Japan: 105–114.

Humes, A. G., and R. F. Cressey. 1958. A new family containing two new genera of cyclopoid copepods parasitic on starfishes. Journal of Parasitology 44: 395–408.

Huq, A., E. B. Small, P. A. West, M. I. Huq, R. Rahman, and R. R. Colwell. 1983. Ecological relationships between *Vibrio cholerae* and planktonic crustacean copepods. Applied Environmental Microbiology 45: 275–283.

Huys, R. 1990a. *Campyloxiphos dineti* gen. et spec. nov. from off Namibia and a redefinition of the Deoterthridae Boxshall & Lincoln (Crustacea: Tantulocarida). Journal of Natural History 24: 415–432.

Huys, R. 1990b. *Coralliotantulus coomansi* gen. et sp. n.: first record of a tantulocaridan (Crustacea: Maxillopoda) from shallow subtidal sands in tropical waters. Stygologia 5: 183–198.

Huys, R. 1990c. Amsterdam expeditions to the West Indian Islands, report 64: a new family of harpacticoid copepods and an analysis of the phylogenetic relationships within the Laophontoidea T. Scott. Bijdragen tot de Dierkunde 60: 79–120.

Huys, R. 1991. Tantulocarida (Crustacea: Maxillopoda): a new taxon from the temporary meiobenthos. P.S.Z.N. [Pubblicazioni della Stazione Zoologica di Napoli] 1, Marine Ecology 12: 1–34.

Huys, R. 1993. Styracothoracidae (Copepoda: Harpacticoida), a new family from the Philippine deep sea. Journal of Crustacean Biology 13: 769–783.

Huys, R. 2001. Splanchnotrophid systematics: a case of polyphyly and taxonomic myopia. Journal of Crustacean Biology 21: 106–156.

Huys, R., and R. Böttger-Schnack. 1994. Taxonomy, biology and phylogeny of Miraciidae (Copepoda: Harpacticoida). Sarsia 79: 207–283.

Huys, R., and G. A. Boxshall. 1991. Copepod Evolution. London: Ray Society.

Huys, R., and S. Conroy-Dalton. 1997. Discovery of hydrothermal vent Tantulocarida on a new genus of Argestidae (Copepoda: Harpacticoida). Cahiers de Biologie Marine 38: 235–249.

Huys, R., P. F. Andersen, and R. M. Kristensen. 1992a. *Tantulacus hoegi* gen. et sp. nov. (Tantulocarida: Deoterthridae) from the meiobenthos of the Faroe Bank, North Atlantic. Sarsia 76: 287–297.

Huys, R., S. Ohtsuka, G. A. Boxshall, and T. Itô. 1992b. *Itoitantulus misophricola* gen. et sp. nov.: first record of Tantulocarida (Crustacea: Maxillopoda) in the North Pacific region. Zoological Science 9: 875–886.

Huys, R., G. A. Boxshall, and R. Böttger-Schnack. 1992c. On the discovery of the male of *Mormonilla* Giesbrecht, 1891. Bulletin of the British Museum (Natural History), Zoology 58: 157–170.

Huys, R., G. A. Boxshall, and R. J. Lincoln. 1993. The tantulocarid life cycle: the circle closed? Journal of Crustacean Biology 13: 432–442.

Huys, R., J. M. Gee, C. G. Moore, and R. Hamond. 1996. Marine and Brackish Water Harpacticoid Copepods: Pt. 1, Keys and Notes for Identification of the Species. Synopses of the British Fauna, n.s., No. 51. Shrewsbury, UK: Field Studies Council.

Huys, R., P. J. López-González, Á. A. Luque, and E. Roldán. 2002. Brooding in cocculiniform limpets (Gastropoda) and familial distinctiveness of the Nucellicolidae (Copepoda): misconceptions reviewed from a chitonophilid perspective. Biological Journal of the Linnean Society 75: 187–217.

Huys, R., J. Llewellyn-Hughes, P. D. Olson, and K. Nagasawa. 2006. Small subunit rDNA and Bayesian inference reveal *Pectenophilus ornatus* (Copepoda *incertae sedis*) as highly transformed Mytilicolidae, and support assignment of Chondracanthidae and Xarifiidae to Lichomolgoidea (Cyclopoida). Biological Journal of the Linnean Society 87: 403–425.

Huys, R., J. Llewellyn-Hughes, S. Conroy-Dalton, P. D. Olson, J. Spinks, and D. A. Johnston. 2007. Extraordinary host switching in siphonostomatoid copepods and the demise of the Monstrilloida: integrating molecular data, ontogeny and antennulary morphology. Molecular Phylogenetics and Evolution 43: 368–378.

Huys, R., F. Fatih, S. Ohtsuka, and J. Llewellyn-Hughes. 2012. Evolution of the bomolochiform superfamily complex (Copepoda: Cyclopoida): new insights from ssrDNA and morphology, and origin of umazuracolids from polychaete-infesting ancestors rejected. International Journal for Parasitology 42: 71–92.

Ikeda, T. 1986. Preliminary observations on the development of the larvae of *Euphausia crystallorophias* Holt and Tattersall in the laboratory. Memoirs of National Institute of Polar Research (Tokyo), Special Issue 40: 183–186.

Ikeda, T. 1992. Laboratory observations on spawning, fecundity and early development of a mesopelagic ostracode, *Conchoecia pseudodiscophora*, from the Japan Sea. Marine Biology 112: 313–318.

Illg, P. 1949. A review of the copepod genus *Paranthessius* Claus. Proceedings of the United States National Museum 99: 391–428.

Illg, P. L., and P. L. Dudley. 1980. The family Ascidicolidae and its subfamilies (Copepoda, Cyclopoida), with descriptions of new species. Mémoires du Muséum d'Histoire Naturelle, A, Zoologie 117: 1–192.

Illig, G. 1930. Die Schizopoden der Deutschen Tiefsee-Expedition. *In* C. Chun, ed. Vol. 22, Wissenschaftliche Ergebnisse Deutschen Tiefsee-Expedition auf dem Dampfer "Valdivia" 1898–1899, 397–625. Jena, Germany: Gustav Fischer Verlag.

Ingle, R. W. 1984. The larval and post-larval development of the thumb-nail crab, *Thia scutellata* (Fabricius), (Decapoda: Brachyura). Bulletin of the British Museum (Natural History), Zoology 47: 53–64.

Ingle, R. W. 1992. Larval Stages of Northeastern Atlantic Crabs: An Illustrated Key. London: Chapman & Hall.

Ingle, R. W., and P. F. Clark. 1983. The larval development of the angular crab, *Goneplax rhomboides* (Linnaeus) (Decapoda: Brachyura). Bulletin of the British Museum (Natural History), Zoology 44: 163–177.

Ingle, R. W., and A. L. Rice. 1971. The larval development of the masked crab, *Corystes cassivelaunus* (Pennant) (Brachyura, Corystidae), reared in the laboratory. Crustaceana 20: 271–284.

Ingólfsson, A., and E. Ólafsson. 1997. Vital role of drift algae in the life history of the pelagic harpacticoid *Parathalestris croni* in the northern North Atlantic. Journal of Plankton Research 19: 15–27.

Inoue, N., H. Minami, and H. Sekiguchi. 2004. Distribution of phyllosoma larvae (Crustacea: Decapoda: Palinuridae, Scyllaridae and Synaxidae) in the western North Pacific. Journal of Oceanography 60: 963–976.

International Commission on Zoological Nomenclature (IZCN). 1964. Opinion 702: *Stereomastis* Bate, 1888 (Crustacea, Decapoda); validated under the plenary powers. Bulletin of Zoological Nomenclature 21: 111–112.

Iorio, M. I., M. A Scelzo, and E. E. Boschi. 1990. Desarrollo larval y post-larval del langostino *Pleoticus muelleri* Bate, 1888 (Crustacea, Decapoda, Solenoceridae). Scientia Marina 54: 329–341.

Ismail, N., S. Ohtsuka, B. A. Venmathi Maran, S. Tasumi, K. Zaleha, and H. Yamashita. 2013. Complete life cycle of a pennellid *Peniculus minuticaudae* Shiino, 1956 (Copepoda: Siphonostomatoida) infecting cultured threadsail filefish, *Stephanolepis cirrhifer*. Parasite 20: 42.

Ito, M., and J. S. Lucas. 1990. The complete larval development of the scyllarid lobster, *Scyllarus demani* Holthuis, 1946 (Decapoda, Scyllaridae), in the laboratory. Crustaceana 58: 144–167.

Itô, T. 1970. The biology of a harpacticoid copepod, *Tigriopus japonicus* Mori. Journal of the Faculty of Science, Hokkaido University, ser. 6, Zoology 17: 474–500.

Itô, T. 1985. Contributions to the knowledge of cypris y (Crustacea: Maxillopoda) with reference to a new genus and three new species from Japan. *In* Special Publication of the Mukaishima Marine Biological Station 1985, 113–122. Hiroshima, Japan: Mukaishima Marine Biological Station, Hiroshima University.

Itô, T. 1990. Naupliar development of *Hansenocaris furcifera* Itô (Crustacea: Maxillopoda: Facetotecta) from Tanabe Bay, Japan. Publications of the Seto Marine Biological Laboratory 34(4/6): 201–224.

Itô, T., and M. J. Grygier. 1990. Description and complete larval development of a new species of *Baccalaureus* (Crustacea: Ascothoracida) parasitic in a zoanthid from Tanabe Bay, Honshu, Japan. Zoological Science 7: 485–515.

Itô, T., and M. Takenaka. 1988. Identification of bifurcate paraocular process and postocular filamentary tuft of facetotectan cyprids (Crustacea: Maxillopoda). Publications of the Seto Marine Biological Laboratory 33(1/3): 19–38.

Itoh, H., and S. Nishida. 1995. Copepodid stages of *Hemicyclops japonicus* Itoh & Nishida (Poecilostomatoida: Clausidiidae) reared in the laboratory. Journal of Crustacean Biology 15: 134–155.

Ivanenko, V. N., P. H. C. Corgosinho, F. Ferrari, P-M. Sarradin, and J. Sarrazin. 2011. Microhabitat distribution of *Smacigastes micheli* (Copepoda: Harpacticoida: Tegastidae) from deep-sea hydrothermal vents at the Mid-Atlantic Ridge, 37° N (Lucky Strike), with a morphological description of its nauplius. Marine Ecology 2011: 1–11.

Iwata, Y., H. Sugita, Y. Deguchi and F. I. Kamemoto. 1991. The larval development of reef lobster, *Enoplometopus occidentalis* Randall (Decapoda, Axiidae) reared in the laboratory. Researches on Crustacea 20: 1–15.

Iwata, Y., H. Sugita, Y. Deguchi and F. I. Kamemoto. 1992. On the larval morphology of *Metanephrops sagamiensis* reared in the laboratory. Suisanzoshoku 40: 183–188.

Izawa, K. 1973. On the development of parasitic Copepoda, 1: *Sarcotaces pacificus* Komai (Cyclopoida: Philichthyidae). Publications of the Seto Marine Biological Laboratory 21: 77–86.

Izawa, K. 1986. On the development of parasitic Copepoda, 4: ten species of poecilostome cyclopoids, belonging to Taeniacanthidae, Tegobomolochidae, Lichomolgidae, Philoblennidae, Myicolidae, and Chondracanthidae. Publications of the Seto Marine Biological Laboratory 31: 81–162.

Izawa, K. 1987. Studies on the phylogenetic implications of ontogenetic features in the poecilostome nauplii (Copepoda: Cyclopoida). Publications of the Seto Marine Biology Laboratory 32(4/6): 151–217.

Izawa, K. 1991. Evolutionary reduction of body segments in the poecilostome Cyclopoida (Crustacea: Copepoda). *In* S-i Uye, S. Nishida, and J.-S. Ho, eds. Proceedings of the Fourth International Conference on Copepoda, Karuizawa, Japan, 16–20 September 1990. Special Volume, Bulletin of the Plankton Society of Japan: 71–86.

Izawa, K. 1997. The copepodid of *Peniculisa shiinoi* Izawa, 1965 (Copepoda, Siphonostomatoida, Pennellidae), a single free-swimming larval stage of the species, Crustaceana 70: 911–919.

Jackson, C. J., P. C. Rothlisberg, R. C. Pendrey, and M. T. Beamish. 1989. A key to genera of the penaeid larvae and early post-larvae of the Indo–West Pacific region, with descriptions of the larval development of *Atypopenaeus formosus* Dall and *Metapenaeopsis palmensis* Haswell (Decapoda: Penaeoidea: Penaeidae) reared in the laboratory. U.S. Fishery Bulletin 87: 703–733.

Jackson, D. L., and D. L. Macmillan. 2000. Tailflick escape behavior in larval and juvenile lobsters (*Homarus americanus*) and crayfish (*Cherax destructor*). Biological Bulletin 198: 307–318.

Jakobi, H. 1954. Biologie, Entwicklungsgeschichte und Systematik von *Bathynella natans* Vejdovsky. Zoologische Jahrbücher, Abteilung für Systematik, Ökologie und Geographie der Tiere 83: 1–62.

Jakobi, H. 1969. Contribuição à ontogenia de *Bathynella* Vejd. e *Brasilibathynella* Jakobi 1958 (Crustacea). Boletim da Universidade Federal do Paraná, Zoologia 3: 131–142.

Jalihal, D. R., K. N. Sankolli, and S. Shenoy. 1993. Evolution of larval developmental patterns and the process of freshwaterization in the prawn genus *Macrobrachium* Bate, 1868 (Decapoda, Palaemonidae). Crustaceana 65: 365–376.

Jamieson, A. J., T. Fujii, D. J. Mayor, M. Solan, and I. G. Priede. 2010. Hadal trenches: the ecology of the deepest places on earth. Trends in Ecology and Evolution 25: 190–197.

Jarman, S. N., S. Nicol, N. G. Elliott, and A. McMinn. 2000. 28S rDNA evolution in the Eumalacostraca and the phylogenetic position of krill. Molecular Phylogenetics and Evolution 17: 26–36.

Jaume, D. 2008. Global diversity of spelaeogriphaceans and thermosbaenaceans (Crustacea: Spelaeogriphacea and Thermosbaenacea) in freshwater. Hydrobiologia 595: 219–224.

Jaume, D., G. A. Boxshall, and R. N. Bamber. 2006. A new genus from the continental slope off Brazil and the discovery of the first males in the Hirsutiidae (Crustacea: Peracarida: Bochusacea). Zoological Journal of the Linnean Society 148: 169–208.

Jenner, R. A. 2010. Higher-level crustacean phylogeny: consensus and conflicting hypotheses. Special Issue, Arthropod Structure and Development 39: 143–153.

Jenner, R. A., C. N. Dhubhghaill, M. P. Ferla, and M. A. Wills. 2009. Eumalacostracan phylogeny and total evidence: limitations of the usual suspects. BMC Evolutionary Biology 9: 21.

Jensen, G. C. 1989. Gregarious settlement by megalopae of the porcelain crabs *Petrolisthes cinctipes* (Randall) and *P. eriomerus* Stimpson. Journal of Experimental Marine Biology and Ecology 131: 223–231.

Jensen, P. G., J. T. Høeg, S. Bower, and A. V. Rybakov. 1994a. Scanning electron microscopy of lattice organs in cyprids of the Rhizocephala Akentrogonida (Crustacea: Cirripedia). Canadian Journal of Zoology 72: 1018–1026.

Jensen, P. G., J. Moyse, J. T. Høeg, and H. Al-Yahya. 1994b. Comparative SEM studies of lattice organs: putative sensory structures on the carapace of larvae from Ascothoracida and Cirripedia (Crustacea Maxillopoda Thecostraca). Acta Zoologica (Stockholm) 75: 124–142.

Jepsen, J. 1965. Marsupial development of *Boreomysis arctica* (Krøyer, 1861). Sarsia 20: 1–8.

Johns, D. G., M. Edwards, W. Greve, and A. W. G. SJohn. 2005. Increasing prevalence of the marine cladoceran *Penilia avirostris* (Dana, 1852) in the North Sea. Helgoländer Meeresuntersuchungen / Helgoland Marine Research 59: 214–218.

Johnson, M. W. 1971a. On palinurid and scyllarid lobster larvae and their distribution in the South China Sea (Decapoda, Palinuridea). Crustaceana 21: 247–282.

Johnson, M. W. 1971b. The phyllosoma larvae of slipper lobster from the Hawaiian Islands and adjacent areas. Crustaceana 21: 77–103.

Johnson, M. W., and M. Knight. 1966. Phyllosoma larvae of the spiny lobster *Panulirus inflatus* (Bouvier). Crustaceana 10: 31–47.

Johnson, M. W., and W. M. Lewis. 1942. Pelagic larval stages of the sand crabs *Emerita analoga* (Stimpson), *Blepharipoda occidentalis* Randall, and *Lepidopa myops* Stimpson. Biological Bulletin 83: 67–87.

Johnson, M. W., and P. B. Robertson. 1970. On the phyllosoma larvae of the genus *Justitia* (Decapoda, Palinuridae). Crustaceana 18: 283–292.

Johnson, W. S., and D. M. Allen. 2012. Zooplankton of the Atlantic and Gulf Coasts: A Guide to Their Identification and Ecology, 2nd ed. Baltimore: Johns Hopkins University Press.

Johnson, W. S., M. Stevens, and L. Watling. 2001. Reproduction and development of marine peracaridans. Advances in Marine Biology 39: 105–260.

Johnston, N. M., and D. A. Ritz. 2005. Kin recognition and adoption in mysids (Crustacea: Mysidacea). Journal of the Marine Biological Association of the United Kingdom 85(6): 1441–1447.

Johnston, N. M., D. A. Ritz, and G. E. Fenton. 1997. Larval development in the Tasmanian mysids *Anisomysis mixta australis*, *Tenagomysis tasmaniae*, and *Paramesopodopsis rufa* (Crustacea: Mysidacea). Marine Biology 130: 93–99.

Joly, N. 1843. Études sur les moeurs, le développement et les métamorphoses d'une petite salicoque d'eau douce (*Caridina desmarestii*) suivies de quelques réflexions sur les métamorphoses des crustacés décapodes en général. Annales des Sciences Naturelles, Zoologie 19: 34–86.

Jones, M. L. 1961. *Lightiella serendipita* gen. nov., sp. nov., a cephalocarid from San Francisco Bay, California. Crustaceana 3: 31–46.

Jorgensen, O. M. 1925. The early stages of *Nephrops norvegicus* from the Northumberland plankton, together with a note on the post-larval development of *Homarus vulgaris*. Journal of the Marine Biological Association of the United Kingdom, n.s., 8: 870–876.

Jungersen, H. F. E. 1914. *Chordeuma obesum*, a new parasitic copepod endoparasitic in *Asteronyx loveni* M. & Tr. Mindeskrift for Japetus Steenstrup 16: 1–18.

Jurasz, W., W. Kittel, and P. Presler. 1983. Life cycle of *Branchinecta gaini* Daday, 1910 (Branchiopoda, Anostraca) from King George Island, South Shetland Islands. Polish Polar Research 4: 143–154.

Just, J., and G. C. B. Poore. 1988. Second record of Hirsutiidae (Peracarida: Mictacea): *Hirsutia sandersetalia*, new species, from southeastern Australia. Journal of Crustacean Biology 8: 483–488.

Jutte, P. A., T. W. Cronin, and R. L. Caldwell. 1998. Photoreception in the planktonic larvae of two species of *Pullosquilla*, a lysiosquilloid stomatopod crustacean. Journal of Experimental Biology 201: 2481–2487.

Kabata, Z. 1958. *Lernaeocera obtusa* n. sp.: its biology and its effects on the haddock. Marine Research (Scottish Home Department) 3: 1–26.

Kabata, Z. 1969. *Phrixocephalus cincinnatus* Wilson, 1908 (Copepoda: Lernaeoceridae): morphology, metamorphosis and host-parasite relationship. Journal of the Fisheries Research Board of Canada 26: 921–934.

Kabata, Z. 1972. Developmental stages of *Caligus clemensi* (Copepoda: Caligidae). Journal of the Fisheries Research Board of Canada 29: 1571–1593.

Kabata, Z. 1976. Early stages of some copepods (Crustacea) parasitic on marine fishes of British Columbia. Journal of the Fisheries Research Board of Canada 33: 2507–2525.

Kabata, Z. 1979. Parasitic Copepoda of British Fishes. London: Ray Society.

Kabata, Z. 1981. Copepoda (Crustacea) parasitic on fishes: problems and perspectives. Advances in Parasitology 19: 1–71.

Kabata, Z., and B. Cousens. 1973. Life cycle of *Salmincola californiensis* (Dana, 1852) (Copepoda: Lernaeopodidae). Journal of the Fisheries Research Board of Canada 30: 881–903.

Kaestner, A. 1967. Lehrbuch der Speziellen Zoologie. Vol. 1, pt. 2: Crustacea. Jena, Germany: Gustav Fischer Verlag.

Kaestner, A. 1970. Invertebrate Zoology. Vol. 3, Crustacea. Trans. and adapted by H. W. Levi and L. R. Levi. New York: Wiley Interscience.

Kaji, T., O. S. Møller, and A. Tsukagoshi. 2011. A bridge between original and novel states: ontogeny and function of "suction discs" in the Branchiura (Crustacea). Evolution & Development 13: 119–126.

Kakati, V. S. 1988. Larval development of the Indian spider crab *Elamenopsis demeloi* (Kemp) (Brachyura, Hymenosomatidae) in the laboratory. Mahasagar 21: 219–227.

Kamenz, C., J. A. Dunlop, G. Scholtz, H. Kerp, and H. Hass. 2008. Microanatomy of Early Devonian book lungs. Biology Letters 4: 212–215.

Kane, J. 1963. Stages in early development of *Parathemisto gaudichaudii* (Guér.) (Crustacea, Amphipoda: Hyperiida), the development of secondary sexual characters and of the ovary. Philosophical Transactions of the Royal Society, New Zealand 3: 37–47.

Kane, W. F. de V. 1904. Further captures of *Mysis relicta* in Ireland. Irish Naturalist 13: 107–109.

Kapler, O. 1960. Ze života škeblovsky male (*Leptestheria dahalacensis* Rüppel). Časopis Slezského Musea. Vědy Přírodní 9: 101–110.

Kawatow, K., K. Muroga, K. Izawa, and S. Kasahara. 1980. Life cycle of *Alella macrotrachelus* (Copepoda) parasitic on cultured Black Sea-bream. Journal of the Faculty of Applied Biological Science, Hiroshima University 19: 199–214.

Kellicott, D. S. 1880. *Argulus stizostethii*, n. s. American Journal of Microscopy and Popular Science 5: 53–58.

Kemp, S. W. 1910a. The Decapoda Natantia of the coast of Ireland. Scientific Investigations (Fisheries Branch, Ireland) 1908: 1–190.

Kemp, S. W. 1910b. The Decapoda collected by the "Huxley" from the north side of the Bay of Biscay in August, 1906. Journal of the Marine Biological Association of the United Kingdom 8: 407–420.

Kern, J. E., N. A. Edwards, and S. S. Bell. 1984. Precocious clasping of early copepodite stages: a common occurrence in *Zausodes arenicolus* Wilson (Copepoda: Harpacticoida). Journal of Crustacean Biology 4: 261–265.

Kesling, R. V. 1951. The morphology of ostracod molt stages. Illinois Biological Monographs 21: 1–324.

Kessler, R. 2011. Crustaceans crave a little quiet. ScienceNow, 7 March 2011. Online at http://news.sciencemag.org/sciencenow/2011/03/crustaceans-crave-a-little-quiet.html?ref=wp/.

Kim, C. H., and I. K. Jang. 1986. The complete larval development of *Cyclograpsus intermedius* Ortmann reared in the laboratory (Decapoda: Brachyura: Grapsidae). Journal of Science, Pusan National University 42: 143–158.

Kim, C. H., and H. S. Ko. 1985. The complete larval development of *Sesarma erythrodactyla* Hess reared in the laboratory (Decapoda: Brachyura: Grapsidae). Journal of Science, Pusan National University 40: 165–178.

Kim, I.-H. 1993. Developmental stages of *Caligus punctatus* Shiino, 1955 (Copepoda: Caligidae). *In* G. A. Boxshall and D. Defaye, eds. Pathogens of Wild and Farmed Fish: Sea Lice, 16–29. New York: Ellis Horwood.

Kim, Y.-H., and S.-C. Chung. 1990. Studies on the growth and molting of the tiger crab, *Orithyia sinica* (Linnaeus). Bulletin of the Korean Fisheries Society 23: 93–108.

King, J. L., and R. Hanner. 1998. Cryptic species in a "living fossil" lineage: taxonomic and phylogenetic relationships within the genus *Lepidurus* (Crustacea: Notostraca) in North America. Molecular Phylogenetics and Evolution 10: 23–36.

Kinne, O. 1955. *Neomysis vulgaris* Thompson: eine autökologisch-biologische Studie. Biologisches Zentralblatt 74: 160–202.

Knight, M. D. 1967. The larval development of the sand crab *Emerita rathbunae* Schmitt (Decapoda, Hippidae). Pacific Science 21: 58–76.

Knight, M. D. 1968. The larval development of *Raninoides benedicti* Rathbun (Brachyura, Raninidae), with notes on the Pacific records of *Raninoides laevis* (Latreille). Crustaceana, Suppl. 2: 145–169.

Knight, M. D. 1970. The larval development of *Lepidopa myops* Stimpson (Decapoda, Albuneidae) reared in the laboratory, and the zoeal stages of another species of the genus from California and the Pacific coast of Baja California, Mexico. Crustaceana 19: 125–156.

Knight, M. D. 1975. The larval development of Pacific *Euphausia gibboides* (Euphausiacea). U.S. Fishery Bulletin 73: 145–168.

Knight, M. D. 1976. Larval development of *Euphausia sanzoi* Torelli (Crustacea: Euphausiacea). Bulletin of Marine Science 26: 583–557.

Knight, M. D. 1978. Larval development of *Euphausia fallax* Hansen (Crustacea: Euphausiacea) with a comparison of larval morphology within the *E. gibboides* species group. Bulletin of Marine Science 28: 255–281.

Knight, M. D. 1980. Larval development of *Euphausia eximia* (Crustacea: Euphausiacea) with notes on its vertical distribution and morphological divergence between populations. U.S. Fishery Bulletin 78: 313–335.

Knight, M. D. 1984. Variation in larval morphogenesis within the Southern California Bight population of *Euphausia pacifica* from winter trough summer, 1977–1978. California Cooperative Oceanic Fisheries Investigation Reports 25: 87–99.

Knight, M. [D.], and M. Omori. 1982. The larval development of *Sergestes similis* Hansen (Crustacea, Decapoda, Sergestidae) reared in the laboratory. U. S. Fishery Bulletin 80: 217–243.

Knudsen, S. W., M. Kirkegaard, and J. Olesen. 2009. The tantulocarid genus *Arcticotantalus* [sic] removed from Basipodellidae into Deoterthridae (Crustacea: Maxillopoda) after the description of a new species from Greenland, with first live photographs and an overview of the class. Zootaxa 2035: 41–68.

Koenemann, S., and T. M. Iliffe. 2013. Class Remipedia. *In* J. Forest and J. C. von Vaupel Klein, eds., F. R. Schram and M. Charmantier-Daures, advisory eds. Vol. 5, The Crustacea: Translated and Updated from the Traité de Zoologie, 125–177. Leiden, Netherlands: Brill.

Koenemann, S., F. R. Schram, A. Bloechl, M. Hoenemann, T. M. Iliffe, and C. Held. 2007a. Post-embryonic development of remipede crustaceans. Evolution & Development 9: 117–121.

Koenemann, S., F. R. Schram, T. M. Iliffe, L. M. Hinderstein, and A. Bloechl. 2007b. Behaviour of Remipedia in the laboratory, with supporting field observations. Journal of Crustacean Biology 27: 534–542.

Koenemann, S., J. Olesen, F. Alwes, T. M. Iliffe, M. Hoenemann, P. Ungerer, C. Wolff, and G. Scholtz. 2009. The post-embryonic development of Remipedia (Crustacea): additional results and new insights. Development Genes and Evolution 219: 131–145.

Kohlhage, K., and J. Yager. 1994. An analysis of swimming in remipede crustaceans. Philosophical Transactions of the Royal Society of London, B 346: 213–221.

Kolbasov, G. A. 2009. Acrothoracica, Burrowing Crustaceans. Moscow: KMK Scientific Press.

Kolbasov, G. A., and J. T. Høeg. 2003. Facetotectan larvae from the White Sea with the description of a new species (Crustacea: Thecostraca). Sarsia 88: 1–15.

Kolbasov, G. A., and J. T. Høeg. 2007. Cypris larvae of the acrothoracican barnacles (Thecostraca, Cirripedia, Acrothoracica). Zoologischer Anzeiger 246: 127–151.

Kolbasov, G. A., and A. S. Savchenko. 2009. *Microdajus tchesunovi* sp. n. (Tantulocarida, Microdajidae): a new crustacean parasite of [sic] from the White Sea. Experimental Parasitology 125: 13–22.

Kolbasov, G. A., J. T. Høeg, and A. S. Elfimov. 1999. Scanning electron microscopy of acrothoracican cypris larvae (Crustacea, Thecostraca, Cirripedia, Acrothoracica, Lithoglyptidae). Contribution to Zoology 68: 143–160.

Kolbasov, G. A., M. J. Grygier, V. N. Ivanenko, and A. A. Vagelli. 2007. A new species of the y-larva genus *Hansenocaris* Itô, 1985 (Crustacea: Thecostraca: Facetotecta) from Indonesia, with a review of y-cyprids and a key to all their described species. Raffles Bulletin of Zoology 55: 343–353.

Kolbasov, G. A., A. Yu. Sinev, and A. V. Tschesunov. 2008a. External morphology of *Arcticotantulus pertzovi* (Tantulocarida, Basipodellidae), a microscopic crustacean parasite from the White Sea. Entomological Review 88: 1192–1207 [original Russian text published in Zoologicheskii Zhurnal 87: 1437–1452].

Kolbasov, G. A., M. J. Grygier, J. T. Høeg, and W. Klepal. 2008b. External morphology of ascothoracid-larvae of the genus *Dendrogaster* (Crustacea, Thecostraca, Ascothoracida), with remarks on the ontogeny of the lattice organs. Zoologischer Anzeiger 247: 159–183.

Komai, T., and Y. M. Tung. 1929. Notes on the larval stages of *Squilla oratoria* with remarks on some other stomatopod larvae found in Japanese seas. Annotationes Zoologicae Japonenses 12: 187–237.

Konishi, K. 2001. First record of larvae of the rare mud shrimp *Naushonia* Kingsley (Crustacea: Decapoda: Laomediidae) from Asian waters. Proceedings of the Biological Society of Washington 114: 611–617.

Konishi, K. and H. Taishaku. 1994. Larval development in *Paralomis hystrix* (De Haan, 1846) (Crustacea, Anomura, Lithodidae) under laboratory conditions. Bulletin of the National Research Institute for Aquaculture 23: 43–54.

Konishi, K., R. Quintana, and Y. Fukuda. 1990. A complete description of larval stages of the ghost shrimp *Callianassa petalura* Stimpson (Crustacea: Thalassinidea: Callianassidae) under laboratory conditions. Bulletin of the National Research Institute of Aquaculture 17: 27–49.

Konishi, K ., Y. Fukuda, and R. Quintana. 1999. The larval development of the mud burrowing shrimp *Callianassa* sp. under laboratory conditions (Decapoda, Thalassinidea, Callianassidae). *In* F. R. Schram and J. C. von Vaupel Klein, eds. Vol. 1, Crustaceans and the Biodiversity Crisis: Proceedings of the Fourth International Crustacean Congress, Amsterdam, The Netherlands, July 20–24, 1998, 781–804. Leiden, Netherlands: Brill.

Korn, M., and A. K. Hundsdoerfer. 2006. Evidence for cryptic species in the tadpole shrimp *Triops granarius* (Lucas, 1864) (Crustacea: Notostraca). *Zootaxa* 1257: 57–68.

Korn, O. M. 1995. Naupliar evidence for cirripede taxonomy and phylogeny. *In* F. R. Schram and J. T. Høeg, eds. New Frontiers in Barnacle Evolution, 87–122. Crustacean Issues No. 10. Rotterdam: A. A. Balkema.

Kornicker, L. S. 1969. Morphology, Ontogeny, and Intraspecific Variation of *Spinacopia*, a New Genus of Myodocopid Ostracod (Sarsiellidae). Smithsonian Contributions to Zoology No. 8. Washington, DC: Smithsonian Institution Press.

Kornicker, L. S. 1981. Revision, Distribution, Ecology, and Ontogeny of the Ostracode Subfamily Cyclasteropinae (Myodocopina: Cylindroleberididae). Smithsonian Contributions to Zoology No. 319. Washington, DC: Smithsonian Institution Press.

Kornicker, L. S. 1983. Rutidermatidae of the Continental Shelf of

Southeastern North America and Gulf of Mexico (Ostracoda: Myodocopina). Smithsonian Contributions to Zoology No. 371. Washington, DC: Smithsonian Institution Press.

Kornicker, L. S. 1992. Myodocopid Ostracoda of the Benthedi Expedition, 1977, to the NE Mozambique Channel, Indian Ocean. Smithsonian Contributions to Zoology No. 531. Washington, DC: Smithsonian Institution Press.

Kornicker, L. S., and E. Harrison-Nelson. 1999. Eumeli Expeditions, Pt. 1: *Tetragonodon rex*, New Species, and General Reproductive Biology of the Myodocopina. Smithsonian Contributions to Zoology No. 602. Washington, DC: Smithsonian Institution Press.

Kornicker, L. S., and E. Harrison-Nelson. 2002. Ontogeny of *Rutiderma darbyi* (Crustacea: Ostracoda: Myodocopida: Rutidermatidae) and comparisons with other Myodocopina. Proceedings of the Biological Society of Washington 115: 426–471.

Kornicker, L. S., and T. M. Iliffe. 1989. Ostracoda (Myodocpina: Cladocopina, Halocypridina) Mainly from Anchialine Caves in Bermuda. Smithsonian Contributions to Zoology No. 475. Washington, DC: Smithsonian Institution Press.

Kornicker, L. S., and T. M. Iliffe. 1995. Ostracoda (Halocypridina, Cladocopina) from an Anchialine Lava Tube in Lanzarote, Canary Islands. Smithsonian Contributions to Zoology No. 568. Washington, DC: Smithsonian Institution Press.

Kornicker, L. S., and G. I. Sohn. 1976. Phylogeny, Ontogeny, and Morphology of Living and Fossil Thaumatocyridacea (Myodocopa: Ostracoda). Smithsonian Contributions to Zoology No. 219. Washington, DC: Smithsonian Institution Press.

Kornienko, E. S., and O. M. Korn. 2010. Illustrated key for the identification of brachyuran larvae in the northwestern Sea of Japan. Vladivostok, Russia: Dalnauka (A. V. Zhirmunsky Institute of Marine Biology, Russian Academy of Sciences) [in Russian].

Kornienko, E. S., and O. M. Korn. 2011. The larval development of the hermit crab *Areopaguristes nigroapiculus* (Decapoda: Anomura: Diogenidae) reared under laboratory conditions. Journal of the Marine Biological Association of the United Kingdom 91: 1031–1039.

Korovchinsky, N. M. 1990. Evolutionary morphological development of the superfamily Sidoidea and life strategies of crustaceans of continental waters. Internationale Revue der Gesamten Hydrobiologie 75: 649–676.

Korovchinsky, N. M. 2009. The genus *Leptodora* Lilljeborg (Crustacea: Branchiopoda: Cladocera) is not monotypic: description of a new species from the Amur River basin (Far East of Russia). Zootaxa 2120: 39–52.

Korovchinsky, N. M., and O. S. Boikova. 1996. The resting eggs of the Ctenopoda (Crustacea: Branchiopoda): a review. Hydrobiologia 320: 131–140.

Korovchinsky, N. M., and N. G. Sergeeva. 2008. A new family of the order Ctenopoda (Crustacea: Cladocera) from the depths of the Black Sea. Zootaxa 1795: 57–66.

Kotov, A. A. 1996. Fate of the second maxilla during embryogenesis in some Anomopoda Crustacea (Branchiopoda). Zoological Journal of the Linnean Society 116: 393–405.

Kotov, A. A., and O. S. Boikova. 1998. Comparative analysis of the late embryogenesis of *Sida crystallina* (O. F. Müller, 1776) and *Diaphanosoma brachyurum* (Lievin, 1848) (Crustacea: Branchiopoda: Ctenopoda). Hydrobiologia 380: 103–125.

Kotov, A. A., and O. S. Boikova. 2001. Study of the late embryogenesis of *Daphnia* (Anomopoda, "Cladocera," Branchiopoda) and a comparison of development in Anomopoda and Ctenopoda. Hydrobiologia 442: 127–143.

Krishnan, T., and T. Kannupandi. 1987. Laboratory larval culture of a spider crab, *Doclea muricata* (Fabricius, 1787) (Decapoda, Majidae). Crustaceana 53: 292–303.

Krøyer, H. 1838. Om snyltekrebsene, issr med hensyn til den danske Fauna [Concerning parasitic Crustacea with special reference to the Danish fauna]. Naturhistorisk Tidsskrift 2(1): 8–52, 131–157.

Krøyer, H. 1841. Fire nye arter af slaegten *Cuma* Edwards. Naturhistorisk Tidsskrift 3(6): 503–534.

Krøyer, H. 1847. Karcinologiske bidrag: *Nebalia bipes* Fabr. Naturhistorisk Tidsskrift, ser. 2, 4: 436–446.

Kuei, H. Y. N., and T. K. S. Björnberg. 2002. Developmental stages of *Oncaea curta* Sars, 1926 (Copepoda, Poecilostomatoida). Nauplius 10: 1–14.

Kühn, A. 1913. Die Sonderung der Keimebezirke in der Entwicklung der Sommereier von *Polyphemus pediculus* de Geer. Zoologische Jahrbücher, Abteilung für Anatomie und Ontogenie der Tiere 35: 243–340.

Kühnert, L. 1934. Beitrag zur Entwicklungsgeschichte von *Alcippe Lampas* Hancock. Zeitschrift für Morphologie und Ökologie der Tiere 29: 45–78.

Kurata, H. 1965. Larvae of decapod Crustacea of Hokkaido, 9: Axiidae, Callianassidae, and Upogebiidae (Anomura). Bulletin of Hokkaido Regional Fisheries Research Laboratory 30: 1–10.

Kurian, C. V. 1956. Larvae of decapod Crustacea from the Adriatic Sea. Acta Adriatica 6: 1–108.

Kutschera, V., A. Maas, D. Waloszek, C. Haug, and J. T. Haug. 2012. Re-study of larval stages of *Amphionides reynaudii* (Eucarida, Malacostraca, Crustacea *s. l.*) with modern imaging techniques. Journal of Crustacean Biology 32: 916–930.

Lagersson, N., and J. T. Høeg. 2002. Settlement behavior and antennulary biomechanics and in cypris larvae of *Balanus amphitrite* (Crustacea: Thecostraca: Cirripedia). Marine Biology 141: 513–526.

Lagersson, N., A. Garm, and J. T. Høeg. 2003. Notes on the ultrastructure of the setae on the fourth antennulary segment of the *Balanus amphitrite* cyprid (Crustacea: Cirripedia: Thoracica). Journal of the Marine Biological Association of the United Kingdom 83: 361–365.

Lai, H.-T., and Shy, J.-Y. 2009. The larval development of *Caridina pseudodenticulata* (Crustacea: Decapoda: Atyidae) reared in the laboratory, with a discussion of larval metamorphosis types. Raffles Bulletin of Zoology, Suppl. 20: 97–107.

Lamb, E. J., G. A. Boxshall, P. J. Mill, and J. Grahame. 1996. Nucellicolidae: a new family of endoparasitic copepods (Poecilostomatoida) from the dog whelk *Nucella lapillus* (Gastropoda). Journal of Crustacean Biology 16: 142–148.

Landeira, J., F., Lozano-Soldevilla, and J. I. González-Gordillo. 2009. Morphology of first seven larval stages of the striped soldier shrimp, *Plesionika edwardsii* (Brandt, 1851) (Crustacea: Decapoda: Pandalidae) from laboratory reared material. Zootaxa 1986: 51–56.

Lang, K. 1948. Monographie der Harpacticiden. Lund, Sweden: Håkan Ohlsson.

Lange, J. C. 1756. Lære om de Naturlige Vande, samt Grundig Undersogning af de Vande, som findes udi Kiøbenhavn og dens Egne, og til allaehaande Nytte af dens Indbyggere meest blive anvendte: Næst Oplosning paa det Sporsmaal; Hvilket Vand der er det bedste? Copenhagen: L. H. Lillie.

Lange, S., C. H. J. Hof, F. R. Schram, and F. A. Steeman. 2001. New genus and species from the Cretaceous of Lebanon links the Thylacocephala to the Crustacea. Palaeontology 44: 905–912.

Larsen, K. 2003. Proposed new standardized anatomical terminology for the Tanaidacea (Peracarida). Journal of Crustacean Biology 23: 644–661.

Latreille, P. A. 1817. Les Crustacés, les Arachnides et les Insectes. Vol. 3, in M. le C. Cuvier, ed. Le Règne Animal Distribué d'après son Organisation, pour Servir de Base à l'Histoire Naturelle des Animaux et d'Introduction à l'Anatomie Comparée. Paris: Deterville.

Laubier, L. 1961. *Phyllodicola petiti* (Delamare et Laubier, 1960) et la famille des Phyllodicolidae, copépodes parasites d'annélides polychètes en Méditerranée occidentale. Crustaceana 2: 228–242.

Laughlin, R. A., P. J. Rodriguez, and J. A. Marval. 1983. Zoeal stages of the coral crab *Carplius corallinus* (Herbst) (Decapoda, Xanthidae) reared in the laboratory. Crustaceana 44: 169–186.

Laval, P. 1963. Sur la biologie et les larves de *Vibilia armata* Boy. et de *V. propinqua* Stebb., amphipodes hyperides. Comptes Rendus Hebdomadaires des Séances de l'Académie des Sciences 257: 1389–1392.

Laval, P. 1980. Hyperiid amphipods as crustacean parasitoids associated with gelatinous zooplankton. *In* M. Barnes, ed. Vol. 18, Oceanography and Marine Biology: An Annual Review, 11–56. Aberdeen, Scotland, UK: Aberdeen University Press.

Laverack, M. S., D. L. MacMillan, G. Ritchie, and S. L. Sandow. 1996. The ultrastructure of the sensory dorsal organ of Crustacea. Crustaceana 69: 636–651.

Lavrov, D. V., W. M. Brown, and J. L. Boore. 2004. Phylogenetic position of the Pentastomida and (pan)crustacean relationships. Proceedings of the Royal Society of London, B 271: 537–544.

Leach, W. E. 1818. Sur quelques genres nouveaux de crustacés. Journal de Physique, de Chimie et d'Histoire Naturelle 86: 304–307.

Lebour, M. V. 1926. A general survey of the larval euphausiids with a scheme for their identification. Journal of the Marine Biological Association of the United Kingdom 14: 519–527.

Lebour, M. V. 1928. The larval stages of the Plymouth Brachyura. Proceedings of the Zoological Society of London 98: 473–560.

Lebour, M. V. 1930. The larval stages of Caridion, with a description of a new species, *C. steveni*. Proceedings of the Zoological Society of London 21: 181–194.

Lebour, M. V. 1931. The larvae of the Plymouth Caridea, 1 and 2: 1, the larvae of the Crangonidae; 2, the larvae of the Hippolytidae. Proceedings of the Zoological Society of London 101: 1–9.

Lebour, M. V. 1932a. The larvae of the Plymouth Caridea, 3: the larvae stages of *Spintocaris cranchii* (Leach). Proceedings of the Zoological Society of London 102: 131–137.

Lebour, M. V. 1932b. The larvae of the Plymouth Caridea, 4: the Alpheidae. Proceedings of the Zoological Society of London 102: 463–469.

Lebour, M. V. 1940. The larvae of the Pandalidae. Journal of the Marine Biological Association of the United Kingdom 24: 239–252.

Lebour, M. V. 1941. The stenopid larvae of Bermuda. Pp. 161–181 *in* R. Gurney and M. V. Lebour. On the Larvae of Certain Crustacea Macrura, Mainly from Bermuda. Journal of the Linnean Society London, Zoology 41: 89–181.

Lebour, M. V. 1944. 11: larval crabs from Bermuda. Zoologica (New York Zoological Society) 29: 113–128.

Lee, C. N., and B. Morton. 2005. Demography of *Nebalia* sp. (Crustacea: Leptostraca) determined by carrion bait trapping in Lobster Bay, Cap d'Aguilar Marine Reserve, Hong Kong. Marine Biology 148: 149–157.

Lereboullet, M. 1866. Observation sur la génération et le développement de la Limnadie de Hermann (*Limnadia hermannni* Ad. Brongn.). Annales des Sciences Naturelles, Zoologie 5: 283–308.

Le Roux, A. 1973. Observation sur le développement larvaire de *Nyctiphanes couchii* (Crustacea: Euphausiacea) au laboratoire. Marine Biology 22: 159–166.

Le Roux, A. 1974. Observation sur le développement larvaire de *Meganyctiphanes norvegica* (Crustacea: Euphausiacea) au laboratoire. Marine Biology 26: 45–56.

Li, C. P., S. De Grave, T. Y. Chan, H. C. Lei, and K. H. Chu. 2011. Molecular systematics of caridean shrimps based on five nuclear genes: implications for superfamily classification. Zoologischer Anzeiger 250: 270–279.

Linder, F. 1943. Über *Nebaliopsis typica* G.O. Sars nebst einigen allgemeinen Bemerkungen über die Leptostraken. Dana Report No. 25. Copenhagen: Reitzel.

Linder, F. 1952. Contributions to the morphology and taxonomy of the Branchiopoda Notostraca, with special reference to the North American species. Proceedings of the United States National Museum 102: 1–69.

Linnaeus, C. von. 1768. Systema Naturae, 12th ed., vol. 3. Stockholm: Laurentius Salvius.

Liu, J., and X. Dong. 2007. *Skara hunanensis* a new species of Skaracarida (Crustacea) from Upper Cambrian (Furongian) of Hunan, south China. Progress in Natural Science 17: 934–942.

Lombardi, J., and E. Ruppert. 1982. Functional morphology of locomotion in *Derocheilocaris typica*. Zoomorphology 100: 1–10.

Longhurst, A. R. 1955. A Review of the Notostraca. Bulletin of the British Museum (Natural History), Zoology 3: 1–57.

López-González, P. J., and J. Bresciani. 2001. New Antarctic records of *Herpyllobius* Steenstrup and Lütken, 1861 (parasitic Copepoda) from the EASIZ-III cruise, with description of two new species. Scientia Marina 65: 357–366.

López-Greco, L. S., V. Viau, M. Lavolpe, G. Bond-Buckup, and E. M. Rodriguez. 2004. Juvenile hatching and maternal care in *Aegla uruguayana* (Anomura, Aeglidae). Journal of Crustacean Biology 24: 309–313.

Löpmann, A. 1937. Die Zweiäugigkeit von *Diaphanosoma* (zugleich ein Beiträg zur Kenntnis des Cladocerenauges). Internationale Revue der Gesamten Hydrobiologie 34: 432–484.

Lorenzen, S. 1969. Harpacticoiden aus dem lenitischen Watt und den Salzwiesen der Nordseeküste. Kieler Meeresforschungen 25: 215–223.

Lucas, J. S. 1971. The larval stages of some Australian species of Halicarcinus (Crustacea, Brachyura, Hymenosomatidae), 1: morphology. Bulletin of Marine Science 21: 471–490.

Lutsch, E., and A. Avenant-Oldewage. 1995. The ultrastructure of the newly hatched *Argulus japonicus* Thiele, 1900 larvae (Branchiura). Crustaceana 68: 329–340.

Lützen, J. 1968. On the biology of the family Herpyllobiidae (parasitic copepods). Ophelia 5: 175–187.

Lützen, J., J. T. Høeg, and Å. Jespersen. 1996. Spermatogonia implantation by antennular penetration in the akentrogonid rhizocephalan *Diplothylacus sinensis* (Keppen, 1877) (Crustacea: Cirripedia). Zoologischer Anzeiger 234: 201–207.

Maas, A., and D. Waloszek. 2001. Cambrian derivatives of the early arthropod stem lineage, Pentastomids, Tardigrades and Lobopodians: an "Orsten" perspective. Zoologischer Anzeiger 240: 451–459.

Maas, A., D. Waloszek, and K. J. Müller. 2003. Morphology, Ontogeny and Phylogeny of the Phosphatocopina (Crustacea) from the Upper Cambrian "Orsten" of Sweden. Fossils and Strata No. 49. Oslo: Taylor & Francis.

Maas, A., D. Waloszek, J. Chen, A. Braun, X. Wang, and D. Huang. 2004. Phylogeny and life habits of early arthropods: predation in the Early Cambrian sea. Progress in Natural Science 14: 158–166.

Maas, A., A. Braun, X. Dong, P. C. J. Donoghue, K. J. Müller, E. Olempska, J. E. Repetski, D. J. Siveter, M. Stein, and D. Waloszek. 2006. The "Orsten": more than a Cambrian Konservat-Lagerstatte yielding exceptional preservation. Palaeoworld 15: 266–282.

Maas, A., C. Haug, J. T. Haug, J. Olesen, X. Zhang, and D. Waloszek. 2009. Early crustacean evolution and the appearance of epipodites and gills. Arthropod Systematics & Phylogeny 67: 255–273.

MacDonald, J. D., R. B. Pike, and D. I. Williamson. 1957. Larvae of the British species of *Diogenes*, *Pagurus*, *Anapagurus*, and *Lithodes* (Crustacea, Decapoda). Proceedings of the Zoological Society of London 128: 209–257.

Macdonald, R. 1927. Irregular development in the larval history of *Meganyctiphanes norvegica*. Journal of the Marine Biological Association of the United Kingdom 14: 785–794.

Magalhães, C. 1988. The larval development of palaemonid shrimps from the Amazon region reared in the laboratory, 2: extremely abbreviated larval development in *Euryrhynchus* Miers, 1877 (Decapoda, Euryrhynchinae). Crustaceana 55: 39–52.

Maisey, J. G., and M. G. P. de Carvalho. 1995. First records of fossil sergestid decapods and fossil brachyuran crab larvae (Arthropoda, Crustacea), with remarks on some supposed palaemonid fossils, from the Santana Formation (Aptian-Albian, NE Brazil). American Museum Novitates 3132: 1–20.

Malaquin, A. 1901. Le parasitisme évolutif des Monstrillides (Crustacés, Copépodes). Archives de Zoologie Expérimentale et Générale 3`: 81–232.

Manning, R. B. 1962. *Alima hyalina* Leach, the pelagic larva of the stomatopod crustacean *Squilla alba* Bigelow. Bulletin of Marine Science of the Gulf and Caribbean 12: 496–507.

Manning, R. B. 1980. The superfamilies, families, and genera of Recent stomatopod Crustacea, with diagnoses of six new families. Proceedings of the Biological Society of Washington 93: 362–372.

Manning, R. B. 1991. Stomatopod Crustacea Collected by the *Galathea* Expedition, 1950–1952, with a List of Stomatopoda Known from Depths below 400 Meters. Smithsonian Contributions to Zoology No. 521. Washington, DC: Smithsonian Institution Press.

Manning, R. B. 1995. Stomatopod Crustacea of Vietnam: The Legacy of Raoul Serène. Crustacean Research, Special No. 4. Tokyo: Carcinological Society of Japan.

Manning, R. B., and A. J. Provenzano Jr. 1963. Studies on development of stomatopod Crustacea, 1: early larval stages of *Gonodactylus oerstedii* Hansen. Bulletin of Marine Science of the Gulf and Caribbean 13: 467–487.

Manton, S. M. 1928. On the embryology of a mysid crustacean,

Hemimysis lamornae. Philosophical Transactions of the Royal Society of London, B 216: 363–463.

Manton, S. M. 1934. On the embryology of the crustacean *Nebalia bipes*. Philosophical Transactions of the Royal Society of London, B 223: 163–238.

Marchenkov, A. V., and G. A. Boxshall. 1995. A new family of copepods associated with ascidiaceans in the White Sea, and an analysis of antennulary segmentation and setation patterns in the order Poecilostomatoida. Zoologischer Anzeiger 234: 133–143.

Marinovic, B., J. Lemmens, and B. Knott. 1994. Larval development of *Ibacus peronii* Leech (Decapoda, Scyllaridae) under laboratory conditions. Journal of Crustacean Biology 14: 80–96.

Marques, F., and G. Pohle. 1995. Phylogenetic analysis of the Pinnotheridae (Crustacea, Brachyura) based on larval morphology, with emphasis on the *Dissodactylus* species complex. Zoologica Scripta 24: 347–364.

Marques, F., and G. Pohle. 1996. Laboratory-reared larval stages of *Dissodactylus mellitae* (Decapoda: Brachyura: Pinnotheridae) and developmental patterns within the *Dissodactylus* species complex. Canadian Journal of Zoology 74: 47–62.

Marr, J. W. S. 1962. The natural history and geography of the Antarctic krill (*Euphausia superba* Dana). Discovery Reports 32: 33–464.

Marshall, H. P., and H. J. Hirche. 1984. Development of eggs and nauplii of *Euphausia superba*. Polar Biology 2: 245–250.

Martin, J. 2001. Les larves de crustacés décapodes des côtes françaises de la Manche: Identification, période, abondance. Brest, France: IFREMER [Institut Français de Recherche pour l'Exploitation de la Mer].

Martin, J. W. 1984. Notes and bibliography on the larvae of xanthid crabs, with a key to the known xanthid zoeas of the western Atlantic and Gulf of Mexico. Bulletin of Marine Science 34: 220–239.

Martin, J. W. 1988. Phylogenetic significance of the brachyuran megalopa: evidence from the Xanthidae. Symposia of the Zoological Society of London 59: 69–102.

Martin, J. W. 1992. Branchiopoda. *In* F. W. Harrison and A. G. Humes, eds. Microscopic Anatomy of Invertebrates. Vol. 9, Crustacea, chapter 3. New York: Wiley-Liss.

Martin, J. W. 2013. Arthropod evolution and phylogeny. McGraw-Hill Yearbook of Science and Technology 2012, 34–37. New York: McGraw-Hill.

Martin, J. W., and D. Belk. 1988. A review of the clam shrimp family Lynceidae Stebbing, 1902 (Branchiopoda, Conchostraca) in the Americas. Journal of Crustacean Biology 8: 451–482.

Martin, J. W., and C. Cash-Clark. 1995. The external morphology of the onychopod "cladoceran" genus *Bythotrephesi* (Crustacea, Branchiopoda, Onychopoda, Cercopagididae), with notes on the morphology and phylogeny of the order Onychopoda. Zoologica Scripta 24: 61–90.

Martin, J. W., and J. C. Christiansen. 1996. A morphological comparison of the phyllopodous thoracic limbs of a leptostracan (*Nebalia* sp.) and a spinicaudate conchostracan (*Leptestheria* sp.), with comments on the use of the term Phyllopoda as a taxonomic category. Canadian Journal of Zoology 73: 2283–2291.

Martin, J. W., and G. E. Davis. 2001. An Updated Classification of the Recent Crustacea. Science Series No. 39. Los Angeles: Natural Museum of Los Angeles County.

Martin, J. W., and G. E. Davis. 2006. Historical trends in crustacean systematics. Crustaceana 79: 1347–1368.

Martin, J. W., and A. Dittel. 2007. The megalopa larval stage of the hydrothermal vent crab genus *Bythograea* (Crustacea, Decapoda, Bythograeidae). Zoosystema 29: 365–379.

Martin, J. W., and J. W. Goy. 2004. The first larval stage of *Microprosthema semilaeve* (von Martens, 1872) (Crustacea: Decapoda: Stenopodidea) obtained in the laboratory. Gulf and Caribbean Research. 16: 19–25.

Martin, J. W., and T. A. Haney. 2009. Leptostraca (Crustacea) of the Gulf of Mexico. *In* D. L. Felder and D. K. Camp, eds. Gulf of Mexico: Origin, Waters, and Biota. Vol. 1, Biodiversity, 895–899. College Station: Texas A&M University Press.

Martin, J. W., and M. S. Laverack. 1992. On the distribution of the crustacean dorsal organ. Acta Zoologica (Stockholm) 73: 357–368.

Martin, J. W., and B. Ormsby. 1991. A large brachyuran-like larva of the Hippidae (Crustacea: Decapoda: Anomura) from the Banda Sea, Indonesia: the largest known zoea. Proceedings of the Biological Society of Washington 104: 561–568.

Martin, J. W., and F. M. Truesdale. 1989. Zoeal development of *Ethusa microphthalma* Smith, 1881 (Brachyura, Dorippidae) reared in the laboratory, with a comparison of other dorippid zoeae. Journal of Natural History 23: 205–217.

Martin, J. W., F. M. Truesdale, and D. L. Felder. 1985. Larval development of *Panopeus bermudensis* Benedict and Rathbun, 1891 (Brachyura, Xanthidae), with notes on zoeal characters in xanthid crabs. Journal of Crustacean Biology 5: 54–105.

Martin, J. W., F. M. Truesdale, and D. L. Felder. 1988. The megalopa stage of the stone crab *Menippe adina* Williams and Felder, with a comparison of megalopae in the genus *Menippe*. U.S. Fishery Bulletin 86: 289–297.

Martin, J. W., E. Vetter, and C. E. Cash-Clark. 1996. Description, external morphology, and natural history observations of *Nebalia hessleri*, new species (Phyllocarida: Leptostraca), from southern California, with a key to the families and genera of the Leptostraca. Journal of Crustacean Biology 16: 347–372.

Martin, J. W., D. B. Cadien, and T. L. Zimmerman. 2002. First record and habitat notes for the genus *Lightiella* (Crustacea, Cephalocarida, Hutchinsoniellidae) from the British Virgin Islands. Gulf and Caribbean Research 14: 75–79.

Martin, J. W., S. L. Boyce, and M. J. Grygier. 2003. New records of *Cyclestheria hislopi* (Baird, 1859) (Crustacea: Branchiopoda: Diplostraca: Cyclestheria) in Southeast Asia. Raffles Bulletin of Zoology 51: 215–218.

Martin, J. W., E. M. Liu, and D. Striley. 2007. Morphological observations on the gills of dendrobranchiate shrimps. Zoologischer Anzeiger 246: 115–125.

Martin, J. W., K. A. Crandall, and D. L. Felder, eds. 2009. Decapod Crustacean Phylogenetics. Crustacean Issues No. 18. Boca Raton, FL: CRC Press.

Martin, M. F. 1932. On the morphology and classification of *Argulus* (Crustacea). Proceedings of the Zoological Society of London 102: 771–806.

Martin, P., N. J. Dorn, T. Kawai, C. van der Heiden, and G. Scholtz. 2010. The enigmatic Marmorkrebs (marbled crayfish) is the parthenogenetic form of *Procambarus fallax* (Hagen, 1870). Contributions to Zoology 79: 107–118.

Martínez Arbizu, P. 2005. Additions to the life cycle of Tantulocarida. Sixth International Crustacean Congress, Glasgow, Scotland, UK, 18–22 July 2005 (abstract).

Maruzzo, D., S. Conlan, N. Aldred, A. S. Clare, and J. T. Høeg. 2011. Video observation of antennulary sensory setae during surface exploration in cyprids of *Balanus amphitrite*. Biofouling 27: 225–239.

Maruzzo D., N. Aldred, A. S. Clare, and J. T. Høeg. 2012. Metamorphosis in the cirripede crustacean *Balanus amphitrite*. PLoS One 7(5): e37408.

Matsudaira, C., T. Kariya, and T. Tsuda. 1952. The study of the biology of a mysid *Gastrosaccus vulgaris* Nakasawa. Tohoku Journal of Agricultural Research 3: 155–174.

Matthews, J. B. L. 1964. On the biology of some bottom-living copepods (Aetideidae and Phaennidae) from western Norway. Sarsia 16: 1–46.

Mattox, N. T. 1937. Studies on the life history of a new species of fairy shrimp, *Eulimnadia diversa*. Transactions of the American Microscopical Society 56: 249–255.

Mattox, N. T. 1950. Notes on the life history and description of a new species of conchostracan phyllopod, *Caenestheriella gynecia*. Transactions of the American Microscopical Society 69: 50–53.

Mauchline, J. 1971. Euphausiacea Larvae. Zooplankton Sheet No. 135/137. [Copenhagen:] Conseil International pour l'Exploration de la Mer.

Mauchline, J. 1972. The biology of bathypelagic organisms, especially Crustacea. Deep-Sea Research 19: 753–780.

Mauchline, J. 1973. The broods of British Mysidacea (Crustacea). Journal of the Marine Biological Association of the United Kingdom 53: 801–817.

Mauchline, J. 1980. The Biology of Mysids and Euphausiids. Vol. 18 *In* J. H. S. Blaxter, F. S. Russell, and M. Younge, eds. Advances in Marine Biology. London: Academic Press.

Mauchline, J. 1985. Growth in mysids and euphausiids. *In* A. M. Wenner, ed. Factors in Adult Growth, 337–354. Crustacean Issues No. 3. Rotterdam: A. A. Balkema.

Mauchline, J. 1998. The biology of calanoid copepods. Advances in Marine Biology 33: 1–710.

Mauchline, J., and L. R. Fisher. 1969. The Biology of Euphausiids. Vol. 7 *in* F. S. Russell and C. M. Yonge, eds. Advances in Marine Biology. London: Academic Press.

Mayrat, A., and M. de Saint Laurent. 1996. Considérations sur la classe des malacostracés (Malacostraca Latreille, 1802). *In* J. Forest, ed. Traité de Zoologie: Anatomie, Systématique, Biologie. Vol. 7, fasc. 2, Généralités (suite) et Systématique (Céphalocarides à Syncarides), 841–863. Paris: Masson.

Mazzocchi, M. G., and G.-A. Paffenhöfer. 1998. First observations on the biology of *Clausocalanus furcatus* (Copepoda, Calanoida). Journal of Plankton Research 20: 331–342.

McLachlan, A. 1977. The larval development and population dynamics of *Derocheilocaris algonensis*. Zoologica Africana 12: 1–14.

McLaughlin, P. A., R. Lemaitre, and C. C. Tudge. 2004. Carcinization in the Anomura—fact or fiction?: 2, evidence from larval, megalopal and early juvenile morphology. Contributions to Zoology 73: 165–205.

McLaughlin, P. A., D. K. Camp, M. Angel, E. L. Bousfield, P. Brunel, R. C. Brusca, D. Cadien, A. C. Cohen, K. Conlan, L. G. Eldredge, D. L. Felder, J. W. Goy, T. A. Haney, B. Hann, R. W. Heard, E. A. Hendrycks, H. H. Hobbs III, J. Holsinger, B. Kensley, D. R. Laubitz, S. E. Lecroy, R. Lemaitre, R. F. Maddocks, J. W. Martin, P. Mikkelsen, E. Nelson, W. A. Newman, R. M. Overstreet, W. J. Poly, W. W. Price, J. W. Reid, A. Robertson, D. C. Rogers, A. Ross, M. Schotte, F. R. Schram, C.-T. Shih, L. Watling, G. D. F. Wilson, and D. D. Turgeon. 2005. Common and Scientific Names of Aquatic Invertebrates from the United

States and Canada: Crustaceans. Special Publication No. 31. Bethesda, MD: American Fisheries Society.

McLaughlin, P. A., R. Lemaitre, and U. Sorhannus. 2007. Hermit crab phylogeny: a reappraisal and its "fall-out." Journal of Crustacean Biology 27: 97–115.

McLaughlin, P. A., C. B. Boyko, K. A. Crandall, T. Komai, R. Lemaitre, M. Osawa, and D. L. Rahayu. 2010. Annotated checklist of anomuran decapod crustaceans of the world (exclusive of the Kiwaoidea and families Chirostylidae and Galatheidae of the Galatheoidea): preamble and scope. Raffles Bulletin of Zoology, Suppl. 23: 1–4.

McLay, C. L., S. S. L. Lim, and P. K. L. Ng. 2001. On the first zoea of Lauridromia indica (Gray, 1831), with an appraisal of the generic classification of the Dromiidae (Decapoda: Brachyura) using larval characters. Journal of Crustacean Biology 21: 733–747.

McWilliam, P. S. 1995. Evolution in the phyllosoma and puerulus phases of the spiny lobster genus Panulirus White. Journal of Crustacean Biology 15: 542–557.

McWilliam, P. S., and B. F. Phillips. 1997. Metamorphosis of the final phyllosoma and secondary lecithotrophy in the puerulus of Panulirus cygnus George: a review. Marine and Freshwater Research 48: 783–790.

McWilliam, P. S., B. F. Phillips, and S. Kelly. 1995. Phyllosoma larvae of Scyllarus species (Decapoda, Scyllaridae) from the shelf waters of Australia. Crustaceana 68: 537–566.

Mehlhorn, H., ed. 2008. Encyclopedia of Parasitology, 3rd ed., vol. 1. New York: Springer.

Meland, K., and E. Willassen. 2007. The disunity of "Mysidacea" (Crustacea). Molecular Phylogenetics and Evolution 44: 1083–1104.

Melo, S. G., and A. L. Brossi-Garcia. 2002. Post-embryonic development of Upogebia paraffinis Williams, 1993 (Decapoda, Thalassinidea), reared in the laboratory. Nauplius 8: 149–168.

Metschnikoff, E. 1868. On the development of Nebalia (Russian). Mémoires de l'Académie Impériale des Sciences de Saint Pétersbourg, ser. 7, 13: 1.

Meyer, R., S. Friedrich, and R. R. Melzer. 2004. Xantho poressa (Olivi, 1792) and Xantho pilipes A. Milne-Edwards, 1867 larvae (Brachyura, Xanthidae): scanning EM diagnosis of zoea I from the Adriatic Sea. Crustaceana 77: 997–1005.

Meyer, R., I. S. Wehrtmann, and R. R. Melzer. 2006. Morphology of the first zoeal stage of Portunus acuminatus Stimpson, 1871 (Decapoda: Portunidae: Portuninae) reared in the laboratory. Scientia Marina 70: 261–270.

Michel, A. 1969. Dernier stade larvaire pélagique et post-larve de Heterosquilla (Heterosquilloides) brazieri (Miers, 1880) (crustacés stomatopodes). Bulletin du Muséum National d'Histoire Naturelle 4: 992–997.

Michel, A. 1970a. Larves pélagiques et post-larves du genre Odontodactylus (Crustacés Stomatopodes) dans le Pacifique tropical sud et équatorial. Cahiers ORSTOM, Océanographie 8: 111–126.

Michel, A. 1970b. Les larves phyllosomes du genre Palinurellus von Martens (Crustacés Décapodes: Palinuridae). Bulletin du Museum National d'Histoire Naturelle 41: 1228–1237.

Michel, A. 1971. Note sur les puerulus de Palinuridae et les larves phyllosomes de Panulirus homarus (L.). Cahiers ORSTOM, Océanographie 9: 459–473.

Michel, A., and R. B. Manning. 1972. The pelagic larvae of Chorisquilla tuberculata (Borradaile, 1907) (Stomatopoda). Crustaceana 22: 113–126.

Michels, J. and M. Büntzow. 2010. Assessment of Congo red as fluorescence marker for the exoskeleton of small crustaceans and the cuticle of polychaetes. Journal of Microscopy 238: 95–101.

Mikheev, V. N., A. F. Pasternak, E. T. Valtonen, and Y. Lankinen. 2001. Spatial distribution and hatching of overwintered eggs of a fish ectoparasite, Argulus coregoni (Crustacea: Branchiura). Diseases of Aquatic Organisms 46: 123–128.

Mileikovsky, S. A. 1971. Types of larval development in marine bottom invertebrates, their distribution and ecological significance: a re-evaluation. Marine Biology 10: 193–213.

Miller, D. G. M. 1986. A late phyllosoma larva of Jasus tristani Holthuis (Decapoda, Palinuridae). Crustaceana 50: 1–6.

Miller, P. E., and H. G. Coffin. 1961. A Laboratory Study of the Developmental Stages of Hapalogaster mertensii [sic] (Brandt), (Crustacea, Decapoda). Walla Walla College Publications of the Dept. of Biological Sciences and the Biological Station No. 30. College Place, WA: Walla Walla College, Dept. of Biological Sciences.

Milne-Edwards, H. 1830a. Description des genres glaucothoé, sicyonie, sergeste et acète, de l'ordre des crustacés décapodes. Annales des Sciences Naturelles 19: 333–351.

Milne-Edwards, H. 1830b. Observations sur le genre Phyllosoma. Bulletin des Sciences Naturelles et de Géologie 23: 1–148.

Milne-Edwards, H. 1832. Note sur un nouveau genre de crustacés de l'ordre des Stomatopodes. Annales de la Société Entomologique de France 1: 336–340.

Milne-Edwards, H. 1835. Note sur le genre Nebalia. Annales des Sciences Naturelles, ser. 2, 3: 309–311.

Miura, Y., and Y. Morimoto. 1953. Larval development of Bathynella morimotoi Uéno. Annotationes Zoologicae Japonenses 2: 238–245.

Modlin, R. F. 1979. Development of Mysis stenolepis (Crustacea: Mysidacea). American Midland Naturalist 101: 250–254.

Modlin, R. F. 1996. Contributions to the ecology of Paranebalia belizensis from the waters off central Belize, Central America. Journal of Crustacean Biology 16: 529–534.

Mohrbeck, I., P. Martínez Arbizu, and T. Glatzel. 2010. Tantulocarida (Crustacea) from the Southern Ocean deep sea, and the description of three new species of Tantulacus Huys, Andersen & Kristensen, 1992. Systematic Parasitology 77: 131–151.

Møller, O. S. 2009. Branchiura (Crustacea): survey of historical literature and taxonomy. Arthropod Systematics & Phylogeny 67: 41–55.

Møller, O. S., and J. Olesen. 2009. The little-known Dipteropeltis hirundo Calman, 1912 (Crustacea, Branchiura): SEM investigations of paratype material in light of recent phylogenetic analyses. Journal of Experimental Parasitology 125: 30–41.

Møller, O. S., and J. Olesen. 2012. First description of larval stage 1 from a non-African fish parasite Dolops (Crustacea: Branchiura). Journal of Crustacean Biology 32: 231–238.

Møller, O. S., J. Olesen, and J. T. Høeg. 2003. SEM studies on the early development of Triops cancriformis (Bosc) (Crustacea: Branchiopoda: Notostraca). Acta Zoologica (Stockholm) 84: 267–284.

Møller, O. S., J. Olesen, and J. T. Høeg. 2004. Study on the larval development of Eubranchipus grubii (Crustacea, Branchiopoda, Anostraca), with notes on the basal phylogeny of the Branchiopoda. Zoomorphology 123: 107–123.

Møller, O. S., J. Olesen, and D. Waloszek. 2007. Swimming and cleaning in the free-swimming phase of Argulus larvae

(Crustacea, Branchiura): appendage adaptation and functional morphology. Journal of Morphology 268: 1–11.

Møller, O. S., J. Olesen, A. Avenant-Oldewage, P. F. Thomsen, and H. Glenner. 2008. First maxillae suction discs in Branchiura (Crustacea): development and evolution in light of the first molecular phylogeny of Branchiura, Pentastomida, and other "Maxillopoda." Arthropod Structure and Development 37: 333–346.

Monakov, A. V., and T. I. Dobrynina. 1977. Post-embryonic development of *Lynceus brachyurus* O. F. Müller (Conchostraca). Zoologicheskii Žhurnal 56: 1877–1880.

Monakov, A. V., E. B. Pavel'eva, and R. Ya. Bratcik. 1980. Postembrionalnoye razvitiye i rost *Leptestheria dahalacensis* (Rüppel) (Branchiopoda, Conchostraca). Trudy Instituta Biologii Vnutrennikh vod Akademii Nauk SSSR 41/44: 53–58.

Monod, T. 1924. Sur un type nouveau de malacostracé: *Thermosbaena mirabilis*, nov. gen. nov. sp. Bulletin de la Société Zoologique de France 49: 58–68.

Montgomery, J. C., A. Jeffs, S. D. Simpson, M. Meekan, and C. Tindle. 2006. Sound as an orientation cue for pelagic larvae of reef fishes and decapod crustaceans. Advances in Marine Biology 51: 143–196.

Montú, M., K. Anger, C. de Bakker, V. Anger, and L. Loureiro Fernandes. 1988. Larval development of the Brazilian mud crab *Panopeus austrobesus* Williams, 1983 (Decapoda: Xanthidae) reared in the laboratory. Journal of Crustacean Biology 8: 594–613.

Montvilo, J. A., M. A. Hegyi, and M. J. Kevin. 1987. Aspects of the jelly coat of *Holopedium* and certain other cladocerans (Crustacea). Transactions of the American Microscopical Society 106: 105–114.

Moore, R. C., and L. McCormick. 1969. General features of Crustacea. *In* R. C. Moore, ed. Treatise on Invertebrate Paleontology, Part R, Arthropoda 4, R57–R120. Lawrence: Geological Society of America and University of Kansas Press.

Mordukhai-Boltovskoi, P. D. 1968. On the taxonomy of the Polyphemidae. Crustaceana 14: 197–209.

Morgan, G. J. 1987. Abbreviated development in *Paguristes frontalis* (Milne-Edwards, 1836) (Anomura: Diogenidae). Journal of Crustacean Biology 7: 536–540.

Morgan, S. G. 1995. Life and death in the plankton: larval mortality and adaptation. *In* L. McEdward, ed. Ecology of Marine Invertebrate Larvae, 279–321. Boca Raton, FL: CRC Press.

Morgan, S. G. 2001. The larval ecology of marine communities. *In* M. Bertness, S. D. Gaines, and M. Hay, eds. Marine Community Ecology, 158–181. Sunderland, MA: Sinauer.

Morgan, S. G., and J. G. Goy 1987. Reproduction and larval development of the mantis shrimp *Gonodactylus bredini* (Crustacea: Stomatopoda) maintained in the laboratory. Journal of Crustacean Biology 7: 595–618.

Morgan, S. G., and A. J. Provenzano Jr. 1979. Development of pelagic larvae and post-larva of *Squilla empusa* (Crustacea, Stomatopoda), with an assessment of larval characters within the Squillidae. U.S. Fishery Bulletin 77: 61–90.

Morgan, S. G., C. S. Manooch III, D. L. Mason, and J. W. Goy. 1985. Pelagic fish predation on *Cerataspis*, a rare larval genus of oceanic penaeoids. Bulletin of Marine Science 36: 249–259.

Morin, J. G., and A. Cohen. 1991. Bioluminescent displays, courtship, and reproduction in ostracods. *In* R. T. Bauer and J. W. Martin, eds. Crustacean Sexual Biology, 1–16. New York: Columbia University Press.

Moura, G. C., and M. L. Christoffersen. 1996. The system of the mandibulate arthropods: Tracheata and Remipedia as sister groups, "Crustacea" non-monophyletic. Journal of Comparative Biology 1: 95–113.

Moyse, J. 1963. A comparison of the value of various flagellates and diatoms as food for barnacle larvae. Journal de Conseil—Conseil Permanent International pour l'Exploration de la Mer 28: 175–187.

Moyse, J. 1987. Larvae of lepadomorph barnacles. *In* A. J. Southward, ed. Barnacle Biology, 329–362. Rotterdam: A. A. Balkema.

Moyse, J., P. G. Jensen, J. T. Høeg, and H. Al-Yahya. 1995. Attachment organs in cypris larvae: using scanning electron microscopy. *In* F. R. Schram and J. T. Høeg, eds. New Frontiers In Barnacle Evolution, 153–178. Crustacean Issues No. 10. Rotterdam: A. A. Balkema.

Müller, A. H. 1962. Pantopoden (Arthrop., Pycnogonida) aus dem Solnhofener Plattenkalk (Malm Zeta) von Süddeutschland. Freiberger Forschungsheft C, Paläontologie 151: 149–157.

Müller, F. 1864. Für Darwin. Leipzig, Germany: Wilhelm Engelmann.

Müller, F. 1871. Bruchstücke zur Naturgeschichte der Bopyriden. Jenaische Zeitschrift für Medicin und Naturwissenschaft 6: 53–73.

Müller, G. W. 1894. Ostracoden. Fauna und Flora des Golfes von Neapel und der Angrenzenden Meeres-Abschnitte, Monographie No. 21. Berlin: R. Friedländer & Sohn.

Müller, K. J. 1983. Crustacea with preserved soft parts from the Upper Cambrian of Sweden. Lethaia 16: 93–109.

Müller, K. J., and D. Walossek. 1985. Skaracarida, a new order of Crustacea from the Upper Cambrian of Västergötland, Sweden. Fossils and Strata 17: 1–65.

Müller, K. J., and D. Walossek. 1986a. Arthropod larvae from the Upper Cambrian of Sweden. Transactions of the Royal Society of Edinburgh, Earth Sciences 77: 157–179.

Müller, K. J., and D. Walossek. 1986b. *Martinssonia elongata* gen. et sp. n., a crustacean-like euarthropod from the Upper Cambrian "Orsten" of Sweden. Zoologica Scripta 15: 73–92.

Müller, K. J., and D. Walossek. 1988. External morphology and larval development of the Upper Cambrian maxillopod *Bredocaris admirabilis*. Fossils and Strata 23: 1–70.

Müller, K. J., and D. Walossek. 1991. Ein Blick durch das "Orsten"-Fenster in die Arthropodenwelt vor 500 Millionen Jahren. Verhandlungen der Deutschen Zoologischen Gesellschaft 84: 281–294.

Müller, O. F. 1785. Entomostraca seu Insecta Testacea, quae in Aquis Daniae et Norvegiae Reperit, Descripsit et Iconibus Illustravit Otho Fridericus Müller. Leipzig, Germany: F. W. Thiele.

Müller, P. E. 1868. Bidrag til Cladocerernes Forplantningshistorie. Naturhistorisk Tidsskrift, ser. 3, 5: 295–354.

Murano, M. 1964. Fisheries biology of a marine relict mysid *Neomysis intermedia* Czerniavsky, 4: life cycle with special reference to growth. Aquiculture 12: 109–117.

Naganawa, H. 2001. Current prospect of the Recent large branchiopodan fauna of East Asia, 3: revised classification of the Recent Spinicaudata. Aquabiology 23: 291–299.

Nagasawa, K., J. Bresciani, and J. Lützen. 1988. Morphology of *Pectenophilus ornatus*, new genus, new species, a copepod parasite of the Japanese scallop *Patinopecten yessoensis*. Journal of Crustacean Biology 8: 31–42.

Nair, K. B. 1939. The reproduction, oogenesis and development of *Mesopodopsis orientalis* Tatt. Proceedings of the Indian Academy of Science 9: 175–223.

Nair, K. K. N. 1968. Observations on the biology of *Cyclestheria hislopi* (Baird), (Conchostraca: Crustacea). Archiv für Hydrobiologie 65: 96–99.

Nates, S. F., and C. L. McKenney Jr. 2000. Ontogenetic changes in biochemical composition during larval and early post-larval development of *Lepidophthalmus louisianensis*, a ghost shrimp with abbreviated development. Comparative Biochemistry and Physiology, B 127: 459–468.

Nates, S. F., D. L. Felder, and R. Lemaitre. 1997. Comparative larval development of two species of the burrowing ghost shrimp genus *Lepidophthalmus* (Decapoda: Callianassidae). Journal of Crustacean Biology 17: 497–519.

Negrea, S., N. Botnariuc, and H. J. Dumont. 1999. Phylogeny, evolution and classification of the Branchiopoda. Hydrobiologia 412: 191–212.

Neil, D. M., D. L. Macmillan, R. M. Robertson and M. S. Laverack. 1976. The structure and function of thoracic exopodites in larvae of the lobster *Homarus gammarus* (L.). Philosophical Transactions of the Royal Society of London, B 294: 53–68.

Newman, W. A., and A. Ross. 2001. Prospectus on larval cirriped setation formulae, revisited. Journal of Crustacean Biology 21: 56–77.

Ng, P. K. L., D. Guinot, and P. J. F. Davie. 2008. Systema Brachyurorum: Pt. 1, an Annotated Checklist of Extant Brachyuran Crabs of the World. Raffles Bulletin of Zoology, Suppl. No. 17. Singapore: Raffles Museum of Biodiversity Research.

Ngoc-Ho, N. 1977. The larval development of *Upogebia darwini* (Crustacea, Thalassinidea) reared in the laboratory, with a redescription of the adult. Journal of Zoology (London) 181: 439–464.

Ngoc-Ho, N. 1981. A taxonomic study of the larvae of four thalassinid species (Decapoda: Thalassinidea) from the Gulf of Mexico. Bulletin of the British Museum (Natural History), Zoology 40: 237–273.

Nielsen, C. 2012. Animal Evolution: Interrelationships of the Living Phyla, 3rd ed. Oxford: Oxford University Press.

Nilsson, D. E., E. Hallberg and R. Elofsson. 1986. The ontogenetic development of refracting superposition eyes in crustaceans: transformation of optical design. Tissue and Cell 18: 509–519.

Noodt, W. 1970. Zur Eidonomie der Stygocaridacea, einer Gruppe interstitieller Syncarida (Malacostraca). Crustaceana 19: 227–244.

Nordmann, A. von. 1832. Mikrographische Beitrage zur Naturgeschichte der Wirbellosen Thiere, vol. 2. Berlin: G. Reimer.

Nott, J. 1969. Settlement of barnacle larvae: surface structure of the antennular attachment disc by scanning electron microscopy. Marine Biology 2: 248–251.

Nott, J., and B. Foster. 1969. On the structure of the antennular attachment organ of the cypris larva of *Balanus balanoides* (L.). Philosophical Transactions of the Royal Society of London, B 256: 115–134.

Nussbaum, J. 1887. L'embryologie de *Mysis chameleo* (Thompson). Archives de Zoologie Expérimentale et Générale, ser. 2, 5: 123–202.

Oakley, T. H., J. M. Wolfe, A. R. Lindgren, and A. K. Zaharoff. 2013. Phylotranscriptomics to bring the understudied into the fold: monophyletic Ostracoda, fossil placement, and pancrustacean phylogeny. Molecular Biology and Evolution 30(1): 215–233.

Oberg, M. 1906. Die Metamorphose der Plankton-Copepoden der Kieler Bucht. Wissenschaftliche Meeresuntersuchungen (Abteilung Kiel), n.s., 9: 39–103.

Obreshkove, V., and A. W. Fraser. 1940. Growth and differentiation of *Daphnia magna* eggs in vitro. Proceedings of the Society for Experimental Biology and Medicine 43: 543–544.

Oehmichen, A. 1921. Wissenschaftliche Mitteilungen, 1: die Entwicklung der äußeren Form des *Branchipus grubei* Dyb. Zoologischer Anzeiger 53: 241–253.

Ogawa, K., K. Matsuzaki, and H. Misaki. 1997. A new species of *Balaenophilus* (Copepoda: Harpacticoida), an ectoparasite of a sea turtle in Japan. Zoological Science 14: 691–700.

Ohtsuka, S. 1993. Note on mature female of *Itoitantulus misophricola* Huys, Ohtsuka and Boxshall, 1992 (Crustacea: Tantulocarida). Publications of the Seto Marine Laboratory 36: 73–77.

Ohtsuka, S., Y. Hanamura, and T. Kase. 2002. A New Species of *Thetispelecaris* (Crustacea: Peracarida) from Submarine Cave on Grand Cayman Island. Zoological Science 19: 611–624.

Ohtsuka, S., G. A. Boxshall, and S. Harada. 2005. A new genus and species of nicothoid copepod (Crustacea: Copepoda: Siphonostomatoida) parasitic on the mysid *Siriella okadai* Ii from off Japan. Systematic Parasitology 62: 65–81.

Ohtsuka, S., S. Harada, M. Shimomura, G. A. Boxshall, R. Yoshizaki, D. Ueno, Y. Nitta, S. Iwasaki, H. Okawachi, and T. Sakakihara. 2007. Temporal partitioning: dynamics of alternating occupancy of a host microhabitat by two different crustacean parasites. Marine Ecology Progress Series 348: 261–272.

Ohtsuka, S., I. Takami, B. A. Venmathi Maran, K. Ogawa, T. Shimono, Y. Fujita, M. Asakawa, and G. A. Boxshall. 2009. Developmental stages and growth of *Pseudocaligus fugu* Yamaguti, 1936 (Copepoda: Siphonostomatoida: Caligidae) host-specific to puffer. Journal of Natural History 43: 1779–1804.

Okada, R., A. Tsukagoshi, R. J. Smith, and D. J. Horne. 2008. The ontogeny of the platycopid *Keijcyoidea infralittoralis* (Ostracoda: Podocopa). Zoological Journal of the Linnean Society 153: 213–237.

Okamoto, K. 2008. Japanese nephropid lobster *Metanephrops japonicus* lacks zoeal stage. Fisheries Science 74: 98–103.

Okomura, T. 2003. Relationship of ovarian and marsupial development to the female molt cycle in *Acanthomysis robusta* (Crustacea: Mysida). Fisheries Science 69: 995–1000.

Olesen, J. 1998. A phylogenetic analysis of the Conchostraca and Cladocera (Crustacea, Branchiopoda, Diplostraca). Zoological Journal of the Linnean Society 122: 491–536.

Olesen, J. 1999. Larval and post-larval development of the branchiopod clam shrimp *Cyclestheria hislopi* (Baird, 1859) (Crustacea, Branchiopoda, Conchostraca, Spinicaudata). Acta Zoologica (Stockholm) 80: 163–184.

Olesen, J. 2000. An updated phylogeny of the Conchostraca–Cladocera clade (Branchiopoda, Diplostraca). Crustaceana 73: 869–886.

Olesen, J. 2001. External morphology and larval development of *Derocheilocaris remanei* Delamare-Deboutteville and Chappuis, 1951 (Crustacea, Mystacocarida), with a comparison of crustacean segmentation and tagmosis patterns. Det Kongelige Danske Videnskabernes Selskab, Biologiske Skrifter 53: 1–59.

Olesen, J. 2004. On the ontogeny of the Branchiopoda (Crustacea): contribution of development to phylogeny and classification. *In* G. Scholtz, ed. Evolutionary Developmental Biology of Crustacea, 217–269. Crustacean Issues No. 15. Lisse, Netherlands: A. A. Balkema.

Olesen, J. 2005. Larval development of *Lynceus brachyurus* (Crustacea, Branchiopoda, Laevicaudata): redescription of unusual crustacean nauplii, with special attention to the molt between last nauplius and first juvenile. Journal of Morphology 264: 131–148.

Olesen, J. 2007. Monophyly and phylogeny of Branchiopoda, with focus on morphology and homologies of branchiopod phyllopodous limbs. Journal of Crustacean Biology 27: 165–183.

Olesen, J. 2009. Phylogeny of Branchiopoda (Crustacea): character evolution and contribution of uniquely preserved fossils. Arthropod Systematics & Phylogeny 67: 3–39.

Olesen, J. 2013. The crustacean carapace: morphology, function, development, and phylogenetic history. *In* M. Thiel and L. Watling, eds. Functional Morphology and Diversity. Vol. 1, Natural History of the Crustacea, 103x-139. New York: Oxford University Press.

Olesen, J., and M. J. Grygier. 2003. Larval development of Japanese "conchostracans": 1, larval development of *Eulimnadia braueriana* (Crustacea, Branchiopoda, Spinicaudata, Limnadiidae) compared to that of other limnadiids. Acta Zoologica (Stockholm) 84: 41–61.

Olesen, J., and M. J. Grygier. 2004. Larval development of Japanese "conchostracans": 2, larval development of *Caenestheriella gifuensis* (Crustacea, Branchiopoda, Spinicaudata, Cyzicidae), with notes on homologies and evolution of certain naupliar appendages within the Branchiopoda. Arthropod Structure and Development 33: 453–469.

Olesen, J., and S. Richter, 2013. Onychocaudata (Branchiopoda: Diplostraca), a new high-level taxon in branchiopod systematics. Journal of Crustacean Biology 33: 62–65.

Olesen, J., and D. Walossek. 2000. Limb ontogeny and trunk segmentation in *Nebalia* species. Zoomorphology 120: 47–64.

Olesen, J., J. W. Martin, and E. W. Roessler. 1996. External morphology of the male of *Cyclestheria hislopi* (Baird, 1859) (Crustacea, Branchiopoda, Spinicaudata), with comparison of male claspers among the Conchostraca and Cladocera and its bearing on phylogeny of the "bivalved" Branchiopoda. Zoologica Scripta 25: 291–316.

Olesen, J., S. Richter, and G. Scholtz. 2001. The evolutionary transformation of phyllopodous to stenopodous limbs in the Branchiopoda (Crustacea): is there a common mechanism for early limb development in arthropods? International Journal of Developmental Biology 45: 869–876.

Olesen, J., S. Richter, and G. Scholtz. 2003. On the ontogeny of *Leptodora kindtii* (Crustacea, Branchiopoda, Cladocera), with notes on the phylogeny of the Cladocera. Journal of Morphology 256: 235–259.

Olesen, J., J. T. Haug, A. Maas, and D. Waloszek. 2011. External morphology of *Lightiella monniotae* (Crustacea, Cephalocarida) in the light of Cambrian "Orsten" crustaceans. Arthropod Structure and Development 40: 449–478.

Olesen, J., M. Fritsch, and M. J. Grygier. 2013. Larval development of Japanese "conchostracans": pt. 3, larval development of *Lynceus biformis* (Crustacea, Branchiopoda, Laevicaudata) based on scanning electron microscopy and fluorescence microscopy. Journal of Morphology 274: 229–242.

Onbé, T. 1978. The life cycle of marine cladocerans. Bulletin of the Plankton Society of Japan 25: 41–54.

Onbé, T. 1984. The developmental stages of *Longipedia americana* (Copepoda: Harpacticoida) reared in the laboratory. Journal of Crustacean Biology 4: 615–631.

O'Neil, J. M., and M. R. Roman. 1994. Ingestion of the cyanobacterium *Trichodesmium* spp. by pelagic harpacticoid copepods *Macrosetella*, *Miracia*, and *Oculosetella*. Hydrobiologia 292/293: 235–240.

Orr, P. J., and D. E. G. Briggs. 1999. Exceptionally preserved conchostracans and other crustaceans from the Upper Carboniferous of Ireland. Special Papers in Palaeontology 62: 1–68.

Ortmann, A. 1893. Decapoden und Schizopoden der Plankton-Expedition. Vol. 2 (G, b), Ergebnisse der Plankton-Expedition der Humboldt-Stiftung. Kiel, Germany: Lipsius & Tischer.

Osawa, M., and P. A. McLaughlin. 2010. Annotated checklist of anomuran decapod crustaceans of the world (exclusive of the Kiwaoidea and families Chirostylidae and Galatheidae of the Galatheoidea), 2: Porcellanidae. Raffles Bulletin of Zoology, Suppl. 23: 109–129.

Overstreet, R. M., I. Dyková, and W. E. Hawkins. 1992. Branchiura. *In* F. W. Harrison and A. G. Humes, eds. Microscopic Anatomy of Invertebrates. Vol. 9, Crustacea, 385–413. New York: Wiley-Liss.

Pabst, T., and S. Richter. 2004. The larval development of an Australian limnadiid clam shrimp (Crustacea, Branchiopoda, Spinicaudata), and a comparison with other Limnadiidae. Zoologischer Anzeiger 243: 99–115.

Pabst, T., and G. Scholtz. 2009. The development of phyllopodous limbs in Leptostraca and Branchiopoda. Journal of Crustacean Biology 29: 1–12.

Paffenhöfer, G.-A., and K. D. Lewis. 1989. Feeding behavior of nauplii of the genus *Eucalanus* (Copepoda, Calanoida). Marine Ecology Progress Series 57: 129–136.

Pai, P. G. 1958. On post-embryonic stages of phyllopod crustaceans, *Triops* (*Apus*), *Streptocephalus*, and *Estheria*. Proceedings of the Indian Academy of Sciences, B, Animal Science 48: 229–250.

Palero, F., and P. Abelló. 2007. The first phyllosoma stage of *Palinurus mauritanicus* (Crustacea: Decapoda: Palinuridae). Zootaxa 1508: 49–59.

Palero, F., G. Guerao, and P. F. Clark. 2008a. *Palinustus mossambicus* Barnard 1926 (Crustacea: Decapoda: Achelata: Palinuridae): morphology of the puerulus stage. Zootaxa 1857: 44–54.

Palero, F., G. Guerao, and P. Abelló. 2008b. Morphology of the final stage phyllosoma larva of *Scyllarus pygmaeus* (Crustacea: Decapoda: Scyllaridae), identified by DNA analysis. Journal of Plankton Research 30: 483–488.

Palero F., K. A. Crandall, P. Abelló, E. Macpherson, and M. Pascual. 2009a. Phylogenetic relationships between spiny, slipper and coral lobsters (Crustacea, Decapoda, Achelata). Molecular Phylogenetics and Evolution 50: 152–162.

Palero, F., G. Guerao, P. F. Clark, and P. Abelló. 2009b. The true identities of the slipper lobsters *Nisto laevis* and *Nisto asper* (Crustacea: Decapoda: Scyllaridae) verified by DNA analysis. Invertebrate Systematics 23: 77–85.

Palero, F., G. Guerao, P. F. Clark, and P. Abelló. 2010a. Final-stage phyllosoma of *Palinustus* A. Milne-Edwards, 1880 (Crustacea: Decapoda: Achelata: Palinuridae): the first complete description. Zootaxa 2403: 42–58.

Palero, F., G. Guerao, P. F. Clark, and P. Abelló. 2010b. *Scyllarus arctus* (Crustacea: Decapoda: Scyllaridae) final stage phyllosoma identified by DNA analysis, with morphological description. Journal of the Marine Biological Association of the United Kingdom 91: 485–492.

Paré, J. A. 2008. An overview of pentastomiasis in reptiles and other vertebrates. Journal of Exotic Pet Medicine 17: 285–294.

Parker, R. R., and L. Margolis. 1964. A new species of parasitic copepod, *Caligus clemensi* sp. nov., (Caligoida: Caligidae) from pelagic fishes in the coastal waters of British Columbia. Journal of the Fisheries Research Board of Canada 21: 873–889.

Pasini, G., and A. Garassino. 2009. A new phyllosoma form (Decapoda, ?Palinuridae) from the Late Cretaceous (Cenomanian) of Lebanon. Atti della Società Italiana di Scienze Naturali e del Museo Civico di Storia Naturale in Milano 150: 21–28.

Paterson, N. F. 1958. External features and life cycle of *Cucumaricola notabilis* nov. gen. et sp., a copepod parasite of the holothurian, *Cucumaria*. Parasitology 48: 269–290.

Patt, D. I. 1947. Some cytological observations on the nährboden of *Polyphemus pediculus* Linn. Transactions of the American Microscopic Society 66: 344–353.

Paul, M. A., and C. K. G. Nayar. 1977. Studies on a natural population of *Cyclestheria hislopi* (Baird) (Conchostraca: Crustacea). Hydrobiologia 53: 173–179.

Paula, J. 1989. Rhythms of larval release of decapod crustaceans in the Mira Estuary, Portugal. Marine Biology 100: 309–312.

Paula, J., and R. G. Hartnoll. 1989. The larval and post-larval development of *Percnon gibbesi* (Crustacea, Brachyura, Grapsidae) and the identity of the larval genus *Pluteocaris*. Journal of Zoology (London) 218: 17–37.

Paulinose, V. T. 1982. Key to the identification of larvae and post-larvae of the penaeid prawns (Decapoda: Penaeidea) of the Indian Ocean. Mahasagar 15: 223–229.

Pearson, J. C. 1939. The early life histories of some American Penaeidae, chiefly the commercial shrimp *Penaeus setiferus* (Linn.). U.S. Fishery Bulletin 49: 1–73.

Pennak, R. W., and D. J. Zinn. 1943. Mystacocarida, a new order of Crustacea from intertidal beaches in Massachusetts and Connecticut. Smithsonian Miscellaneous Collections No. 103. Washington, DC: Smithsonian Institution.

Pereyra Lago, R. 1988. Larval development of *Spiroplax spiralils* (Barnard, 1950) (Brachyura: Hexapodidae) in the laboratory: the systematic position of the family on the basis of larval morphology. Journal of Crustacean Biology 8: 576–593.

Pérez-Farfante, I., and B. F. Kensley. 1997. Penaeoid and sergestoid shrimps and prawns of the world: keys and diagnoses for the families and genera. Mémoires du Muséum National d'Histoire Naturelle 175: 1–233.

Pérez-Losada, M., J. T. Høeg, and K. A. Crandall. 2009. Remarkable convergent evolution in specialized parasitic Thecostraca (Crustacea). BMC Biology 2009 7: 15.

Perkins, H. C. 1973. The larval stages of the deep sea red crab, *Geryon quinquedens* Smith, reared under laboratory conditions (Decapoda: Brachyrhyncha). U.S. Fishery Bulletin 71: 69–82.

Perkins, P. S. 1983. The life history of *Cardiodectes medusaeus* (Wilson), a copepod parasite of lanternfishes (Myctophidae). Journal of Crustacean Biology 3: 70–87.

Peterson, K. J. 2005. Macroevolutionary interplay between planktic larvae and benthic predators. Geology 33: 929–932.

Petrov, B. 1992. Larval development of *Leptestheria saetosa* Marinček and Petrov, 1992 (Leptestheriidae, Conchostraca, Crustacea). Archives of Biological Sciences (Belgrade) 44: 229–241.

Petrunina, A. S., and G. A. Kolbasov. 2012. Morphology and ultrastructure of definitive males of *Arcticotantulus pertzovi* and *Microdajus tchesunovi* (Crustacea: Tantulocarida). Zoologischer Anzeiger 251: 223–236.

Petrunina, A. S., T. V. Neretina, N. S. Mugue, and G. A. Kolbasov. 2013. Tantulocarida versus Thecostraca: inside or outside? First attempts to resolve phylogenetic position of Tantulocarida using gene sequences. Journal of Zoological Systematics and Evolutionary Research: online 21 October 2013. DOI: 10.1111 /jzs.12045.

Phillips, B. F., and D. L. Macmillan. 1987. Antennal receptors in puerulus and postpuerulus stages of the rock lobster *Panulirus cygnus* (Decapoda: Palinuridae) and their potential role in puerulus navigation. Journal of Crustacean Biology 7: 122–135.

Piasecki, W., and A. Avenant-Oldewage. 2008. Diseases caused by Crustacea. *In* J. C. Eiras, H. Segner, T. Wahli, and B. G. Kapoor, eds. Fish Diseases, vol. 2, 1115–1200. Enfield, NH: Science Publishers.

Pike, R. B., and R. G. Wear. 1969. Newly hatched larvae of the genera *Gastroptychus* and *Uroptychus* (Crustacea, Decapoda, Galatheidea) from New Zealand waters. Transactions of the Royal Society of New Zealand 11: 189–195.

Pires, A. M. S. 1987. *Potiicoara brasiliensis*: a new genus and species of Spelaeogriphacea (Crustacea: Peracarida) from Brazil with a phylogenetic analysis of the Peracarida. Journal of Natural History 21: 225–238.

Platt, T., and N. Yamamura. 1986. Prenatal mortality in a marine cladoceran, *Evadne nordmanni*. Marine Ecology Progress Series 29: 127–139.

Pohle, G., and M. Telford. 1981. Morphology and classification of decapod crustacean larval setae: a scanning electron microscope study of *Dissodactylus crinitichelis* Moreira, 1901 (Brachyura: Pinnotheridae). Bulletin of Marine Science 31: 736–752.

Pohle, G., F. L. Mantelatto, M. L. Negreiros-Fransozo, and A. Fransozo. 1999. Larval Decapoda (Brachyura). *In* D. Boltovskoy, ed. South Atlantic Zooplankton, vol. 2, 1281–1351. Leiden, Netherlands: Backhuys.

Pohle, G., W. Santana, G. Jansen, and M. Greenlaw. 2011. Plankton-caught zoeal stages and megalopa of the lobster shrimp *Axius serratus* (Decapoda: Axiidae) from the Bay of Fundy, Canada, with a summary of axiidean and gebiidean literature on larval descriptions. Journal of Crustacean Biology 31: 82–99.

Polz, H. 1971. Eine weitere Phyllosoma-Larve aus den Solnhofener Plattenkalken. Neues Jahrbuch für Geologie und Paläontologie, Monatshefte 1971(8): 474–488.

Polz, H. 1972. Entwicklungsstadien bei fossilen Phyllosomen (Form A) aus den Solnhofener Plattenkalken. Neues Jahrbuch für Geologie und Paläontologie, Monatshefte 1972(11): 678–689.

Polz, H. 1973. Entwicklungsstadien bei fossilen Phyllosomen (Form B) aus den Solnhofener Plattenkalken. Neues Jahrbuch für Geologie und Paläontologie, Monatshefte 1973(5): 284–296.

Polz, H. 1984. Krebslarven aus den Solnhofener Plattenkalken. Archaeopteryx 2: 30–40.

Polz, H. 1987. Zur Differenzierung der fossilen Phyllosomen (Crustacea, Decapoda) aus den Solnhofener Plattenkalken. Archaeopteryx 5: 23–32.

Polz, H. 1995. Ein außergewöhnliches Jugendstadium eines palinuriden Krebses aus den Solnhofener Plattenkalken. Archaeopteryx 13: 67–74.

Polz, H. 1996. Eine Form-C-Krebslarve mit erhaltenem Kopfschild (Crustacea, Decapoda, Palinuroidea) aus den Solnhofener Plattenkalken. Archaeopteryx 14: 43–50.

Polz, H. 1997. Der Carapax von *Mayrocaris bucculata* (Thylacocephala, Conchyliocarida). Archaeopteryx 15: 59–71.

Poore, G. C. B. 2005. Peracarida: monophyly, relationships and evolutionary success. Nauplius 13: 1–27.

Poore, G. C. B. 2012. The nomenclature of the Recent Pentastomida (Crustacea), with a list of species and available names. Systematic Parasitology 82: 211–240.

Poore, G. C. B., and N. L. Bruce. 2012. Global diversity of marine isopods (except Asellota and crustacean symbionts). PLoS One 7: e43529. doi:10.1371/journal.pone.0043529.

Poore, G. C. B., and W. F. Humphreys. 1998. First record of Spelaeogriphacea from Australasia: a new genus and species from an aquifer in the arid Pilbara of Western Australia. Crustaceana 71: 721–742.

Poore, G. C. B., and W. F. Humphreys. 2003. Second species of *Mangkurtu* (Spelaeogriphacea) from north-western Australia. Records of the Western Australia Museum 22: 67–74.

Poore, G. C. B., and D. M. Spratt. 2011. Pentastomida. Australian Faunal Directory, Australian Biological Resources Study. Online at www.environment.gov.au/biodiversity/abrs/online -resources/fauna/afd/taxa/PENTASTOMIDA.

Porter, M., Y. Zhang, S. Desai, R. L. Caldwell, and T. W. Cronin. 2010. Evolution of anatomical and physiological specialization in the compound eyes of stomatopod crustaceans. Journal of Experimental Biology 213: 3473–3486.

Potts, W. T. W., and C. T. Durning. 1980. Physiological evolution in the branchiopods. Comparative Biochemistry and Physiology 67B: 475–485.

Poulsen, E. M. 1962. Ostracoda–Myodocopa, 1: Cypridiniformes–Cyprididae. Dana Report No. 57. Copenhagen: Høst.

Poulsen, E. M. 1965. Ostracoda–Myodocopa, 2: Cypridiniformes–Rutidermatidae, Sarsiellidae, and Asteropidae. Dana Report 65: 1–484. Copenhagen: Høst.

Pretus, J. L. 1991. Morfologia de la zoe I de *Stenopus spinosus* Risso (Crustacea, Stenopodidea del litoral de Menorca. Bolletí de la Societat d'Història Natural de les Balears 34: 51–60.

Price, J. O., and J. F. Payne. 1984. Post-embryonic to adult growth in the crayfish *Orconectes neglectus chaenodactylus* Williams, 1952 (Decapoda, Astacidea). Crustaceana 46: 176–194.

Provenzano, A. J., Jr. 1962. Pagurid crabs (Decapoda Anomura) from St. John, Virgin Islands, with descriptions of three new species. Crustaceana 3: 151–166.

Provenzano, A. J., Jr. 1971. Biological results of the University of Miami deep-sea expeditions, 73: zoeal development of *Pylopaguropsis atlantica* Wass, 1963, and evidence from larval characters of some generic relationships within the Paguridae. Bulletin of Marine Science (University of Miami) 21: 237–255.

Provenzano, A. J., Jr., and R. B. Manning. 1978. Studies on development of stomatopod Crustacea, 2: the later larval stages of *Gonodactylus oerstedii* Hansen reared in the laboratory. Bulletin of Marine Science 28: 297–315.

Purcell, E. M. 1977. Life at low Reynolds number. American Journal of Physics 45: 3–11.

Pyne, R. 1972. Larval development and behaviour of the mantis shrimp *Squilla armata* Milne-Edwards (Crustacea: Stomatopoda). Journal of the Royal Society of New Zealand 2: 121–146.

Quetin, L. B., and R. M. Ross. 1984. Depth distribution of develop-ing *Euphausia superba* embryos, predicted from sinking rates. Marine Biology 79: 47–53.

Quintana, R. 1984a. On the prezoeal stage of three species of the genus *Cancer* L. from Chilean waters (Cancridae, Brachyura). Researches on Crustacea 13, 14: 143–151.

Quintana, R. 1984b. Observations on the early post-larval stages of *Leucosia craniolaris* (L., 1758) (Brachyura: Leucosiidae). Reports of the Usa Marine Biological Institute, Kochi University 6: 7–21.

Quintana, R. 1986a. The megalopal stage in the Leucosiidae (Decapoda, Brachyura). Zoological Science (Tokyo) 3: 533–542.

Quintana, R. 1986b. On the early post-larval stages of some leucosiid crabs from Tosa Bay, Japan (Decapoda: Brachyura, Leucosiidae). Journal of the Faculty of Science, Hokkaido University, ser. 6, Zoology 24: 227–266.

Quintana, R. 1987. Later zoeal stages and early post-larval stages of three dorippid species from Japan (Brachyura: Dorippidae: Dorippinae). Publications of the Seto Marine Biological Laboratory 32: 233–274.

Quintana, R., and K. Konishi. 1986. On the prezoeal stage: observations on three *Pagurus* species (Decapoda, Anomura). Journal of Natural History 20: 837–844.

Quintana, R., and K. Konishi. 1988. Ultrastructure of the prezoeal cuticle and integument of first zoea of *Pagurus brachiomastus* (Thallwitz) (Anomura: Paguridae) just after hatching. Journal of Natural History 22: 1077–1084.

Quintana, R., and H. Saelzer. 1986. The complete larval development of the edible crab, *Cancer setosus* Molina and observations on the prezoeal and first zoeal stages of *C. coronatus* Molina (Decapoda: Brachyura, Cancridae). Journal of the Faculty of Science, Hokkaido University, ser. 6, Zoology 24: 267–303.

Quintero, R. C., and E. Z. de Roa. 1973. Notas bioecológicas sobre *Metamysidopsis insularis* Brattegard (Crustacea–Mysidacea) en una laguna littoral de Venezuela. Acta Biologica Venezuela 8: 245–278.

Rabalais, N. N., and R. H. Gore. 1985. Abbreviated development in decapods. *In* A. M. Wenner, ed. Larval Growth, 67–126. Crustacean Issues No. 2. Rotterdam: A. A. Balkema.

Rabet, N. 2010. Revision of the egg morphology of *Eulimnadia* (Crustacea, Branchiopoda, Spinicaudata). Zoosystema 33: 373–391.

Raibaut, A. 1985. Les cycles évolutifs des copépodes parasites et les modalités de l'infestation. L'Année Biologique, ser. 4, 24: 233–274.

Raje, P. C., and M. R. Ranade. 1975. Early life history of a stenopodid shrimp, *Microprosthema semilaeve* (Decapoda: Macrura). Journal of the Marine Biological Association of India 17: 213–222.

Ramdohr, K. A. 1805. Beitrage zur Naturgeschichte einiger Deutschen Monoculusarten. Mikrographische Beitrage zur Entomologie und Helminthologie No. 1. Halle, Germany: Hemmerde und Schwetschke.

Rathke, H. 1840. Zur Entwicklungsgeschichte der Decapoden. Archiv für Naturgeschichte 6: 214–249.

Rathke, H. 1842. Beitrage zur vergleichenden Anatomie und Physiologie: reisebemerkungen aus Skandinavien, nebst einem Anhange über die rückschreitender Metamorphose der Thiere, 2; zur Entwicklungsgeschichte der Dekapoden. Neueste Schriften der Naturforschenden Gesellschaft zu Danzig 3: 23–55.

Reaka, M. L. 1986. Biogeographic patterns of body size in stomatopod Crustacea: ecological and evolutionary consequences. *In* R. H. Gore and K. H. Heck, eds. Crustacean Biogeography, 209–235. Crustacean Issues No. 4. Rotterdam: A. A. Balkema.

Reese, E. S., and R. A. I. Kinzie. 1968. The larval development of the coconut or robber crab *Birgus latro* (L.) in the laboratory (Anomura, Paguridea). Crustaceana, Suppl. 2: 117–144.

Regier, J. C., J. W. Shultz, and R. E. Kambic. 2005. Pancrustacean phylogeny: hexapods are terrestrial crustaceans and maxillopods are not monophyletic. Proceedings of the Royal Society of London, B 272: 395–401.

Regier, J. C., J. W. Shultz, A. R. D. Ganley, A. Hussey, D. Shi, B. Ball, A. Zwick, J. E. Stajich, M. P. Cummings, J. W. Martin, and C. W. Cunningham. 2008. Resolving arthropod phylogeny: exploring phylogenetic signal within 41 kb of protein-coding nuclear gene sequence. Systematic Biology 57: 920–938.

Regier, J. C., J. W. Shultz, A. Zwick, A. Hussey, B. Ball, R. Wetzer, J. W. Martin, and C. W. Cunningham. 2010. Arthropod relationships revealed by phylogenomic analysis of nuclear protein-coding sequences. Nature 463: 1079–1084.

Reid, J. W. 2001. A human challenge: discovering and understanding continental copepod habitats. *In* R. M. Lopes, J. W. Reid, and C. E. F. Rocha, eds. Copepoda: Developments in Ecology, Biology and Systematics. Hydrobiologia 453/454: 201–226.

Reinhardt, J. 1849. Phyllamphion, en ny Slaegt af Stomatopodernes Orden. Videnskabelige Meddelelser Naturhistorisk Forening i København, 1849–1850: 2–6.

Reverberi, G., and M. Pittoti. 1943. Il ciclo biologico e la determinazione fenotipica del sesso di *Ione thoracica* Montagu, bopiride parassita di *Callianassa laticauda* Otto. Pubblicazioni della Stazione Zoologica di Napoli 19: 111–184.

Rice, A. L. 1964. The metamorphosis of a species of *Homola* (Crustacea, Decapoda, Dromiacea). Bulletin of Marine Science of the Gulf and Caribbean 14: 221–238.

Rice, A. L. 1968. Growth "rules" and the larvae of decapod crustaceans. Journal of Natural History 2: 525–530.

Rice, A. L. 1980. Crab zoeal morphology and its bearing on the classification of the Brachyura. Transactions of the Zoological Society of London 35: 271–424.

Rice, A. L. 1981a. Crab zoeae and brachyuran classification: a re-appraisal. Bulletin of the British Museum (Natural History), Zoology 40: 287–296.

Rice, A. L. 1981b. The megalopa stage in brachyuran crabs: the Podotremata Guinot. Journal of Natural History 15: 1003–1011.

Rice, A. L. 1982. The megalopa stage of *Latreillia elegans* Roux (Decapoda, Brachyura, Homoloidea). Crustaceana 43: 205–210.

Rice, A. L. 1983. Zoeal evidence for brachyuran phylogeny. *In* F. R. Schram, ed. Crustacean Phylogeny, 313–329. Crustacean Issues No. 1. Rotterdam: A. A. Balkema.

Rice, A. L. 1988. The megalopa stage in majid crabs, with a review of spider crab relationships based on larval characters. Symposia of the Zoological Society of London 59: 27–46.

Rice, A. L., and A. J. Provenzano Jr. 1966. The larval development of the West Indian sponge crab *Dromidia antillensis* (Decapoda: Dromiidae). Journal of Zoology 149: 297–319.

Rice, A. L., and A. J. Provenzano Jr. 1970. The larval stages of *Homola barbata* (Fabricius) (Crustacea, Decapoda, Homolidae) reared in the laboratory. Bulletin of Marine Science 20: 446–471.

Rice, A. L., and K. G. von Levetzow. 1967. Larvae of *Homola* (Crustacea, Dromiacea) from South Africa. Journal of Natural History 1: 435–453.

Rice, A. L., and D. I. Williamson. 1970. Methods for rearing larval decapod Crustacea. Helgoländer Wissenschaftliche Meeresuntersuchungen / Helgoland Marine Research 20: 417–434.

Rice, A. L., and D. I. Williamson. 1977. Planktonic stages of

Crustacea Malacostraca from Atlantic seamounts. "Meteor" Forsch-Ergebnisse D 26: 28–64.

Richardson, A. J. 2008. In hot water: zooplankton and climate change. ICES Journal of Marine Science 65: 279–295.

Richter, S., and G. Scholtz. 2001. Phylogenetic analysis of the Malacostraca (Crustacea). Journal of Zoological Systematics and Evolutionary Research 39: 113.-136.

Richter, S., A. Braband, N. Aladin, and G. Scholtz. 2001. The phylogenetic relationships of "predatory water-fleas" (Cladocera: Onychopoda, Haplopoda) inferred from 12S rDNA. Molecular Phylogenetics and Evolution 19: 105–113.

Richter, S., J. Olesen, and W. C. Wheeler. 2007. Phylogeny of Branchiopoda (Crustacea) based on a combined analysis of morphological data and six molecular loci. Cladistics 23: 301–336.

Richter, S., R. Loesel, G. Purschke, A. Schmidt-Rhaesa, G. Scholtz, T. Stach, L. Vogt, A. Wanninger, G. Brenneis, C. Döring, S. Faller, M. Fritsch, P. Grobe, C. M. Heuer, S. Kaul, O. S. Møller, C. H. Müller, V. Rieger, B. H. Rothe, M. E. Stegner, and S. Harzsch. 2010. Invertebrate neurophylogeny: suggested terms and definitions for a neuroanatomical glossary. Frontiers in Zoology 7: 1–49. doi:10.1186/1742-9994-7-29.

Richters, F. 1873. Die Phyllosomen: ein Beitrag zur Entwicklungsgeschichte der Loricaten. Zeitschrift für Wissenschaftliche Zoologie 23: 623–646.

Riley, J. 1981. An experimental investigation of the development of *Porocephalus crotali* (Pentastomida: Porocephalida) in the western diamondback rattlesnake (*Crotalus atrox*). International Journal of Parasitology 11: 127–131.

Riley, J. 1983. Recent advances in our understanding of pentastomid reproductive biology. Systematic Parasitology 86: 59–83.

Riley, J. 1986. The biology of pentastomids. Advances in Parasitology 25: 45–128.

Riley, J., and J. T. Self. 1979. On the systematics of the pentastomid genus *Porocephalus* (Humboldt, 1811) with descriptions of new species. Systematic Parasitology 1: 25–42.

Riley, J., A. A. Banaja, and J. L. James. 1978. The phylogenetic relationship of the Pentastomida: the case for their inclusion within the Crustacea. International Journal of Parasitology 8: 245–254.

Risso, A. 1816. Histoire Naturelle des Crustacés des Environs de Nice. Paris: Librairie Grecque-Latine-Allemande.

Ritchie, L. E., and J. T. Høeg. 1981. The life history of *Lernaeodiscus porcellanae* (Cirripedia: Rhizocephala) and co-evolution with its porcellanid host. Journal of Crustacean Biology 1: 334–347.

Rivera, J., and G. Guzmán. 2002. Descripción de larvas mysis de tres especies de camarones mesopelágicos del género *Gennadas* (Decapoda: Aristeidae) en aguas del Pacífico sudoriental. Investigaciones Marina 30: 33–44.

Rivier, I. K. 1998. The Predatory Cladocera (Onychopoda: Podonidae, Polyphemidae, Cercopagidae) and Leptodoridae of the World. Leiden, Netherlands: Backhuys.

Robertson, P. B. 1968. The complete larval development of *Scyllarus americanus* (Smith), (Decapoda, Scyllaridae) in the laboratory with notes on larvae from the plankton. Bulletin of Marine Science 18: 294–342.

Robertson, P. B. 1969. Biological investigations of the deep sea, 48: phyllosoma larvae of a scyllarid lobster, *Arctides guineensis*, from the western Atlantic. Marine Biology 4: 143–151.

Robinson, M. 1906. On the development of *Nebalia*. Quarterly Journal of Microscopical Science 50: 383–378.

Robles, R., C. C. Tudge, P. C. Dworschak, G. C. B. Poore, and D. L. Felder. 2009. Molecular phylogeny of the Thalassinidea based on nuclear and mitochondrial genes. *In* J. W. Martin, K. A. Crandall, and D. L. Felder, eds. Decapod Crustacean Phylogenetics, 309–326. Crustacean Issues No. 18. Boca Raton, FL: CRC Press.

Rode, A. L., and B. S. Lieberman. 2002. Phylogenetic and biogeographic analysis of Devonian phyllocarid crustaceans. Journal of Paleontology 76: 271–286.

Rodrigues, S. de A. 1984. Desenvolvimento pós-embryonário de *Callichirus mirim* (Rodrigues, 1971) obtido em condições artificias (Crustacea, Decapoda, Thalassinidea). Boletim de Zoologia, Universidade de São Paulo 8: 239–256.

Rodrigues, S. de A. 1994. First stage larva of *Axiopsis serratifrons* (A. Milne-Edwards, 1873) reared in the laboratory (Decapoda: Thalassinidea: Axiidae). Journal of Crustacean Biology 14: 314–318.

Rodrigues, S. [de] A., and R. B. Manning. 1992. First stage larva of *Coronis scolopendra* Latreille (Stomatopoda: Nannosquillidae). Journal of Crustacean Biology 12: 79–82.

Rodrigues, W. and N. J. Hebling. 1978. Estudos biológicos em *Aegla perobae* Hebling and Rodrigues, 1977 (Decapoda, Anomura). Revista Brasileira de Biologia, 38: 383–390.

Rodríguez, A., and J. A. Cuesta. 2011. Morphology of larval and first juvenile stages of the kangaroo shrimp *Dugastella valentina* (Crustacea, Decapoda, Caridea), a freshwater atyid with abbreviated development and parental care. Zootaxa 2867: 43–58.

Rodríguez, A., and J. W. Martin. 1997. Larval development of the crab *Xantho poressa* (Decapoda: Xanthidae) reared in the laboratory. Journal of Crustacean Biology 17: 98–110.

Roessler, E. W. 1983. Estudios taxónomicos, ontogenéticos, ecológicos y etológicos sobre los ostrácodos de agua dulce en Colombia, 4: desarrollo postembrionario de *Heterocypris bogotensis* Roessler (Ostracoda, Podocopa, Cyprididae). Caldasia 13: 755–776.

Roessler, E. W. 1995. Review of Colombian Conchostraca (Crustacea): ecological aspects and life cycles—family Cyclestheriidae. Hydrobiologia 298: 113–124.

Roessler, E. W. 1998. On crucial developmental stages in podocopid ostracod ontogeny and their prenauplius as a missing link in crustacean phylogeny. Advances in Limnology, Special Issue, Archiv für Hydrobiologie 52: 535–547.

Roger, J. 1944. *Eryoneicus? sahel-almae* n. sp., crustacé décapode du Sénonien du Liban. Bulletin du Muséum National d'Histoire Naturelle 16: 191–194.

Roger, J. 1946. Les invertébrés des couches à poissons du Crétacé Supérieur du Liban. Mémoires de la Société Géologique de France 12: 1–92.

Rogers, D. C. 2001. Revision of the Nearctic *Lepidurus* (Notostraca). Journal of Crustacean Biology 21: 991–1006.

Rogers, D. C. 2009. Branchiopoda (Anostraca, Notostraca, Laevicaudata, Spinicaudata, and Cyclestherida). *In* G. F. Likens, ed. Encyclopedia of Inland Waters, 242–249. Boston: Elsevier.

Rogers, D. C., D. L. Quinney, J. Weaver, and J. Olesen. 2006. A new giant species of predatory fairy shrimp from Idaho, USA (Branchiopoda: Anostraca). Journal of Crustacean Biology 26: 1–12.

Rogers, D. C., N. Rabet, and S. C. Weeks. 2012. Revision of the extant genera of Limnadiidae (Branchiopoda: Spinicaudata). Journal of Crustacean Biology 32: 827–842.

Ross, R., and L. Quetin. 2000. Reproduction in Euphausiacea. *In* I. Everson, ed. Krill: Biology, Ecology and Fisheries, 150–181. Fish and Aquatic Resources No. 6. London: Blackwell Science.

Rossi, F. 1980. Comparative observations on the female reproductive system and parthenogenetic oogenesis in Cladocera. Bollettino di Zoologia 47: 21–38.

Rothlisberg, P. C., J. A. Church, and C. Fandry. 1995. A mechanism for near-shore density and estuarine recruitment of post-larval *Penaeus plebejus* Hess (Decapoda, Penaeidae). Estuarine, Coastal and Shelf Science 40: 115–138.

Roy, K., and L. E. Fåhræus. 1989. Tremadocian (Early Ordovician) nauplius-like larvae from the Middle Arm Point Formation, Bay of Islands, western Newfoundland. Canadian Journal of Earth Sciences 26: 1802–1806.

Rudolph, E. H., and C. S. Rojas. 2003. Embryonic and early post-embryonic development in the burrowing crayfish *Virilastacus araucanius* (Faxon, 1914) (Decapoda, Parastacidae) under laboratory conditions. Crustaceana 76: 835–850.

Rushton-Mellor, S. K., and G. A. Boxshall. 1994. The developmental sequence of *Argulus foliaceus* (Crustacea: Branchiura). Journal of Natural History 28: 763–785.

Rybakov, A. V., O. M. Korn, J. T. Høeg, and D. Walossek. 2002. Larval development in *Peltogasterella* studied by scanning electron microscopy (Crustacea: Cirripedia: Rhizocephala). Zoologischer Anzeiger 241: 199–221.

Rybakov, A. V., J. T. Høeg, P. G. Jensen, and G. A. Kolbasov. 2003. The chemoreceptive lattice organs in cypris larvae develop from naupliar setae (Thecostraca: Cirripedia, Ascothoracida And Facetotecta. Zoologischer Anzeiger 242: 1–20.

Saba, M., M. Takeda, and Y. Nakasone. 1978. Larval development of *Epixanthus dentatus* (White) (Brachyura, Xanthidae). Bulletin of the National Science Museum, A, Zoology 4: 151–161.

Saint Laurent, M. de. 1979. Vers une nouvelle classification des crustacés décapodes Reptantia. Bulletin de l'Office National des Pêches, République Tunisienne, Ministère de l'Agriculture 3: 15–31.

Saint Laurent, M. de. 1988. Enoplometopoidea, nouvelle superfamille de crustacés décapodes Astacidea. Comptes Rendus Hebdomadaires des Séances de l'Académie des Sciences, ser. 3, 307: 59–62.

Saint Laurent, M. de, and E. Macpherson. 1990. Crustacea Decapoda: le genre *Eumunida* Smith, 1883 (Chirostylidae) dans les eaux neo-caledoniennes. Mémoires du Museum National d'Histoire Naturelle, A, Zoologie 145: 227–288.

Saito, T., and K. Konishi. 1999. Direct development in the sponge-associated deep-sea shrimp *Spongicola japonica* (Decapoda: Spongicolidae). Journal of Crustacean Biology 19: 46–52.

Saito, T., and K. Konishi. 2002. Description of the first stage zoea of the symmetrical hermit crab *Pylocheles mortensenii* (Boas, 1926) (Anomura, Paguridea, Pylochelidae). Crustaceana 75: 621–628.

Sakai, K. 1971. The larval stages of *Ranina ranina* (Linnaeus) (Crustacea, Decapoda, Raninidae) reared in the laboratory, with a review of uncertain zoeal larvae attributed to *Ranina*. Publications of the Seto Marine Biological Laboratory 19: 123–156.

Samassa, P. 1893. Die Keimblätterbildung bei den Cladoceren, 2: *Daphnelle* und *Daphnia*. Archiv für Mikroskopische Anatomie 41: 650–688.

Sambon, L. W. 1922. A synopsis of the family Linguatulidae. Journal of Tropical Medicine and Hygiene 12: 188–206, 391–428.

Samia, B. 1993. Contribution à l'étude de *Peroderma cylindricum* Heller, 1865, parasite de la sardine, *Sardina pilchardus* (Walbaum, 1792) des côtes tunisiennes. PhD diss., University of Tunis.

Samter, M. 1895. Die Veränderung der Form und Lage der Schale von *Leptodora hyalina* Lillj. während der Entwicklung. Zoologischer Anzeiger 483/484: 334–338, 341–344.

Samter, M. 1900. Studien zur Entwicklungsgeschichte der *Leptodora hyalina* Lillj. Zeitschrift für Wissenschaftliche Zoologie 68: 169–260.

Samuelsen, T. J. 1972. Larvae of *Munidopsis tridentata* (Esmark) (Decapoda, Anomura) reared in the laboratory. Sarsia 48: 91–98.

Sanchez, G., and P. G. Aguilar. 1975. Notas sobre crustáceos del Mar Peruano, 1: desarrollo larvario de *Lepidopa chilensis* Lenz (Decapoda, Anomura: Albuneidae). Anales Cientificos UNA (Lima) 13: 1–11.

Sanders, H. L. 1955. The Cephalocarida, a new subclass of Crustacea from Long Island Sound. Proceedings of the National Academy of Sciences 41: 61–66.

Sanders, H. L. 1957. The Cephalocarida and crustacean phylogeny. Systematic Zoology 6: 112–128.

Sanders, H. L. 1963. The Cephalocarida: Functional Morphology, Larval Development, Comparative External Anatomy. Memoirs of the Connecticut Academy of Arts and Sciences No. 15. New Haven: Connecticut Academy of Arts and Sciences.

Sanders, H. L., and R. R. Hessler. 1964. The larval development of *Lightiella incisa* Gooding (Cephalocarida). Crustaceana 7: 81–97.

Sanders, H. L., R. R. Hessler, and S. P. Garner. 1985. *Hirsutia bathyalis*, a new unusual deep-sea benthic peracaridan crustacean from the tropical Atlantic. Journal of Crustacean Biology 5: 30–57.

Sanders, K. L., and M. S. Y. Lee. 2010. Arthropod molecular divergence times and the Cambrian origin of pentastomids. Systematics and Biodiversity 8: 63–74.

Sankolli, K. N. 1967. Studies on larval development in Anomura (Crustacea, Decapoda). Proceedings of the Symposium on Crustacea, Held at Ernakulam from January 12–15, 1965, pt. 1, 744–776. Mandapam Camp: Marine Biological Association of India.

San Vicente, C., and J. C. Sorbe. 1990. Biología del misidáceo suprabentónico *Schistomysis kervillei* (Sars, 1885) en la plataforma continental Aquitana (suroeste de Francia). Actas del 6 Simposio Ibérico de Estudios del Bentos Marino: 245–267. Palma de Mallorca, Spain: Bilbilis.

San Vicente, C., and J. C. Sorbe. 1993. Biologie du mysidacé suprabenthique *Schistomysis parkeri* Norman, 1892, dans la zone sud du Golfe de Gascogne (Plage d'Hendaye). Crustaceana 65: 222–252.

San Vicente, C., and J. C. Sorbe. 1995. Biology of the suprabenthic mysid *Schistomysis spiritus* (Norman, 1860) in the southeastern part of the Bay of Biscay. Scientia Marina 59, Suppl. 1: 71–86.

San Vicente, C., and J. C. Sorbe. 2003. Biology of the suprabenthic mysid *Schistomysis assimilis* (Sars, 1877) on Creixell Beach, Tarragona (northwestern Mediterranean). Boletín del Instituto Español de Oceanografía 19: 391–120.

Sars, G. O. 1873. Om en dimorph udvikling samt generationsvexel hos *Leptodora*. Forhandlinger i Videnskabsselskabet i Kristiania 1873: 1–15.

Sars, G. O. 1885. Report on the Schizopoda collected by H.M.S. *Challenger*, during the years 1873–1876. Vol. 37, *in* C. W. Thomson and J. Murray, eds. Reports on the Scientific Results of the Voyage of H.M.S. *Challenger* during the Years 1873–76, Zoology. London.

Sars, G. O. 1887. On *Cyclestheria hislopi* (Baird), a new generic type of bivalve Phyllopoda, raised from dried Australian mud. Forhandlinger i Videnskabsselskabet i Kristiania, 1887: 1–65.

Sars, G. O. 1896a. Fauna Norvegiæ. Vol. 1, Descriptions of the Norwegian Species at Present Known Belonging to the Suborders Phyllocarida and Phyllopoda. Christiania, Norway: Joint-Stock.

Sars, G. O. 1896b. Development of *Estheria packardi* as shown by artificial hatching from dried mud. Archiv for Mathematik og Naturvidenskab 18: 1–27.

Sars, G. O. 1896–1899. An Account of the Crustacea of Norway, with Short Descriptions and Figures of All the Species. Vol. 2, Isopoda. Bergen, Norway: Bergen Museum. [Bopyrid and dajid text and plates published in 1898.]

Sars, G. O. 1900a. An Account of the Crustacea of Norway, with Short Descriptions and Figures of All the Species. Vol. 3, Cumacea. Bergen, Norway: Bergen Museum.

Sars, G. O. 1900b. Account of the post-embryonal development of *Pandalus borealis* Krøyer, with remarks on the development of other Pandali, and descriptions of the adult *Pandalus borealis*. Reports on Norwegian Fishery and Marine Investigations Vol. 1, No. 3. Christiania, Norway: O. Andersons Bogtrykkeri.

Sars, G. O. 1906. Post-embryonal development of *Athanas nitescens* Leach. Archiv for Mathematik og Naturvidenskab 27: 1–29.

Sars, G. O. 1912. Account of the post-embryonal development of *Hippolyte varians* Leach. Archiv for Mathematik og Naturvidenskab 32: 1–25.

Sars, G. O. 1921. An Account of the Crustacea of Norway, with Short Descriptions and Figures of All the Species. Vol. 8, Copepoda Monstrilloida and Notodelphyoida. Bergen, Norway: Bergen Museum.

Sasaki, J., and Y. Mihara. 1993. Early larval stages of the hair crab *Erimacrus isenbeckii* (Brandt) (Brachyura: Atelecyclidae), with special reference to its hatching process. Journal of Crustacean Biology 13: 511–522.

Sassaman, C. 1995. Sex determination and evolution of unisexuality in the Conchostraca. Hydrobiologia 298: 45–65.

Savchenko, A. S., and G. A. Kolbasov. 2009. *Serratotantulus chertoprudae* gen. et sp. n. (Crustacea, Tantulocarida, Basipodellidae): a new tantulocaridan from the abyssal depths of the Indian Ocean. Integrative and Comparative Biology 49: 106–113.

Schaeffer, J. C. 1756. Der Krebsartige Kiefenfuss mit der Kurzen und Langen Schwanzklappe. Regensburg, Germany.

Scheerer-Ostermeyer, E. 1940. Beitrag zur Entwicklungsgeschichte der Süßwasserostrakoden. Zoologische Jahrbücher, Abteilung für Anatomie und Ontogenie der Tiere 66: 349–370.

Schmalfuss, H. 2003. World catalog of terrestrial isopods (Isopoda:Oniscidea). Stuttgarter Beiträge zur Naturkunde, Serie A, Biologie No. 654. Stuttgart: Staatliches Museum für Naturkunde.

Schminke, H. K. 1978. Larval development of Syncarida (Crustacea). American Zoologist 18: 58.

Schminke, H. K. 1981. Adaptation of Bathynellacea (Crustacea, Syncarida) to life in the interstitial ("Zoea Theory"). Internationale Revue der Gesamten Hydrobiologie 66: 575–637.

Schminke, H. K. 1982. Die Nauplius-Stadien von *Parastenocaris vicesima* Klie, 1935 (Copepoda, Parastenocarididae). Drosera 82: 101–108.

Schminke, H. K. 2007. Entomology for the copepodologist. Journal of Plankton Research 29, Suppl. 1: i149–i162.

Schnabel, K. E., and S. T. Ahyong. 2010. A new classification of the Chirostyloidea (Crustacea: Decapoda: Anomura). Zootaxa 2687: 56–64.

Schnabel, K. E., S. T. Ahyong, and E. W. Maas. 2011. Galatheoidea are not monophyletic—molecular and morphological phylogeny of the squat lobsters (Decapoda: Anomura) with recognition of a new superfamily. Molecular Phylogenetics and Evolution 58: 157–168.

Scholtz, G. 1995. The attachment of the young in the New Zealand freshwater crayfish *Paranephrops zealandicus* (White, 1847) (Decapoda, Astacida, Parastacidae). New Zealand Natural Sciences 22: 81–89.

Scholtz, G. 2000. Evolution of the nauplius stage in malacostracan crustaceans. Journal of Zoological Systematics and Evolutionary Research 38: 175–187.

Scholtz, G., ed. 2004. Evolutionary Developmental Biology of Crustacea. Crustacean Issues No. 15. Rotterdam: A. A. Balkema.

Scholtz, G. 2008. Zoological detective stories: the case of the face-totectan crustacean life cycle. Journal of Biology 7: 16.

Scholtz, G., and T. Kawai. 2002. Aspects of embryonic and post-embryonic development of the Japanese freshwater crayfish *Cambaroides japonicus* (Crustacea, Decapoda) including a hypothesis on the evolution of maternal care in the Astacida. Acta Zoologica (Stockholm) 83: 23–212.

Scholtz, G., and S. Richter. 1995. Phylogenetic systematics of the reptantian Decapoda (Crustacea, Malacostraca). Zoological Journal of the Linnean Society 113: 289–328.

Schram, F. R. 1978. Arthropods: a convergent phenomenon. Fieldiana, Geology 39:61–106.

Schram, F. R. 1981. On the classification of Eumalacostraca. Journal of Crustacean Biology 1: 1–10.

Schram, F. R. 1983. Remipedia and crustacean phylogeny. *In* F. R. Schram, ed. Crustacean Phylogeny. Crustacean Issues No. 1. Rotterdam: A. A. Balkema.

Schram, F. R. 1984. Relationships within eumalacostracan Crustacea. Transactions of the San Diego Society of Natural History 20: 301–312.

Schram, F. R. 1986. Crustacea. New York: Oxford University Press.

Schram, F. R. 2007. Paleozoic proto-mantis shrimp revisited. Journal of Paleontology 81: 895–916.

Schram, F. R., and C. H. J. Hof. 1998. Fossils and the interrelationships of major crustacean groups. *In* G. D. Edgecombe, ed. Arthropod Fossils and Phylogeny, 232–302. New York: Columbia University Press.

Schram, F. R., and S. Koenemann. 2001. Developmental genetics and arthropod evolution, 1: on legs. Evolution & Development 3: 343–354.

Schram, F. R., and H. G. Müller. 2004. Catalog and Bibliography of the Fossil and Recent Stomatopoda. Leiden, Netherlands: Backhuys.

Schram, T. A. 1979. The life history of the eye-maggot of the sprat, *Lernaeenicus sprattae* (Sowerby) (Copepoda, Lernaeoceridae). Sarsia 64: 279–316.

Schram, T. A., and P. E. Aspholm. 1997. Redescription of male *Hatschekia hippoglossi* (Guérin-Méneville, [1837]) (Copepoda: Siphonostomatoida) and additional information on the female. Sarsia 82: 1–18.

Schrehardt, A. 1987. A scanning electron-microscope study of the post-embryonic development of *Artemia*. *In* P. Sorgeloos, D. A. Bengtson, W. Declair, and E. Jasper, eds. *Artemia* Research and Its Applications. Vol. 1, Morphology, Genetics, Strain Characterization, Toxicology, 5–32. Wetteren, Belgium: Universa Press.

Schreiber, E. 1922. Beitrage zur Kenntnis der Morphologie, Entwicklung und Lebensweise der Sußwasser-Ostracoden. Zoologische Jahrbücher, Abteilung für Anatomie und Ontogenie der Tiere 43: 485–539.

Schutze, M. L. M., C. E. F. Rocha, and G. A. Boxshall. 2000. Antennulary development during the copepodid phase in the family Cyclopidae (Copepoda, Cyclopoida). Zoosystema 22: 749–806.

Schweitzer, C. E., R. M. Feldmann, A. Garassino, H. Karasawa, and G. Schweigert. 2010. Systematic list of fossil decapod crustacean species. Crustaceana Monographs No. 10. Leiden, Netherlands: Brill.

Schwentner, M., B. V. Timms, R. Bastrop, and S. Richter. 2009. Phylogeny of Spinicaudata (Branchiopoda, Crustacea) based on three molecular markers—an Australian origin for *Limnadopsis*. Molecular Phylogenetics and Evolution 53: 716–725.

Schwentner, M., S. Clavier, M. Fritsch, J. Olesen, S. Padhye, B. Timms, S. Richter. 2013. *Cyclestheria hislopi* (Crustacea: Branchiopoda): a group of morphologically cryptic species with origins in the Cretaceous. Molecular Phylogenetics and Evolution 66: 800–810.

Scott, T., and A. Scott. 1913. The British Parasitic Copepoda. Copepoda Parasitic on Fishes, 2 vols. London: Ray Society.

Scotto, L. E. 1979. Larval development of the Cuban stone crab, *Menippe nodifrons* (Brachyura, Xanthidae), under laboratory conditions with notes on the status of the family Menippidae. U.S. Fishery Bulletin 77: 359–386.

Scotto, L. E., and R. H. Gore. 1981. Studies on decapod Crustacea from the Indian River region of Florida, 23: the laboratory cultured zoeal stages of the coral gall-forming crab *Troglocarcinus corallicola* Verrill, 1908 (Brachyura: Hapalocarcinidae) and its familial position. Journal of Crustacean Biology 1: 486–505.

Scourfield, D. J. 1926. On a new type of crustacean from the Old Red Sandstone (Rhynie chert bed, Aberdeenshire)—*Lepidocaris rhyniensis* gen. et sp. nov. Philosophical Transactions of the Royal Society of London, B 214: 153–187.

Scourfield, D. J. 1940. Two new and nearly complete specimens of young stages of the Devonian fossil crustacean *Lepidocaris rhyniensis*. Proceedings of the Linnean Society 152: 290–298.

Sebestyén, O. 1931. Contribution to the biology and morphology of *Leptodora kindtii* (Focke) (Crustacea, Cladocera). A Magyar Biológiai Kutatóintézet Munkai 4: 151–170.

Sebestyén, O. 1949. On the life-method of the larva of *Leptodora kindtii* (Focke), (Cladocera, Crustacea). Acta Biologica Hungarica 1: 71–81.

Secretan, S. 1985. Conchyliocarida, a class of fossil crustaceans: relationships to Malacostraca and postulated behaviour. Transactions of the Royal Society of Edinburgh 76: 381–389.

Sekiguchi, H. 1974. Relation between the ontogenetic vertical migration and mandibular gnathobase in pelagic copepods. Bulletin of the Faculty of Fisheries, Mie University 1: 1–10.

Sekiguchi, H., J. D. Booth, and J. Kittaka. 1996. Phyllosoma larva of *Puerulus angulatus* (Bate, 1888) (Decapoda: Palinuridae) from Tongan waters. New Zealand Journal of Marine and Freshwater Research 30: 407–411.

Selbie, C. M. 1914. The Decapoda Reptantia of the Coasts of Ireland. Vol. 1, Palinura, Astacura and Anomura (except Paguridea). Ireland, Fisheries Branch, Report on the Sea and Inland Fisheries of Ireland, Scientific Investigations 1914, No. 1. London: His Majesty's Stationery Office.

Self, J. T. 1969. Biological relationships of the Pentastomida: a bibliography on the Pentastomida. Experimental Parasitology 21: 63–119.

Semmler, H., A. Wanninger, J. T. Høeg, and G. Scholtz. 2008. Immunocytochemical studies on the nceaupliar nervous system of *Balanus improvisus* (Crustacea, Cirripedia, Thecostraca). Arthropod Structure & Development 37: 383–395.

Semmler, H., J. T. Høeg, G. Scholtz, and A. Wanninger. 2009. Three-dimensional reconstruction of the naupliar musculature and a scanning electron microscopy atlas of nauplius development of *Balanus improvisus*. (Crustacea: Cirripedia: Thoracica). Arthropod Structure & Development 38: 135–145.

Serban, E. 1972. *Bathynella* (Podophallocarida, Bathynellacea). Travaux de l'Institut de Spéologie "Emile Racovitza" 11: 11–224.

Serban, E. 1985. Le développement post-embryonnaire chez *Gallobathynella coiffaiti* (Delamare) (Gallobathynellinae, Bathynellidae, Bathynellacea). Travaux de l'Institut de Spéologie "Emile Racovitza" 24: 47–61.

Serban, E., and N. Coineau. 1990. Données concernant le développement post-embryonnaire dans la famille des Parabathynellidae Noodt (Bathynellacea, Podophallocarida, Malacostraca). Travaux de l'Institut de Spéologie "Emile Racovitza" 29: 3–24.

Seridji, R. 1985. Larves de crustacés décapodes des eaux jordaniennes du Golfe d'Aqaba: note préliminaire. Vie et Milieu, A, Biologie Marine 7: 1–13.

Seridji, R. 1990. Description of some planktonic larval stages of *Stenopus spinosus* Risso, 1826: notes on the genus and the systematic position of the Stenopodidae as revealed by larval characters. Scientia Marina 54: 293–303.

Servais, T., O. Lehnert, J. Li, G. L. Mullins, A. Munnecke, A. Nützel, and M. Vecoli. 2008. The Ordovician biodiversification: revolution in the oceanic trophic chain. Lethaia 41: 99–109.

Shafir, A., and J. G. van As. 1986. Laying, development and hatching of eggs of the fish ectoparasite *Argulus japonicus* (Crustacea: Branchiura). Journal of Zoology (London) 210: 401–414.

Shaw, G. 1791. Description of *Cancer stagnalis* of Linnaeus. Transactions of the Royal Society of London 1: 103–110.

Shen, H., A. Braband, and G. Scholtz. 2013. Mitogenomic analysis of decapod crustacean phylogeny corroborates traditional views on their relationships. Molecular Phylogenetics and Evolution 66: 776–789.

Shen, Y., and D. Huang 2008. Extant clam shrimp egg morphology: taxonomy and comparison with other fossil branchiopod eggs. Journal of Crustacean Biology 28: 352–360.

Shenoy, S. 1967. Studies on larval development in Anomura (Crustacea, Decapoda). Proceedings of the Symposium on Crustacea, Held at Ernakulam from January 12–15, 1965, pt. 2, 777–803. Mandapam Camp: Marine Biological Association of India.

Shenoy, S., and K. N. Sankolli. 1967. Studies on larval development in Anomura (Crustacea, Decapoda), Proceedings of the Symposium on Crustacea, Held at Ernakulam from January 12–15, 1965, pt. 2, 805–814. Mandapam Camp: Marine Biological Association of India.

Shimono, T., N. Iwasaki, and H. Kawai. 2004. A new species of *Dactylopusioides* (Copepoda: Harpacticoida: Thalestridae) infesting a brown alga, *Dictyota dichotoma* in Japan. Hydrobiologia 523: 9–15.

Shimono, T., N. Iwasaki, and H. Kawai. 2007. A new species of *Dactylopusioides* (Copepoda: Harpacticoida: Thalestridae) infesting brown algae, and its life history. Zootaxa 1582: 59–68.

Shimura, S. 1981. The larval development of *Argulus coregoni*

Thorell (Crustacea: Branchiura). Journal of Natural History 15: 331–348.

Shimura, S. 1983. SEM observations on the mouth tube and pre-oral sting of *Argulus coregoni* Thorell and *Argulus japonicus* Thiele (Crustacea: Branchiura). Fish Pathology 18: 151–156.

Shimura, S., and K. Inoue. 1984. Toxic effects of extract from the mouth-parts of *Argulus coregoni* Thorell (Crustacea: Branchiura). Bulletin of the Japanese Society of Scientific Fisheries 50: 729.

Shojima, Y. 1963. Scyllarid phyllosomas' habit of accompanying the jelly-fish. Bulletin of the Japanese Society of Scientific Fisheries 29: 349–353.

Shotter, R. A. 1971. The biology of *Clavella uncinata* (Müller) (Crustacea: Copepoda). Parasitology 63: 419–430.

Shu, D. 1990. Cambrian and Early Ordovician "Ostracoda" (Bradoriida) in China. Courier Forschungsintitut Senckenberg 123: 315–330.

Shy, J-Y., and T-Y. Chan. 1996. Complete larval development of the edible mud shrimp *Upogebia edulis* Ngoc-Ho & Chan, 1992 (Decapoda, Thalassinidea, Upogebiidae) reared in the laboratory. Crustaceana 69: 175–186.

Siddiqui, F. A., and F. S. Ghory. 2006. Complete larval development of *Emerita holthuisi* Sankolli, 1965 (Crustacea: Decapoda: Hippidae) reared in the laboratory. Turkish Journal of Zoology 30: 121–135.

Siddiqui, F. A., and N. M. Tirmizi. 1995. Laboratory rearing of *Upogebia quddusiae* Tirmizi & Ghani, 1978 (Decapoda, Thalassinidea) from ovigerous female to post-larva. Crustaceana 68: 445–460.

Siewing, R. 1956. Untersuchungen zur morphologie der Malacostraca (Crustacea). Zoologischer Jahrbücher, Anatomie 75: 39–176.

Siewing, R. 1958. Anatomie und Histologie von *Thermosbaena mirabilis*: ein Beitrag zur Phylogenie der Reihe Pancarida (Thermosbaenacea). Abhandlungen der Mathematisch-Naturwissenschaftlichen Klasse, Akademie der Wissenschaften und der Literatur zu Mainz 7: 195–270.

Siewing, R. 1963. Studies in malacostracan morphology: results and problems. *In* H. B. Whittington and W. D. I. Rolfe, eds. Phylogeny and Evolution of Crustacea: Proceedings of a Conference Held at Cambridge, Massachusetts, March 6–8, 1962, 85–103. Museum of Comparative Zoology Special Publication No. 13. Cambridge, MA: Harvard University, Museum of Comparative Zoology.

Silas, E. G., and K. J. Mathew. 1977. A critique to the study of larval development in Euphausiacea. Proceedings of the Symposium on Warm Water Zooplankton, Held in Goa, 14–19 October 1976, 571–582. Special Publication, National Institute of Oceanography. Goa, India: National Institute of Oceanography. Online at http://eprints.cmfri.org.in/6024/.

Silén, L. 1963. *Clionophilus vermicularis* n. gen., n. sp., a copepod infecting the burrowing sponge, *Cliona*. Zoologisk Bidrag från Uppsala 35: 269–288.

Simpson, S. D., A. N. Radford, E. J. Tickle, M. G. Meekan, and A. G. Jeffs. 2011. Adaptive avoidance of reef noise. PLoS One 6: e16625. doi:10.1371/journal.pone.0016625.

Sims, H. W. 1966. The phyllosoma larvae of the spiny lobster *Palinurellus gundlachi* von Martens (Decapoda, Palinuridae). Crustaceana 11: 205–215.

Siveter, D. J. 2008. Ostracods in the Palaeozoic? Senckenbergiana Lethaea 88: 1–9.

Siveter, D. J., and G. B. Curry. 1984. Lower Ordovician (Arenig) ostracods from the Highland Border Complex. *In* G. B. Curry, B. J. Bluck, C. J. Burton, J. K. Ingham, D. J. Siveter, and A. Williams, eds. Age, evolution and tectonic history of the Highland Border Complex, Scotland. Transactions of the Royal Society of Edinburgh, Earth Sciences 75: 113–133.

Siveter, D. J., M. Williams, and D. Waloszek. 2001. A phosphatocopid crustacean with appendages from the Lower Cambrian. Science 293: 479–481.

Siveter, D .J., M. D. Sutton, D. E. G. Briggs, and D. J. Siveter. 2003a. An ostracod crustacean with soft parts from the Lower Silurian. Science 300: 1749–1751.

Siveter, D. J., D. Waloszek and M. Williams. 2003b. An Early Cambrian phosphatocopid crustacean with three-dimensionally preserved soft parts from Shropshire, England. Special Papers in Palaeontology 70: 9–30.

Skogsberg, T. 1920. Studies on Marine Ostracods: Pt. 1, Cypridinids, Calocyprids, and Polycopids. Zoologiske Bidrag från Uppsala, Suppl. 1. Uppsala, Sweden: Almqvist & Wiksell.

Slinn, D. J. 1970. An infestation of adult *Lernaeocera* (Copepoda) on wild sole, *Solea solea*, kept under hatchery conditions. Journal of the Marine Biological Association of the United Kingdom 50: 787–800.

Smit, N. J., and A. J. Davies. 2004. The curious life-style of the parasitic stages of gnathiid isopods. Advances in Parasitology 58: 289–391.

Smith, D. L. 1977. A Guide to Marine Coastal Plankton and Invertebrate Larvae. Dubuque, IA: Kendall/Hunt.

Smith, J. A., and P. J. Whitfield. 1988. Ultrastructural studies on the early cuticular metamorphosis of adult female *Lernaeocera branchialis* (L.) (Copepoda, Pennellidae). *In* G. A. Boxshall and H. K. Schminke, eds. Biology of Copepods: Proceedings of the Third International Conference on Copepoda, held at the British Museum (Natural History), London, August 10–August 14, 1987. Hydrobiologia 167/168: 607–616.

Smith, R. J. 2000. Morphology and ontogeny of Cretaceous ostracods with preserved appendages from Brazil. Palaeontology 43: 63–98.

Smith, R. J., and T. Kamiya. 2002. The ontogeny of *Neonesidea oligodentata* (Bairdioidea, Ostracoda, Crustacea). Hydrobiologia 489: 245–275.

Smith, R. J., and T. Kamiya. 2003. The ontogeny of *Loxoconcha japonica* Ishizaki, 1968 (Cytheroidea, Ostracoda, Crustacea). Hydrobiologia 490: 31–52.

Smith, R. J., and T. Kamiya. 2005. The ontogeny of the entocytherid ostracod *Uncinocythere occidentalis* (Kozloff & Whitman, 1954) Hart, 1962 (Crustacea). Hydrobiologia 538: 217–229.

Smith, R. J., and T. Kamiya. 2008. The ontogeny of two species of Darwinuloidea (Ostracoda, Crustacea). Zoologischer Anzeiger 247: 275–302.

Smith, R. J., and K. Martens. 2000. The ontogeny of *Eucypris virens* (Cyprididae, Ostracoda). Hydrobiologia 419: 31–63.

Smith, R. J., and A. Tsukagoshi. 2005. The ontogeny and musculature of the antennae of podocope ostracods (Crustacea). Journal of Zoology 265: 157–177.

Smith, S. I. 1873. The early stages of the American lobster (*Homarus americanus* Edwards). Transactions of the Connecticut Academy of Arts and Sciences 2: 351–381.

Spears, T., and L. G. Abele. 2000. Branchiopod monophyly and interordinal phylogeny inferred from 18S ribosomal DNA. Journal of Crustacean Biology 20: 1–24.

Spears, T., R. W. Debry, L. G. Abele, and K. Chodyla. 2005. Peracarid monophyly and interordinal phylogeny inferred from nuclear small-subunit ribosomal DNA sequences (Crustacea: Malacostraca: Peracarida). Proceedings of the Biological Society of Washington 118: 117–157.

Sproston, N. G. 1942. The developmental stages of *Lernaeocera branchialis* (Linn.). Journal of the Marine Biological Association of the United Kingdom, n.s., 25: 441–466.

Stanley, J. A., C. A. Radford, and A. G. Jeffs. 2010. Induction of settlement in crab megalopae by ambient underwater reef sound. Behavioral Ecology 21: 112–120.

Stanley, J. A., C. A. Radford, and A. G. Jeffs. 2011. Behavioural response thresholds in New Zealand crab megalopae to ambient underwater sound. PLoS One 6: e28572. doi:10.1371/journal.pone.0028572.

Stanley, J. A., C. A. Radford, and A. G. Jeffs. 2012. Location, location, location: finding a suitable home among the noise. Proceedings of the Royal Society of London, B, 279: 3622–3631.

Stein, M., D. Waloszek, and A. Maas. 2005. *Oelandocaris oelandica* and the stem lineage of Crustacea. *In* S. Koenemann and R. A. Jenner, eds. Crustacea and Arthropod Relationships, 55–71. Crustacean Issues No. 16. Boca Raton, FL: CRC Press.

Stein, M., D. Waloszek, A. Maas, J. T. Haug, and K. J. Müller. 2008. The stem crustacean *Oelandocaris oelandica* re-visited. Acta Palaeontologica Polonica 53: 461–484.

Stella, E. 1959. Ulteriori osservazioni sulla riproduzione e lo sviluppo di *Monodella argentarii* (Pancarida: Thermosbaenacea). Rivista di Biologia (Perugia) 51: 121–144.

Stenderup, J. T., J. Olesen, and H. Glenner. 2006. Molecular phylogeny of the Branchiopoda (Crustacea): multiple approaches suggest a "diplostracan" ancestry of the Notostraca. Molecular Phylogenetics and Evolution 41: 182–194.

Stephensen, K. 1935. Two Crustaceans (a Cirriped and a Copepod) Endoparasitic in Ophiurids. Danish Ingolf Expedition, 1895–1896, Report No. 3, pt. 12. Copenhagen: Bianco Luno.

Stock, J. H. 1968. The Calvocheridae, a family of copepods inducing galls in sea-urchin spines. Bijdragen tot de Dierkunde 38: 85–90.

Stone, C. J. 1989. A comparison of algal diets for cirripede nauplii. Journal of Experimental Marine Biology and Ecology 132: 17–40.

Storch, V., and B. G. M. Jamieson. 1992. Further spermatological evidence for including Pentastomida (tongue worms) in the Crustacea. International Journal of Parasitology 22: 95–108.

Strasser, K. M., and D. L. Felder. 1999. Larval development of two populations of the ghost shrimp *Callichirus major* (Decapoda: Thalassinidea) under laboratory conditions. Journal of Crustacean Biology 19: 844–878.

Strasser, K. M., and D. L. Felder. 2000. Larval development of the ghost shrimp *Callichirus islagrande* (Decapoda: Thalassinidea: Callianassidae) under laboratory conditions. Journal of Crustacean Biology 20: 100–117.

Strasser, K. M., and D. L. Felder. 2005. Larval development of the mud shrimp *Axianassa australis* (Decapoda: Thalassinidea) under laboratory conditions. Journal of Natural History 39: 2289–2306.

Strathmann, R. R. 1993. Hypotheses on the origins of marine larvae. Annual Review of Ecology and Systematics 24: 89–117.

Strathmann, R. R. 2007. Three functionally different kinds of pelagic development. Bulletin of Marine Science 81: 167–179.

Strathmann, R. R. 2012. Early and diverse origins of feeding larvae.

Tenth International Larval Biology Symposium, Berkeley, California: 11 (abstract).

Strenth, N. E., and S. L. Sissom. 1975. A morphological study of the post-embryonic stages of *Eulimnadia texana* Packard (Conchostraca, Crustacea). Texas Journal of Science 26: 137–154.

Strickler, J. R. 1984. Sticky water: a selective force in copepod evolution. *In* D. G. Meyers and J. R. Strickler, eds. Trophic Interactions within Aquatic Ecosystems, 1–40. Boulder, CO: Westview Press.

Strömberg, J.-O. 1971. Contribution to the embryology of bopyrid isopods, with special reference to *Bopyroides*, *Hemiarthrus* and *Pseudione* (Isopoda, Epicaridea). Sarsia 47: 1–47.

Stuck, K. C., and F. M. Truesdale. 1986. Larval and early post-larval development of *Lepidopa benedicti* Schmitt, 1935 (Anomura: Albuneidae) reared in the laboratory. Journal of Crustacean Biology 6: 89–110.

Sudler, M. T. 1899. The development of *Penilia schmaeckeri* Richard. Proceedings of the Boston Society of Natural History 29: 107–133.

Sulkin, S. D. 1984. Behavioral basis of depth regulation in the larvae of brachyuran crabs. Marine Ecology Progress Series 15: 181–205.

Sund, O. 1920. The *Challenger* Eryonidea (Crustacea). Annals and Magazine of Natural History, ser. 9, 6: 220–226.

Surriray, J. S. A. 1819. Sur les oeufs de *Calyge*. Journal de Physique, de Chimie, d'Histoire Naturelle et des Arts 89: 400.

Suter, P. J. 1977. The biology of two species of *Engaeus* (Decapoda: Parastacidae) in Tasmania, 2: life history and larval development, with particular reference to *E. cisternarius*. Australian Journal of Marine and Freshwater Research 28: 85–93.

Swain, T. D., and D. J. Taylor. 2003. Structural rRNA characters support monophyly of raptorial limbs and paraphyly of limb specialization in water fleas. Proceedings of the Royal Society of London, B 270: 887–896.

Swanepoel, J. H., and A. Avenant-Oldewage. 1992. Comments on the morphology of the pre-oral spine in *Argulus* (Crustacea: Branchiura). Journal of Morphology 212: 155–162.

Swanson, K. M. 1989. *Manawa staceyi* n. sp. (Punciidae, Ostracoda): soft anatomy and ontogeny. Courier Forschungsinstitut Senckenberg 113: 235–249.

Swanson, K. M. 1991. Distribution, affinities and origin of the Punciidae (Crustacea: Ostracoda). Memoirs of the Queensland Museum 31: 77–92.

Taishaku, H., and K. Konishi. 1995. Zoeas of *Calappa* species with special reference to larval characters of the family Calappidae (Crustacea, Brachyura). Zoological Science 12: 649–654.

Taishaku, H., and K. Konishi. 2001. Lecithotrophic larval development of the spider crab *Goniopugettia sagamiensis* (Gordon, 1931) (Decapoda, Brachyura, Majidae) collected from the continental shelf break. Journal of Crustacean Biology 21: 748–759.

Takahashi, K., and K. Kawaguchi. 2004. Reproductive biology of the intertidal and infralittoral mysids *Archaeomysis kokuboi* and *A. japonica* on a sandy beach in NE Japan. Marine Ecology Progress Series 283: 219–231.

Tamaki, A., H. Tanoue, J. Itoh, and Y. Fukuda. 1996. Brooding and larval developmental periods of the callianassid ghost shrimp, *Callianassa japonica* (Decapoda: Thalassinidea). Journal of the Marine Biological Association of the United Kingdom 76: 675–689.

Tanaka, G., R. J. Smith, D. J. Siveter, and A. R. Parker. 2009. Three-dimensionally preserved decapod larval compound eyes from the Cretaceous Santana Formation of Brazil. Zoological Science 26: 846–850.

Tanaka, K. 2007. Life history of gnathiid isopods: current knowledge and future directions. Plankton Benthos Research 2: 1–11.

Tasch, P. 1969. Branchiopoda. *In* R. C. Moore, ed. Treatise on Invertebrate Paleontology, Pt. R, Arthropoda 4, vol. 1: Crustacea (except Ostracoda), Myriapoda–Hexapoda, R128–R191. Lawrence: University of Kansas Press and Geological Society of America.

Taton, H. 1934. Contribution à l'étude du copépode gallicole *Mesoglicola Delagei* Quidor. Travaux de la Station Biologique de Roscoff 12: 51–68.

Tavares, C., and J. W. Martin. 2010. Suborder Dendrobranchiata Bate, 1888. *In* J. Forest and J. C. von Vaupel Klein, eds., F. R. Schram and M. Charmantier-Daures, advisory eds. The Crustacea: Complementary to the Volumes Translated from the French of the Traité de Zoologie (Founded by P.-P. Grassé). Vol. 9, pt. A, Eucarida: Euphausiacea, Amphionidacea, and Decapoda, 99–164. Leiden, Netherlands: Brill.

Tavares, C., C. Serejo, and J. W. Martin. 2009. A preliminary phylogenetic analysis of the Dendrobranchiata based on morphological characters. *In* J. W. Martin, K. A. Crandall, and D. L. Felder, eds. Decapod Crustacean Phylogenetics, 255–273. Crustacean Issues No. 18. Boca Raton, FL: CRC Press.

Taylor, D. J., T. J. Crease, and W. M. Brown. 1999. Phylogenetic evidence for a single long-lived clade of crustacean cyclic parthenogens and its implications for the evolution of sex. Proceedings of the Royal Society of London, B 266: 791–797.

Telford, M. J., S. J. Bourlat, A. Economou, D. Papillon, and O. Rota-Stabelli. 2008. The evolution of the Ecdysozoa. Philosophical Transactions of the Royal Society of London, B 363: 1529–1537.

Teodósio, É. A. F. M. O., and S. Masunari. 2007. Description of first two juvenile stages of *Aegla schmitti* Hobbs III, 1979 (Anomura: Aeglidae). Nauplius 15: 73–80.

Terossi, M., J. A. Cuesta, I. S. Wehrtmann, and F. l. Mantelatto. 2010 Revision of the larval morphology (zoea I) of the family Hippolytidae Bate (Caridea), with a description of the first stage of the shrimp *Hippolyte obliquimanus* Dana, 1852. Zootaxa 2624: 49–66.

Thatje, S., R. Bacardit, M. C. Romero, F. Tapella, and G. A. Lovrich. 2001. Description and key to the zoeal stages of the Campylonotidae (Decapoda, Caridea) from the Magellan region. Journal of Crustacean Biology 21: 492–505.

Thessalou-Legaki, M., A. Peppa, and M. Zacharaki. 1999. Facultative lecithotrophy during larval development of the burrowing shrimp *Callianassa tyrrhena* (Decapoda: Callianassidae). Marine Biology 133: 635–642.

Thiele, J. 1904. Beiträge zur Morphologie der Arguliden. Mitteilungen aus der Zoologischen Sammlung des Museums für Naturkunde in Berlin 2: 5–51.

Thiéry, A., and C. Gasc. 1991. Resting eggs of Anostraca, Notostraca and Spinicaudata (Crustacea, Branchiopoda) occurring in France: identification and taxonomical value. Hydrobiologia 212: 245–259.

Thomas, W. J. 1973. The hatchling setae of *Austropotamobius pallipes* (Lereboullet) (Decapoda, Astacidae). Crustaceana 24: 77–89.

Thompson, J. V. 1828. On the metamorphosis of the Crustacea, and on zoea. Zoological Researches and Illustrations; or, Natural History of Nondescript or Imperfectly Known Animals, in a Series of Memoirs, vol. 1, pt. 1, memoir 1, 1–11. Cork, Ireland: King & Ridings.

Thompson, J. V. 1830. On the cirripedes or barnacles. *In* Zoological

Researches and Illustrations; or, Natural History of Nondescript or Imperfectly Known Animals, in a Series of Memoirs, vol. 1, pt. 1, memoir 4, 69–88. Cork, Ireland: King & Ridings.

Thompson, J. V. 1836. Natural history and metamorphosis of an anomalous crustaceous parasite of *Carcinus maenas*, the *Sacculina carcini*. Entomological Magazine (London) 3: 452–456.

Thompson, M. T. 1903. The metamorphoses of the hermit crab. Proceedings of the Boston Society of Natural History 31: 147–209.

Thorell, T. 1859. Bidrag till Kännedomen om Krustaceer, som lefva i Arter af Slägtet *Ascidia* L. Kungliga Svenska Vetenskapsakademiens Handlingar, n.s., 3: 1–84.

Tiemann, H. 1984. Is the taxon Harpacticoida a monophyletic one? *In* W. Vervoort and J. C. von Vaupel Klein, eds. Studies on Copepoda 2: Proceedings of the First International Conference on Copepoda, Amsterdam, The Netherlands, 24–28 August 1981. Crustaceana, Suppl. 7: 47–59.

Tokioka, T. 1936. Larval development and metamorphosis of *Argulus japonicus*. Memoirs of the College of Science, Kyoto Imperial University, B 12: 93–114.

Tokioka, T., and R. Bieri. 1966. Juveniles of *Macrosetella gracilis* (Dana) from clumps of *Trichodesmium* in the vicinity of Seto. Publications of the Seto Marine Biological Laboratory 14: 177–184.

Tomescu, N., and C. Craciun. 1987. Post-embryonic development in *Porcellio scaber* (Crustacea: Isopoda). Pedobiologia 30: 345–350.

Tomlinson, J. T. 1969. The burrowing barnacles (Cirripedia: order Acrothoracica). Bulletin of the United States National Museum No. 296. Washington, DC: Smithsonian Institution Press.

Tommasini, S., F. S. Sabelli, and M. Trentini. 1989. Scanning electron microscopy of eggshell development in *Triops cancriformis* (Bosc) (Crustacea, Notostraca). Vie et Milieu 39: 29–32.

Townsley, S. J. 1953. Adult and larval stomatopod crustaceans occurring in Hawaiian waters. Pacific Science 7: 399–437.

Tseng, W. Y. 1975. Biology of the pelagic ostracod, *Euconchoecia elongata* Müller. Report of the Laboratory of Fishery Biology Report No. 27. Keelung, Taiwan, R.O.C.: Taiwan Fisheries Research Institute.

Tsukagoshi, A. 1990. Ontogenetic change of distributional patterns of pore systems in *Cythere* species and its phylogenetic significance. Lethaia 23: 225–241.

Tudge, C. C., and C. W. Cunningham. 2002. Molecular phylogeny of the mud lobsters and mud shrimps (Crustacea: Decapoda: Thalassinidea) using nuclear 18S rDNA and mitochondrial 16S rDNA. Invertebrate Systematics 16: 839–847.

Turquier, Y. 1967. Le développement larvaire de *Trypetesa nassarioides* Turquier, cirripède acrothoracique. Archives de Zoologie Expérimentale et Générale 108: 33–47.

Turquier, Y. 1985. Cirripèdes acrothoraciques des côtes occidentales de la Mediterranée et de l'Afrique du Nord, 1: Cryptophialidae. Bulletin de la Société Zoologique de France 110: 151–168.

Uchida, T., and Y. Dotsu. 1973. Collection of the T. S. *Nagasaki Maru* of Nagasaki University, 4: on the larval hatching and larval development of the lobster *Nephrops thomasoni*. Bulletin of the Faculty of Fisheries, Nagasaki University 36: 23–35.

Ungerer, P., and C. Wolff. 2005. External morphology of limb development in the amphipod *Orchestia cavimana* (Crustacea, Malacostraca, Peracarida). Zoomorphology 124: 89–99.

Urawa, S., K. Muroga, and S. Kasahara. 1980a. Naupliar development of *Neoergasilus japonicus* (Copepoda: Ergasilidae). Bulletin of the Japanese Society of Scientific Fisheries 46: 941–947.

Urawa, S., K. Muroga, and S. Kasahara. 1980b. Studies on *Neoergasilus japonicus* (Copepoda: Ergasilidae), a parasite of freshwater fishes, 2: development in copepodid stage. Journal of the Faculty of Applied Biological Science, Hiroshima University 19: 21–38.

Utinomi, H. 1961. Studies on the Cirripedia Acrothoracica, 3: development of the female and male of *Berndtia purpurea* Utinomi. Publications of the Seto Marine Biology Laboratory 9: 413–446.

Uye, S.-i. 1980. Development of neritic copepods *Acartia clausi* and *A. steueri*, 1: some environmental factors affecting egg development and the nature of resting eggs. Bulletin of the Plankton Society of Japan 27: 1–9.

Uye, S.-i., Y. Iwai, and S. Kasahara. 1983. Growth and production of the inshore marine copepod *Pseudodiaptomus marinus* in the central part of the Inland Sea of Japan. Marine Biology 73: 91–98.

Valenti, W. C., W. H. Daniels, M. B. New, and E. S. Correia. 2009. Hatchery systems and management. *In* M. B. New, W. C. Valenti, J. H. Tidwell, L. R. D'Abramo, and M. N. Kutty, eds. Freshwater Prawns: Biology and Farming, 55–85. Oxford: Wiley-Blackwell.

van As, L. L., and J. G. van As. 1996. A new species of *Chonopeltis* (Crustacea: Branchiura) from the southern Rift Valley, with notes on larval development. Systematic Parasitology 35: 69–77.

Van Beneden, P. J. 1861. Recherches sur la Faune Littorale de Belgique. Brussels: Hayez.

Van Dover, C. L., J. R. Factor, and R. H. Gore. 1982. Developmental patterns of larval scaphognathites: an aid to the classification of Anomuran and Brachyuran Crustacea. Journal of Crustacean Biology 2: 48–53.

Van Dover, C. L., A. B. Williams, and J. R. Factor. 1984. The first zoeal stage of a hydrothermal vent crab (Decapoda: Brachyura: Bythograeidae). Proceedings of the Biological Society of Washington 97: 413–418.

van Niekerk, J. P., and D. J. Kok. 1989. *Chonopeltis australis* (Branchiura): structural, developmental and functional aspects of the trophic appendages. Crustaceana 57: 51–56.

Vannier, J., P. Boissy, and P. Racheboeuf. 1997. Locomotion in *Nebalia bipes*: a possible model for Palaeozoic phyllocarid crustaceans. Lethaia 30: 89–104.

van Straelen, V. 1938. Sur une forme larvaire nouvelle de stomatopodes du Cenomanien du Liban. Palaeobiologica: Archiv für die Erforschung des Lebens der Vorzeit und Seiner Geschichte 6: 394–400.

Vaugelas, J. de, B. Delesalle, and C. Monier. 1986. Aspects of the biology of *Callichirus armatus* (A. Milne-Edwards, 1870) (Decapoda, Thalassinidea) from French Polynesia. Crustaceana 50: 204–216.

Venmathi Maran, B. A., S. Y. Moon, S. Ohtsuka, S.-Y. Oh, H. Y. Soh, J.-G. Myoung, A. Iglikowska, and G. A. Boxshall. 2013. The caligid life cycle: new evidence from *Lepeophtheirus elegans* reconciles the cycles of *Caligus* and *Lepeophtheirus* (Copepoda: Caligidae). Parasite 20: 15.

Vetter, E. W. 1996. Life-history patterns of two Southern Californian *Nebalia* species (Crustacea: Leptostraca): the failure of form to predict function. Marine Biology 127: 131–141.

Villamar, D. F., and G. J. Brusca. 1988. Variation on the larval development of *Crangon nigricauda* (Decapoda: Caridea), with notes on larval morphology and behaviour. Journal of Crustacean Biology 8: 410–419.

Vinogradov, G. M. 1999. Amphipoda. *In* D. Boltovskoy, ed. Vol. 2, South Atlantic Zooplankton, 1141–1240. Leiden, Netherlands: Backhuys.

Vlasblom, A. G., and J. H. B. W. Elgershuizen. 1977. Survival and oxygen consumption of *Praunus flexuosus* and *Neomysis integer*, and embryonic development of the latter species, in different temperature and chlorinity combinations. Netherlands Journal of Sea Research 11: 305–315.

Vogt, G., and L. Tolley. 2004. Brood care in freshwater crayfish and relationship with the offspring's sensory deficiencies. Journal of Morphology 262: 566–582.

Wagin, V. L. 1976. Ascothoracida. Kazan, Russia: Kazan University Press [in Russian].

Wagner, H. P. 1994. A Monographic Review of the Thermosbaenacea (Crustacea: Peracarida). Zoologische Verhandelingen No. 291. Leiden, Netherlands: Nationaal Natuurhistorisch Museum.

Wagner, W. 1896. Einige Beobachtungen über die embryonale Entwicklung von *Neomysis vulgaris* var. *baltica* Czern. Travaux de la Société Impériale des Naturalistes de St. Pétersbourg, Zoologie et Physiologie 26: 177–221.

Wakayama, N. 2007. Embryonic development clarifies polyphyly in ostracod crustaceans. Journal of Zoology 273: 406–413.

Walker, G. 1973. Frontal horns and associated gland cells of the nauplii of the barnacles *Balanus hameri*, *Balanus balanoides* and *Elminius modestus* (Crustacea Cirripedia). Journal of the Marine Biological Association of the United Kingdom 53: 455–463.

Walker, G. 1985. The cypris larvae of Thompson (Crustacea: Cirripedia: Rhizocephala). Journal of Experimental Marine Biology and Ecology 93: 131–145.

Walker, G. 1995. Larval settlement: historical and future perspectives. *In* F. R. Schram and J. T. Høeg, eds. New Frontiers In Barnacle Evolution, 69–85. Crustacean Issues No. 10. Rotterdam: A. A. Balkema.

Walker, G., A. B. Yule, and J. A. Nott. 1987. Structure and function of balanomorph larvae. *In* A. J. Southward, ed. Barnacle Biology, 307–328. Crustacean Issues No. 5. Rotterdam: A. A. Balkema.

Walker-Smith, G. K., and G. C. B. Poore. 2001. A phylogeny of the Leptostraca (Crustacea) with keys to families and genera. Memoirs of Museum Victoria 58: 383–410.

Walley, J. 1969. Studies on the larval structure and metamorphosis of *Balanus balanoides* (L.) Philosophical Transactions of the Royal Society of London, B 256: 237–280.

Walossek, D. 1993. The Upper Cambrian *Rehbachiella* and the phylogeny of Branchiopoda and Crustacea. Fossils and Strata 32: 1–202.

Walossek, D. 1995. The Upper Cambrian *Rehbachiella*, its larval development, morphology and significance for the phylogeny of Branchiopoda and Crustacea. Hydrobiologia 298: 1–13.

Walossek, D. 1996. *Rehbachiella*, der bisher älteste Branchiopode. Stapfia 42: 21–28.

Walossek, D. 1999. On the Cambrian diversity of Crustacea. *In* F. R. Schram and J. C. von Vaupel Klein, eds. Vol. 1, Crustaceans and the Biodiversity Crisis: Proceedings of the Fourth International Crustacean Congress, Amsterdam, the Netherlands, July 20–24, 1998, 3–27. Leiden, Netherlands: Brill.

Walossek, D., and K. J. Müller. 1989. A second type A-nauplius from the Upper Cambrian "Orsten" of Sweden. Lethaia 22: 301–306.

Walossek, D., and K. J. Müller. 1990. Upper Cambrian stem-lineage crustaceans and their bearing on the monophyletic origin of Crustacea and the position of *Agnostus*. Lethaia 23: 409–427.

Walossek, D., and K. J. Müller. 1994. Pentastomid parasites from the lower Palaeozoic of Sweden. Transactions of the Royal Society of Edinburgh, Earth Sciences 85: 1–37.

Walossek, D., and K. J. Müller. 1998a. Early arthropod phylogeny in the light of Cambrian "Orsten" fossils. *In* G. D. Edgecombe, ed. Arthropod Fossils and Phylogeny, 185–231. New York: Colombia University Press.

Walossek, D., and K. J. Müller. 1998b. Cambrian "Orsten"-type arthropods and the phylogeny of Crustacea. *In* R. A. Fortey and R. H. Thomas, eds. Arthropod Relationships, 139–153. Systematics Association Special Volume No. 55. London: Chapman & Hall.

Walossek, D., I. Hinz-Schallreuter, J. H. Shergold, and K. J. Müller. 1993. Three-dimensional preservation of arthropod integument from the Middle Cambrian of Australia. Lethaia 26: 7–15.

Walossek, D., J. E. Repetski, and K. J. Müller. 1994. An exceptionally preserved parasitic arthropod, *Heymonsicambria taylori* n. sp. (Arthropoda *incertae sedis*: Pentastomida) from the Cambrian-Ordovician boundary beds of Newfoundland, Canada. Canadian Journal of Earth Sciences 31: 1664–1671.

Walossek, D., J. T. Høeg, and T. C. Shirley. 1996. Larval development of the rhizocephalan cirripede *Briarosaccus tenellus* (Maxillopoda: Thecostraca) reared in the laboratory: a scanning electron microscopy study. Hydrobiologia 328: 9–47.

Waloszek, D. 2003a. The "Orsten" window: a three-dimensionally preserved Upper Cambrian meiofauna and its contribution to our understanding of the evolution of Arthropoda. Paleontological Research 7: 71–88.

Waloszek, D. 2003b. Cambrian "Orsten"-type preserved arthropods and the phylogeny of Crustacea. *In* A. Legakis, S. Sfenthourakis, R. Polymeni, and M. Thessalou-Legaki, eds. The New Panorama of Animal Evolution: Proceedings of the Eighteenth International Congress of Zoology, Athens, Greece, August 8–September 2, 2000, 69–87. Sofia, Greece: Pensoft.

Waloszek, D., J. E. Repetski, and A. Maas. 2006. A new Late Cambrian pentastomid and a review of the relationships of this parasitic group. Transactions of the Royal Society of Edinburgh, Earth Sciences 96: 163–176.

Waloszek, D., A. Maas, J. Chen, and M. Stein. 2007. Evolution of cephalic feeding structures and the phylogeny of Arthropoda. Palaeogeography, Palaeoclimatology, Palaeoecology 254: 273–287.

Warren, E. 1901. A preliminary account of the development of the free-living nauplius of *Leptodora hyalina* (Lillj). Proceedings of the Royal Society of London, B 68: 210–218.

Watling, L. 1983. Peracaridan disunity and its bearing on eumalacostracan phylogeny with redefinition of eumalacostracan superorders. *In* F. R. Schram, ed. Crustacean Phylogeny, 214–227. Crustacean Issues No. 1. Rotterdam: A. A. Balkema.

Watling, L. 1999. Toward understanding the relationship of the peracaridan orders: the necessity of determining exact homologies. *In* F. R. Schram and J. C. von Vaupel Klein, eds. Crustaceans and the Biodiversity Crisis: Proceedings of the Fourth International Crustacean Congress, Amsterdam, the Netherlands, July 20–24, 1998, 73–89. Leiden, Netherlands: Brill.

Watling, L. 2005. Cumacea World Database. Online at www .marinespecies.org/cumacea/.

Wear, R. G. 1965. Zooplankton of Wellington Harbour, New Zealand. Zoology Publications from Victoria University of Wellington 38: 1–31.

Wear, R. G. 1967. Life-history studies on New Zealand Brachyura, 1: embryonic and post-embryonic development of *Pilumnus novaezealandiae* Filhol, 1886, and of *P. lumpinus* Bennett, 1964 (Xanthidae, Pilumninae). New Zealand Journal of Marine and Freshwater Research 1: 482–535.

Wear, R. G. 1968. Life-history studies on New Zealand Brachyura, 2: family Xanthidae—larvae of *Heterozius rotundifrons* A. Milne-Edwards, 1867, *Ozius truncatus* H. Milne-Edwards, 1834, and *Heteropanope (Pilumnopeus) serratifrons* (Kinahan, 1856). New Zealand Journal of Marine and Freshwater Research 2: 293–332.

Wear, R. G. 1970. Some larval stages of *Petalomera wilsoni* (Fulton & Grant, 1902) (Decapoda, Dromiidae). Crustaceana 18: 1–12.

Wear, R. G. 1976. Studies on the larval development of *Metanephrops challengeri* (Balss, 1914) (Decapoda, Nephropidae). Crustaceana 30: 113–122.

Wear, R. G., and E. J. Batham. 1975. Larvae of the deep sea crab *Cymonomus bathamae* Dell, 1971 (Decapoda, Dorippidae), with observations on larval affinities of the Tymolinae. Crustaceana 28: 113–120.

Wear, R. G., and D. R. Fielder. 1985. The Marine Fauna of New Zealand: Larvae of the Brachyura (Crustacea, Decapoda). New Zealand Oceanographic Institute Memoir No. 92. Wellington: New Zealand Department of Scientific and Industrial Research.

Wear, R. G., and J. C. Yaldwyn. 1966. Studies on thalassinid Crustacea (Decapoda, Macrura, Reptantia) with description of a new *Jaxea* from New Zealand and an account of its larval development. Zoology Publications from Victoria University of Wellington 41: 1–27.

Webber, W. R., and J. D. Booth. 2001. Larval stages, developmental ecology, and distribution of *Scyllarus* sp. Z (probably *Scyllarus aoteanus* Powell, 1949) (Decapoda: Scyllaridae). New Zealand Journal of Marine and Freshwater Research 35: 1025–1056.

Webber, W. R., and R. G. Wear. 1981. Life history studies on New Zealand Brachyura, 5: larvae of the family Majidae. Journal of Marine and Freshwater Research 15: 331–383.

Weeks, S. C., C. Benvenuto, and S. K. Reed. 2006. When males and hermaphrodites coexist: a review of androdioecy in animals. Integrative and Comparative Biology 46: 449–464.

Weeks, S. C., E. G. Chapman, D. C. Rogers, D. M. Senyo, and W. R. Hoeh. 2009. Evolutionary transitions among dioecy, androdioecy and hermaphroditism in limnadiid clam shrimp (Branchiopoda: Spinicaudata). Journal of Evolutionary Biology 22: 1781–1799.

Wehrtmann, I., L. Albornoz, D. Véliz, and L. M. Pardo. 1996. Early developmental stages, including the first crab, of *Allopetrolisthes angulosus* (Decapoda: Anomura: Porcellanidae) from Chile, reared in the laboratory. Journal of Crustacean Biology 16: 730–747.

Weismann, A. 1876–1879. Beiträge zur Naturgeschichte der Daphnoiden. Zeitschrift für Wissenschaftliche Zoologie, vols. 27–33: 1–486.

Weisz, P. B. 1947. The histological pattern of metameric development in *Artemia salina*. Journal of Morphology 81: 45–95.

Wenner, E. L. 1979. Some aspects of the biology of deep-sea lobsters of the family Polychelidae (Crustacea, Decapoda) from the western North Atlantic. U.S. Fishery Bulletin 77: 435–444.

Weygoldt, P. 1960. Embryologische Untersuchungen an Ostrakoden: die Entwicklung von *Cyprideis litoralis* (G. S. Brady) (Ostracoda, Podocopa, Cytheridae). Zoologische Jahrbücher, Abteilung für Anatomie und Ontogenie der Tiere 78: 367–426.

Wicksten, M. K. 1981. New records of *Stereomastis sculpta pacifica* (Faxon) (Decapoda: Polychelidae) in the eastern Pacific Ocean. Proceedings of the Biological Society of Washington 93: 914–919.

Wilkens, H., J. Parzefall, and A. Ribowski. 1990. Population biology and larvae of the anchialine crab *Munidopsis polymorpha* (Galatheidae) from Lanzarote (Canary Islands). Journal of Crustacean Biology 10: 667–675.

Willen, E. 2000. Phylogeny of the Thalestridimorpha Lang, 1944 (Crustacea, Copepoda). Göttingen, Germany: Cuvilier Verlag.

Williams, B. G., B. G. Greenwood, and J. Jillett. 1985. Seasonality and duration of the developmental stages of *Heterosquilla tricarinata* (Claus, 1871) (Crustacea: Stomatopoda) and the replacement of the larval eye at metamorphosis. Bulletin of Marine Science 36: 104–114.

Williams, J. D., and C. B. Boyko. 2012. The global diversity of parasitic isopods associated with crustacean hosts (Isopoda: Bopyroidea and Cryptoniscoidea). PLoS One 7: e35350.

Williams, T. A. 1994a. A model of rowing propulsion and the ontogeny of locomotion in *Artemia* larvae. Biological Bulletin 187: 164–173.

Williams, T. A. 1994b. The nauplius larva of crustaceans: functional diversity and the phylotypic stage. American Zoologist 34: 562–569.

Williams, T. A. 1998. Distalless expression in crustaceans and the patterning of branched limbs. Development Genes and Evolution 207: 427–434.

Williams, T. A. 2007. Limb morphogenesis in the branchiopod crustacean, *Thamnocephalus platyurus*, and the evolution of proximal limb lobes within Anostraca. Journal of Zoological Systematics and Evolutionary Research 45: 191–201.

Williams-Howze, J. 1997. Dormancy in the free-living copepod orders Cyclopoida, Calanoida, and Harpacticoida. *In* A. D. Ansell, R. N. Gibson, and M. Barnes, eds. Vol. 35, Oceanography and Marine Biology: An Annual Review, 257–321. London: UCL Press.

Williams-Howze, J., and J. W. Fleeger. 1987. Pore pattern: a possible indicator of tube-building in *Stenhelia* and *Pseudostenhelia* (Copepoda: Harpacticoida). Journal of Crustacean Biology 7: 148–157.

Williamson, D. I. 1957. Crustacea Decapoda: Larvae, 5; Caridea, Family Hippolytidae. Fiches d'Identification du Plancton, Leaflet No. 68. Copenhagen: International Council for the Exploration of the Sea.

Williamson, D. I. 1960. Crustacea, Decapoda: Larvae, 7; Caridea, Family Crangonidae, Stenopodidea. Fiches d'Identification du Plancton, Leaflet No. 90. Copenhagen: International Council for the Exploration of the Sea.

Williamson, D. I. 1969. Names of larvae in the Decapoda and Euphausiacea. Crustaceana 16: 210–213.

Williamson, D. I. 1970. On a collection of planktonic Decapoda and Stomatopoda (Crustacea) from the east coast of the Sinai Peninsula, northern Red Sea. No. 45, Contributions to the Knowledge of the Red Sea. Bulletin No. 56. Haifa, Israel: Sea Fisheries Research Station.

Williamson, D. I. 1973. *Amphionides reynaudii*, representative of a proposed new order of eucaridan Malacostraca. Crustaceana 25: 35–50.

Williamson, D. I. 1976a. Larval characters and the origin of crabs. Thalassia Jugoslavia 10: 401–414.

Williamson, D. I. 1976b. Larvae of Stenopodidae (Crustacea: Decapoda) from the Indian Ocean. Journal of Natural History 10: 497–509.

Williamson, D. I. 1982a. Larval morphology and diversity. In L. G. Abele, ed. The Biology of Crustacea. Vol. 2, Embryology, Morphology, and Genetics, 43–110. New York: Academic Press.

Williamson, D. I. 1982b. The larval characters of Dorhynchus thomsoni Thomson (Crustacea, Brachyura, Majoidea) and their evolution. Journal of Natural History 16: 727–744.

Williamson, D. I. 1983. Crustacea, Decapoda: Larvae, 8; Nephropidea, Palinuridea, and Eryonidea. Fiches d'Identification du Zooplancton No. 167/168. Copenhagen: Conseil International pour l'Exploration de la Mer.

Williamson, D. I. 1988a. Evolutionary trends in larval form. In A. A. Fincham and P. S. Rainbow, eds. Aspects of Decapod Crustacean Biology: Proceedings of a Symposium Held at the Zoological Society of London, 8–9 April 1987, 11–25. Symposia of the Zoological Society of London No. 59. Oxford: Clarendon Press.

Williamson, D. I. 1988b. Incongruous larvae and the origin of some invertebrate life-histories. Progress in Oceanography 19: 87–116.

Williamson, D. I. 1992. Larvae and Evolution: Toward a New Zoology. New York: Chapman & Hall.

Williamson, D. I., and S. E. Vickers. 2007. The origins of larvae. American Scientist 95: 509–517.

Williamson, D. I., and K. G. Von Levetzow. 1967. Larvae of Parapagurus diogenes (Whitelegge) and some related species (Decapoda, Anomura). Crustaceana 12: 179–192.

Williamson, H. C. 1901. On the larval stages of decapod Crustacea: the Shrimp (Crangon vulgaris Fabr.). Annual Report of the Fisheries Board for Scotland 19: 92–120.

Williamson, H. C. 1905. A contribution to the life history of the lobster (Homarus vulgaris). Annual Report of the Fisheries Board for Scotland 23: 65–155.

Williamson, H. C. 1915. Decapoden-Larven. In K. Brandt and C. Apstein, eds. Nordisches Plankton No. 18, fasc. 6, pt. 1, 315–588. Kiel, Germany: Lipsius & Tischer.

Wills, M. 1998. A phylogeny of the Recent and fossil Crustacea derived from morphological characters. In R. A. Fortey and R. H. Thomas, eds. Arthropod Relationships, 189–209. Systematics Association Special Volume No. 55. London: Chapman & Hall.

Wilson, C. B. 1902. North American parasitic copepods of the family Argulidae, with a bibliography of the group and a systematic review of all known species. Proceedings of the United States National Museum 25: 635–742.

Wilson, C. B. 1904. The fish parasites of the genus Argulus found in the Woods Hole region. Bulletin of the Bureau of Fisheries 24: 115–131.

Wilson, C. B. 1905. North American parasitic copepods belonging to the family Caligidae, pt. 1: the Caliginae. Proceedings of the United States National Museum 28: 479–672.

Wilson, C. B. 1907a. North American parasitic copepods belonging to the family Caligidae, pts. 3 and 4: a revision of the Pandarinae and the Cecropinae. Proceedings of the United States National Museum 33: 323–490.

Wilson, C. B. 1907b. Additional notes on the development of the Argulidae, with description of a new species. Proceedings of the United States National Museum 32: 411–424.

Wilson, C. B. 1912. Descriptions of new species of parasitic copepods in the collections of the United States National Museum. Proceedings of the United States National Museum 42: 233–243.

Wilson, G. D. F. 2008. Global diversity of isopod crustaceans (Crustacea: Isopoda) in freshwater. Hydrobiologia 595: 231–240.

Wilson, G. D. F., C. A. Sims, and A. S. Grutter. 2011. Toward a taxonomy of the Gnathiidae (Isopoda) using juveniles: the external anatomy of Gnathia aureamaculosa zuphea stages using scanning electron microscopy. Journal of Crustacean Biology 31: 509–522.

Wilson, K. A., and R. H. Gore. 1980. Studies on decapod Crustacea from the Indian River region of Florida, 17: larval stages of Plagusia depressa (Fabricius, 1775) cultured under laboratory conditions (Brachyura: Grapsidae). Bulletin of Marine Science 30: 776–798.

Winch, J. M., and J. Riley. 1986. Studies on the behaviour, and development in fish, of Subtriquetra subtriquetra, a uniquely free-living pentastomid larva from a crocodilian. Parasitology 93: 81–98.

Wingstrand, K. G. 1972. Comparative spermatology of a pentastomid, Raillietiella hemidactyli, and a branchiuran crustacean, Argulus foliaceus, with a discussion of pentastomid relationships. Det Kongelige Danske Videnskabernes Selskab, Biologiske Skrifter 19: 1–72.Wingstrand, K. G. 1988. Comparative spermatology of the Crustacea, Entomostraca, 2: subclass Ostracoda. Det Kongelige Danske Videnskabernes Selskab, Biologiske Skrifter 32: 1–149.

Winsor, M. P. 1969. Barnacle larvae in the nineteenth century: a case study in taxonomic theory. Journal of the History of Medicine and Allied Sciences 24: 294–309.

Wittmann, K. J. 1978. Adoption, replacement, and identification of young in marine Mysidacea (Crustacea). Journal of Experimental Marine Biology and Ecology 32: 259–274.

Wittmann, K. J. 1981a. Comparative biology and morphology of marsupial development in Leptomysis and other Mediterranean Mysidacea (Crustacea). Journal of Experimental Marine Biology and Ecology 52: 243–270.

Wittmann, K. J. 1981b. On the breeding biology and physiology of marsupial development in Mediterranean Leptomysis (Mysidacea: Crustacea), with special reference to the effects of temperature and egg size. Journal of Experimental Marine Biology and Ecology 53: 261–279.

Wittmann, K. J. 1984. Ecophysiology of marsupial development and reproduction in Mysidacea (Crustacea). In H. Barnes, ed. Vol. 22, Oceanography and Marine Biology: An Annual Review, 293–428. London: Routledge.

Wolfe, A. F. 1980. A light and electron microscopic study of the frontal knob of Artemia (Crustacea, Branchiopoda). In G. Persoone, P. Sorgeloos, O. Roels, and E. Jaspers, eds. The Brine Shrimp Artemia. Vol. 1, Morphology, Genetics, Radiobiology, Toxicology, 117–130. Wetteren, Belgium: Universa Press.

Wolff, C. 2009. The embryonic development of the malacostracan crustacean Porcellio scaber (Isopoda, Oniscidea). Development Genes and Evolution 219: 545–564.

Woltereck, R. 1904. Zweite Mitteilung über die Hyperiden der Deutschen Tiefsee-Expedition: "Physosoma", ein neuer pelagischer Larventypus; nebst Bemerkungen zur Biologie von Thaumatops und Phronima. Zoologischer Anzeiger 27: 553–563.

Woltereck, R. 1905. Mitteilungen über Hyperiden der Valdivia- (Nr. 4), der Gauss- (Nr. 2) und der Schwedischen Südpolarexpedition: a. Scypholanceola, eine neue Hyperidengattung mit Reflektororganen; b. die Physosoma-Larve der Lanceoliden. Zoologischer Anzeiger 29: 413–417.

Wood-Mason, J. 1876. On the mode in which the young of the New Zealand Astacidae attach themselves to the mother. Annals and Magazine of Natural History, ser. 8, 18: 306–307.

Wooldridge, T. H. 1981. Zonation and distribution of the beach mysid, *Gastrosaccus psammodytes* (Crustacea: Mysidacea). Journal of Zoology 193: 183–189.

Wortham-Neal, J. L., and W. W. Price. 2002. Marsupial developmental stages in *Americamysis bahia* (Mysida: Mysidae). Journal of Crustacean Biology 22: 98–112.

Wray, G. A. 1995. Evolution of larvae and developmental modes. *In* L. McEdward, ed. Ecology of Marine Invertebrate Larvae, 413–447. Boca Raton, FL: CRC Press.

Xu, L., B.-P. Han, K. V. Damme, A. Vierstraete, J. R. Vanfleteren, and H. J. Dumont. 2010. Biogeography and evolution of the Holarctic zooplankton genus *Leptodora* (Crustacea: Branchiopoda: Haplopoda). Journal of Biogeography 38: 359–370.

Yager, J. 1981. Remipedia, a new class of Crustacea from a marine cave in the Bahamas. Journal of Crustacean Biology 1: 328–333.

Yamanaka, K., R. Kuwabara and T. Shio. 1997. Larval development of a Japanese crayfish, *Cambaroides japonicus* (De Haan). Bulletin of Marine Science 61: 165–175.

Yanagimachi, R. 1961. The life-cycle of *Peltogasterella* (Cirripedia, Rhizocephala). Crustaceana 2: 183–186.

Yang, W. T. 1971. The larval and post-larval development of *Parthenope serrata* reared in the laboratory and the systematic position of the Parthenopidae (Crustacea, Brachyura). Biological Bulletin 140: 166–189.

Young, C. M. 1999. Marine invertebrate larvae. *In* E. Knobil and J. D. Neil, eds. Vol. 3, M–Pri, Encyclopedia of Reproduction, 89–97. London: Academic Press.

Young, C. M. 2001. A brief history and some fundamentals. *In* C. Young, M. Sewell, and M. Rice, eds. Atlas of Marine Invertebrate Larvae, 1–19. London: Academic Press.

Zaddach, E. G. 1841. De apodis cancriformis Schaeff. anatome et historia evolutionis. Dissertatio inauguralis zootomica [thesis], Academia Fridericia Guilelmia Rhenana, Bonnae [Bonn].

Zaffagnini, F. 1971. Alcune precisazioni sullo sviluppo post-embrionale del concostraco *Limnadia lenticularis* (L.). Memorie dell'Instituto Italiano di Idrobiologia 27: 45–60.

Zaffagnini, F. 1987. Reproduction in *Daphnia*. Memorie dell' Istituto Italiano di Idrobiologia 45: 245–284.

Zaffagnini, F., and M. Trentini. 1980. The distribution and reproduction of *Triops cancriformis* (Bosc) in Europe (Crustacea, Notostraca). Monitore Zoologico Italiano / Italian Journal of Zoology 14: 1–8.

Zhang, D., J. Lin, and R. L. Creswell. 1997. Larviculture and effect of food on larval survival and development in golden coral shrimp *Stenopus scutellatus*. Journal of Shellfish Research 16: 367–369.

Zhang, W., Y. Shen, and S. Niu. 1990. Discovery of Jurassic clam shrimps with well-preserved soft parts and notes on its biological significance. Palaeontologia Cathayana 5: 311–352.

Zhang, X., D. J. Siveter, D. Waloszek, and A. Maas. 2007. An epipodite-bearing crown-group crustacean from the Lower Cambrian. Nature 449: 595–598.

Zhang, X., A. Maas, J. T. Haug, D. J. Siveter, and D. Waloszek. 2010. A eucrustacean metanauplius from the Lower Cambrian. Current Biology 20: 1075–1079.

Zhao, J. 1989. Cladistics and classification of Cambrian bradoriids from north China. Acta Palaeontologica Sinica 28: 463–473.

Zierold, T., J. Montero-Pau, B. Hänfling, and A. Gómez. 2009. Sex ratio, reproductive mode and genetic diversity in *Triops cancriformis*. Freshwater Biology 54: 1392–1405.

Zilch, R. 1972. Beitrag zur Verbreitung und Entwicklungbiologie der Thermosbaenacea. Internationale Revue der Gesamten Hydrobiologie 57: 75–107.

Zilch, R. 1974. Die Embryonalentwicklung von *Thermosbaena mirabilis* Monod (Crustacea, Malacostraca, Pancarida). Zoologischer Jahrbücher, Anatomie 93: 462–576.

Zilch, R. 1975. Etappen der Frühontogenese von *Thermosbaena mirabilis* Monod (Crustacea, Malacostraca, Pancarida). Verhandlungen der Deutschen Zoologischen Gesellschaft 121–126.

Zimmer, C. 1904. *Amphionides valdiviae*, n. g., n. sp. Zoologischer Anzeiger 28: 225–228.

Zrzavý, J. 2001. The interrelationships of metazoan parasites: a review of phylum- and higher-level hypotheses from recent morphological and molecular phylogenetic analyses. Folia Parasitologica 48: 81–103.